NATURALLY OCCURRING
QUINONES

NATURALLY OCCURRING QUINONES

by

R. H. Thomson

Department of Chemistry, University of Aberdeen, Scotland

Second Edition

1971

ACADEMIC PRESS
London and New York

ACADEMIC PRESS INC. (LONDON) LTD
Berkeley Square House
Berkeley Square,
London, W1X 6BA

U.S. Edition published by

ACADEMIC PRESS INC.
111 Fifth Avenue,
New York, New York 10003

Library of Congress Catalog Card Number: 74–141723

ISBN: 0-12-689650-X

PRINTED IN GREAT BRITAIN BY
WILLIAM CLOWES & SONS LIMITED
LONDON, COLCHESTER AND BECCLES

Preface

Since this book first appeared in 1957 the quinones have become recognised as a major group of natural pigments. In number the natural quinones have multiplied nearly three times and continue to increase. While the majority belong to well-known structural patterns several novel features have appeared including new heterocyclic quinones, tri- and even tetra-meric quinones, and the important families of "bioquinones" with lengthy polyprenoid side chains. Knowledge of the biogenesis of quinones has greatly advanced in recent years and most of Chapter 1 is devoted to that topic. Chapter 2 is also new and is concerned mainly with the use of spectroscopic data in structural elucidation. Inevitably the bulk of the book is a mixture of the new and the old, since recent investigations are based primarily on spectroscopic evidence, while the older work was done by classical methods.

The manuscript was completed in April 1970 and an Appendix covers the literature up to early October.

I am greatly indebted to many colleagues in various parts of the world who supplied samples, spectra, and information in advance of publication. I am grateful also to Professor D. W. Cameron who read the section on aphin pigments and to Dr P. Margaret Brown who ran numerous spectra for me. Finally I wish to thank Miss Mary Smith for her impeccable typing and for help with indexing and proof-reading.

University of Aberdeen   R. H. THOMSON
October, 1970

Acknowledgements

The author would like to thank copyright holders and authors who gave permission to reproduce some of the figures in this book. Detailed references are given in the legends.

Contents

Chapter 1

Distribution and Biogenesis

The natural quinones range from pale yellow to almost black in colour, but although they are widely distributed, and exist in large numbers and great structural variety, they make relatively little contribution to natural colouring compared to, say, the carotenoids or anthocyanins. They are most evident in sea urchins and related marine animals, and in certain fungi and lichens, but many are present in bark or roots, or in animal tissues, where they are not normally seen, and frequently they occur only in very low concentration and are masked by other pigments. Nevertheless the colouring power of some of the plant and animal pigments was discovered at a very early date, and natural quinone dyestuffs, such as kermes and madder, were of great importance until the nineteenth century and are not entirely obsolete even now.

Distribution

Quinones are found chiefly in higher plants, fungi and bacteria, and, in the animal kingdom, in arthropods and echinoderms. Their appearance in other phyla is rare apart from the widely distributed "bioquinones" which are involved in cellular respiration and photosynthesis. The ubiquinones occur fairly generally in animals, including man,[39] and are probably present in most plants although they have not yet been reported in bryophytes or pteridophytes. Plastoquinones, α-tocopherol-quinone and phylloquinone are probably present in all photosynthetic tissue. At the other extreme quinones have been obtained, on occasion, from mineral sources, and they appear in the fossil record.

The anthraquinones are by far the largest group (ca. 170) nearly half of which have been found in fungi and lichens, and a similar number in higher plants. Their first appearance in bacteria has been reported recently.[1] In animals, a few occur in insects (Coccidae only) and in feather stars (Crinoidea). Recent advances in chromatography have revealed the existence of quite complex mixtures of anthraquinones in some species, and these pigments may be formed in remarkable abundance. More than twenty anthraquinones and closely related compounds have been isolated from cultures of *Penicillium islandicum*[2] and nineteen

1

from the root of *Rubia tinctorum*[3] (madder); the bark of *Coprosma australis*[4] has been known to contain 17% of its dry weight of anthraquinones and about 30% of the dried mycelium of *Pyrenophora graminea*[5] (= *Helminthosporium gramineum*) may consist of helminthosporin and catenarin.

Anthraquinones are distributed fairly widely in moulds, especially in *Aspergillus* and *Penicillium* spp.; they are uncommon in higher fungi but are found more frequently in lichens. In higher plants they are located chiefly in heartwood, bark and roots (often as glycosides), occasionally in stems, seeds and fruit. As yet, they appear to be confined to about a dozen families. The Rubiaceae accounts for half the total number, with up to a dozen or so pigments in each of the families Rhamnaceae, Leguminosae (especially *Cassia*), Polygonaceae (particularly *Rumex* and *Rheum* spp.), Bignoniaceae, Verbenaceae (teak), Scrophulariaceae (*Digitalis* spp.) and Liliaceae. Emodin (1) is probably the most widely distributed anthraquinone, being found in moulds,

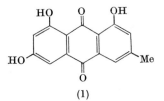

(1)

higher fungi and lichens, flowering plants and insects, and a large proportion of the natural anthraquinones are biosynthetic variations of this basic structure.

The distribution of the naphthaquinones (>120) is sporadic. Nearly half of them occur in higher plants, scattered through some twenty families.† They have been found in leaves, flowers, wood, bark, roots and fruit. There is some overlapping with the distribution of anthraquinones but this is only of significance in the Bignoniaceae and Verbenaceae where related naphthaquinones and anthraquinones co-exist in the same species. On the information at present available, naphthaquinones seem to be of little taxonomic value but juglone (2) is characteristic of the Juglandaceae, plumbagin (2-methyljuglone) occurs regularly in Plumbaginaceae (tribe Plumbagineae[6]) and alkannin (shikonin) (3) in Boraginaceae. On the other hand, the same quinone may be found in botanically diverse families as shown, for example, by the occurrence of lawsone (4) in both Lythraceae and Balsaminaceae, and of mansonones [e.g. (5)] in Sterculiaceae and Ulmaceae. Both

† Phylloquinone (vitamin K_1) probably occurs generally in all families.

2- and 7-methyljuglone and related pigments are found in the Ebenaceae (*Diospyros* spp.) and Droseraceae, the fourteen *Diospyros* quinones being the largest group in any one botanical family. However, as the

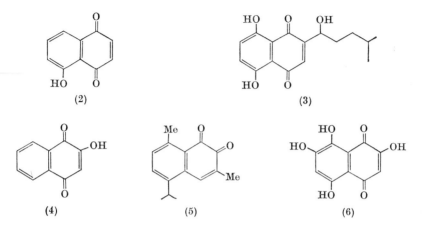

distribution of naphthaquinones has received scant attention such numbers are probably a reflection of the amount of study[7] given rather than the natural abundance. Naphthaquinones seem to occur frequently as individuals rather than in mixtures (cf. the anthraquinones) but the pattern may change on further investigation. They do not occur as glycosides but may exist *in vivo* in a reduced form which may be glycosidic.

The fungal sources of naphthaquinones are also scattered and, in contrast to the anthraquinones, there are none in the Penicillia and only one representative in the Aspergilli. *Fusarium* spp. elaborate a small group of naphthaquinones but there is only one example (rhodocladonic acid) in lichens. The menaquinones (vitamins K_2) occur widely in bacteria (except menaquinone-1 which has been found in a higher plant) while Streptomycetes produce another dozen quinones of varied structure.

In the animal kingdom, some twenty closely related naphthaquinones have been found in echinoderms,[8] chiefly sea urchins with occasional examples in brittle stars, sea stars and starfish. One of the sea urchin pigments, mompain (6) is also a fungal product. Two complex stereoisomeric naphthaquinones occur in many aphid species and give rise, *postmortem*, to a series of more highly condensed pigments.[9]

About ninety benzoquinones have been obtained from natural sources, of which approximately one-third occur in flowering plants and a similar number in fungi, in both cases distributed among twenty or so families.† A small group, related to thymoquinone, appears in the Gymnospermae. Benzoquinones have been found in all parts of higher plants and usually as single entities, but a few alkylated quinones and biquinones, of which rapanone (7) is representative, frequently occur

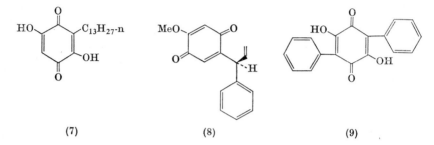

(7)   (8)   (9)

together and are characteristic of the Myrsinaceae. However rapanone is also found in the Connaraceae and Geraniaceae, and polygonaquinone appears in the Liliaceae. The dalbergiones (e.g. (8)) comprise the only other benzoquinone group in flowering plants, and they are restricted to *Dalbergia* and *Machaerium* spp. (Leguminosae). The heartwood of *Dalbergia melanoxylon* contains in addition 2,5-dimethoxybenzoquinone, as does the fungus *Polyporus fumosus* and 2,6-dimethoxybenzoquinone, another fungal product, is also present in several heartwoods (oak, elm, beech). Several higher fungi elaborate benzoquinones including a small group of terphenyl derivatives, for example polyporic acid (9), two of which have been found in lichens. Among the lower fungi *Penicillium* and *Aspergillus* spp. are again responsible for several quinones, and *Helicobasidium mompa* is unique in producing both a terpenoid benzoquinone, helicobasidin (10), and the polyhydroxy-naphthaquinone, mompain (6), mentioned above.

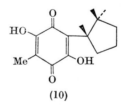

(10)

In animals, benzoquinone itself and several simple derivatives are synthesised by numerous beetles, millipedes and other arthropods, and one isolated benzoquinone has been obtained from an echinoid.

The anthracyclinones, numbering about thirty, form the only other sizeable group of natural quinones, all of which are produced by *Streptomycetes*. These bacteria also elaborate all of the nitrogeneous heterocyclic

---

† As already mentioned, the ubiquinones and plastoquinones are widely distributed.

quinones discussed in Chapter 8, the remaining "miscellaneous quinones" being of terpenoid origin and found in higher plants. The relatively large molecules, known as extended quinones, are mainly of fungal origin; they are surprisingly rare in higher plants (*Hypericaceae* and *Polygonaceae* only) and common in aphids. Only about fifteen have been recognised but they are of great chemical interest. It is curious that arthropods elaborate both the simplest and some of the most complex quinones.

BIOGENESIS

Like other secondary metabolites, the quinones are derived from a few key intermediates, principally acetate, shikimate and mevalonate, by a series of reactions which lead to the formation of benzenoid compounds. It is assumed that the last stages involve the oxidation of a phenol, as in many laboratory syntheses. As the biosynthetic routes to aromatic compounds via acetate[10] and shikimate[11] are now understood in considerable detail, and as the formation of terpenes from mevalonate[10] is well established, these topics will not be discussed here in detail. Probably most quinones arise by the acetate–malonate pathway, particularly those elaborated by fungi, but the extent to which shikimate is involved is still a matter for conjecture. It does, however, appear to be an important intermediate in the formation of many quinones in higher plants.[117] Wholly terpenoid quinones are relatively few but some quinones of mixed origin possess a side chain or ring derived from mevalonate. However, it must be emphasised that the available experimental data are very limited, and a molehill of fact is supporting a mountain of speculation.

Anthraquinones

It is appropriate to consider the anthraquinones first as they form a large and compact group, nearly all of which are polyhydroxy (methoxy) derivatives with little variation in skeletal structure. Nevertheless they arise by at least two biosynthetic routes. This was implicit in an early analysis of the natural anthraquinones by Birch and Donovan[12] which revealed that while many, like emodin, had structures in accord with the "acetate hypothesis" a group related to alizarin seemed to be formed in some other way. Subsequently, labelling experiments with [14C]-acetate carried out by Birch and his colleagues, and by Gatenbeck, established the acetate derivation of helminthosporin,[13] emodin,[14] islandicin[15] and cynodontin.[16] Additional confirmation was obtained using [14C, 18O]-acetate as precursor.[15] Later investigations[17] showed that aromatic polyketides were actually built up from a starter unit (usually acetate) and a chain of malonate units (formed by carboxylation of acetyl co-enzyme A), and this was confirmed in the case of islandicin[18] and the "bianthraquinone", rugulosin.[19] All these results were obtained using moulds in laboratory culture, and since all the fungal anthraquinones are structurally consistent with their formation by the acetate–malonate pathway it seems reasonable to conclude that this is so. It is of interest too that the fungal component of the lichen *Xanthoria parietina* is known to be capable of elaborating anthraquinones in pure culture.[20] As such typical fungal anthraquinones as emodin and chrysophanol are also found in higher plants it was tempting to assume that they are formed in the same way, and the co-occurrence of anthrones† and bianthrones, on occasion, gave support to this view. Recent work by Leistner and Zenk[124] has now shown that this is the case, the incorporation of labelled acetate into chrysophanol by both *Rumex alpinus* (Poly-gonaceae) and *Rhamnus frangula* (Rhamnaceae) being in complete agreement with polyketide biogenesis.

The majority of the anthraquinones which are assumed to be elaborated by the acetate–malonate pathway conform to the emodin (1) pattern and this is considered to arise by suitable folding and condensation of a polyketide chain derived from eight acetate units [as (11)]. Numerous variations of this basic structure exist resulting from O-methylation, side-chain oxidation, chlorination, dimerisation and the

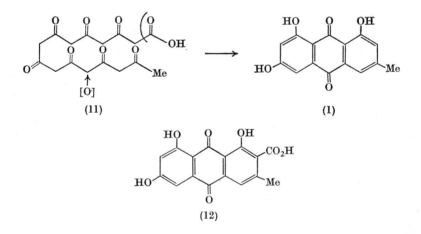

(11)   (1)

(12)

introduction or omission of nuclear hydroxyl groups, while in endo-crocin (12) the terminal carboxyl group is retained. In a few cases

† The isolation of *both* anthrones of physcion (emodin-6-methyl ether) from cultures of several *Aspergilli*[128] and from the root bark of *Ventilago maderaspatana*[129] (Rhamnaceae) has not been explained.

(nalgiovensin, ptilometric acid) the usual β-methyl group is replaced by a propyl substituent (from a C_{18} polyketide, presumably) while madagascin (13) provides the only example of prenylation in the anthraquinone series. It occurs[22] in the bark of *Harungana madagascariensis* together with the 9-anthrone and polyprenylated analogues, [e.g. (14)]. The pigment (15) (from *Curvularia* spp.) is exceptional in having two β-methyl substituents, the origin of the second being a

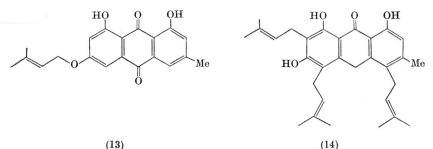

(13)   (14)

(15)

matter for conjecture.[25] Other variations in the carbon skeleton can be attributed to folding of the polyketide chain in different ways prior to cyclisation. Thus condensation of (16) could lead to the formation of 1,3,6,8-tetrahydroxyanthraquinones (17) such as the lichen pigment,

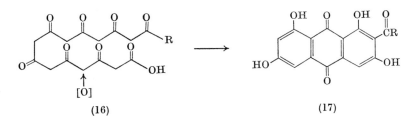

(16)   (17)

solorinic acid (18) and the remarkable group of quinones having C_4 and C_6 side chains elaborated by the mould *Aspergillus versicolor*, for example versicolorin A (19). Another possibility, indicated by (20) → (21), could account for the rhodocomatulin pigments found in crinoids [e.g. (22)] which possess an α-side chain. These are products of animal metabolism and although this derivation is still speculative there are now indications[23] that the phenolic compounds found in marine

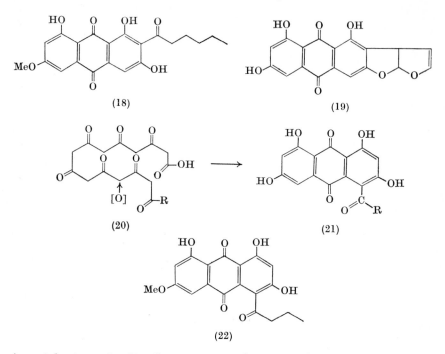

(18)   (19)

(20)   (21)

(22)

invertebrates arise by the acetate–malonate pathway (although biosynthesis by microbial flora cannot be excluded). The presence of an α-side-chain is a feature of anthraquinones of animal origin (exceptionally emodin and 7-hydroxyemodin occur in *Eriococcus confusus*) and may

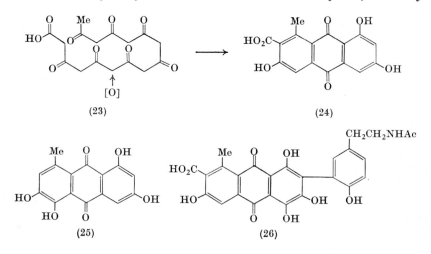

(23)   (24)

(25)   (26)

arise from yet another mode of cyclisation (23) → (24). An example is erythrolaccin (25) but in most cases the β-carboxyl group is retained as in laccaic acid D (24). Laccaic acid A (26) also contains the very unusual *p*-hydroxyphenylethylamine residue.[24] Presumably this originates from tyrosine and is linked to the main chromophore at some stage by phenolic coupling.

In contrast to the anthraquinones of the emodin type about half of those found in higher plants are substituted in only one benzenoid ring and may be totally devoid of a carbon side chain [e.g. (27)] or hydroxyl groups [e.g. (28)]. The majority of these occur in the Rubiaceae sub-family Rubioideae) and, to a lesser extent, in the Bignoniaceae and Verbenaceae,

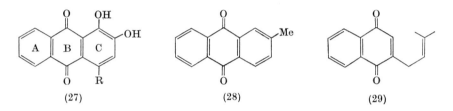

(27)   (28)   (29)

tectoquinone (28) being present in all three. Significantly,[30] the anthraquinones present in Bignoniaceae[26, 27] and Verbenaceae[28] (teak) heartwoods are accompanied by C_{15} naphthaquinones, notably deoxylapachol (29), while the Rubiaceous plants[29] contain a number of C_{15} naphthalenic compounds represented by (30), (31) and (32). 4-Methoxy-1-naphthol has also been found in Rubiaceae.[29] These findings suggested

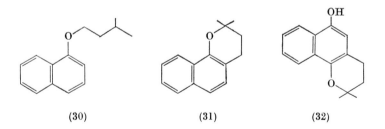

(30)   (31)   (32)

that (29) is synthesised *in vivo* by prenylation of a naphthol precursor followed by oxidation, and since (29) can also be converted into (28) *in vitro*,[26] either by boron trifluoride catalysis or by irradiation, it seems likely that the substituted (C) ring in this group of anthraquinones is derived from mevalonate. This was established[3, 31, 32] by feeding *Rubia tinctorum* (madder) plants with [2-^{14}C]-mevalonate. Four radioactive pigments were isolated,[3] the specific activity of rubiadin (35) and pseudopurpurin (36) being twice that of alizarin (27; R = H) and purpurin

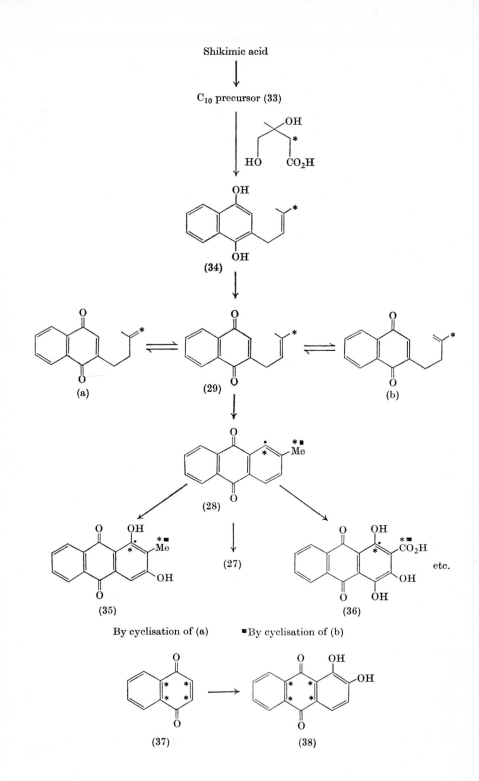

Shikimic acid

C_{10} precursor (33)

(34)

(a)   (29)   (b)

(28)

(35)   (27)   (36)   etc.

By cyclisation of (a)   ■By cyclisation of (b)

(37)   (38)

(27; R = OH). Appropriate degradations of pseudopurpurin established that the carbon-14 was distributed between the side chain and C_1 in ring (C).† It seems therefore that ring (C) in the Rubiaceous anthraquinones is formed as shown below, and presumably the same pathway is followed in the Bignoniaceae and Verbenaceae. The intermediates (28), (29) and (34) have not been detected in *R. tinctorum* but the pyran (32) [equivalent to (29) and (34)] has been isolated,[3] and a low incorporation of deoxylapachol (29) into alizarin has been effected.[35] The "C_{10} precursor" (33) has still to be identified; it could be α-naphthol but a search for this compound (or a glycoside thereof) was unsuccessful.[33] It is significant that labelled 1,4-naphthaquinone was efficiently incorporated[34] into alizarin when fed to the same plant, (37) → (38), presumably by reduction and prenylation of the quinol. The "C_{10} precursor" is known to originate from shikimic acid as Leistner and Zenk[36] have shown that uniformly labelled [^{14}C]-shikimic acid (but not [U-^{14}C]-phenylalanine) is incorporated *in toto* into alizarin and purpurin by *R. tinctorum*, and thereby provides the whole of ring A and one of the quinone carbonyl groups. This leaves three carbon atoms in the centre of these molecules whose origin is still unknown.

It should not be thought that all the anthraquinones in the Rubiaceae are substituted only in one ring. Thus juzunal (39; R = OH) accompanies damnacanthal (39; R = H) in *Damnacanthus major*, lucidin (40; $R_1 = R_2 = H$) occurs with colucidin (40; $R_1 = H$, $R_2 = OH$)[149] in

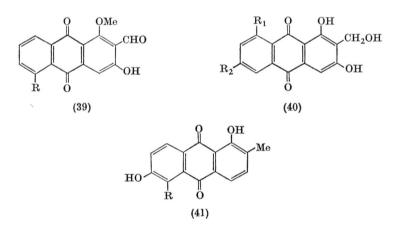

(39)   (40)

(41)

Coprosma lucida and with coelulatin (40; $R_1 = OH$, $R_2 = H$) in *Coelospermum* spp., and *Morinda* spp. elaborate morindone (41; R = OH) and soranjidiol (41; R = H) together with numerous quinones of the alizarin

† Feeding experiment of fifteen days; if the operation is stopped after a few hours no equilibration of label is observed and the radioactivity is confined to the carboxyl group.[32]

type substituted in one ring only. These may all be derived from shiki-mate and mevalonate but there is no experimental evidence, and the possibility that these plants may utilise more than one biosynthetic pathway must be kept in mind. Mere inspection of formulae is of little value as may be illustrated by reference to recent experience with *Digitalis* (Scrophulariaceae), the only other higher plant genus known to form anthraquinones substituted in one ring only. The presence of digitolutein (42; R = Me) in *Digitalis* spp. has long been known, and 3-methylalizarin (42; R = H) and other very simple anthraquinones have also been found,[37] in rather low concentration, all of which appear

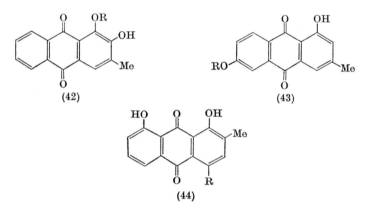

(42)   (43)

(44)

to belong to the shikimate–mevalonate group. However, in a recent investigation[38] of *D. purpurea* (42; R = H and Me) were absent or barely detectable among a large number of trace components, of which (43; R = H and Me) and (44; R = H and HO) could be identified (see also ref. 110). The origin of the *Digitalis* quinones is, therefore, quite un-certain and can only be established by direct experiment. Another pigment of uncertain origin is pachybasin (43; H in place of OR), which very probably is synthesised *in vivo* by two routes since it occurs in both fungi and in teak wood (Verbenaceae).

The most mysterious quinone in this series is anthraquinone itself which has been isolated from *Quebrachia lorentzii* (Anacardiaceae) and *Acacia* spp. (Leguminosae), and from tobacco leaves, and another, possibly related, enigma is the occurrence of 10,10′-bianthrone in cultures of *Ustilago zeae*.[40]

Until recently anthraquinones were regarded as metabolic end products and they are accumulated by certain organisms in relatively large amounts. However in some fungi they are metabolised further, notably in ergot (*Claviceps purpurea*).[132] The ergot pigments include anthraquinones and ergochromes [e.g. (45)], and it has been conclusively established that the latter are derived from the former by feeding experiments with labelled emodin (1a; R = H).[133] The essential steps

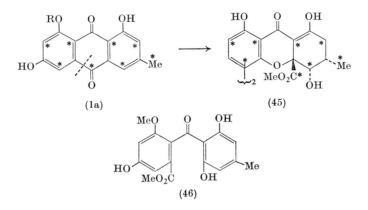

(1a)   (45)

(46)

in this transformation are oxidative cleavage of the central ring to form an *o*-benzoylbenzoic acid, followed by cyclisation to form a xanthone derivative, and dimerisation; thus the ergochromes may be regarded as *seco*-anthraquinones. Similarly, it has been shown[151] that sulochrin (46) is derived from its co-metabolite questin (1a; R = Me) in cultures of *Penicillium frequentans*, and several related examples are under investigation.

Naphthaquinones

Turning now to the naphthaquinones, until very recently this area of biogenesis had been neglected although the structures of the mould products, flaviolin (48; R = H), javanicin (52) and fusarubin (53) had been correctly predicted[41] on the basis of acetate biogenesis. This assumed that flaviolin, for example, was derived from the polyketide (47) and obviously further oxidation could lead to another fungal metabolite, mompain (48; R = OH). An attempt[42] to establish this gave disappointing results. [2-^{14}C]-Acetate was successfully incorporated (0·53%) and after C-methylation with acetyl peroxide the derived 2,7-dimethylmompain was subjected to Kuhn-Roth degradation. However the acetic acid obtained showed only half the activity expected on the basis of normal acetate biogenesis and indicated a high degree of randomisation of the label. This may be connected with the rather lengthy period (43 days) of incubation;[42] the incorporation of malonate was only 0·12%. Mompain itself and numerous other polyhydroxynaphthaquinones (spinochromes) occur in echinoderms and rather similar results were obtained by Lederer and his co-workers[23a] in their study of the biogenesis of the spinochrome (49) in *Arbacia pustulosa.* Incorporation of radioactivity from [2-^{14}C]-acetate was low, again there was considerable randomisation and, unexpectedly, the

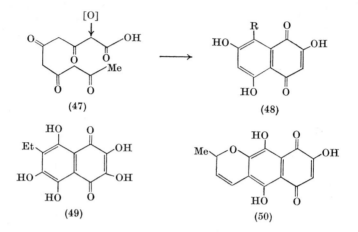

(47)   (48)

(49)   (50)

activity in the ethyl side chain was only 7·1% of the total. If the pigment was formed by cyclisation of a single polyketide derived from six acetate residues the side chain should contain one-sixth (16·6%) of the total activity. Lederer[23a] therefore suggests that biogenesis may proceed in two stages, first cyclisation of (47) and then introduction of the side chain. This would accord with the observation that spino-chromes both with and without a two-carbon side chain (ethyl or acetyl) frequently occur together in the same animal. The recent isolation[43] of the pyranospinochrome (50) is also indicative of acetate biogenesis. The French work does establish that spinochrome (49) is synthesised *de novo* in the animal but does not exclude the participation of microbial flora.

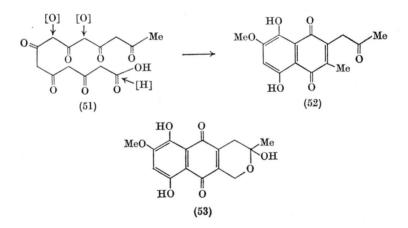

(51)   (52)

(53)

More definite results were obtained by Gatenbeck and Bentley[44] from their study of the biosynthesis of javanicin (52) by *Fusarium javanicum*. Feeding experiments with acetate and malonate established its origin from a hepta-acetyl polyketide (51), and revealed a rare example of the reduction of a terminal carboxyl group to methyl which provided 11% of the total activity from [1-^{14}C]-acetate. Partial reduction of the terminal carboxyl group is observed in fusarubin (53), a co-metabolite. The *O*-methyl group of javanicin was derived from methionine. The same workers[45] have also investigated the origin of a more unusual quinone, mollisin (55), found in cultures of *Mollisia caesia*.

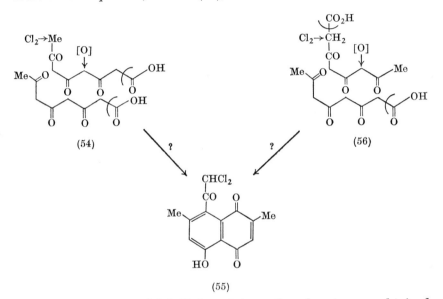

(54)   (56)

(55)

Good incorporation of labelled acetate and malonate was obtained, one-third of the total activity being definitely located at C-2 and C-7. This is consistent with "polyacetate" theory but if the pigment originated from a single polyketide chain then one of the two ring-methyl groups would have to be introduced from the C_1-pool. The very low incorporation of labelled methionine makes this unlikely. Similarly, the failure to incorporate mevalonate excludes the possibility that one ring and its methyl side chain might originate from a C_5-unit (cf. rubiadin, p. 10 and chimaphilin, p.18). Bentley and Gatenbeck[45] therefore suggest that two polyketide chains are involved. Two possibilities are shown in (54) and (56); the latter, which requires chlorination of a β-ketoacid and then decarboxylation having preference, since it has been shown that a chloroperoxidase system catalyses the chlorination of β-diketones and β-ketoacids.[46] Labelled chloroacetate was not utilised by *M. caesia* for the biosynthesis of mollisin.

Probably all the other fungal naphthaquinones are formed by the acetate–malonate pathway but there can be no certainty about 6-methyl-1,4-naphthaquinone (57), and the branched side chain in lambertellin (58) is anomalous. The structures of the bacterial pigments frenolicin

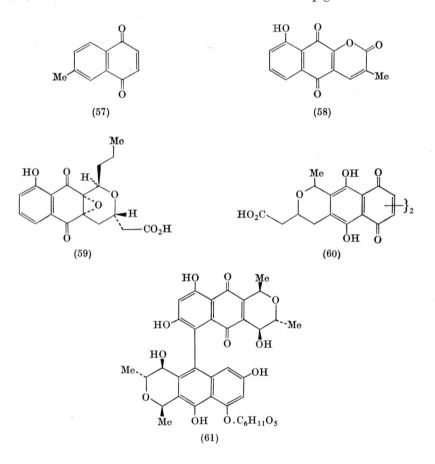

(57)   (58)

(59)   (60)

(61)

(59) and actinorhodin (60) are also consistent with this theory, and protoaphin (61) and the related pigments found in aphids appear to belong to the same category.

The most widely distributed bacterial naphthaquinones are the menaquinones (vitamins K_2) (62). Their biogenesis is completely different and much of the work in this field has been plagued by very low incorporations and contamination with radioactive impurities. Incorporation of shikimic acid into ring A was first reported by Cox and Gibson,[47] and fully confirmed by Bentley[48] and Zenk[49] and their co-workers. *B. megaterium*[49] incorporated 1·14% [U-14C]-D-shikimic acid into MK-7,† and *E. coli*[48] 0·5% into MK-8,† the acid being incorporated *in toto* with the carboxyl group providing one (or part of both) quinone carbonyl groups. (Using *M. phlei*[48] as the experimental organism only 0·02% incorporation into MK-9(H$_2$) was found and about 30% of the activity appeared in the polyprenyl side chain. This suggests some degree of interaction between acetate and shikimate metabolism, and appreciable incorporation of acetate into ring A of MK-9(H$_2$) has also been observed.[48] The origin of three nuclear carbon atoms of ring B has

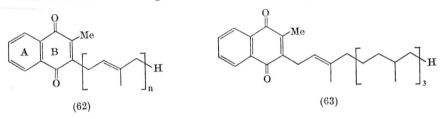

(62)   (63)

still to be determined but there is now evidence that these originate from glutamate as in the biosynthesis of lawsone (see p. 19). It is possible that the side chains are introduced by alkylation of a naphthol precursor, but although Zenk and his colleagues[49] have recently reported the incorporation of [1-^{14}C]-α-naphthol into MK-7 by *B. megaterium*, this could not be confirmed by Threlfall and his co-workers.[99] They suggest that Zenk's menaquinone was contaminated with [1-^{14}C]-α-naphthol (or its breakdown products) as their separation is difficult. The methyl side chain originates from methionine[50, 51, 118] but experimental evidence that the polyprenyl side chain comes from mevalonate is still lacking.[52] The failure of various bacteria to incorporate mevalonate into menaquinone is apparently not attributable to permeability difficulties. It is generally accepted that the isoprenoid side chain in "bioquinones" is introduced by way of a polyprenyl pyrophosphate and it is known that certain bacteria (as well as higher plants) are able to incorporate mevalonic acid into polyprenyl alcohols.[53] In higher plants, which contain phylloquinone (63), incorporation of mevalonate has been reported in maize shoots,[54, 137] and the incorporation of methionine[55] and shikimic acid[55] has also been demonstrated. The C$_{20}$ side chain of phylloquinone could be attached either by way of phytylpyrophosphate or by alkylation with geranylgeranyl pyrophosphate followed by partial saturation. Another uncertainty concerns the order in which the two side chains are introduced, and quite possibly both alternatives are used by different organisms. Martius and Leuzinger[57] have shown that cultures of the heterotrophic anaerobe *Fusiformis nigrescens* can convert 2-methyl-1,4-naphthaquinone into MK-9, and indeed 1,4-naphthaquinone into its 2-methyl

† For nomenclature see p. 340.

homologue and then into MK-9. The same organism can also detach the C_{20} side chain from phylloquinone to form 2-methyl-1,4-naphthaquinone which is then converted into MK-9. A corresponding exchange may be effected in the animal body by the intestinal flora which transform exogenous phylloquinone into MK-4, which Martius[58] considers to be the specific menaquinone exhibiting vitamin activity *in vivo*. However, MK-4 has been found only after feeding 2-methyl-1,4-naphthaquinone. These experiments indicate successive methylation and prenylation, and similarly, Lederer and his co-workers[59] observed that 2-methyl-1,4-naphthaquinone could be incorporated into MK-9(H_2) by *M. phlei*. On the other hand it has been found[51] that cell-free extracts of *M. phlei* will methylate DMK-9 to form MK-9 by transfer of a methyl group from methionine, that is, prenylation is followed by methylation. A detailed mechanism has been suggested.[148] It is of interest that demethyl-menaquinol-monomethyl ether has been isolated from cultures of photo-synthetic bacteria.[154]

The origin of several simple naphthaquinones in higher plants has been investigated recently by Zenk.[117] The co-existence of chimaphilin (69) with glucosides of (64) (homo-arbutin), (65) (renifolin) and (66)

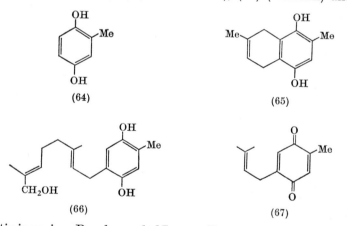

(64)   (65)

(66)   (67)

pyrolatin in various Pyrolaceae led Inouye[62] to suggest that the naphtha-quinone was formed *in vivo* by prenylation of (64) and cyclisation. This idea was strongly reinforced by the isolation[61] of the benzoquinone (67) along with chimaphilin in *Pyrola media*, and by a simple conversion of (67), as its quinol into (69) *in vitro*.[61] Experimental proof has been provided[60] by feeding *Chimaphila umbellata* with labelled tyrosine (68) and mevalonic acid. Appropriate degradations of the radioactive quinone to methylphthalic acid and acetic acid revealed that the 7-methyl group comes from the C-2 carbon of mevalonate and the 2-methyl group derives from the β-carbon atom of tyrosine, the tyrosine ring providing the entire quinone ring of chimaphilin (69) (see below). Evidence that the incorporation of tyrosine involves a migration of the

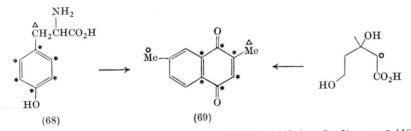

(68)   (69)

side chain to the *ortho*-position was obtained[142] by feeding D-2-^{14}C-shikimic acid when it was found that 48% of the total activity was located at C-2 of the derived chimaphilin. This is reminiscent of homogentisic acid biosynthesis and indeed further experiments showed that both *p*-hydroxyphenylpyruvate and homogentisic acid were specifically incorporated into chimaphilin. Thus the route to chimaphilin proposed by Bolkart and Zenk[142] is shikimic acid → *p*-hydroxyphenylpyruvic acid → homogentisic acid (glucoside) → (64) (glucoside) → (67) (quinol glucoside) → chimaphilin quinol (glucoside) → chimaphilin. After *p*-hydroxyphenylpyruvic acid, all the intermediate compounds in this sequence (except 67) have been isolated from *Chimaphila*.[142] The formation of homogentisic acid may be important in other biosyntheses (see plastoquinone, p. 29) and possibly occurs, as indicated below ($R = CH_2COCO_2H$ or CH_2CO_2H), for which there are *in vitro* analogies.

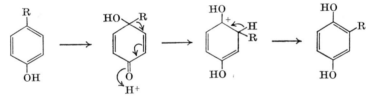

Shikimic acid is a good precursor of lawsone (4) when fed to *Impatiens balsamina*[63, 64] and is incorporated as a C_7 unit to form the benzenoid ring† and contribute 50% of the two carbonyl groups.[64] Recent efforts[143] to ascertain the origin of the three remaining carbon atoms have shown

$HO_2C\overset{*}{C}H(NH_2)CH_2CH_2CO_2H \longrightarrow HO_2\overset{*}{C}COCH_2CH_2CO_2H$

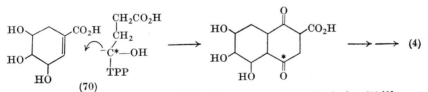

(70)

† The conflicting results of Chen and Bohm[63] are now considered to be invalid.[123]

that both alanine and aspartate contribute to carbon atoms 1 to 4, but neither appears to be a direct source of the C_3 unit. Campbell[143] has suggested that the latter may originate from a succinate semialdehyde-thiamine pyrophosphate complex derived from acetate and oxalo-acetate (and hence from alanine and aspartate) via the Krebs cycle and found, in support, that $2\text{-}^{14}C$-glutamate was well incorporated by *I. balsamina* into lawsone, the radio-carbon being specifically located at C-1 or C-4. The proposed mechanism [see (70)], involving reactions formally analogous to a Michael addition and Claisen condensation, are consistent with the observation (see below) that in juglone biosynthesis, C-1 and C-6 of shikimic acid become C-9 and C-10 of the quinone.

Several prenylated derivatives of lawsone are known, for example, lapachol (71), and in Bignoniaceae and Verbenaceae they occur with

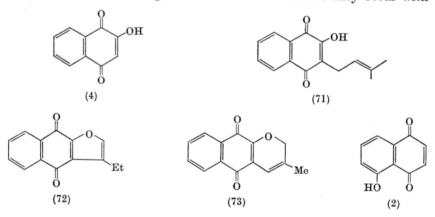

deoxylapachol and related anthraquinones which suggests that prenyla-tion is effected before oxidation at C-2, but this is uncertain. The naphthaquinones in Bignoniaceae with branched C_4 side chains, for example (72) and (73) are anomalous, (73) having the same carbon skeleton as lambertellin (58).

Shikimate is also a good precursor of juglone (2) (in *Juglans regia*)[65] and is again incorporated *in toto* to form the benzenoid ring and 50% of each of the quinone carbonyl groups. This was established by degrada-tion of the radioactive juglone to 3-hydroxyphthalic acid with alkaline hydrogen peroxide, followed by decarboxylation in refluxing 50% sulphuric acid to give *m*-hydroxybenzoic acid, and carbon dioxide which contained 6·3% (theory 7·2%) of the activity of the juglone.[65] This suggests that biosynthesis proceeds by way of a symmetrical inter-mediate, most likely 1,4-naphtha-quinone or -quinol, and in fact $[2,3,9,10\text{-}^{14}C]$-1,4-naphthaquinone was found to be incorporated effici-ently into juglone (and also into lawsone by *I. balsamina*).[65] Zenk[65] regards 1,4-naphthaquinone (or quinol) as an important intermediate in the biosynthesis of naphthaquinones in higher plants. The occurrence of 4-methoxy-1-naphthol in *Galium mollugo* and *Asperula odorata* (Rubiaceae) has already been mentioned, and it may be distributed more widely. The origin of the remaining three carbon atoms in the quinone ring of juglone was explored using a number of potential precursors. Some incorporation of acetate, malonate, fumarate and succinate was observed which suggests a close analogy with the biosynthesis of lawsone (see p. 19). It was also shown that the final cyclisation to form the naphthalene nucleus takes place at the carbon atom derived from C-6 in shikimic acid. This was determined by feeding experiments using

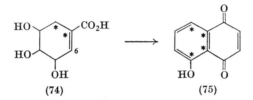

(74)   (75)

[1,2-^{14}C]-shikimic acid (74). The resulting radioactive juglone (75) was then degraded, stepwise, to *p*-nitrophenol, and finally to radioactive bromopicrin containing the C-8 carbon atom of the juglone and 25·4% of the activity of the *p*-nitrophenol.[65] 7-Methyljuglone and 2-methyl-juglone (plumbagin), and numerous related compounds occur in higher plants but whether they are of shikimate-mevalonate origin, or arise by the acetate pathway, is not known.

Benzoquinones

In the benzoquinone series nearly all investigations have been carried out with moulds, for obvious reasons, and very little study of their biogenesis in higher plants has been made. Cultures of *Aspergillus fumigatus*, *Penicillium patulum*, *P. spinulosum*, *Gliocladium roseum* and other organisms produce a number of closely related phenols, including quinols which may become oxidised to the corresponding quinone as the culture ages. On present evidence, a majority of the fungal benzoquinones appear to be formed by the acetate-malonate pathway, O-methyl groups being derived from methionine. The ubiquinones are important exceptions and others will probably appear as some moulds are known to elaborate phenols from shikimic acid. A single organism may utilize both the shikimate and the acetate–malonate routes for the biosynthesis of quinones. Inspection of formulae is a poor guide to the mode of biogenesis. Gentisic acid is elaborated from acetate by *P. urticae*[80] and so, presumably, are its co-metabolites gentisaldehyde, gentisyl alcohol, gentisylquinone and toluquinol. However, in higher plants gentisic acid is derived from shikimic acid by way of cinnamic, benzoic and salicylic acids.[81] Labelling studies have firmly established the acetate–malonate origin of fumigatin (76; R = H),[67, 68] spinulosin (76; R = OH),[69] aurantiogliocladin (77)[70] and its reduced form gliorosein (78),[71] terreic acid (79)[72] and the dimers phoenicin (80; R = H)[73] and oosporein (80; R = OH).[74] Orsellinic acid (83) or 6-methylsalicylic acid are frequent co-metabolites of the quinones (76)–(80); they have been incorporated, after labelling, into several of the pigments, and in

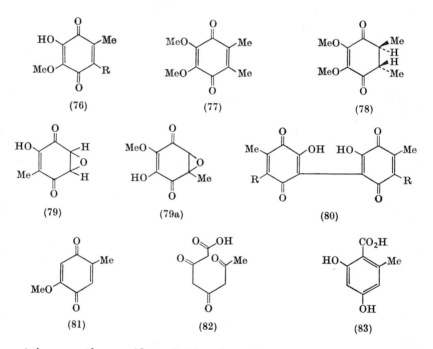

(76)   (77)   (78)

(79)   (79a)   (80)

(81)   (82)   (83)

certain cases the specific activities show that the latter are formed by the same pathway as orsellinic acid, that is (82) → (83). The route from orsellinic acid appears to proceed by decarboxylation to orcinol, and then dimerisation and subsequent oxidation to form the dimeric quinones (80),[73, 74] or sequential hydroxylation via (84)[77] and O-methylation to

(84)   (85)

produce fumigatin (76; R = H) and spinulosin (76; R = OH). In terreic acid (79) biogenesis,[72] epoxidation is probably the final step although, conversely, incorporation of the epoxide (79a) into fumigatin and fumigatol (the quinol) has been reported.[78] Aurantiogliocladin (77) and gliorosein (78) possess an additional C-methyl group which is introduced (from methionine[125]) at the polyketide level.[71] 5-Methylorcylaldehyde (85) is a good precursor but not orsellinic acid nor 5-methylorsellinic acid. It is suggested[71] that (85) is actually formed *in vivo* as an enzyme-bound intermediate but exogenous acid is lost by rapid decarboxylation. Conversion of (85) into the final products involves further oxidation and decarboxylation to form aurantiogliocladin quinol which is tauto-merised under enzymic control to give optically active gliorosein (78). *G. roseum* will also convert aurantiogliocladin into gliorosein. Specific activities (after feeding [2-^{14}C]-acetate) always increased in the order gliorosein, quinol, quinhydrone and quinone, indicating that gliorosein is the actual metabolite released into the medium from which the other compounds are derived.[71] Incorporation of [U-^{14}C]-tyrosine into glio-rosein was also observed[71] but as radioactivity was found in the un-saponifiable lipids (e.g. ergosterol) as well it suggested that the tyrosine was being metabolised to homogentisic acid[79] and then degraded[111] to fumarate and aceto-acetate (and thus acetate). However (U-^{14}C)-tyrosine is incorporated effectively into coprinin (81) quinol with virtually no incorporation of acetate,[75, 76] and degradation studies[75] have established that the ring carbon atoms and the C-methyl group are derived from the ring and β-carbon atom of tyrosine, respectively, whereas the O-methyl group comes from methionine.

The other fungal benzoquinones, apart from ubiquinones, which have been studied experimentally originate from shikimic acid. Volucrisporin (88), found in cultures of *Volucrispora aurantiaca*,[82] differs from the other terphenylquinones, otherwise found in higher fungi and lichens, in the absence of hydroxyl groups attached to the quinone ring and the unusual hydroxylation pattern in the phenyl side chains. Other members of this small group of pigments possess either unsubstituted phenyl substituents (9) or they are hydroxylated at the 4', or 3',4'-positions, reminiscent of the B ring of flavanoids and other shikimate-derived compounds. Feeding experiments[83, 119] with labelled precursors showed that shikimic acid, phenylalanine, phenyllactic acid, phenylserine and *m*-tyrosine were incorporated but acetate and cinnamate were not. Assuming a reversible interconversion of phenyllactic acid, phenyl-alanine and phenyl-pyruvic acid it is suggested[119] that biogenesis most probably proceeds via phenylpyruvic acid which is *m*-hydroxylated in some way before condensation to form volucrisporin. The sequence shown below accounts for the incorporation of the labelled carboxyl carbon atom of phenylalanine and *m*-tyrosine, and the α-carbon atom of phenyllactic acid into the quinone ring of volucrisporin, and an attractive feature is the intermediate (87) which is at the oxidation

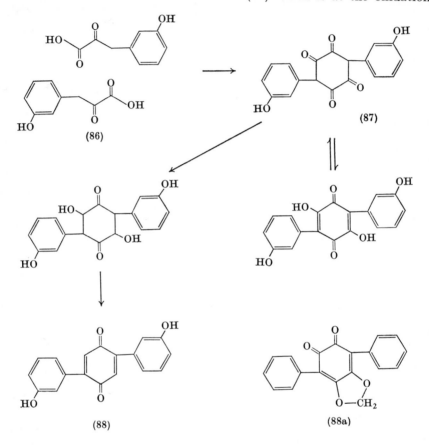

level found in the other terphenylquinones. It seems likely that these are formed by a similar condensation of appropriate C_6–C_3 precursors, and it has been found recently that phlebiarubrone (88a) is derived from two moles of phenylalanine, with formate providing the methylene group. This was demonstrated[139] using ^{13}C-labelled precursors, the site

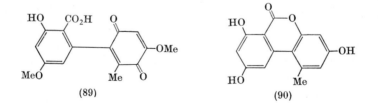

and extent of incorporation being measured by n.m.r. and mass spectrometry. On the other hand the close relationship of the biphenylquinone, botrallin (89)[101] to alternariol(90)[102] strongly suggests that this is of acetate–malonate origin.

The terphenylquinones provide further examples of quinone metabolism. Polyporic acid (9a)† occurs with pigments of the pulvic acid group (e.g. 90a) in certain lichens, and labelling experiments have shown that the latter are derived from the former. *Pseudocyphellaria crocata* elaborates pulvic anhydride (90a) and calycin (90b) (quinones have not been detected in this case), and both incorporated L-phenylalanine, L-phenyllactic acid and polyporic acid efficiently.[134] Rate studies[134]

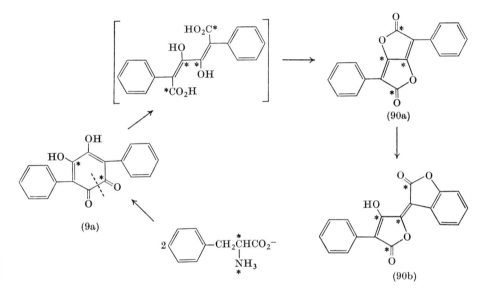

support the reaction sequence phenylalanine → phenylpyruvic acid → polyporic acid → pulvic anhydride → calycin, that is, formation of polyporic acid by self-condensation of phenylpyruvic acid, as in volucrisporin biogenesis, followed by oxidative fission and lactonisation. The conversion of polyporic acid into pulvic anhydride by oxidation with lead tetra-acetate[135] or dimethyl sulphoxide/acetic anhydride[127] provides chemical analogy. (See also the oxidation of atromentin, p. 159.) It is of interest that the fungal component of the lichen *Candellariella vitellina* produces four pulvic acid pigments, including (90a) and (90b), in pure culture.[21]

Concerning the ubiquinones (91)[152] elaborated by micro-organisms, the polyprenyl side chains are derived from mevalonate,[52b, 70b, 84] as

† Tautomeric form of (9).

expected, and the nuclear methyl group from methionine.[85, 118] It is of interest that certain moulds[52] can utilize both the D and L forms of mevalonic acid and Stone and Hemming[86] have demonstrated, using the stereospecifically labelled substrates [2-^{14}C, 4R-^{3}H$_1$]- and [2-^{14}C, 4S-^{3}H$_1$]-mevalonate, that the polyprenyl side chain in Q-10,† elaborated by *Aspergillus fumigatus*, is biogenetically *trans* throughout. From the structural relationship of the ubiquinones and aurantiogliocladin (77)

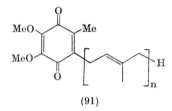

(91)

it would appear that their nuclei are also acetate-derived but in fact acetate can only be incorporated into the polyprenyl chain.‡ In searching for possible aromatic precursors of the ubiquinone nucleus Rudney and Parson[88] made the important discovery that the radioactivity incorporated when *Rhodospirillum rubrum* was fed with [U-^{14}C]-tyrosine was actually derived from an impurity in the commercial tyrosine preparation. The impurity was identified as *p*-hydroxybenzaldehyde which was shown to undergo a highly specific incorporation into the ring of Q-10 during which the carbonyl group was lost. *p*-Hydroxybenzoic acid (92) is equally effective in *R. rubrum*[85] and *Azotobacter vinelandii*,[89] and it has been shown that this metabolite arises from the shikimic acid pathway in *E. coli* and *Aerobacter aerogenes* at a branch point (chorismic acid) preceding the formation of phenylalanine and tyrosine.[90] Incorporation of labelled shikimic acid into Q-8 by *E. coli* has been reported[47a] but similar experiments with *R. rubrum* were unsuccessful.[85] In further work Rudney and Parson[87] detected a lipophilic "Compound X" which rapidly incorporated *p*-hydroxybenzoate and was in turn converted into Q-10. Experiments with carboxyl-labelled *p*-hydroxybenzoic acid, [^{14}C]-acetate and [methyl-^{14}C]-methionine demonstrated that the formation of "Compound X" entailed decarboxylation of the phenolic acid and addition of a prenyl side chain, but there was no evidence of C-methylation. Following a large scale extraction of the lipids of *R. rubrum* "Compound X" was

† For nomenclature see p. 173.

‡ Except in the photosynthetic bacterium *Rhodospirillum rubrum* where only 35% of the activity from [2-^{14}C]-acetate incorporated into Q was in the side chain, the remainder being in the nucleus.[87] Carboxylation of the acetate to form a three-carbon unit which then enters the shikimic acid pathway may account for this anomaly.[84]

isolated and identified as 2-decaprenylphenol (94; n = 10),[91] and chromatographic evidence has since been obtained for the presence of (93)[153] (see also ref. 147). Subsequent exhaustive fractionation of the lipids of *R. rubrum* by Folkers, Rudney and their co-workers[92, 93, 94] led to the discovery of the intermediates (94; n = 4, 9 and 10), (95; R = Me, n = 10), (96; n = 10), (97; n = 8, 9 and 10) and (98; R = H, n = 10), and it has been shown that radioactivity from [U-14C]-*p*-hydroxybenzoic acid is incorporated into all of these by *R. rubrum*. The biogenesis of

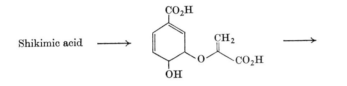

 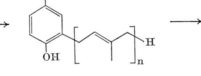

(92)   (93)

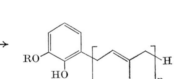

(94)   (95)

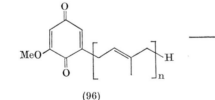

(96)

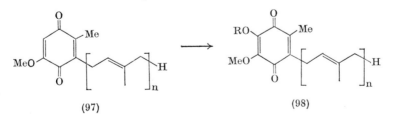

(97)   (98)

ubiquinones in this organism [92, 152] clearly follows the sequence shown on p. 27 (Q-8, Q-9 and Q-10 were found but Q-4 has not been detected). Evidently prenylation with different pyrophosphates takes place in *R. rubrum*†[147] and the whole process may be general for ubiquinone biosynthesis in other forms of life. This looks very probable in animals,[95, 97, 146, 152] but a recent search[150] for prenylated ubiquinone precursors in several higher plants was unsuccessful which may mean that prenylation occurs at a very late stage in plant biosynthesis. The incorporation of *p*-hydroxybenzoate into Q-9 or its precursors in the rat has been demonstrated[89, 96, 145, 146] but here it is derived from phenylalanine or tyrosine for which two biosynthetic routes have been suggested.[95]

Insects also appear to elaborate benzoquinones by two biosynthetic routes. In a study of the beetle *Eleodes longicollis*, which secretes benzoquinone, toluquinone and ethylbenzoquinone, Meinwald and co-workers[98] concluded, after injecting labelled precursors, that while all three incorporated acetate, propionate, malonate, phenylalanine and tyrosine, the relative activities indicated that phenylalanine and tyrosine were the principal precursors of benzoquinone whereas the alkylated quinones were formed by the acetate–malonate pathway. Degradation showed that propionate provided the side chain of ethylbenzoquinone. These results are not clear-cut and do not exclude the possibility that the biosyntheses were effected by symbiotic microorganisms. It is of interest that the beetle *Zophobas rugipes* secretes benzoquinone, toluquinone and ethylbenzoquinone from one pair of defensive glands, and phenol, *m*-cresol and *m*-ethylphenol from another pair.[126]

Very little is known about the biogenesis of benzoquinones in higher plants. Pear leaves (*Pyrus communis*) incorporate phenylalanine into the ring of arbutin (quinol-β-D-glucoside),[114] and *p*-hydroxybenzoic acid (but not gentisic acid) was found to be a precursor in *Bergenia crassifolia*.[120] Thus the biosynthetic pathway proceeds presumably by way of cinnamic acid and *p*-coumaric acid to *p*-hydroxybenzoic[121] acid, followed by a process of oxidative decarboxylation (see also ref. 138). Similar feeding experiments with *Triticum vulgare* (wheat) indicate that the 2-methoxy- and 2,6-dimethoxyquinols, which are present, are derived by oxidative decarboxylation of vanillic and syringic acid, respectively.[122] 2,6-Dimethoxybenzoquinone is fairly widely distributed, especially in bark and heartwoods, where it probably arises from syringyl derivatives as a by-product of lignin formation or it may be a lignin degradation product.[100]

† Also in many other bacteria which appear to elaborate ubiquinones in the same way.[130, 136, 144]

Maize and barley shoots incorporate mevalonic acid[54, 137] into the side chains of ubiquinone, plastoquinone and α-tocopherolquinone, and the O-methyl and nuclear methyl groups of ubiquinone and one nuclear methyl group in plastoquinone and α-tocopherolquinone are derived from methionine.[55] In maize also, shikimic acid,[56] phenylalanine[56] and tyrosine[56] are incorporated into the nucleus of ubiquinone, plasto-quinone and α-tocopherolquinone but p-hydroxybenzoic acid is utilised only for the biosynthesis of ubiquinone. Goodwin and his co-workers[56] also detected an unidentified compound, possibly a polyprenylphenol, which may be a precursor of ubiquinone. The biosynthesis of plasto-quinone and α-tocopherolquinone however, proceeds by a different pathway. Incorporation experiments[115] with labelled tyrosine, p-hydroxyphenylpyruvic acid, and homogentisic acid strongly suggest the following sequence: prephenic acid → p-hydroxyphenylpyruvic acid → homogentisic acid → plastoquinone and α-tocopherolquinone. The details of the final stages are not yet known; it is established that the β-carbon atom of tyrosine provides one of the nuclear methyl groups of plastoquinone and α-tocopherolquinone by migration of the side chain to form homogentisic acid, although toluquinol does not appear to be an intermediate[116] (cf. p. 19). More or less plausible speculations can be made regarding the origin of many of the higher plant benzoquinones. The dalbergiones,[112] [e.g. (8)], clearly contain a C_6–C_3 moiety of, presumably, shikimate origin, while alkylated benzoquinones of the rapanone (7) type seem obviously derived from a polyketide precursor.

Other Quinones

Many of the larger natural quinones appear to be formed by the acetate–malonate pathway. This has been established[103] for the bacterial pigment ε-pyrromycinone (100) by tracer studies which show that it arises from nine acetate and one propionate unit, the latter being used as the starter (99), and similarly, the phenanthrene derivative, pilo-quinone (102), another bacterial colouring matter, originates from a polyketide (101) in which isobutyrate (fed as such or as valine) is the starter.[104]

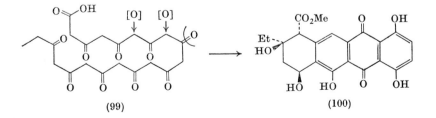

(99)   (100)

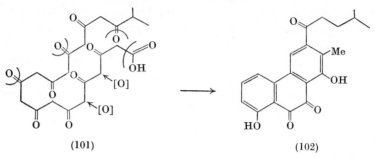

(101)   (102)

Phenolic coupling[109] is a biosynthetic process frequently encountered in organisms which elaborate phenols and all the main quinone groups

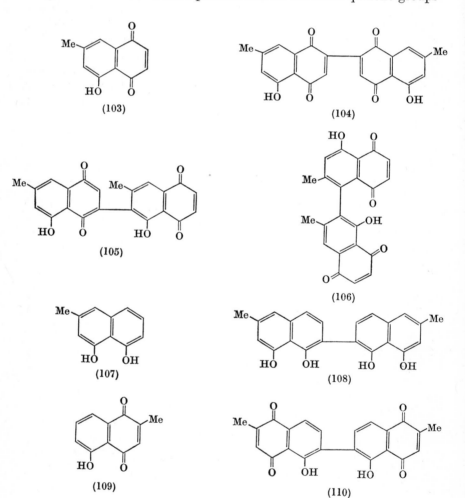

(103)

(104)

(105)

(106)

(107)

(108)

(109)

(110)

include dimers. Sometimes the monomers occur as co-metabolites but the actual coupling step probably takes place before oxidation to the quinone level. The presence of 7-methyljuglone (103) and the dimers (104)–(106) in *Drosera*[105] and *Diospyros*[7] spp. is illustrative, and the existence of (107) and (108) in the fruit of *D. ehretioides* exemplifies phenolic coupling prior to quinone formation.† 2-Methyljuglone (plumbagin) (109) and the dimer (110) are also found in *Diospyros* and the isolation of many more biquinones can be expected. Tri- and tetra-naphthaquinones have been discovered recently (pp. 235 and 242).

If phenolic coupling results in the formation of a 1,1′-binaphthyl then, assuming that suitably disposed phenolic groups are available, an intramolecular repetition of the process will lead to the formation of perylene derivatives such as elsinochrome A (111) whose biosynthesis by the acetate pathway has been established by feeding experiments.[106] Isotopic analysis of the radioactive pigment was consistent with the labelling pattern shown in (111) which in turn implies that the biosynthesis includes a dimerisation step, an unusual feature being the apparent coupling of two benzylic groups to form an additional six-membered ring. By similar steps more highly condensed compounds

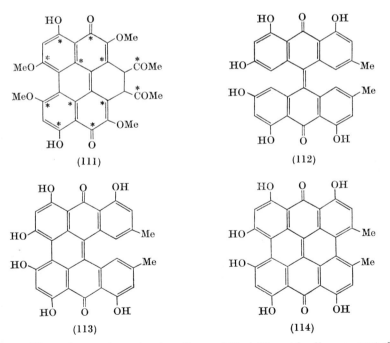

(111)

(112)

(113)

(114)

† Studies[131] on pigment formation in cultures of *Penicillium islandicum* suggest that monomer-dimer pairs are formed independently from a common precursor at enzyme level.

may be formed starting from an anthracene unit, and not only are 10,10′-bianthrones and 1,1′-bianthraquinones known in Nature but in the Hypericaceae[107] we find (112), (113) and (114) (hypericin), evidently formed by sequential phenolic coupling.

Finally, there are a number of quinones which are simply oxidised terpenes. The details are not known but terpenoid phenols are not uncommon and conversion to a quinone is unexceptional. Thymoquinone is the simplest example and the formation of thymol from mevalonate has been confirmed[108] in *Orthodon japonicum* (Labiatae). In some cases the oxidation level of the quinone may be no higher than that of accompanying phenols [e.g. (115) and (5)] occur together in

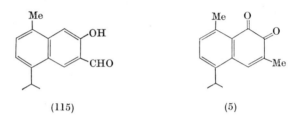

(115)   (5)

elm heartwood (*Ulmus glabra*).[105] Helicobasidin (10), a fungal metabolite from *Helicobasidium mompa*, is the only terpenoid quinone whose origin from mevalonate has been established.[140, 141] Several pathways from farnesyl pyrophosphate to helicobasidin are consistent with the labelling pattern determined but the direct concerted cyclisation of the *trans-cis* isomer, shown below, is favoured[141] on the grounds of simplicity.

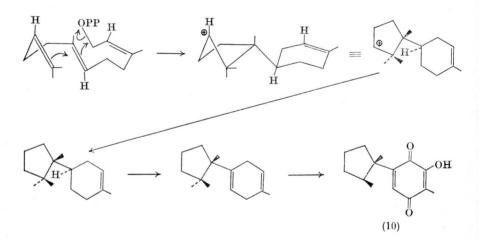

(10)

Most of the terpenoid quinones discovered so far are sesquiterpenes like helicobasidin and biflorin (116), or diterpenes like cryptotanshinone (117), but the possible existence of triterpenoid and tetra-terpenoid quinones cannot be excluded. There are also one or two quinones in the flavanoid series, and the appearance of quinones in other groups of natural phenolic compounds may be expected.

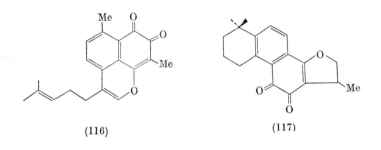

(116)   (117)

REFERENCES

1. H. Brockmann, *Angew. Chem.* (1968), **7**, 481.
2. S. Shibata, *Chem. Brit.* (1967), **3**, 110; Y. Ogihara, N. Kobayashi and S. Shibata, *Tetrahedron Lett.* (1967), 1881; J. D. Bu'Lock and J. R. Smith, *J. Chem. Soc. (C)* (1968), 1941.
3. A. R. Burnett and R. H. Thomson, *J. Chem. Soc. (C)* (1968), 2437.
4. L. H. Briggs and J. C. Dacre, *J. Chem. Soc.* (1948), 564.
5. J. H. V. Charles, H. Raistrick, R. Robinson and A. R. Todd, *Biochem. J.* (1933), **27**, 499 but cf. ref. 13.
6. J. B. Harborne, *Phytochemistry* (1967), **6**, 1415.
7. *Inter alia* G. S. Sidhu and A. V. B. Sankaram, *Annalen* (1966), **691**, 172; G. S. Sidhu, A. V. B. Sankaram and S. M. Ali, *Indian J. Chem.* (1968), **6**, 681; A. L. Fallas and R. H. Thomson, *J. Chem. Soc. (C)* (1968), 2279, K. Yoshihira, M. Tezuka and S. Natori, *Tetrahedron Lett.* (1970), 7.
8. H. Singh, R. E. Moore and P. J. Scheuer, *Experientia* (1967), **23**, 624; H. A. Anderson, J. W. Mathieson and R. H. Thomson, *Comp. Biochem. Physiol.* (1969), **28**, 333.
9. D. W. Cameron and Lord Todd, ref. 109, p. 203.
10. J. H. Richardson and J. B. Hendrickson. "The Biosynthesis of Steroids, Terpenes, and Acetogenins" (1964). Benjamin Inc., New York.
11. D. B. Sprinson, *Adv. Carbohydrate Chem.* (1960), **15**, 235.
12. A. J. Birch and F. W. Donovan, *Aust. J. Chem.* (1953), **6**, 360; (1955), **8**, 529.
13. A. J. Birch, A. J. Ryan and H. Smith, *J. Chem. Soc.* (1958), 4773.
14. S. Gatenbeck, *Acta Chem. Scand.* (1958), **12**, 1211.
15. S. Gatenbeck, *Acta Chem. Scand.* (1960), **14**, 296.
16. A. J. Birch, R. I. Fryer, P. J. Thomson and H. Smith, *Nature (Lond.)* (1961), **190**, 441.
17. R. Bentley and J. G. Keil, *Proc. Chem. Soc.* (1961), 111; J. D. Bu'Lock and H. M. Smalley, *Proc. Chem. Soc.* (1961), 209; J. D. Bu'Lock, H. M. Smalley and G. N. Smith, *J. Biol. Chem.* (1962), **237**, 1778.
18. S. Gatenbeck, *Acta Chem. Scand.* (1962), **16**, 1053.
19. S. Shibata and T. Ikekawa, *Chem. Ind. (London)* (1962), 360; *Chem. Pharm. Bull. (Tokyo)* (1963), **11**, 368.

2

20. R. Tomaselli, *Arch. Bot. Biogeog. Ital.* (1963), **39**, 4th Ser. Vol. VIII, 1; G. D. Scott, personal communication.
21. K. Mosbach, *Acta Chem. Scand.* (1967), **21**, 2331.
22. E. Ritchie and W. C. Taylor, *Tetrahedron Lett.* (1964), 1431.
23. (a) A. Salaque, M. Barbier and E. Lederer, *Bull. Soc. Chim. Biol* (1967), **49**, 841; (b) M D. Sutherland, personal communication (1969).
24. E. D. Pandhare, A. V. Rama Rao, R. Srinivasan and K. Venkataraman, *Tetrahedron* (1966), Suppl. **8**, 229.
25. R. H. Thomson *In* "Chemistry and Biochemistry of Plant Pigments" (ed. T. W. Goodwin) (1965). Academic Press, New York.
26. A. R. Burnett and R. H. Thomson, *J. Chem. Soc. (C)* (1967), 1261.
27. A. R. Burnett and R. H. Thomson, *J. Chem. Soc. (C)* (1968), 850.
28. W. Sandermann and M. H. Simatupang, *Holz Roh Werkst.* (1966), **24**, 190.
29. A. R. Burnett and R. H. Thomson, *J. Chem. Soc. (C)* (1968), 854.
30. W. Sandermann and H.-H. Dietrichs, *Holz Roh Werkst.* (1957), **15**, 281.
31. A. R. Burnett and R. H. Thomson, *Chem. Comm.* (1967), 1125.
32. E. Leistner and M. H. Zenk, *Tetrahedron Lett.* (1968), 1395; E. Leistner and M. H. Zenk, *Abstr. IUPAC 5th Int. Sym. Chem. Nat. Prods. (London)* (1968), p. 113.
33. A. R. Burnett, E. J. C. Brew and R. H. Thomson, unpublished.
34. E. Leistner and M. H. Zenk, *Tetrahedron Lett.* (1968), 861.
35. M. H. Zenk, personal communication.
36. E. Leistner and M. H. Zenk, *Z. Naturforsch.* (1967), **22b**, 865.
37. A. R. Burnett and R H Thomson, *Phytochemistry* (1968), **7**, 1423.
38. E. Brew and R. H. Thomson, *J. Chem. Soc.* (c). In Press.
39. P. H. Gale, F. R. Koniuszy, A. C. Page and K. Folkers, *Arch. Biochem. Biophys.* (1961), **93**, 211; K. Folkers, C. H. Shunk, B. O. Linn, N. R. Trenner, D. E. Wolf, C. H. Hoffman, A. C. Page and F. R. Koniuszy. *In* "Quinones in Electron Transport" (eds. G. E. W. Wolstenholme and C. M. O'Connor) (1961), p. 100. Churchill, London.
40. R. H. Haskins, J. A. Thorn and B. Boothroyd, *Can. J. Microbiol.* (1955), **1**, 749.
41. A. J. Birch and F. W. Donovan, *Chem. Ind. (London)* (1954), 1047.
42. S. Natori, Y. Inouye and H. Nishikawa, *Chem. Pharm. Bull. (Tokyo)* (1967), **15**, 380.
43. R. E. Moore, H. Singh and P. J. Scheuer, *Tetrahedron Lett.* (1968), 4581.
44. S. Gatenbeck and R. Bentley, *Biochem. J.* (1965), **94**, 478.
45. R. Bentley and S. Gatenbeck, *Biochemistry* (1965), **4**, 1150.
46. J. R. Beckwith and L. P. Hager, *J. Biol. Chem.* (1963), **238**, 3091.
47. (a) G. B. Cox and F. Gibson, *Biochim. Biophys. Acta* (1964), **93**, 204; (b) *Biochem. J.* (1966), **100**, 1.
48. I. M. Campbell, C. J. Coscia, M. Kelsey and R. Bentley, *Biochem. Biophys. Res. Comm.* (1967), **28**, 25.
49. E. Leistner, J. H. Schmitt and M. H. Zenk, *Biochem. Biophys. Res. Comm.* (1967), **28**, 845.
50. M. Guerin, R. Azerad and E. Lederer, *Bull. Soc. Chim. Biol.* (1965), **47**, 2105; R. Azerad, R. Bleiler-Hill and E. Lederer, *Biochem. Biophys. Res. Comm.* (1965), **19**, 194; G. Jauréguiberry, M. Lenfant, B. C. Das and E. Lederer, *Tetrahedron* (1966), Suppl. 8, 27.
51. R. Azerad, R. Bleiler-Hill, F. Catala, O. Samuel and E. Lederer, *Biochem. Biophys. Res. Comm.* (1967), **27**, 253; O. Samuel and R. Azerad, *FEBS Letters* (1969), **2**, 336.

52. (a) T. Ramasarma and T. Ramakrishnan, *Biochem. J.* (1961), **81**, 303; (b) T. S. Raman, B. V. S. Sharma, J. Jayaraman and T. Ramasarma, *Arch. Biochem. Biophys.* (1965), **110**, 75.

53. A. A. Kandutsch, H. Paulus, E. Levin and K. Bloch, *J. Biol. Chem.* (1964), **239**, 2507; K. J. I. Thorne and E. Kodicek, *Biochem. J.* (1966), **99**, 123.

54. D. R. Threlfall, W. T. Griffiths and T. W. Goodwin, *Biochem. J.* (1967), **103**, 831.

55. D. R. Threlfall, G. R. Whistance and T. W. Goodwin, *Biochem. J.* (1968), **106**, 107.

56. G. R. Whistance, D. R. Threlfall and T. W. Goodwin, *Biochem. J.* (1967), **105**, 145.

57. C. Martius and W. Leuzinger, *Biochem. Z.* (1964), **340**, 304.

58. C. Martius and H. Esser, *Biochem. Z.* (1959), **331**, 1; M. Billeter and C. Martius, *Biochem. Z.* (1960), **333**, 430; M. Billeter, W. Bolliger and C. Martius, *Biochem. Z.* (1964), **340**, 290; C. Martius, E. G. Semadeni and C. Alvino, *Biochem. Z.* (1965), **342**, 492.

59. R. Azerad, R. Bleiler-Hill and E. Lederer. *In* "Biosynthesis of Aromatic Compounds" (ed. G. Billek) (1966), p. 75. Pergamon, London.

60. K. H. Bolkart and M. H. Zenk, *Naturwiss.* (1968), **55**, 444; K. H. Bolkart, M. Knobloch and M. H. Zenk, *Naturwiss.* (1968), **55**, 445.

61. A. R. Burnett and R. H. Thomson, *J. Chem. Soc. (C)* (1968), 857.

62. H. Inouye, *J. Pharm. Soc. Japan* (1956), **76**, 976.

63. D. Chen and B. A. Bohm, *Canad. J. Biochem.* (1966), **44**, 1389; B. A. Bohm, *Biochem. Biophys. Res. Comm.* (1967), **26**, 621.

64. M. H. Zenk and E. Leistner, *Z. Naturforsch.* (1967), **22b**, 460.

65. E. Leistner and M. H. Zenk, *Z. Naturforsch.* (1968), **23b**, 259.

66. S. Gatenbeck, *Acta Chem. Scand.* (1957), **11**, 555.

67. N. M. Packter, *Biochem. J.* (1965), **97**, 321; (1966), **98**, 353.

68. G. Pettersson, *Acta Chem. Scand.* (1963), **17**, 1323.

69. G. Pettersson, *Acta Chem. Scand.* (1964), **18**, 335; (1965), **19**, 1016.

70. (a) A. J. Birch, R. I. Fryer and H. Smith, *Proc. Chem. Soc.* (1958), 343; (b) R. Bentley and W. V. Lavate, *J. Biol. Chem.* (1965), **240**, 532.

71. N. M. Packter and M. W. Steward, *Biochem. J.* (1967), **102**, 122; M. W. Steward and N. M. Packter (1968), **109**, 1.

72. G. Read and L. C. Vining, *Chem. Comm.* (1968), 935; G. Read, D. W. S. Westlake and L. C. Vining, *Canad. J. Biochem.* (1969), **47**, 1071.

73. E. Charollais, S. Fliszár and T. Posternak, *Arch. Sci. Soc. Phys. Hist. Nat. Genève* (1963), **16**, 474.

74. S. H. El Basyouni and L. C. Vining, *Canad. J. Biochem.* (1966), **44**, 557; A. J. Birch and R. I. Fryer, *Austral. J. Chem.* (1969), **22**, 1319.

75. N. M. Packter, *Biochem. J.* (1969), **114**, 369.

76. A. J. Birch. *In* "Quinones in Electron Transport" (eds. G. E. W. Wolstenholme and C. M. O'Connor) (1961), p. 233. Churchill, London.

77. P. Simonart and H. Verachtert, *Bull. Soc. Chim. Biol.* (1966), **48**, 943; (1967), **49**, 543, 919; N. M. Packter, *Abstr. IUPAC 5th Int. Sym. Chem. Nat. Prods., London* (1968), p. 126.

78. Y. Yamamoto, K. Nitta and A. Jinbo, *Chem. Pharm. Bull. (Tokyo)* (1967). **15**, 427.

79. M. Suda and Y. Takeda, *J. Biochem. (Tokyo)* (1950), **37**, 375; S. Dagley, M. E. Fewster and F. C. Happold, *J. Gen. Microbiol.* (1953), **8**, 1.

80. S. Gatenbeck and I. Lönnroth, *Acta Chem. Scand.* (1962), **16**, 2298.

81. G. Billek and F. P. Schmook, *Österr. Chem.-Zeit.* (1966), **67**, 401; *Monatsh.* (1967), **98**, 1651.
82. P. V. Divekar, G. Read, L. C. Vining and R. H. Haskins, *Canad. J. Chem.* (1959), **37**, 1970.
83. G. Read, L. C. Vining and R. H. Haskins, *Canad. J. Chem.* (1962), **40**, 2357.
84. D. R. Threlfall. *In* "Terpenoids in Plants" (ed. J. B. Pridham), (1967), p. 191. Academic Press, London.
85. W. W. Parson and H. Rudney, *J. Biol. Chem.* (1965), **240**, 1855.
86. K. J. Stone and F. W. Hemming, *Biochem. J.* (1967), **104**, 43.
87. W. W. Parson and H. Rudney, *Proc. Nat. Acad. Sci. U.S.A.* (1965), **53**, 599; see also D. R. Threlfall and J. Glover, *Biochem. J.* (1962), **82**, 14P.
88. H. Rudney and W. W. Parson, *J. Biol. Chem.* (1963), **238**, PC3137.
89. W. W. Parson and H. Rudney, *Proc. Nat. Acad. Sci. U.S.A.* (1964), **51**, 444.
90. B. D. Davis, *Nature* (1950), **166**, 1120; C. H. Doy and F. Gibson, *Biochim. Biophys. Acta* (1961), **50**, 495; M. I. Gibson and F. Gibson, *Biochem. J.* (1964), **90**, 248.
91. R. K. Olsen, J. L. Smith, G. D. Daves, H. W. Moore, K. Folkers, W. W. Parson and H. Rudney, *J. Amer. Chem. Soc.* (1965), **87**, 2298.
92. (a) R. K. Olsen, G. D. Daves, H. W. Moore, K. Folkers and H. Rudney, *J. Amer. Chem. Soc.* (1966), **88**, 2346; (b) P. Friis, G. D. Daves and K. Folkers, *J. Amer. Chem. Soc.* (1966), **88**, 4754.
93. P. Friis, J. L. G. Nilsson, G. D. Daves and K. Folkers, *Biochem. Biophys. Res. Comm.* (1967), **28**, 324.
94. R. K. Olsen, G. D. Daves, H. W. Moore, K. Folkers, W. W. Parson and H. Rudney, *J. Amer. Chem. Soc.* (1966), **88**, 5919.
95. R. E. Olson, *Vitamins and Hormones* (1966), **24**, 551.
96. R. E. Olson, R. Bentley, A. S. Aiyar, G. H. Dialameh, P. H. Gold, V. G. Ramsey and C. M. Springer, *J. Biol. Chem.* (1963), **238**, PC3146.
97. J. E. Miller, *Biochem. Biophys. Res. Comm.* (1965), **19**, 335.
98. J. Meinwald, K. F. Koch, J. E. Rogers and T. Eisner, *J. Amer. Chem. Soc.* (1966), **88**, 1590.
99. B. S. Brown, G. R. Whistance and D. R. Threlfall, *FEBS Letters*, (1968), **1**, 323; cf. J. R. S. Ellis and J. Glover, *Biochem. J.* (1968), **110**, 22P.
100. E. S. Caldwell and C. Steelink, *Biochim. Biophys. Acta* (1969), **184**, 420, and references therein.
101. J. C. Overeem and A. van Dijkman, *Rec. Trav. Chim.* (1968), **87**, 940.
102. R. Thomas, *Biochem. J.* (1961), **78**, 748; S. Gatenbeck and S. Hermodsson, *Acta Chem. Scand.* (1965), **19**, 65.
103. W. D. Ollis, I. O. Sutherland, R. C. Codner, J. J. Gordon and G. A. Miller, *Proc. Chem. Soc.* (1960), 347.
104. J. Zylber, E. Zissmann, J. Polonsky, E. Lederer and M. A. Merrien, *Europ. J. Biochem.* (1969), **10**, 278.
105. V. Krishnamoorthy and R. H. Thomson, *Tetrahedron Lett.* In Press; J. W. Rowe, personal communication.
106. C.-T. Chen, K. Nakanishi and S. Natori, *Chem. Pharm. Bull. (Tokyo)* (1966), **14**, 1434.
107. H. Brockmann, *Prog. Chem. Org. Nat. Prods.* (1957), **14**, 142.
108. M. Yamazaki, T. Usui and S. Shibata, *Chem. Pharm. Bull. (Tokyo)* (1963), **11**, 363.
109. W. I. Taylor and A. R. Battersby. "Oxidative Coupling of Phenols" (1967). Arnold, London.
110. S. Imre, *Phytochemistry* (1969), **8**, 315.

111. R. G. Ravdin and D. I. Crandall, *J. Biol. Chem.* (1951), **189**, 137; W. E. Knox and S. W. Edwards, *J. Biol. Chem.* (1955), **216**, 479.
112. W. D. Ollis, *Experientia* (1966), **22**, 777.
113. W. Parker and J. S. Roberts, *Quart. Revs.* (1967), **21**, 331.
114. S. K. Grisdale and G. H. N. Towers, *Nature (Lond.)* (1960), **188**, 1130.
115. G. R. Whistance and D. R. Threlfall, *Biochem. Biophys. Res. Comm.* (1967), **28**, 295; *Biochem. J.* (1968), **109**, 482.
116. G. R. Whistance and D. R. Threlfall, *Biochem. J.* (1968), **109**, 577.
117. M. H. Zenk and E. Leistner, *Lloydia* (1968), **31**, 275.
118. L. M. Jackman, I. G. O'Brien, G. B. Cox and F. Gibson, *Biochim. Biophys. Acta* (1967), **141**, 1.
119. P. Chandra, G. Read and L. C. Vining, *Can. J. Biochem.* (1966), **44**, 403.
120. M. H. Zenk, *Z. Naturforsch.* (1964), **19b**, 856.
121. M. H. Zenk. *In* "Biosynthesis of Aromatic Compounds" (ed. G. Billek) (1966), p. 45. Pergamon, London.
122. K. H. Bolkart and M. H. Zenk, *Z. Pflanzenphysiol.* (1968), **59**, 439.
123. B. A. Bohm, personal communication.
124. E. Leistner and M. H. Zenk, *Chem. Comm.* (1969), 210; cf. A. Meynaud, A. Ville and H. Pacheco, *C. R. H. Acad. Sci.* (1968), **266D**, 1783.
125. M. Lenfant, G. Farrugia and E. Lederer, *C. R. Acad. Sci.* (1969), **268D**, 1986.
126. W. R. Tschinkel, *J. Insect Physiol.* (1969), **15**, 191.
127. H. W. Moore and R. J. Wikholm, *Tetrahedron Lett.* (1968), 5049.
128. J. N. Ashley, H. Raistrick and T. Richards, *Biochem. J.* (1939), **33**, 1291.
129. A. G. Perkin and J. J. Hummel, *J. Chem. Soc.* (1894), **65**, 923.
130. S. Imamoto and S. Senoh, *Tetrahedron Lett.* (1967), 1237; *J. Chem. Soc. Japan (Pure Chem. Sect.)* (1968), **89**, 316.
131. S. Gatenbeck, *Acta Chem. Scand.* (1960), **14**, 102; S. Gatenbeck and P. Barbesgärd, *Acta Chem. Scand.* (1960), **14**, 230; M. Kikuchi, *Bot. Mag. Tokyo* (1962), **75**, 158.
132. B. Franck, *Angew. Chem.* (1969), **8**, 251.
133. D. Gröger, D. Erge, B. Franck, U. Ohnsorge, H. Flasch and F. Hüper, *Chem. Ber.* (1968), **101**, 1970.
134. W. S. G. Maass and A. C. Neish, *Canad. J. Bot.* (1967), **45**, 59.
135. R. L. Frank, G. R. Clark and J. N. Coker, *J. Amer. Chem. Soc.* (1950), **72**, 1824.
136. G. R. Whistance, J. F. Dillon and D. R. Threlfall, *Biochem. J.* (1969), **111**, 461.
137. O. A. Dada, D. R. Threlfall and G. R. Whistance, *Europ. J. Biochem.* (1968), **4**, 329.
138. H. Kindl, *Z. Physiol. Chem.* (1969), **350**, 1289.
139. A. K. Bose, K. S. Khanchandani, P. T. Funke and M. Anchel, *Chem. Comm.* (1969), 1347; (1970), 252.
140. S. Natori, Y. Inouye and H. Nishikawa, *Chem. Pharm. Bull. (Tokyo)* (1967), **15**, 380.
141. R. Bentley and D. Chen, *Phytochemistry* (1969), **8**, 2171.
142. K. H. Bolkart and M. H. Zenk, *Z. Pflanzenphysiol.*, (1969), **61**, 356.
143. I. M. Campbell, *Tetrahedron Lett.* (1969), 4777.
144. G. B. Cox, I. G. Young, L. M. McCann and F. Gibson, *J. Bacteriol.* (1969), **99**, 450.
145. M. J. Winrow and H. Rudney, *Biochem. Biophys. Res. Comm.* (1969), **37**, 833.

146. H. G. Nowicki, G. H. Dialameh, B. L. Trumpower and R. E. Olson, *Fed. Proc.* (1969), **28**, 884.
147. T. S. Raman, H. Rudney and N. H. Buzzelli, *Arch. Biochem. Biophys.* (1969), **130**, 164.
148. E. Lederer, *Quart. Revs.* (1969), **23**, 453.
149. L. H. Briggs, M. Kingsford and R. N. Seelye, unpublished work.
150. D. R. Threlfall and G. R. Whistance, *Phytochemistry* (1970), **9**, 355.
151. S. Gatenbeck and L. Malmström, *Acta Chem. Scand.* (1970), **23**, 3493.
152. H. Rudney. *In* "Natural substances formed biologically from mevalonic acid" (ed. T. W. Goodwin) (1970), p. 89. Academic Press, London.
153. J. L. G. Nilsson, T. M. Farley and K. Folkers, *Anal. Biochem.* (1968), **23**, 422.
154. R. Powls, *FEBS Letters* (1970), **6**, 40.
155. D. J. Robins, I. M. Campbell and R. Bentley, *Biochem. Biophys. Res. Comm.* (1970), **39**, 10811

CHAPTER 2

Identification

The evidence for the structures of the naturally occurring quinones is given in the succeeding chapters of this book, and this is necessarily a mixture of the old and the new. Modern investigations rely very heavily on the interpretation of spectral information whereas older work, going back to the structural elucidation of alizarin a hundred years ago, was mainly dependent on the preparation of derivatives and on degradative experiments. This chapter, in consequence, is mainly concerned with spectra but other methods of identification still in use are considered briefly.

Isolation and purification

The methods employed are common to natural products chemistry and not peculiar to quinones. Normally these pigments are isolated by sequential extraction with solvents of increasing polarity, and the appropriate fractions are purified by column and/or thin layer chromatography (t.l.c.) or preparative layer chromatography (p.l.c.). Ultrasonic extraction may offer some advantage.[79] As some quinones are photolabile, unknown pigments should preferably be stored out of light and occasionally they require handling in the dark. It is not possible to recommend any particular procedure as quinones vary so much in polarity and solubility, for example, the bioquinones are easily taken into non-polar solvents whereas xylindein is extracted with phenol, and obviously the chromatographic behaviour of polyhydroxyquinones may be very different from that of hydroxyl-free analogues. Original papers should be consulted for details. As the majority of quinones are relatively involatile, and t.l.c. is so efficient, preparative g.l.c. has rarely been used but undoubtedly many quinones could be separated in this way, using, where appropriate, trimethylsilyl ethers.[1] Quinones usually occur in the free state but hydroxyquinones, especially in the anthraquinone series, are frequently present in the plant in glycosidic combination, some naphthaquinones occur as glycosides of the corresponding quinols, and on occasion hydroxyquinones may exist *in vivo* as metal salts (spinochromes), O-sulphates (fusarubinogen) or in association with protein (namakochrome). Appropriate treatment is required in each case.

Chemical methods are avoided as far as possible but may be useful in the separation of hydroxylated quinones. Hydroxybenzoquinones and 2(3)-hydroxynaphthaquinones are vinylogous carboxylic acids and hence can be extracted into aqueous sodium bicarbonate; naphthaquinones and anthraquinones β-hydroxylated in a benzenoid ring dissolve in aqueous sodium carbonate (some pass into bicarbonate solution), whereas the chelated α-isomers require aqueous sodium hydroxide. Similar considerations apply to other types of quinones and alkaline extraction, used with caution, can be very helpful in the separation of complex mixtures.[2]

Colour reactions

Although colour reactions are much less important than formerly they are still useful, particularly at the beginning of an investigation when crude extracts or even natural tissue may yield information of value. Very little material is required and the reactions may be carried out, if desired, by spraying chromatograms.[82] These rough tests should preferably be supplemented later by spectrophotometric measurements on purified material. The most useful diagnostic tests depend upon the redox properties of quinones and the presence of hydroxyl groups. Leucomethylene blue[78] is a useful spray for the detection of benzoquinones and naphthaquinones on paper or thin layer chromatograms, the quinones appearing as blue spots on a white background. Reduction to a colourless (or much less highly coloured) product, and easy restoration of the original colour on oxidation, is characteristic and distinguishes quinones from nearly all other natural compounds. Re-oxidation can usually be effected simply by shaking the solution in air, but the leuco compounds of non-hydroxylated benzoquinones and naphthaquinones do not oxidise so readily. Reduction is easily effected with neutral or alkaline sodium dithionite but many other reducing agents may be used. Catalytic hydrogenation can be employed quantitatively and sodium borohydride is convenient when reductions are to be followed spectrophotometrically. For hydroxyquinones the colour changes are more striking in alkaline solution and re-oxidation (by air) is more rapid. Anthraquinones can be distinguished from benzoquinones and naphthaquinones as they usually give red solutions on reduction in alkaline solution (zinc or dithionite in aqueous sodium hydroxide). A yellow-brown colour is ambiguous however, as some hydroxynaphthaquinones (e.g. lawsone) may behave in the same way, but generally anthraquinols, in contrast to benzoquinols and naphthaquinols, absorb >400 nm in alkaline solution. Reductions in aqueous alkaline dithionite can be followed spectrophotometrically if a layer of dioxan is added to the solution to prevent aerial oxidation, or if tetraethylenepentamine is used in place of sodium hydroxide.[3]

The characteristic colours given by hydroxyquinones in alkaline solution are useful aids to structure determination. Those listed in Table I are obtained with excess sodium hydroxide although compounds

TABLE I

Colour of hydroxyquinones in alkaline solution

Quinone	Colour	$\lambda_{max.}^{EtOH/HO^-}$ nm
Benzoquinones		
2-hydroxy-5-methyl-	Red	493
2,3-dihydroxy-5,6-dimethyl-	Bluish–purple	522
2,5-dihydroxy-	Bluish–Red	505
Naphthaquinones		
2-hydroxy-	Orange	459
5-hydroxy-	Violet	538
6-hydroxy-	Violet–red	520
2,3-dihydroxy-	Blue	650
2,5-dihydroxy-	Violet–red	490
3,5-dihydroxy-	Red	435
5,6-dihydroxy-	Blue	571
5,7-dihydroxy-	Violet	542
5,8-dihydroxy-	Blue	655
Anthraquinones		
1-hydroxy-	Red	500
2-hydroxy-	Orange–red	478
1,2-dihydroxy-	Violet–blue	576
1,3-dihydroxy-	Red	485
1,4-dihydroxy-	Violet	560
1,5-dihydroxy-	Red	496
1,8-dihydroxy-	Red	513
1,2,3-trihydroxy-	Green	668
1,2,4-trihydroxy-	Violet–red	544
1,4,5-trihydroxy-	Violet	561
1,4,5,8-tetrahydroxy-	Blue	630

containing more than one hydroxyl group may show more than one indicator change. Obviously too much reliance should not be placed on these colour changes which may be modified by cross-conjugation and other structural features. Many older tests for hydroxyquinones are now obsolescent but the zirconium nitrate test[4] (a red–violet precipitate in acid solution) for vicinal hydroxyl groups (as in alizarin) is useful, and so is Shibata's methanolic magnesium acetate reagent. This was originally introduced[5] for hydroxyanthraquinones, the colour obtained being indicative of the orientation of the hydroxyl groups, but it appears to be of wider utility and a positive colour reaction with magnesium acetate is a general test for a hydroxyquinone. *Peri*-hydroxyquinones form boro-acetates (1) with boro-acetic anhydride[6, 67] resulting in a significant bathochromic shift in the visible region.

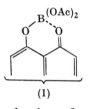

(1)

Non-hydroxylated quinones having a free quinonoid position, that is, principally benzoquinones and naphthaquinones, can be recognized by the Craven[7] or Kesting[8] test. In this reaction the quinone is treated in alcoholic solution with a reagent containing a reactive methylene group

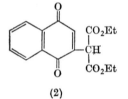

(2)

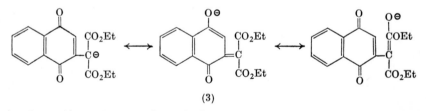

(3)

(aceto-acetic ester, malononitrile, etc., also nitromethane[9]) and ammonia; the anion so formed then undergoes Michael addition.[62] The blue–green or violet–blue colour which appears is that of the mesomeric anion [e.g. (3)] derived from the addition product [e.g. (2)]. If a hydroxyl group is present the Kesting–Craven colour may be masked or suppressed, and a free quinonoid position is not, in fact, essential, as alkoxyl and halogen groups can be displaced by the reagent.[63] The mesomeric

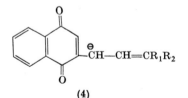

(4)

anion (3) is similar to that (4) given by allylquinones in the Dam–Karrer test[64] which is responsible for a transient blue–violet colour obtained when such quinones are treated with alcoholic potassium hydroxide. Other colour reactions with bases (e.g. indole[10]) designed to detect free quinonoid positions are limited in scope, and again nucleophilic displacement of substituents can occur.

Derivatives

The only derivatives routinely prepared are leuco-acetates, and the acetates, methyl ethers and trimethylsilyl ethers of hydroxyquinones, and frequently these are made solely in order to obtain a more soluble or more volatile form of the parent compound for spectroscopic study (their n.m.r. spectra give a convenient "count" of the number of hydroxyl groups). Some indication of the orientation of hydroxyl groups can be gained from R_F values, and by the use of selective reagents. Only β-hydroxyl groups are readily methylated with diazomethane; chelated α-hydroxyl groups are normally resistant but they succumb to methyl iodide-silver oxide-chloroform or methyl sulphate-potassium carbonate-acetone. It is sometimes possible to methylate one hydroxyl group in a 1,8-dihydroxyanthraquinone using diazomethane. All nuclear hydroxyl groups can be esterified by reaction with acetic anhydride but selective β-acetylation can be achieved using ketene[65] or acetic anhydride in the presence of boro-acetic anhydride, followed by hydrolysis of chelated(α) boric esters with cold water.[66] Reductive acetylation is extremely useful as the electronic spectrum of the leuco-acetate is very similar to that of the parent hydrocarbon, bathochromically shifted.[11] The main group to which the quinone belongs can thus be determined in most cases. Quinoxaline derivatives are frequently prepared from o-quinones but it should be remembered that some p-quinones also react with o-phenylenediamine.[12]

Degradation reactions

Most new quinones can be identified without recourse to degradation but it is often necessary when novel structures are encountered. If the basic polycyclic system is not recognizable from the ultraviolet spectrum of the leuco-acetate or in other ways, zinc dust distillation or the less drastic zinc dust fusion[13] can be used to obtain the parent hydrocarbon (or heterocycle).[14] The disadvantages are well-known but milder methods[15] of reduction eliminate only the quinone oxygen functions leaving substituents intact. However, the yields obtained, for example by reduction of anthraquinones with diborane,[15] are vastly superior to those available by degradation with zinc and such methods might repay further study. Oxidative degradation is used mainly to establish the structure of a side chain attached to a quinone ring or to obtain an identifiable fragment containing a benzenoid ring. In the former case, usually effected with alkaline hydrogen peroxide,[16] the group R in (5) is isolated as the acid (7). Under similar, or more vigorous, conditions, naphtha-

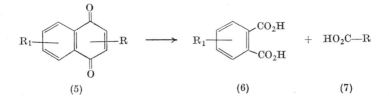

(5)   (6)   (7)

quinones yield phthalic acids (6), but anthraquinones and more highly condensed compounds where the quinone ring is protected on both sides by benzene rings, are much more difficult to degrade. However, methoxylated anthraquinones can be cleaved to form benzoic acids by heating with potassium t-butoxide and water (molecular ratio 3:1) and this recent method[31] may be useful.

SPECTRA

Ultraviolet-visible Spectra

Benzoquinones

The spectrum of p-benzoquinone is characterized by intense absorption near 240 nm ($\epsilon_{max.}$ 26,000), a medium band $\sim$285 nm ($\epsilon_{max.}$ $\sim$300) attributed to an electron-transfer (E.T.)[17] transition, and much weaker absorption (n → π^*) in the visible region (Fig. 1) (Table II). The spectrum is modified by the introduction of substituents, and spectroscopically nearly all the natural compounds can be regarded as alkyl, hydroxyl, and alkoxyl derivatives of the parent quinone [see structures (8) to (21)]. Exceptionally, aryl groups may also be present. The electronic spectra of p-benzoquinones have been discussed and analysed in a number of papers[18] but for organic chemists concerned only with empirical interpretations, that of Flaig *et al.*[19] is the most useful. Much of the information in Table II is drawn from their work while the spectra of a representative group of natural quinones are listed in Table IV. The great majority of natural quinone spectra have been determined in ethanol solution although this is not the ideal spectroscopic solvent. The possible presence of basic impurities must be borne in mind since these will ionise hydroxyquinones resulting in misleading spectra;[18, 21] this

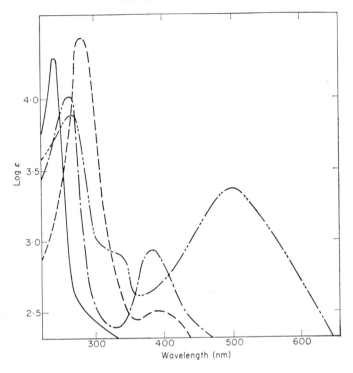

FIG. 1. Electronic absorption spectrum (in EtOH) of: 1,4-benzoquinone (———);
2,5-dihydroxy-1,4-benzoquinone (– – –); 2-hydroxy-5-methyl-1,4-benzoquinone
(·—·—·); 2-hydroxy-5-methyl-1,4-benzoquinone (EtOH/HO) (··—··—··).

can be avoided by running spectra in acidified ethanol. For simple
synthetic compounds most information is available for chloroform
solutions (Table II)† and when comparing spectra run in different media
the solvent shifts recorded by Flaig[19] should be consulted. Chloroform
is not suitable, however, for measuring alkali shifts, etc., and much of
the information in Tables IV–X, and the spectra in Figs 1–6, were obtained in ethanolic solution.

Introduction of a substituent of the usual type into the benzoquinone
nucleus produces a small bathochromic displacement of the first band
in the spectrum (<10 nm) but the second band undergoes a more
significant red shift in the order Me (27 nm), MeO (69 nm), HO (81 nm)
in chloroform solution. The visible band is little affected and is frequently
obscured under the envelope of band 2. Introduction of a second

† For a short list of spectra measured in ethanol see ref. 20.

TABLE II

Ultraviolet-visible absorption of 1,4-benzoquinones

Quinone	$\lambda_{max.}^{CHCl_3}$ nm ($\log \epsilon$)		
Parent	246 (4·42)	288 (2·50)	439 (1·35)
	242[a] (4·26)	285[a](sh) (2·6)	434[a] (1·26)
			454[a] (1·22)
2-Methyl-	249 (4·33)	315 (2·80)	436 (1·38)
	255(sh) (4·27)		
2-Ethyl-	248 (4·30)	318 (2·95)	437 (1·53)
	256(sh) (4·29)		
2,3-Dimethyl-	250 (4·26)	337 (3·05)	425 (1·55)
	257(sh) (4·20)		
2,5-Dimethyl-	254 (4·37)	316 (2·42)	430 (1·43)
	261(sh) (4·32)		
2,6-Dimethyl-	255 (4·29)	319 (2·54)	429 (1·45)
Trimethyl-	258 (4·30)	340 (2·61)	425 (1·53)
	263 (4·27)		430(sh)
Tetramethyl-	262 (4·30)	342 (2·34)	430(sh) (1·5)
	269 (4·31)		
Methoxy-	254 (4·26)	357 (3·21)	
2,3-Dimethoxy-	254 (4·17)	398 (3·17)	
2,5-Dimethoxy-	278 (4·37)	370 (2·48)	
	284 (4·38)		
2,6-Dimethoxy-	287 (4·38)	377 (2·78)	
Trimethoxy-	291 (4·29)	418 (2·65)	
Hydroxy-	256 (4·14)	369 (3·07)	
2,5-Dihydroxy-	279 (4·35)	393 (2·43)	
	286(sh) (4·34)		
2-Hydroxy-3-methyl-	255 (4·17)	396 (3·16)	
2-Methoxy-3-methyl[b]	254 (4·12)	374 (3·13)	
2-Hydroxy-5-methyl-	264 (4·28)	382 (2·82)	
2-Methoxy-5-methyl-	264 (4·32)	365 (2·88)	
2-Hydroxy-6-methyl-	268 (4·21)	380 (2·87)	
2-Methoxy-6-methyl-[b]	263 (4·23)	355 (2·97)	
2-Hydroxy-3,6-dimethyl-	266 (4·24)	402 (2·94)	
2-Methoxy-3,6-dimethyl-[b]	265 (4·15)	376 (2·89)	
2-Hydroxy-3,5,6-trimethyl-	272 (4·29)	409 (2·65)	
	280 (4·30)		
2-Methyl-5,6-dimethoxy-	264 (4·16)	402 (2·94)	
3-Hydroxythymoquinone[a]	267 (4·16)	404 (3·01)	
3,6-Dihydroxythymoquinone[a]	293 (4·31)	435 (2·36)	

[a] In EtOH.
[b] In CCl$_4$.

substituent has much less effect on band 2 than the first and may be negligible; it is always greatest in 2,3-disubstituted derivatives. Further substitution results in further bathochromic displacements but the

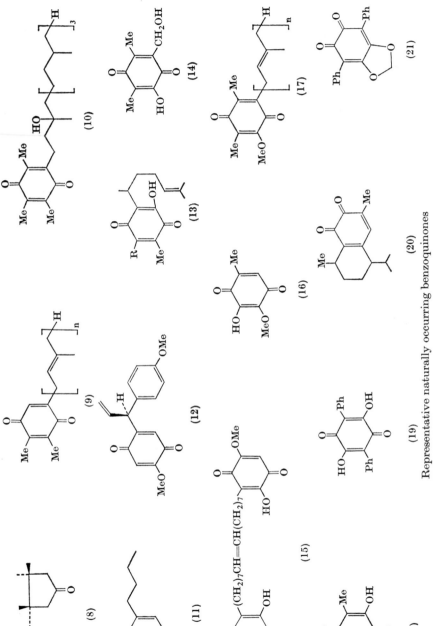

Representative naturally occurring benzoquinones

shifts are not particularly useful for empirical calculations of λ_{max}. Since model compounds are available for most of the natural structures which are likely to appear, an unknown may be best identified by spectral comparison with known compounds. For example shanorellin (14), a recent addition to the list of fungal benzoquinones, shows λ_{max}. 272 and 406 nm, and may be regarded as a trialkylhydroxybenzoquinone; 2-hydroxy-3,5,6-trimethylbenzoquinone absorbs at 272 and 409 nm. Note, however, that plastoquinones and ubiquinones show only one maximum.

In alkaline alcoholic solution there is a marked shift of the visible absorption (Fig. 1) (Table I). In this respect only instantaneous shifts, reversible on acidification, are important; on keeping, other changes may take place more slowly, notably hydroxylation of the ring or nucleophilic replacement of methoxyl by hydroxyl. Addition of sodium borohydride to an ethanolic solution of a benzoquinone affords the quinol accompanied by a sharp change to benzenoid absorption with λ_{max}. ca. 290 nm and marked reduction in ϵ_{max}. Spectrophotometric assays of plastoquinone and ubiquinone have been developed on this basis.[18]

Only two natural o-benzoquinones have been found, both highly substituted (Table IV); relatively few o-benzoquinone spectra (Table III) have been recorded.[18, 22] They show triple absorption peaks and are easily distinguished from the p-isomers by the relatively low intensity and marked bathochromic displacements. Band 2 is frequently shifted

TABLE III

Ultraviolet-visible absorption of 1,2-benzoquinones

Quinone	λ_{max}. nm (log ϵ)		
Parent		375 (3·23)	568 (1·48)
4-Methyl-[b]	249(sh) (3·32)	387 (3·23)	544(sh) (1·48)
			570 (1·52)
3,5-Dimethyl-[a]	260 (3·19)	410 (3·20)	558 (1·66)
4,5-Dimethyl-[a]	260 (3·46)	400 (3·12)	555 (1·48)
			570 (1·48)
3,4,5-Trimethyl-[a]	265 (3·33)	425 (3·11)	545 (1·75)
3-Methoxy-[b]	269(sh) (2·66)	465 (3·26)	545(sh) (1·78)
			575(sh) (1·56)
4-Methoxy-[b]	255(sh) (3·75)	406 (3·21)	538(sh) (1·66)
4,5-Dimethoxy-[b]	283 (4·09)	406 (2·82)	504(sh) (1·60)
5-Ethyl-3-methoxy-[a]		470 (3·19)	570(sh) (2·60)

[a] In $CHCl_3$.
[b] In CH_2Cl_2.

into the visible region and the low intensity n → π* absorption may extend as far as 600 nm.

TABLE IV

Ultraviolet-visible absorption of some naturally occurring benzoquinones

Quinone	$\lambda_{max.}^{EtOH}$ nm (log ϵ)		
Lagopodin A (8)	257 (4·25)	310 (2·55)	435 (1·55)
Plastoquinones (9)	254		
	262(sh)		
α-Tocopherolquinone (10)	261 (4·38)		
	269 (4·29)		
Primin (11)	267 (4·33)	365 (2·54)	
4,4'-Dimethoxydalbergione (12)	258 (4·14)	333 (3·23)	
Perezone (13; R = H)	266 (4·05)	412 (3·01)	
Hydroxyperezone (13; R = OH)	295 (4·26)	425 (2·41)	
Shanorellin (14)	272 (4·05)	406 (2·07)	
Ardisiaquinone A (15)	289 (4.60)	420 (2.83)	
Fumigatin (16)	265 (4·14)	450 (2·96)	
Ubiquinones (17)	275	405	
Spinulosin[a] (18)	297 (4·35)	460 (2·29)	
Polyporic acid (19)	256 (4·63)	330(sh) (4·06)	465 (2·60)
	262 (4·63)		
o-Benzoquinones			
Mansonone A (20)	260(sh) (3·5)	432 (3·2)	
	280(sh) (3·1)		
Phlebiarubrone (21)	268 (4·48)	332 (3·64)	465 (3·54)

[a] In CHCl$_3$.

Naphthaquinones

These spectra are inevitably more complex than those of benzoquinones since both benzenoid and quinonoid absorption is involved, and either or both rings may be substituted. In the natural compounds the principal substituents can be regarded as alkyl, hydroxyl, and alkoxyl groups, but in addition acyl groups and conjugated double bonds may be present and the quinonoid ring may be fused to a furan or pyrone ring system [see structures (22) to (34)]. The spectra of numerous simple 1,4-naphthaquinones have been measured and analysed,[18, 23, 69] the most useful compilation being that of Scheuer and his co-workers[24] from which many of the data in Table V are taken. Illustrative spectra for some natural quinones are given in Table VII.

The spectrum of the parent compound (Fig. 2) comprises intense benzenoid and quinonoid electron-transfer absorption in the region

TABLE V

Ultraviolet-visible absorption of 1,4-naphthaquinones

Quinone	$\lambda_{max.}^{CHCl_3}$ nm (log ϵ)			
Parent	245 (4·34)	257(sh) (4·12)	335 (3·48)	
	251 (4·37)			
2-Methyl-	245·5 (4·27)	258·5 (4·22)	335·5 (3·43)	
	251 (4·30)			
5-Methyl-	252 (4·29)	260(sh) (4·07)	355 (3·59)	
2,3-Dimethyl-	243 (4·26)	260 (4·28)	330 (3·8)	
	249 (4·26)	269 (4·28)		
2-Hydroxy-	249 (4·20)	275·5 (4·23)	339 (3·53)	
		282 (4·24)		
2-Methoxy-	242 (4·22)	274 (4·21)	333 (3·47)	
	248 (4·25)	280 (4·21)		
5-Hydroxy-	253·5 (4·19)		337 (3·15)	429 (3·64)
5-Methoxy-	247 (4·24)		324(sh) (3·10)	396 (3·52)
6-Hydroxy-	254 (4·30)		344(sh) (3·24)	388 (3·38)
	261 (4·29)			
2-Methoxy-3-isopentyl	252 (4·41)		333 (3·66)	382(sh) (3·10)
	279 (4·32)			

Substituent					
2-Hydroxy-3-iso-α-pentenyl					
2,3-Dihydroxy-	262 (4·25)	265 (4·41), 274(sh) (4·23), 288(sh) (4·16)	316 (3·51), 335 (3·36)	419 (3·30), 439 (3·17)	
2,5-Dihydroxy-	240 (4·00)	286 (4·10)		418 (3·56), 430 (3·56)	
3,5-Dihydroxy-	240 (3·89)	283 (4·15)		419 (3·64)	
5,6-Dihydroxy-[a]		263 (4·01)		461 (3·51)	
5,7-Dihydroxy-		249(sh) (4·04), 263 (4·10)	371(sh) (3·35)	436 (3·57)	
5,8-Dihydroxy-		269 (3·87)	338 (3·00)	490 (3·73), 547 (3·56)	524 (3·78), 564 (3·57)
2,5,7-Trihydroxy-[b]	262 (4·12)	308 (3·93)	403 (3·30)	452 (3·35)	
2,5,8-Trihydroxy-	292 (3·91)	390(sh) (3·06)	481 (3·74)	496(sh) (3·79), 528 (3·67)	506 (3·80), 543 (3·60)
2,3,5,7-Tetrahydroxy-	270 (4·20)	320 (3·89)	387 (3·45)	470 (3·16)	
2,3,5,8-Tetrahydroxy-	249 (4·19)	288 (3·66), 302 (3·66)	463 (3·74), 488 (3·80)	478(sh) (3·77), 511 (3·65)	524 (3·64)

[a] In EtOH.
[b] In MeOH.

240–290 nm and a medium intensity benzenoid E.T. band at 335 nm. A broad weak local excitation (L.E.)[17] band at 425 nm (ϵ 32) is discernible in iso-octane solution but not in more polar solvents. The shoulder at 257 nm is ascribed to a quinonoid E.T. transition, and is shifted bathochromically by +I and +M substituents in the quinone ring whereas the benzenoid absorption at 245 and 251 nm is usually

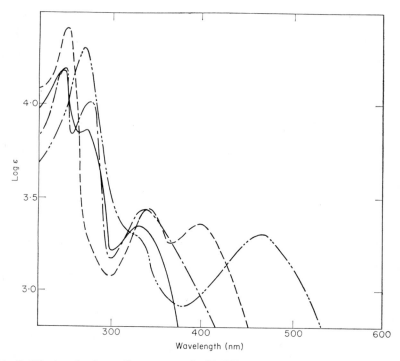

Fig. 2. Electronic absorption spectra (in EtOH) of: 1,4-naphthaquinone (———); 1,2-naphthaquinone (– – – –) ;2-hydroxy-1,4-naphthaquinone (·—·—·; 2-hydroxyl,4-naphthaquinone (EtOH/HO⁻) (··—··—··).

scarcely affected. A visible band in the spectrum of, for example, 2,3-dihydroxy-1,4-naphthaquinone is attributed to a quinonoid E.T. transition usually obscured by the broad benzenoid band at ∼335 nm except in derivatives which have powerful electron-donating groups at positions 2 and 3. In bz-substituted 1,4-naphthaquinones the benzenoid and quinonoid E.T. bands in the region 240–290 nm frequently coalesce, and the prominent benzenoid absorption near 335 nm is shifted towards the red.

Peri-substitution by a hydroxyl group is exceptional, the benzenoid band shifting almost 100 nm into the visible to give a peak at 429 nm

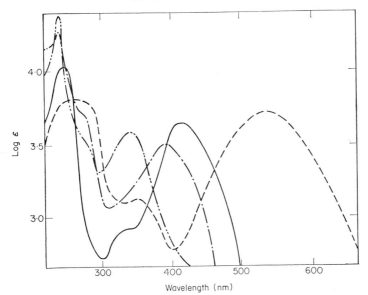

FIG. 3. Electronic absorption spectra (in EtOH) of: 5-hydroxy-1,4-naphtha-quinone (——); 5-hydroxy-1,4-naphthaquinone (EtOH/HO⁻) (– – – –); 5-methoxy 1,4-naphthaquinone (·—·—·); 5-acetoxy-1,4-naphthaquinone (··—··—··).

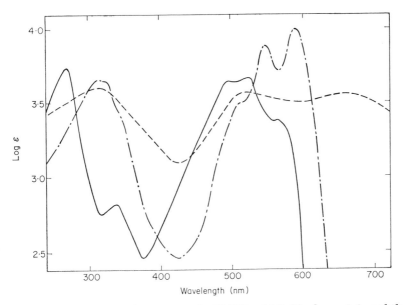

FIG. 4. Electronic absorption spectra (in EtOH) of 5,8-dihydroxy-1,4-naphtha-quinone. —— (neutral); – – – – (+HO⁻); ·—·—· (+AlCl₃).

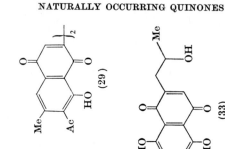

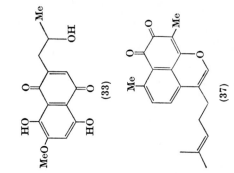

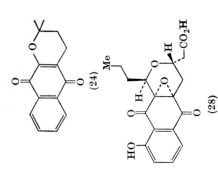

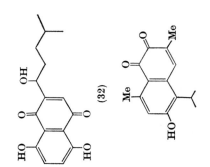

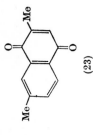

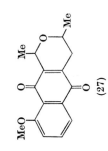

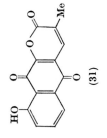

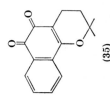

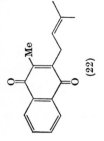

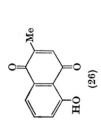

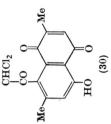

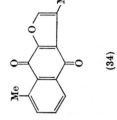

Representative naturally occurring naphthaquinones

(422 nm in MeOH) characteristic of simple juglones (Fig. 3). The effect is less marked in juglone methyl ether (λ_{max}. 396 nm), while in the acetate (5-acetoxy-1,4-naphthaquinone) the —M effect of the ester carbonyl group inhibits transfer of an oxygen p electron into the juglone π system and the spectrum reverts to that of the parent 1,4-naphthaquinone. Acetylation is thus a useful way of observing the spectrum of a hydroxy- or polyhydroxyquinone uninfluenced by the hydroxyl group(s). An

TABLE VI

Ultraviolet-visible absorption of 1,2-naphthaquinones

Quinone	λ_{max}. nm (log ϵ)			
Parent[c]	255 (4·36)	343 (3·37)	406 (3·33)	
3-Methyl-[a]	253 (4·40)	340 (3·30)		
4-Methyl-[a]	252 (4·40)	345 (3·40)		490(sh) (2·20)
5-Hydroxy-[b]	250 (4·26)	350(sh) (3·23)	430 (3·67)	
6-Hydroxy-[c]	276 (4·18)	375 (3·79)	425(sh) (3·20)	
7-Hydroxy-[c]	265 (4·47)	335 (3·11)	455 (3·36)	
4-Methoxy-[c]	250 (4·41)	335 (3·30)	400 (3·35)	490(sh) (2·20)
8-Methoxy-[a]	242 (4·22)		422 (3·82)	
8-Methoxy-6-methyl-[a]	244 (4·27)		420 (3·87)	
5-Methoxy-7-methyl-[a]	262 (4·26)	376 (3·24)	472 (3·64)	
3,8-Dimethoxy-[c]	261 (4·16)	290(sh) (3·46)	450 (3·55)	

[a] In ethanol.
[b] In dioxan.
[c] In chloroform.

acetyl group directly attached to a ring has little effect on the spectrum. Scheuer et al.[24] tentatively ascribe the weaker band at ca. 340 nm in juglone and some related compounds to a quinonoid E.T. transition. The benzenoid E.T. band at 429 nm in the juglone spectrum is not significantly changed by further benzenoid substitution with the exception of further o- and p-hydroxylation. Naphthazarins, with two peri-hydroxyl groups, are characterised by a combined benzenoid and quinonoid E.T. band in the region 270–350 nm (ϵ <10,000) which is to be expected from such a tautomeric system, and multibanded benzenoid absorption centred around 525 nm (ϵ 6000–9000). These bands often engulf a weak quinonoid E.T. band observed in some derivatives in the region 330–500 nm. The combination band is susceptible to substitution, and it has been noted in polyhydroxynaphthazarins that each β-hydroxyl group produces a bathochromic shift of ca. 20 nm. The substitution pattern in such compounds is associated with the fine structure of the visible absorption band.

TABLE VII

Ultraviolet-visible absorption of some naturally occurring naphthaquinones

Quinone	$\lambda_{max.}^{EtOH}$ nm (log ϵ)			
Menaquinone-1 (22)	245 (4·46) 249 (4·44)	262 (4·34) 273 (4·31)	331 (3·58)	
Chimaphilin (23)	248 (4·19) 254·5 (4·19)	265(sh) (4·02)	338 (3·19)	
α-Lapachone (24)	251 (4·45)	282 (4·22)	332 (3·44) 375(sh) (3·15)	
Dehydro-α-lapachone (25)		267 (4·35) 276(sh) (4·17)	333 (3·40)	434 (3·21)
Plumbagin (26)		266 (4·12)		418 (3·61)
Eleutherin (27)	246 (4·13)	269(sh) (4·07)	393 (3·58)	
Frenolicin (28)	234 (4·26)	284(sh) (3·54)	362 (3·71)	
Dianellinone (29)[a]		275(sh) (4·29)		429 (3·89)
Mollisin (30)	259 (4·26)	280(sh) (3·9)		420 (3·52)
Lambertellin (31)		284(sh) (4·08) 290 (4·10)		430 (3·68)
Alkannin (32)		280 (3·84)		480 (3·74) 510 (3·78) 546 (3·60)
Solaniol (33)[a]			304 (3·97)	472(sh) (3·84) 500 (3·91) 536(sh) (3·72)
Maturinone (34)	251 (4·39)	266 (4·00) 287(sh) (3·74)	355 (3·58)	
o-Naphthaquinones				
β-lapachone (35)	256 (4·44)	282 (4·01)	330 (3·28)	431 (3·26)
Mansonone G (36)	244 (4·25)	274 (4·40)		407 (3·95)
Biflorin (37)	234 (4·50)		340 (3·75)	555 (3·76)
8-Methoxy-3-methyl-	241 (4·29)			427 (3·78)

[a] In dioxan.

The alkali shifts shown by hydroxynaphthaquinones are of diagnostic value (Figs 2–4). The anion of 2-hydroxy-1,4-naphthaquinone is orange, that of 5-hydroxy-1,4-naphthaquinone is violet, while those of 2,3-, 5,6- and 5,8-dihydroxy-1,4-naphthaquinones are blue (see Table I). Addition of anhydrous aluminium chloride to an ethanolic solution of a naphthazarin gives a visible spectrum showing characteristic triplet absorption (Fig. 4);[68] this is not shown by juglones.

1,2-Naphthaquinones are comparatively rare in Nature and, with two exceptions, are at least partly terpenoid in origin, structures (35)–(37) being representative (see Table VII). Only two possess phenolic groups. No extensive study of 1,2-naphthaquinone spectra has been published[18, 25] and the selection listed in Table VI is drawn from various sources. As the parent quinone itself contains two different carbonyl groups and a substituent in any position will be conjugated with one of these, a detailed analysis of a larger number of 1,2-naphthaquinones might yield some useful structural correlations. The parent quinone shows intense absorption near 250 nm and bands of medium intensity near 340 and 400 nm (Fig. 2). A weak longwave band >500 nm is only observed in non-polar solvents. Substitution in the quinonoid ring has relatively little effect on the spectrum whereas substitution in the benzenoid ring produces marked changes which clearly depend upon the position of substitution.

Anthraquinones

This large group of pigments consists almost entirely of polyhydroxy or alkoxy derivatives, and the influence of these substituents dominates the spectra. Table VIII lists the spectra of some of the more common synthetic compounds while formulae (38)–(56) illustrate the diversity of hydroxyl substitution found among the natural pigments (spectral data are given in Table IX). Several surveys have been published,[18, 26-28, 76] the most useful analyses being those of Morton and Earlam,[29] and Peters and Sumner[30] although unfortunately these are limited to mono- and disubstituted derivatives. Anthraquinone shows intense benzenoid absorption at ca. 250 nm and medium absorption at 322 nm, strong quinonoid E.T. bands are seen at 263 and 272 nm and there is weak quinonoid absorption at 405 nm. These areas of selective absorption are characteristic and the pattern in the ultraviolet region is not seriously affected by substitution, the benzenoid bands appearing fairly regularly within the range 240–260 and 320–330 nm and the quinonoid band(s) at 260–290 nm (see Table VIII). In addition hydroxy-anthraquinones show an absorption band in the region 220–240 nm, not shown by the parent compound. This is frequently ignored although Ikeda *et al.*[27] consider that the $\lambda_{max.}$ and $\epsilon_{max.}$ values depend upon the number and orientation of the hydroxyl groups. By regarding the spectrum of anthraquinone as a combination of the absorptions arising from partial acetophenone and benzoquinone chromophores,[29] Scott[81] has shown that the ultraviolet spectra of simple derivatives can be calculated empirically by the addition of increments, appropriate to the substituent and its position, to the relevant absorption bands of the parent quinone. Rough agreement with the observed values for $\lambda_{max.}$ can be obtained but unfortunately such predictions are not applicable

TABLE VIII

Ultraviolet-visible absorption of anthraquinones

Quinone	$\lambda_{max.}^{EtOH}$ nm (log ϵ)			
Parent	243 (4·52)	263 (4·31)	322 (3·75)	405 (1·95)
1-Methyl-	252 (4·71) 252 (4·66)	272 (4·31) 263 (4·26) 272 (4·16) 265 (4·32)	331 (3·68)	415 (2·18)
2-Methyl-	255 (4·65)	274 (4·24) 266 (4·18)	324 (3·66)	
1-Hydroxy-	252 (4·46)	277 (4·44) 262 (4·36)	327 (3·52)	402 (3·74)
1-Methoxy-	254 (4·52)	270 (4·18)	328 (3·46)	378 (3·72)
2-Hydroxy-	241 (4·31)	271 (4·55) 283 (4·46)	330 (3·55)	378 (3·55)
2-Methoxy-	246 (4·21)	267 (4·51) 280 (4·38)	329 (3·56)	363 (3·60)
1,2-Dihydroxy-	247 (4·45)	278 (4·13)	330 (3·46)	434 (3·70)
1,2-Dimethoxy-	251 (4·46)	270 (4·30) 280 (4·28)	330 (3·50)	374 (3·71)

1,3-Dihydroxy-	246 (4·43)	284 (4·36)	330 (3·30)	420 (3·65)	465(sh) (3·89), 499·5 (3·76), 512 (3·67)
1,4-Dihydroxy-	248 (4·23), 254 (4·26)	279 (3·95)	325 (3·36)	452(sh) (3·87), 480 (3·91)	
1,5-Dihydroxy-	253 (4·33)	275(sh) (4·05), 284·5 (4·03)		418 (4·00), 432 (4·00)	
1,8-Dihydroxy-	251 (4·27)	273(sh) (4·00), 283 (3·99)		429 (3·98)	
1,2,3-Trihydroxy-	241 (4·28), 245 (4·30)	287 (4·49)		414 (3·81)	
1,2,4-Trihydroxy-	255 (4·46)	295(sh) (3·98)		452(sh) (3·83)	483 (3·93), 518(sh) (3·81)
1,4,5-Trihydroxy-	250 (4·21)	284 (3·94)		459(sh) (4·02), 489 (4·12)	478(sh) (4·08), 510(sh) (4·00), 523 (3·93)
1,3,8-Trihydroxy-6-methyl-	253 (4·31)	266 (4·29), 289 (4·36)		436 (4·14)	
1,4,5,8-Tetrahydroxy-[a]		287 (3·81), 299(sh) (3·79)		477(sh) (3·81), 509 (4·07), 546·5 (4·14), 601(sh) (3·25)	487(sh) (3·91), 521 (4·14), 560 (4·19)
1,3,6,8-Tetrahydroxy-	253 (4·17) 262 (4·17)	291 (4·48)	318 (3·96)	372 (3·47) 452 (4·01)	

[a] In CHCl₃.

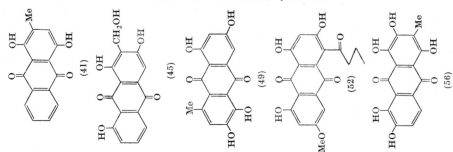

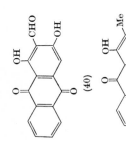

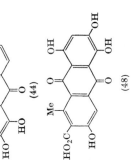

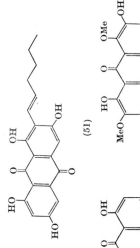

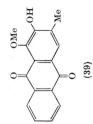

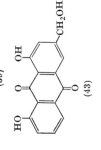

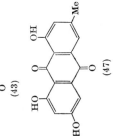

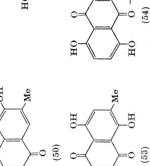

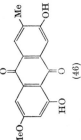

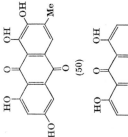

Representative naturally occurring anthraquinones

to the longwave quinonoid absorption band which is of most value in the structural elucidation of the natural anthraquinones. Substitution by hydroxyl or alkoxyl invariably intensifies this band, and in general there is a bathochromic shift. However, there is no shift in the case of 1-

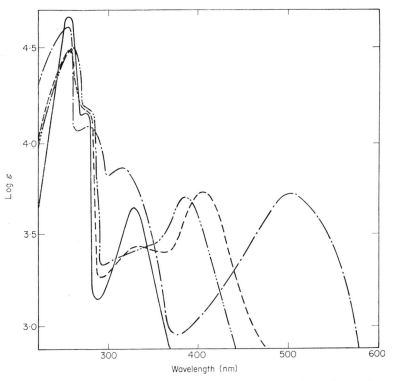

Fig. 5. Electronic absorption spectra (in EtOH) of: anthraquinone (——); hydroxyanthraquinone (————); 1-hydroxyanthraquinone (EtOH/HO⁻) (—·—·—); 1-methoxyanthraquinone (··—··—··).

hydroxyanthraquinone and on methylation the hydrogen bond contribution is eliminated and the peak at 402 nm is moved hypsochromically to 378 nm (Fig. 5). Exceptionally 2-hydroxyanthraquinone absorbs at 378 nm and its methyl ether at 363 nm.

Absorption above 360 nm is dominated by the number of α-hydroxyl groups, the influence of β-hydroxyls being much weaker except when adjacent to an α-hydroxyl[28] (Fig. 6). The visible absorption is of medium intensity ($\epsilon \sim 10,000$) and may obscure the benzenoid band at 320–330 nm; tri- and tetra-α-hydroxylated derivatives frequently display fine structure in the longwave band. Quinones possessing one α-hydroxyl group normally absorb in the range 400–420 nm but this may extend as far as 436 nm if an adjacent hydroxyl or a *peri*-methoxyl group is also present. The spectra of 1,8-dihydroxyanthraquinones show a peak at ca. 430–450 nm (still higher in the 1,3,6,8-tetrahydroxy series), 1,5-dihydroxy compounds usually display two maxima in the region 418–440 nm (again shifted bathochromically by an adjacent β-hydroxyl), while

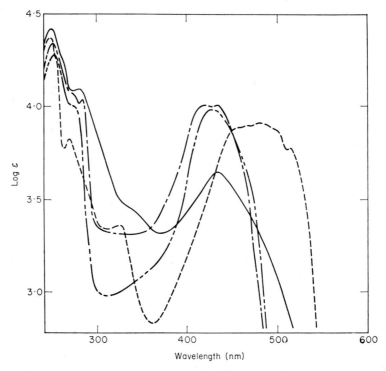

Fig. 6. Electronic absorption spectra (in EtOH) of: 1,2-dihydroxyanthraquinone (——); 1,4-dihydroxyanthraquinone (– – – – –); 1,5-dihydroxyanthraquinone (·—·—·); 1,8-dihydroxyanthraquinone (··—··—··).

the 1,4-dihydroxy quinones absorb at 470–500 nm with a peak (or shoulder) above 500 nm (Fig. 6). Additional α-hydroxylation results in a further red shift of the longwave absorption. 1,4,5-Trihydroxyanthraquinones show two or more maxima in the range 485–530 nm, and the 1,4,5,8-tetrahydroxy compounds show multibanded absorption in the 540–560 nm region. Excellent examples of poly-α-hydroxylated anthraquinones can be found in the anthracyclinone series (Chapter 6). Most of the natural anthraquinones contain at least four substituents which modify the absorption spectra and irregularities will be found which do not fit in with the above generalisations (Table IX). Further information

TABLE IX

Ultraviolet-visible absorption of some naturally occurring anthraquinones

Quinone	$\lambda_{max.}^{EtOH}$ nm (log ε)							
Pachybasin (38)	224 (3·74)	252 (4·01)	281 (3·65)		403 (3·02)			
Digitolutein (39)		242·5 (4·27)	274 (4·40)		386 (3·52)			
Nordamnacanthal (40)	219 (4·56)	262 (4·73)	292 (4·57)			421 (4·09)		
2-Methylquinizarin (41)		250 (4·59)	286 (4·06)			482 (4·29)		
Soranjidiol (42)	220 (4·45)	245 (4·04)	271·5 (4·46)	322 (3·48)		430 (4·04)		
Aloe-emodin (43)	225 (4·59)	254 (4·34)	276·5 (4·01), 287 (4·03)	337·5 (3·17)	411 (3·80)			
Morindone (44)	232 (4·50)	259 (4·53)	292 (4·16)	301 (4·18)		448 (4·07)		
Coelulatin (45)		247 (4·23)	270 (4·28), 285 (4·30)	320 (4·30)		437 (3·54)		
Macrosporin (46)	225 (3·60)		284 (4·80)	305 (4·40)	381 (4·15)	436 (4·14)		
Emodin (47)	222 (4·46)	253 (4·31)	266 (4·29), 289 (4·36)			436 (4·14)		
Kermesic acid (48)			266 (4·18), 292 (3·73)			498 (3·04)	538 (2·94)	
Erythrolaccin (49)	228 (4·26)		295 (4·13)			466 (3·74)		
Alaternin (50)	229 (4·32)	252 (4·11)	283·5 (4·40)	317·5 (4·06)		430 (4·01)		
Averythrin (51)	223 (4·46)	255 (4·12)	266 (4·18), 294 (4·45)	324 (4·02)		453 (3·95)		
Rhodocomatulin 6-methyl ether (52)		256 (4·24), 263 (4·25)	293 (4·42)	317 (3·96)	366 (3·46)	456 (4·05)		
Catenarin (53)	231 (4·19)	257 (4·19)	281 (4·21)	300 (4·01)		491 (4·13)	512 (4·05)	525 (3·99)
Cynodontin (54)		241 (4·56)	295 (4·06)			471 (4·06), 514 (4·38)	483 (4·14), 539 (4·37)	503 (4·31), 552 (4·42)
Aurantio-obtusin (55)	226 (4·20)		286 (4·58)	314 (3·94)	388 (3·82)	418 (3·62)	468 (3·95)	495 (4·06)
Ventimalin (56)	227 (4·23)		272 (4·39)	312 (4·03)				

concerning the number and arrangement of α-hydroxyl groups can be obtained by measuring the bathochromic shift of the longwave absorption which occurs on ionization, accompanied always by an increase in intensity (see Table I).

Infrared spectra

The carbonyl frequencies of quinones are useful diagnostic aids in structure determination and have been studied extensively. Ideally spectra should be run in dilute solution but many natural quinones are poorly soluble in suitable solvents and in practice most spectra are now measured in potassium bromide discs. It should be noted that many symmetric *p*-quinones show multiple absorption in the carbonyl region, even in dilute solution, which is attributed mainly to Fermi resonance effects.[37] In published spectra minor peaks or shoulders in the 6 μ region are frequently ignored, only the principal band(s) being cited. Useful compilations are available for benzoquinones[20, 32, 33] and anthraquinones,[34, 35] but naphthaquinones[36] are less well served and the information is more scattered.

The carbonyl absorption of *p*-benzoquinone falls at 1669 cm^{-1} (in solution) which is normal for an $\alpha\beta:\alpha'\beta'$ di-unsaturated ketone, and the frequency rises as the number of linear fused rings increases (1,4-naphthaquinone 1675 cm^{-1}, 9,10-anthraquinone 1678 cm^{-1}, naphthacene-5,12-quinone 1682 cm^{-1}).[38] The carbonyl frequency is lowered by hydrogen bonding, by substitution either in the quinonoid ring or an adjacent benzenoid ring with +I or +M groups, and by separation of the carbonyl functions so that quinonoid conjugation extends through more than one ring (extended quinones). The carbonyl frequency is raised by −M substituents and by steric strain, both of which are relatively rare in natural quinones. The commonest substituents are alkyl, hydroxyl and alkoxyl, and the shift to lower frequencies can be appreciable in highly substituted quinones [for example, α-tocopherolquinone (10) shows ν_{CO} 1641 cm^{-1} (CCl$_4$)], particularly if they are chelated (see below). Consequently it is difficult to give a meaningful range of ν_{CO} values for any large group of quinones (Tables X–XII). It follows that ν_{CO} alone is insufficient to classify a quinone and additional information is required. *o*-Quinones normally exhibit two carbonyl bands, usually at somewhat higher frequency than expected for the *p*-isomers. There is very little information available for *o*-benzoquinones[39]† but *o*-naphthaquinones can often be distinguished from the *p*-quinones by the presence of a medium band (or shoulder) in the range 1700–1680 cm^{-1} (Table X).

---

† *o*-Benzoquinone samples should not be prepared in KBr disc as the heat generated causes decomposition. Even in Nujol the spectra may show several bands in the carbonyl region.[85]

TABLE X

Carbonyl absorption of various quinones

	$\nu_{CO}{}^e$ cm^{-1}
1,4-Benzoquinones	
Parent[a]	**1669**, 1653
Methyl-[a]	1661
Ethyl-[d]	1661
2,3-Dimethyl-[a]	**1656**, 1626
2,5-Dimethyl-[a]	**1661**, 1637
2,6-Dimethyl-[a]	1656
Trimethyl-[a]	1647
Tetramethyl-[a]	1653, **1639**
Methoxy-[b]	**1678**, 1645
2,3-Dimethoxy-[b]	**1664**, 1637
2,5-Dimethoxy-[b]	1658
2,6-Dimethoxy-[b]	1695, **1645**
2-Methyl-3-methoxy-[a]	1669, **1653**
2-Methyl-5-methoxy-[a]	1681, **1653**
1,2-Benzoquinones	
Parent[c]	1680, 1658
3,4-Dimethyl-[c]	1675, 1646
1,4-Naphthaquinones	
Parent[a]	1675
2-Methyl-[a]	1670
5-Methyl-[b]	1669, 1662
6-Methyl-[b]	1661
2,3-Dimethyl-[b]	1660
2,7-Dimethyl-[a]	1673
2-Methoxy-[b]	**1678**, 1645
5-Methoxy-[b]	1667(sh), 1657
2,3-Dimethoxy-[b]	**1680**, 1650
5-Acetoxy-[b]	1760, 1669(sh), 1661
6-Hydroxy-[b]	1664
1,2-Naphthaquinones	
Parent[b]	1678, 1661
3-Methoxy-[b']	1700, 1665, 1645
4-Methoxy-[b]	1700, 1667, 1648
6-Methoxy-[b']	1685, 1655
7-Methoxy-[b']	1695, 1660
8-Methoxy-[b]	1683(sh), 1653
5-Hydroxy-[b']	1689, 1637
6-Hydroxy-[b']	1680, 1645
7-Hydroxy-[b']	1685, 1645
5-Methoxy-7-methyl-[b]	1695, 1653
8-Methoxy-6-methyl-[b]	1681, 1659
3,5-Dimethoxy-[b']	1690, 1670
4,7-Dimethoxy-[b]	1696, 1661(sh), 1644
3,8-Dimethoxy-[b']	1690, 1670, 1660

TABLE X—*continued*

	$\nu_{CO}{}^e$ cm^{-1}
9,10-Anthraquinones	
Parent [c]	1675
2-Methyl- [c]	1675
1-Methoxy- [a]	1675
1-Acetoxy- [c]	1764, 1678
2-Hydroxy- [c]	1667
2-Hydroxy-3-methyl- [c]	1658
2,3-Dihydroxy- [c]	1675
2,6-Dihydroxy- [c]	1656
1,2,3-Trimethoxy- [c]	1664

[a] CCl$_4$.   [c] Nujol.
[b] KBr.   [d] CS$_2$.
[b'] KCl.   [e] Bold numerals = principal band.

Chelated quinones are numerous, and can be recognised by their displaced carbonyl absorption (see Tables XI and XII) together with the downward shift in hydroxyl frequency (*o*-hydroxyquinones) or the complete absence of hydroxyl absorption in the 3 μ region (*peri*-hydroxyquinones).† In all cases acetylation eliminates the effect of the chelating group and ν_{CO} is restored to approximately that of the parent quinone. Thus 2- and 5-hydroxy-1,4-naphthaquinones can be differentiated from the 6-isomer, and likewise 1- from 2-hydroxyanthraquinones. 2,5- and 3,5-Dihydroxy-1,4-naphthaquinones can be distinguished (in CHCl$_3$) as the former show doublet carbonyl absorption near 1660 and 1620 cm^{-1} whereas the spectra of the latter have only one carbonyl peak near 1630 cm^{-1} (ignoring unimportant shoulders). Juglones generally show two carbonyl peaks in the regions 1675–1650 and 1645–1620 cm^{-1} while naphthazarins absorb strongly at ~1620–1590 cm^{-1}. From a study of fifty-nine anthraquinones Briggs and his co-workers[35] found the following correlations (see also ref. 34). Some

Number of α-HO groups	ν_{CO} (Nujol) cm^{-1}
None	1678–1653
1	1675–1647 and 1637–1621
2 (1,4- and 1,5-)	1645–1608
2 (1,8-)	1678–1661 and 1626–1616
3	1616–1592
4	1592–1572

† Obviously this only applies in the absence of other non-chelated hydroxyl groups.

TABLE XI

Carbonyl absorption of chelated hydroxyquinones

	ν_{CO} cm^{-1}
Benzoquinones	
Monohydroxy-[a]	1669, 1658
2,5-Dihydroxy-[b]	1646(sh), 1620(br)
3,5-Dimethyl-2,6-dihydroxy-[b]	1660, 1641
Naphthaquinones	
2-Hydroxy-[b]	1674, 1640
5-Hydroxy-[b]	1666, 1645
2,3-Dihydroxy-[b]	1672, 1638
2,5-Dihydroxy-[b]	1658, 1614
3,5-Dihydroxy-[b]	1645, 1625
5,8-Dihydroxy-[a]	1623 (1613[b])
Anthraquinones	
1-Hydroxy-[c]	1667, 1631
1,2-Dihydroxy-[c]	1658, 1634
1,3-Dihydroxy-[c]	1675, 1637
1,4-Dihydroxy-[c]	1626
1,5-Dihydroxy-[c]	1634
1,6-Dihydroxy-[c]	1664, 1634
1,8-Dihydroxy-[c]	1678, 1621
1,2,3-Trihydroxy-[c]	1650, 1626
1,2,4-Trihydroxy-[c]	1621
1,4,5-Trihydroxy-[c]	1605
1,2,8-Trihydroxy-[c]	1661, 1621
1,3,5,7-Tetrahydroxy-[c]	1608
1,3,6,8-Tetrahydroxy-[c]	1669, 1624
1,4,5,8-Tetrahydroxy-[c]	1592

[a] CCl$_4$.
[b] KBr.
[c] Nujol.

exceptions were noted and the ranges quoted should now be slightly extended for KBr measurements. 1,4,5-Trihydroxyanthraquinones can be further distinguished from the 1,4- and 1,5-dihydroxy derivatives as the latter normally show a strong doublet in the 6 μ region not seen in the spectra of the former. Anthraquinones having no α-hydroxyl groups† show only one carbonyl peak and its position is little affected by substitution except (usually) in the case of β-hydroxylation but the maximum shift is <20 cm^{-1}. All the naturally occurring extended quinones are chelated. The simplest example, diphenoquinone (no natural

† Aminoanthraquinones are ignored in this discussion.

TABLE XII

Carbonyl absorption of some naturally occurring quinones

	ν_{CO} cm^{-1}
Benzoquinones	
Plastoquinone (9)	1647
Lagopodin A (8)[a]	1745, 1658
Perezone (13; R = H)[b]	1650, 1628
Hydroxyperezone (13; R = OH)[c]	1640(sh), 1615
Ardisiaquinone A (15)[c]	1655(sh), 1633
Spinulosin (18)[c]	1655(sh), 1637
Mansonone A (20)[c]	1685, 1670
Naphthaquinones	
α-Lapachone (24)[c]	1678, 1640
Plumbagin (26)[c]	1659, 1637
Eleutherin (27)[c]	1662
Dianellinone (29)[c]	1695, 1667, 1623
Alkannin (32)[c]	1603
Solaniol (33)[c]	1602
β-Lapachone (35)[c]	1690, 1640
Anthraquinones	
Digitolutein (36)[c]	1666
Soranjidiol (39)[c]	1664, 1634
Kermesic acid (47)	1670, 1623
Morindone (41)[c]	1634
Emodin (46)[c]	1675, 1631
Catenarin (52)	1598
Cynodontin (53)[b]	1572

[a] In CCl$_4$.
[b] In Nujol.
[c] In KBr.

derivatives), shows ν_{CO} 1639 cm^{-1} (Nujol) which shifts to 1650 cm^{-1} in perylene-3,10-quinone, and to 1631 cm^{-1} in its 4,9-dihydroxy derivative, the simplest natural quinone in this class.

In other regions of the spectrum substituent groups reveal their characteristic absorption independently of the quinone chromophore but one or two correlations have been noted which are of some diagnostic value for certain types of quinone. Most anthraquinones have an intense absorption band in the region 1600–1575 cm^{-1} but this overlaps the carbonyl peak in highly chelated derivatives as already noted.[35] In benzoquinones and naphthaquinones a corresponding band, attributed to a C=C vibration, usually appears at higher wave numbers (~1620–1590 cm^{-1}). Benzoquinones which have an isolated ring hydrogen atom usually show a strong or medium band between 910 and 880 cm^{-1}, and if two adjacent ring hydrogen atoms are present, medium to strong absorption occurs between 840 and 805 cm^{-1}.[32] Bands at 1299 and 714 cm^{-1} are considered[40] to be characteristic for menaquinones but are not found regularly in other naphthaquinones with an unsubstituted benzenoid ring. Whiffen and co-workers[41] have published useful correlation tables relating the in-plane and out-of-plane C—H deformation frequencies of a large number of naphthalene derivatives with their substitution pattern, and these correlations appear to be equally valid for naphthaquinones. Little attention has been paid to these empirical data but it could be helpful in cases where insufficient material is available for n.m.r. determination.

N.m.r. spectra

The quinonoid protons in *p*-benzoquinone resonate at τ 3·28 and in 1,4-naphthaquinone at τ 3·03. The effect of substitution (Tables XIII and XIV)[42, 47, 48, 83] is analogous to that observed in comparable *cis*-vinyl compounds, and for benzoquinones the chemical shift of Q—H

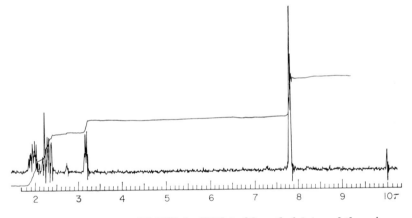

FIG. 7. N.m.r. spectrum (60 MHz in CDCl$_3$) of 2-methyl-1,4-naphthaquinone.

is very similar to that found in cyclohex-2-ene-1,4-diones.[42] On reduction to a quinol (leuco-acetate formation is usually most convenient) the signal shifts downfield to the aromatic region, frequently with a reduction in multiplicity, and this is a useful criterion for a quinonoid structure. The signal from an alkyl substituent undergoes a corresponding shift and reduction in multiplicity. Thus in 2-methyl-1,4-naphthaquinone the quinonoid proton at C-3 gives rise to a quartet at τ 3·16 (J, 1·5 Hz) coupled to a doublet from the allylic methyl protons at τ 7·81 (Fig. 7).

TABLE XIII

N.m.r. spectra (60 MHz) of some p-benzoquinones (τ values)

Benzoquinone	H-2	H-3	H-5	H-6	Others
Parent[b]	3·28	3·28	3·28	3·28	
2-Methyl-[b]	—	3·42(m)	3·30[a]	3·30[a]	Me, 7·93(d)
2,6-Dimethyl-[b]	—	3·50(m)	3·50(m)	—	Me, 7·97(d)
2-Methyl-5-isopropyl-[b]	—	3·45(q)	—	3·53(d)	Me, 7·98(d); Pr, 7·00(m)
2,6-Dimethoxy-[c]	—	4·14(m)	4·14(m)	—	MeO, 6·16
2-Methoxy-5-methyl-[b]	—	4·26	—	3·60(q)	Me, 8·02(d); MeO, 6·26
2,3-Dimethoxy-5-methyl-[b]	—	—	—	3·74(q)	Me, 8·05(d); MeO, 6·10, 6·12

[a] Narrow multiplet.
[b] In CCl$_4$.
[c] In CHCl$_3$.

TABLE XIV

N.m.r. spectra (60 MHz) of some naphthaquinones (τ values)

1,4-Naphthaquinone[a]	H-2	H-3	H-5	H-6	H-7	H-8	Others
Parent	3·05	3·05	1·93(m)	2·23(m)	2·23(m)	1·93(m)	
2-Methyl-	—	3·21(q)	—	—	—	—	Me, 7·87(d)
2-Hydroxy-	—	3·63	—	—	—	—	
2-Methoxy-	—	3·83	—	—	—	—	MeO, 6·11
2-Acetoxy-	—	3·24	—	—	—	—	
2-Acetyl-	—	2·94	—	—	—	—	
5-Hydroxy-	3·03	3·03	—	2·75(m)	2·40(m)	2·30(m)	HO, −1·93
5-Hydroxy-7-methyl-	3·09	3·09	—	2·92(d)	—	2·59(d)	HO, −1·83; Me, 7·58
5-Hydroxy-3,7-dimethoxy-	3·92	—	—	3·40(d)	—	2·92(d)	HO, −1·97; MeO, 3·91
5,8-Dihydroxy-	2·87	2·87	—	2·87	2·87	—	HO, −2·43
5,8-Dihydroxy-2-methoxy-	—	3·83	—	2·77	2·77	—	HO, −2·63, −2·17; MeO, 6·08
5,8-Dihydroxy-2-ethyl-	—	3·16(t)	—	2·80	2·80	—	HO, −2·45, −2·60
5,8-Dihydroxy-2,7-dimethoxy-	—	3·60	—	3·60	—	—	HO, −3·12, −2·70; MeO, 6·06

1,2-Naphthaquinone[a]	H-3	H-4	H-5	H-7	Others
Parent	2·60[b]	3·49[b]			
	2·71	3·65			
8-Methoxy-6-methyl-	2·62[c]	3·56[c]	3·14(d)	3·27(d)	MeO, 6·04; Me, 7·60
	2·78	3·73			

[a] In CDCl₃.
[b] J, 7 Hz.
[c] J, 10 Hz.

In the spectrum of the leuco-acetate (Fig. 8) the C-3 proton and the ring-methyl protons are revealed as singlets at τ 2·87 and 7·70, respectively. Note that in unsymmetrical bz-substituted 1,4-naphthaquinones (Table XIV) the adjacent protons at C-2 and C-3 frequently give a singlet rather than an AB quartet, and this is also true for some mono-substituted benzoquinones. The nuclear protons in simple benzoquinones give rise to multiplets which originate from long range interactions. The following coupling constants have been observed:[42, 43]

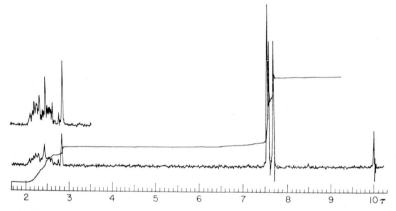

Fig. 8. N.m.r. spectrum (60 MHz in $CDCl_3$) of 2-methylnaphthaquinone leuco-diacetate.

J_{allylic} (CH_3—C=C—H) 1·5–1·7 Hz, $J_{\text{homo-allylic}}$ (CH_3—C=C—CH_3) 1·3 Hz, J (H—C—C—C—H) 2·2–2·5 Hz. Other long range spin-spin couplings, if present, are much smaller, and the coupling constants for allylic and homo-allylic systems vary with the angle between the C=C double bond and the relevant C—H bonds.[44] Very little information is available on the n.m.r. spectra of o-quinones but 1,2-naphthaquinones with no quinonoid substituents can be distinguished from 1,4-isomers by virtue of the AB quartet given by the non-equivalent quinonoid protons (Table XIV and Fig. 9).

    The other signals observed in the n.m.r. spectra of quinones, arising from aromatic and side chain protons, are not peculiar to these compounds and little comment is needed. In 1,4-naphthaquinone and 9,10-anthraquinone the α- and β-protons give A_2B_2 multiplets centred at τ 1·93 and 2·33 respectively, and these are modified by substitution in the normal way. In naphthaquinones the benzenoid substitution pattern can usually be deduced from the aromatic proton signals without difficulty but the situation is more complex in anthraquinones.[49, 50] In practice, however, certain types of substitution predominate and some representative examples are given in Table XV. In this connection, as many quinones are phenolic, the substituent constants compiled by Ballantine and Pillinger[45] for phenols can be used to predict the chemical shift of the aromatic protons and hence the orientation of substituents. Chelated α-hydroxyl groups are easily recognized as their protons resonate at very low field ($\sim\tau$ −2 to −3). For methyl and methoxyl groups, which are common substituents, additional information can be obtained from solvent shifts. Relative to deuterochloroform solution,

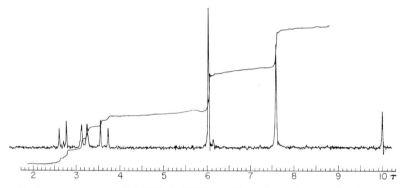

FIG. 9. N.m.r. spectrum (60 MHz in CDCl$_3$) of 8-methoxy-6-methyl-1,2-naphthaquinone.

the chemical shift of anthraquinonoid β-methyl protons moves upfield 0·52–0·60 ppm in benzene solution, that of an α-methyl group is little affected (+0·06–0·17 ppm)† while for methyl groups attached to a quinone ring the upfield shift is 0·29–0·51 ppm.[46]‡ For methoxyl substituents the solvent shifts $\Delta_{C_6H_6}^{CDCl_3}$ are again upfield being $\sim$1 ppm when the group is attached to a quinone ring,‡ 0·65–0·74 ppm for *peri*-methoxyls, and $\sim$0·8 ppm for β-methoxyl groups (in anthraquinones‡). However when two methoxyls occupy *ortho*- or *meta*-positions the shifts may be considerably modified and difficult to interpret. Data of this sort are still limited and the use of other solvents might be explored with advantage. It has been shown[47] recently that isomeric compounds of the rhodoquinone type can be distinguished by their solvent shifts in benzene and pyridine (relative to carbon tetrachloride), the protons of the methyl and methylene groups attached to the quinone ring being chiefly affected.

The location of O-methyl and O-glycosyl substituents in 1,8-dihydroxyanthraquinones is a common structural problem. Steglich and Lösel[80] have found a solution by measuring the acylation shift (Δ_{Ac}) of

† $\Delta_{C_6H_6}^{CDCl_3} = \delta_{CDCl_3} - \delta_{C_6H_6}$ ppm.
‡ Only a few compounds have been studied.

TABLE XV

N.m.r. spectra (60 MHz) of some anthraquinones (τ values)

Anthraquinone	H-2	H-3	H-4	H-5	H-6	H-7	H-8	Me
1,2-Dihydroxy-[a]	—	2·79(d)	2·44(d)		1·7-2·3(m)			—
1,2,4-Trihydroxy-[a]	—	3·44	—		1·7-2·3(m)			
1,2,4-Triacetoxy-[b]	—	2·60	—	1·85(m)	2·23(m)	2·23(m)	1·85(m)	
1,2,4-Trimethoxy-	2·97	—			1·9-2·9(m)			
1,8-Dihydroxy-3-methyl-[b]	3·02[d]	—	2·70[d]	3·02(d)		3·54(d)		7·58
1,6,8-Trihydroxy-3-methyl-[a]	2·92	—	2·40	2·65(d)		3·33(d)		7·65
1,8-Dihydroxy-3-methyl-6-methoxy-[b]	2·90	—	2·36	2·67(d)		3·23(d)		7·52
1,6,8-Trimethoxy-3-methyl-[b]	3·49(d)	—	2·96(d)	—	3·49(d)	—	2·96(d)	7·63
1,3,5,7-Tetrahydroxy-[a]	3·58(d)	—	3·08(d)	—		3·17		
1,3,5,6-Tetrahydroxy-8-methyl-[a]	3·30(d)	—	2·76(d)	—		3·03		7·28[c]
1,3,5,6-Tetramethoxy-8-methyl-[b]	—	—	2·12	2·61(d)		3·28(d)		7·25
1,2,6,8-Tetramethoxy-3-methyl-[b]	3·03	—	2·71	2·84		—		7·63
1,6,7,8-Tetrahydroxy-3-methyl-[a]	2·83	—	—	2·75(d)		3·26(d)		7·62
1,4,6,8-Tetramethoxy-3-methyl-[b]	—	—	2·56	2·83(d)		3·30(d)		7·63
2-n-Hexanoyl-1,3,6,8-tetramethoxy-[b]	3·20	—	—	2·76(d)		3·25(d)		
4-n-Butyryl-1,3,6,8-tetramethoxy-[b]	—	—	—	—		—		

[a] In DMSO.

[b] In CDCl3.

[c] In C5H5N.

[d] Meta-protons at H-2 and H-4 in a 1-hydroxy(methoxy)-3-methyl system may show broadened but unsplit singlets. The methyl signal is also broadened.[49]

nuclear protons, this being the difference in chemical shift observed on comparison of pertrimethylsilyl ethers and per-acetates. Significant differences are found for protons in different environments (see examples below) which are of diagnostic value. Glycosidic substituents may be

Acylation Shifts (Δ_{A_c})†

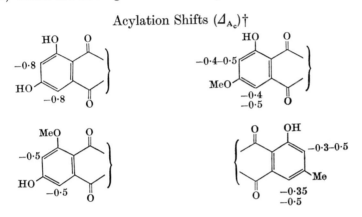

similarly located using Δ'_{Ac} values; figures for emodin 1- and 8-glucoside are shown below.

Acylation Shifts (Δ'_{A_c})‡

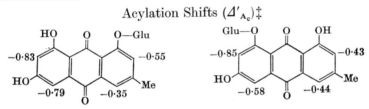

The tautomeric naphthazarin system is a special case. Hitherto redox potentials have been used to assess the equilibrium position since electron releasing substituents attached to a quinone ring lower the potential, and vice versa. Thus [(57a); R = Me] has a lower redox potential than [(57c); R = Me)] and hence a lower energy, and consequently will predominate in solution. (It does not necessarily follow that this is the tautomeric form adopted in the crystalline state, nor that which preferentially takes part in chemical reactions.) In naphthazarin systems 1,5-quinone structures [see (57)] must also be considered although their contribution to the equilibrium is normally small. However, the parent compound has the 1,5-quinone structure [(57b); R = H)] in the crystalline state.[77] The predominant tautomeric form in solution can now be conveniently determined by n.m.r. spectroscopy, and Moore and Scheuer[73] have done this for a large number of

† $\Delta_{Ac} = \delta_{Harom}$ (per-trimethylsilylanthraquinone—δ_{Harom} (per-acetylanthraquinone.
‡ $\Delta'_{Ac} = \delta_{Harom}$ (per-trimethylsilyl-aglycone)—δ_{Harom} (per-acetyl-glycoside).

naphthazarins carrying up to four of the substituents normally encountered in natural quinones of this type.

The n.m.r. spectrum of naphthazarin consists of two singlets; one at $\tau -2\cdot43$ arising from the *peri*-hydroxyl protons and the other, at $\tau\ 2\cdot87$,† is the signal from *both* the aromatic and the quinonoid protons which results from the rapid interconversion of the tautomers. In naphthazarins (57) where R = OH, OMe, OAc, Et, structure (a) predominates.

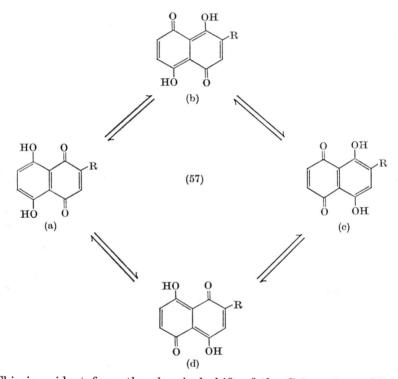

This is evident from the chemical shift of the C-3 proton, which is essentially the same as that for the corresponding 2-substituted-1,4-naphthaquinone or -juglone (cf. Figs 7 and 10), and from the downfield shift of the C-6 and C-7 proton signals. (In juglone the C-6 proton resonates at $\tau\ 2\cdot75$.) Moreover, in [(57); R = Et], the C-3 proton signal is a sharp triplet (J, $1\cdot5$ Hz) clearly coupled with the methylene protons, confirming that a double bond is localized between C-2 and C-3. On the other hand where R = Ac (in 57) the predominant tautomer is (c) as the singlet signal for the two vinyl protons has shifted upfield to $\tau\ 2\cdot92$, and the C-3 proton resonates at $\tau\ 2\cdot44$ (whereas the C-3 proton in 2-acetyl-1,4-naphthaquinone resonates at $\tau\ 2\cdot94$). Similar data for

† Unchanged at $-60°$.[72]

naphthazarins with one substituent in each ring lead to the conclusion that substituents promote quinonoid properties in the ring to which they are attached in the order OH > OMe ≫ OAc > Et ≫ H ≫ Ac, and for two adjacent substituents the order is OH, Ac > OH, OMe > OMe, OMe >

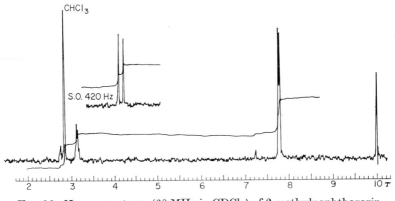

FIG. 10. N.m.r. spectrum (60 MHz in CDCl₃) of 2-methylnaphthazarin.

OMe, Et > OMe, Ac. The position of the pair OH, Ac, at the head of the second list, is attributed to increased stabilisation arising from the tautomeric forms (58) and (59) which involve enolic structures.

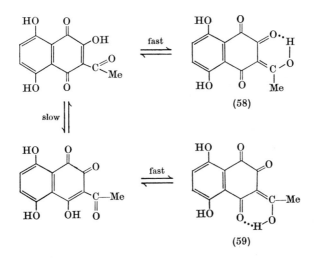

Mass spectra

Features common to the mass spectra of all quinones are peaks corresponding to the loss of one and two molecules of carbon monoxide.[51,75] Benzoquinones and naphthaquinones also eliminate an acetylenic fragment from the quinone ring, and if the latter is hydroxylated breakdown is accompanied by a characteristic hydrogen rearrangement.

Benzoquinones

These compounds normally form abundant molecular ions which frequently give rise to the base peak. The principal fragmentation processes are shown in (60) for the parent compound.[52] In addition loss

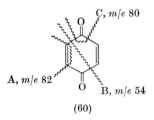

(60)

of two molecules of carbon monoxide gives an important peak at m/e 52. This necessarily requires the formation of at least one carbon–carbon bond and the fragment is most simply represented as ionised

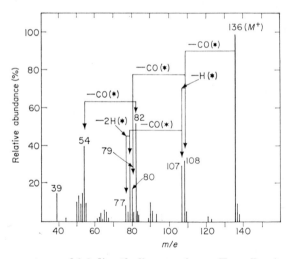

FIG. 11. Mass spectrum of 2,3-dimethylbenzoquinone. From Bowie *et al.* *J. Chem. Soc.* (*B*) (1966), 335.

cyclobutadiene. For unsymmetrical quinones fragmentation A always leads to the elimination of the most highly substituted acetylene, for example, 2,3-dimethylbenzoquinone gives an ion m/e 82 (M—MeC≡CMe) but no M—HC≡CH fragment (Fig. 11). Similarly in fragmentation B, the most highly substituted neutral moiety is eliminated. Appropriate metastable peaks show that this breakdown is frequently a two-step process, elimination of an acetylene being followed by loss of carbon monoxide. Scheme 1 shows the postulated[52] fragmentation path for tetramethylbenzoquinone (Fig. 12). The formation of a hydroxy-tropylium ion (a) is supported by evidence from other spectra which

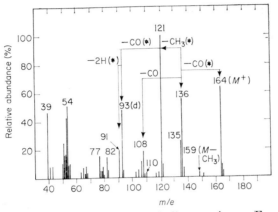

FIG. 12. Mass spectrum of 2,3,5,6-tetramethylbenzoquinone. From Bowie et al. J. Chem. Soc. (B) (1966), 335.

indicates that the presence of at least two methyl (or larger) groups is necessary to allow an M—CO ion to break down by loss of a radical with formation of a relatively stable carbonium ion, and also by the subsequent decomposition of (a) into (b) and (c) which is exactly analogous

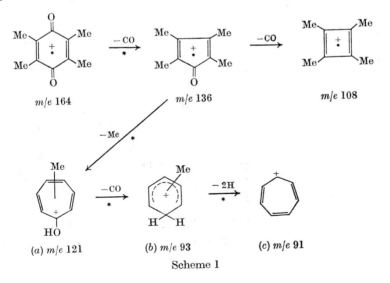

Scheme 1

to the fragmentation of the hydroxytropylium ion in the spectrum of benzyl alcohol.

A distinguishing feature of the spectra of hydroxybenzoquinones arises from the fact that fragmentation C is accompanied by hydrogen

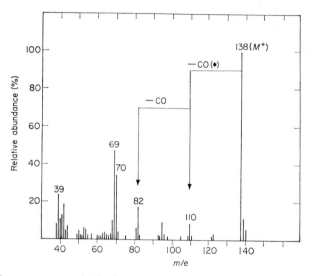

FIG. 13. Mass spectrum of 2-hydroxy-5-methylbenzoquinone. From Bowie *et al.* *J. Chem. Soc. (B)* (1966), 335.

rearrangement. This has been established by deuteriation studies[52] which show, for example, in the breakdown of 2-hydroxy-5-methyl-benzoquinone (Fig. 13), that the ion $C_4H_5O^+$ (60% of m/e 69) shifts to

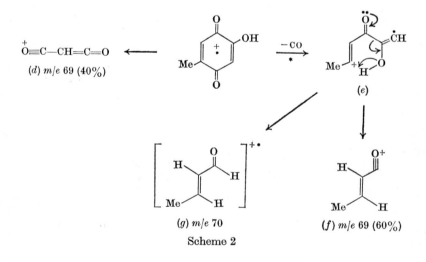

Scheme 2

m/e 70 in the spectrum of the monodeuteriated derivative which means that the hydroxyl group is the principal source of the rearranged hydrogen. The transfer most probably occurs in the M—CO ion (e) (Scheme 2) to give (f) while the $C_4H_6O^+$ peak at m/e 70 (Fig. 13) must be associated with a double hydrogen transfer and is plausibly represented by (g). The remaining part (40%) of m/e 69 is due to $C_3HO_2^+$ (d) which is commonly encountered in the spectra of compounds containing

the $—O—\overset{|}{C}{=}CH—\overset{|}{C}{=}O$ system. The mass spectra of methoxybenzo-quinones are rather more complicated but ions (d) appear regularly, and

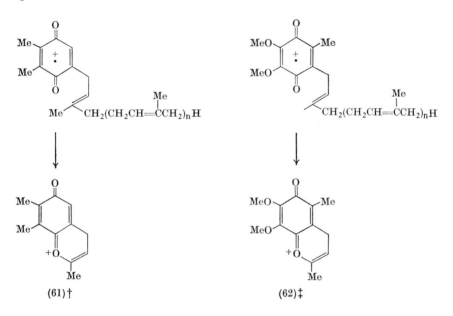

(61)†   (62)‡

M-71 peaks are also prominent. They correspond to the formal loss of two molecules of carbon monoxide and a methyl group from the molecular ion, and further decompose by loss of carbon monoxide to give M-99. These fragments may be represented by (h) and (i), respectively.

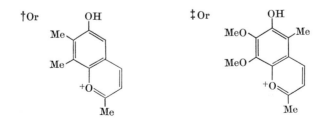

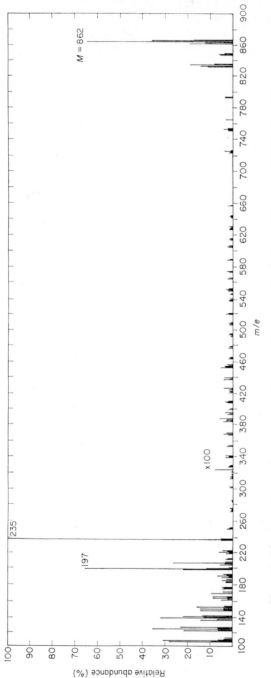

FIG. 14. Mass spectrum of ubiquinone 10. From Muraca *et al. J. Amer. Chem. Soc.* (1967), **89**, 1505.

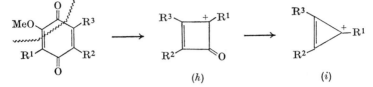

Somewhat different but very characteristic behaviour is shown by the plastoquinones[53] and ubiquinones[54] under electron bombardment. Both groups show strong molecular ion peaks, and base peaks at m/e 189 and 235, respectively, attributed to the pyrylium ions (61) and (62), formed by cleavage of the bond δ to the ring and cyclisation. Between the molecular ion peak and the base peak, a series of weak peaks appears at $[M\text{-}69\text{-}(68)_n]$, due to successive loss of prenyl units (Fig. 14).

Significant M + 2 peaks are frequently observed in the mass spectra of benzoquinones, sometimes increasing in intensity with time. Introduction of deuterium oxide into the ionisation chamber along with a quinone results in the appearance of M + 3 and M + 4 peaks in the mass spectrum, indicating that the M + 2 peaks originate from a reaction involving water.[55, 56] The intensity of the M + 2 peak can be correlated with the redox potential of the quinone.[84] However in some cases the M + 2 ion may arise by hydrogen transfer from a side chain.

No systematic study of the mass spectra of o-benzoquinones has been reported but in three examples they gave M + 2 peaks comparable in intensity with the molecular ion.[56]

Naphthaquinones

The fragmentation processes of 1,4-naphthaquinone[51] and its simple derivatives are essentially similar to those of the benzoquinones. Scheme 3 is illustrative.[57] Usually the molecular ion forms the base peak. The appearance of an abundant ion m/e 104 (j) and its decomposition products m/e 76 (k) and m/e 50 are characteristic for naphthaquinones with no benzenoid substituents. On the other hand substitution in the benzenoid ring causes these peaks to shift to the appropriate higher m/e values. As in the polyprenylbenzoquinones, phylloquinone (63) shows a base peak at m/e 225 ascribed to a pyrylium ion (64) while

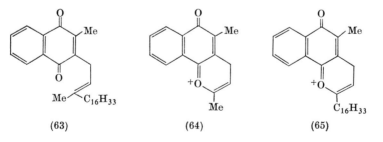

(63) (64) (65)

an intense M-15 peak may be attributed to (65) arising by cleavage of the C—Me bond δ to the quinone ring.[59] Again, as in the benzoquinone series, a 2-hydroxynaphthaquinone undergoes a characteristic hydrogen rearrangement which results in a partial or almost complete replacement

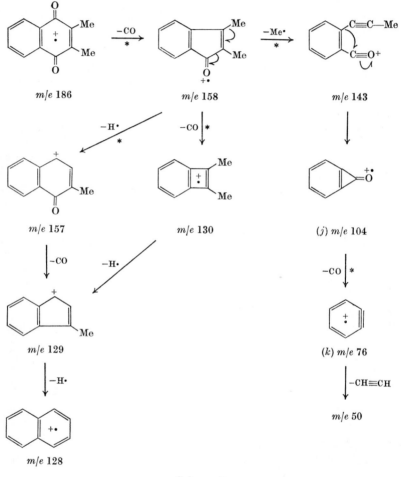

Scheme 3

of the ion m/e 104 (j) by the benzoyl ion (l) at m/e 105 in the case of lawsone (see Scheme 4).[57] However, in hydroxynaphthazarins this rearrangement may be suppressed.[58] Naphthazarin itself is very stable and undergoes remarkably little fragmentation. Comparison of the spectra of 2,5- and 5,7-dihydroxy-1,4-naphthaquinone (Figs 15 and 16) illustrates the differences which arise from substitution in different rings,

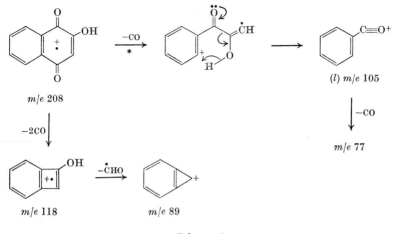

Scheme 4

the "doublet" at M-54 (m/e 136) and M-56 (m/e 134) (Fig. 16) being characteristic for a 1,4-naphthaquinone with no quinonoid substituents. In methoxynaphthaquinones the initial fragmentation tends to involve

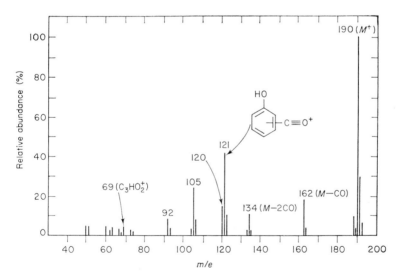

FIG. 15. Mass spectrum of 2,5-dihydroxy-1,4-naphthaquinone. From Bowie *et al. J. Amer. Chem. Soc.* (1965), **87**, 5094.

the methoxyl groups, the naphthaquinone skeleton being broken subsequently. 2-Methoxy-1,4-naphthaquinone affords abundant M—Me and M—CH_2O ions while the 5-methoxy isomer shows an M—CHO peak at m/e 159 represented as (m). Elimination of a second formyl radical leads to m/e 130 (n) which breaks down in the usual way. The same spectrum contains an M—H_2O peak (11%) which is characteristic

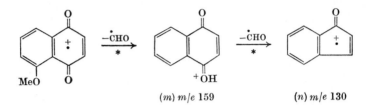

$(m)\ m/e\ 159$ $(n)\ m/e\ 130$

of *peri*-methoxyquinones, the water originating from the carbonyl oxygen and the hydrogen atoms of the methoxyl group.[70] (A *peri*-ethoxyquinone fragments initially by loss of Me· followed by ·CHO.[71])

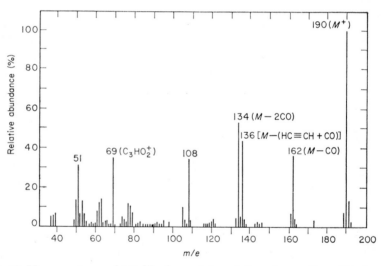

FIG. 16. Mass spectrum of 5,7-dihydroxy-1,4-naphthaquinone. From Bowie *et al.* *J. Amer. Chem. Soc.* (1965), **87**, 5094.

The presence of a C-acetyl group has a marked effect on the fragmentation pattern.[58] 2-Acetyl-1,4-naphthaquinone breaks down by loss of a methyl radical followed by carbon monoxide to give a peak at M-43, and also by elimination of ketene to give M-42. Further losses of carbon monoxide and acetylene from these two peaks gives a characteristic series of "doublets" (Fig. 17). This behaviour is considerably modified if a hydroxyl group is adjacent to the acetyl group (Fig. 18). Elimination of ketene is now largely suppressed, the major decomposition being loss of carbon monoxide followed by a methyl radical to give the base peak at m/e 173 (p). A minor fragmentation involving loss of water from (o)

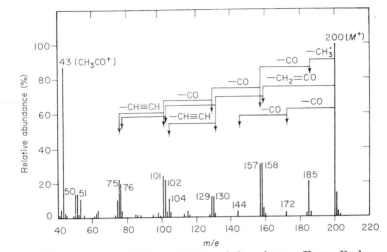

Fig. 17. Mass spectrum of 2-acetyl-1,4-naphthaquinone. From Becher et al. J. Org. Chem. (1966), **31**, 3650.

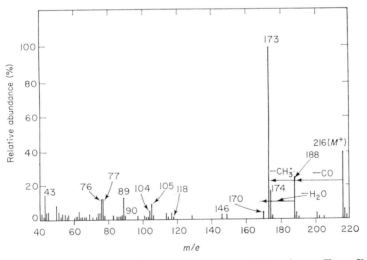

Fig. 18. Mass spectrum of 3-acetyl-2-hydroxy-1,4-naphthaquinone. From Becher et al. J. Org. Chem. (1966), **31**, 3650.

to give (q) is diagnostic for vicinal acetyl and hydroxyl groups. These and related fragmentations have been applied extensively by Djerassi and his co-workers[58] to the highly substituted spinochrome pigments.

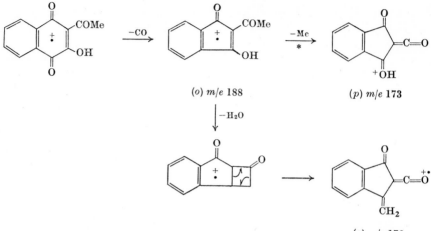

(o) m/e 188

(p) m/e 173

(q) m/e 170

The behaviour of 1,2-naphthaquinones under electron impact is similar to that of the 1,4-isomers except that the ion m/e 104 (j) does not arise. The o-isomers can usually be distinguished by the fact that their molecular ions are weak and the M + 2 peaks are of similar or greater intensity (cf. ref. 86).[56, 60]

Anthraquinones

In this series the molecular ion almost invariably forms the base peak. Anthraquinone itself[61] undergoes successive elimination of two molecules of carbon monoxide to give strong peaks at m/e 180 (M—CO) and 152

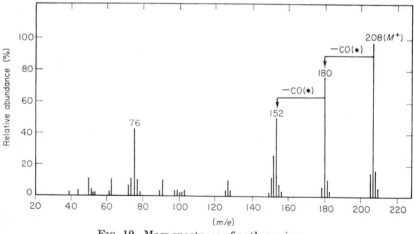

FIG. 19. Mass spectrum of anthraquinone.

(M—2CO) (and strong doubly charged ions at m/e 90 and 76) which correspond to the molecular ions of fluorenone and biphenylene, respectively. Otherwise there is very little fragmentation (Fig. 19). The spectra of derivatives follow the same pattern with additional peaks appropriate to the substituents and to their α- or β-orientation.[51,74] 2-Hydroxyanthraquinone shows more intense M—CO and M—HO peaks than does the chelated 1-isomer and both spectra have a peak at m/e 140 corresponding to the loss of three molecules of carbon monoxide, the third arising from the phenolic group in the normal manner. A very stable ion at m/e 139 (M—2CO—CHO) may be (66) or (67). Dihydroxy-

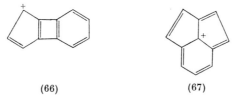

(66)   (67)

anthraquinones behave similarly and a peak at M—4CO (m/e 128) may be the molecular ion of naphthalene. As in the naphthaquinone series, a $peri$-methoxy group gives rise to M—HO and M—H$_2$O peaks which are not observed in β-methoxy derivatives,[70] and 1,8-dimethoxyanthraquinone can be distinguished from the 1,5-isomer by virtue of its abundant M—Me ion.[71] Both α- and β-methoxy compounds eliminate a formyl radical (M—CHO) but again this fragmentation is only significant for $peri$-isomers.[51]

REFERENCES

1. T. Furuya, S. Shibata and H. Iizuka, *J. Chromatog.* (1966), **21**, 116; M. Kotakemori and M. Ando, *J. Agric. Chem. Soc. Japan* (1968), **42**, 726; G. H. Dialameh and R. E. Olson, *Anal. Biochem.* (1969), **32**, 263.
2. A. R. Burnett and R. H. Thomson, *J. Chem. Soc.* (C) (1967) 2100; (1968), 2437.
3. N. R. Rao, K. H. Shah and K. Venkataraman, *Curr. Sci.* (1950), **19**, 149; (1951), **20**, 66; M. R. Padhye, N. R. Rao and K. Venkataraman, *J. Sci. Ind. Res.* (*India*) (1954), **13B**, 759; N. S. Bhide, B. S. Joshi, V. Patwardhan, R. Srinivasan and K. Venkataraman, *Bull. Nat. Inst. Sci. India* (1965), No. 28, 114.
4. F. Feigl and V. Anger, "Spot Tests in Organic Analysis", 7th English edit. (1966). Elsevier, Amsterdam.
5. S. Shibata, M. Takito and O. Tanaka, *J. Amer. Chem. Soc.* (1950), **72**, 2789.
6. H. Brockmann and W. Müller, *Chem. Ber.* (1958), **91**, 1920.
7. R. Craven, *J. Chem. Soc.* (1931), 1605.
8. W. Kesting, *Ber.* (1929), **62**, 1422.
9. W. Brackman and E. Havinga, *Rec. Trav. Chim.* (1955), **74**, 1021.
10. H. Karius and G. E. Mapstone, *Chem. Ind.* (*London*) (1956), 266.
11. H. Brockmann and G. Budde, *Chem. Ber.* (1953), **86**, 432.
12. R. Ott, *Monatsh.* (1959), **90**, 827; H. Immer, G. Kunesch, J. Polonsky and E. Wenkert, *Bull. Soc. Chim. France* (1968), 2420.
13. E. Clar, *Ber.* (1939), **72**, 1645.
14. R. L. Edwards and N. Kale, *Tetrahedron* (1965), **21**, 2095.

15. C. J. Sanchorawala, B. C. Subba Rao, M. K. Unni and K. Venkataraman, *Indian J. Chem.* (1963), **1**, 19 and references therein.
16. H. Ogawa and S. Natori, *Chem. Pharm. Bull.* (*Tokyo*) (1968), **16**, 1709; D. W. Cameron, R. I. T. Cromartie, D. G. I. Kingston and Lord Todd, *J. Chem. Soc.* (1964), 51.
17. H. C. Longuet-Higgins and J. N. Murrell, *Proc. Phys. Soc.* (1955), **68A**, 60; J. N. Murrell, *J. Chem. Soc.* (1956), 3779.
18. R. A. Morton (ed.), "Biochemistry of Quinones" (1965). Academic Press, New York, and references therein.
19. W. Flaig, J.-C. Salfeld and E. Baume, *Annalen* (1958), **618**, 117.
20. S. Natori, H. Nishikawa and H. Ogawa, *Chem. Pharm. Bull.* (*Tokyo*) (1964), **12**, 236.
21. V. C. Farmer and R. H. Thomson, *Chem. Ind.* (*London*) (1957), 112.
22. H.-J. Teuber and G. Staiger, *Chem. Ber.* (1955), **88**, 802; W. Flaig, Th. Ploetz and A. Küllmer, *Z. Naturforschung* (1955), **10b**, 668.
23. C. J. P. Spruit, *Rec. Trav. Chim.* (1949), **68**, 309.
24. I. Singh, R. T. Ogata, R. E. Moore, C. W. J. Chang and P. J. Scheuer, *Tetrahedron* (1968), **24**, 6053.
25. R. G. Cooke, A. K. Macbeth and F. L. Winzor, *J. Chem. Soc.* (1939), 878; H.-J. Teuber and N. Götz, *Chem. Ber.* (1954), **87**, 1236.
26. C. J. P. Spruit, *Rec. Trav. Chim.* (1949), **68**, 325; Z. Yoshida and F. Takabayashi, *Tetrahedron* (1968), **24**, 933.
27. T. Ikeda, Y. Yamamoto, K. Tsukida and S. Kanatomo, *J. Pharm. Soc. Japan* (1956), **76**, 217.
28. L. H. Briggs, G. A. Nicholls and R. M. L. Paterson, *J. Chem. Soc.* (1952), 1718; J. H. Birkinshaw, *Biochem. J.* (1955), **59**, 485; J. H. Birkinshaw and R. Gourlay, *Biochem. J.* (1961), **80**, 387.
29. R. A. Morton and W. T. Earlam, *J. Chem. Soc.* (1941), 159.
30. R. H. Peters and H. H. Sumner, *J. Chem. Soc.* (1953), 2101.
31. G. A. Swan, *J. Chem. Soc.*, 1948, 1408; D. G. Davies and P. Hodge, *Chem. Comm.* (1968), 953.
32. P. Yates, M. I. Ardao and L. F. Fieser, *J. Amer. Chem. Soc.* (1956), **78**, 650; W. Flaig and J.-C. Salfeld, *Annalen* (1959), **626**, 215.
33. H. Ogawa and S. Natori, *Chem. Pharm. Bull.* (*Tokyo*) (1968), **16**, 1709.
34. M. St. C. Flett, *J. Chem. Soc.* (1948), 1441; O. Tanaka, *Pharm. Bull.* (*Tokyo*) (1958), **6**, 18.
35. H. Bloom, L. H. Briggs and B. Cleverley, *J. Chem. Soc.* (1959), 178.
36. M.-L. Josien, N. Fuson, J.-M. Lebas and T. M. Gregory, *J. Chem. Phys.* (1953), **21**, 331.
37. T. L. Brown, *Spectrochim. Acta* (1962), **18**, 1065; E. D. Becker, H. Ziffer and E. Charney, *Spectrochim. Acta* (1963), **19**, 1871.
38. R. A. Durie, R. E. Lack and J. S. Shannon, *Austral. J. Chem.* (1957), **10**, 429.
39. W. Otting and G. Staiger, *Chem. Ber.* (1955), **88**, 828.
40. O. Isler and O. Wiss, *Vitamins and Hormones* (1959), **17**, 53.
41. J. G. Hawkins, E. R. Ward and D. H. Whiffen, *Spectrochim. Acta* (1957), **10**, 105.
42. R. K. Norris and S. Sternhell, *Austral. J. Chem.* (1966), **19**, 617.
43. E. W. Garbisch, *Chem. Ind.* (*London*) (1964), 1715; C. M. Orlando and A. K. Bose, *J. Amer. Chem. Soc.* (1965), **87**, 3782.
44. J. T. Pinhey and S. Sternhell, *Tetrahedron Lett.* (1963), 275.
45. J. A. Ballantine and C. T. Pillinger, *Tetrahedron* (1967), **23**, 1691.
46. J. H. Bowie, D. W. Cameron, P. E. Schultz, D. H. Williams and N. S. Bhacca, *Tetrahedron* (1966), **22**, 1771.

47. J. J. Wilczynski, G. D. Daves and K. Folkers, *J. Amer. Chem. Soc.* (1968), **90**, 5593.
48. E. R. Wagner, R. D. Moss and R. M. Brooker, *Tetrahedron Lett.* (1965), 4233.
49. A. W. K. Chan and W. D. Crow, *Aust. J. Chem.* (1966), **19**, 1704.
50. K. Venkataraman, *J. Sci. Ind. Res. (India)* (1966), **25**, 97.
51. J. H. Beynon and A. E. Williams, *Appl. Spectroscopy* (1960), **14**, 156.
52. J. H. Bowie, D. W. Cameron, R. G. F. Giles and D. H. Williams, *J. Chem. Soc. (B)* (1966), 335.
53. D. Misiti, H. W. Moore and K. Folkers, *J. Amer. Chem. Soc.* (1965), **87**, 1402; B. C. Das, M. Lounasmaa, C. Tendille and E. Lederer, *Biochem. Biophys. Res. Comm.* (1965), **21**, 318; W. T. Griffiths, *Biochem. Biophys. Res. Comm.* (1966), **25**, 596.
54. R. F. Muraca, J. S. Whittick, G. D. Daves, P. Friis and K. Folkers, *J. Amer. Chem. Soc.* (1967), **89**, 1505; H. Morimoto, T. Shima, I. Imada, M. Sasaki and A. Onchida, *Annalen* (1967), **702**, 137.
55. R. T. Aplin and W. T. Pike, *Chem. Ind. (London)* (1966), 2009.
56. S. Ukai, K. Hirose, A. Tatematsu and T. Goto, *Tetrahedron Lett.* (1967), 4999.
57. J. H. Bowie, D. W. Cameron and D. H. Williams, *J. Amer. Chem. Soc.* (1965), **87**, 5094.
58. D. Becher, C. Djerassi, R. E. Moore, H. Singh and P. J. Scheuer, *J. Org. Chem.* (1966), **31**, 3650.
59. S. J. Di Mari, J. H. Supple and H. Rapoport, *J. Amer. Chem. Soc.* (1966), **88**, 1226.
60. R. W. A. Oliver and R. M. Rashman, *J. Chem. Soc. (B)* (1968), 1141.
61. J. H. Beynon, G. R. Lester and A. E. Williams, *J. Phys. Chem.* (1959), **63**, 1861; J. H. Beynon. In "Advances in Mass Spectrometry" (ed. J. D. Waldron) (1959), p. 328. Pergamon Press, London.
62. J. A. D. Jeffreys, *J. Chem. Soc.* (1959), 2153; T. J. King and C. E. Newall, *J. Chem. Soc.* (1965), 974.
63. C. H. Shunk, J. F. McPherson and K. Folkers, *J. Org. Chem.* (1960), **25**, 1053; M. Akatsuka, *J. Pharm. Soc. Japan* (1970), **90**, 160.
64. H. Dam, A. Geiger, J. Glavind, P. Karrer, W. Karrer, E. Rothschild and H. Salomon, *Helv. Chim. Acta* (1939), **22**, 310; P. Karrer, *Helv. Chim. Acta* (1939), **22**, 1146; L. F. Fieser, W. P. Campbell and E. M. Fry, *J. Amer. Chem. Soc.* (1939), **61**, 2206.
65. H. Brockmann, E. H. Frhr. von Falkenhausen, R. Neeff, A. Dorlars and G. Budde, *Chem. Ber.* (1951), **84**, 865.
66. E. T. Jones and A. Robertson, *J. Chem. Soc.* (1930), 1699.
67. O. Dimroth, *Annalen* (1925), **446**, 97; O. Dimroth and Th. Faust, *Ber.* (1921), **54**, 3020.
68. H. A. Anderson, J. W. Mathieson and R. H. Thomson, *Comp. Biochem. Physiol.* (1969), **28**, 333.
69. R. R. Hill and G. H. Mitchell, *J. Chem. Soc. (B)* (1969), 61.
70. J. H. Bowie and P. Y. White, *J. Chem. Soc. (B)* (1969), 89.
71. J. H. Bowie, P. J. Hoffmann and P. Y. White, *Tetrahedron* (1970), **26**, 1163.
72. H. Brockmann and A. Zeeck, *Chem. Ber.* (1968), **101**, 4221.
73. R. E. Moore and P. J. Scheuer, *J. Org. Chem.* (1966), **31**, 3272.
74. J. H. Beynon, R. A. Saunders and A. E. Williams, "The Mass Spectra of Organic Molecules" (1968), p. 206. Elsevier, Amsterdam.
75. H. Budzikiewicz, C. Djerassi and D. H. Williams, "Mass Spectrometry of Organic Compounds" (1967). Holden-Day, San Francisco.

76. T. Yoshimoto, *J. Chem. Soc. Japan* (*Pure Chem. Sect.*) (1963), **84**, 733.
77. C. Pascard-Billy, *Bull. Soc. Chim. France* (1962), 2282, 2293, 2299; *Acta Cryst.* (1962), **15**, 519.
78. B. O. Linn, A. C. Page, E. L. Wong, P. H. Gale, C. H. Shunk and K. Folkers, *J. Amer. Chem. Soc.* (1959), **81**, 4007.
79. I. C. Patel and D. M. Skauen, *J. Pharm. Sci.* (1969), **58**, 1135.
80. W. Steglich and W. Lösel, *Tetrahedron* (1969), **25**, 4391.
81. A. I. Scott, "Interpretation of the Ultraviolet Spectra of Natural Products" (1964). Pergamon Press, London.
82. M. H. Simatupang and B. M. Hausen, *J. Chromatog.* (1970), **52**, 180.
83. R. W. Crecely, K. M. Crecely and J. H. Goldstein, *J. Mol. Spectrosc.* (1969), **32**, 407.
84. J. Heiss, K.-P. Zeller and A. Rieker, *Org. Mass Spec.* (1969), **2**, 1325.
85. E. Geyer, personal communication, 1970.
86. T. A. Elwood, K. H. Dudley, J. M. Tesarek, P. F. Rogerson and M. M. Busley, *Org. Mass Spec.* (1970), **3**, 841.

CHAPTER 3

Benzoquinones

Very simple benzoquinones are highly reactive substances and their absence from natural sources would occasion no surprise. However they do occur, although rarely, in higher plants, and in fact there are two natural o-benzoquinones† which, not surprisingly, are highly substituted. p-Benzoquinone is an insect metabolite but only the reduced form is found in plants, usually as the mono-β-D-glucoside, arbutin,[25, 151] or its monomethyl ether (methylarbutin). The resistance of pear trees (*Pyrus communis*) to fire blight is associated with the relatively high concentration of arbutin, and the free quinol, in the leaves.[252] Quinol is present in the heartwood of *Pinus resinosa* Ait.,[26] and the sapwood of *P. radiata* D. Don[27] (Pinaceae), and it is the toxic principle in the burs of *Xanthium canadense* Mill.[33] (Compositae) which are poisonous to animals. Similarly, toluquinone, a common arthropod metabolite, appears in the plant kingdom as the corresponding quinol, again usually bound to glucose (homo-arbutin, isohomo-arbutin),[25] and a series of long-chain alkylquinols is present in the exudate of *Campnosperma*

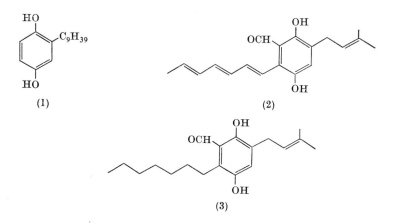

(1)　　　　　　　　(2)

(3)

† Mansonone A is discussed in Chapter 4.

auriculata Hook. f. (Anacardiaceae), of which the nonadecyl homologue (1) has been identified.[28] The mould products auroglaucin (2)[29] and flavoglaucin (3)[29] (from *Aspergillus* spp.) are more elaborate quinols, and these also appear to exist entirely in the reduced form. Catechols occur frequently and some of these take part, by way of their o-quinones, in the formation of complex products such as melanins,[30] and tanned protein in insect cuticle.[31]

Arthropods produce a variety of substances for use as defensive agents against predators including formic and other fatty acids, hydrogen cyanide, phenols, terpenes and quinones.[32, 454] Quinones are much favoured and eight (4–11)† have been found in arachnids, millipedes, beetles, earwigs and termites (see Table I). Indeed Schildknecht[23] suggests that black beetles (Tenebrionidae) could more appropriately be described as "quinone beetles" as quinones are invariably present. The secretions are often discharged as a fine spray, sometimes with great accuracy, and are undoubtedly effective in repelling predators. Remarkable photographic proof of this has been obtained recently by Eisner and his colleagues.[469] By inducing bombardier beetles to discharge onto a thermocouple which acted as the trigger to an electronic flash unit they were able to photograph the spray at the moment of emergence (see Fig. 1). Quinols are

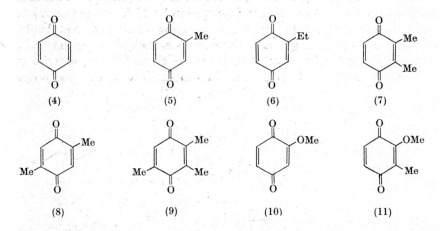

usually present in the defensive glands, both free and as monogluco-sides,[453] and sometimes other substances. In bombardier beetles (*Brachynus* spp.) an aqueous solution of quinols and hydrogen peroxide is passed to an outer compartment where an explosive reaction is triggered off by enzymic action.[407, 461, 474] When attacked, these carabids *audibly* discharge a mixture of quinone vapours and oxygen,

† The quinol of (10) also occurs in wheat germ (see p. 107).

(a)

(b)

FIG. 1. (a) Bombardier beetle discharging. The beetle is fastened with wax to a hook on its back. The arrow points to the thermocouple used to prod the beetle and to trigger the electronic flash unit. (b), (c) and (d) Bombardier beetle under attack by an ant. (b) Approaching.

(c)

(d)

FIG. 1. (c) Poised for attack. (d) Ant bites the beetle's left leg and is accurately sprayed in return. From D. J. Aneshansley, T. Eisner, J. M. Widom, and B. Widom, *Science N.Y.* (1969). **165** (no. 3888), 61-3. (Copyright 1969 by the American Association for the Advancement of Science.)

TABLE I

Benzoquinones[e] in Arthropoda

Arthropod	Quinone				Reference
	Benzoquinone (m.p. 116°)	Methyl-benzoquinone (m.p. 69°)	Ethyl-benzoquinone (m.p. 39·5°)	2-Methoxy-3-methyl-benzoquinone (m.p. 19–20°)	
ARACHNIDA					
Phalangida					
Heteropachyloidellus robustus Roewer[a]					1, 2
DIPLOPODA					
Archiulus sabulosus (L.)		+		+	12
Aulonopygus aculeatus Attems[d]		+			20
Aulonopygus aculeatus barbieri Demange		+			20
Brachyiulus unilineatus Koch		+		+	6
Cambala hubrichti Hoffman		+		+	22
Chicobolus spinigerus (Wood)		+		+	3
Cylindroiulus teutonicus Pocock		+		+	6
Doratogonus annulipes Carl		+		+	22
Floridobolus penneri Causey		+			3
Julus terrestris L.	+?				24
Narceus annularis (Rafinesque)		+		+	3
Narceus gordanus (Chamberlin)		+		+	3
Orthoporus conifer (Attems)		+		+	22
Orthoporus flavior Chamberlin and Mulaik		+		+	22
Orthoporus punctilliger Chamberlin		+			22
Pachybolus laminatus Cook		+			14
Peridontopyge aberrans Attems		+			20

TABLE I—continued

Arthropod	Quinone				Reference
	Benzoquinone (m.p. 116°)	Methyl-benzoquinone (m.p. 69°)	Ethyl-benzoquinone (m.p. 39·5°)	2-Methoxy-3-methyl-benzoquinone (m.p. 19–20°)	
DIPLOPODA—continued					
Peridontopyge vachoni Demange		+			20
Rhinocricus insulatus (Chamberlin)		+			21
Spirostreptus castaneus Attems	+				14
Spirostreptus multisulcatus Demange		+			20
Spirostreptus virgator Silvestri		+			14
Trigoniulus lumbricinus (Gerstäcker)				+	3
INSECTA					
Dictyoptera					
Diploptera punctata (Eschscholtz)	+	+	+		13
Isoptera					
Mastotermes darwiniensis Frogg.	+	+			405
Dermaptera					
Forficula auricularia L.		+	+		9
Coleoptera					
Arthropterus sp.		+	+		492
Blaps gigas Fischer	+	+	+		23
Blaps lethifera Marsham	+	+	+		10
Blaps mortisaga L.		+	+		8, 10
Blaps mucronata Latreille		+	+		10
Blaps requienii Solier		+			10
Brachinus crepitans L.	+	+			4, 15
Brachinus explodens Duft.	+	+			4
Brachinus sclopeta Fabr.	+	+			4

Species	1	2	3	4	Ref.[e]
Callistus lunatus Fab.		+	+		406
Chlaenius vestitus Paykull	+	+	+		406
Clivina basalis Chaudoir	+	+			492
Clivina fossor L.	+	+	+		406
Diaperis boleti L.		+	+		23
Diaperis maculata Olivier		+	+	+	13
Eleodes hispilabris Say		+	+		5
Eleodes longicollis Leconte	+	+	+		7
Gnaptor spinimanus Pallas[b]		+	+		10
Helops quisquilius Strm.		+	+		23
Leichenum caniculatum variegatum Klug	+	+			455
Morica planata tingitana Baudi[b]	+	+			10
Mystropomus regularis Bänn.		+	+		492
Opatroides punctulatus Brullé		+	+		23
Opatrum sabulosum L.		+	+		23
Paussus favieri Fairmaire	+	+	+		456
Pheropsophus catoirei Dejean[b]	+	+	+		4
Pheropsophus verticalis Dejean		+	+		492
Pimelia confusa Senac[b]		+			10
Prionychus ater Fabr.		+	+		23
Stenomax aeneus Scop.		+	+		23
Tenebrio molitor L.		+			11
Tenebrio obscurus Fabr.[b]	+		+		10
Tribolium castaneum (Herbst.)[c]		+	+		13, 16, 17, 18
Tribolium confusum J. du Val		+			16, 17, 19
Zophobas rugipes Kirsch	+	+			38

[a] Secretes gonylyleptidine, a 71:11:15 mixture of 2,3-dimethyl- (m.p. 55°), 2,5-dimethyl- (m.p. 125°) and 2,3,5-trimethylbenzoquinone (m.p. 32°).

[b] Museum specimen.

[c] A trace of methoxy-1,4-benzoquinone (m.p. 145°) was also found.[18]

[d] Identified by t.l.c. in one system only.

[e] For ultraviolet and infrared data see Tables II and XI in Chapter 2.

the pressure of the latter providing the driving force. The discharge is not only chemically but also thermally repellent, the spray reaching a temperature of 100°C.[469] The quinone/quinol content varies being 20 μg per insect in *Blaps mortisaga*,[23] and as high as 329 μg per insect in the flour beetle *Tribolium brevicornis*.[34] The relatively high quinone content of the common grain and flour beetles is of some concern as stored food in mills and warehouses is frequently infested by these insects.[35] Flour reacts rapidly with simple quinones becoming pink and unpalatable, owing, presumably, to combination with protein by reaction with amino and thiol groups, and all the quinones secreted by these insects are potentially harmful, and possibly carcinogenic.[36] The problem is world-wide and some indication of its magnitude is seen in a recent report[37] on infested sunflower seeds in Yugoslavia which disclosed that 95,000 tons of seeds were harbouring 18 thousand million *Tribolium castaneum*, equivalent to almost a ton of quinones.[34]

These simple quinones are usually identified by their spectral and chromatographic properties and those of the derived 2,4-dinitrophenyl-hydrazones, or by g.l.c., but in an earlier investigation[1, 2] of gonylepti-dine, an antibiotic secretion from a South American phalangid, suf-ficient material was available to permit separation and identification by chemical means. The major component, 2,3-dimethylbenzoquinone (7), was isolated as its Diels-Alder adduct with 2,3-dimethylbutadiene. The unreacted quinones from the addition reaction were then subjected to Thiele acetylation which gave the leucodiacetate of (8), and un-changed (9).

Thymoquinone (12), $C_{10}H_{12}O_2$

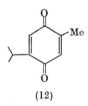

(12)

Occurrence: essential oil of *Seseli hippomarathrum* Jacq.[44] (Umbelliferae), *Monarda fistulosa* L.,[46, 47] (= *M. menthaefolia* R. Grah.) and *Nepeta leucophylla* Benth.[103] (Labiatae), the seeds of *Carum roxburghianum* Benth.[53] (Umbelliferae) and *Nigella sativa* L.[45] (Ranunculaceae), leaves of *Eupatorium japonicum* Thunb.[72] (Compositae), heartwood of *Tetraclinis articulata* (Vahl) Masters,[48, 49] *Juniperus chinensis* (Sab.),[50] *J. cedrus* L.[51] and *Heyderia decurrens* (Torr.) K. Koch[52] (= *Libocedrus decurrens* Torr.) (Coniferae).

Physical properties: yellow tablets, m.p. 45° (49–50°), $\lambda_{max.}$ (EtOH) 276, 282(sh) nm (log ε 3·41, 3·39), $\nu_{max.}$ (CCl₄) 1656, 1610 cm⁻¹.

Nearly all the sources listed above contain thymoquinol which has usually been isolated by steam-distillation during which it is partly oxidised to the quinone. In many cases it may be an artefact rather than a true natural product although it exists as such in the heartwood of the incense cedar *Heyderia decurrens*.[52] In some Compositae the quinol is present as its dimethyl ether.[79] Both the quinone and the quinol are toxic to wood-rotting fungi and no doubt they contribute to the durability of Coniferae heartwoods.

Thymoquinone is easily identified by its n.m.r. spectrum. It forms a photodimer (probably 13) on exposure to light and has been isolated in this form.[45]

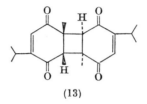

(13)

γ,γ-Dimethylallyl-1,4-benzoquinone (14), $C_{11}H_{12}O_2$

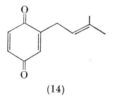

(14)

Occurrence: aerial parts of *Phagnalon saxitale* Cass.[55] (Compositae).
Physical properties: yellow crystals, m.p. 30·5°, $\lambda_{max.}$ (ether) 243, 315, 440 nm (log ε 4·28, 2·87, 1·52), $\nu_{max.}$ (CCl$_4$) 3320,† 3280,† 1665, 1603 cm^{-1}.

The n.m.r. spectrum of this simple quinone is complicated by extensive long-range coupling but that of the quinol, also present in *P. saxitale*, shows clearly the characteristic signals for a γ,γ-dimethylallyl group. As the quinol has three aromatic protons this means that the natural quinone, $C_{11}H_{12}O_2$, has structure (14). This was supported by the mass spectral fragmentation pattern and by the formation of acetone on ozonolysis, and was confirmed by synthesis. Condensation of quinol with 2-methylbut-3-en-2-ol in the presence of boron trifluoride gave 2-(γ,γ-dimethylallyl)quinol which was converted into the quinone with silver oxide.[55]

† Impurity?

5-(γ,γ-Dimethylallyl)-2-methylbenzoquinone (15), $C_{12}H_{14}O_2$

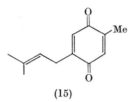

(15)

Occurrence: leaves of *Pyrola media* Sw.[56] (Pyrolaceae).
Physical properties: yellow oil, λ_{max}. (EtOH) 251, 258(sh), 310 nm (log ϵ 4·28, 4·23, 2·50), ν_{max}. (film) 2960, 2920, 1659, 1610 cm^{-1}.

The leaves of *P. media* contain several quinones and phenolic compounds, the monocyclic products being quinol mono- and dimethyl ethers, toluquinol, and the quinones (15) and (16). The C_{12} quinone is a light-sensitive oil showing ultraviolet absorption similar to that of 2,5-dimethyl-1,4-benzoquinone; the n.m.r. spectrum shows the expected signals for methyl and γ,γ-dimethylallyl groups each adjacent to a free position on a quinone ring, and two ring protons which appear as a multiplet. The orientation of the two substituents was shown by an oxidative cyclisation, catalysed by boron trifluoride, in an atmosphere of oxygen which gave 2,7-dimethyl-1,4-naphthaquinone (chimaphilin, p. 199), another constituent of *P. media*. The C_{12} quinone is therefore (15) which was confirmed by a conventional condensation of toluquinol with 2-methylbut-3-en-2-ol, followed by oxidation with silver oxide.[56]

5-Geranyl-2-methyl-1,4-benzoquinone (16), $C_{17}H_{22}O_2$

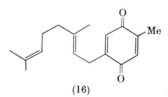

(16)

Occurrence: leaves of *Pyrola media* Sw.[56] (Pyrolaceae).
Physical properties: yellow oil, λ_{max}. (hexane) 250, 257, 308 nm (log ϵ 4·35, 4·28, 2·43), ν_{max}. (film) 2959, 2918, 1660, 1612 cm^{-1}.

The structure of this quinone was immediately suggested by its molecular formula, light-sensitivity and similar spectral properties to those of (15), and by the fact that the quinol of (16) occurs as a glucoside, pyrolatin,† in *P. japonica*.[57] The tetrahydro derivative of the aglycone

† It has been shown recently[408] that the terminal grouping is actually

Me
$\diagup$
$|$
CH$_2$OH

is, in fact, obtained when the quinone is hydrogenated over palladised charcoal, three mols. being absorbed. The n.m.r. spectrum of the quinone is fully consistent with structure (16), the identical splitting pattern of the ring protons in both (15) and (16) indicating that they have the same orientation. It was synthesised by condensing toluquinol with geraniol in the usual way, and subsequent oxidation with silver oxide.[56]

Gentisylquinone (17), $C_7H_6O_3$

(17)

Occurrence: as a quinhydrone complex in cultures of *Penicillium urticae* Bain.[39] (= *P. patulum* Bain.).

Physical properties: yellow needles, m.p. 75–76°, $\lambda_{max.}$ (EtOH) 247, 432 nm (log ε 4·38, 1·32), $\nu_{max.}$ (Nujol) 3350(sh), 3250, 1670(sh), 1660, 1600(sh) cm⁻¹.

The free quinone has not been isolated but several micro-organisms elaborate the quinol (gentisyl alcohol), and a deep violet quinhydrone complex, $C_7H_6O_3.3C_7H_8O_3$, m.p. 86–89°, has been obtained from *P. urticae*. It can be oxidised to gentisylquinone, reduced to gentisyl alcohol, and reconstituted by mixing ethereal solutions of both. The quinone–quinol ratio was determined polarographically and by titration with iodine.

Gentisyl alcohol is also a metabolite of certain unidentified *Phoma* spp.[400] together with the related product, epoxydon (18) (cf. terremutin p. 116) and is found in the bark of *Populus balsamifera* (Salicaceae).[450] Gentisaldehyde and gentisic acid are present in cultures of *P. urticae*,[40] and the acid occurs in other *Penicillia*[41] and is very widespread in leaf tissue.[42]

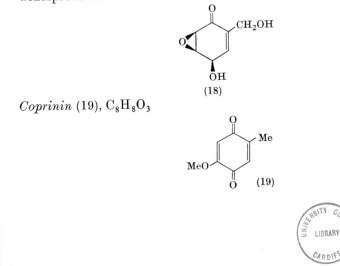

(18)

Coprinin (19), $C_8H_8O_3$

(19)

Occurrence: cultures of *Lentinus degener* Kalchbr.[58, 59] and *Coprinus radians* (Desm.) Fr. (= *C. similis* Berk. & Br.).[58]

Physical properties: yellow spangles, m.p. 175°, $\lambda_{max.}$ (CHCl$_3$) 264, 365 nm (log ϵ 4·32, 2·88), $\nu_{max.}$ (CCl$_4$) 1681, 1653 cm^{-1}.

This pigment is one of two toluquinones produced by *L. degener*. Only the quinols are present in the early stages of growth and the subsequent autoxidation can be accelerated by removal of the mycelium, and by aeration.[59] The compound is most conveniently identified by its n.m.r. spectrum, and can be prepared by the addition of methanol to toluquinone, catalysed by zinc chloride.[60] However, the yield is poor, and it is more satisfactory to methylate the hydroxyquinone which can be obtained by Thiele acetylation of toluquinone followed by oxidative hydrolysis.[61]

Phoenicin (20), C$_{14}$H$_{10}$O$_6$

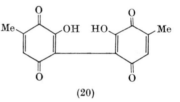

(20)

Occurrence: cultures of *Penicillium phoeniceum* van Beyma,[64, 65] *P. rubrum* O. Stoll,[65, 66] and *P. chermisinum* Biourge.[67]

Physical properties: golden-yellow plates, m.p. 230–231° (dec.), $\lambda_{max.}$ (CHCl$_3$) 268, 406 nm (log ϵ 4·52, 3·36), $\nu_{max.}$ (KBr) 3236, 1662, 1653(sh), 1647, 1611 cm^{-1}.

The structure of phoenicin, the first naturally occurring biquinone† to be discovered, was determined by Posternak.[68] It contains two C-methyl groups and gives a diacetate, and was recognised as a biquinone by titration with acidified potassium iodide, the catalytic absorption of two mols. of hydrogen, and the formation of a leucohexa-acetate. Two series of salts are formed, exemplified by the violet–red monopotassium salt and the violet diammonium salt, and the quinone shows two indicator changes from yellow to red at pH 1·8–3·4, and from red to violet at pH 5·4–6·4. With quinol it forms a red quinhydrone, C$_{14}$H$_{10}$O$_6$·2C$_6$H$_6$O$_2$. Phoenicin therefore appeared to be a dihydroxy-dimethyl-bibenzoquinone, and this was confirmed by Thiele acetylation of the biquinone (21) (easily obtained from toluquinone[74]) which gave two hexa-acetates, the major component being identical with leuco-phoenicin hexa-acetate. Although oxidative hydrolysis of the latter gave phoenicin, an unambiguous synthesis was necessary to establish

† The monomer 6-hydroxy-2-methylbenzoquinone has been identified chromatographically in the metabolism solution of *Aspergillus fumigatus* Fr.[62] and the quinol has been isolated from submerged cultures.[63]

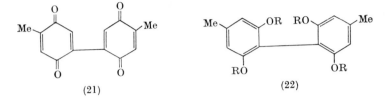

(21) (22)

the positions of the hydroxyl groups. This was done initially starting from the biphenyl (22; R = Me); dinitration and reduction afforded a diamino derivative which gave phoenicin dimethyl ether when oxidised with chromic acid,[69] but it is much simpler to demethylate (22; R = Me) with pyridine hydrochloride to give the bi-orcinol (22; R = H) which yields phoenicin almost quantitatively on oxidation with Fremy's salt.[70] It is interesting that when orcinol is exposed to air in alkaline solution[71] oxidative coupling occurs *ortho* to the methyl group, and the product[63] ("Henrich's quinone") is a mixture of the monoquinone (23) and the biquinone (24) which is an isomer of phoenicin. The biquinone

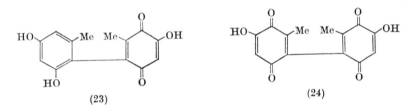

(23) (24)

(24) has been separated[72] into its optical isomers by chromatography on a starch column in phosphate buffer at pH 7 but resolution of phoenicin, or a derivative, has not been reported.

Primin (25), $C_{12}H_{16}O_3$

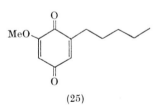

(25)

Occurrence: glandular hairs on the leaves of *Primula obconica* Hance[75,76] (Primulaceae).

Physical properties: golden-yellow crystals, m.p. 62–63°,† $\lambda_{max.}$ (EtOH) 267, 365 nm (log ε 4·33, 2·54), $\nu_{max.}$ (KBr) 2924, 2857, 1685, 1655, 1634, 1603, 1240 cm⁻¹.

† After normal crystallisation and sublimation. Raised to 66·7–66·8° after zone melting.

Primula obconica is an attractive ornamental plant frequently used for household decoration. Unfortunately some people are allergic to it, the cause of the trouble being the secretion of a simple quinone, primin, from the glandular hairs on the leaves which causes dermatitis.[344] The active principle was first isolated by Bloch and Karrer,[75] and was identified recently by Schildknecht and co-workers.[76] It was recognised as a methoxybenzoquinone by analysis, colour reactions, and its ultra-violet absorption, which suggested, by comparison with model compounds, that it was a 2,5- or 2,6-disubstituted quinone. Strong infrared absorption in the region 2850–3000 cm^{-1} indicated the presence of an alkyl side chain, and both the n.m.r. and mass spectra showed that this was n-pentyl. In the mass spectrum the base peak falls at m/e 153 (M-55) attributed to a dihydroxymethoxytropylium ion formed by cleavage at the "benzyl" position and rearrangement. The orientation of the two substituents follows from the n.m.r. spectrum which shows a multiplet at τ 3·54 and a doublet at τ 4·22 arising from the two ring protons. These are obviously coupled indicating that they occupy the 3,5-positions, so that primin must be 6-methoxy-2-n-pentyl-1,4-benzoquinone (25).[76] This structure was further confirmed by incubation with tritiated water ("electron pyrolysis") when primin fragmented into 6-methoxy-2-methylbenzoquinone and n-butane.[77]

Primin has been synthesised[78] by the reaction of 2,3-dimethoxy-benzaldehyde with n-butyl magnesium bromide, followed by oxidation

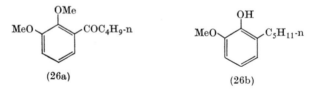

(26a)   (26b)

to the ketone (26a). Partial demethylation with aluminium chloride in toluene and then Clemmensen reduction led to the phenol (26b), from which primin was obtained by oxidation with Fremy's salt.

Sarcodontic Acid (27), $C_{22}H_{32}O_5$

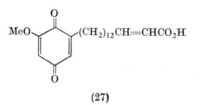

(27)

Occurrence: in the fungus *Sarcodontia setosa* (Pers.) Donk.[459]

Physical properties: yellow crystals, m.p. 122–124°, $\lambda_{max.}$ (EtOH) 264, 366 nm (log ϵ 4·82, 3·12), $\nu_{max.}$ (CHCl$_3$) 3200–2500, 1685, 1645, 1638(sh), 1610 cm^{-1}.

Sarcodontic acid is responsible for the intense sulphur-yellow colour of the wood-rotting fungus *Sarcodontia setosa* found on old apple trees. It contains a methoxyl (n.m.r.) and a carboxyl (i.r.) group, and shows a marked resemblance to methoxybenzoquinone in its ultraviolet absorption. Its quinonoid character was verified by polarographic reduction. On catalytic hydrogenation a benzenoid tetrahydro derivative is formed having a non-conjugated carboxyl group ($\nu_{max.}$ 2600–3400, 1716 cm^{-1}) which afforded a methyl ester on brief treatment with diazomethane, and was reoxidised to a quinone (dihydrosarcodontic acid) with oxygen and a palladium catalyst. The n.m.r. spectrum of tetrahydrosarcodontic acid revealed two *meta*-coupled aromatic protons, a complex multiplet (4H) at τ 7·37–8·10, and a broad signal (24H) centred at τ 8·60. As the methyl ester failed to form a lead salt on treatment with lead acetate, it cannot be a catechol, and hence the natural pigment is a 2,6-disubstituted *p*-benzoquinone, the substituents being methoxyl (probably) and a long aliphatic side chain containing an α,β-unsaturated carboxyl group. The nature of the side chain was determined by oxidative degradation; ozonolysis of sarcodontic acid gave an acid HO$_2$C(CH$_2$)$_{12}$CO$_2$H together with smaller amounts of lower α,ω-dicarboxylic acids, while the same treatment of dihydrosarcodontic acid afforded thapsic acid, HO$_2$C(CH$_2$)$_{14}$CO$_2$H, in high yield. These results establish that the natural compound contains an unbranched C$_{15}$ side chain terminated by an α,β-unsaturated carboxyl group. Sarcodontic acid is therefore (27).[459]

Shanorellin (28), C$_9$H$_{10}$O$_4$

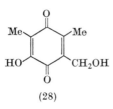

(28)

Occurrence: cultures of *Shanorella spirotricha* (Benjamin).[412]

Physical properties: orange-yellow needles, m.p. 121°, $\lambda_{max.}$ (CHCl$_3$) 272, 406 nm (log ϵ 4·05, 2·07), $\nu_{max.}$ (KBr) 3450, 1660, 1640, 1620 cm^{-1}.

The fungal metabolite shanorellin shows the spectroscopic and redox properties of a benzoquinone, and forms a diacetate and a leucotetraacetate. The n.m.r. spectrum revealed two methyl groups attached to the quinone ring, one of which gives a triplet centred at τ 8·075 (J, 0·5 Hz) coupled to a quartet (2H) at τ 5·475. This indicates homo-allylic coupling between a methyl and a methylene group, and hence structures (28) and (29) can be written for shanorellin. The former was shown to be

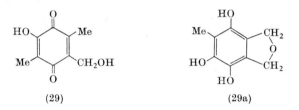

(29)   (29a)

correct by X-ray crystallographic analysis[491] of the chloromethyl derivative obtained by treating shanorellin with thionyl chloride in chloroform-pyridine.[412] The quinol (29a), from *Aspergillus terreus*,[494] is closely related.

2,5-Dimethoxybenzoquinone (30), $C_8H_8O_4$

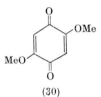

(30)

Occurrence: cultures of *Polyporus fumosus* Pers. ex Fr.[80] and the wood of *Dalbergia melanoxylon* Guill. & Perr.[81] (Leguminosae).

Physical properties: yellow prisms, dec. ca. 250°, $\lambda_{max.}$ (CHCl₃) 278, 284, 370 nm (log ϵ 4·37, 4·38, 2·48), $\nu_{max.}$ (KBr) 1658 cm⁻¹.

This quinone was obtained from *P. fumosus* when grown on corn-steep liquor which may contain a precursor. Other *Polyporus* spp. give rise to substances having similar light absorption when grown on the same medium. The quinone is easily identified from its spectral properties and can be prepared by addition of methanol to benzoquinone.[82]

2,6-Dimethoxybenzoquinone (31), $C_8H_8O_4$

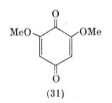

(31)

Occurrence: in wheat grains (*Triticis vulgare* Vill.[83, 84, 95]), the aerial parts of *Adonis vernalis* L.[85] (Ranunculaceae), the root of *Rauwolfia vomitoria* Afz.[86] (Apocynaceae), heartwood of *Fagus sylvatica* L.,[87] *Quercus rubra* L.[402] (Fagaceae), *Ulmus thomasii* (Sarg.)[401] (Ulmacea), wood of *Kielmeyera rupestris* Duarte[464] (Guttiferae), bark of *Populus balsamifera* L.,[403] *P. trichocarpa* Torr. & Gray[404] (Salicaceae), and *Khaya senegalensis* A. Juss.[88] (Meliaceae), sap of *Acer saccharum* Marshall[451] (Aceraceae), and in the following Simarubaceae: *Ailanthus altissima* (Mill.) Swingle (fruits),[89] *A. excelsa* Roxb. (bark),[90] *Eurycoma longifolia* Jack.,[120] *Picrasma ailanthoides* Planchon (wood),[91] *P. crenata* Engl. (wood),[88] *Samadera indica* Gärtn. (bark),[92] *Simaruba amara* Aubl. (bark)[88] and possibly in *Quassia amara* L. (wood).[93]

Physical properties: yellow needles, m.p. 254–255°, $\lambda_{max.}$ (CHCl$_3$) 287, 377 nm (log ϵ 4·28, 2·78), $\nu_{max.}$ (Nujol) 1703, 1647, 1631, 1597 cm^{-1}.

Wheat germ has an adverse effect on the baking quality of flour which is attributed to the presence of glutathione.[94] This can be removed by prefermentation with baker's yeast as the products of fermentation include methoxybenzoquinone and 2,6-dimethoxybenzoquinone which combine with the glutathione by addition of the thiol group. Methoxyquinone is present in wheat germ as a monoglucoside of the quinol and probably the dimethoxyquinone exists in a similar form; the glycosides appear to play a significant part in the germination process.[96] Both methoxy-quinone and methoxyquinol are effective flour improvers[97] but are of no technical value as they discolour bread, a pink tinge arising by combination of the quinone with protein, similar to the effect produced by flour beetle secretions (p. 98). The trace of methoxyquinone obtained from *Tribolium castaneum*[18] (Table I) possibly originated from the flour on which the insects were cultured.

2,6-Dimethoxybenzoquinone can be obtained by the action of nitrous[467] or nitric[468] acid on 1,2,3-trimethoxybenzene.

Most nucleophilic reactions of simple quinones lead to ring-substituted derivatives (1,4-addition) but direct attack on a carbonyl group is also possible (1,2-addition). 2,6-Dimethoxybenzoquinone provides a simple example, the more electrophilic carbonyl group at C-1 being the reactive centre. Prolonged reaction with acetic anhydride in the presence of sodium acetate leads to the formation of the esters (33) and (35) which is interpreted[98] as a Perkin reaction[339] followed, either by addition of acetic anhydride to the quinone-methide (32), and hydrolysis, or by hydrolysis and decarboxylation of (32) and subsequent addition of acetic anhydride to (34). 2,5-Dimethoxybenzoquinone and alkylated benzoquinones behave similarly.[99] In the presence of potassium carbonate,[100] or alumina,[101] acetone will add to the C-1 carbonyl group of 2,6-dimethoxybenzoquinone to give (36), but there is no reaction with the 2,5-isomer. *o*-Benzoquinones undergo a similar aldol reaction with acetone, even in the cold, using aluminium oxide as catalyst.[102]

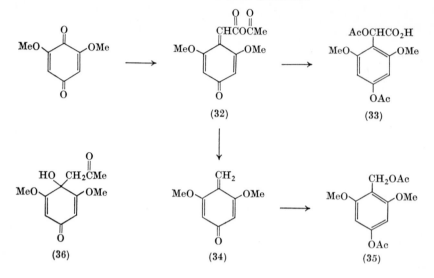

(32)   (33)

(36)   (34)   (35)

3-Hydroxythymoquinone (37), $C_{10}H_{12}O_3$

3,6-Dihydroxythymoquinone (38), $C_{10}H_{12}O_4$

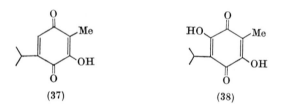

(37)   (38)

Occurrence: heartwood of *Juniperus chinensis* (Sabina)[50] (Coniferae).
Physical properties: (37) orange needles, m.p. 168–170°, $\lambda_{max.}$ (EtOH) 267, 404 nm (log ε 4·16, 3·01), $\nu_{max.}$ (CCl$_4$) 3425, 1657, 1625 cm^{-1}.
(38) red needles, m.p. 223–224°, $\lambda_{max.}$ (EtOH) 293, 435 nm (log ε 4·31, 2·36), $\nu_{max.}$ (CHCl$_3$) 3327, 1640 cm^{-1}, $\nu_{max.}$ (KBr) 3319, 1616 cm^{-1}.

The status of these quinones is uncertain as they were isolated from a tree which was probably killed by fungal infection, and they may not be present in sound wood. Furthermore the alkaline treatment used[50] as part of the isolation procedure suffices to convert thymoquinol (also present in *J. chinensis*) partly into the two hydroxylated thymoquinones which are thus possibly artefacts. The monohydroxy compound (37) is prepared from 2,4-dinitrothymol by reduction to the diamine and oxidation with ferric chloride,[104] and the dihydroxyquinone can be obtained by addition of methylamine to thymoquinone and hydrolysis of the bis-methylamino derivative.[105]

3-Libocedroxythymoquinone (39), $C_{32}H_{40}O_6$

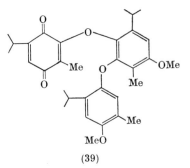

(39)

Occurrence: heartwood of *Heyderia decurrens* (Torr.) K. Koch[106] (= *Libocedrus decurrens* Torr.) (Coniferae).

Physical properties: dark red crystals, m.p. 155°, λ_{max}. (?) 267, 285(sh), 389, 480(sh), nm (log ε 4·22, 3·88, 3·17, 2·55), ν_{max}. (KBr) 1660, 1645, 1620 cm^{-1}.

The heartwood of the incense cedar (*H. decurrens*) contains a group of aromatic monoterpenoids including carvacrol, *p*-methoxycarvacrol, *p*-methoxythymol, and thymoquinone, the dimers libocedrol (40)[107] and heyderiol (41),[108] and a trimeric red quinone, $C_{32}H_{40}O_6$.[106] The close relationship of this more complex product to the others was recognised[106] from the similarity of its ultraviolet-visible and infrared spectra to those of libocedrone (42), the quinone obtained by oxidation of libocedrol with ferric sulphate, and by the observation that oxidation of libocedrol with alkaline ferricyanide gave the new red quinone as the major product, together with *p*-methoxythymol and libocedrone. This synthetic reaction suggests that the C_{32} quinone is derived from three *p*-methoxythymol units which must be joined by ether linkages as the compound contains two methoxyl groups but no hydroxyls. Assuming that the libocedrol moiety is contained intact in the trimeric structure, the remaining problem concerns its point of attachment to the quinonoid ring. That this is as shown in (39) was supported by the formation of 3-hydroxythymoquinone (37) when the red pigment was reduced to the quinol, refluxed with hydrogen bromide and reoxidised with ferric chloride, and also by the (poorly resolved) doublet signal from the quinonoid proton in the n.m.r. spectrum. The rest of the n.m.r. spectrum is consistent with structure (39), 3-libocedroxythymoquinone, and excludes other possibilities bearing in mind reasonable modes of formation.

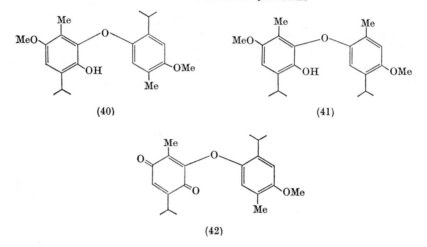

(40)   (41)

(42)

The decay resistance of incense cedar heartwood is attributed largely to the fungicidal phenols, p-methoxythymol and p-methoxycarvacrol.[110] As the timber ages the resistance declines accompanied by a decrease in the concentration of these phenols and a concomitant increase in the concentration of the less toxic dimers libocedrol and heyderiol, and the quinones thymoquinone and (39). This is consistent with a process of oxidative coupling via intermediate aroxyl radicals, all of which have been observed (in solution) and characterized *in vitro*.[109] The radicals are remarkably stable,[409] particularly those from the dimeric phenols which survive after exposure to air for 24 hours. A plausible reaction path *in vivo* proceeds by coupling of two p-methoxythymol radicals to give libocedrol (40), followed by coupling with a third p-methoxythymol radical to give the trimer (43), and hence (39). In the *in vitro* synthesis from libocedrol the radicals formed initially dimerise to (45; RO = libocedroxy) which can then dissociate to form the radicals (46) and (47)

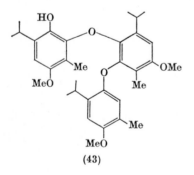

(43)

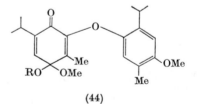

(44)

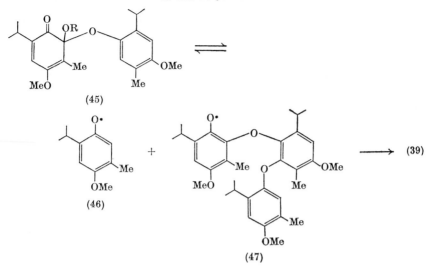

(45)

+ (46) (47) ⟶ (39)

leading to the products (cf. ref. 410). An alternative coupling of the libocedroxy radicals to give the dimer (44; RO = libocedroxy) followed by hydrolysis of the ketal would account for the formation of libocedrone (42). That such dimers are formed is supported by the oxidation of libocedrol with lead dioxide which gave a yellow oil showing the spectral characteristics of a cyclohexadienone-ether, and a weak e.s.r. signal which intensified on dilution.

Fumigatin (48), $C_8H_8O_4$

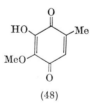

(48)

Occurrence: cultures of *Aspergillus fumigatus* Fres.[111] and *Penicillium spinulosum* Thom.[112, 113]

Physical properties: maroon needles, m.p. 116°, $\lambda_{max.}$ (EtOH) 265, 450 nm (log ϵ 4·14, 2·96), $\nu_{max.}$ (KBr) 3310, 1663, 1635, 1610 cm^{-1}.

Fumigatin was one of the first quinones identified by Raistrick[411] in his pioneer work on fungal metabolites. From the molecular formula, group analyses and the formation of an acetate and methyl ether, fumigatin was clearly[111] a hydroxymethoxytoluquinone, and synthesis[115] of the three possible dimethoxytoluquinones showed that 5,6-dimethoxy-2-methylbenzoquinone was identical with fumigatin methyl ether. Of the two possible structures for the natural pigment, (48) was shown[115] to be correct when the ethyl ether was found to be different from synthetic 5-ethoxy-6-methoxy-2-methylbenzoquinone. This has been confirmed by several syntheses,[116−119] initially by Baker and Raistrick[116] who prepared the phenolic ketone (50) by Friedel-Crafts acylation of 3,4,5-trimethoxytoluene (49) and converted it into the

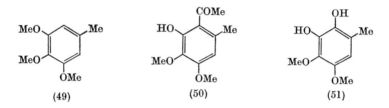

(49)   (50)   (51)

catechol (51) by Dakin oxidation, from which fumigatin was derived by treatment with ferric chloride.

Aspergillus fumigatus elaborates several toluquinones (they are all present as quinols in the early stages of growth[114]) which have not been completely characterised. Two of these, which give fumigatin methyl ether on methylation, appear to be 5,6-dihydroxy-2-methyl- and 5-hydroxy-6-methoxy-2-methylbenzoquinone.[62]

3-Hydroxy-5-methoxy-2-methylbenzoquinone (52), $C_8H_8O_4$

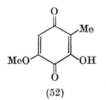

(52)

Occurrence: cultures of *Lentinus degener* Kalchbr.[59]

Physical properties: brownish-orange prisms, m.p. 202–203°(dec.), $\lambda_{max.}$ (EtOH) 290, 425 nm (log ε 4·20, 2·72), $\nu_{max.}$ (KBr) 3240, 1681, 1664, 1620, 1608 cm^{-1}.

This copigment of coprinin is an isomer of fumigatin. It titrates as a monobasic acid (pK = 4·8) giving a purple colour in alkaline solution. As it contains one O–Me and one C–Me group it must be a hydroxy-methoxytoluquinone and the methyl ether was found to be identical with 3,5-dimethoxy-2-methylbenzoquinone.[111] This leaves two possible structures for the natural pigment and it proved to be identical with (52) which was already known. Compound (52) was synthesised by way of Thiele acetylation of 5-methoxy-2-methylbenzoquinone, hydrolysis and oxidation with ferric chloride.[121]

Oosporein (iso-oosporein, chaetomidin) (53), $C_{14}H_{10}O_8$

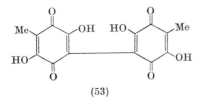

(53)

Occurrence: cultures of *Oospora colorans* van Beyma,[124] *Chaetomium aureum* Chivers[125] and unidentified *Chaetomium* spp.,[126] *Verticillium psalliotae* Treschow,[127, 128] an unidentified *Acremonium* sp.,[129] *Beauveria bassiana* (Bals.) Vuill.[130] and *B. tenella* (Delacroix) Siem.,[457] *Phlebia mellea* Overholts[128] and *P. albida* Fries.[128]

Physical properties: bronze plates, m.p. 290–295°,† $\lambda_{max.}$ (EtOH) 287, 425–450 nm (log ε 4·60, 2·85), $\nu_{max.}$ (KBr) 3300, 1621, 1610 cm^{-1}.

The biquinone‡ oosporein enjoys a relatively wide fungal distribution occurring in Ascomycetes, Basidiomycetes, and several Fungi Imperfecti. Like phoenicin (20) it passes through a series of colour changes as the pH is raised. Oosporein forms a tetra-acetate and a leuco-octa-acetate, and absorbs two mols. of hydrogen to form a colourless leuco compound which is easily reoxidised in air. It is therefore a biquinone and evidently a dihydroxyphoenicin as it gives *p,p*-bitolyl on zinc dust distillation. This was confirmed by Kögl and Van Wessem[124] who converted phoenicin, by reaction with excess methylamine, into the tetra-aminobiquinone (53a) which yielded oosporein on acid hydrolysis.

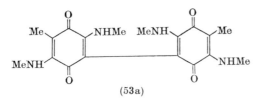

(53a)

Oosporein undergoes a remarkable transformation when fused[126a, 131] with potash to give a naphthaquinone, tomichaedin (56) in 10–15%

† Several different m.p.s have been reported for oosporein, its tetra-acetate and leuco-octa-acetate.
‡ The monomer, 3,6-dihydroxy-2-methylbenzoquinone has been found in cultures of *Aspergillus fumigatus* Fres. It was identified by ultraviolet spectroscopy, and by chromatographic comparison, after methylation and reduction, with 3,6-dimethoxy-2-methylbenzoquinone and 2,3,5,6-tetrahydroxytoluene.[62]

yield. A speculative interpretation[131b] of the reaction suggests that oxalic acid is split off from the tautomeric form (54) by β-keto fission, as indicated, followed by a benzilic acid rearrangement of (55), dehydration, aldol condensation and final oxidation.

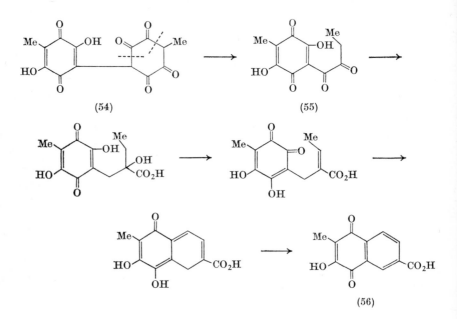

5,6-Dihydroxy-2,3-dimethyl-1,4-benzoquinone (57), $C_8H_8O_4$
5-Hydroxy-6-methoxy-2,3-dimethyl-1,4-benzoquinone (58), $C_9H_{10}O_4$
Aurantiogliocladin (59), $C_{10}H_{12}O_4$

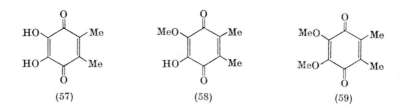

Occurrence: cultures of *Gliocladium roseum* Bain.[122, 132]

Physical properties: (57) orange needles, m.p. 182° (dec.), $\lambda_{max.}$ (EtOH) 277, 445 nm (log ε 4·07, 2·61), $\nu_{max.}$ (CCl₄) 3378, 1642, 1618 cm⁻¹.

(58) reddish-brown prisms, m.p. 70°, $\lambda_{max.}$ (EtOH) 278, 442 nm (log ε 4·21, 2·81), $\nu_{max.}$ (CCl₄) 3390, 1639, 1618 cm⁻¹.

(59) orange needles, m.p. 63–64°, $\lambda_{max.}$ (EtOH) 275, 414 nm (log ε 4·20, 2·73), $\nu_{max.}$ (CCl₄) 1653, 1616 cm⁻¹.

In the original study of the metabolites of *G. roseum* by Brian and co-workers[132] three compounds were isolated, the quinone aurantio-gliocladin (59), its dark red quinhydrone rubrogliocladin, and gliorosein, a tautomeric form of the quinol. Two other closely related quinones, (57) and (58), formed in small amounts, have been isolated recently by Pettersson.[122]

Aurantiogliocladin, $C_{10}H_{12}O_4$, contains two methoxyl groups and its structure was ascertained[134] by reduction and methylation of the quinol, followed by permanganate oxidation which gave tetramethoxyphthalic acid. As the latter gave 1,2,3,4-tetramethoxybenzene on decarboxyl-ation aurantiogliocladin must be 5,6-dimethoxy-2,3-dimethyl-1,4-benzo-quinone (59). This has been confirmed by several syntheses.[119, 133, 135, 136] Initially the aldehyde (60), obtained from 3,4-dimethylphenol by a Duff reaction, was nitrated and then subjected to a Dakin reaction to yield the catechol (61). Conversion to the amine (62) and then to the

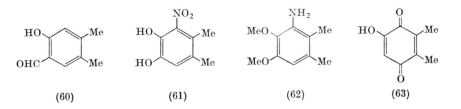

(60)   (61)   (62)   (63)

p-diamine was effected by standard methods, and finally oxidation with ferric chloride provided the desired quinone.[135] Pettersson's syn-thesis,[136] involving Thiele acetylation of (63), is shorter but difficult to reproduce.[137] The two metabolites (57) and (58) can both be obtained from aurantiogliocladin by mild alkaline hydrolysis.[136]

Gliorosein has been isolated in good yield from surface cultures of *G. roseum*.[138] It is optically active and has the unusual structure (64),[134, 138, 140] being the diketo tautomer of aurantiogliocladin quinol, into which it is converted in the presence of base. It is biosynthesised by way of the quinol.[138] This is effected by a tautomerase enzyme but efforts to reproduce this *in vitro* with the demethylated quinol failed[137] although 1,2,3,4-tetrahydroxybenzene has been successfully rearranged to (65) simply by acidification of an alkaline solution.[139]

Another example of an ene-dione is the optically active antibiotic, terreic acid (66), elaborated by *Aspergillus terreus* Thom.[143] It shows the expected chemical and spectroscopic properties of its functional groups and is converted, under mild, basic conditions, into a mixture of quinones.[144] The racemic form has been prepared[145] by carefully controlled epoxidation of 3-hydroxy-2-methylbenzoquinone using sodium perborate. A mutant of *A. terreus* elaborates terremutin (67).[413] A closely related compound is the optically active epoxide (68) found in a

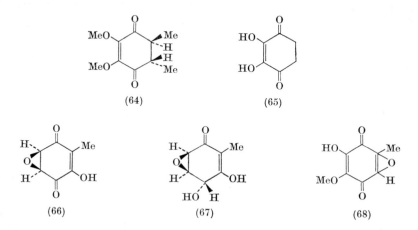

(64)   (65)

(66)   (67)   (68)

strain of *Aspergillus fumigatus*.[54] In base it is rapidly converted into spinulosin (68) which was isolated from the same cultures but appeared to be an artefact derived from (70).

2-Ethyl-3,6-dihydroxy-1,4-benzoquinone (69), $C_8H_8O_3$

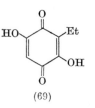

(69)

Occurrence: spines of *Echinothrix diadema* L.[141]

Physical properties: orange prisms, subl. 130–145°, $\lambda_{max.}$ (CHCl₃) 282(sh), 288, 422 nm (log ε 4·12, 4·13, 2·62), $\nu_{max.}$ (?) 1613 cm⁻¹.

Spine pigments in sea urchins are usually naphthaquinones. This benzoquinone is exceptional and could, conceivably, be a spinochrome degradation product. As there is only one carbonyl absorption band (at 1613 cm⁻¹) both carbonyl groups must be chelated to hydroxyls, and the presence of an ethyl group was obvious from the n.m.r. spectrum. The compound is thus 2,3- or 2,5-dihydroxyethylbenzoquinone, and must be the latter as the ring proton shows a singlet peak at τ 4·18. It was synthesised[141] by oxidation of ethylquinol with alkaline hydrogen peroxide by which method quinol itself yields 2,5-dihydroxy-1,4-benzo-quinone.[142]

Spinulosin (70), $C_8H_8O_5$

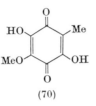

(70)

Occurrence: cultures of *Penicillium spinulosum* Thom,[112] *P. fellutanum* Biourge (= *P. cinerascens* Biourge)[146] and *Aspergillus fumigatus* Fres.[147]
Physical properties: purple-black plates, m.p. 203°, $\lambda_{max.}$ (CHCl$_3$) 297, 460 nm (log ϵ 4·35, 2·29), $\nu_{max.}$ (KBr) 3310, 1655(sh), 1637, 1608 cm^{-1}.

Spinulosin forms a diacetate and titrates as a dibasic acid. It has one methoxyl group and hence, from the molecular formula, it must be a dihydroxymethoxytoluquinone,[112] and the orientation of the substituents became apparent when it was found that fumigatin (48), a cometabolite of spinulosin in *P. spinulosum* and *A. fumigatus*, could be converted into spinulosin.[148] Thiele acetylation of fumigatin gave a tetra-acetate identical with spinulosin leucotetra-acetate from which the dihydroxyquinone was obtained by hydrolysis and oxidation. Spinulosin is thus (70) which was confirmed synthetically[149] by treatment

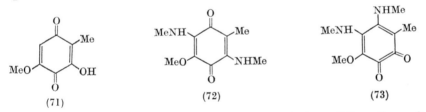

(71)   (72)   (73)

of the quinone (71), obtained by Thiele acetylation of 5-methoxy-2-methylbenzoquinone, with excess methylamine and hydrolysis of the bis-methylamino derivative. The latter, however, is not the diamino-quinone (72), which was prepared later in other ways, and may be the *o*-quinone (73) or a tautomer.[150] A number of tautomeric polyhydroxy-(methoxy)benzoquinones give similar anomalous results with methyl-amine.[150]

(*R*)-*4-methoxydalbergione*† (74), $C_{16}H_{14}O_3$
(*S*)-*4-methoxydalbergione* (75; R = H), $C_{16}H_{14}O_3$
(*S*)-*4'-hydroxy-4-methoxydalbergione*‡ (75; R = OH), $C_{16}H_{14}O_4$
(*S*)-*4,4'-dimethoxydalbergione*§ (75; R = OMe), $C_{17}H_{16}O_4$

† Originally named[153] dalbergione and also dalbergenone.[156a, 161]
‡ Originally named[153] hydroxydalbergione.
§ Originally named[153] methoxydalbergione.

(*R*)-*3,4-dimethoxydalbergione* (76; R = H), $C_{17}H_{16}O_4$
(*R*)-*4'-hydroxy-3,4-dimethoxydalbergione* (76; R = OH), $C_{17}H_{16}O_5$

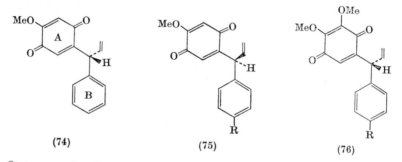

(74)   (75)   (76)

Occurrence: (74) heartwood of *Dalbergia nigra* Fr. Allem,[153,154] *D. latifolia* Roxb.,[155,156,159] *D. spruceana* Benth.,[157] *D. cochinchinensis* Pierre,[356] *D. sissoo* Roxb.[457]† and *D. obtusa* Lecomte[399] (Leguminosae).

(75; R = H) heartwood and sapwood of *D. miscolobium* Benth.[153,158] [= *D. violacea* (Vog.) Malme] and wood of *D. baroni* Baker.[159]

(75; R = OH) heartwood of *D. nigra*[153] and *D. miscolobium*.[158]

(75; R = OMe) heartwood of *D. nigra*.[153,154]

(76; R = H) heartwood of *Machaerium scleroxylon* Tull.,[160] *M. kuhlmannii* Hoehne[399] and *M. nictitans* (Vell.) Benth.[399] (Leguminosae).

(76; R = OH) heartwood of *M. nictitans*.[399]

Physical properties: (74) yellow needles, m.p. 114–116°, $[\alpha]_D$ −51° (dioxan), +13° (CHCl₃), $\lambda_{max.}$ (EtOH) 260 nm (log ε 4·18), $\nu_{max.}$ (CHCl₃) 1672, 1650, 1605, 992, 928, 904 cm⁻¹.

(75; R = H) yellow needles, m.p. 118–120°, $\lambda_{max.}$ and $\nu_{max.}$ as for (74).

(75; R = OH) orange rhombs, m.p. 172–178° (dec.), $[\alpha]_D$ −52° (dioxan), $\lambda_{max.}$ (EtOH) 228, 262, 330 nm (log ε 4·10, 4·12, 3·20), $\nu_{max.}$ (Nujol) 3530, 1664, 1639, 1600, 987, 928 cm⁻¹.

(75; R = OMe) orange needles, m.p. 109·5–111°, $[\alpha]_D$ −32° (dioxan), −139° (CHCl₃), $\lambda_{max.}$ (EtOH) 228, 258, 333 nm (log ε 4·16, 4·14, 3·23), $\nu_{max.}$ (CHCl₃) 1672, 1645, 1600, 992, 928, 905 cm⁻¹.

(76; R = H) red needles, m.p. 41–42°, $[\alpha]_D$ 60° (CHCl₃), $\lambda_{max.}$ (EtOH) 260, 405 nm (log ε 4·06, 3·00), $\nu_{max.}$ (CHCl₃) 1660, 1605, 995, 930 cm⁻¹.

(76; R = OH) red oil, $[\alpha]_D$ +68° (CHCl₃), $\lambda_{max.}$ (MeOH) 230, 264 nm (log ε 4·19, 4·02), $\nu_{max.}$ (CHCl₃) 3500, 3250, 1650, 1595 cm⁻¹.

The woods of several *Dalbergia* spp. are of commercial value (for example, *D. nigra*, Brazilian rosewood, *D. latifolia*, Indian rosewood) but some have the disadvantage that they contain inhibitors[162] which retard the hardening of polyester varnishes, now widely used, or toxic constituents[163] which cause eczema among workmen in the wood-working industry. These problems, and the observation that extracts of *D. nigra* had antibiotic activity, led to the isolation[162b, 164] of the active principles which were regarded, initially, as naphthaquinones.

† (74), or (75), or a racemate.

The subsequent structural determination and the recognition of a new type of natural benzoquinone belonging to the neoflavonoid group are due to Ollis and his co-workers.[153, 160]

The dalbergiones contain either one or two O-methyl groups, show normal redox properties [the quinone (76; R = H) is accompanied by its quinol in *M. scleroxylon*[160]] and form leucodi- or tri-acetates. On catalytic hydrogenation they absorb two mols. of hydrogen and can be reoxidised to dihydrodalbergiones which are still optically active and show virtually the same ultraviolet absorption as the parent compounds. Their ultra-violet absorption is similar to that of 2-methyl-5-methoxy-1,4-benzo-quinone (λ_{max}. 262 nm) and in certain cases (see above) there is an additional peak at 228 nm also seen in the spectra of *p*-cresol and *p*-methoxytoluene; the absorption spectrum of, for example, 4,4'-di-methoxydalbergione corresponds closely to the summation curve of 2-methoxy-5-methylbenzoquinone and *p*-methoxytoluene. The presence of a benzenoid moiety was confirmed by permanganate oxidation when methoxy- and 3,4-dimethoxydalbergione gave benzoic acid and 4,4'-dimethoxydalbergione gave *p*-anisic acid. The latter can therefore be represented by the partial structure (77; R = OMe). It is now evident that the C_3H_4 fragment must be responsible for the optical activity of

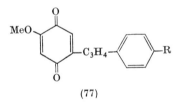

(77)

these compounds and is therefore either a 1,2-disubstituted cyclopropane or a 3,3-disubstituted propene. The former is excluded by the absence of signals at high field in the n.m.r. spectra of these compounds whereas the spectra are in full accord with the ABCX system of a 3,3-disubstituted propene structure.[165] Infrared bands at ca. 910 and 990 cm⁻¹, which disappear on hydrogenation, can also be attributed to a vinyl group. Other n.m.r. data are consistent with structures (74) and (75; R = H), the orientation of the substituents in ring A being confirmed by singlets at ca. τ 4·1 (Q–*H* adjacent to OMe) and doublets at ca. τ 3·5 (Q–*H* coupled to the benzylic proton). The structure of the dimethoxyquinone (76; R = H) is also defined by its n.m.r. spectrum which is similar to that of (74) except for the singlet peak at ca. τ 4·1 and the additional methoxyl resonance.[160] Although a benzoic acid could not be isolated from the permanganate oxidation of the hydroxymethoxydalbergione (75; R = OH) the position of the hydroxyl group is established by the n.m.r. spectrum of the quinol triacetate in which the four aromatic protons of the B ring form an A_2B_2 system. Surprisingly, (75; R = OH) was never converted into (75; R = OMe). The n.m.r. spectrum of (76; R = OH) shows the expected correlations with other members of the group.[399]

The dalbergione structures have been confirmed by the synthesis[165] of (±)-4-methoxydalbergione and (±)-3,4-dimethoxydalbergione. Claisen rearrangement of the cinnamyl ethers (76; R = H and OMe) yielded the

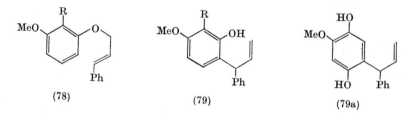

(78)   (79)   (79a)

o-allylphenols (79; R = H and OMe) (two isomers from 78; R = H) which were oxidised with Fremy's salt to the corresponding p-quinones identical with the racemates of 4-methoxy- and 3,4-dimethoxydalber-giones, respectively. Alternatively, methoxyquinol may be condensed with cinnamyl alcohol[470] or 1-phenylallyl alcohol[471] in aqueous citric (+ascorbic)[470] or formic[471] acid to give (79a). Oxidation to the quinone was effected inter alia by t.l.c. on silica gel impregnated with silver nitrate.[471] Racemic dihydrodalbergiones have also been synthesised; for example, reduction of 4-methoxypropiophenone with sodium boro-hydride gives an alcohol which, without isolation, can be condensed with methoxyquinol in acetic acid, in the presence of air, to give (±)-dihydro-4,4'-dimethoxydalbergione.[444]

The absolute stereochemistry of this group of pigments is of some

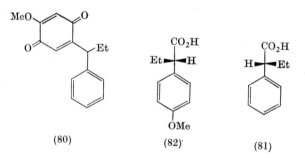

(80)   (82)   (81)

interest.[153] Ozonolysis of the dihydro-4-methoxydalbergione (80) yielded (−)-α-ethylphenylacetic acid of established absolute configuration (81)[166] which has a negative plain o.r.d. curve. However, similar treatment

of 4,4'-dimethoxydalbergione gave (+)-α-ethyl-*p*-methoxyphenylacetic
acid having a *positive* plain o.r.d. curve. It follows that this acid has the

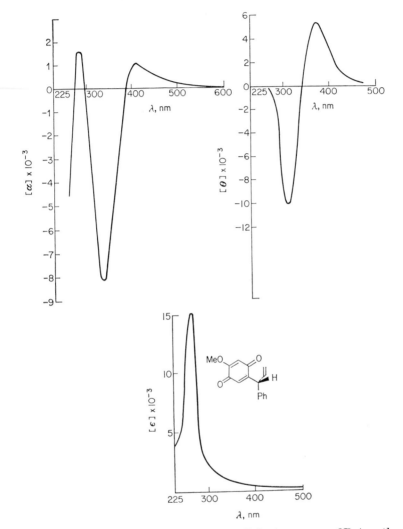

Fig. 2. Optical rotatory dispersion and circular dichroism curves of R-4-methoxy-
dalbergione (74) and its ultraviolet spectrum. From Eyton *et al.* *Tetrahedron*
(1965), **21**, 2707.

absolute configuration (82) which leads to the conclusion that the
4-methoxydalbergione in *D. nigra* has the R-configuration (74) whereas
the 4,4'-dimethoxydalbergione, in the same plant, must have the

S-configuration (75; R = OMe). This antipodal relationship is fully supported by the c.d. curves of (74) and (75; R = OMe) which are of

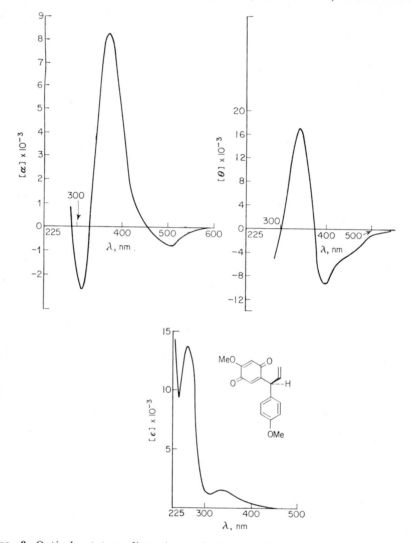

FIG. 3. Optical rotatory dispersion and circular dichroism curves of S-4, 4'-dimethoxydalbergione (75) and its ultraviolet spectrum. From Eyton *et al.* *Tetrahedron* (1965), **21**, 2707.

opposite type (see Figs 2 and 3), and since the 4'-hydroxy-4-methoxy-dalbergione shows an o.r.d. curve similar to that of the 4,4'-dimethoxy analogue it also has the S-configuration (75; R = OH). The 3,4-dimethoxy isomer and also (76; R = OH) were found to have the R-configuration by comparison of their o.r.d. curves with that of R-4-methoxydalbergione. It is of interest that the o.r.d.[170] and c.d. curves of the dalbergiones are difficult to correlate with their ultraviolet spectra because the optically active absorption bands are either not detectable (e.g. 74) or very weak (e.g. 75; R = OMe, λ_{max} 333 nm, ϵ 1700) (see Figs 1 and 2). Very weak electronic transitions characteristic of the p-quinone chromophore can thus be detected by c.d. studies.

These and other[155, 159] investigations reveal the unusual situation that *D. nigra* and *D. latifolia* elaborate R-4-methoxydalbergione but *D. miscolobium* and *D. baroni* produce S-4-methoxydalbergione, and whereas both dalbergiones (75; R = H and OH) in *D. miscolobium* have the S-configuration, *D. nigra* has three dalbergiones of which one (74) has the R- and two (75; R = OH and OMe) have the S-configuration. In an earlier examination[167] of *D. nigra* two pigments, J-1 and J-2 were reported without comment. J-1 proved to be hydroxymethoxydalbergione (75; R = OH) but J-2, m.p. 111–112°, which appeared to behave as a pure compound of formula $C_{30}H_{21}O_4(OMe)_3$, was later separated[153] by repeated chromatography into R-4-methoxydalbergione and S-4,4′-dimethoxydalbergione. When equimolecular amounts of these two quinones were crystallised together, J-2 was obtained and was unchanged by further crystallisation. The compound J-2 is thus a quasiracemate[168] of (74) and (75; R = OMe) and appears to be the first to be isolated from natural sources. (A quasiracemic alkaloid has been described since.[169])

The neoflavanoids[157a] include the dalbergiones, 4-arylchromenes (neoflavenes) and 4-arylcoumarins (dalbergins), and these are accompanied in *Dalbergia* and *Machaerium* spp. by a variety of related C_6–C_3–C_6 type compounds. The origin of these compounds has been the subject of considerable speculation (see ref. 157b and references therein).

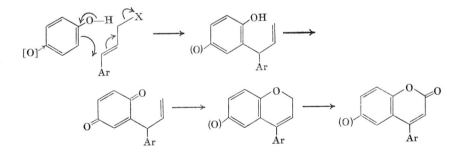

the most satisfying proposal[157b] at present involving the alkylation of a phenolic precursor by a cinnamyl pyrophosphate or its equivalent. This is illustrated on p. 123 for compounds of the neoflavonoid type but alternative modes of interaction could account for the formation of the cinnamylphenols[152] and other products found in these hardwoods (see also ref. 444).

(*S*)-*Mucroquinone* (83), $C_{17}H_{16}O_6$

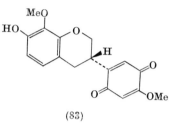

(83)

Occurrence: wood of *Machaerium mucronulatum* Mart.[414] (Leguminosae).
Physical properties: orange needles, m.p. 192°, [α] −15·4° (acetone), $\lambda_{max.}$ (MeOH) 230(sh), 264, 360 nm (log ε 4·01, 4·13, 2·00), $\nu_{max.}$ (CHCl₃) 3500, 1680, 1650, 1595 cm⁻¹.

Until recently isoflavans were unknown in plants but a group of five has been found recently in *Dalbergia* and *Machaerium* woods. They differ in the hydroxylation patterns in rings A and B and in one case ring B is quinonoid. This is evident from the ultraviolet absorption and from signals at τ 3·54 (doublet, 1H) and 4·02 (singlet, 1H) which correspond to the Q–H resonances in 4-methoxydalbergione (74). Other signals were readily assigned to a hydroxyl and two methoxyl groups, and two *ortho*-aromatic protons, while the 220 MHz n.m.r. spectrum could be analysed in terms of an ABMXY system comprising the protons at the C-2, C-3 and C-4 positions of an isoflavan system. On this basis structure (83) (or the alternative with 7-OH and 8-OMe interchanged) seemed very probable bearing in mind the structures of the other *Machaerium* isoflavans. That mucroquinone is in fact (83) was proved by synthesis of the racemic form which was derived from the synthetic isoflavan (84) by oxidation with Fremy's salt.[414] The stereochemistry

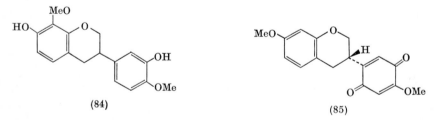

(84)   (85)

shown in (83) was assigned[415] by comparison of the o.r.d. curve with that of the quinone (85) of known configuration. However, these curves are somewhat atypical and this must be associated with the unusual conformation adopted by this system. This is evident from the coupling

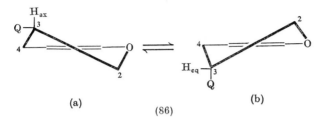

(a)   (b)
(86)

constants of the ABMXY system of the dihydropyran ring which show
that it adopts the half-chair form (86), conformation (b) in which the
quinonyl substituent is *axial*, being predominant.[415]

(R)-*perezone* (pipitzahoic acid) (87), $C_{15}H_{20}O_3$
(R)-*hydroxyperezone* (88), $C_{15}H_{20}O_4$

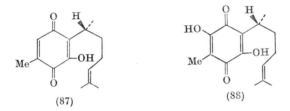

(87)   (88)

Occurrence: (87) roots of *Perezia adnata* A. Gray[171,172] (= *Trixis pipitzahuac*
Schaffn. ex Herrera), *P. alamani* Hemsl. var. *adnata*,[173] *P. cuernavacana* Robinson
& Greenm.,[175] *P. reticulata* A. Gray[180] and *Trixis cacaloides* D. Don[174] (Compositae).
   (88) Roots of *P. alamani* Hemsl. var. *adnata*.[173]
   Physical properties: (87) orange leaflets, m.p. 102–103°, $[\alpha]_D^{20}$ −17° (ether),
$\lambda_{max.}$ (EtOH) 206, 266, 412 nm (log ε 4·15, 4·05, 3·01), $\nu_{max.}$ (Nujol) 3308, 1650,
1628, 1610 cm⁻¹.
   (88) red plates, m.p. 143°, $[\alpha]_D$ −15° (?), $\lambda_{max.}$ (EtOH) 295, 425 nm (log ε 4·26,
2·41), $\nu_{max.}$ (KBr) 3320, 1640(sh), 1615 cm⁻¹.

Investigations on perezone date back to 1852[171] but the structure
was not finally determined until the advent of n.m.r. It occurs in the
roots of several Mexican Compositae and was used as a mild purgative
("Aurum vegetabile"). A hydroxyquinone structure is evident from its
violet colour in alkaline solution and the formation of various derivatives,
but an early suggestion[176] that perezone might be an anthraquinone was
abandoned when the molecular formula, $C_{15}H_{20}O_4$, was established and
several workers[177] then favoured an alkylated hydroxybenzoquinone
structure. Reaction with aniline and other amines gave a series of mono-
aminoperezones from which Anschütz and Leather[177] concluded that
one quinonoid position was unsubstituted, and they drew attention to

the similarity between perezone and hydroxythymoquinone. On this basis the alkyl group(s) (C_9H_{18}) must include one double bond or ring The former was confirmed[178] by hydrogenation, two mols. of hydrogen being taken up to give, after aeration, dihydroperezone, the position of the double bond being established by the formation of acetone on ozonolysis of leucoperezone triacetate.

The nature of the side chains was determined by further oxidative degradations. Fichter *et al.*[179] obtained $\alpha\beta$-diketobutyric acid by ozonolysis of perezone while Kögl and Boer[178] obtained acetic acid and a nonenoic acid by treatment with alkaline hydrogen peroxide. Reduction of the C_9 acid gave 2,6-dimethylheptoic acid identified by comparison with synthetic material. The alkyl substituents are thus methyl and —CH(Me)CH$_2$CH$_2$CH=CMe$_2$, and the orientation of the side chains was found by zinc dust distillation of perezone which gave a liquid hydrocarbon oxidisable to terephthalic acid. The natural pigment

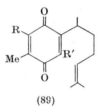

(89)

is therefore (89; R = H, R' = OH or R = OH, R' = H). Hydroxyperezone (89; R = R' = OH), obtained by hydrolysis of anilinoperezone, undergoes an intramolecular condensation on warming in sulphuric acid, with the elimination of water, whereas perezone, itself, does not give a definite product (see also ref. 180). This implies that the hydroxyl group in perezone is not on the same side of the ring as the C_8 side chain, and consequently Kögl and Boer[178] reasonably concluded that perezone had structure (89; R = OH, R' = H).

This structure was accepted for thirty years until n.m.r. studies in four different laboratories[175, 180, 181, 182] showed that the signal from the methyl group was a doublet (τ 7·94) coupled to a quartet (τ 3·53) from a ring proton. Similar signals are given by perezone methyl ether and dihydroperezone, and it is clear that the position next to the ring methyl group is unsubstituted. Perezone is therefore (89; R = H, R' = OH) (= 87). Supporting evidence came from the dihydro derivative (90) of (89; R = OH, R' = H) which was synthesised[181, 183, 184] and found to be different from dihydroperezone, and the chemical and physical properties of the adducts (91) and (92) were only consistent with structures based on the revised formula (87). For example, heating the diazomethane adduct (92) gives the methylperezone methyl ether (93).[175] Finally,

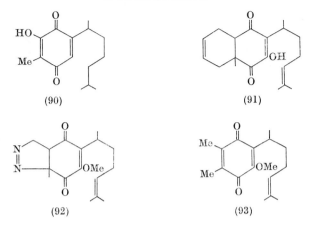

(90)   (91)

(92)   (93)

structure (87) was established by a synthesis of racemic perezone effected by the Mexican group.[185] The carbon skeleton was assembled by interaction of the lithium derivative (94) of 3,5-dimethoxytoluene with the ketone (95). Dehydration of the resulting tertiary alcohol,

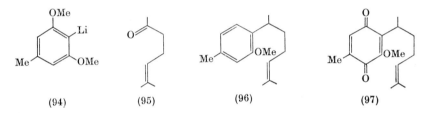

(94)   (95)   (96)   (97)

followed by Birch reduction of the derived styrene afforded (96) which was converted into the quinone (97) in low yield by oxidation with the Jones reagent in cold acetone, and finally hydrolysed with acid.

Kögl and Boer[178] found that degradation of hydroxydihydroperezone with alkaline hydrogen peroxide gave D(−)-2,6-dimethylheptanoic acid (98) which had previously been prepared[186] from D(+)-citronellal (99).

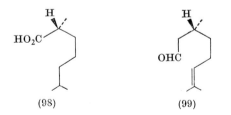

(98)   (99)

It follows that perezone has the absolute configuration shown in (87)[187] which was confirmed later by o.r.d. and c.d. studies.[188]

Hydroxyperezone (88) occurs in the roots of *P. alamani* var. *adnata*[173] as a mono-isovalerate. The synthetic compound is prepared by acid hydrolysis of anilinoperezone but under appropriate conditions the "hydrate" (100) is formed by addition of water to the isopropylidene

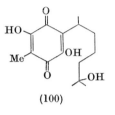

(100)

double bond.[178, 183]

Perezinone. As already mentioned, when perezone is warmed in sulphuric acid the elements of water are eliminated and a new yellow compound, perezinone, is formed. As neither perezone nor dihydro-hydroxyperezone undergo a similar change Kögl and Boer[178] concluded that both the side chain double bond and the neighbouring hydroxyl group were involved in the reaction, and they regarded perezinone as 101). The n.m.r. spectrum fits this structure satisfactorily except for a

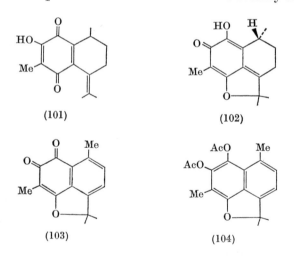

(101)   (102)

(103)   (104)

singlet (6H) at τ 8·48 assigned to two equivalent methyl groups. The *gem*-dimethyl groups in (101) are clearly non-equivalent (a space-filling model cannot be constructed) and perezinone has, in fact, the quinone-methide structure (102) which is consistent with all the spectral data.[183, 189] On oxidation with silver oxide it gives the *o*-naphthaquinone (103), reminiscent of biflorin (p. 322), and reductive acetylation, followed by

dehydrogenation with DDQ leads to the furonaphthalene (104).[189]
The conversion of hydroxyperezone into perezinone is essentially an
intramolecular electrophilic attack on the side chain double bond by a
protonated form of the hydroxyquinone system (see 105), formation of
the furan ring being assisted by electron release from the hydroxyl group
at C-6 which may explain why perezone itself does not undergo a similar
cyclisation.

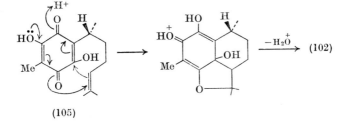

(105)

Perezinone slowly autoxidizes when exposed to air and the surface
becomes coated with the yellow ketoquinone-methide (106).[183, 189, 416]
This occurs more rapidly in solution, when a second compound is formed

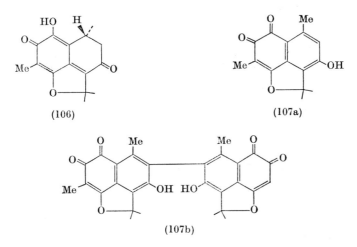

(106)   (107a)

(107b)

which may be the dimeric quinone (107b) since its n.m.r. spectrum
consists only of four singlets (OH, Ar–CH_3, Q–CH_3 and >C(CH_3)$_2$)
and no aromatic protons are present.[183] On the other hand Mexican
workers[416] have found that (106) is converted into the monomeric
o-quinone (107a) under various conditions, including exposure to air.
This appears to be very similar to (107b) but contains an aromatic proton
resonating at τ 3·30.

5

α- and *β-Pipitzols*. It has long been known[178, 190] that perezone undergoes rearrangement on heating (conveniently in boiling tetralin[189]) to give an isomeric product which is actually a mixture of two very similar isomers, α-pipitzol, m.p. 146–147°, $[α]_D$ +192°, and β-pipitzol, m.p. 131–132°, $[α]_D$ −172°. The mixture has been isolated from the roots of *P. cuernavacana*.[189]† In an elegant spectroscopic and chemical study Romo and his collaborators[188, 189] have shown that the pipitzols have the cedrenoid structures (108a) (α) and (108b) (β) and their formation is an example of a thermal cyclo-addition of the (4 + 2) type involving the pentadienyl cation system represented by (87a).[417]

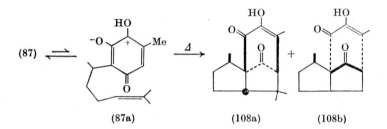

(87) ⇌   (87a)   (108a)   (108b)

(*S*)-*Helicobasidin* (109), $C_{15}H_{20}O_4$
(*S*)-*Deoxyhelicobasidin* (110), $C_{15}H_{20}O_3$

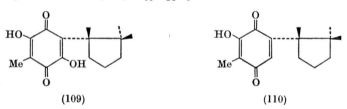

(109)   (110)

Occurrence: cultures of *Helicobasidium mompa* Tanaka.[191, 192, 194]

Physical properties: (109) orange-red needles, m.p. 190–192°, $[α]_D^{25}$ −123·1° (CHCl₃), $λ_{max.}$ (EtOH) 297, 377, 430 nm (log ε 4·15, 2·61, 2·47), $ν_{max.}$ (CHCl₃) 3327, 1684(w), 1638 cm⁻¹, $ν_{max.}$ (KBr) 3300, 1609 cm⁻¹.

(110) yellow needles, m.p. 194–195°, $[α]_D^{25}$ −186·5° (CHCl₃), $λ_{max.}$ (EtOH) 274, 404 nm (log ε 4·11, 3·02), $ν_{max.}$ (Nujol) 3275, 1664, 1633, 1619(sh), 1594 cm⁻¹.

The potent plant pathogen *H. mompa*[192] produces an unusual mixture of pigments. One, mompain (p. 256), is a polyhydroxynaphthaquinone while two others are sesquiterpenoid benzoquinones. Helicobasidin gives a violet solution in aqueous sodium carbonate, shows typical redox properties, and forms a diacetate and a leucotetra-acetate. The ultraviolet and infrared spectra are consistent with a 2,5-dihydroxy-3,6-dialkylbenzoquinone structure, confirmed by the absence of vinylic

† From plants collected in June. In November only perezone was obtained.

proton signals in the n.m.r. spectrum which also revealed the presence of a quinonoid methyl group and three other quaternary methyls. Degradation with alkaline hydrogen peroxide yielded, besides acetic acid, a laevorotatory acid, $C_9H_{16}O_2$, which was identified as the enantiomer of (+)-camphonanic acid prepared from (+)-camphor. This degradation product is therefore (111) and as zinc dust fusion of helicobasidin gave

(111)   (112)

a product containing cuparene (112) (identified by g.l.c.) Natori et al.[193] consider the natural pigment to have structure (109). Thus helicobasidin is a cyclic isomer of hydroxyperezone. An attempt to synthesise helicobasidin by peroxide alkylation of 3-methyl-2,5-dihydroxybenzoquinone using camphonanoyl peroxide was not successful.

A minor metabolite from *H. mompa*, $C_{15}H_{20}O_3$, contains one atom of oxygen less than helicobasidin, and its ultraviolet and infrared agree well with a dialkylhydroxybenzoquinone structure. The n.m.r. spectrum is virtually the same as that of helicobasidin except for a signal (1H) at τ 3·42 due to a ring proton, and as this, and the signal from the ring methyl group are both singlets, the position next to the methyl group must be occupied. On this evidence deoxyhelicobasidin is clearly (110).[194]

(S)-*Lagopodin A* (113), $C_{15}H_{18}O_3$

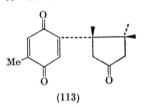

(113)

Occurrence: cultures of *Coprinus lagopus* Fr.[195]

Physical properties: yellow needles, m.p. 96–97°, $[\alpha]_D^{20}$ −10° (CHCl₃), $\lambda_{max.}$ (EtOH) 257, 310, 435 nm (log ϵ 4·25, 2·55, 1·55), $\nu_{max.}$ (CHCl₃) 1745, 1658, 1597 cm⁻¹.

The lagopodins are further examples of fungal sesquiterpenoid quinones. The ultraviolet-visible absorption of lagopodin A is that of a 2,5- or 2,6-dialkylbenzoquinone but in the infrared there is an additional carbonyl band at 1745 cm⁻¹. As there are only three atoms of oxygen in the molecule this must be attributed to a five-membered ring ketone.

The n.m.r. spectrum revealed the presence of a quinonoid methyl group coupled to a vinylic proton at τ 3·45 (quartet), and a second ring proton resonates at τ 3·33. As this latter signal is a singlet it follows that the quinone ring is 2,5-disubstituted (2,6-disubstitution would be disclosed by long-range coupling between the protons at C-3 and C-5), and also that the second substituent is linked to the quinone ring at a quaternary carbon atom (absence of allylic coupling). The n.m.r. spectrum also includes signals from three quaternary methyl groups and these must be attached to the cyclopentanone ring, the nature of which was established by ozonolysis. This gave an optically active C_9 acid identified as

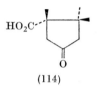

(114)

(114), the enantiomer of the (1R)-acid derived from (+)-camphor. Accordingly, lagopodin A must have structure (113).[195]

(7S)-*Lagopodin B* (115), $C_{15}H_{18}O_4$

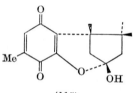

(115)

Occurrence: cultures of *Coprinus lagopus* Fr.[195]

Physical properties: yellow crystals, m.p. 113–115°, $[\alpha]_D^{20}$ +17° (CHCl$_3$), $\lambda_{max.}$ (EtOH) 268, 412 nm (log ϵ 4·10, 2·98), $\nu_{max.}$ (CCl$_4$) 3550–3200, 1675, 1635 cm^{-1}.

Lagopodin B is similar to lagopodin A yet it differs in various ways; *inter alia* there is an additional atom of oxygen, a hydroxyl group, the ultraviolet-visible absorption is like that of a 2,5 (or 2,6)-dialkylhydroxy-benzoquinone, and there is only one vinylic proton which is disclosed by a quartet at τ 3·60 coupled to a ring methyl group at τ 7·98 (doublet).

(116)   (117)

On ozonolysis, or oxidation with alkaline hydrogen peroxide, it gives the same acid (114) as that obtained by degradation of lagopodin A. These facts suggest that lagopodin B has structure (116). (The n.m.r. evidence advanced[195] in support of a 2,5- as opposed to a 2,6-dialkylated quinone structure is not convincing but the biogenetical argument is strong.)

However, pK measurements showed that lagopodin B was a relatively weak acid compared to the hydroxythymoquinones, and the position of the hydroxyl proton resonance (τ 5·12) is more characteristic of an aliphatic hydroxyl group than a hydroxyquinone, and from the infrared spectrum this is hydrogen-bonded. Furthermore, the cyclopentanone carbonyl band can only be discerned in the infrared as a weak shoulder at 1745 cm^{-1} when measured as a 5% solution in carbon tetrachloride, but on dilution (0·16% solution) a distinct band appears at 1745 cm^{-1} and simultaneously a sharp hydroxyl peak is seen at 3580 cm^{-1} and another at 3475 cm^{-1}. Consideration of these observations leads to the conclusion that lagopodin B has the hemiketal structure (115), but exists partly in the open form (116) in solution. This view is supported by an unusual reaction which lagopodin B undergoes on keeping with ferric chloride in absolute methanol. From its n.m.r. spectrum, which is very similar to that of the natural pigment, there is little doubt that the product is the methyl ketal derivative (117). It is also of interest that lagopodin B reacts with aniline to give two products; one is a normal anilino derivative (118) and the other is the dianilino compound (119).

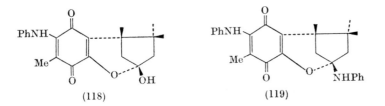

(118)   (119)

It is not immediately obvious why structure (115) should rearrange to (116) on dilution, and the explanation may be that lagopodin B (115) is in equilibrium with a small proportion of the open form (116) and partly associated by intermolecular hydrogen bonding. This second equilibrium would be concentration-dependent.[195]

When lagopodin B is treated with aqueous sodium hydroxide under very mild conditions it is partly converted into the dimer lagopodin C (120).† As lagopodin C has not been positively detected in culture

---

† *Physical properties*: $C_{30}H_{34}O_8$, golden yellow crystals, m.p. 239–243° (dec.), $[\alpha]_D^{22} + 87°$ (CHCl$_3$—MeOH), λ_{max} (EtOH) 211·5, 275, 412 nm, ν_{max} (Nujol) 3480, 1665, 1630 cm^{-1}.

filtrates of *C. lagopus* and has only been isolated, together with lagopodin B, from an alkaline extract, it is almost certainly an artefact of the isolation procedure.[195] Lagopodin C is a biquinone from its molecular weight, the presence of an M + 4 peak in the mass spectrum, and the absorption of two mols. of hydrogen on catalytic hydrogenation. Its spectral properties are very similar to those of lagopodin B except for the absence of vinylic proton absorption in the n.m.r. spectrum and the singlet signal from the quinonoid methyl groups. The n.m.r. spectrum also establishes that the dimer is symmetrical. Chemically it can be degraded to the acid (114) with alkaline hydrogen peroxide, and on treatment with a catalytic amount of ferric chloride in methanol it forms a bis-methyl ketal. These facts can all be accounted for if lagopodin C has structure (120). On pyrolysis it yields a little of the cyclopentenone

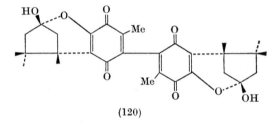

(120)

(122) (molecular weight 124), a reaction which may be regarded as a reversed Michael reaction involving the tautomeric form (121), and it may be noted that a fragment *m/e* 124 forms the molecular ion in the mass spectrum of lagopodin C.

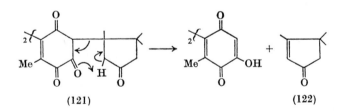

(121)   (122)

The formation of lagopodin C is probably initiated by addition of a nucleophile (e.g. OH⁻) to a molecule of lagopodin B and thereafter may be autocatalytic. Addition to lagopodin B (as 123) would give the quinol (124) which, by a redox reaction with (123), would provide the quinol of lagopodin B (125). Dimerisation can then occur[196] by nucleophilic addition of (125) to (123) leading to the biquinol (126), and by a further redox reaction with lagopodin B this would be oxidised to lagopodin C

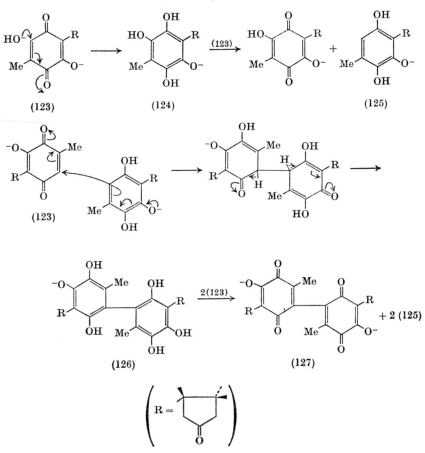

(123) (124) (123) (125)

(126) 2(123) (127) + 2 (125)

$$R = $$

(as 127) with concomitant formation of two molecules of the quinol of lagopodin B (125). Further dimerization can then ensue.[195]

Tauranin† (128), $C_{22}H_{30}O_4$

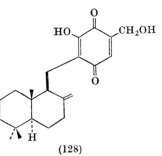

(128)

† Originally known in impure form as aurantin, m.p. 175–180° (dec.).[198]

Occurrence: mycelium of *Oospora aurantia* (Cooke) Sacc. & Vogl.[197]

Physical properties: orange-yellow prisms, m.p. 150–160° (dec.), $[\alpha]_D^{21}$ −148° (MeOH), $\lambda_{max.}$ (MeOH), 266, 415 nm (log ϵ 4·07, 3·07), $\nu_{max.}$ (CHCl₃) 3640, 3415, 1662(sh), 1644, 1622 cm⁻¹.

The ultraviolet-visible absorption of tauranin is indicative of a 2,5(or 2,6)-dialkylhydroxybenzoquinone. On catalytic hydrogenation three mols. of hydrogen are taken up to give, after reoxidation, two epimeric dihydrodeoxytauranins, $C_{22}H_{30}O_3$, in which a hydroxymethyl group attached to the quinone ring has been reduced to methyl. This is clear from the n.m.r. spectra; in the spectrum of tauranin the quinonoid ring proton shows a triplet at τ 3·35 coupled to a two proton doublet at τ 5·47, which changes in the spectrum of deoxytauranin to a quartet at τ 3·42 coupled to a doublet (3H) at τ 7·96. Both dihydrodeoxytauranins gave monoacetates, and from one of these a leucotriacetate was obtained. The third mol. of hydrogen which is absorbed catalytically saturates a terminal methylene group. This is observed as a singlet (2H) at τ 5·33 in the spectrum of tauranin, and on reduction is replaced by a doublet (3H) at τ 9·04 and 9·25, best seen in the spectra of the monoacetates just mentioned.

In order to determine the orientation of the substituents in the quinone ring, *epi*-dihydrodeoxytauranin was converted into an anilino

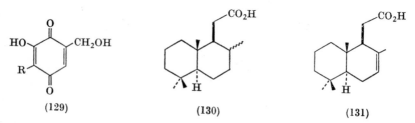

(129)   (130)   (131)

derivative and then hydrolysed to give a dihydroxyquinone. As this compound shows only a single carbonyl absorption band [at 1631 cm⁻¹ (CHCl₃)] it appears to be a 2,5-dihydroxybenzoquinone,[193,200] and consequently tauranin may be represented by the part structure (129; $R = C_{15}H_{25}$). The nature of the group R was determined by ozonolysis

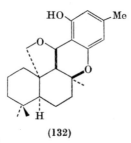

(132)

of the dihydrodeoxytauranins which gave the epimeric acids (130) identical with the product previously obtained[201] by hydrogenation of the acid (131) derived from ambrein and sclareol. As the stereochemistry of (131) is known, it follows that tauranin has structure (128).[199]

The combination of a sesquiterpenoid moiety with a benzoquinone group of non-terpenoid origin is unusual,† and is comparable with siccanin (132), a metabolite of *Helminthosporium siccans* Drechsler.[202]

Embelin (embelic acid) (133), $C_{17}H_{26}O_4$
Rapanone (134), $C_{19}H_{30}O_4$

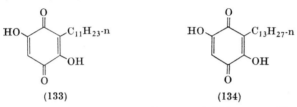

(133)   (134)

Occurrence: (133) in the following Myrsinaceae: *Ardisia crenata* Sims. (roots),[203] *A. japonica* (Thunb.) Blume (rhizome),[203] *A. sieboldii* Miq. (bark and root bark),[203] *Embelia ribes* Burm. (berries),[204] *E. robusta* Roxb. (berries),[205] *E. kilimandscharica* Gilg. (7·58% of the dried berries),[206] *E. barbeyana* Mez. (roots),[207] *Myrsine africana* L. (berries),[208] *M. capitellata* Wall. (berries),[205] *M. seguinii* Lév. (= *Rapanea neriifolia* Mez.) (bark and root bark),[203] *M. semiserrata* Wall. (berries),[205] *M. stolonifera* (Koidz.) Walker (berries),[203] *M. australis* Allan (berries, bark and heartwood),[452] *Rapanea pulchra* Gilg. & Schellenb. (berries)[209] and *R. neurophylla* Mez. (berries).[206]

(134) roots of *Connarus monocarpus* L.[210] (Connaraceae), bulbs of *Oxalis purpurata* (Sonder) var. *jacquinii*[211] (Geraniaceae), and in the following Myrsinaceae: *Aegiceras corniculatum* (Stickm.) Blanco (bark),[213] *Ardisia crenata* Sims. (roots),[203] *A. colorata* Roxb. (bark and heartwood),[462] *A. japonica* (Thunb.) Blume (rhizome),[203] *A. macrocarpa* Wall. (berries, bark and wood),[214] *A. quinquegona* Blume [= *Bladhia quinquegona* (Blume) Nakai] (bark),[215] *A. sieboldii* Miq. (bark and root bark),[203,215] *Myrsine seguinii* Lév. (bark and root bark),[203,215] *M. stolonifera* (Koidz.) Walker (berries),[203] *Rapanea maximowiczii* Koidz. (bark and wood)[215] and *R. pulchra* Gilg. & Schellenb. (bark and root bark).[209]

† Another example is pleurotin (132a), $C_{21}H_{22}O_5$, orange needles, dec. 200–215°, $[\alpha]_D^{23}$ −20° (CHCl$_3$) from the basidiomycete, *Pleurotus griseus*.[487] Details are not available.[488]

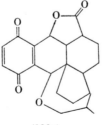

(132a)

Physical properties: (133) lustrous orange plates, m.p. 142–143°, λ_{max}. (EtOH) 292·5, 426 nm (log ϵ 4·24, 2·53), ν_{max}. (KBr) 3305, 1637(sh), 1605 cm^{-1}.

    (134) lustrous orange plates, m.p. 142–143°, λ_{max}. (EtOH) 292·5, 425 nm (log ϵ 4·24, 2·43), ν_{max}. (KBr) 3305, 1605 cm^{-1}.

Embelin and rapanone are characteristic of the Myrsinaceae although rapanone is not confined to this family. The two homologues have virtually the same appearance, electronic absorption, R_F value (t.l.c.) and melting point, and a mixture of the two shows no melting point depression. This is also true for some of their derivatives which led to some confusion in early investigations. However, embelin and rapanone can now be distinguished[203] by significant differences in their infrared spectra in the region 850–1000 cm^{-1}, by their mass spectra (parent peaks at m/e 294 and 322, respectively; M—CO peaks are *not* observed), by integrated proton counts of their n.m.r. spectra,[452] and by g.l.c. separation of the methyl esters of the fatty acids obtained by degradation with hydrogen peroxide (or ozone[213]). From a recent survey of *Myrsine* and *Ardisia* spp. Ogawa and Natori[203] concluded that both homologues were normally present, although in different proportions, and it seems likely that many samples of embelin and rapanone isolated in early work were contaminated with the other homologue. Higher or lower homologues do not seem to be present although closely related pigments with C_9, C_{15} and longer side chains will be described later.

    Embelin has anthelmintic properties and was formerly listed in the U.S. Dispensatory.[216] It has the spectral properties of an alkylated 2,5-dihydroxybenzoquinone but was first recognised as such by its red-violet colour in alkaline solution, redox properties and the formation of a dibenzoate and a leucotetra-acetate.[217] At first it was regarded as a C_9 compound,[204b] and later as a C_{18} molecule[217, 218] which was "proved"[219] by a synthesis of the homologue (135) which was "identical" (by mixed melting point) with embelin, as were the respective dibenzoates and leucotetra-acetates. The presence of an actual C_{11} side chain is

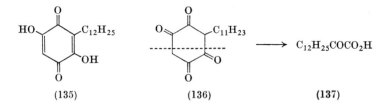

    (135)   (136)   (137)

shown by the formation of lauric acid[217] on oxidation with alkaline permanganate, and of α-ketomyristic acid (137)[218] on alkaline hydrolysis which should be regarded as a β-diketone cleavage of the tautomeric form (136).

Asano and Yamaguti[220] established structure (133) by a lengthy synthesis, the natural and synthetic pigments being identified by comparison of their X-ray powder photographs. A shorter route[221] starts from the ketone (138) obtained by Friedel-Crafts acylation of quinol dimethyl ether with undecoyl chloride. Clemmensen reduction

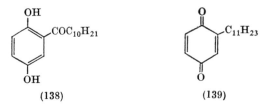

(138)   (139)

and oxidation to the quinone (139) was followed by reaction with dimethylamine, and final hydrolysis of the 3,6-diamino derivative. By far the most convenient procedure, despite the low yield, is Fieser's synthesis[222] of embelin by decomposition of lauroyl peroxide in a hot acetic acid solution of 2,5-dihydroxybenzoquinone.

The structure of the higher homologue, rapanone, follows from the formation of myristic acid and α-ketopalmitic acid on alkaline oxidation and hydrolysis, respectively.[223] It was first synthesised by the Japanese workers[223] and later obtained by peroxide alkylation.[222]

Vilangin (140), $C_{35}H_{52}O_8$

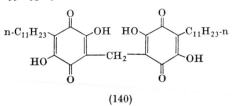

(140)

Occurrence: dried berries of *Embelia ribes* Burm.[224] and *E. robusta* Roxb.,[224] flowers and berries of *Myrsine australis* Allan[452] (Myrsinaceae).

Physical properties: orange-yellow prisms, m.p. 264–265° (dec.), $\lambda_{max.}$ (dioxan) 290·5, 298, 416 nm (log ϵ 5·55, 5·56, —), $\nu_{max.}$ (KBr) 3282, 1610 cm⁻¹.

Vilangin† is a minor constituent of *E. ribes* discovered fairly recently.[224] It is relatively insoluble which may account for its being overlooked in earlier investigations; on the other hand it was not detected in the most recent study.[203] Like embelin it affords lauric acid (and lower acids) when oxidised with alkaline permanganate, and yields α-ketomyristic acid on alkaline hydrolysis; indeed, embelin can be

† The name is derived from the Telugu vernacular name for *E. ribes*.

obtained by thermal decomposition of vilangin. The presence of two embelin moieties is indicated by its molecular weight, and by the formation of a tetra-acetate, leuco-octa-acetate and numerous other derivatives. Structure (140) satisfies the molecular formula, and is consistent with the tendency to form anhydrovilangin (141) (and derivatives thereof) under acid conditions. The structure of vilangin was

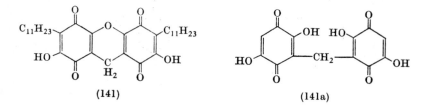

(141)   (141a)

confirmed by synthesis[224] from embelin and formaldehyde (a general reaction[225] with aldehydes). It has also been obtained by condensing formaldehyde with 2,5-dihydroxybenzoquinone to give (141a), followed by alkylation with lauroyl peroxide.[226]

As vilangin was obtained from dried *E. ribes* berries it may be an artefact.[465]

Bovinone (142), $C_{26}H_{36}O_4$

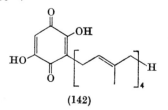

(142)

Occurrence: sporophore of *Boletus* (*Suillus*) *bovinus* (Linn. ex Fr.).[465]
Physical properties: orange needles, m.p. 84–85°, $\lambda_{max.}$ (EtOH) 287 nm (log ϵ 4·31), $\nu_{max.}$ (KBr) 3300, 1610 cm^{-1}, $\nu_{max.}$ (CHCl$_3$) 3360, 1640 cm^{-1}.

Boletus (*Suillus*) *bovinus* elaborates two novel polyprenylbenzo-quinones and also atromentin (192). These account for the pink colour which develops when the sporophore is moistened with alcohol whereas variegatic acid[462] is responsible for the blueing reaction which occurs when the flesh is bruised. From its acidity, redox behaviour and ultra-violet/infrared spectra, bovinone is a dihydroxybenzoquinone, and it yields a diacetate and leucotetra-acetate. A singlet (1H) at τ 4·02 in the n.m.r. spectrum shows that one quinonoid position is free, and the arrangement of the substituents was established *inter alia* by oxidation of the leucotetramethyl ether with permanganate which gave 2,3,5,6-tetramethoxyphenylacetic acid. Bovinone is thus a 2,5-dihydroxy-

benzoquinone alkylated at C-3. The formation of an octahydro derivative by catalytic hydrogenation (and reoxidation), and laevulinic
aldehyde by ozonolysis of bovinone leucotetra-acetate, suggested the
presence of a tetraprenyl side chain, which was confirmed by the mass
spectrum. This included strong peaks at M-69, M-69-68 and M-69-68-68,
while cleavage of the fourth isoprene unit gave the benzylium ions

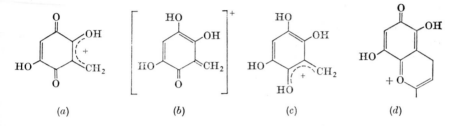

| (a) | (b) | (c) | (d) |

(a), (b) and (c) at m/e 153, 154 and 155. The peak at m/e 193 could be
attributed to (d) while the base peak occurred at m/e 69. From the
n.m.r. spectrum the side chain has an all-*trans* structure. Hence bovinone
is (142),[465] the monomeric quinone corresponding to amitenone (143)
reported [420] a little earlier from the same fungus.

Amitenone (143), $C_{53}H_{72}O_8$

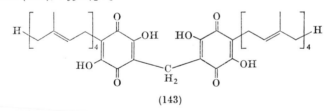

(143)

Occurrence: dried sporophores of *Boletus (Suillus) bovinus* (Linn. ex Fr.).[419,465]
Physical properties: yellow-orange crystals, m.p. 187–188°, $\lambda_{max.}$ (EtOH),
288 nm (log ε 4·45), $\lambda_{max.}$ (CHCl$_3$) 410(sh), nm (log ε 2·67), $\nu_{max.}$ (KBr) 3290, 1615
cm^{-1}, $\nu_{max.}$ (CHCl$_3$) 3358, 1646 cm^{-1}.

Amitenone is the second prenylated quinone in the (Japanese) edible
fungus *S. bovinus*.[419]† The structure [420] is analogous in type to vilangin
but, as in bovinone, the geranylgeranyl side chains are reminiscent of the
plastoquinones and ubiquinones and may presage a new family of
terpenoid benzoquinones. Amitenone shows a marked tendency to form
gels in non-polar solvents. The n.m.r. spectrum shows the characteristic
resonances of a *trans*-polyprenyl side chain containing four isoprene

† The structure previously advanced [419] for bovinin, another quinone from this fungus
has now been withdrawn.[421]

units, and the mass spectrum provides evidence for the successive loss of three C_5 units, the base peak appearing at m/e 69 ($C_5H_9^+$). Further support for a tetraprenyl side chain was obtained by degradation with alkaline hydrogen peroxide which gave a C_{21} unsaturated acid (methyl ester, M^+ 332). The molecular weight suggests that two such side chains are present and from the n.m.r. spectrum they must be symmetrical. A four-proton signal at τ 6·93 (J, 7 Hz) could be attributed to two terminal methylene groups adjacent to a hydroxyquinone ring and a singlet (2H) at τ 6·48 was assigned to a methylene bridge between two quinone nuclei. That two quinone rings are present can be deduced from the results of hydrogenation. Ten mols. of hydrogen were absorbed but the product, after aeration, was a hexadecahydro derivative. This corresponds to the saturation of eight side chain double bonds and the reduction of two quinone systems which were reoxidised. Oxidative degradation of hexadecahydro-amitenone gave a saturated acid (methyl ester, M^+ 340). The remaining functional groups are four hydroxyls and from the ultraviolet and infrared data these are evidently arranged as two 2,5-dihydroxybenzoquinone systems. If these are now linked by a methylene bridge and two geranyl-geranyl side chains are attached to the vacant ring positions, structure (143) emerges for amitenone.[420] Beaumont and Edwards[465] suggest that amitenone, which has only been isolated from the dried mushrooms, may be an artefact derived from bovinone (142) which is present in the fresh fungus but is very difficult to extract from dried material. Amitenone has been synthesised from bovinone by condensation with formaldehyde.[465]

Amitenone probably decomposes on heating (cf. vilangin) and it is not surprising that no molecular ion appears in the mass spectrum. Cleavage at the central C—C bonds results in formation of the ions e and

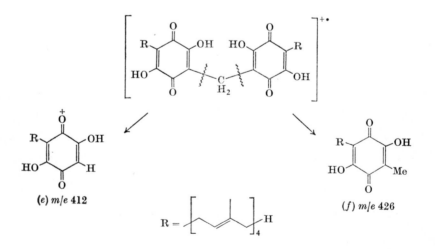

(e) m/e 412   (f) m/e 426

f followed by side chain fragmentation to give e-69, e-69-68, e-69-68-68 and (d); m/e 69 is the base peak.

2-Hydroxy-5-methoxy-3-pentadecenyl-1,4-benzoquinone (144), $C_{22}H_{34}O_4$

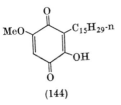

(144)

Occurrence:† *Ardisia japonica* (Thunb.) Blume (fruit and rhizome),[203] *A. japonica* (Thunb.) Blume var. *angusta* (rhizome),[203] *A. crenata* Sims. (fruit),[203] *A. montana* (Miq.) Sieb. (rhizome),[203] and *A. quinquegona* Blume (fruit).[203]

Physical properties: yellow plates, m.p. 66·5–67°, λ_{max}. (EtOH) 289, 420 nm (log ϵ 4·21, 2·65), ν_{max}. (KBr) 3318, 1655(sh), 1632, 1595 cm^{-1}.

Ardisia japonica contains several alkylated benzoquinones including this pentadecenyl derivative which could not be separated from the tridecenyl and tridecyl analogues. The physical properties above refer to a product containing ca. 13% of each. The structure[424] includes a methoxyl group, an acidic hydroxyl group (violet solution in aqueous sodium carbonate), and a double bond which yields a dihydro derivative on hydrogenation. Demethylation of the latter with aluminium chloride afforded an orange dihydroxyquinone very similar in spectral properties (λ_{max}. 292 nm, ν_{max}. 1605 cm^{-1}) to embelin and rapanone, and hence a 2,5-dihydroxybenzoquinone. Only one side chain is present as the n.m.r. spectrum includes a singlet (1H) at τ 4·27 due to a quinonoid proton, and since the spectrum of the dihydro compound is almost superimposable on that of 2-hydroxy-5-methoxy-3-tridecylbenzo-quinone (rapanone methyl ether) (144; $C_{13}H_{27}$ in place of $C_{15}H_{29}$) it was concluded that the ring substituents were arranged as in (144). The nature of the side chain was established by hydrogen peroxide oxidations and gas chromatography of the methyl esters of the resulting fatty acids. The dihydro compound gave mainly methyl palmitate and some methyl myristate, while the natural pigment yielded chiefly the methyl ester of a hexadecenoic acid (presumably) and lesser amounts of methyl myristate and a methyl tetradecenoate. These results agreed with the molecular ion intensities seen in the mass spectrum of the mixture and hence the principal pigment has a C_{15} side chain (144) and this is con-taminated by the tridecyl- and tridecenyl-quinones (144; $R = C_{13}H_{27}$ and $C_{13}H_{25}$, respectively).[424] The position of the double bond has not been ascertained.

† In all cases contaminated with the tridecenyl and tridecyl analogues.

Bhogatin (145), $C_{16}H_{24}O_4$

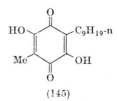

(145)

Occurrence: leaves of *Maesa macrophylla* Wall.[422] (Myrsinaceae).
Physical properties: orange-red plates, m.p. 156–157°, $\lambda_{max.}$ (EtOH) 295 nm (log ϵ 3·15), $\nu_{max.}$ (CHCl$_3$) 3385, 1635 cm^{-1}.

This new quinone forms a diacetate and a leucotetra-acetate, and its spectral properties suggest a dialkylated 2,5-dihydroxybenzoquinone structure. As permanganate degradation affords both acetic and n-capric acids the side chains must contain one and nine carbon atoms, respectively, and hence the pigment must be (145).[422] This was confirmed synthetically by alkylating 2,5-dihydroxy-3-methylbenzoquinone with capryl peroxide. The compound was prepared earlier[423] by a conventional route.

Polygonaquinone (146), $C_{28}H_{48}O_4$

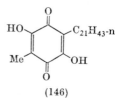

(146)

Occurrence: roots of *Polygonatum falcatum* A. Gray[231] (Liliaceae).
Physical properties: orange prisms, m.p. 133–134°, $\lambda_{max.}$ (CHCl$_3$) 295, 417 nm (log ϵ 4·32, 2·33), $\nu_{max.}$ (KBr) 1615 cm^{-1}, $\nu_{max.}$ (CHCl$_3$) 1635 cm^{-1}.

The general nature of this quinone is apparent from its ultraviolet and infrared spectra, colour reactions and derivatives. The n.m.r. spectrum reveals the presence of a methyl group and a long, unbranched alkyl side chain attached to a quinone ring; vinylic protons are absent. The leucotetra-acetate shows benzenoid light absorption, and the 2,5-orientation of the two hydroxyl groups is indicated by the intense carbonyl absorption at 1615 cm^{-1} (KBr) which shifts to 1655 cm^{-1} in the dimethyl ether and 1672 cm^{-1} in the diacetate. The nature of the side chain was determined by oxidation with alkaline hydrogen peroxide, and with permanganate in pyridine, both of which gave behenic acid (C_{22}). This means that polygonaquinone has a C_{21} side chain and is therefore (146).[231] The structure has been confirmed by synthesis.[232]

Friedel-Crafts acylation of toluquinol dimethyl ether with heneicosanoyl chloride gave the ketone (147) and a little of the more unusual keto-ester

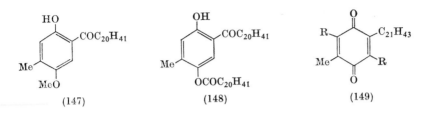

(147)   (148)   (149)

(148). Hydrolysis of the latter, followed by Clemmensen reduction and oxidation of the resulting quinol afforded the quinone (149; R = H), from which polygonaquinone was derived by conversion to the bis-methylamino derivative (149; R = NHMe) and hydrolysis.

Maesaquinone (150), $C_{26}H_{42}O_4$
Maesaquinone acetate, $C_{28}H_{44}O_5$

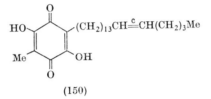

(150)

Occurrence: *Maesa emirnensis* Arn. (fruit),[207] *M. japonica* (Thunb.) Moritzii (bark and fruit)[203,227] and *M. tenera* Mez. (fruit)[203] (Myrsinaceae).

Physical properties: orange plates, m.p. 123°, $\lambda_{max.}$ (EtOH) 295, 440 nm (log ϵ 4·36, 2·58), $\nu_{max.}$ (CHCl$_3$) 3340, 1635 cm^{-1}, $\nu_{max.}$ (KBr) 3320, 1605 cm^{-1}.

Mono-acetate, yellow solid, m.p. 28–41°, $\lambda_{max.}$ (EtOH) 275, 415 nm (log ϵ 4·18, 2·85), $\nu_{max.}$ (KBr) 3315, 1782, 1655, 1639(sh), 1615 cm^{-1}.

Maesaquinone gives a violet alkaline solution and yields a diacetate and leucotetra-acetate. The latter absorbs one mol. of hydrogen catalytically whereas the natural pigment absorbs two giving a product which reoxidises in air to form dihydromaesaquinone. Oxidation[227,228] of this new quinone with alkaline hydrogen peroxide afforded acetic acid and a C_{20} acid identified as arachidic. Dihydromaesaquinone was therefore regarded as (151), the orientation of the substituents being assumed by analogy with embelin and the similarity of a synthetic homologue. Synthetic confirmation was secured[228] by alkylation of the keto-ester (152) with heptadecyl iodide, followed by hydrolysis to the ketone (153), Clemmensen reduction, and oxidation with chromic acid to give the quinone (154). Dihydromaesaquinone (151) was then derived

by addition of methylamine and final hydrolysis, the synthetic product
being identical (mixed melting point) with the "natural" material. The

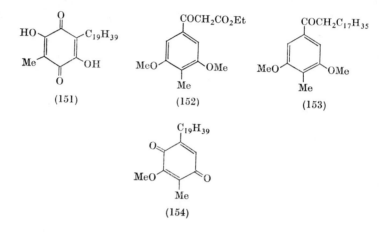

(151)   (152)   (153)

(154)

position of the side chain double bond in maesaquinone was not established but as ozonolysis gave a dibasic acid which corresponded (in
melting point) with pentadecan-1,13-di-oic acid, Hiramoto[228] proposed
structure (150) for the natural pigment.

Owing to the unreliability of mixed melting point determinations in
this series of compounds a check on structure (150) by modern methods
was desirable, and this was recently carried out by Ogawa and Natori.[229]
Their findings fully confirmed Hiramoto's earlier work. Maesaquinone
has the expected ultraviolet and infrared spectral properties, and its
n.m.r. spectrum confirms the presence of a methyl group and a normal
alkenyl chain attached to the quinone ring. Oxidation of maesaquinone
with alkaline hydrogen peroxide yielded an unsaturated acid, $C_{20}H_{38}O_2$,
which afforded arachidic acid on hydrogenation and it was cleaved by
ozonolysis to give, after further oxidation, a mixture of valeric and
pentadecan-1,15-di-oic acids. This confirmed the position of the double
bond, and both the lack of C–H absorption in the region 960–980 cm^{-1}
and the vicinal coupling constant[424] of the vinyl protons (J, 5 Hz)
indicate that it has the cis-configuration. The acids were identified by
conversion to their methyl esters and direct comparison (g.l.c.) with
authentic samples. Finally, dihydromaesaquinone was again synthesised by alkylation of 3,6-dihydroxytoluquinone with arachidoyl
peroxide.

In the fresh fruits of M. japonica (and M. tenera[203]) maesaquinone
exists mainly as a mono-acetate (it is not known which hydroxyl group
is esterified), the content being as high as 9·5%.[229]

Ardisiaquinone† *A* (155), $C_{30}H_{40}O_8$
Ardisiaquinone B (156), $C_{29}H_{38}O_8$
Ardisiaquinone C $C_{32}H_{42}O_9$ (mono-acetate of *B*)

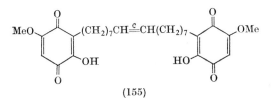

(155)

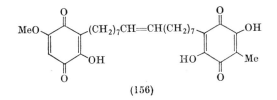

(156)

Occurrence: root bark of *Ardisia sieboldii* Miq.[203] (Myrsinaceae).
Physical properties: *A* (155), yellow crystals, m.p. 154°, $\lambda_{max.}$ (EtOH) 289, 415–425 nm (log ϵ 4·60, 2·83), $\nu_{max.}$ (KBr) 3320, 1655(sh), 1633, 1598 cm^{-1}.
B (156), red crystals, m.p. 119°, $\lambda_{max.}$ (EtOH) 291, 420–428 nm (log ϵ 4·56, 2·82), $\nu_{max.}$ (KBr) 3300, 1605 cm^{-1}.
C, yellow crystals, m.p. 69–70°, $\lambda_{max.}$ (EtOH) 281, 420 nm (log ϵ 4·41, 2·87), $\nu_{max.}$ (KBr) 3328, 1774, 1657(sh), 1635, 1603 cm^{-1}.

These quinones accompany embelin and rapanone in *A. sieboldii*, and their ultraviolet-visible and infrared spectra indicate that they are alkylated dihydroxybenzoquinones of the same general type.[230] Hydrogenation of *A* and *B*, followed by aeration, afforded dihydro derivatives indicating the presence of a double bond in addition to quinonoid unsaturation. In contrast to the quinones discussed previously, oxidation of the dihydro derivatives with alkaline hydrogen peroxide gave a dibasic acid, octadecan-1,18-di-oic acid, while similar degradations of ardisiaquinones *A* and *B* yielded *cis*-octadec-9-en-1,18-di-oic acid identified by direct comparison with authentic material, and by further degradation with ozone which gave azelaic acid as the sole product. These findings, together with the absence of terminal methyl groups (n.m.r.), the molecular formulae and molecular extinction values, indicate that the new pigments are biquinones with the two rings linked as in (157).[424] It could be shown by study of the n.m.r. spectrum and the formation of

† The unidentified ardisiols (from *A. fuliginosa* Blume[466]) may be related: α-ardisiol, $C_{35}H_{46}O_{10}$, orange, m.p. 107°, β-ardisiol, $C_{35}H_{46}O_{10}$, pale brown, m.p. 183°, oxyardisiol, $C_{35}H_{46}O_{11}$, brownish-yellow, m.p. 191°.

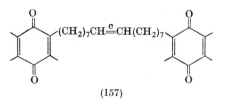

(157)

derivatives, that ardisiaquinone A contained two hydroxyl and two methoxyl groups, and two symmetrical protons attached to the quinone rings. As the chemical shift of these protons does not change appreciably when the compound is acetylated, it may be deduced that they lie adjacent to methoxyl and not hydroxyl groups. This was confirmed by observing the nuclear Overhauser effect on irradiating the methoxyl protons at τ 6·14 which resulted in a 14% increase in the intensity of the ring proton signal at τ 4·18.[425] Alkaline hydrolysis of A affords a tetrahydroxy derivative showing λ_{max}. 293–4 nm and ν_{max}. (KBr) 1605 cm^{-1} which is clearly a bis-2,5-dihydroxybenzoquinone. All this leads logically to structure (155) for A,[424] and the general substitution pattern was confirmed by the following synthetic scheme.[425] The diketone (158), prepared by Friedel-Crafts acylation of quinol dimethyl ether with thapsyl dichloride, was reduced (CO → CH$_2$), demethylated, and the biquinol oxidised to the unstable biquinone (159). Treatment with

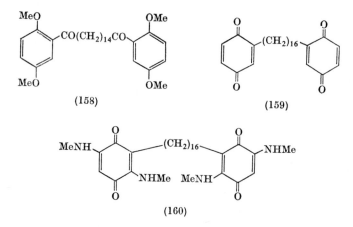

(158)　　　　　　　　(159)

(160)

methylamine then gave the tetrakisamino derivative (160) which was identical with that derived from dihydro-ardisiaquinone A dimethyl ether with the same reagent. An unexpected product of the reaction with (159) was bis-2,5-methylaminobenzoquinone which is also formed in the reaction of toluquinone with methylamine.[425]

The n.m.r. spectrum of ardisiaquinone B shows that it has the same structural features as *A* except that one methoxyl group is replaced by hydroxyl and one of the quinonoid protons is replaced by a methyl group. It must therefore have structure (156) and in ardisiaquinone *C* one of its hydroxyl groups is acetylated.[230, 424] No precursors of these quinones have been found as yet but it is interesting to note the occurrence of striatol (160a) in the wood of *Grevillea striata* (Proteaceae).[446]

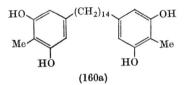

(160a)

The mass spectra[424] of ardisiaquinones *A* and *B* and their dihydro derivatives show strong M + 2 peaks and a major cleavage occurs at the "benzyl" position forming significant ions at *m/e* 167 (*g*), 168 (*h*) and 169 (*i*) (base peak). In the spectra of embelin and rapanone the base

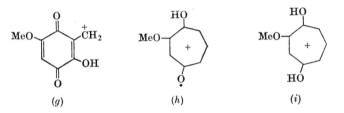

(*g*)   (*h*)   (*i*)

peak is at *m/e* 154 (*f*, HO in place of OMe).

Carthamone (161), $C_{25}H_{20}O_{11}$

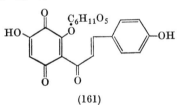

(161)

Occurrence: coloured flowers of *Carthamus tinctorius* L.[233] (Compositae).

Physical properties: scarlet needles, m.p. 228–230° (dec.), $\lambda_{max.}$ (?) 242, 368, 510 nm, $\nu_{max.}$ (?) 1672, 1621 cm^{-1}.

Safflower consists of the dried, yellow to red, flowers of *C. tinctorius*, and was formerly cultivated in Asia, and to some extent in Europe, as a natural dyestuff for cotton and silk. It contains a mixture of flavonoids, one of which happens to possess a quinonoid ring. This red compound, formerly known as carthamin or carthamic acid, which is the dyeing principle, was first obtained pure by Perkin.[233] Subsequently Kuroda[234] isolated a very similar yellow pigment which she regarded as the chalcone glucoside (162). This, in fact, is the main pigment component, but Seshadri[235] pointed out that structure (162) would be unstable[236] and cyclise readily to a colourless flavanone. However the required stability

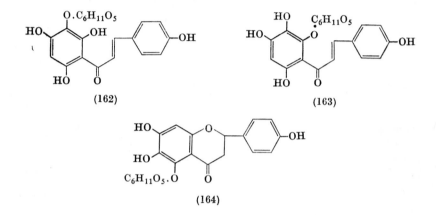

(162)   (163)

(164)

is attained if the sugar residue is relocated as in (163). Both the chalcone and the corresponding flavanone (164) are present in ivory-white flowers, but coloured varieties, especially orange-red flowers, contain an additional red component. This pigment can be reduced to the yellow chalcone (163) by sulphur dioxide and reoxidised to the original material by peroxidase. It must therefore be the quinone (161) which was renamed carthamone, the chalcone (163) being called carthamin.[237] The conversion of (163) into (161) can easily be observed; flowers of *C. tinctorius* which have just opened are yellow and they change via orange to red in about ten hours, and if fresh yellow flowers are homogenised they redden almost at once (but *not* in the absence of oxygen).[238]

Pedicinin (165; R = H), $C_{16}H_{12}O_6$
Methylpedicinin (165; R = Me), $C_{17}H_{14}O_6$

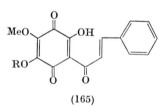

(165)

Occurrence: leaves of *Didymocarpus pedicellata* R. Br.[239, 240] (Gesneriaceae).
Physical properties:[261] (165; R = H), red needles, m.p. 202–204° and 206–207°,

$\lambda_{max.}$ (EtOH) 246, 300, 385 nm (log ϵ 3·85, 4·09, 4·47), $\nu_{max.}$ (KBr) 3272, 1670 1643(sh), 1635(sh), 1620 cm^{-1}.

(165; R = Me), orange-yellow needles, m.p. 110–112°.

As in safflower, the leaves of *D. pedicellata* (and probably other Gesneriaceae[241]) contain a group of flavonoid compounds (for a review see Seshadri[242]), the most abundant being a chalcone-quinol, pedicin (166), which itself is an orange-red pigment. Its structure has been established by synthesis.[240] Oxidation of (166) with silver oxide gives

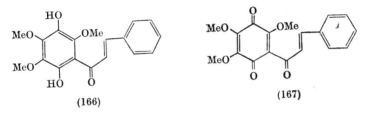

(166)   (167)

the quinone (167); this has not been found in leaves possibly because of the exceptional ease of hydrolysis. The methoxyl group at C-2, which is conjugated with two carbonyl groups, is very susceptible to nucleophilic attack and treatment with cold aqueous sodium bicarbonate suffices to convert (167) into methylpedicinin (165; R = Me).† In the presence of cold aqueous sodium hydroxide a second methoxyl group is hydrolysed to give the other natural quinone, pedicinin (165; R = H). Methylation of pedicinin with diazomethane gives methylpedicinin confirming that the free (that is, unreacted) hydroxyl group is at C-2 where it is chelated with the chalcone carbonyl group.[240] The remaining uncertainty[243] is the position of the methoxyl group in pedicinin, and this was eliminated by *ethylation* of leucopedicinin to give 2,3,5,6-tetra-ethoxy-4-methoxy-chalcone identical with synthetic material. Pedicinin is therefore methylated at C-6 as in (165; R = H).[244]

Botrallin (168), $C_{16}H_{14}O_7$

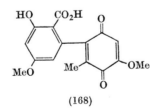

(168)

Occurrence: cultures of *Botrytis allii* Munn.[426]

Physical properties: yellow prisms, m.p. 165–185° (dec.), $\lambda_{max.}$ (EtOH) 238, 273, 370 nm (log ϵ 4·36, 4·20, 3·95), $\nu_{max.}$ (KBr) 1656, 1629, 1580 cm^{-1}.

† This could have occurred during the isolation procedure.[240]

Botrallin is elaborated by *B. allii* when grown on malt-agar but not on oatmeal-agar or potato-dextrose-agar. It is the only example of its type (cf. the biphenylbiquinones phoenicin and oosporein), and bears a close resemblance to the benzocoumarin alternariol (169) isolated from *Alternaria tenuis*.[427] Quinone character was revealed by redox properties and a positive Craven test, and the compound gives a violet–red solution in aqueous sodium bicarbonate but is unstable in dilute ammonia. The n.m.r. spectrum revealed the presence of a methyl group and a proton attached to a quinone ring, two methoxyls, a hydroxyl and a carboxyl group, and two *meta*-coupled aromatic protons. On reduction with zinc and acetic acid it gave a neutral product, closely resembling alternariol (169) in its ultraviolet absorption. This compound, $C_{16}H_{14}O_6$, is an anhydro derivative of the expected quinol; it contained only two hydroxyl groups and no carboxyl group, and showed ν_{CO} 1656 cm^{-1} which shifted to 1724 cm^{-1} on acetylation. This is also analogous to alternariol and indicates that the reduction product is a benzocoumarin with a

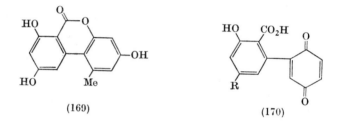

(169)   (170)

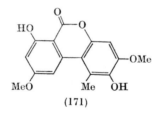

(171)

hydroxyl substituent adjacent to the carbonyl group. It follows that botrallin is of the form (170) with a substituent R as shown to allow for two unsubstituted *meta*-positions. The group R must be methoxyl as the methyl group is attached to the quinone ring (τ 8·23 shifting to τ 7·30 in the reduction product). Botrallin is therefore assigned structure (168),[426] the positions of the methyl and methoxyl groups in the quinone ring being located by analogy with alternariol. This is favoured, but not decisively, by the relatively high field resonance of the quinonoid proton (τ 3·86). Again the spectroscopic data is not decisively in favour of the *para*-quinone structure although this is most probable; an attempted reaction with *o*-phenylenediamine was inconclusive. On this basis the reduction product is (171).

Polyporic Acid (172), $C_{18}H_{12}O_4$

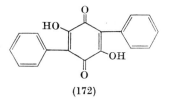

(172)

Occurrence: in the fungi *Polyporus nidulans* Fr. [= *P. rutilans* (Pers.) Fr.],[245, 246] *Lopharia papyracea* (Jungh.) Reid[247] and *Peniophora filamentosa* (B. & C.) Burt,[248] and the lichens *Sticta coronata* Muell. Arg.[248] [= *S. orygmaea* Ach. (? Hook.)] and *S. colensoi* Bab.[248]

Physical properties: bronze leaflets,† m.p. 310–312°,‡ $\lambda_{max.}$ (EtOH) 256, 262, 330(sh), 465 nm (log ε 4·63, 4·63, 4·06, 2·60), $\nu_{max.}$ (KBr) 3310, 1637, 1613, 1597 cm^{-1}.

The original source of this colouring matter was an unidentified *Polyporus* sp. which Stahlschmidt[249] found on decaying oak trees but a later search[250] of the same area failed to locate it (a common hazard of chemical research on higher fungi). The pigment was recognisable by its deep violet colour with ammonia and is probably the same as Zopf's lichen pigment, orygmaeic acid.[248, 251] The pigment content may be high—11·4% of the dry weight of *L. papyracea*[247] and 23% of *P. nidulans*.[246]

The structure was elucidated by Kögl.[245] It forms a yellow diacetate and dimethyl ether, and gives *p*-terphenyl on zinc dust distillation. This hydrocarbon, which accounts for all the carbon atoms in polyporic acid, had not previously been derived from a natural product. As a chromic acid oxidation of the diacetate gave only benzoic acid the two hydroxyl groups must be attached to the quinone ring and the structure is therefore (172).[245] This was confirmed by comparison with an authentic specimen previously obtained by Fichter[253] as a by-product in the reaction of ethyl oxalate with ethyl phenylacetate in the presence of sodium. This was recognized by Kögl[254] as an acyloin reaction in which the intermediate diketone (173) underwent Claisen condensation with ethyl oxalate to give (174) which is tautomeric with polyporic acid. However the yield is only ca. 1%. A more satisfactory preparation[246, 255] proceeds by way of 2,5-diphenylbenzoquinone which can be obtained

† Red solvate from pyridine, yellow solvate from dioxan.
‡ M.p.s from 305° to 318° have been reported.

from benzoquinone by reaction with benzene in the presence of aluminium chloride. The crude product contains a mixture of quinols and quinhydrones which can be converted entirely to quinones by careful

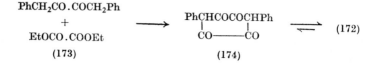

$$PhCH_2CO.COCH_2Ph \quad + \quad EtOCO.COOEt \quad (173) \longrightarrow PhCHCOCOCHPh \atop CO{\rule{0.6cm}{0.4pt}}CO \quad (174) \rightleftharpoons \quad (172)$$

oxidation with chromic or nitrous acids. Bromination then gives 3,6-dibromo-2,5-diphenylbenzoquinone from which polyporic acid is derived by brief treatment with alkali. Polyporic acid dimethyl ether can be obtained by addition of methanol to 2,5-diphenylbenzoquinone but the yield is low.[254] Another approach starts from 2,5-dihydroxy(or dichloro)benzoquinone, the aryl groups being introduced by free radical phenylation using either *N*-nitrosoacetanilide or the Meerwein procedure.[256] The method is general[258, 259] and can be adapted to prepare unsymmetrical derivatives of polyporic acid. A number of these show limited anti-tumour activity,[258] and some are effective anti-oxidants, the t-butyl derivative (175) being outstanding.[259]

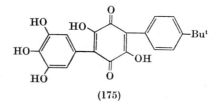

(175)

Hydroxybenzoquinones are prone to undergo ring opening and ring contraction reactions, especially under basic conditions. This behaviour is exemplified by polyporic acid; on prolonged boiling in alkaline solution the intense violet colour completely disappears, the products from polyporic acid being 1-benzyl-2-phenylsuccinic acid (183), *cis* and *trans*-α-benzylcinnamic acid (184), and oxalic acid.[260] It is suggested[447] that these may be formed by way of hydrolytic fission of the tautomers (176)–(177) to give (178), followed by a benzilic acid rearrangement to form (179) and its lactone (180). This could collapse in two ways, as indicated, to give either the succinic acid (183), or the cinnamic acids (184) and oxalic acid. Fichter[253] has shown that alkylated lactones of type (180) yield carbon dioxide and dialkylsuccinic acids when treated with alkali under the same conditions, and Kögl *et al.*[260] were able to isolate the lactone (180; *p*-HO–C_6H_4 in place of Ph) from the alkaline decomposition of atromentin (p. 160) and show that it gave the corresponding cinnamic acids on further treatment. However, as the lactones would not exist as such in alkaline solution the decompositions represented by (181) and (182) are invalid, and it is difficult to see how the succinic acid (183) is derived from the ring-opened form of (181).

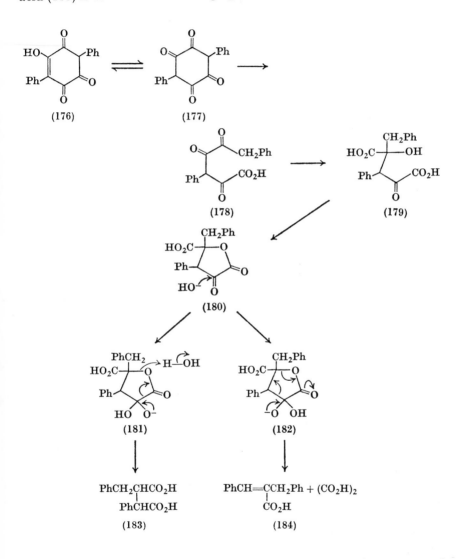

As an example of ring contraction of a hydroxybenzoquinone, a recent reaction of the pedicinin derivative (185) may be quoted.[261] When this was kept in cold aqueous sodium hydroxide it rearranged to (186).

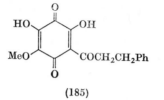

(185)

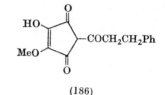

(186)

Phlebiarubrone (187), $C_{19}H_{12}O_4$

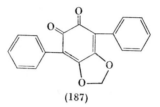

(187)

Occurrence: cultures of *Phlebia strigoso-zonata* (Schw.) Lloyd.[262]

Physical properties: red needles, m.p. 248–250°, λ_{max}. (EtOH) 268, 332, 465 nm (log ϵ 4·47, 3·64, 3·54), ν_{max}. (KBr) 1653, 1640 cm^{-1}.

It is of interest that this unique *o*-benzoquinone is found in the mycelium of *P. strigoso-zonata* when cultured artificially but under natural conditions another terphenyl quinone, thelephoric acid (p. 161) is found in the fruiting bodies.[284] The key to the elucidation of the structure of phlebiarubrone was found in the action of alkali. The quinone is insoluble in alkali but if dissolved in acetic acid and then added to dilute sodium hydroxide solution an immediate purple colour appears, and acidification then liberates polyporic acid. Thus all four oxygen atoms are attached to the central ring and two of these form a methylene-dioxy group. This is evident from the n.m.r. spectrum of the pigment (2H singlet at τ 3·75) and its leucodiacetate (2H singlet at τ 4·03), and from the formation of formaldehyde in the alkaline hydrolysis. Phlebia-rubrone is therefore an *o*-quinone of structure (187).[262] It reacted with *o*-phenylenediamine to give a phenazine which, unexpectedly, was the dihydroxy derivative (188), and can be obtained equally well from polyporic acid. In this latter reaction polyporic acid is behaving as a tautomeric *o*-quinone, and similarly, if the potassium salt is stirred in acetone in the presence of excess methylene sulphate and sodium hydrogen carbonate it is converted, in high yield, into phlebiarubrone.[263]

As noted earlier (p. 107) benzoquinones are prone to nucleophilic attack at a carbonyl group and phlebiarubrone provides a further ex-ample. Refluxing with acetic anhydride and sodium acetate gave the ester (189) whose structure was established by hydrogenolysis of the benzyl ace-tate, followed by hydrolysis with hydrogen bromide, and oxidation with

ferric chloride to give the known quinone (190).[262]† Using completely
anhydrous sodium acetate the reaction takes another course giving a

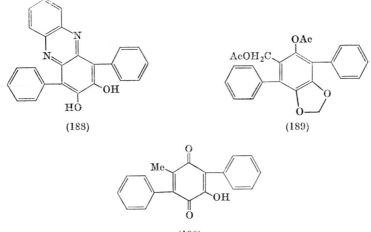

(188)   (189)

(190)

number of products, including the normal leuco-acetate. (Polyporic acid
gives the normal leuco-acetate under similar conditions.)

Volucrisporin (191), $C_{18}H_{12}O_4$

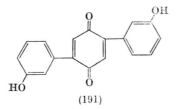

(191)

Occurrence: cultures of *Volucrispora aurantiaca* Haskins.[265, 266]
Physical properties: red, rhombic plates, m.p. >300°, $\lambda_{max.}$ (dioxan) 234(sh),
281·5, 329, 377(sh), nm (log ε 4·47, 4·00, 3·90, 3·45), $\nu_{max.}$ (KBr) 3280, 1637, 1630(sh),
1590 cm⁻¹.

Volucrispora is a new Hyphomycete genus and this pigment has
certain novel features. It forms a dimethyl ether, diacetate and leucotetra-acetate, and the structure was revealed by two degradative
experiments; zinc dust distillation yielded *p*-terphenyl while oxidation
with alkaline hydrogen peroxide gave *m*-hydroxybenzoic acid. These
results point to structure (191) which was confirmed by synthesis of the
dimethyl ether. Arylation of benzoquinone with diazotised *m*-anisidine

† See also ref. 264.

gave 2-(*m*-methoxyphenyl)benzoquinone, and the process was then repeated to give, with some difficulty, 2,5-bis(*m*-methoxyphenyl)benzoquinone identical with volucrisporin dimethyl ether.[265] Volucrisporin is present in cultures of *V. aurantiaca* partly as the quinol.

This pigment differs from all the other natural terphenylquinones (from Basidiomycetes and lichens) in the absence of hydroxyl groups attached to the quinone ring and the presence of the unusual *m*-hydroxyphenyl side chains. The reported[265] failure to react with diazomethane is curious as is the purple colour in aqueous sodium hydroxide. However, the initial colour is actually a brownish-yellow ($\lambda_{max.}$ 365 nm)[295] which changes to violet in air, evidently due to hydroxylation of the quinone ring as a mixture of products is rapidly formed, one of which, 3-hydroxy-2,5-bis(*m*-hydroxyphenyl)benzoquinone, gives a purple solution in aqueous bicarbonate.

Atromentin (192). $C_{18}H_{12}O_6$
Aurantiacin (193), $C_{32}H_{20}O_8$

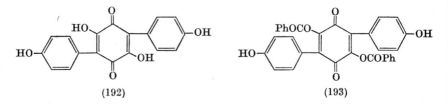

(192)   (193)

Occurrence: (192) in the fungi *Paxillus atrotomentosus* (Batsch) Fr.,[267, 268] *Hydnellum diabolus* Banker,[269] *Clitocybe subilludens* Murr.,[458] *Xerocomus chrysenteron* (Bull. ex St. Amans)[489] and *Boletus* (*Suillus*) *bovinus* (Linn. ex Fr.) Kuntze.[465]

(193) in the fungi *Hydnum aurantiacum* Batsch,[270] *Hydnellum caeruleum* (Hornem ex Pers.) Karst.,[271] *H. pseudocaeruleum* ined.[272] and *H. scrobiculatum* (Fr. ex Secr.) Karst. var. *zonatum* (Batsch ex Fr.) Harrison.[272]

Physical properties: (192) bronze leaflets, no m.p., $\lambda_{max.}$ (dioxan) 269, 366 nm (log ε 4·40, 3·64), $\nu_{max.}$ (KBr) 3448, 3350, 1650, 1600 cm⁻¹.

(193) dark red needles, m.p. 305–308° (285–295°), $\lambda_{max.}$ (dioxan) 234, 402 nm (log ε 4·71, 3·70), $\nu_{max.}$ (KBr) 3480, 1739, 1668, 1608 cm⁻¹.

P. atrotomentosus occurs frequently on old tree stumps, and is reddish-brown externally but the inner parts are not pigmented. The quinone is present mainly in the leuco form (3·6% dry weight) and is oxidised during the extraction with alkali, or better, cold acetone.[259] From the early work of Thörner[267] it appeared to be a hydroxyquinone but nearly fifty years elapsed before the structure was determined by Kögl and Postowsky.[268] It gives a di- and tetramethyl ether, a tetra-acetate and a hexaleuco-acetate; reduction with zinc gives *p*-terphenyl and oxidation with alkaline peroxide yields p-hydroxybenzoic acid. Atromentin is therefore 4′,4″-dihydroxypolyporic acid which is consistent with the facile hydrolysis of the dimethyl ether obtained by reaction with diazomethane. It has been synthesised by all the methods used for polyporic acid.[254, 255, 256b, 257, 259]

Aurantiacin is the 3,6-dibenzoate of atromentin. The leuco compound has been isolated from *Hydnellum caeruleum*[272] and *H. pseudocaeruleum*,[272] and the leucodibenzoate (194) from *Hydnum aurantiacum*.[273] The structure[270] follows from its facile hydrolysis to atromentin and benzoic acid, and by the formation (with difficulty) of a dimethyl ether identical with synthetic 4′,4″-dimethoxypolyporic acid dibenzoate.

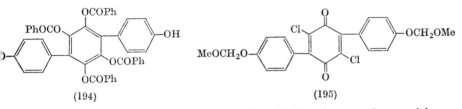

(194)   (195)

It has been synthesised[274] by arylation of 2,5-dichlorobenzoquinone with diazotized p-methoxymethoxyaniline to form (195); hydrolysis with dilute alkali then gave the 3,6-dihydroxy derivative which was benzoylated, and finally the protecting acetal groups were detached by hydrolysis with dilute acid.

Vigorous hydrolysis[260] of atromentin opens the ring and the lactone (196) can then be *isolated* (cf. the behaviour of polyporic acid, p. 154); or more vigorous treatment with hot 50% aqueous potassium hydroxide this decomposes to give a mixture of the *cis* and *trans* forms of the acid (197). Whereas oxidation of atromentin with alkaline hydrogen peroxide affords p-hydroxybenzoic acid, under acid conditions and with a limited amount of peroxide,[260] a yellow dilactone (200; R = OH) results by oxidative cleavage of the o-quinonoid form (198) leading to the intermediate (199) which lactonises. A similar oxidation of polyporic acid gives only a trace of dilactone (200; R = H) but this can be accomplished smoothly with lead tetra-acetate,[246, 275] and using dimethylsulphoxide-acetic anhydride[212] the yield is 95%. The dilactones behave as vinylogous acid anhydrides and on warming with methanol are easily converted into monolactone-esters for example, vulpinic acid (201; R = H, R′ = Me) a natural product. These reactions provide a chemical link between the terphenylquinones and the yellow tetronic acid pigments† which occur frequently in lichens,[276] sometimes accompanied by the quinones. *Sticta coronata*[248] for example, elaborates pulvinic (pulvic) acid (201; R = R′ = H), pulvinic anhydride (200; R = H) and

† See also ref. 463.

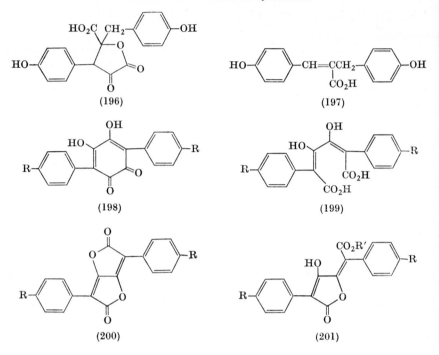

(196)   (197)

(198)   (199)

(200)   (201)

polyporic acid, and there is evidence[43] that the pulvinic acid pigments are derived from the quinones *in vivo* (see p. 25).

Leucomelone (202), $C_{18}H_{12}O_7$

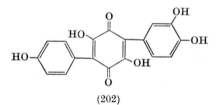

(202)

Occurrence: in the fungus *Polyporus leucomelas* Pers. ex Fr.[277]
Physical properties: brown leaflets, m.p. 320° (dec.).

From the Japanese black edible mushroom *P. leucomelas*, Akagi[277] extracted a quinonoid pigment, leucomelone, and a colourless substance, protoleucomelone. As the pigment formed a penta-acetate and a leuco-hepta-acetate, and gave *p*-terphenyl on zinc dust distillation, it appeared to be a pentahydroxydiphenylbenzoquinone, and since a mixture of *p*-hydroxybenzoic acid and protocatechuic acid was obtained by oxidation with alkaline hydrogen peroxide, structure (202) is clearly indicated. This was confirmed synthetically[278] by way of a two-stage arylation of

2,5-dichlorobenzoquinone. Treatment with N-nitroso-p-methoxyacet-anilide followed by reaction with diazotised 4-aminoveratrole gave (203),

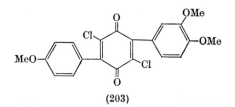

(203)

from which leucomelone was obtained by demethylation and hydrolysis (cf. ref. 259). The other natural product, protoleucomelone, proved to be the leucohepta-acetate [277] (cf. leuco-aurantiacin dibenzoate, p. 159).

Leucomelone is clearly related to the cyclopentenone, involutin (204), present in *Paxillus involutus* (Oeder ex Fr.),[279] and analogous

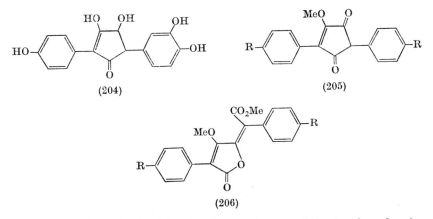

(204)   (205)

(206)

ketones have been derived *in vitro* from quinones of the terphenyl series. Kögl,[254] for example, obtained the cyclopentenones (205; R = H and OMe) by treating the tetronic esters (206; R = H and OMe; from polyporic acid and atromentin, respectively) with cold methanolic potash, the reaction proceeding by solvolysis of the lactone followed by an intramolecular Claisen condensation.

Thelephoric Acid (207), $C_{18}H_8O_8$

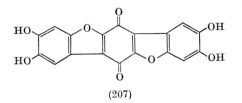

(207)

Occurrence: in the higher fungi, *Thelephora*,[280, 281, 297] *Hydnum*[273, 282–285, 297] and *Hydnellum*[272] spp., *Polystictus versicolor* (L.) Fr.,[283, 286] *Polyozellus multiplex* (Underw.) Murr.[283] (= *Cantharellus multiplex* Underw.), *Phlebia strigoso-zonata* (Schw.) Lloyd,[284] and in the lichens *Lobaria retigera* Trév.[287] and other *Lobaria* spp.[287, 288, 398]

Physical properties: very dark violet prisms, no m.p., λ_{max}. (EtOH) 217, 264, 305, 390(sh), 483 nm (log ε 4·33, 4·27, 4·30, 3·48, 3·86), ν_{max}. (KBr) 3550, 3260, 1645, 1610 cm^{-1}.

The original chemical work,[281] on the basis of which a bizarre phen-anthraquinone structure† was advanced, has been completely dis-credited.[285, 289–291] Nearly thirty years later new structures were tentatively advanced based *inter alia* on the isolation of *p*-terphenyl[291] from a zinc dust distillation, and the similarity of the ultraviolet absorp-tion of the leuco-acetate to that of a benzo*bis*benzofuran.[290] Shortly afterwards the correct structure was proposed[292] by Gripenberg and confirmed[285] by synthesis. The following summary is based on his work. Thelephoric acid forms a tetra-acetate and a leucohexa-acetate whose light absorption is similar to that of (208). This was supported by the

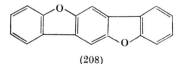

(208)

observation that zinc dust distillation gives several products besides *p*-terphenyl, one of which also showed the very characteristic ultraviolet absorption of (208). These facts suggest that thelephoric acid is a tetrahydroxybenzobis[1,2-b,4,5-b′]benzofuranquinone, and by analogy with the terphenylquinone group, (207) seemed the most probable structure. This was confirmed by condensing chloranil with 3,4-di-methoxyphenol which gave a product identical with thelephoric acid tetramethyl ether (209; R = Me), whence demethylation with pyridine hydrochloride yielded thelephoric acid. Two other syntheses are of interest: in one[294] thelephoric acid was obtained (in 3% yield) in one step by oxidising a mixture of 2,5-dihydroxybenzoquinone and catechol with ferricyanide, and in the other,[293] which is completely unambiguous, the terphenyl (211) (prepared by a standard Ullmann condensation) was converted to the somewhat unstable triquinone (210) which cyclised to thelephoric acid tetra-acetate (209; R = Ac) when boiled in acetic anhydride. This is a novel extension of the well-known cyclisation of 2,2′-biquinones to dibenzofuranquinones.[295]

† For a summary see ref. 29.

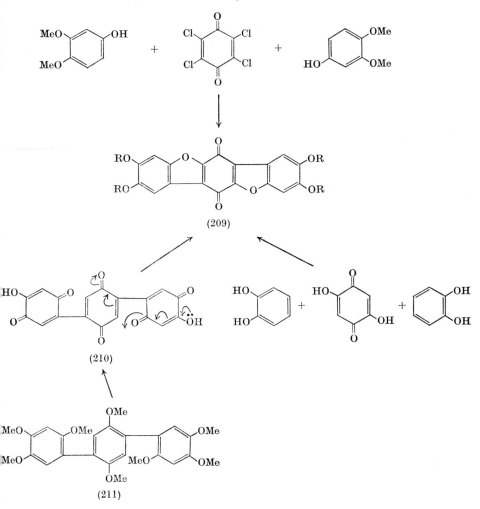

(209)

(210)

(211)

Thelephoric acid is intensely coloured in the solid state and forms red solutions which become blue on addition of base. The alkaline solution absorbs strongly throughout the visible region showing a maximum at ca. 740 nm scarcely diminished in intensity at 1000 nm. This may be

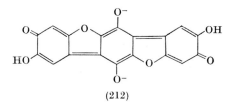

(212)

attributed to poly-anions having the extended conjugated system of (212).[295]

On prolonged methylation using dimethyl sulphate and potassium carbonate in acetone, containing a little pyridine, the insoluble tetramethyl ether, first formed, gradually dissolves and the solution becomes colourless. The product is the leucohexamethyl ether (213) formed, it is

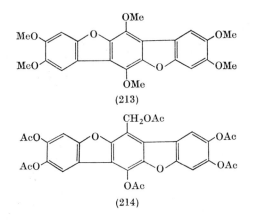

(213)

(214)

suggested, by oxidation of pyridinium hydroxide to α-pyridone at the expense of the quinone.[285] Prolonged reaction of thelephoric acid with acetic anhydride/sodium acetate gives the hexa-acetate (214),[296] (cf. pp. 107 and 156).

Before leaving this group of terphenylquinones† attention should be drawn to the closely related quinone methide, xylerythrin (215), which occurs in the wood-rotting fungus *Peniophora sanguinea* Bres.[298]

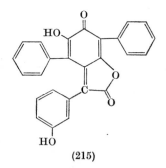

(215)

† The bright red caps of the toadstool *Amanita muscaria* (L. ex Fr.) S. F. Gray are alleged[430] to contain a pigment, "muscarufin", which has an unusual terphenylquinone structure. (For a summary, see ref. 29.) In fact a mixture of unstable pigments is present which bears little resemblance to "muscarufin" and attempts to repeat the original work[430] have failed.[431]

(Thelephoraceae) together with other pigments, one of which is a mono-methyl ether of (215), and another possibly a hydroxy derivative. The functional groups were detected by their spectral properties, derivative formation, and the rather easy decomposition in alkaline solution which gives benzoic acid and an unknown yellow acid, $C_{19}H_{14}O_6$. The complete structure was determined by X-ray analysis of the bis-bromoacetate[299] and a synthesis has been described.[298(b), 429]

Cyperaquinone (216), $C_{14}H_{10}O_4$
Hydroxycyperaquinone (217), $C_{14}H_{10}O_5$
Demethylcyperaquinone (218), $C_{13}H_8O_4$
Dihydrocyperaquinone (219), $C_{14}H_{12}O_4$
Tetrahydrocyperaquinone (220), $C_{14}H_{14}O_4$

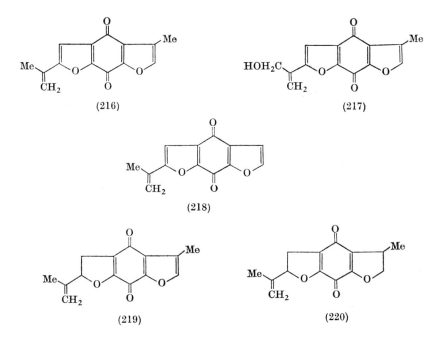

(216) (217) (218) (219) (220)

Occurrence: roots and rhizomes of the following Cyperaceae:[472] (216) *Cyperus haspan* L. and other *Cyperus* spp.,[490] *Remirea maritima* Aubl. and *Fimbristylis dichotoma* L., (217) *C. haspan*, (218) *C. compressus* L. and other *Cyperus* spp.,[490] (219) *R. maritima* and *F. dichotoma*, (220) *R. maritima*.

Physical properties: (216) carmine needles, m.p. 182–183° (dec.), $\lambda_{max.}$ (EtOH) 259, 347, 473 nm (log ε 4·46, 3·50, 3·63), $\nu_{max.}$ (Nujol) 1665, 1650(sh) cm⁻¹.

(217) russet plates, m.p. 166–168° (dec.), $\lambda_{max.}$ (EtOH) 262, 347, 473 nm (log ε 4·62, 3·49, 3·62), $\nu_{max.}$ (Nujol) 3420, 1673, 1660(sh) cm⁻¹.

(218) bright red crystals, m.p. 136–138°.

(219) orange needles, m.p. 113–114°, $[\alpha]_D$ −35° $(CHCl_3)$, $\lambda_{max.}$ (EtOH) 275, 334, 463 nm $(\log \epsilon$ 4·28, 3·93, 2·75), $\nu_{max.}$ (Nujol) 1675, 1640 cm^{-1}.

(220) maroon prisms, m.p. 138–140°, $[\alpha]_D$ +210° $(CHCl_3)$, $\lambda_{max.}$ (EtOH) 264, 323, 480 nm $(\log \epsilon$ 3·66, 4·36, 2·50), $\nu_{max.}$ (Nujol) 1690, 1642 cm^{-1}.

The cyperaquinones are the first quinones to be obtained from sedges and were found recently by Allan et al.[472] in four of the 35 species (in ten genera of Cyperaceae) which were examined. Structural determinations were based entirely on spectroscopic data, apart from the observation that the compounds showed redox properties and failed to react with o-phenylenediamine.

Cyperaquinone. The molecular ion of this pigment forms the base peak of the mass spectrum at m/e 242 and there are no fragment ions of relative abundance greater than 3% which suggests a fully conjugated system. The n.m.r. spectrum showed signals at τ 7·71 (3H, d) and 2·57 (1H, finely split singlet) consistent with a 3-methylfuran system, and also at τ 7·92 (3H, bs), 4·72 (1H, bs), 4·15 (1H, bs) and 3·30 (1H, s) ascribed to a 2-isopropenylfuran structure. On this basis, together with the data on the di- and tetrahydro compounds (see below), cyperaquinone was regarded as (216).[472] A 2,6-dioxygenated quinone system was preferred on biogenetic grounds as cyperaquinone and two other

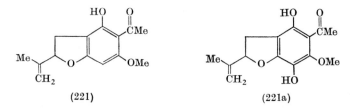

(221)   (221a)

members of the group occur in *Remirea maritima* together with remirol (221)[473] and (221a)[490] which, conceivably, could be the precursors of all the quinones.

Hydroxycyperaquinone has virtually the same electronic spectrum as (216) but the infrared spectrum shows that a hydroxyl group is present. This is confirmed by the n.m.r. spectrum which differs from that of (216) only insofar as the C_3 side chain shows signals at τ 5·34 (2H, bs), 4·7 (1H, bs), 3·3 (1H, s) and 6·7 (1H, s, disappearing on deuteriation) consistent with a HO—CH$_2$—$\overset{|}{C}$=CH$_2$ structure. Hydroxycyperaquinone is therefore (217).[472]

Demethylcyperaquinone has almost the same ultraviolet-visible and n.m.r. spectra as (216), but the methyl resonance at τ 7·71 is replaced by a signal (1H) at τ 3·3 corresponding to a furanoid β-proton. The structure is thus (218).[472]

Dihydrocyperaquinone is optically active. The n.m.r. spectrum shows signals for (a) a β-methylfuran system as in (216), and (b), a 2,3-dihydro 2-isopropenylfuran structure as in remirol (221), consisting of the following: τ 8·20 (3H, s), 6·9 (2H, octet), 5·03 (1H, bs), 4·91 (1H, bs) and 4·62 (1H, t). Structure (219) was therefore proposed which is consistent with the mass spectrum. Dihydrocyperaquinone is highly toxic to fish.

Tetrahydrocyperaquinone. The structure (220) was again deduced[472] principally from the n.m.r. spectrum which includes signals arising from a 2,3-dihydro-2-isopropenylfuran system as in (219), and also the following at τ 8·67 (3H, d), 6·5 (1H, m), 5·75 (1H, q) and 5·23 (1H, t) corresponding to a 2,3-dihydro-3-methylfuran structure. The tetrahydro compound is therefore regarded[472] as (220) but the visible absorption at 480 nm is anomalous.

Plastoquinones

Plastoquinone-9 (PQ-9)† (222; n = 9), $C_{53}H_{80}O_2$
Plastoquinone-8 (PQ-8) (222; n = 8), $C_{48}H_{72}O_2$
Plastoquinone-4 (PQ-4) (222; n = 4), $C_{28}H_{40}O_2$
Plastoquinone-3 (PQ-3) (222; n = 3), $C_{23}H_{32}O_2$

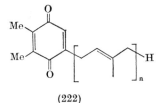

(222)

Occurrence: PQ-9 occurs generally in green tissues throughout the plant kingdom;[300-304, 310, 428, 432, 443, 445] PQ-8 has been found in the leaves of *Aesculus hippocastanum* L.[493] and *Ficus elastica* Roxb.,[493] *Zea mays*[493] seedlings and cultures of *Euglena gracilis*,[493] PQ-4 in the leaves of *A. hippocastanum* L.[305,cf433] and *A. glabra* Willd.,[432] and PQ-3 in *Spinacia oleracea* L.[306]

Physical properties: PQ-9 has m.p. 48–49°, PQ-4 and PQ-3 are yellow oils (PQ-3 very unstable), all show λ_{max}. (EtOH) 255, 262(sh), nm, ν_{max}. (?) 1647, 1623 cm^{-1} (PQ-9).

Plastoquinone was first isolated by Kofler[300] from alfalfa (*Medicago sativa* L.). It remained in obscurity for some years but, following the

† Abbreviation proposed by IUPAC-IUB Commission on Biochemical Nomenclature.[437]

discovery of ubiquinone, interest revived and it became known for a time as "Kofler's quinone". PQ-9 occurs generally in green plants, chiefly in the chloroplasts[307, 308] (hence the name[311]), although it is not confined to photosynthetic tissue.[301, 307, 322] Its function in photosynthetic electron transport[309, 393] has been shown by solvent extraction of chloroplasts leading to loss of activity which could be restored by addition of plastoquinone. Reduction of plastoquinone in light and reoxidation in the dark, has also been demonstrated in chloroplasts, and its place in the electron transport chain seems to lie between light reactions I and II.[434]

Kofler[300] deduced from the ultraviolet absorption that the new quinone was an alkylated benzoquinone and later investigations confirmed this. The n.m.r. spectrum showed[312] that there were two methyl groups attached to the quinone ring and one proton (best "seen" in the spectrum of the quinone after saturation of the side chain). As the signal from the latter is a triplet it follows that the adjacent substituent is a methylene group which must be the terminus of a long side chain. Other spectral data indicated that the side chain was polyprenyl and very similar to that of ubiquinone-10, and the length of the chain was deduced from the intensity of absorption at 254 and 262 nm in comparison with that of synthetic 2,3-dimethyl-5-farnesyl-benzoquinone (PQ-3). Folkers and his co-workers[313] confirmed their conclusions by a synthesis in which 2,3-dimethylquinol was condensed with the C_{45}-alcohol, all-*trans*-solanesol (223), in the presence of boron trifluoride, followed by oxidation.

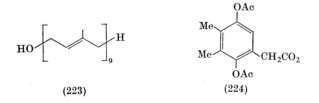

(223)   (224)

PQ-4[326] and PQ-3[312] were synthesised in the same way. Isler and his group also effected the same synthesis,[314] and provided chemical evidence for the nature of the side chain by ozonolysis of the leucodiacetate which yielded the acid (224), levulinic aldehyde and acetone.[310] N.m.r. studies show that pure PQ-9 has an all-*trans* side chain.[323] Slight differences between the R_F values of natural and synthetic PQ-9 have been observed[315] which may have been due to impurities in the natural sample, or possibly differences in the stereochemistry of the side chain The same comment applies to PQ-3, isolated[306] from spinach chloroplasts, which contained "dimers" of PQ-3 and PQ-9. Such "dimers" were first encountered as yellow oils by Eck and Trebst[305] in their isolation of PQ-4 and PQ-9 from horse chestnut leaves. They are probably artefacts as it was shown that the "dimer" of PQ-9 could be formed by ultraviolet-irradiation of the monomer. Photo-dimers of simple quinones are well-known[316] but the structures of the plastoquinone derivatives have not yet been resolved.

Other Plastoquinones

Although only four members of the plastoquinone group (222) have been discovered so far numerous other compounds have been reported, all of which appear to be derivatives of PQ-9 (= PQ-A). Most of these are still inadequately characterised and it is difficult to compare reports from different laboratories; possibly there are more plastoquinones in the literature than can be found in leaves. Plastoquinone C,[317, 334] which occurs in many species, is a hydroxylated PQ-9. This is evident from its chromatographic behaviour (more polar than PQ-9),[317] hydroxyl absorption[318, 319] at 3545 and 3620 cm^{-1}, and the mass spectrum[319] which confirms the molecular formula, $C_{53}H_{80}O_3$, and shows a peak at M-18. Reductive acetylation gave a product showing a small molecular ion peak at m/e 892 corresponding to a leucotriacetate, a stronger peak at m/e 850 corresponding to the diacetate of a triol, and a very strong peak at m/e 832 corresponding to an anhydro-diacetate, from which it appears that the hydroxyl group is secondary and/or tertiary. In the lower mass region, a series of peaks at m/e 832-69-(68)$_n$ (n = 0 to 6) arises from successive loss of seven isoprene units while the base peak at m/e 189 corresponds to the ion (j) common to all the plastoquinones. This shows that PQ-C contains the structures (225) and (226), and hence

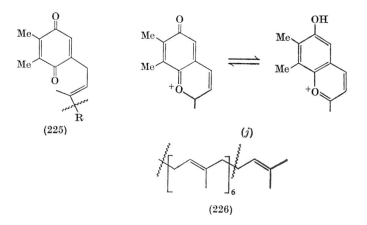

(225)   (j)

(226)

the hydroxyl group must be located in the second isoprene unit counting from the quinone ring. In recent work, Pennock[320] and Crane[432, 436] and their co-workers have shown that the leaves of several species contain a family of quinones showing the normal ultraviolet absorption of plastoquinones and similar polarity to PQ-C and PQ-D.† In view of this, the work just described on "plastoquinone C" may have been carried out on a mixture of isomers. From *Polygonum cuspidatum*[435] a crude plastoquinone fraction was separated into a relatively polar (PQ-C) group, and a less polar group (PQ-Z) consisting of eight components each containing hydroxyl and ester groups on infrared evidence. The PQ-C group was further separated by t.l.c. on silica gel/AgNO$_3$ into (at least) ten quinones, of which six were considered to possess an allylic tertiary hydroxyl group and the others were thought to have an allylic secondary alcohol group, again on infrared evidence. PQ-C in algae has been separated into seven components.[428]

A further group of plastoquinones (PQ-B) has been found in the leaves of spinach,[317, 436] alfalfa[319] and many other species,[320, 432, 436] and in algae,[428] and again there are at least six of these.[320] Their infrared spectra are essentially the same as that of PQ-9 with additional bands at 1730, 1253 and 722 cm^{-1} which are typical of fatty acid esters. On hydrolysis, or reduction with lithium aluminium hydride, they yield long chain alcohols and quinones having the same polarity and infrared spectra as the PQ-C group which has a secondary allylic alcohol function. Evidently the PQ-B group are esters of PQ-C. The mass spectrum of PQ-B$_2$ shows a molecular peak at m/e 1002, which corresponds to PQ-9 [748] plus a palmitic acid residue [256], and there is a strong peak at m/e 746 (loss of CH$_3$(CH$_2$)$_{14}$CO$_2$H) at first thought[319] to be the molecular ion peak. The mass spectra of PQ-B$_5$ and PQ-B$_6$ are similar.[321]

Plastochromanols

Like other prenyl-quinones and -quinols the plastoquinones can be converted into cyclic derivatives and this occurs *in vivo*. The plasto-chromenol-8 (227) (solanachromene), isolated[325] from tobacco, was racemic[326] and probably an artefact of the drying and curing processes,

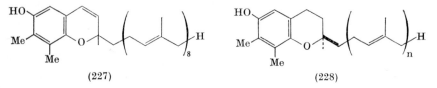

(227)   (228)

† Plastoquinone D is an isomer of PQ–C.[319, 436]

but plastochromanol-8 (228; n = 8) from the leaves [327] of *Hevea brasiliensis* and plastochromanol-3 (228; n = 3) from the latex,[328] both have the 2R-configuration,[326] and are evidently formed under ensymic control. *In vitro* the quinones can be cyclised to chromenols by refluxing in pyridine and then reduced to chromanols using sodium in ethanol. Plastochromanol-3 is otherwise known as γ-tocotrienol, and relatively large amounts of tocotrienols are present in palm oil [327] and the latex of *H. brasiliensis*.[328]

Phytylplastoquinone (229), $C_{28}H_{66}O_2$

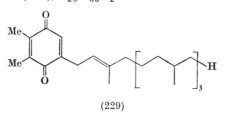

(229)

Occurrence: two strains of *Euglena gracilis*,[478] *E. gracilis* var. *saccarophila*[478] and probably in the alga *Tribonema*.[478,479]

Physical properties: an oil, $\lambda_{max.}$ (C_6H_{12}) 254, 261 nm, $\nu_{max.}$ (film) 1647 cm^{-1}.

The observation that this new quinone was eluted from an alumina column just before PQ-9 recalled the parallel behaviour of phylloquinone and menaquinone-9, and suggested to Whistance and Threlfall [478] that it might be phytylplastoquinone. Direct comparison with authentic [310] material showed that this was the case and by cyclisation in pyridine to the chromenol, followed by hydrogenation, it was converted into γ-tocopherol. Phytylplastoquinone (229) is possibly a natural precursor of γ-tocopherol. A second compound isolated from *E. gracilis* proved to be a monomethyl ether of phytylplastoquinol.[478]

α-*Tocopherolquinone* (230), $C_{29}H_{50}O_3$

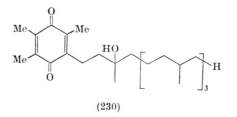

(230)

Occurrence: in green leaves,[320,324,332,335,337] algae,[304,324,428,445] pteridophytes,[324] a bryophyte,[324] in various animal (chordata) tissues[329] and in shrimps and limpets.[330]

Physical properties: unstable yellow oil, $\lambda_{max.}$ (EtOH) 261, 269 nm (log ε 4·38, 4·29), $\nu_{max.}$ (film) 3559, 1641 cm^{-1}.

α-Tocopherolquinone (230) can be prepared by oxidation[336] of α-tocopherol and was known as a synthetic compound long before it was found *in vivo*. The reverse process, conversion of the quinone into α-tocopherol, can be effected by reduction and treatment with acid.[336, 438] Not surprisingly, when first detected in Nature it was regarded as an artefact of isolation but it now appears to be a genuine constituent of animal tissues and is normally present in photosynthetic tissue in plants. In *Vicia faba* leaves up to 60% exists in the leuco form.[324] It has been detected in chloroplast preparations[331] and was first isolated by Morton and his co-workers[333] from holly leaves (*Ilex aquifolium* L.) and characterized as the quinol di-*p*-bromobenzoate. Trace amounts of *β*-, *γ*- and *δ*-tocopherolquinones have been reported in spinach chloroplasts,[332, 334] algae,[428] senescent tobacco[475] and maple[475] leaves, green ivy leaves[475] and cactus[475] tissues.

The chemistry of α-tocopherolquinone has been reviewed recently.[336]

Ubiquinones

Ubiquinone precursors

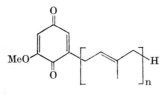

(231)

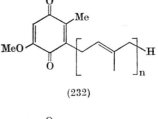

(232)

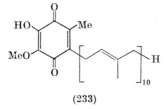

(233)

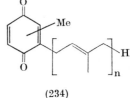

(234)

Occurrence: (231; $n = 10$), (232; n = 10), (233) and (234; n = 9 and 10) in cultures of *Rhodospirillum rubrum*;[340, 341] (231; n = 9)[484] and (232; n = 5–9)[343, 484] in *Pseudomonas ovalis*, (232) in *Ps. fluorescens*[338] and *Esch. coli*,[338] (232; n = 8 and 9) in *Ps. alkanolytica*,[485] (231; n = 9)[484] and (232; n = 4–8)[484] in *Proteus mirabilis*, (232; n = 9) in Vibrio 01,[484] and (232; n = 9) in *Euglena gracilis*.[481]

Physical properties: (231), $\lambda_{max.}$ (C_6H_{12}) 264·5 nm; (232), $\lambda_{max.}$ (CCl_4) 269, 276(sh) nm; (233), $\lambda_{max.}$ (hexane) 271, 277 nm.

These quinones are important intermediates in the biogenesis of the ubiquinones. Most have been detected in only trace amounts and few have been fully characterised. The structures of the quinones (232) were deduced from their ultraviolet, n.m.r. and mass spectra, and comparison with ubiquinones and synthetic models. The position of the methoxyl group was initially inferred[340] from its relationship to a precursor [342] in *R. rubrum* (see p. 27), and this was supported[343] by the observation that neither (232; n = 9) nor 5-methoxy-2-methyl-3-phytylbenzoquinone could be cyclised to a chromenol, whereas the 6-methoxy isomer was readily cyclised by treatment with sodium hydride or triethylamine (cf. ref. 439). The quinones (232) can be synthesised by alkylation of 5-methoxy-2-methylquinol.[439]

The quinones (231) were identified mainly on the basis of their ultraviolet and mass spectra[340, 484] while the presence of (234; n = 9 and 10) as major components of a complex lipid fraction is based solely on the mass spectrum (the location of the quinone methyl group is not known).† The only direct evidence for the structure of (233) is the light absorption [λ_{max}. (EtOH) 282, 430 nm, λ_{max}. (EtOH/OH⁻) 283, 535 nm] which is very similar to that of 6-hydroxy-5-methoxy-2-methyl-3-phytylbenzoquinone.[341] It has also been obtained[439] by mild hydrolysis of rhodoquinone-10.

Ubiquinones (Qn)‡ (235; n = 1 to 12)

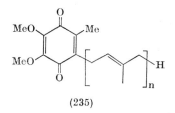

(235)

Occurrence: very widespread[345] in bacteria and fungi, higher plants and algae, vertebrates and invertebrates, although not ubiquitous.

Physical properties:[346] λ_{max}. (EtOH) 275, 405 nm, ν_{max}. 1653, 1608 cm⁻¹, Q-1 to Q-5 are yellow oils, Q-6, m.p. 19–20°, Q-7, m.p. 31–32°, Q-8, m.p. 37–38°, Q-9, m.p. 44–45°, Q-10, m.p. 49°.

About fifteen years ago, independent work by Green and his collaborators at Wisconsin, and by Morton and his co-workers in Liverpool, led to the isolation of a new substance which opened a new chapter in quinone chemistry.[347] (The history of these events has been described by Morton.[352]) It was found in liver, kidney, heart, spleen, pancreas and other tissues, in various species, including man, and was named[353]

† A 2-polyprenylbenzoquinone has been tentatively identified in *Ps. ovalis* and *Esch. coli*.[338]

‡ Abbreviation proposed by IUPAC-IUB Commission on Biochemical Nomenclature.[437] Ubiquinone-50, ubiquinone-10, Q-10, co-enzyme Q_{10}, all refer to (235; n = 10).

ubiquinone to indicate its wide occurrence in Nature. (For many years American workers preferred the name Co-enzyme Q but ubiquinone is now recommended.†) "It" subsequently turned into a family of quinones (235; n = 1–12). Nearly all vertebrates contain Q-10 (235; n = 10) although Q-9 has been found in fishes, and Q-11 and Q-12 in the rat.[480] Q-9 and Q-10 occur in insects and marine invertebrates. From chromatographic behaviour a ubiquinone found in the crab, *Carcinas maenas*, was stated[350] to be Q-10 "or possibly a higher homologue." In higher plants (most tissues) the principal representatives are again Q-9 and Q-10, but a series of ubiquinones may occur together, for example, Q-7 to Q-10 in the fruit of *Cucumis melo*,[481] and Q-8 to Q-12 in the fruit of *Capsicum annuum*.[481, 482] The lower prenylogues down to Q-5 have been isolated from micro-organisms. The discovery of Q-5 in *Escherichia coli*[348] was of interest as C_{25} terpenoids are still extremely rare. However, recent work in Folkers' laboratory[349] has revealed that micro-organisms may contain the whole family of ubiquinones from Q-1 to Q-10.‡ Mass spectrometry of a sample of Q-6 from *Saccharomyces cerevisiae*, purified by paper chromatography, disclosed the presence of Q-1 to Q-6 which was confirmed by reverse-phase t.l.c. Similarly, "pure" Q-8 from *E. coli* was found to contain Q-1 to Q-8, and "pure" Q-10 from *Rhodospirillum rubrum* contained all the ubiquinones from Q-1 to Q-10. In each species the concentration of individual quinones diminished with chain length and the amounts of Q-1 to Q-4 were extremely small. It seems likely that many other species contain a similar multiplicity of ubiquinones. The distribution of ubiquinones has been reviewed by Crane[345] (see also ref. 330).

Ubiquinones play a major role in respiratory electron transport systems.[392, 449] They are localised mainly in mitochondria and have been shown to undergo cyclic oxidation-reduction during the oxidation of substrates of the citric acid cycle. Extraction of mitochondrial particles with acetone removes the ubiquinone and results in loss of enzymic activity which can be restored on addition of ubiquinone. However, recent work shows that electron transport can still occur in mitochondrial particles from which all ubiquinone has been removed indicating that electron transfer proceeds through a branched chain system; one pathway depends on ubiquinone but there is at least one other in which ubiquinone does not participate. Ubiquinone also appears to function as an electron carrier in photosynthetic bacteria, taking the place occupied by plastoquinone in higher plants.

---

† Abbreviation proposed by IUPAC-IUB Commission on Biochemical Nomenclature.[437]
‡ Q-11 and Q-12 have since been detected in photosynthetic bacteria.[481]

The structure of Q-10 was determined independently by two groups; in Europe the Liverpool group collaborated with Hoffmann-La Roche[354] at Basel, and in America the investigations begun at Wisconsin were continued by Merck.[355] By large-scale extractions as much as 37 g of crystalline Q-10 was isolated from 750 kg of pig heart.[354] The redox properties and ultraviolet absorption indicated a benzoquinone structure, the latter showing a very close resemblance to that of aurantiogliocladin (p. 114) and 2,3-dimethoxy-5-methyl-6-phytylbenzoquinone. Infrared comparisons with synthetic compounds and members of the vitamin K series pointed to the existence of a polyisoprenoid chain, the length of which was deduced from the molecular weight, the hydrogen uptake, titration with perbenzoic acid and the n.m.r. spectrum. The arrangement of the substituents on the quinone ring was established by oxidative degradation; ozonolysis of the leucodiacetate afforded the acid (236) as

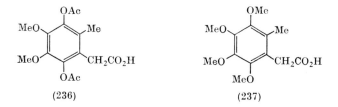

(236)   (237)

well as levulinic aldehyde and acetone, and similarly, oxidation of the leucodimethyl ether with permanganate gave (237) and tetramethoxy-phthalic anhydride. These results show that Q-10 has structure (235; n = 10) which was confirmed by synthesis. Several members of the series were synthesised before they were recognised *in vivo*, the general procedure[359, 360] being to condense the quinol (238) with the appropriate

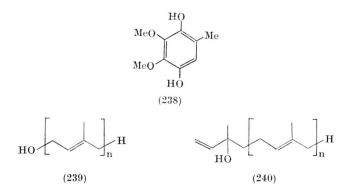

(238)

(239)   (240)

all-*trans* primary (239) or tertiary (240) allylic alcohol in the presence of boron trifluoride or zinc chloride (or both), followed by oxidation of the intermediate quinol with silver oxide.[357, 358, 460]

In early investigations on ubiquinone in animal tissue two substances, designated SA and SC, were detected; SA was later recognised as Q-10 while SC is the cyclic isomer, ubichromenol-9 (241; n = 9). It was first

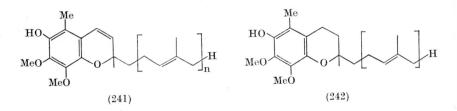

(241)   (242)

isolated[381] from human kidney and later ubichromenols-5 and -6 (241; n = 5 and 6) were found[382] in *Torula* yeast but they appear to be arte-facts of storage.[383] Ubichromenol-9 has been discussed as an artefact of isolation and its natural status is still controversial. Base-catalysed cyclisation of ubiquinone (as shown) can be effected easily with the aid

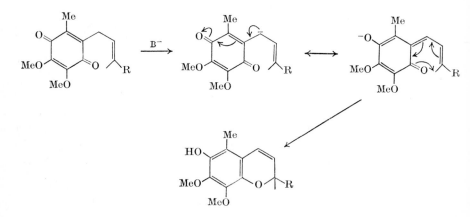

of pyridine,[385] or sodium hydride,[384] and may occur to some extent in ethanolic potassium hydroxide[386, cf. 387] and during chromatography on alumina columns.[387, 390] Ubichromenol formation can also be induced photochemically (with visible light) in boiling ethanol.[448] Careful re-examination[388] of the isolation of ubichromenol-9 from human kidney showed that part of it was an artefact of isolation but the remainder was considered to be of natural origin. Morton and his co-workers[388] also demonstrated that ubichromenol could be isolated from kidney by solvent extraction alone, avoiding alkaline saponification and chromato-graphy on alumina, and it was stated[347] to be optically active. Further examination of this point might settle the issue which has not been clarified by biogenetical studies.[391]

The quinol of Q-10 can be cyclised under acid conditions (potassium hydrogen sulphate in refluxing acetic acid[389]) to the chromanol (242; n = 9) but heating the quinone with stannous chloride in acetic acid, gave a product whose n.m.r. spectrum showed that the side chain was now essentially saturated, most of the proton signals appearing at τ 8·7–9·2. It was suggested[389] that the side chain had cyclised, more or less completely, to form structures such as (243). This possibility does

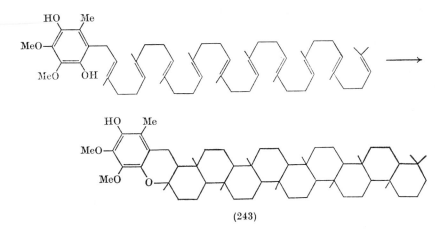

(243)

not exist in the case of hexahydro-Q-4 (having a phytyl side chain) which gives the normal chromanol under these conditions.

X-*Dihydro-ubiquinone*-10 [Q-10(X H_2)] (244), $C_{59}H_{92}O_4$

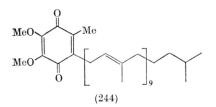

(244)

Occurrence: cultures of *Penicillium stipitatum* Thom,[361] *Gibberella fujikuroi* (Saw.) Wr.,[362,363] *Aspergillus niger* v. Tieghem,[363] *Alternaria porri* (Ell.) Saw. (= *A. solani*),[363] *Curvularia lunata* (Wakker) Boedijn,[363] *Neurospora crassa* Shear & B. O. Dodge[363] and other moulds.[363]

Physical properties: orange-yellow crystals, m.p. 29°, $\lambda_{max.}$ (EtOH) 275 nm (log ε 4·08), $\nu_{max.}$ (?) 1653 cm⁻¹.

This modified ubiquinone is widely distributed in moulds but was not discovered until 1962. It is essential to use a chromatographic system[364] which will separate it from other members of the series; in certain reverse-phase paper chromatographic systems it is likely that

Q-9 (IX H$_2$) will run like Q-10, and Q-8 (VIII H$_2$) like Q-9, and such compounds may exist but would be easily overlooked. From the ultraviolet and infrared spectra it was clearly very similar to Q-9 and Q-10, and was originally regarded as a tetrahydro-Q-10. More accurate microhydrogenation figures (the product was perhydro-Q-10), and determination of the molecular weight of the leuco-acetate (prepared with ^{14}C-acetic anhydride)[377] indicated a dihydro-Q-10 structure, the remaining problem being the location of the "extra" hydrogen atoms. The n.m.r. spectrum includes all the bands shown by Q-9 and Q-10 but there are only nine side chain vinylic protons and a doublet (6H) at τ 9·13 (J, 6 Hz) is clearly due to an isopropyl group. It follows that the isoprene unit furthest from the ring is saturated, and the quinone is therefore (244).[362, 364]

*Epoxyubiquinones-*10 (245)

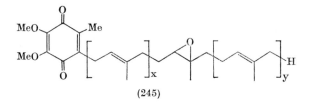

(245)

Occurrence: (245; x + y = 7 and 8) in *Pseudomonas alkanolytica*;[485] (245; x + y = 9) in *Rhodospirillum rubrum*,[394] beef heart[394, 395, 483] and whale heart muscle.[485]

Physical properties: λ_{max}. (EtOH) 275 nm, ν_{max}. (film) 1660, 1650 cm^{-1}.

These compounds may be artefacts. Five were found initially in the lipids of *R. rubrum* in low concentration (1–2% of Q-10) and probably all are mixtures. Structural evidence is mainly spectroscopic.[394, 485] The *R. rubrum* epoxides[394] show essentially the same ultraviolet, infrared and n.m.r. spectra, the ultraviolet and infrared curves being indistinguishable from those of Q-10. The molecular ions correspond to Q + 16, that is, an additional atom of oxygen is present and must be in a side chain epoxide as hydroxyl and additional carbonyl groups are

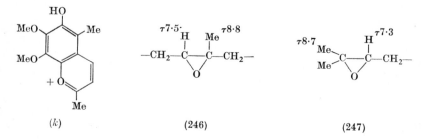

(k)   (246)   (247)

absent (infrared), and there is no M-18 peak in the mass spectrum. An intense peak at m/e 235, attributable to the ion k,[396] shows that the ring structure, and the first isoprene unit are unaltered. Positive evidence for the presence of an epoxide comes from the triplet signal (1H) at τ 7·52 and the singlet (3H) at τ 8·84 which are characteristic of isoprenoid epoxides (246). This also shows that the epoxy group is not located in the terminal isoprene unit as this would give rise to a six-proton doublet at τ 8·7 (see 247). These assignments were supported by studies on synthetic epoxyubiquinones. Reaction of Q-10 with perphthalic acid in boiling ether gave a non-terminal epoxide having the same ultraviolet, infrared and n.m.r. spectra as two of the epoxides from *R. rubrum*; similar epoxidation of Q-2 afforded the terminal epoxide (248) while

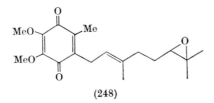

(248)

Q-3 gave the corresponding terminal epoxide and the in-chain epoxide (245; x = y = 1). Cleavage of the epoxides by heating with acetic acid yielded the expected hydroxyacetates (ν_{max}. 3500, 1735 cm^{-1}), the infrared spectra of the products from two of the *R. rubrum* epoxides and the synthetic epoxyubiquinone-10 being indistinguishable.

By reductive acetylation followed by perhydrogenation, Morimoto and co-workers[485] were able to convert an epoxyubiquinone-9 fraction (from *Ps. alkanolytica*) into a product, still apparently an epoxide, which could be cleaved with periodate to give aldehydic and ketonic fragments. These were mixtures and by mass spectrometry it could be shown that they must have been derived from nine epoxides in the original epoxy-Q-9 fraction, indicating that, to some extent, each side chain double bond was epoxidised.

Rhodoquinone-10 (249; n = 10), $C_{58}H_{89}NO_3$
Rhodoquinone-9 (249; n = 9), $C_{53}H_{81}NO_3$

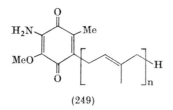

(249)

Occurrence: (249; n = 10) cultures of *Rhodospirillum rubrum*[365] and other *Athiorhodaceae*;[372] (249; n = 9) also in *R. rubrum*[351] and in the alga *Euglena gracilis* var. *bacillaris*[351] and the nematodes *Metastrongylus elongatus*[476] and *Ascaris lumbricoides* var. *suis*.[476]

Physical properties: (249; n = 10) violet needles, m.p. 69–70°, $\lambda_{max.}$ (EtOH) 283, 500 nm (log ε 3·99, 3·34), $\nu_{max.}$ (KBr) 3580, 3480, 1640, 1600 cm^{-1}.

(249; n = 9) magenta needles, m.p. 66·5–67°, $\lambda_{max.}$ (EtOH) 283, 506 nm (log ε 4·04, 3·14), $\nu_{max.}$ (KBr) 3478, 3362, 1650, 1610 cm^{-1}.

On the basis of analyses, and ultraviolet and infrared spectra, rhodo-quinone-10 was recognised[365] as another member of the ubiquinone group, and was originally regarded as a hydroxyquinone (Q-10, OH in place of OMe) as it showed two peaks in the 3μ region of the infrared. Unexpectedly however, it contains nitrogen which is present as a primary amino group attached to the quinone ring. This was demonstrated[366] by the disappearance of a two-proton peak in the n.m.r. spectrum on addition of a trace of formic acid, the presence of amide absorption bands (3250, 1655 cm^{-1}) in the mono-acetyl derivative, and by conversion into ubiquinone-10 by replacement of NH_2 by OH, and methylation. (To avoid cyclisation of the side chain the hydrolysis was effected in hot acetic acid containing hydrated cupric chloride.) The reverse process was then carried out, converting Q-10 into rhodoquinone-10 by treatment with ammonia at room temperature, the isomeric isorhodoquinone-10 (249; NH_2 and OMe interchanged) being formed simultaneously.[367, 439] These very similar isomers can be distinguished by their n.m.r. solvent shifts in benzene and pyridine.[440]

The orientation of the amino and methoxyl groups in natural rhodo-quinone-10 was determined as follows.[367] Conversion to the hydroxy-methoxy analogue, as above, followed by methylation with deuteriated methyl iodide and cyclisation with the aid of triethylamine, gave the chromenol (250) showing singlet methoxyl absorption at τ 6·25. This

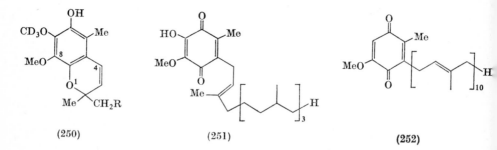

(250)   (251)   (252)

showed that the methoxyl group was at C-8 by comparison with the corresponding chromenol previously synthesised[368] from phytyl-fumigatin (251). When a methoxyl group is at C-7 in (250) the signal appears at τ 6·13. It follows that in rhodoquinone-10 the amino group is at C-6 as in (242). This was confirmed synthetically by addition of ammonia to the decaprenylquinone (252).[439] However, the sample of (252) used was a mixture of *cis* and *trans* isomers at the double bond nearest to the quinone ring and the resulting rhodoquinone-10 was a similar mixture of m.p. 47–49°.

Rhodoquinone-9 was recognised initially from its ultraviolet-visible absorption and the detailed structure was elucidated from further spectroscopic data.[351] The relative positions of the amino and methoxyl groups were determined by comparison with the two synthetic isomers derived from Q-9 which were differentiated from each other by n.m.r. solvent shifts.[477] Rhodoquinone-9 has been synthesised by 1,4-addition of ammonia to (232; n = 9).[477]

A methoxyl group attached to a quinone ring behaves as a vinylogous ester and it is not surprising to find[369, 441] that ubiquinones-9 and -10 are the precursors of the rhodoquinones *in vivo*. However as only one isomer is formed in each case, the replacement of a methoxyl group by amino must be under enzymic control. Homologous ubiquinones in which one, or both, methoxyl groups are replaced by ethoxyl, have been obtained[370] as artefacts of isolation when ethanolic potassium hydroxide was used to remove saponifiable material, and other homologues have been prepared by reaction with appropriate alkoxides. The methoxyl groups of ubiquinone can also be displaced by anions of ethyl cyanoacetate when treated with the ester in ethanolic solution containing ammonia.[379] This is a modification of the Craven test and estimation of the resulting blue colour has been developed into a quantitative method for the determination of Q-10 in urine.[380]

A different type of non-selective attack on ubiquinone methoxyl groups was discovered in the Takeda laboratories. When Q-7, in ethanol, was exposed to sunlight, a mixture of products was obtained of which acetaldehyde,[371, cf. 316] the corresponding ubichromenol,[373] "isoubiquinone-7",[374] and "demethylubiquinone-7",[375] have been identified. "Demethylubiquinone-7" was originally regarded as the monohydroxy-monomethoxy analogue (253),[375, 378] but methylation with deuteriated

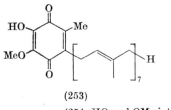

(253)

(254; HO and OMe interchanged)

methyl iodide and conversion to the chromenol gave a product showing two methoxyl signals of equal intensity in its n.m.r. spectrum. Hence "demethylubiquinone-7" is a 50:50 mixture of (253) and (254).[368] A mixture of (253) and (254) is also obtained on reduction of Q-7 with

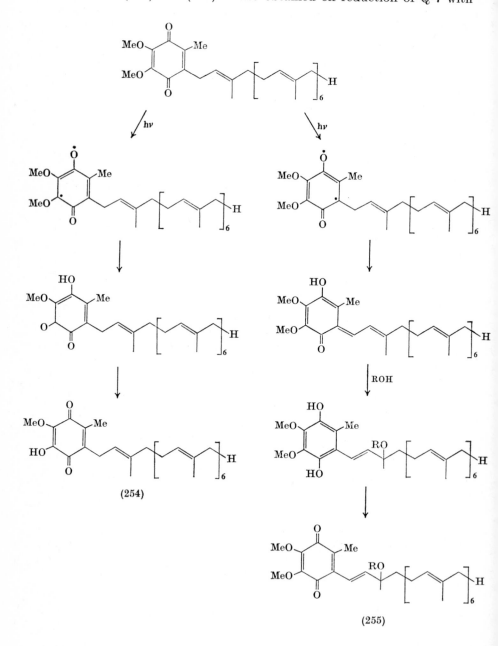

(254)

(255)

lithium aluminium hydride or sodium borohydride (followed by re-oxidation) and, more surprisingly, by similar treatment (LiAlH$_4$) of the leucodiacetate of Q-7.[397] "Isoubiquinone-7", as the name implies, was at first[374] thought to be an isomer of Q-7 but further investigation[486] established the structure (255; R = Et). If the photolysis is carried out in methanol the corresponding methoxy analogue (255; R = Me) is obtained. To account for the formation of (253), (254) and (255) the scheme on p. 182 has been proposed,[486] the first step in each case being an n → π^* transition of the appropriate enone system.

REFERENCES

1. C. Estable, M. I. Ardao, N. P. Brasil and L. F. Fieser, *J. Amer. Chem. Soc.* (1955), **77**, 4942.
2. L. F. Fieser and M. I. Ardao, *J. Amer. Chem. Soc.* (1956), **78**, 774.
3. A. Monro, M. Chadha, J. Meinwald and T. Eisner, *Ann. Entomol. Soc. Amer.* (1962), **55**, 261.
4. H. Schildknecht and K. Holoubek, *Angew. Chem.* (1961), **73**, 1.
5. M. S. Blum and R. D. Crain, *Ann. Entomol. Soc. Amer.* (1961), **54**, 474.
6. H. Schildknecht and K. H. Weis, *Z. Naturforsch.* (1961), **16b**, 810.
7. M. S. Chadha, T. Eisner and J. Meinwald, *J. Insect Physiol.* (1961), **7**, 46.
8. H. Schildknecht and H. Weis, *Z. Naturforsch.* (1960), **15b**, 200.
9. H. Schildknecht and K. H. Weis, *Z. Naturforsch.* (1960), **15b**, 755.
10. H. Schildknecht and K. H. Weis, *Z. Naturforsch.* (1960), **15b**, 757.
11. H. Schildknecht, *Angew. Chem.* (1959), **71**, 524
12. R. Trave, L. Garanti and M. Pavan, *Chim. e Ind.* (1959), **41**, 19.
13. L. M. Roth and B. Stay, *J. Insect Physiol.* (1958), **1**, 305.
14. M. Barbier and E. Lederer, *Biokhimiya* (1957), **22**, 236.
15. H. Schildknecht, *Angew. Chem.* (1957), **69**, 62.
16. P. Alexander and D. H. R. Barton, *Biochem. J.* (1943), **37**, 463.
17. R. H. Hackman, M. G. M. Pryor and A. R. Todd, *Biochem. J.* (1948), **43**, 474.
18. J. D. Loconti and L. M. Roth, *Ann. Entomol. Soc. Amer.* (1953), **46**, 281.
19. M. Engelhardt, H. Rapoport and A. Sokoloff, *Science, N.Y.* (1965), **150**, 632.
20. M. Barbier, *J. Chromatog.* (1959), **2**, 649.
21. J. W. Wheeler, J. Meinwald, J. J. Hurst and T. Eisner, *Science, N.Y.* (1964), **144**, 540.
22. T. Eisner, J. J. Hurst, W. T. Keeton and Y. Meinwald, *Ann. Entomol. Soc. Amer.* (1965), **58**, 247.
23. H. Schildknecht, K. Holoubek, K. H. Weis and H. Krämer, *Angew. Chem.* (1964), **3**, 73.
24. A. Béhal and M. Phisalix, *Bull. Mus. Natl. Hist. nat. (Paris)* (1900), **6**, 388.
25. R. Hegnauer. "Chemotaxonomie der Pflanzen", Vols. 4, 5, (1966, 1969). Birkhäuser, Basel.
26. E. von Rudloff, *Chem. Ind. (London)* (1965), 180.
27. W. E. Hillis and T. Inoue, *Phytochemistry* (1968), **7**, 13.
28. J. A. Lamberton, *Aust. J. Chem.* (1959), **12**, 224.
29. R. H. Thomson, "Naturally Occurring Quinones", 1st Edit. (1957). Butterworths, London.
30. R. A. Nicolaus, "Melanins" (1968). Hermann, Paris.

31. D. Gilmour. "The Biochemistry of Insects" (1961). Academic Press, New York. M. G. M. Pryor. *In* "Comparative Biochemistry" (1962), Vol. 4, p. 371. Academic Press, New York.

32. L. M. Roth and T. Eisner, *Ann. Revs. Entomol.* (1962), **7**, 107; J. Weatherston, *Quart. Rev.* (*London*) (1967), **21**, 287.

33. N. R. Kuzel and C. E. Miller, *J. Am. Pharm. Assoc.* (1950), **39**, 202.

34. R. K. Ladisch, S. K. Ladisch and P. M. Howe, *Nature* (*London*) (1967), **215**, 939.

35. R. K. Ladisch, *Penn. Acad. Sci.* (1963), **37**, 127.

36. R. K. Ladisch, *Penn. Acad. Sci.* (1965), **38**, 144.

37. P. Vukasović, T. Stojanović and V. Kosovac,*J.Stored Prod. Res.* (1966), **2**, 69.

38. W. R. Tschinkel, *J. Insect Physiol.* (1969), **15**, 191.

39. B. G. Engel and W. Brzeski, *Helv. Chim. Acta* (1947), **30**, 1472.

40. A. Brack,*Helv.Chim. Acta* (1947), **30**, 1; S. W. Tanenbaum and E. W. Bassett, *Biochim. Biophys. Acta* (1958), **28**, 21.

41. H. Raistrick and P. Simonart, *Biochem. J.* (1933), **27**, 628; J. Barta and R. Mečir, *Experientia* (1948), **4**, 277; R. K. Crowden, *Canad. J. Microbiol.* (1967), **13**, 181.

42. L. A. Griffiths, *J. Exp. Bot.* (1959), **10**, 437.

43. K. Mosbach, *Biochem. Biophys. Res. Comm.* (1964), **17**, 363; W. S. G. Maass, G. H. N. Towers and A. C. Neish, *Ber. Deut. Bot. Ges.* (1964), **77**, 157; W. S. G. Maass and A. C. Neish, *Canad. J. Bot.* (1967), **45**, 59.

44. E. Tardy, *Bull. Soc. Chim. France* (1902), **27**, 994.

45. M. El-Dakhakny, *Planta Med.* (1963), **11**, 465.

46. Brandel and Kremers, *Chem. Z.* (1901), **II**, 1007.

47. R. S. Justice, *J. Am. Pharm. Assoc.* (1936), **25**, 850.

48. E. Grimal, *Compt. Rend.* (1904), **139**, 927.

49. H. Erdtman and E. Rennerfelt, *Acta Chem. Scand.* (1949), **3**, 906; Y.-L. Chow and H. Erdtman, *Acta Chem. Scand.* (1962), **16**, 1291.

50. C. Pilo and J. Runeberg, *Acta Chem. Scand.* (1960), **14**, 353.

51. J. Runeberg, *Acta Chem. Scand.* (1960), **14**, 1991.

52. E. Zavarin and A. B. Anderson, *J. Org Chem.* (1955), **20**, 82.

53. B. K. Malavya and S. Dutt, *Proc. Indian Acad. Sci.* (1942), **16A**, 157.

54. Y. Yamamoto. K. Nitta. K. Tango, T. Saito and M. Tsuchimuro, *Chem. Pharm. Bull.* (*Tokyo*) (1965), **13**, 935.

55. F. Bohlmann and K.-M. Kleine, *Chem. Ber.* (1966), **99**, 885.

56. A. R. Burnett and R. H. Thomson, *J. Chem. Soc.* (*C*) (1968), 857.

57. M. Tomita, H. Inouye, Y. Miura, S. Moriguchi, N. Inaba, K. Nakagawa and Y. Tomita, *J. Pharm. Soc. Japan* (1952), **72**, 223; H. Inouye, *Pharm. Bull.* (*Tokyo*) (1954), **2**, 359 and other papers.

58. M. Anchel, A. Hervey, F. Kavanagh, J. Polatnick and W. J. Robbins, *Proc. Natl. Acad. Sci. U.S.* (1948), **34**, 498.

59. G. Pettersson, *Acta Chem. Scand.* (1966), **20**, 45; N. M. Packter, *Biochem J.* (1969), **114**, 369.

60. J. N. Ashley, *J. Chem. Soc.* (1937), 1471.

61. R. B. Woodward, F. Sondheimer, D. Taub, K. Heusler and W. M. McLamore, *J. Amer. Chem. Soc.* (1952), **74**, 4234.

62. G. Pettersson, *Acta Chem. Scand.* (1963), **17**, 1771.

63. N. M. Packter, *Abstr. IUPAC 5th Int. Sym. Chem. Nat. Prods.* (*London*), 1968, p. 126.

64. E. A. H. Friedheim, *C. R. Soc. Biol.* (1933), **112**, 1030; *Helv. Chim. Acta* (1938), **21**, 1464.

65. T. Curtin, G. Fitzgerald and J. Reilly, *Biochem. J.* (1940), **34**, 1605.
66. T. Posternak, *C. R. Soc. Phys. Hist. nat. Genève* (1939), **56**, 28.
67. T. Rosett, Ph.D. Thesis, University of London (1955).
68. T. Posternak, *Helv. Chim. Acta* (1938), **21**, 1326.
69. T. Posternak, H. W. Ruelius and J. Tcherniak, *Helv. Chim. Acta* (1943), **26**, 2031.
70. H. Musso and H. Beecken, *Chem. Ber.* (1959), **92**, 1416.
71. F. Henrich, W. Schmidt and F. Rossteutscher, *Ber.* (1915), **48**, 483.
72. H. Shimada and T. Sawada, *J. Pharm. Soc. Japan* (1957), **77**, 1246.
73. H. Krebs, J. A. Wagner and J. Diewald, *Chem. Ber.* (1956), **89**, 1875.
74. T. Posternak, W. Alcalay and R. Huguenin, *Helv. Chim. Acta* (1956), **39**, 1556
75. B. Bloch and P. Karrer, *Vjschr. naturforsch. Ges. Zürich* (1927), **72**, Suppl. 1.
76. H. Schildknecht, I. Bayer and H. Schmidt, *Z. Naturforsch.* (1967), **22b**, 36.
77. H. Schildknecht, F. Enzmann, K. Gessner, K. Penzien, F. Römer and O. Volkert, *Angew. Chem.* (1966), **5**, 751.
78. H. Schildknecht and H. Schmidt, *Z. Naturforsch.* (1967), **22b**, 287.
79. Ref. 25, Vol. 3.
80. J. D. Bu'Lock, *J. Chem. Soc.* (1955), 575.
81. B. J. Donnelly, D. M. X. Donnelly, A. O'Sullivan and J. P. Prendergast, *Tetrahedron* (1969), **25**, 4409.
82. E. Knoevenagel and C. Bückel, *Ber.* (1901), **34**, 3993.
83. L. Vuataz, *Helv. Chim. Acta* (1950), **33**, 433.
84. D. J. Cosgrove, D. G. H. Daniels, E. N. Greer, J. B. Hutchinson, T. Moran and J. K. Whitehead, *Nature (London)* (1952), **169**, 966; D. J. Cosgrove, D. G. H. Daniels, J. K. Whitehead and J. D. S. Goulden, *J. Chem. Soc.* (1952), 4821; D. G. H. Daniels, *Cereal Chem.* (1959), **36**, 32.
85. W. Karrer, *Helv. Chim. Acta* (1930), **13**, 1424.
86. S. M. Kupchan and M. E. Obasi, *J. Amer. Pharm. Assoc.* (1960), **49**, 257.
87. K. Freudenberg and G. B. Sidhu, *Holzforschung* (1961), **15**, 33.
88. J. Polonsky and E. Lederer, *Bull. Soc. Chim. France* (1959), 1157.
89. J. Polonsky and J.-L. Fourrey, *Tetrahedron Lett.* (1964), 3983.
90. M. K. Jain, *Indian J. Chem.* (1964), **2**, 40.
91. N. Inamoto, S. Masuda, O. Simamura and T. Tsuyuki, *Bull. Chem. Soc. Japan* (1961), **34**, 888
92. J. Polonsky, J. Zylber and R. O. B. Wijesekera, *Bull. Soc. Chim. France* (1962), 1715.
93. E. London, A. Robertson and A. Worthington, *J. Chem. Soc.* (1950), 3431.
94. E. W. Hullett and R. Stern, *Cereal Chem.* (1941), **18**, 561.
95. H. L. Bungenberg de Jong, W. J. Klaar and J. A. Vliegenhart, *Nature (London)* (1953), **172**, 402.
96. H. L. Bungenberg de Jong, W. J. Klaar and J. A. Vliegenhart, *3rd Internationaler Brotkongress, Hamburg*, 1955.
97. E. N. Greer, P. Halton, J. B. Hutchinson and T. Moran, *J. Sci. Fd. Agric.* (1953), **4**, 34; R. W. Cawley and E. W. Hullett, *Chem. Ind. (London)* (1955), 1424.
98. M. Lounasmaa, *Acta Chem. Scand.* (1968), **22**, 70.
99. M. Lounasmaa, *Acta Chem. Scand.* (1967), **21**, 2807.
100. K. Aghoramurthy, K. V. Rao and T. R. Seshadri, *Proc. Indian Acad. Sci.* (1953), **37A**, 798; M. G. Sarngadharan and T. R. Seshadri, *Tetrahedron* (1966), **22**, 739.
101. R. Magnusson, *Acta Chem. Scand.* (1958), **12**, 791.

102. R. Magnusson, *Acta Chem. Scand.* (1960), **14**, 1643.
103. G. N. Gupta, Y. P. Talwar, M. C. Nigam and K. L. Handa, *Soap, Perfumery and Cosmetics* (1964), **37**, 45.
104. E. Carstanjen, *J. prakt. Chem.* (1877), [2], **15**, 399.
105. Th. Zincke, *Ber.* (1881), **14**, 95.
106. E. Zavarin, *J. Org.* (1958), **23**, 1198.
107. E. Zavarin and A. B. Anderson, *J. Org. Chem.* (1955), **20**, 788.
108. E. Zavarin, *J. Org. Chem.* (1958), **23**, 1264.
109. J. D. Fitzpatrick, C. Steelink and R. E. Hansen, *J. Org. Chem.* (1967), **32**, 625.
110. A. B. Anderson, T. C. Scheffer and C. G. Duncan, *Holzforschung* (1963), **17**, 1.
111. W. K. Anslow and H. Raistrick, *Biochem. J.* (1938), **32**, 687.
112. J. H. Birkinshaw and H. Raistrick, *Trans. Roy. Soc.* (1931), **220B**, 245.
113. G. Pettersson, *Acta Chem. Scand.* (1965), **19**, 1016.
114. G. Pettersson, *Acta Chem. Scand.* (1964), **18**, 1428.
115. W. K. Anslow, J. N. Ashley and H. Raistrick, *J. Chem. Soc.* (1938), 439.
116. W. Baker and H. Raistrick, *J. Chem. Soc.* (1941), 670.
117. T. Posternak and H. W. Ruelius, *Helv. Chim. Acta* (1943), **26**, 2045.
118. K. Aghoramurthy, M K. Ramanathan and T. R. Seshadri, *Chem. Ind. (London)* (1954), 1327; K. Aghoramurthy and T. R. Seshadri, *Proc. Indian Acad. Sci.* (1952), **35A**, 331.
119. T. R. Seshadri and G. B. Venkatasubramanian, *J. Chem. Soc.* (1959), 1660.
120. Le-Van-Thoi and Nguyen-Ngoc-Suong, *Ann. Fac. Sci. Univ. Saigon* (1963), 43. *J. Org. Chem.* (1970), **35**, 1104.
121. G. Aulin and H. Erdtman, *Svensk Kem. Tidskr.* (1938), **50**, 42.
122. G. Pettersson, *Acta Chem. Scand* (1964), **18**, 2303.
123. Fr. Fichter, *Annalen* (1908), **361**, 363.
124. F. Kögl and G. C. van Wessem, *Rec. Trav. Chim.* (1944), **63**, 5.
125. G. Lloyd, A. Robertson, G. B. Sankey and W. B. Whalley, *J. Chem. Soc.* (1955), 2163.
126. H. Nishikawa, *J. Fac. Agric. Tottori Univ.* (1952), **1**, 71.
127. H. W. Reading, Ph.D. Thesis, University of London (1955).
128. H. Takeshita and M. Anchel, *Science, N.Y.* (1965), **147**, 152.
129. P. V. Divekar, R. H. Haskins and L. C. Vining, *Canad J. Chem.* (1959), **37**, 2097
130. L. C. Vining, W. J. Kelleher and A. E. Schwarting, *Canad. J. Microbiol.* (1962), **8**, 931.
131. (a) N. Shigematsu, *J. Inst. Polytechnics. Osaka City Univ. Ser. C* (1956), **5**, 100; (b) J. Smith and R. H. Thomson, *Tetrahedron* (1960), **10**, 148.
132. P. W. Brian, P. J. Curtis, S. R. Howland, E. G. Jefferys and H. Raudnitz, *Experientia* (1951), **7**, 266.
133. H. Aquila, *Annalen* (1969), **721**, 220
134. E. B. Visher, *J. Chem. Soc.* (1953), 815.
135. W. Baker, J. F. W. McOmie and D. Miles, *J. Chem. Soc.* (1953), 820.
136. G. Pettersson, *Acta Chem. Scand.* (1964), **18**, 2309.
137. H. A. Anderson, Ph.D. Thesis, University of Aberdeen (1966).
138. N. M. Packter and M. W. Steward, *Biochem. J.* (1967), **102**, 122.
139. W. Mayer and R. Weiss, *Angew. Chem.* (1956), **68**, 680.
140. J. F. Grove, *J. Chem. Soc. (C)* (1966), 985.
141. R. E. Moore, H. Singh and P. J. Scheuer, *J. Org. Chem.* (1966), **31**, 3645.

142. R. G. Jones and H. A. Shonle, *J. Amer. Chem. Soc* (1945), **67**, 1034.
143. E. P. Abraham and H. W. Florey. "Antibiotics" (eds. H. W. Florey *et al.*), (1969), Vol. I, p. 337. Oxford University Press, London; M. A. Kaplan, I. R Hooper and B. Heinemann, *Antibiot. and Chemother.* (1954), **4**, 746.
144. J. C. Sheehan, W. B. Lawson and R. J. Gaul, *J. Amer. Chem. Soc.* (1958), **80**, 5536.
145. A. Rashid and G. Read, *J. Chem. Soc.* (*C*) (1967), 1323.
146. A. Bracken and H. Raistrick, *Biochem. J.* (1947), **41**, 569.
147. W. K. Anslow and H. Raistrick, *Biochem. J.* (1938), **32**, 2288.
148. W. K. Anslow and H. Raistrick, *Biochem. J.* (1938), **32**, 687.
149. W. K. Anslow and H. Raistrick, *Biochem. J.* (1938), **32**, 803.
150. W. K. Anslow and H. Raistrick, *J. Chem. Soc* (1939), 1446
151. M. C. B. van Rheede van Oudtshoorn, *Planta Med.* (1963), **11**, 399; R. Hegnauer, *Pure Appl. Chem.* (1967), **14**, 173.
152. M. Gregson, K. Kurosawa, W. D. Ollis, B. T. Redman, R. J. Roberts, I. O. Sutherland, A. Braga de Oliveira, W. B. Eyton, O. R. Gottlieb and H. H. Dietrichs, *Chem. Comm.* (1968), 1390.
153. W. B. Eyton, W. D. Ollis, I. O. Sutherland, L. M. Jackman, O. R. Gottlieb and M. Taveira Magalhães, *Proc. Chem. Soc.* (1962), 301; W. B. Eyton, W. D. Ollis, I. O. Sutherland, O. R. Gottlieb, M. Taveira Magalhães and L. M. Jackman, *Tetrahedron* (1965), **21**, 2683.
154. G. B. Marini-Bettolo, C. G. Casinovi, O. Gonçalves Da Lima, M. H. Dalia Maia and I. L. D'Albuquerque, *Ann. Chim.* (*Roma*) (1962), **52**, 1190.
155. C. B. Dempsey, D. M. X. Donnelly and R. A. Laidlaw, *Chem. Ind.* (*London*) (1963), 491.
156. (a) M. M. Rao and T. R. Seshadri, *Tetrahedron Lett.* (1963), 211; (b) G. D. Bhatia, S. K. Mukerjee and T. R. Seshadri, *Indian J. Chem.* (1965), **3**, 422.
157. (a) W. D. Ollis, *Experientia* (1966), **22**, 777; (b) W. D. Ollis and O. R. Gottlieb, *Chem. Comm.* (1968), 1396.
158. W. D. Ollis, H. J. P. E. M. Landgraf, O. R. Gottlieb and M. Taviera Magalhães, *An. Acad. brasil Ciênc.* (1964), **36**, 31.
159. B. J. Donelly, D. M. X. Donnelly and C. B. Sharkey, *Phytochemistry* (1965), **4**, 337.
160. O. R. Gottlieb, M. Fineberg, I. Salignac de Souza Guimarães, M. Taveira Magalhães, W. D. Ollis and W. B. Eyton, *An. Acad. brasil Ciênc.* (1964), **36**, 33; W. B. Eyton, W. D. Ollis, M. Fineberg, O. R. Gottlieb, I. Salignac de Souza Guimarães and M. Taveira Magalh es, *Tetrahedron* (1965), **21**, 2697.
161. D. Kumari, S. K. Mukerjee and T. R. Seshadri, *Tetrahedron* (1966), **22**, 3491.
162. (a) W. Sandermann and E. Schwarz, *Farbe u. Lack.* (1956), **62**, 134; (b) W. Sandermann, H. H. Dietrichs and M. Puth, *Holz Roh Werkst.* (1960), **18**, 63.
163. K. H. Schultz and H. H. Dietrichs, *Allergie u. Asthma* (1962), **8**, 125.
164. See refs. in 153.
165. M. F. Barnes, W. D. Ollis, I. O. Sutherland, O. R. Gottlieb and M. Taviera Magalhães, *Tetrahedron* (1965), **21**, 2707.
166. B. Sjoberg, *Acta Chem. Scand.* (1960), **14**, 273.
167. O. R. Gottlieb and M. Taveira Magalhães, *J. Org. Chem.* (1961), **26**, 2449.
168. A. Fredga, *Tetrahedron* (1960), **8**, 126.
169. S. M. Faiho and H. M. Fales, *J. Amer. Chem. Soc.* (1964), **86**, 4434.

170. D. M. X. Donnelly, B. J. Nangle, P. B. Hulbert, W. Klyne and R. J. Swan, *J. Chem. Soc.* (*C*) (1967), 2450.

171. Rio de la Loza, Disertación presentada a la Academia de Medecina, México, (1852).

172. M. C. Weld, *Annalen* (1855), **95**, 188.

173. T. Garcia, E. Dominguez and J. Romo, *Bol. Inst. Quím. Univ. nac. auton. México* (1965), **17**, 16.

174. D. I. del Valle L., *An. Fac. Farm. Bioquím.* (1951), **2**, 80.

175. F. Walls, M. Salmon, J. Padilla, P. Joseph-Nathan and J. Romo, *Bol. Inst. Quím. Univ. nac. auton. México* (1965), **17**, 3.

176. S. B. Vigener, *Sitz. Ber. niederrhein. Ges. Bonn* (1884), 86.

177. F. Mylius, *Ber.* (1885), **18**, 480, 936; R. Anschütz, *Ber.* (1885), **18**, 709; R. Anschütz and W. Leather, *Ber.* (1885), **18**, 715; *Annalen* (1887), **237**, 90; M. McC. Sanders, *Proc. Chem. Soc.* (1906), **22**, 134.

178. F. Kögl and A. G. Boer, *Rec. Trav. Chim.* (1935), **54**, 779.

179. F. Fichter, M. Jetzer and R. Leepin, *Annalen* (1913), **395**, 15.

180. E. R. Wagner, R. D. Moss, R. M. Brooker, J. P. Heeschen, W. J. Potts and M. L. Dilling, *Tetrahedron Lett.* (1965), 4233.

181. D. A. Archer and R. H. Thomson, *Chem. Comm.* (1965), 354.

182. R. B. Bates, S. K. Paknikar and V. P. Thalacker, *Chem. Ind.* (*London*) (1965), 1793.

183. D. A. Archer and R. H. Thomson, *J. Chem. Soc.* (*C*) (1967), 1710.

184. K. Yamaguchi, *J. Pharm. Soc. Japan* (1942), **62**, 491.

185. E. Cortés, M. Salmon and F. Walls, *Bol. Inst. Quím. Univ. nac. auton. México* (1965), **17**, 19.

186. J. von Braun and W. Teuffert, *Ber.* (1929), **62**, 235.

187. D. Arigoni and O. Jeger, *Helv. Chim. Acta* (1954), **37**, 881.

188. J. Padilla, J. Romo, F. Walls and P. Crabbé, *Rev. Soc. Quím. Méx.* (1967), **11**, 7.

189. F. Walls, J. Padilla, P. Joseph-Nathan, F. Giral, M. Escobar and J. Romo, *Tetrahedron* (1966), **22**, 2387.

190. F. G. P. Remfry, *J. Chem. Soc.* (1913), **103**, 1076.

191. H. Nishikawa, *Agr. Biol. Chem.* (*Tokyo*) (1962), **26**, 696.

192. S. Takai, *Phytopathol. Z.* (1961), **43**, 175; *Bull. Govt. Forest Exp. Stat. Tokyo* (1966), No. 195, 1.

193. S. Natori, H. Ogawa, K. Yamaguchi and H. Nishikawa, *Chem. Pharm. Bull.* (*Tokyo*) (1963), **11**, 1343; S. Natori, H. Nishikawa and H. Ogawa, *Chem. Pharm. Bull.* (*Tokyo*) (1964), **12**, 236.

194. S. Natori, Y. Inouye and H. Nishikawa, *Chem. Pharm. Bull.* (*Tokyo*) (1967), **15**, 380.

195. P. Bollinger, Thesis, Eidgenössischen Technischen Hochschule, Zürich (1965).

196. H. Musso, *Angew. Chem.* (1963), **2**, 723.

197. H. Nishikawa, *Trans. Tottori Soc. Agr. Sci.* (1949), **9**, 1.

198. H. Nishikawa, *Proc. Imp. Acad.* (*Tokyo*) (1934), **10**, 414.

199. K. Kawashima, K. Nakanishi, M. Tada and H. Nishikawa, *Tetrahedron Lett.*, 1964, 1227; K. Kawashima, K. Nakanishi and H. Nishikawa, *Chem. Pharm. Bull. Tokyo* (1964), **12**, 796.

200. B. W. Bycroft and J. C. Roberts, *J. Org. Chem.* (1963), **28**, 1429.

201. M. Stoll and M. Hinder, *Helv. Chim. Acta* (1954), **37**, 1859.

202. K. Hirai, S. Nozoe, K. Tsuda, Y. Iitaka, K. Ishibashi and M. Shirasaka, *Tetrahedron Lett.* (1967), 2177.

203. H. Ogawa and S. Natori, *Phytochemistry* (1968), **7**, 773.
204. (a) L. Scott, *Chem. and Drugg.* (1888), 163; (b) C. J. H. Warden, *Pharm. J.* (1888), **18** [3], 601; (1888), **19** [3], 305.
205. S. Krishna and B. S. Varma, *For. Bull. Dehra Dun* (1941), **102**.
206. R. Merian and E. Schlittler, *Helv. Chim. Acta* (1948), **31**, 2237.
207. R. Paris and C. Rabenoro, *Ann. pharm. franç.* (1950), **8**, 380.
208. S. Krishna and B. S. Varma, *J. Indian Chem. Soc.* (1936), **13**, 115; Anon., *Bull. Imp. Inst. London* (1938), **36**, 319.
209. S. Wilkinson, *Planta Med.* (1961), **9**, 121.
210. S. N. Aiyan, M. K. Jain, M. Krishnamurti and T. R. Seshadri, *Phytochemistry* (1964), **3**, 335.
211. O. Fernandez and A. Pizarroso, *Farm. Nueva (Madrid)* (1946), **11**, 1; *Chem. Abs.* (1948), **42**, 8888.
212. H. W. Moore and R. J. Wikholm, *Tetrahedron Lett.* (1968), 5049.
213. O. D. Hensens and K. G. Lewis, *Aust. J. Chem.* (1966), **19**, 169.
214. V. K. Murthy, T. V. P. Rao and V. Venkateswarlu, *Tetrahedron* (1965), **21**, 1445; *Curr. Sci.* (1965), **34**, 16; V. K. Murty, K. R. Prabhu and V. Venkateswarlu, *Curr. Sci.* (1969), **38**, 90.
215. J. Kawamura, *Repts. Japan Sci. Assoc.* (1937), **12**, 377.
216. U.S. Dispensatory, 1950 Edition.
217. A. Heffter and W. Feuerstein, *Arch. Pharm.* (1900), **238**, 15.
218. K. S. Nargund and B. W. Bhide, *J. Indian Chem. Soc.* (1931), **8**, 237.
219. K. H. Hasan and E. Stedman, *J. Chem. Soc.* (1931), 2112.
220. M. Asano and K. Yamaguti, *J. Pharm. Soc. Japan* (1940), **60**, 105; *Proc. Imp. Acad. Japan* (1940), **16**, 36.
221. M. Asano and Z. Hase, *J. Pharm. Soc. Japan* (1940), **60**, 650.
222. L. F. Fieser and E. M. Chamberlin, *J. Amer. Chem. Soc.* (1948), **70**, 71.
223. M. Asano and K. Yamaguti, *J. Pharm. Soc. Japan* (1940), **60**, 585.
224. Ch. B. Rao and V. Venkateswarlu, *Curr. Sci.* (1961), **30**, 259; *J. Org. Chem.* (1961), **26**, 4529.
225. Ch. B. Rao and V. Venkateswarlu, *Tetrahedron* (1962), **18**, 361.
226. Ch. B. Rao and V. Venkateswarlu, *Tetrahedron* (1962), **18**, 951.
227. M. Hiramoto, *Proc. Imp. Acad. (Tokyo)* (1939), **15**, 220.
228. M. Hiramoto, *J. Pharm. Soc. Japan* (1942), **62**, 460, 464.
229. H. Ogawa and S. Natori, *Chem. Pharm. Bull. (Tokyo)* (1965), **13**, 511.
230. H. Ogawa, S. Sakaki, K. Yoshihira and S. Natori, *Tetrahedron Lett.* (1968), 1387.
231. H. Nakata, K. Sasaki, I. Morimoto and Y. Hirata, *Tetrahedron* (1964), **20**, 2319.
232. K. Yoshihara and S. Natori, *Chem. Pharm. Bull. (Tokyo)* (1966), **14**, 1052.
233. T. Kametaka and A. G. Perkin, *J. Chem. Soc.* (1910), **97**, 1415.
234. C. Kuroda, *Sci. Papers Inst. Phys. Chem. Res. (Tokyo)* (1930), **13**, 59; *J. Chem. Soc.* (1930), 752.
235. T. R. Seshadri, *Sci. Proc. Roy. Dublin Soc.* (1956), 77.
236. N. Nasarimhachari and T. R. Seshadri, *Proc. Indian Acad. Sci.* (1948), **27A**, 223.
237. T. R. Seshadri and R. S. Thakur, *Curr. Sci.* (1960), **29**, 54.
238. M. Shimokoriyama and S. Hattori, *Arch. Biochem. Biophys.* (1955), **54**, 93.
239. S. Siddiqui, *J. Indian Chem. Soc.* (1937), **14**, 703.
240. K. V. Rao and T. R. Seshadri, *Proc. Indian Acad. Sci.* (1948), **27A**, 375.
241. J. B. Harborne, *Phytochemistry* (1967), **6**, 1643.
242. T. R. Seshadri, *Revs. Pure Appl. Chem. (Australia)* (1951), **1**, 186.

243. P. K. Bose and P. Dutt, *J. Indian Chem. Soc.* (1940), **17**, 499.
244. G. S. Rao, K. V. Rao and T. R. Seshadri, *Proc. Indian Acad. Sci.* (1948), **28A**, 103.
245. F. Kögl, *Annalen* (1926), **447**, 78.
246. R. L. Frank, G. R. Clark and J. N. Coker, *J. Amer. Chem. Soc.* (1950), **72**, 1824.
247. V. Jirawongse, E. Ramstad and J. Wolinsky, *J. Pharm. Sci.* (1962), **51**, 1108.
248. J. Murray, *J. Chem. Soc.* (1952), 1345.
249. C. Stahlschmidt, *Annalen* (1877), **187**, 177; (1879), **195**, 365.
250. F. Klingemann, *Annalen* (1893), **275**, 89.
251. W. Zopf, *Annalen* (1901), **317**, 110, 124.
252. B. C. Smale and H. L. Keil, *Phytochemistry* (1966), **5**, 1120.
253. Fr. Fichter, *Annalen* (1908), **361**, 363.
254. F. Kögl, H. Becker, G. de Voss and E. Wirth, *Annalen* (1928), **465**, 243.
255. P. R. Shildneck and R. Adams, *J. Amer. Chem. Soc.* (1931), **53**, 2373.
256. (a) D. E. Kvalnes, *J. Amer. Chem. Soc.* (1934), **56**, 2478; (b) E. Ghigi, *Boll. Sci. Fac. Chim. Bologna* (1944–47), **5**, 38.
257. M. Akagi, *J. Pharm. Soc. Japan* (1942), **62**, 195.
258. B. F. Cain, *J. Chem. Soc.* (1961), 936; (1963), 356; (1966), 1041.
259. G. J. Bennett and N. Uri, *J. Chem. Soc.* (1962), 2753.
260. F. Kögl, H. Becker, A. Detzel and G. de Voss, *Annalen* (1928), **465**, 211.
261. H. H. Lee and C. H. Tan, *J. Chem. Soc. (C)* (1967), 1583.
262. T. C. McMorris and M. Anchel, *Tetrahedron Lett.* (1963), 335; *Tetrahedron* (1967), **23**, 3985.
263. J. Gripenberg, *Tetrahedron Lett.* (1966), 697.
264. M. Lounasmaa, *Acta Chem. Scand.* (1966), **20**, 2304.
265. P. V. Divekar, G. Read and L. C. Vining, *Chem. Ind. (London)* (1959), 731; P. V. Divekar, G. Read, L. C. Vining and R. H. Haskins, *Canad. J. Chem.* (1959), **37**, 1970.
266. L. C. Vining, L. R. Nesbitt and R. H. Haskins, *Canad. J. Microbiol.* (1966), **12**, 409.
267. W. Thörner, *Ber.* (1878), **11**, 533; (1879), **12**, 1630.
268. F. Kögl and J. J. Postowsky, *Annalen* (1924), **440**, 19; (1925), **445**, 159.
269. K. L. Euler, V. E. Tyler, L. R. Brady and M. H. Malone, *Lloydia* (1965), **28**, 203; J. M. Khanna, M. H. Malone, K. L. Euler and L. R. Brady, *J. Pharm. Sci.* (1965), **54**, 1016.
270. J. Gripenberg, *Acta Chem. Scand.* (1956), **10**, 1111.
271. M. L. Montfort, V. E. Tyler and L. R. Brady, *J. Pharm. Sci.* (1966), **55**, 1300.
272. G. Sullivan, *Diss. Abs.* (1967), **27B**, 3054; G. Sullivan, L. R. Brady and V. E. Tyler, *Lloydia* (1967), **30**, 84.
273. J. Gripenberg, *Acta Chem. Scand.* (1958), **12**, 1411.
274. R. L. Edwards and N. Kale, *J. Chem. Soc.* (1964), 4084.
275. O. P. Mittal and T. R. Seshadri, *Curr. Sci.* (1957), **26**, 4.
276. F. M. Dean, "Naturally Occurring Oxygen Ring Compounds" (1963), p. 74. Butterworths, London.
277. M. Akagi, *J. Pharm. Soc. Japan* (1942), **62**, 129.
278. M. Akagi, *J. Pharm. Soc. Japan* (1942), **62**, 202.
279. R. L. Edwards, G. C. Elsworthy and N. Kale, *J. Chem. Soc. (C)* (1967), 405.
280. W. Zopf, *Bot. Zeit.* (1889), **47**, 69.
281. F. Kögl, H. Erxleben and L. Jänecke, *Annalen* (1930), **482**, 105.
282. J. Zellner, *Monatsh.* (1915), **36**, 615.

283. M. Sawada, *Nippon Ringaku Kaishi* (1952), **34**, 110; *Chem. Abs.* (1953), **47**, 4436.

284. M. Sawada, *Nippon Ringaku Kaishi* (1958), **40**, 195; *Chem. Abs.* (1958), **52**, 18638.

285. J. Gripenberg, *Acta Chem. Scand.* (1960), **10**, 135.

286. T. Shimano, K. Taki and K. Goto, *Ann. Proc. Gifu Coll. Pharm.* (1953), No. 3, 43.

287. Y. Asahina and S. Shibata, *Ber.* (1939), **72**, 1531.

288. T. R. Seshadri, *Indian J. Pharm.* (1953), **15**, 286.

289. B. B. Millward and M. C. Whiting, *J. Chem. Soc.* (1958), 903.

290. G. Read and L. C. Vining, *Canad. J. Chem.* (1959), **37**, 1442.

291. K. Aghoramurthy, K. G. Sarma and T. R. Seshadri, *Tetrahedron Lett.* (1959), No. 8, 20; (1960), No. 16, 4.

292. J. Gripenberg, *Suomen Kemistilehti* (1960), **B33**, 72.

293. M. Lounasmaa, *Acta Chem. Scand.* (1965), **19**, 540.

294. H. W. Wanzlick, *Angew. Chem.* (1964), **3**, 401; H. Heidepriem, Doktorarbeit, Technischen Universität, Berlin, 1964.

295. A. J. Shand and R. H. Thomson, *Tetrahedron* (1963), **19**, 1919 and refs. therein.

296. J. Gripenberg and M. Lounasmaa, *Acta Chem. Scand.* (1966), **20**, 2202.

297. M. Sawada, *Bull. Tokyo Univ. For.* (1965), No. 59, 33; *Chem. Abs.* (1966), **65**, 6197.

298. (a) J. Gripenberg, *Acta Chem. Scand.* (1965), **19**, 2242; (b) J. Gripenberg and J. Martikkala, *Acta Chem. Scand.* (1969), **23**, 2583.

299. S. Abrahamson and M. Innes, *Acta Cryst.* (1966), **21**, 948.

300. M. Kofler, "Festschrift E. C. Barell" (1946). Hoffmann-La Roche, Basle.

301. R. L. Lester and F. L. Crane, *J. Biol. Chem.* (1959), **234**, 2169.

302. K. Egger, *Planta* (1965), **64**, 41.

303. C. Bucke, R. M. Leech, M. Hallaway and R. A. Morton, *Biochem. Biophys. Acta* (1965), **112**, 19.

304. N. G. Carr, G. Exell, V. Flynn, M. Hallaway and S. Talukdar, *Arch. Biochem. Biophys.* (1967), **120**, 503.

305. H. Eck and A. Trebst, *Z. Naturforsch.* (1963), **18b**, 446.

306. D. Misiti, H. W. Moore and K. Folkers, *J. Amer. Chem. Soc.* (1965), **87**, 1402.

307. F. L. Crane, *Plant Physiol.* (1959), **34**, 128.

308. E. R. Redfearn and J. Friend, *Phytochemistry* (1962), **1**, 147.

309. R. A. Dilley and F. L. Crane, *Plant Physiol.* (1964), **39**, 33.

310. M. Kofler, A. Langemann, R. Rüegg, L. H. Chopard-dit-Jean, A. Rayroud and O. Isler, *Helv. Chim. Acta* (1959), **42**, 1283.

311. F. L. Crane, *Plant Physiol.* (1959), **34**, 546.

312. N. R. Trenner, B. H. Arison, R. E. Erickson, C. H. Shunk, D. E. Wolf and K. Folkers, *J. Amer. Chem. Soc.* (1959), **81**, 2026.

313. C. H. Shunk, R. E. Erickson, E. L. Wong and K. Folkers, *J. Amer. Chem. Soc.* (1959), **81**, 5000.

314. M. Kofler, A. Langemann, R. Rüegg, U. Gloor, U. Schwieter, J. Würsch, O. Wiss and O. Isler, *Helv. Chim. Acta* (1959), **42**, 2252.

315. R. A. Dilley, *Anal. Biochem.* (1964), **7**, 240.

316. J. M. Bruce, *Quart. Rev. (London)* (1967), **21**, 405; A. Schönberg, G. O. Schenck and O.-A. Neumüller, "Preparative Organic Photochemistry" (1968). Springer-Verlag, Berlin.

317. L. P. Kegel, M. D. Henninger and F. L. Crane, *Biochem. Biophys. Res. Comm.* (1962), **8**, 294.

318. D. R. Threlfall, W. T. Griffiths and T. W. Goodwin, *Biochim. Biophys. Acta* (1965), **102**, 614.

319. B. C. Das, M. Lounasmaa, C. Tendille and E. Lederer, *Biochem. Biophys. Res. Comm.* (1965), **21**, 318.

320. W. T. Griffiths, J. C. Wallwork and J. F. Pennock, *Nature (London)* (1966), **211**, 1037.

321. W. T. Griffiths, *Biochem. Biophys. Res. Comm.* (1966), **25**, 596.

322. P. J. Dunphy, K. J. Whittle and J. F. Pennock. *In* "Biochemistry of Chloroplasts" (ed. T. W. Goodwin) (1966), Vol. 1, p. 165. Academic Press, New York.

323. O. Isler, A. Langemann, H. Mayer, R. Rüegg and P. Schudel, *Bull. Nat. Inst. Sci. India* (1965), **28**, 132.

324. C. Bucke, R. M. Leech, M. Hallaway and R. A. Morton, *Biochim. Biophys. Acta* (1966), **112**, 19.

325. R. L. Rowland, *J. Amer. Chem. Soc.* (1958), **80**, 6130; R. L. Rowland and J. A. Giles, *Tobacco Sci.* (1960), **4**, 29.

326. H. Mayer, J. Metzger and O. Isler, *Helv. Chim. Acta* (1967), **50**, 1376.

327. J. F. Pennock, F. W. Hemming and J. D. Kerr, *Biochem. Biophys. Res. Comm.* (1964), **17**, 542.

328. P. J. Dunphy, K. J. Whittle, J. F. Pennock and R. A. Morton, *Nature (London)* (1965), **207**, 521; K. J. Whittle, P. J. Dunphy and J. F. Pennock, *Biochem. J.* (1966), **100**, 138.

329. R. A. Morton and W. E. J. Phillips, *Biochem. J.* (1959), **73**, 427; A. J. Diplock and G. A. D. Hazelwood, *Proc. 6th Internat. Congr. Biochem. New York* (1964), Vol. 8, p. 26.

330. J. F. Pennock, *Vitamins and Hormones* (1966), **24**, 307.

331. H. K. Lichtenthaler and R. B. Park, *Nature (London)* (1963), **198**, 1070.

332. R. A. Dilley and F. L. Crane, *Plant Physiol.* (1963), **38**, 452.

333. C. Bucke, M. Hallaway and R. A. Morton, *Biochem. J.* (1964), **90**, 41P.

334. M. D. Henninger and F. L. Crane, *Biochemistry* (1963). **2**, 1168.

335. I. Hindberg and H. Dam, *Physiol. Plant.* (1965), **18**, 838.

336. P. Schudel, H. Mayer and O. Isler. *Vitamins.* In press.

337. W. A. Skinner and P. A. Sturm, *Phytochemistry* (1968), **7**, 1893.

338. G. R. Whistance, J. F. Dillon and D. R. Threlfall, *Biochem. J.* (1969), **111**, 461.

339. S. M. Bloom, *J. Amer. Chem. Soc.* (1961), **83**, 3808.

340. P. Friis, G. D. Daves and K. Folkers, *J. Amer. Chem. Soc.* (1966), **88**, 4754.

341. P. Friis, J. L. G. Nilsson, G. D. Daves and K. Folkers, *Biochem. Biophys. Res. Comm.* (1967), **28**, 324.

342. R. K. Olsen, G. D. Daves, H. W. Moore and K. Folkers, *J. Amer. Chem. Soc.* (1966), **88**, 2346.

343. S. Imamoto and S. Senoh, *Tetrahedron Lett.* (1967), 1237; *J. Chem. Soc. Japan (Pure Chem. Sect.)* (1968), **89**, 316.

344. O. Gessner, "Die Gift- und Arzneipflanzen von Mitteleuropa", (1953), p. 538. Carl Winter, Heidelberg.

345. F. L. Crane. *In* "Biochemistry of Quinones" (ed. R. A. Morton) (1965), p. 183. Academic Press, New York and "Progress in the Chemistry of Fats and Other Lipids" (ed. R. T. Holman) (1964), Vol. VII, Part 2, p. 267. Pergamon Press, Oxford.

346. P. Sommer and M. Kofler, *Vitamins and Hormones* (1966), **24**, 349.

347. R. A. Morton, *Vitamins and Hormones* (1961), **19**, 1; R. A. Morton (ed.) "Biochemistry of Quinones" (1965). Academic Press, New York.

348. P. Friis, G. D. Daves and K. Folkers, *Biochem. Biophys. Res. Comm.* (1966), **24**, 252.

349. G. D. Daves, R. F. Muraca, J. S. Whittick, P. Friis and K. Folkers, *Biochemistry* (1967), **6**, 2861.

350. D. H. Burrin and R. B. Beechey, *Biochem. J.* (1963), **89**, 111P.

351. R. Powls and F. W. Hemming, *Phytochemistry* (1966), **5**, 1235.

352. R. A. Morton, *Nature (London)* (1958), **182**, 1764.

353. R. A. Morton, G. M. Wilson, J. S. Lowe and W. M. F. Leat, *Chem. Ind. (London)* (1957), 1649.

354. R. A. Morton, U. Gloor, O. Schindler, G. M. Wilson, L. H. Chopard-dit-Jean, F. W. Hemming, O. Isler, W. M. F. Leat, J. F. Pennock, R. Rüegg, U. Schwieter and O. Wiss, *Helv. Chim. Acta* (1958), **41**, 2343.

355. R. L. Lester, F. L. Crane and Y. Hatefi, *J. Amer. Chem. Soc.* (1958), **80**, 4751; D. E. Wolf, C. H. Hoffman, N. R. Trenner, B. G. Arison, C. H. Shunk, B. O. Linn, J. F. McPherson and K. Folkers, *J. Amer. Chem. Soc.* (1958), **80**, 4752.

356. D. M. X. Donnelly, B. J. Nangle, J. P. Prendergast and A. M. O'Sullivan, *Phytochemistry* (1968), **7**, 647.

357. U. Gloor, O. Isler, R. A. Morton, R. Rüegg and O. Wiss, *Helv. Chim. Acta* (1958), **41**, 2357; R. Rüegg, U. Gloor, R. N. Goel, G. Ryser, O. Wiss and O. Isler, *Helv. Chim. Acta* (1959), **42**, 2616.

358. C. H. Shunk, B. O. Linn, E. L. Wong, P. E. Wittreich, F. M. Robinson and K. Folkers, *J. Amer. Chem. Soc.* (1958), **80**, 4753.

359. O. Isler, H. Mayer, R. Rüegg and J. Würsch, *Vitamins and Hormones* (1966), **24**, 331.

360. O. Schindler, *Prog. Chem. Org. Nat. Prods.* (1962), **20**, 73.

361. W. V. Lavate, J. R. Dyer, C. M. Springer and R. Bentley, *J. Biol. Chem.* (1962), **237**, PC2715.

362. P. H. Gale, B. H. Arison, N. R. Trenner, A. C. Page and K. Folkers, *Biochemistry* (1963), **2**, 196.

363. W. V. Lavate and R. Bentley, *Arch. Biochem. Biophys.* (1964), **108**, 287.

364. W. V. Lavate, J. R. Dyer, C. M. Springer and R. Bentley, *J. Biol. Chem.* (1965), **240**, 524.

365. J. Glover and D. R. Threlfall, *Biochem. J.* (1962), **85**, 14P.

366. H. W. Moore and K. Folkers, *J. Amer. Chem. Soc.* (1965), **87**, 1409.

367. H. W. Moore and K. Folkers, *J. Amer. Chem. Soc.* (1966), **88**, 567.

368. H. W. Moore and K. Folkers, *J. Amer. Chem. Soc.* (1966), **88**, 564.

369. W. W. Parson and H. Rudney, *J. Biol. Chem.* (1965), **240**, 1855.

370. B. O. Linn, N. R. Trenner, C. H. Shunk and K. Folkers, *J. Amer. Chem. Soc.* (1959), **81**, 1263; B. O. Linn, N. R. Trenner, B. H. Arison, R. G. Weston, C. H. Shunk and K. Folkers, *J. Amer. Chem. Soc.* (1960), **82**, 1647; C. H. Shunk, D. E. Wolf, J. F. McPherson, B. O. Linn and K. Folkers, *J. Amer. Chem. Soc.* (1960), **82**, 5914.

371. I. Imada, Y. Sanno and H. Morimoto, *Chem. Pharm. Bull. (Tokyo)* (1964), **12**, 1042.

372. N. G. Carr, *Biochem. J.* (1964), **91**, 28P.

373. I. Imada and H. Morimoto, *Chem. Pharm. Bull. (Tokyo)* (1964), **12**, 1047.

374. I. Imada and H. Morimoto, *Chem. Pharm. Bull. (Tokyo)* (1964), **12**, 1051.

375. I. Imada, Y. Sanno and H. Morimoto, *Chem. Pharm. Bull. (Tokyo)* (1964), **12**, 1056.

376. B. Frydman and H. Rapoport, *J. Amer. Chem. Soc.* (1963), **85**, 823.

377. O. Isler, R. Rüegg and A. Langemann, *Chem. Weekblad.* (1960), **56**, 613.

378. I. Imada and H. Morimoto, *Chem. Pharm. Bull.* (*Tokyo*) (1965), **13**, 130.
379. C. H. Shunk, J. F. McPherson and K. Folkers, *J. Org. Chem.* (1960), **25**, 1053; cf. J. A. D. Jeffreys, *J. Chem. Soc.* (1959), 2153.
380. F. R. Koniuszy, P. H. Gale, A. C. Page and K. Folkers, *Arch. Biochem. Biophys.* (1960), **87**, 298.
381. D. L. Laidman, R. A. Morton, J. Y. F. Paterson and J. F. Pennock, *Biochem. J.* (1960), **74**, 541.
382. A. T. Diplock, J. Green, E. E. Edwin and J. Bunyan, *Nature* (*London*) (1961), **189**, 749.
383. J. Stevenson, F. W. Hemming and R. A. Morton, *Biochem. J.* (1963), **89**, 58P.
384. B. O. Linn, C. H. Shunk, E. L. Wong and K. Folkers, *J. Amer. Chem. Soc.* (1963), **85**, 239.
385. D. McHale and J. Green, *Chem. Ind.* (*London*) (1962), 1867.
386. H. H. Draper and A. S. Csallany, *Biochem. Biophys. Res. Comm.* (1960), **2**, 307.
387. J. Links, *Biochim. Biophys. Acta* (1960), **38**, 193; J. Links and O. Tol, *Biochim. Biophys. Acta* (1963), **73**, 349; J. Green, E. E. Edwin, A. T. Diplock and D. McHale, *Biochem. Biophys. Res. Comm.* (1960), **2**, 269.
388. F. W. Hemming, D. L. Laidman, R. A. Morton and J. F. Pennock, *Biochem. Biophys. Res. Comm.* (1961), **4**, 393.
389. C. H. Shunk, N. R. Trenner, C. H. Hoffman, D. E. Wolf and K. Folkers, *Biochem. Biophys. Res. Comm.* (1960), **2**, 427.
390. F. W. Hemming, R. A. Morton and J. F. Pennock, *Biochem. J.* (1961), **80**, 445.
391. T. S. Raman and H. Rudney, *Arch. Biochem. Biophys.* (1966), **116**, 75.
392. E. R. Redfearn, *Vitamins and Hormones* (1966), **24**, 465.
393. F. L. Crane and M. D. Henninger, *Vitamins and Hormones* (1966), **24**, 489.
394. P. Friis, G. D. Daves and K. Folkers, *Biochemistry* (1967), **6**, 3618.
395. G. L. Sottocasa and F. L. Crane, *Biochemistry* (1965), **4**, 305.
396. R. F. Muraca, J. S. Whittick, G. D. Daves, P. Friis and K. Folkers, *J. Amer. Chem. Soc.* (1967), **89**, 1505.
397. H. Morimoto, I. Imada and M. Sasaki, *Annalen* (1965), **690**, 115.
398. P. S. Rao, K. G. Sarma and T. R. Seshadri, *Curr. Sci.* (1966), **35**, 147.
399. W. D. Ollis, B. T. Redman, R. J. Roberts and I. O. Sutherland, *Chem. Comm.* (1968), 1392.
400. A. Closse, R. Mauli and H. P. Sigg, *Helv. Chim. Acta* (1966), **49**, 204.
401. J. W. Rowe, J. K. Toda and M. Fracheboud, *Abstr. IUPAC 5th Int. Sym. Chem. Nat. Prods.* (*London*) (1968), p. 309; F. D. Hostettler and M. K. Seikel, *Tetrahedron* (1969), **25**, 2325.
402. J. W. Rowe, personal communication, 1968.
403. I. A. Pearl and S. F. Darling, *Phytochemistry* (1968), **7**, 1851.
404. I. A. Pearl and S. F. Darling, *Tappi* (1968), **51**, 537.
405. B. P. Moore, *J. Insect Physiol.* (1968), **14**, 33.
406. H. Schildknecht, U. Maschwitz and H. Winkler, *Naturwissenschaften* (1968), **55**, 112; H. Schildknecht, H. Winkler and U. Maschwitz, *Z. Naturforsch.* (1968), **23b**, 637.
407. H. Schildknecht, E. Maschwitz and U. Maschwitz, *Z. Naturforsch.* (1968), **23b**, 1213.
408. H. Inouye, K. Tokaru and S. Tobita, *Chem. Ber.* (1968), **101**, 4057.
409. A. R. Forrester, J. M. Hay and R. H. Thomson. "Organic Chemistry of Stable Free Radicals", (1968). Academic Press, London.

410. C. J. R. Adderley and F. R. Hewgill, *J. Chem. Soc. (C)* (1968), 1434.
411. H. Raistrick, *Proc. Roy. Soc.* (1949), **A199**, 141.
412. C.-K. Wat, A. Tse, R. J. Bandoni and G. H. N. Towers, *Phytochemistry* (1968), **7**, 2177.
413. M. W. Miller, *Tetrahedron* (1968), **24**, 4839.
414. K. Kurosawa, W. D. Ollis, B. T. Redman, I. O. Sutherland, A. Braga de Oliveira, O. R. Gottlieb and H. Magalhães Alves, *Chem. Comm.* (1968), 1263.
415. K. Kurosawa, W. D. Ollis, B. T. Redman, I. O. Sutherland, O. R. Gottlieb and H. Magalhães Alves, *Chem. Comm.* (1968), 1265.
416. P. Joseph-Nathan, J. Reyes and Ma. P. González, *Tetrahedron* (1968), **24**, 4007.
417. R. B. Woodward in "Aromaticity", *Chem. Soc. Special Pub.*, No. 21, (1967), p. 217; G. B. Gill, *Quart. Rev. (London)* (1968), **22**, 338.
418. R. C. Cookson, D. A. Cox and J. Hudec, *J. Chem. Soc.* (1961), 4499.
419. M. Sawada, *Bull. Tokyo Univ. For.* (1965), No. 59, 54.
420. K. Minami, K. Asawa and M. Sawada, *Tetrahedron Lett.* (1968), 5067.
421. M. Sawada, personal communication, 1968.
422. K. R. Prabhu, Ch. B. Rao and V. Venkateswarlu, *Curr. Sci.* (1969), **38**, 15; C. Chandrasekhar, K. R. Prabhu and V. Venkateswarlu, *Phytochemistry* (1970), **9**, 414.
423. K. H. Hasan and E. Stedman, *J. Chem. Soc.* (1931), 2112.
424. H. Ogawa and S. Natori, *Chem. Pharm. Bull. (Tokyo)* (1968), **16**, 1709.
425. K. Yoshihira, S. Sakaki, H. Ogawa and S. Natori, *Chem. Pharm. Bull. (Tokyo)* (1968), **16**, 2383.
426. J. C. Overeem and A. van Dijkman, *Rec. Trav. Chim.* (1968), **87**, 940.
427. R. Thomas, *Biochem. J.* (1961), **80**, 234.
428. E. Sun, R. Barr and F. L. Crane, *Plant Physiol.* (1968), **43**, 1935.
429. H.-W. Wanzlick and U. Jahnke, *Chem. Ber.* (1968), **101**, 3753.
430. F. Kögl and H. Erxleben, *Annalen* (1930), **479**, 11.
431. Personal communications from C. H. Eugster and J. Gripenberg.
432. R. Barr and F. L. Crane, *Plant Physiol.* (1967), **42**, 1255.
433. K. Egger and H. Kleinig, *Z. Pflanzenphysiol.* (1967), **56**, 113.
434. B. Rumberg, P. Schmidt-Mende and H. T. Witt, *Nature (London)* (1964), **201**, 466; H. H. Stiehl and H. T. Witt, *Z. Naturforsch.* (1968), **23b**, 220; (1969), **24b**, 1588; P. Schmidt-Mende and B. Rumberg, *Z. Naturforsch.* (1968), **23b**, 225.
435. J. C. Wallwork and J. F. Pennock, *Chem. Ind. (London)* (1968), 1571.
436. R. Barr, M. D. Henninger and F. L. Crane, *Plant Physiol.* (1967), **42**, 1246.
437. *Biochem. J.* (1967), **102**, 15.
438. W. John, E. Dietzel and W. Emte, *Z. Physiol. Chem.* (1939), **257**, 173.
439. G. D. Daves, J. J. Wilcznski, P. Friis and K. Folkers, *J. Amer. Chem. Soc.* (1968), **90**, 5587.
440. J. J. Wilcznski, G. D. Daves and K. Folkers, *J. Amer. Chem. Soc.* (1968), **90**, 5593.
441. R. Powls and F. W. Hemming, *Phytochemistry* (1966), **5**, 1269.
442. R. Powls, E. Redfearn and S. Trippett, *Biochem. Biophys. Res. Comm.* (1968), **33**, 408.
443. H. K. Lichtenthaler, *Z. Pflanzenphysiol.* (1968), **59**, 195; *Planta* (1968), **81**, 140.
444. L. Jurd, *Tetrahedron* (1969), **25**, 1407.
445. W. A. Skinner and P. A. Sturm, *Phytochemistry* (1968), **7**, 1893.

446. M. Rasmussen, D. D. Ridley, E. Ritchie and W. C. Taylor, *Aust. J. Chem.* (1968), **21**, 2989. See also D. D. Ridley, E. Ritchie and W. C. Taylor, *Aust. J. Chem.*, 1970, **23**, 147.
447. J. F. Corbett and A. G. Fooks, *J. Chem. Soc.* (*C*), 1967, 1909.
448. H. W. Moore and K. Folkers, *Annalen*, 1965, **684**, 212.
449. F. L. Crane. *In* "Biological Oxidations" (ed. T. P. Singer) (1968), p. 533. Interscience, New York; F. L. Crane and H. Low, *Physiol. Revs.* (1966), **46**, 662.
450. I. A. Pearl and S. F. Darling, *Phytochemistry* (1968), **7**, 1855.
451. V. J. Filipic and J. C. Underwood, *J. Food Sci.* (1964), **29**, 464.
452. R. C. Cambie and R. A. F. Couch, *New Zealand J. Sci.* (1967), **10**, 1020.
453. G. M. Happ, *J. Insect Physiol.* (1968), **14**, 1821.
454. H. Schildknecht, *Angew. Chem.* (1970), **82**, 17.
455. G. M. Happ, *Ann. Entomol. Soc. Amer.* (1967), **60**, 279.
456. H. Schildknecht and K. Koob, *Naturwissenschaften* (1969), **56**, 328.
457. S. H. El Basyouni, D. Brewer and L. C. Vining, *Canad. J. Bot.* (1968), **46**, 441.
458. G. Sullivan and W. L. Guess, *Lloydia* (1969), **32**, 72.
459. F. Kotlaba, J. Křepinský, V. Herout, R. Prokeš and A. Vystrčil, *Naturwissenschaften* (1965), **52**, 591; J. Křepinský, V. Herout, F. Šorm, A. Vystrčil, R. Prokeš and G. Jommi, *Coll. Czech. Chem. Comm.* (1965), **30**, 2626.
460. E. A. Obol'nikova, O. I. Volkova and G. I. Samokhvalov, *Zh. Obshch. Khim.* (1968), **38**, 459.
461. E. Schnepf, W. Wenneis and H. Schildknecht, *Z. Zellforsch.* (1969), **96**, 582.
462. Panida Kanchanapee, H. Ogawa and S. Natori, *Japan J. Pharmacog.* (1967), **21**, 68.
463. P. C. Beaumont and R. L. Edwards, *J. Chem. Soc.* (*C*) (1968), 2968.
464. A. P. Duarte, D. de Barros Corrêa, L. G. Fonseca e Silva, S. Janot and O. R. Gottlieb, *Acad. Brasil Cienc.* (1968), **40**, 307; D. de Barros Corrêa, L. G. Fonseca e Silva, O. R. Gottlieb and S. Janot Gonçalves, *Phytochemistry* (1970), **9**, 447.
465. P. C. Beaumont and R. L. Edwards, *J. Chem. Soc.* (*C*) (1969), 2398.
466. M. Greshoff and J. Sack, *Pharm. Weekblad.* (1903), **40**, 127.
467. C. Graebe and H. Hess, *Annalen* (1905), **340**, 237; H. I. Bolker and F. L. Kung, *J. Chem. Soc.* (*C*) (1969), 2298.
468. W. Will, *Ber.* (1888), **21**, 608; W. Baker, *J. Chem. Soc.* (1941), 665.
469. D. J. Aneshansley, T. Eisner, J. M. Widom and B. Widom, *Science* (1969), **165**, 61.
470. L. Jurd, *Tetrahedron Lett.* (1969), 2863.
471. S. Mageswaran, W. D. Ollis, R. J. Roberts and I. O. Sutherland, *Tetrahedron Lett.* (1969), 2897.
472. R. D. Allan, R. L. Correll and R. J. Wells, *Tetrahedron Lett.* (1969), 4669.
473. R. D. Allan, R. L. Correll and R. J. Wells, *Tetrahedron Lett.* (1969), 4673.
474. H. Schildknecht, *Angew. Chem.* (1970), **82**, 17.
475. R. Barr and C. J. Arntzen, *Plant. Physiol.* (1969), **44**, 591.
476. M. Sato and H. Ozawa, *J. Biochem.* (*Tokyo*) (1969), **65**, 861.
477. H. Ozawa, M. Sato, S. Natori and H. Ogawa, *Experientia* (1969), **25**, 484; H. Ogawa and S. Natori, *Tetrahedron Lett.* (1969), 1969; H. Ogawa, M. Sato, S. Natori and H. Ogawa, *Chem. Pharm. Bull* (*Tokyo*) (1970), **18**, 1099.
478. G. R. Whistance and D. R. Threlfall, *Phytochemistry* (1970), **9**, 213.

479. W. T. Griffiths, Ph.D. Thesis, University College of Wales, Aberystwyth (1965).

480. F. E. Field, B.Sc. Thesis, University College of Wales, Aberystwyth (1969).

481. D. R. Threlfall and G. R. Whistance, *Phytochemistry* (1970), **9**, 355.

482. D. R. Threlfall and G. R. Whistance, *Biochem. J.* (1969), **113**, 38P.

483. T. M. Farley, J. Blake and K. Folkers, *Internat. J. Vit. Res.* (1969), **39**, 168.

484. G. R. Whistance, B. S. Brown and D. R. Threlfall, *Biochim. Biophys. Acta* (1969), **176**, 895.

485. H. Morimoto, I. Imada, M. Watanabe, Y. Nakao, M. Kuno and N. Matsumoto, *Annalen* (1969), **729**, 158.

486. H. Morimoto, I. Imada and G. Goto, *Annalen* (1969), **729**, 184.

487. W. J. Robbins, F. Kavanagh and A. Hervey, *Proc. Nat. Acad. Sci. U.S.A.*, 1947, **33**, 171; F. Kavanagh, *Arch. Biochem.* (1947), **15**, 95.

488. D. Arigoni, *Pure Appl. Chem.* (1968), **17**, 331.

489. W. Steglich, W. Furtner and A. Prox, *Z. Naturforsch.* (1968), **23b**, 1044.

490. R. J. Wells, personal communication (1970).

491. E. Subramanian and J. Trotter, *J. Chem. Soc. (A)* (1970), 622.

492. B. P. Moore and B. E. Wallbank, *Proc. Roy. Ent. Soc. Lond.* (1968), **(B)37**, 62.

493. G. R. Whistance and D. R. Threlfall, *Phytochemistry* (1970), **9**, 737.

494. S. Naito and Y. Kaneko, *Tetrahedron Lett.* (1969), 4675.

CHAPTER 4

Naphthaquinones†

The vast majority of naphthalene derivatives found in Nature are
quinones, and the others are mainly related naphthols or naphthyl
ethers. Increasing numbers of *o*-naphthaquinones (mainly of terpenoid
origin) and binaphthaquinones have been found in recent years, and
such compounds can no longer be considered rare. Another by no means
uncommon structural feature is an oxygen heterocycle fused to the
quinonoid or benzenoid ring, and the apparent distinction between
naphthaquinones of plant and animal origin has now disappeared. In
the animal kingdom only echinoderms are known to elaborate naphthaquinones but they are found throughout the plant kingdom, and have
turned up in artefacts ranging from golf clubs to tobacco smoke.

6-Methyl-1,4-naphthaquinone (1), $C_{11}H_8O_2$

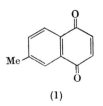

(1)

Occurrence: *Marasmius graminum* (Libert) Berk.[3]
Physical properties: golden-yellow needles, m.p. 90–91°, $\lambda_{max.}$ (EtOH) 249,
255·5, 342·5 nm (log ϵ 4·34, 4·29, 3·49), $\nu_{max.}$ (KBr) 1660, 1599 cm^{-1}.

During a search for new antibiotics several *Marasmius* spp. were
found[1] to produce antibacterial substances which led to subsequent
chemical investigations.[2] One of these compounds, isolated by steam
distillation of the culture filtrate of *M. graminum*, showed the light
absorption of a typical 1,4-naphthaquinone and formed a leucodiacetate.
A Kuhn–Roth estimation indicated the presence of one C-Me group,
evidently located at C-6, since oxidation with neutral permanganate
yielded trimellitic anhydride. The metabolite is therefore 6-methyl-1,4-
naphthaquinone which was confirmed by direct comparison with a
synthetic specimen prepared via Diels–Alder addition of isoprene to

† The protoaphins are discussed in Chapter 7.

benzoquinone.[3] It can also be obtained by cyclisation[4] of γ,γ-dimethyl-allylbenzoquinone (p. 99) either by irradiation or using boron trifluoride as catalyst. This is the only fungal naphthaquinone lacking a hydroxyl or methoxyl function but it is accompanied in *M. graminum* by a trace of red material which gave a red-violet colour in aqueous sodium hydroxide, possibly indicative of a hydroxynaphthaquinone.

Chimaphilin (2), $C_{12}H_{10}O_2$

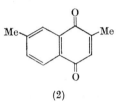

(2)

Occurrence: in the aerial parts of the following Pyrolaceae; *Chimaphila umbellata* Nutt.,[5, 24] *C. japonica* Miq.,[6b] and *C. maculata* Pursh.,[7] *Moneses grandiflora* S. F. Gray,[8] *Pyrola japonica* Klenze,[6b] *P. incarnata* Fisch. (rhizomes),[6a] *P. media* Sw.,[13] and other *Pyrola* spp.[6b]

Physical properties: yellow needles, m.p. 114°, $\lambda_{max.}$ (EtOH) 248, 254·5, 265(sh), 338 nm (log ϵ 4·19, 4·19, 4·02, 3·19), $\nu_{max.}$ (KBr) 1680(sh), 1662, 1622, 1600 cm^{-1}.

Chimaphilin was first isolated[5] from *C. umbellata* in 1860 by steam distillation, and subsequently examined by several investigators[7, 9] but its structure was not established until 1956.[10] The light absorption and the formation of a leucodiacetate suggested a 1,4-naphthaquinone structure, and the substitution pattern was revealed by (a) oxidative degradation with hot alkaline hydrogen peroxide to give 4-methyl-phthalic acid and with nitric acid to give trimellitic acid, and (b), Clemmensen reduction, followed by dehydrogenation of the resulting tetralin, which afforded 2,7-dimethylnaphthalene. Reoxidation of the latter[11] with chromic acid yielded 2,7-dimethyl-1,4-naphthaquinone identical with chimaphilin. Chimaphilin has been prepared recently[13] by cyclisation of 2-methyl-5-(γ,γ-dimethylallyl)-benzoquinone (4) which is probably analogous to the *in vivo* synthesis.[13, 14] Chimaphilin occurs with (4) and glucosides of (3) and (5) in *Pyrola media*.[13] Renifolin, the β-D-

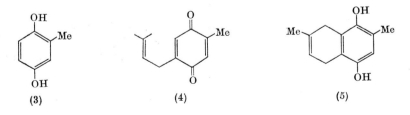

(3)   (4)   (5)

glucoside of (5), is also present with chimaphilin in *P. renifolia*[12] and *P. secunda*.[12] Unpublished work by Bohm[27] suggests that chimaphilin itself exists in *C. umbellata* in an "acid and β-glucosidase sensitive form" which may refer to renifolin or to a glucoside of chimaphilin quinol.

Deoxylapachol (DMK-1)† (6), $C_{15}H_{14}O_2$

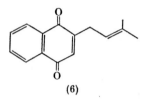

(6)

Occurrence: heartwood of *Tectona grandis* L. fil.[29] (Verbenaceae), *Tabebuia avellanedae* Lor. ex Griseb.[28] [syn. *T. ipé* (Mart. ex Schum.) Standl.] (Bignoniaceae) and an unidentified *Tabebuia* sp.[30]

Physical properties: yellow leaflets, m.p. 62°, λ_{max}. (EtOH) 246, 251, 276, 332 nm (log ϵ 4·41, 4·32, 3·54, 3·21), ν_{max}. (KBr) 1660, 1610 cm^{-1}.

This is one of a group of related quinones found in teak[31] (*T. grandis*) and "lapacho" wood (*T. avellanedae*),[28] and can lead to severe skin irritation and eczema among workmen handling such wood. Veneers made from "toxic" varieties of teak become covered with a surface layer of deoxylapachol after storage in a warm room.[29] In contrast to all the other quinones in teak and "lapacho" wood it has a free quinonoid position, and this structural feature may be associated with its physiological properties, and, in part, with the termite resistance of such woods, these activities arising, perhaps, from its ability to interfere with enzyme systems by reaction with thiol or amino groups at C-3. Lapachenole (7), another skin irritant, is also present in these timbers and may

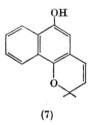

(7)

owe its activity to conversion into deoxylapachol; alternatively the latter may function by cyclisation to lapachenole. Lapachol (11) is a much less active skin irritant[120] than deoxylapachol but has roughly the same toxicity towards termites.

† Abbreviation proposed by IUPAC-IUB Commission on Biochemical Nomenclature.[395]

Deoxylapachol gives a positive Craven test,[32] forms a leucodiacetate and yields phthalic acid on oxidation with permanganate. This indicates a 1,4-naphthaquinone structure with a free quinonoid position, and since it also gives a positive Dam–Karrer test the remaining C_5H_9 fragment must be a substituted allyl side chain in which the double bond is situated $\beta\gamma$ to the quinone ring. The ultraviolet spectrum of deoxylapachol is, in fact, identical with that of 2-allyl-1,4-naphthaquinone. Cleavage of the double bond by oxidation of the leucodiacetate either with periodate-permanganate,[35] or by ozonolysis, gave acetone, and hence deoxylapachol must be (6).[29] This was confirmed synthetically by condensing 1,4-naphthaquinol with γ,γ-dimethylallyl alcohol in the presence of boron trifluoride, followed by oxidation with silver oxide.[29, 33]

2-Methyl-3-(γ,γ-dimethylallyl)-1,4-naphthaquinone (MK-1)† (8), $C_{16}H_{16}O_2$

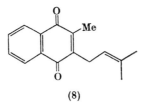

(8)

Occurrence: heartwood of *Tabebuia avellanedae* Lor. ex Griseb.[28] (Bignoniaceae).
Physical properties: yellow oil, $\lambda_{max.}$ (EtOH) 245, 249, 262, 273, 331 nm (logϵ 4·46, 4·44, 4·34, 4·31, 3·58), $\nu_{max.}$ (film) 1660, 1623, 1599 cm^{-1}.

This quinone, and the preceding one, are the simplest members of the menaquinone (vitamin K_2) series, all the others being bacterial metabolites. They are included at this point along with the various isopentylnaphthaquinones with which they occur but the more important members of the series are discussed at the end of this chapter. The synthetic compound was known[34] before its recent discovery in Nature and was easily recognised[28] by its n.m.r. spectrum. Steam distillation of "lapacho" wood (*T. avellanedae*) affords two artefacts, phthiocol (3-hydroxy-2-methyl-1,4-naphthaquinone) and 2,3-dimethyl-1,4-naphthaquinone, neither of which could be detected in the wood itself. Both were probably derived from MK-1, the former via autoxidation on the allylic methylene group (phthiocol is a common oxidative artefact of K vitamins[447]) and the latter, possibly by migration of the side chain double bond into the $\alpha\beta$-position in conjugation with the ring, followed by hydration to give (9) and a retro-aldol reaction.

† Abbreviation proposed by IUPAC-IUB Commission on Biochemical Nomenclature.[395]

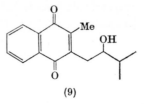

(9)

2,3,6-Trimethyl-1,4-naphthaquinone has been isolated from tobacco smoke but its origin is unknown.[527]

Lawsone (henna, naphthalenic acid, isojuglone) (10), $C_{10}H_6O_3$
Lawsone methyl ether, $C_{11}H_8O_2$

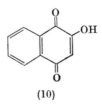

(10)

Occurrence: (10) leaves of *Lawsonia alba* (Lam.)[15] (= *L. inermis*) and *L. spinosa* L.[19] (Lythraceae), *Impatiens balsamina* L., *I. capensis* Meerb., *I. pallida* Nutt. and other *Impatiens* spp.[16,17,18] (Balsaminaceae); the methyl ether occurs in aerial parts of *Impatiens* spp.[16,25]
Physical properties: (10) yellow needles, m.p. 192° (dec.), λ_{max}. (EtOH) 242·5, 248, 274, 334 nm (log ϵ 4·17, 4·21, 4·14, 3·40), ν_{max}. (KBr) 3150, 1674, 1640 cm^{-1}; methyl ether, yellow needles, m.p. 183·5°, λ_{max}. (EtOH) 241·5, 247, 276, 333 nm (log ϵ 4·23, 4·24, 4·17, 4·45), ν_{max}. (KBr) 1658, 1645, 1602 cm^{-1}.

Lawsone is the dyeing principle of the ancient colouring matter, henna, prepared from the leaves of *L. alba*, which is cultivated in Africa and India for medicinal[149] and dyeing[19] purposes. The plant has many pseudonyms and is probably identical with camphire mentioned in the Song of Solomon.[20] The pigment is readily extracted[15] from the leaves with aqueous sodium carbonate, and is used as a cosmetic in African and Eastern countries, not only for dyeing hair[91] (Mohammed is said to have dyed his beard with henna) but also for staining the hands and feet. Henna can be found on the nails of Egyptian mummies and must be one of the oldest cosmetics still in use.

The structure of lawsone is established[15,22] by the formation of an acetate and a leucotriacetate, and by oxidation to phthalic acid. It is conveniently prepared by hydrolysis and oxidation of the leucotriacetate, obtained by Thiele acetylation of 1,4-naphthaquinone,[23] or by autoxidation of α-tetralone, 1,2- or 1,3-dihydroxynaphthalene in a basic medium.[205] The occurrence of the 1,3-diol has been reported[19] in *Lawsonia* leaves but requires authentication.

The presence of lawsone[18] and its methyl ether[25] in the flowers of the garden balsam (*I. balsamina*) is unusual. In the leaves it occurs, at least in part, in a colourless reduced form.[16, 17] Ruwet[533] isolated a fraction which appeared to contain a naphthol glucoside since on treatment with a glycosidase (extracted from the cotyledons), in the presence of air, it was converted into lawsone and glucose was liberated.

Lapachol (lapachic acid, taiguic acid, greenhartin, tecomin) (11), $C_{15}H_{14}O_3$
Lapachol methyl ether, $C_{16}H_{16}O_3$

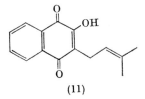

(11)

Occurrence: (11) heartwood of several genera of the Bignoniaceae[41] (especially the tribe *Tecomeae*); *Tabebuia flavescens* Benth. & Hook. f. ex Griseb.,[36] *T. guayacan* Hemsl.,[37] *T. avellanedae* Lor. ex Griseb.,[28] *Paratecoma peroba* (Record) Kuhlm.,[39] *Tecoma araliaceae* DC.[40] and several other *Tecoma* spp.,[41, 45] *Stereospermum suaveolens* DC.;[38] also in the heartwood of *Tectona grandis* L. fil.[38] (Verbenaceae) and *Avicennia tomentosa* Jacq.[182] (Verbenaceae), and in the root of *Conospermum teretifolium* R. Br.[564] (Proteaceae); the methyl ether occurs in the heartwood of *Tabebuia avellanedae* Lor. ex Griseb.[28] (Bignoniaceae).

Physical properties: (11) yellow prisms or plates, m.p. 140°, $\lambda_{max.}$ (EtOH) 251, 278, 333 nm (logϵ 4·23, 4·08, 3·34), $\nu_{max.}$ (KBr) 3322, 1660, 1637, 1591 cm^{-1}; methyl ether, yellow crystals, m.p. 54–55°, $\lambda_{max.}$ (EtOH) 252, 278, 333, 381(sh) nm (logϵ 4·41, 4·32, 3·67, 3·11), $\nu_{max.}$ (KBr) 1660, 1645, 1598 cm^{-1}.

Lapachol has been known since 1858,[64] and is the most abundant quinone in the wood of various Bignoniaceae found mainly in tropical America, for example 7·64% in *Tecoma araliaceae*,[40] and is sometimes visible in yellow deposits.[43] A number of early records are probably chemically sound but insecure botanically, owing to the unfortunate practice of adopting names used in the timber trade without proper botanical identification, but some are definitely wrong (see ref. 41). *Bignonia tecoma*, and no doubt other species, was used as a dyewood in Brazil, a crude extract for dyeing cotton, known as tecomin, being obtained by heating sawdust and shavings with lime and water.[44] At the present time the anti-tumour properties of lapachol are of considerable interest and the compound is undergoing clinical trials.[418]

Our knowledge of the chemistry of lapachol and related compounds is largely due to a long series of careful researches carried out by Hooker and his co-workers during the periods 1889–96 and 1915–35.[46] Prior to these investigations Paterno[47] had shown that his "lapachic acid" was probably identical with Arnaudon's "taiguic acid"[64] and Stein's "greenhartin",[65] and had proposed structure (12) for lapachol since it

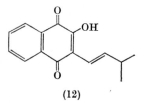

(12)

could be reversibly reduced to a colourless compound, possessed an acidic hydroxyl group, was oxidised to phthalic acid by nitric acid, and on zinc dust distillation yielded, *inter alia*, naphthalene and a gaseous hydrocarbon which absorbed bromine to give isobutylene dibromide. This structure was adopted initially by Hooker[48] who prepared addition products with bromine and with hydrochloric acid, and noted the resemblance of lapachol to 2-hydroxy-1,4-naphthaquinone ("naphthalenic acid"), but subsequently[49, 56] he condensed the latter with isovaleraldehyde and obtained (12) (isolapachol) which proved to be isomeric with the natural quinone. Lapachol was therefore regarded as (11) consistent with the isolation of acetone, later,[50] by degradation with alkaline hydrogen peroxide.

2-Hydroxy-1,4-naphthaquinone is tautomeric with 4-hydroxy-1,2-naphthaquinone and ethers of both forms can be obtained by reaction of the silver salt with alkyl halides. The ambident anion can also react on carbon (C-3) if an allyl or benzyl halide is used,[51] and this was exploited by Fieser[52] in his synthesis of lapachol. Treatment of a suspension of the silver salt of 2-hydroxy-1,4-naphthaquinone in ether at 0° (or better, the potassium salt in hot acetone[53]) with γ,γ-dimethylallyl bromide gave a little 3-(γ,γ-dimethylallyl)-2-hydroxy-1,4-naphthaquinone (11), identical with lapachol, and some of the normal ether (13). The latter rearranges to the C-allyl isomer (14) in refluxing ethanol and at higher temperatures (boiling xylene) to both (14) and (15) (isodunniol).[53] The best yield of lapachol is gained by alkylation of 2-methoxy-1,4-dihydroxynaphthalene with 2-methylbut-3-en-2-ol in the presence of boron trifluoride, followed by oxidation with silver oxide and alkaline hydrolysis.[193] Other syntheses have been described.[54, 55]

The dominant feature of the chemistry of lapachol and related compounds is the ease with which the side chain cyclises onto an oxygen function to give a variety of pyrano- and furano-naphthaquinones. Some

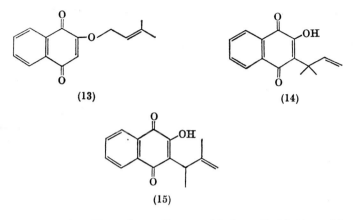

(13)

(14)

(15)

of these occur naturally and are discussed below. Oxidation of lapachol with lead dioxide in acetic acid gives a neutral orange-yellow compound, $C_{30}H_{26}O_6$ ("lapachol peroxide"),[59] evidently a dehydro-dimer. Structure (17) has been suggested[60] which is consistent with most of the available evidence and its formation can be regarded as a C—O coupling of the radical (16). The compound is remarkably sensitive to base and reverts to lapachol on warming in alkaline solution.

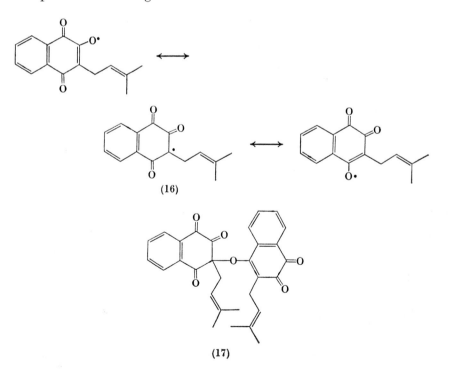

(16)

(17)

α-*Lapachone* (18), $C_{15}H_{14}O_3$
β-*Lapachone* (19), $C_{15}H_{14}O_3$

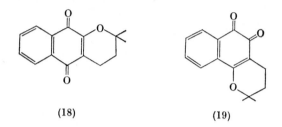

(18)   (19)

Occurrence: heartwood of *Tabebuia avellanedae* Lor. ex Griseb.[28, 62] (Bignoniaceae).

Physical properties: (18) yellow needles, m.p. 119°, λ_{max}. (EtOH) 251, 282, 332, 375(infl.) nm (log ε 4·45, 4·22, 3·44, 3·15), ν_{max}. (KBr) 1678, 1640, 1610, 1595 cm⁻¹; (19) orange needles, m.p. 155–156°, λ_{max}. (EtOH) 256, 282, 330, 431 nm (log ε 4·44, 4·01, 3·28, 3·26), ν_{max}. (KBr) 1690, 1640, 1632, 1598 cm⁻¹.

Paternò[47] first obtained these two isomers in 1882 by acid treatment of lapachol, and subsequently Hooker[48, 49] determined their structure and the precise conditions necessary for their formation. They have recently been found,[28] with lapachol, in a *Tabebuia* species and there is no doubt that they exist as such in the wood and are not artefacts. The lapachones, which are isomeric with lapachol, show the normal properties of α- and β-naphthaquinones; the β-isomer is orange-red, soluble in sodium bisulphite solution, involatile in steam, and forms a quinoxaline derivative, whereas the α-quinone is yellow, steam-volatile and does not condense with *o*-phenylene-diamines. The α-isomer has a lower redox potential than the β-quinone, and both show characteristic ultraviolet-visible and infrared absorption (see above). All these properties are readily explained if α- and β-lapachone have structures (18) and (19), respectively, which are consistent with their n.m.r. spectra and the interconversions described below.

If lapachol is dissolved in concentrated sulphuric acid and then poured into water, β-lapachone is precipitated, but when lapachol is treated with hydrochloric acid (in acetic acid) α-lapachone is formed. Both isomers are produced when lapachol is dissolved in concentrated nitric acid. Similarly, concentrated sulphuric acid converts α-lapachone into the β-isomer and the process is reversed in concentrated hydrochloric acid. All these reactions are virtually quantitative. In the conversion of β-lapachone into α-lapachone, chlorohydrolapachol (21) can be isolated under appropriate conditions showing that the isomerisation proceeds by the opening and closing of the heterocyclic ring. On further heating with hydrochloric acid, (21) is transformed into α-lapachone but

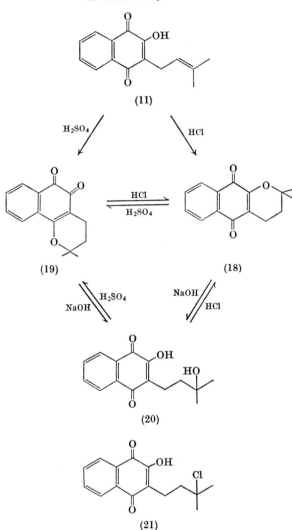

(11)

H₂SO₄   HCl

HCl
H₂SO₄

(19)   (18)

H₂SO₄   NaOH
NaOH   HCl

(20)

(21)

concentrated sulphuric acid reconverts it into the β-isomer. The di-hydropyran ring of both lapachones can be opened by heating with dilute sodium hydroxide (hydrolysis of a vinylogous ester) to give hydroxyhydrolapachol (20) which can be recyclised to either (18) or (19) with the appropriate acid.

These isomerisations, which seemed to depend upon the specific acid employed, were found later by Ettlinger[63] to be related to the basicities of the two lapachones. β-Lapachone is much more basic than the α-isomer (by ~ two pK_a units), and consequently in concentrated sulphuric acid the equilibrium favours the β-cation (23) and β-lapachone

(19) is liberated on rapid dilution with water. However, in concentrated hydrochloric acid ionisation is insufficient to shift the equilibrium from the side of the unchanged quinone, and the more stable *p*-quinone (18) separates on gradual addition of water, the equilibrium being continuously displaced. The product obtained is not dependent on the acid employed. If either lapachone is treated with aqueous sulphuric acid

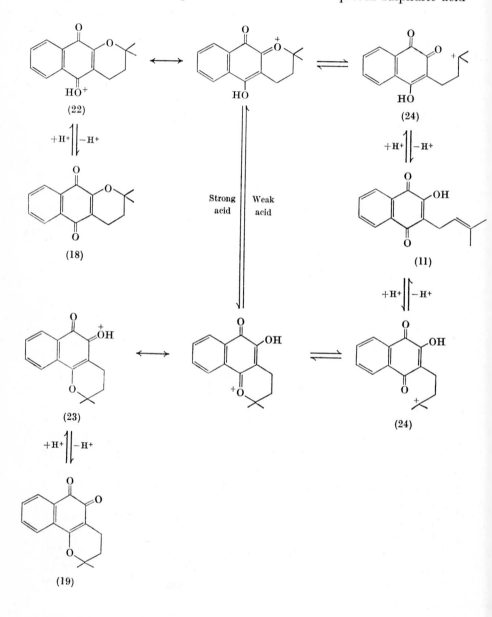

(<62%) and then rapidly diluted, α-lapachone is formed, but if the acid concentration is >75%, β-lapachone results; with 67% sulphuric acid a mixture (~1:1) of both isomers is obtained. Aluminium chloride converts lapachol into β-lapachone exclusively.[45]

Isomeric changes of the type discussed are not confined to lapachol and apply generally to 3-alkenyl-2-hydroxy-1,4-naphthaquinones in which the double bond is situated αβ or βγ to the quinone ring, although the rearrangements are not usually so clear-cut as in the lapachol-lapachone examples. The formation and rearrangement of cyclic ethers proceeds most easily when a tertiary carbonium ion (e.g. 24) is involved.

Dehydro-α-lapachone (xyloidone) (25), $C_{15}H_{12}O_3$

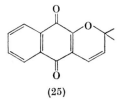

(25)

Occurrence: heartwood of *T. avellanedae* Lr. ex Griseb.[28, 62, 206](Bignoniaceae).
Physical properties: orange needles, m.p. 143°, $\lambda_{max.}$ (EtOH) 267, 276(sh), 333, 434 nm (log ε 4·35, 4·17, 3·40, 3·21), $\nu_{max.}$ (KBr) 1681, 1640, 1590 cm^{-1}.

Although one of the products obtained in the early investigations[47, 67] on lapachol, the structure of this compound was only established[66] recently following its isolation[28] from lapacho wood. As lapachol appears to undergo partial conversion to dehydro-α-lapachone on prolonged storage,[61] or on exposure to light and air for a few days on a layer of silica,[66] the possibility that the "natural" material is an artefact cannot be excluded. The transformation of lapachol into dehydro-α-lapachone is more effective in boiling pyridine.[206c]

Lapachol undergoes another cyclisation reaction with acetic anhydride in the presence of sodium acetate† to give a mixture[48] of two colourless isomeric diacetates, *both* of which give dehydro-α-lapachone after hydrolysis and oxidation. The two diacetates take up one mol. of hydrogen, catalytically, to give the leucodiacetates of α-(26) and β-lapachone (27), and must therefore have structures (28) and (29), respectively, that is, the leucodiacetates of dehydro-α- and β-lapachone; thus the quinone derived from (28) and (29) could be either (25) or (30) but the evidence was conflicting.[61, 68, 69]

The problem was resolved by synthesis.[66] The α-isomer (25) was obtained by dehydrogenation of α-lapachone with D.D.Q. and also by

† The normal acetate is obtained using zinc chloride as catalyst.[69]

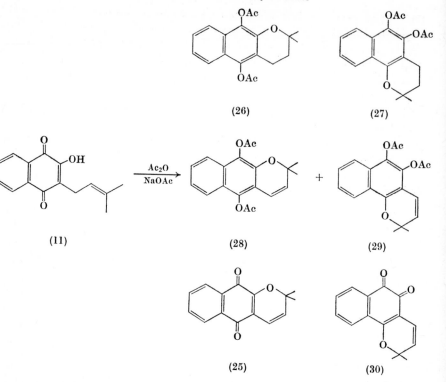

(26)　　　　　　　　　　　　(27)

(11)　　　　　　　(28)　　　　　　　(29)

(25)　　　　　　　　(30)

converting the diacetate (28) into its quinol using a Grignard reagent, followed by oxidation. The β-isomer (30),† a dark red compound which looks purple in the mass, was obtained in similar fashion but is extremely prone to rearrange to the α-isomer. Both (25) and (30) are formed

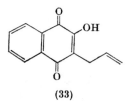

(33)

when isolapachol (12) is oxidised with 2,3-dichloro-5,6-dicyanobenzo-quinone;[557] the reaction can be reversed by Birch reduction, lapachol and hydrolapachol also being formed, according to the conditions.[557] Dehydro-β-lapachone is easily distinguished from the α-isomer by its ultraviolet-visible absorption (λ_{max}. (EtOH) 265, 281(sh), 334, 484), and

† The compound, m.p. 152°, isolated from teak, cannot be dehydro-β-lapachone as suggested.[31]

an infrared band at 1697 cm⁻¹ characteristic of β-naphthaquinones, and is much less stable. Comparison of the ultraviolet-visible spectra of both α- and β-lapachones with the corresponding dehydro compounds shows the marked bathochromic shift which results when a double bond is introduced into an alkyl side chain in conjugation with the quinone chromophore.

The preparation of the acetates (28) and (29) from lapachol was the first example of the formation of a chromenol by base-catalysed cyclisation of an allylquinone. The reaction proceeds by normal acetylation, apparently giving both isomers (31) and (32), followed by proton abstraction from the activated methylene group and rearrangement of the dienone system as indicated. 3-Allyl-2-hydroxy-1,4-naphthaquinone (33) also gives a similar pair of isomeric diacetates when treated with acetic anhydride and sodium acetate and fails to form a normal acetate.[61, 70]

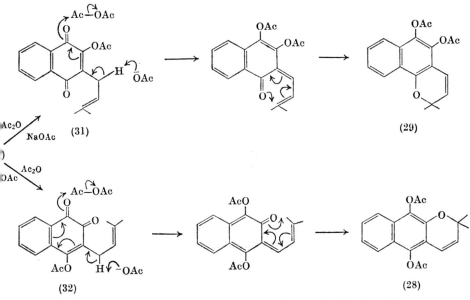

In more recent examples[71, 72] of chromenol formation hydroxyl groups have been absent and only one product was obtained.

Lomatiol (34), $C_{15}H_{14}O_4$

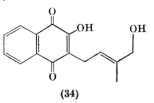

(34)

Occurrence: seeds of *Lomatia ilicifolia* R. Br.,[73] *L. longifolia* R. Br.,[73] *L. silaifolia* R. Br.[75] and other *Lomatia* spp.[75] (Proteaceae); heartwood of *Paratecoma peroba* (Record) Kuhlm.[74] (Bignoniaceae).

Physical properties: yellow needles, m.p. 127°, λ_{max}. (EtOH) 251, 280, 333 nm (log ε 4·35, 4·21, 3·48), ν_{max}. (KBr) 3443, 1668, 1635, 1590 cm^{-1}.

Whereas lomatiol occurs with lapachol and other related quinones in *Paratecoma peroba*, and no doubt in other Bignoniaceae, it is recorded as the sole colouring matter in the yellow outer layer of the seeds of various Australian *Lomatia*† (12% of the dry weight in *L. longifolia*[75]) although apparently not present in three Chilean spp.[75] Lomatiol may also occur in the bark of *L. ilicifolia* which has been used as a dyewood.

In an early paper Rennie[73] showed that lomatiol was a hydroxy-lapachol since it formed a diacetate and could be converted by treatment with sulphuric acid into the known hydroxy-β-lapachone (42). As phthalic acid can be obtained by oxidation of lomatiol, the second hydroxyl group must be located in the side chain. It proved to be allylic which complicates the chemistry of lomatiol, relative to lapachol, and Hooker's initial structural proposal[81] was later abandoned.[75] Hydrogenation over platinum gave (after reoxidation) both hydrolomatiol and hydrolapachol which confirmed that lomatiol had a lapachol carbon skeleton and a double bond in the side chain. Hydrogenolysis of a hydroxyl group suggests that it occupies an allylic or benzylic position, and its location on an ω-carbon atom was deduced by Hooker[75] as follows. Of the four possible structures (a–d) for the side chain of hydrolomatiol, (b) and (c) could be eliminated as the relevant compounds

(a) —CHCH₂CHMe₂   (b) —CH₂CHCHMe₂   (c) —CH₂CH₂CMe₂   (d) —CH₂CH₂CHMe
$\quad\ \ $|$\qquad\qquad\qquad\qquad$|$\qquad\qquad\qquad\qquad$|$\qquad\qquad\qquad\qquad$|
$\quad\ \ OH\qquad\qquad\qquad\ \ OH\qquad\qquad\qquad\ \ OH\qquad\qquad\qquad$CH₂OH

were known and were different, and an isomer having structure (a) should be convertible into isolapachol (12) on treatment with acid which was not the case. This leaves structure (d) and the double bond in lomatiol must occupy the βγ position shown in (34) as no other structure could account for the colour of the compound, its behaviour on hydrogenation and the rearrangement reactions described below. Additional support was provided[76] by a permanganate oxidation of lomatiol which gave the deep orange quinone (47) (norlomatiol), a

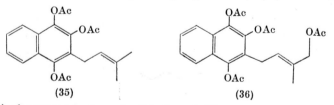

(35)$\qquad\qquad\qquad\qquad\qquad\qquad$(36)

† Juglone is also present in the seeds of *L. tinctoria*.[355]

remarkable example of the Hooker oxidation in which both the double bond and the allylic hydroxyl group are preserved. A synthesis of

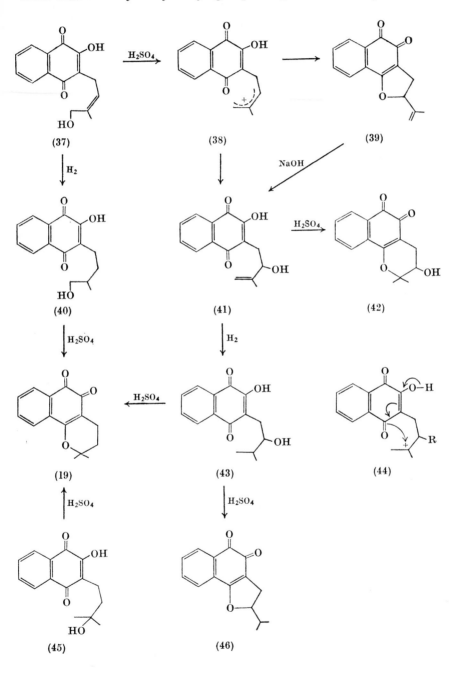

lomatiol has been effected[77] by allylic oxidation of leucolapachol triacetate (35) with a limited amount of selenium dioxide in acetic anhydride, followed by hydrolysis and oxidation of the crude tetraacetate (36).

Some of the rearrangement reactions of lomatiol and its derivatives are set out on p. 213.[75] If the quinone is dissolved in sulphuric acid for a short time and then diluted with water, it can be converted, by judicious selection of conditions, into either dehydro-iso-β-lapachone (39) or hydroxy-β-lapachone (42). In the former case, the reaction proceeds by cyclisation of an oxygen function onto an allylic carbonium ion (38) and in the latter, the isomeric allyl alcohol (41) is evidently formed followed by cyclisation onto the tertiary carbonium ion (44; R = OH). The structure of (39) is supported by hydrogenation to give iso-β-lapachone (46) and by the formation of isolomatiol (41) on alkaline hydrolysis. At the hydrolomatiol level all three quinones (40), (43) and (45) cyclise in sulphuric acid to form β-lapachone (19) via the tertiary carbonium ion (44; R = H), hydro-isolomatiol (43) giving, in addition, some iso-β-lapachone (46) by cyclisation of the initial secondary carbonium ion.

Norlomatiol (47), the Hooker oxidation product of lomatiol, undergoes cyclisation on heating in dilute hydrochloric acid forming the red pyrano-α-naphthaquinone (48) which can be transformed, by dissolution in alkali, into the isomeric violet β-quinone (49), similar to dehydro-β-

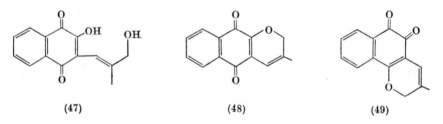

| (47) | (48) | (49) |

lapachone but somewhat more stable.[78] In both acid and alkaline solutions it tends to revert to the red α-quinone (48) but redox reactions, which are not clearly understood, also take place.[78]

Dunnione (50), $C_{15}H_{14}O_3$

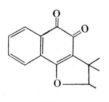

(50)

Occurrence: on the leaves, stems and flowers of *Streptocarpus dunnii* Mast.[79] and on the leaves of *S. pole-evansii* Verd.[80] (Gesneraceae).

Physical properties: orange-red needles, m.p. 98–99°, $[\alpha]_D^{18}$ +310° (CHCl$_3$), λ_{max}. (EtOH) 261, 270(sh), 347(sh), 445 nm (log ε 4·47, 4·41, 3·08, 3·36), ν_{max}. (KCl) 1698, 1638, 1611, 1590, 1570 cm^{-1}.

The leaves of *Streptocarpus dunnii* are large (approximately 60 × 45 cm) and by extraction with cold benzene < 2 g of pigment per leaf can be obtained.[79, 82] It shows the typical ultraviolet-visible and infrared absorption of a β-naphthaquinone and the usual chemical properties, namely reversible reduction, solubility in aqueous sodium bisulphite and formation of a leucodiacetate and a quinoxaline. Dunnione is isomeric with lapachol and there are obvious similarities in the chemistry of the two compounds.[50, 82] Thus dunnione rearranges in hot hydrochloric acid to a yellow isomer, α-dunnione [λ_{max}. (EtOH) 253, 289, 336 nm, ν_{max}. (KCl) 1675 cm^{-1}], and if this is dissolved in alkali and then acidified, dunnione is regenerated. These reactions are reminiscent of β-lapachone and the behaviour of the latter in cold alkaline solution is strikingly similar to dunnione. This led Price and Robinson[50] to conclude that the third oxygen atom in dunnione was present in a dihydrofuran or dihydropyran ring, that is (a) or (b) below. The preferred structure (a)

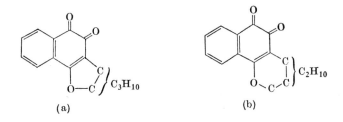

(a)   (b)

(50) was based on the results of several oxidative degradations which yielded phthalic acid, methyl isopropyl ketone and ca. 1·3 mol. of acetic acid. This is confirmed by the n.m.r. spectrum[209] which shows two singlets at τ 8·52 and 8·72 (>CMe$_2$), the third methyl group appearing as a doublet at τ 8·55 coupled to a quartet (1H) centred at τ 5·33 (J, 3 Hz). It will be seen that the isoprenoid chain is attached at the tertiary carbon to the quinone ring, which is comparatively rare (a similar example is found in atrovenetin[85]). The structure was established by Cooke's synthesis[83] of (±)-dunnione effected by cyclisation of the lapachol isomer (14) in cold sulphuric acid. It follows that α-dunnione is the isomer (51).

In hot sulphuric acid solution α-dunnione,[50] and to a much lesser extent dunnione[82] itself, rearranges to another orange-red isomer,

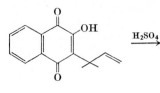

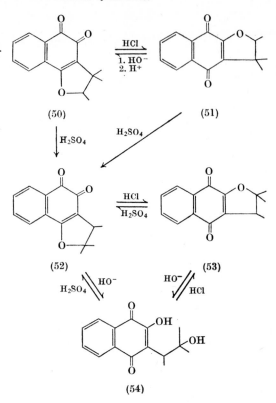

β-isodunnione (52), which in turn can be converted into another yellow isomer, α-isodunnione (53). This last isomerisation can be reversed in concentrated sulphuric acid. The isodunniones closely resemble the lapachones, and from their properties and easy interconversion they are obviously isomeric o- and p-quinones. The heterocyclic rings can be opened in alkaline solution, and by careful neutralisation the intermediate compound, hydroxyhydro-isodunniol (54) may be isolated; recyclisation to α- or β-isodunnione is easily brought about by appropriate treatment with strong acids. (The corresponding intermediate, hydroxyhydrodunniol, the product expected on ring opening of either dunnione or α-dunnione, could not be isolated because of its sensitivity to alkali which leads to yet another rearrangement product, allodunnione, and to acids, which bring about rapid ring closure.) The structures of the isodunniones were deduced[82] from the observation that oxidation of β-isodunnione with chromic acid or alkaline hydrogen peroxide yielded acetone and were confirmed later by synthesis.[58] It is clear that the isodunniones are derived from dunnione and α-dunnione by the carbonium ion rearrangements (51 → 52) shown below.

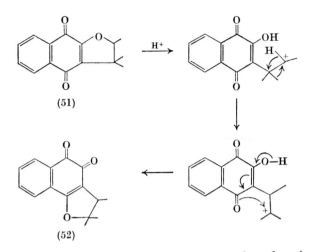

(51)

(52)

Allodunnione is a yellow isomer which arises when dunnione is kept in cold dilute alkali or, more rapidly, on boiling, the initial red solution changing to yellow in a few minutes.[50] Its colour in alkaline solution indicates that it is not a naphthaquinone. The formation of acetaldehyde, acetone, and acetic acid (1·23 mol.) in oxidation reactions, implied that the isoprenoid chain was still intact, and the solubility of this neutral compound in warm aqueous sodium hydroxide (from which it could be recovered on acidification) suggested the presence of a dihydrofuran, dihydropyran or lactone ring. On this basis Price and Robinson[50] proposed structure (55) for allodunnione and tentatively suggested the following reaction sequence. It was realised later that ring contraction

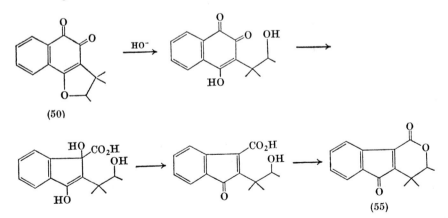

(50)

(55)

of 3-alkyl-2-hydroxy-1,4-naphthaquinones to indenone carboxylic acids is a general rearrangement[84, 86, 87, 88] which is particularly easy if the side chain is attached to the ring by a tertiary carbon atom.[88, 89] The reaction was utilised to provide synthetic confirmation of the structure of allodunnione. Rearrangement of the quinone (14) in boiling alkaline solution gave the indenone acid (56) which was cyclised in cold

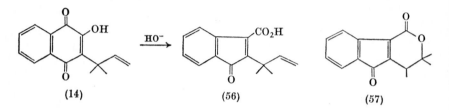

(14)   (56)   (57)

hydrochloric acid to give (±)-allodunnione (55).[84] The same lactone was also obtained on brief treatment of the acid (56) with concentrated sulphuric acid but on longer contact the isomer (57) appeared, resulting from carbonium ion rearrangement.

Oxidation of dunnione or allodunnione gives acetaldehyde and a compound, $C_{12}H_{12}O_4$, originally formulated[50] as a lactonic acid. Reinvestigation[90] later showed the absence of lactone carbonyl absorption in the infrared and the compound was identified as the known acid (60).

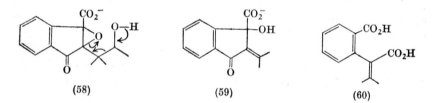

(58)   (59)   (60)

The reaction may proceed by way of (58), (59) and a 1,3-indandione, but other pathways are possible.[90]

Ring contraction of a 2-hydroxy-1,4-naphthaquinone as in the formation of allodunnione, occurs during the early stages of the Hooker oxidation.[59, 92, 93] In this reaction, originally carried out with cold alkaline permanganate, a side chain methylene group adjacent to the quinone ring is eliminated and the positions of the quinonoid substituents are interchanged (61 → 62).[94] The intermediate dihydroxy-indanone acid (64), which can be isolated, is thought[89, 95, 121] to result from a benzilic acid rearrangement of (63) and is oxidised further in the tautomeric ketol form to a triketo-acid (65) and this, by aldol condensation, is cyclised to (66) so leading to the final product by decarboxylation and further oxidation. In an improved procedure[95a] the reaction is carried out in two stages, oxidation to the indanone acid being effected with hydrogen peroxide and sodium carbonate in aqueous dioxan under

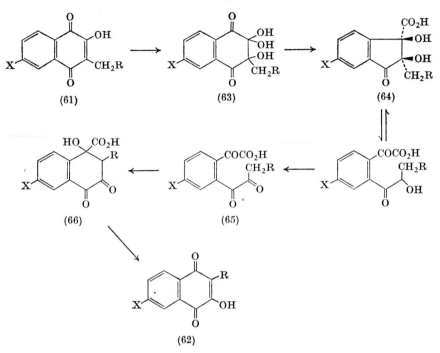

(61) → (63) → (64)

(66) ← (65) ←

(62)

nitrogen, and the further oxidation of the ketol is effected with copper sulphate in alkaline solution. Hydrates analogous to (63) are known[121] and being non-planar, they assist the facile ring contraction of 2-hydroxy-1,4-naphthaquinones with bulky alkyl substituents at C-3 by relief of steric strain.[89]

Dehydro-iso-α-lapachone (67), $C_{15}H_{12}O_3$

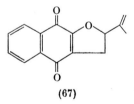

(67)

Occurrence: heartwood of *Paratecoma peroba* (Record) Kuhlm.[490] (Bignoniaceae).
Physical properties: golden-yellow leaflets, m.p. 104–105·5°, $[\alpha]^{21}_{589}$ −31·61° (C_6H_6), $\lambda_{max.}$ (EtOH) 253, 295, 342 nm (logϵ 4·45, 3·86, 3·18), $\nu_{max.}$ (KBr) 1667 cm^{-1}.

Sandermann and co-workers[490] have recently found four more quinones in the wood of *Paratecoma peroba* in addition to those already known. They all belong to the lapachol group and in three of them the side chain has been degraded to four carbon atoms. Full information is not yet available but in all cases fragmentation under electron impact leads to the ions m/e 104 and 76 showing that the benzenoid ring is unsubstituted. Structure (67) is consistent with the n.m.r. spectrum. It is isomeric with dehydro-iso-β-lapachone (39) derived from lomatiol (p. 213) and interconversion of these two compounds should be possible.

"α-Ethylfurano-1,4-naphthaquinone" (68), $C_{14}H_{10}O_3$

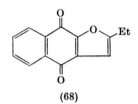

(68)

Occurrence: heartwood of *Paratecoma peroba* (Record) Kuhlm.[490] (Bignoniaceae).
Physical properties: lemon-yellow needles, m.p. 143°, $\lambda_{max.}$ (EtOH) 250, 294, 388 nm (log ϵ 4·68, 3·97, 3·59), $\nu_{max.}$ (KBr) 1667 cm⁻¹.

The structure of this quinone has been deduced from its spectral properties; the presence of the ethyl group and an isolated aromatic proton was revealed by the n.m.r. spectrum, and the ultraviolet-visible absorption is very similar to that of the known α-isopropyl analogue. It has been synthesised by addition of bromine to (69) to give a di-

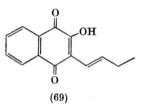

(69)

bromide which cyclises to (68) with loss of hydrogen bromide on keeping in ethanol.[490]

"β-Methylpyrano-1,4-naphthaquinone" (48), $C_{14}H_{10}O_3$

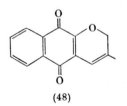

(48)

Occurrence: heartwood of *Paratecoma peroba* (Record) Kuhlm.[490] (Bignoniaceae). *Physical properties*: red needles, m.p. 196·5–197·5° (192–194°), $\lambda_{max.}$ (EtOH) 260, 285, 333, 460 nm (log ϵ 4·25, 4·27, 3·51, 2·99), $\nu_{max.}$ (KBr) 1670, 1645 cm⁻¹.

This pyranonaphthaquinone was recognised[490] from its spectroscopic properties; the presence of a methylallyl system could be deduced from the n.m.r. spectrum while the ultraviolet-visible absorption is similar to that of dehydro-α-lapachone. The pigment was found to be identical[491] with the red quinone Hooker and Steyermark[76] prepared many years ago by cyclisation of norlomatiol in hydrochloric acid (see p. 214).

The fourth quinone in this group from *P. peroba*, $C_{14}H_{10}O_4$, brown-yellow needles, m.p. 183–186°, $[\alpha]_D$ +45·42° (pyridine), has not yet been identified.†

Juglone (nucin, regianin) (70), $C_{10}H_6O_3$

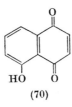

(70)

Occurrence: green parts of *Juglansr egia* Linn.,[42,96,97] *J. nigra* Linn.,[97] *J. cinerea* Linn.,[97] *J. sieboldiana* Max.,[42,98] and *J. mandshurica* Max.,[99] *Carya olivaeformis* Nutt.[97] and *Pterocarya caucasica* C.A.,[97] bark of *C. alba* Nutt.[26] and *C. ovata* (Mill.) Koch.[538] (all Juglandaceae).

Physical properties: orange needles, m.p. 164–165°,‡ $\lambda_{max.}$ (EtOH) 249, 345(sh), 422 nm (log ϵ 4·09, 3·08, 3·56), $\nu_{max.}$ (KBr) 1666, 1645, 1603 cm⁻¹.

Juglone is most conveniently isolated from green walnut shells but it seems to occur, in a reduced form, in the green parts of the tree generally.[101] The structure was indicated by the formation of an acetate and by degradation to 3-hydroxyphthalic acid and naphthalene, with appropriate reagents, and was confirmed synthetically by oxidation of 1,5-dihydroxynaphthalene with chromic acid.[104] Oxidation of 1,5- or 1,8-dihydroxynaphthalene with Fremy's salt is more efficient.[106] A curious reaction is the formation of juglone 4-oxime by diazotizing 8-nitro-1-naphthylamine followed by boiling with sulphuric acid.[116] Like similar *peri*-hydroxynaphthaquinones, juglone forms intense violet complexes with copper(II) and nickel(II) salts, and has been suggested as a spot reagent for nickel.[107]

In *J. regia*, juglone occurs[101–103, 109] as the 4-β-D-glucoside of 1,4,5-trihydroxynaphthalene (α-hydrojuglone). Methylation and hydrolysis

† The structure suggested in ref. 490 has been withdrawn.[491]
‡ In Pyrex glass,[100] in soda glass it has m.p. 153–154° (dec.). A number of quinones have higher and sharper m.p.s in Pyrex.

gave a dimethoxynaphthol identical with that obtained by catalytic reduction of juglone acetate, followed by methylation of (71) with diazomethane, and hydrolysis. This was reasonably assumed[103] to be (72) but in fact it is (73) as the acetyl group in (71) migrates to the

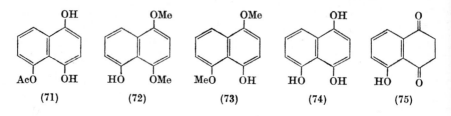

(71)   (72)   (73)   (74)   (75)

adjacent *peri*-hydroxyl group during the methylation.[109] Similar migrations have been observed in *o*-dihydroxyanthraquinones.[110] It was thought,[111] at an early stage, that juglone existed *in vivo* as a mixture of α- (74) and β- (75) -hydrojuglones. Both tautomers are present when juglone (and certain derivatives[113]) is reduced in warm acid solution and they can be separated.[112, 115] β-Hydrojuglone can also be obtained, more directly, by hydrogenation of juglone using a homogeneous catalyst.[105] The addition reactions of juglone have been studied in some detail.[114]

Finely powdered juglone is an effective sternutator and like many other quinones has weak fungicidal and bactericidal properties. It stains the skin a dark yellow-brown and will dye wool; the husks and inner bark of the walnut tree appear to have been used for this latter purpose whilst its poisonous character has been known for centuries. Juglone is generally regarded as the toxic principle and this property has been exploited in various ways from the catching of fish to the treatment of ringworm. Pliny recorded that walnut trees were poisonous to surrounding plants and it has usually been assumed that the toxic agent is transferred to the roots of, for example, tomatoes when in direct contact with those of the tree.[539] However, recent work by Bode[119] indicates that the wilting agent is a secretion from the leaves, rather than the roots, which is carried down by rain. Juglone was identified in the run-off.

Diosquinone (76), $C_{10}H_6O_3$

(76)

Occurrence: bark of *Diospyros tricolor* Hiern[492] (Ebenaceae).
Physical properties: orange-red needles, m.p. 198–199°, $\lambda_{max.}$ (EtOH) 250, 325, 430 nm (log ϵ 4·30, 3·62, 3·57), $\nu_{max.}$ (KBr) 1698, 1642, 1610 cm^{-1}.

Diosquinone yields naphthalene on zinc dust distillation and shows the chemical and spectroscopic properties of a β-naphthaquinone.[492, 493] It contains one active hydrogen atom, and its colour reactions with sodium hydroxide, copper(II) and nickel(II) acetate, and boroacetic anhydride suggest the presence of a *peri*-hydroxyl group. Diosquinone therefore appears to have structure (76) and in the absence of a C-methyl group it differs from all the other *Diospyros* quinones. Attempts to synthesise this simple quinone have so far failed.[494]

8-Methoxy-3-methyl-1,2-naphthaquinone (77), $C_{12}H_{10}O_3$

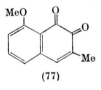

(77)

Occurrence: heartwood of *Diospyros melanoxylon* Roxb.[497] (Ebenaceae).
Physical properties: red crystals, m.p. 144°, $\lambda_{max.}$ (95% EtOH) 217, 241, 427 nm (log ϵ 4·36, 4·29, 3·78), $\nu_{max.}$ (KBr) 1675, 1650, 1580 cm^{-1}.

Like diosquinone this pigment forms a quinoxaline and it was noted that the ultraviolet-vis. absorption was very similar to that of 8-methoxy-1,2-naphthaquinone. The n.m.r. spectrum established the presence of one C–Me and one O–Me group, the former being located in the quinone ring as the signal is split and coupled to a vinylic proton at τ 2·86. The chemical shift of the vinylic quartet at relatively low field ($< \tau$ 3·0) suggests that it is situated at C-4, the methyl group being at C-3. The three aromatic protons show an ABX pattern indicating a *peri*-location for the methoxyl group in the benzenoid ring. If this was at C-5 it would be expected that the signal at lowest field would be that of the C-8 proton adjacent to a quinone carbonyl group, and this should be singly *ortho*-coupled. In fact the proton signal at lowest field is doubly *ortho*-coupled suggesting that the methoxyl group must be at C-8. The new quinone should therefore be (77) and this was confirmed by exposing it to air in alkaline solution when it was converted by nucleophilic hydroxylation, and oxidation, into droserone 5-methyl ether (p. 236). This is also present in *D. melanoxylon*.[497]

Mycochrysone (78), $C_{20}H_{12}O_7$

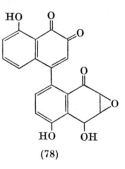

(78)

Occurrence: cultures of an unidentified inoperculate discomycete.[480]

Physical properties: fine red needles (α) or yellow-brown prisms (β), dec. >195°,† $[\alpha]_D^{22}$ −331° (acetone), $\lambda_{max.}$ (EtOH) 236, 303, 438 nm (log ϵ 4·49, 3·75, 3·80), $\nu_{max.}$ (KBr) 3400, 1666, 1637 cm^{-1}.

Mycochrysone is dimorphic, thermolabile and photosensitive in solution, and is best crystallised from cold aqueous acetone or by concentration of an ethereal solution at room temperature. It is unstable in both acid and alkaline solution but can be recovered after a short interval in sodium bicarbonate. Redox properties and various colour reactions indicated that the pigment was a quinone probably possessing two phenolic groups, one of which is *peri* to the quinone system (positive boroacetate test), the other being more acidic (pK_a ∼7·7). On treatment with methyl iodide-silver oxide the more acidic group reacts first giving a red monomethyl ether ($\lambda_{max.}$ 443 nm), insoluble in sodium bicarbonate solution, which can be converted into a yellow dimethyl ether ($\lambda_{max.}$ 419 nm). Acetylation was not very informative (but see ref. 481b), treatment with acetic anhydride-silver carbonate affording principally two amorphous diacetates, while reductive acetylation gave mainly an amorphous hexa-acetate showing naphthalenic absorption. This suggested that the reduction product contains six hydroxyl groups, and if so, the parent quinone would have four; however a Zerewitenoff determination indicated only three (2·66) active hydrogen atoms per molecule. Another oxygen atom is evidently contained in a carbonyl group as mycochrysone and all the derivatives mentioned show ν_{CO} near 1690 cm^{-1}. Both the pigment and its methyl ethers readily form quinoxaline derivatives and hence they are *o*-quinones.[480]

An unusual feature of mycochrysone is the ease with which it can be reduced to perylene derivatives. Zinc dust fusion afforded perylene itself (13%) together with the hexahydroperylene (79) and the decahydro derivative (80), and catalytic hydrogenation of mycochrysone

† α form dec. 175–195° or >195°, β form dec. 205–220°.

dimethyl ether yielded, *inter alia*, tetradecaperylene (81); (79), (80) and (81) show, respectively, ultraviolet absorption characteristic of phenanthrene, naphthalene and benzene. These results imply that the

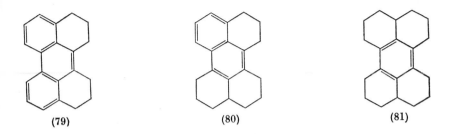

(79)   (80)   (81)

pigment is either a perylene or a 1,1'-binaphthyl derivative although it is of interest that zinc dust fusion of β-naphthaquinone also produced a trace of perylene.[480] The binaphthyl structure was preferred[481] as it is otherwise difficult to accommodate the multiplicity of functional groups into a perylene system.

Structure (78) was advanced by Read and Vining[481] on the following grounds. The *peri*-hydroxy-*o*-naphthaquinone moiety is consistent with the reactions already mentioned and with the general similarity of the ultraviolet-visible absorption of mycochrysone to that of diosquinone. Regarding the other half of the molecule, oxidation with manganese dioxide in chloroform afforded dehydromycochrysone, $C_{20}H_{10}O_7$, which contains a new ketonic group showing an additional ultraviolet band at 365 nm, and chelated carbonyl absorption near 1660 cm^{-1}. In the monomethyl ether the 365 nm band shifts to 350 nm and from this, and various colour reactions, it seemed that a second *peri*-hydroxycarbonyl system was present in dehydromycochrysone. As another aryl ketonic group has to be accommodated, the structure of mycochrysone now appears to be (78) to accord with the molecular formula, the dehydro derivative (82) being formed by oxidation of the benzylic alcohol function. In support of structure (78), it is significant that summation of the light absorption of 8-methoxy-1,2-naphthaquinone and 5-methoxy-1,4-naphthaquinone-2,3-epoxide gives a curve showing excellent agreement with that of dehydromycochrysone dimethyl ether. The presence of an epoxide group is consistent with the active hydrogen determination, and direct evidence for its existence was obtained from mycochrysone quinoxaline. When this was treated with hydrochloric acid in aqueous acetone a chlorohydrin, $C_{26}H_{17}ClN_2O_5$, was formed which reverted to the epoxide on brief treatment with sodium bicarbonate. On longer treatment with hydrochloric acid the chlorohydrin loses water and is then oxidised to another product, $C_{26}H_{13}ClN_2O_4$,

8

formulated as the biquinone (83). As dehydromycochrysone can be selectively reduced with sodium borohydride to mycochrysone it was argued, by analogy with juglone epoxide,[571] that the epoxide and the

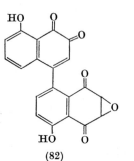

(82)

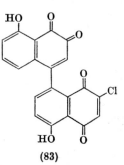

(83)

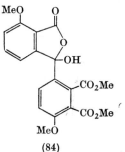

(84)

secondary alcohol function must have a *cis* relationship. Finally, the position of the internuclear link is indicated by the failure of myco-chrysone dimethyl ether to undergo the usual 1,4-addition reactions of a quinone, and was confirmed by the formation of the lactol (84) on permanganate oxidation of the dimethyl ether followed by reaction with diazomethane.

The n.m.r. spectrum of mycochrysone is complex[481b] owing to rotation about the central carbon–carbon bond. For example, the quinonoid proton gives rise to three signals in the region τ 3–4, regarded as a sharp doublet centred at τ 3·78 and a second broad overlapping doublet centred at τ 3·57, each corresponding to one proton. This is attributed to the different degrees of shielding experienced by this proton in two principal conformations of mycochrysone resulting from slow rotation about the central bond. Slow interconversion of conformers may also account for the dimorphism shown by this pigment.

The inoperculate discomycete which produces mycochrysone elabor-ates other pigments known as mycorubrones[481b] which may be 1,1′-binaphthylbiquinones.

7-Methyljuglone (ramentacéone) (85), $C_{11}H_8O_3$

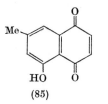

(85)

Occurrence: leaves of *Diospyros ebenum* Koen.[122] (= *D. hebecarpa* A. Cunn.), heartwood of *D. chloroxylon* Roxb.,[123] root and root-bark of *D. lotus* L.[156] (Ebenaceae), whole plants of *Drosera ramentacea* Burch.[138] (= *D. madagascariensis* DC.), *D. intermedia* Hayne,[495] leaves of *D. aliciae* Hamet.,[57] *D. capensis* L.[57] and other *Drosera* spp.[57, 496] (Droseraceae).

Physical properties: orange-red needles, m.p. 125·5–126·5°, λ_{max}. (EtOH) 218, 253, 424 nm (log ε 4·19, 4·13, 3·60), ν_{max}. (KBr) 1663, 1635, 1585 cm⁻¹.

Quinones of this type, including several dimers, a trimer and a tetramer, occur frequently in *Diospyros* spp.[128] as do related naphthols.[124, 125, 497, 573] Some of the latter are sensitive to oxidation and are easily converted into black polymeric materials, probably similar to the black pigments in the heartwood, i.e. ebony, and senescent leaves of many *Diospyros* spp.[126] In *D. ebenum*, 7-methyljuglone is accompanied by the isomer 2-methyljuglone (plumbagin) and by the reduction

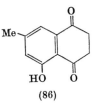

(86)

product (86).[122] This closely resembles β-hydrojuglone in physical and chemical properties, and on treatment with ferric chloride it is converted into 7-methyljuglone. The quinone was recognised from its spectral properties and colour reactions, and the location of the C-methyl group was established by a synthesis of 1,4,5-trimethoxy-7-methylnaphthalene which was identical with the trimethyl ether obtained (via enolisation) from the diketone. The natural quinone was synthesised later[127] by condensation of 4-chloro-3-methylphenol with maleic anhydride in an aluminium chloride-sodium chloride melt to give 8-chloro-5-hydroxy-7-methyl-1,4-naphthaquinone from which the chlorine atom was removed by hydrogenolysis of the leucotriacetate.† Hydrolysis and oxidation then gave the quinone (85), and the quinol, on fusion *in vacuo*, isomerised to form the diketone (86) identical with the natural product.

† See also ref. 496.

Plumbagin (87), $C_{11}H_8O_3$
Plumbagin methyl ether, $C_{12}H_{10}O_3$

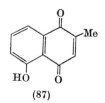

(87)

Occurrence: Plumbaginaceae†—roots and leaves of *Plumbago pulchella* Boiss.,[130, 131] *P. scandens* L.,[130] and *P. europea* L.[129, 130, 144] (also in the flowers[130]) roots of *P. rosea* L.,[130, 132, 143] *P. zeylanica* L.,[130, 133, 143] *P. capensis* Thunb.[130] and *P. coerulea* H.B. & K.;[130] roots of *Ceratostigma plumbaginoides* Bunge,[130] *C. willmottianum* Stapf.,[130] and roots and leaves of *Plumbagella micrantha* Spach.[130] Droseraceae—*Drosera rotundifolia* L.,[134, 135]‡ *D. intermedia* Hayne,[136] *D. peltata* Sm.,[137] *D. binata* Labill.,[135]§ *D. auriculata* Backh. ex Planch.,[138] *D. indica* L.,[138] *D. anglica* Huds.,[495] *D. ramentacea* Burch,[138b] and other *Drosera* spp.,[57] *Aldrovanda vesiculosa* L.,[57] *Dionaea muscipula* Ellis[57] and *Drosophyllum lusitanicum* Link.[57] Ebenaceae—bark of *Diospyros mespiliformis* Hochst.,[141] *D. maritima* Blume[142] and *D. siamang* Bakh.[208] (= *D. elliptifolia* Merr.), bark and leaves of *D. xanthoclamys* Gürke[141] and *D. ebenum* Koen[122] (= *D. hebecarpa* A. Cunn.) Euphorbiaceae—root bark of *Pera ferruginea* Muell-Arg.;[574] the methyl ether occurs in the heartwood of *Diospyros melanoxylon* Roxb.[497]

Physical properties: (87) orange needles, m.p. 77°, $\nu_{max.}$ (EtOH) 220, 266, 418 nm (log ε 3·87, 4·12, 3·61), $\lambda_{max.}$ (KBr) 1659, 1637, 1602 cm^{-1}; methyl ether, yellow needles, m.p. 94°, $\lambda_{max.}$ (95% EtOH) 209, 250, 394 nm (log ε 4·62, 4·24, 3·66), $\nu_{max.}$ (KBr) 1650, 1620 cm^{-1}.

This simple quinone was known for a hundred years before structural investigations were started and some of the initial findings were confusing. It was identified[143, 144, 145] as a methyljuglone on the basis of its molecular formula and colour reactions, the formation of an acetate and an oxime, oxidative degradation with alkaline hydrogen peroxide to 3-hydroxyphthalic acid,[135b] and the formation of 2-methylnaphthalene by zinc dust distillation. This last reaction was reversed[145] by oxidation of 2-methylnaphthalene with Caro's acid which implied that plumbagin was most probably 2- or 3-methyljuglone, and the location of the methyl group was finally settled by an unambiguous synthesis.[146] Condensing *m*-toluoyl chloride with diethyl acetylsuccinate gave (88) which was hydrolysed to the keto-acid (89) and cyclised, after Clemmensen reduction, to the tetralone (90). Plumbagin was then obtained by

† The quinone occurs partly in a reduced form as a glycoside.[130, 540]
‡ The original name[134] for this quinone, "droserone", was later abandoned. An alleged "carboxy-oxy-naphthochinon"[161] was actually impure quercetin.[162]
§ Dieterle[139] reported the isolation of a second quinone, $C_{10}H_8O_3$, m.p. 107–108° from this plant but it could not be detected in a later investigation.[140]

dehydrogenation to the naphthol and oxidation of the acetate with
chromium trioxide. Other syntheses have been reported.[135b, 147, 496]

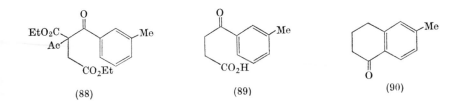

(88)   (89)   (90)

Plumbagin has physiological properties and *Plumbago* roots have been
credited with some remarkable powers. Like juglone it has an irritating
odour and affects the mucous membrane; it stains the skin and produces
blisters[143] readily which may account for the vesicant properties of
the sap of *D. maritima*.[148] Extracts of *Drosera* spp. still find a place in
modern pharmacopeias for the treatment of whooping cough.[149]
Chewing *Plumbago* root is said to be effective both for relieving tooth-
ache[150] (in France) and producing abortion[151] (in Malaya), and in
India, where the root is known as Chita or Chitraka, the drug is reported
to "increase digestive power, to promote the appetite, and to be useful
in dyspepsia, piles, anasarca, diarrhoea, skin diseases, etc.".[153] The
"etc." was apparently broad enough to include leprosy and plague. A
novel exploitation of the vesicant properties of plumbagin has been
reported from Turkey.[152] If a paste of Babink or Babini root (unidentified botanically) is applied to the skin, a deep purple stain results,
indistinguishable from a bruise. By this means false, though painful,
evidence of assault and battery can be produced. Following cases of
alleged violence, an investigation of the root was carried out at the
instigation of the local police, which led to the isolation and identification
of plumbagin. The same purple mark is produced by the crystalline
material accompanied by considerable pain and followed later by
vesication.[152]

Plumbagin methyl ether, one of five quinones in *D. melanoxylon*, was
identified by its spectroscopic properties, particularly the n.m.r.
spectrum, which revealed that a methyl group was located in the quinone
ring next to a vinylic proton and the splitting pattern of the three
aromatic protons showed, by comparison with the spectrum of juglone,
that a methoxyl group occupied a *peri*-position. The structure is therefore
2-methyl-5- or 8-methoxy-1,4-naphthaquinone, both of which were
synthesised. Plumbagin methyl ether was obtained from the tetralone
(91), prepared by standard procedures from β-(*o*-methoxyphenyl)ethyl
bromide and diethyl methylmalonate, by oxidation with selenium

dioxide and by dehydrogenation to the naphthol followed by oxidation with Fremy's salt.[497]

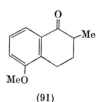

(91)

3-Chloroplumbagin (92), $C_{11}H_7ClO_3$

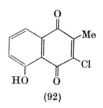

(92)

Occurrence: whole plants of *Drosera intermedia* Hayne,[495] and *D. anglica* Huds.[495] (Droseraceae).

Physical properties: orange-yellow leaflets, m.p. 125°, $\lambda_{max.}$ (EtOH) 213, 245, 278, 423 nm (log ε 4·37, 3·79, 4·02, 3·52), $\nu_{max.}$ (KBr) 1654, 1629, 1600 cm^{-1}.

Chlorinated naphthaquinones are extremely rare and this one is unique in higher plants. It accompanies plumbagin in several Droseraceae and was identified[495] initially by its mass spectrum, and then by direct comparison with authentic material synthesised earlier by C-methylation of 3-chlorojuglone with acetyl peroxide.[168]

Diospyrin† (93), $C_{22}H_{14}O_6$

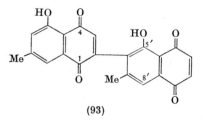

(93)

Occurrence: bark of *Diospyros montana* Roxb.,[154] *D. chloroxylon* Roxb.[155] and *D. mespiliformis* Hochst.[208] (Ebenaceae).

† Not to be confused with a leucodelphinidin glucoside in *D. kaki*, also named diospyrin.[157]

Physical properties: orange-red cubes, m.p. 258°, λ_{max}. (EtOH) 223, 254, 438 nm (log ϵ 4·59, 4·43, 3·99), ν_{max}. (KBr) 1664, 1634, 1590 cm^{-1}.

Diospyrin has two hydroxyl groups (diacetate, dimethyl ether), two C-methyl groups (Kuhn-Roth, n.m.r.), and two quinone systems (formation of a leucohexa-acetate), and from its violet colour in alkaline solution, the ultraviolet-visible spectrum, and the absence of a normal hydroxyl band in the infrared, it is evidently a bis-methyljuglone.[154, 155, 158] Zinc dust distillation gave a crystalline hydrocarbon, $C_{22}H_{18}$, which appeared to be (MeC$_{10}$H$_6$)$_2$, and from its ultraviolet absorption was most probably a β,β'-binaphthyl. The relative positions of the hydroxyl and methyl groups was revealed by the formation of 3-methoxy-5-methyl-phthalic acid[158] on oxidation of the "dimethyl ether" (see below) of diospyrin, the implication being that the natural pigment is a dimer of 7-methyl-juglone. This indeed is the case, but the dimer is not symmetrical and the biquinone did not correspond to diospyrol (94), the binaphthyl derivative already known in *D. ehretioides*.[124, 159] The

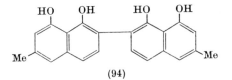

(94)

unsymmetrical nature of diospyrin is clearly evident both from n.m.r. spectral data[155] and the finding that oxidation of diospyrin itself, with alkaline hydogen peroxide, gives a mixture of isocochenillic acid (95) and

(95)   (96)

3-hydroxy-5-methylphthalic acid (96).[155] The n.m.r. spectra of diospyrin and of its more soluble dimethyl ether reveal the presence of *three* vinyl protons, seen in the spectrum of the dimethyl ether as singlets at τ 3·0 (2H) and 3·13 (H). Moreover the two C-methyl groups and the two O-methyl groups have different chemical shifts appearing at τ 7·44, 7·66, and τ 5·91, 6·24, respectively. This implies, *inter alia*, that one half of the molecule contains an unsubstituted quinone ring which was confirmed by evidence for two *ortho*-coupled protons in the n.m.r. spectrum of the leucohexa-acetate. These data can only be accounted for if diospyrin is (93) or the 3,6'-coupled isomer.

To distinguish between these two possibilities diospyrin dimethyl ether was reduced with dithionite to a biquinol which was then methylated with diazomethane. The product was an orange-yellow compound, $C_{27}H_{24}N_2O_6$, one quinol moiety having been oxidised to a quinone followed by addition of diazomethane to form a pyrazole ring (cf. ref. 550). It contained one N-methyl and three O-methyl groups, two of

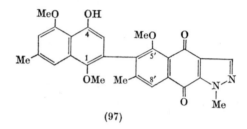

(97)

which were deshielded and gave n.m.r. signals at τ 6·30 and 6·47. It follows that these two methoxyls are both *ortho* to the central C–C bond which therefore links C_2 and C_6', as shown in (97).† A C_3–C_6' linkage is also excluded by the fact that (97) forms an azo dye with diazotised sulphanilic acid. Diospyrin is therefore (93).[551]

The "dimethyl ether" referred to above is a red compound, m.p. 330°, M.W. 416, obtained by methylation with dimethyl sulphate and anhydrous potassium carbonate in refluxing dioxan, whereas the true dimethyl ether, yellow needles, m.p. 256°, M.W. 402, is obtained by cold methylation with methyl iodide-silver oxide.[155] The red compound, which gave 3-methoxy-5-methylphthalic acid on oxidation, is a re-

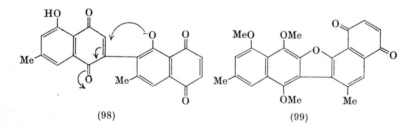

(98)   (99)

arrangement product (99), actually containing three methoxyl groups, which is formed by a nucleophilic intramolecular addition of the anion (98), followed by enolisation and methylation.[469]

† The pyrazole ring could have an isomeric structure.

Isodiospyrin (100), $C_{22}H_{14}O_6$

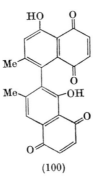

(100)

Occurrence: bark of *Diospyros mespiliformis* Hochst.,[208] bark and wood of *D. virginiana* L.,[208] wood of *D. chloroxylon* Roxb.,[160] root and root-bark of *D. lotus* L.,[156] root of *D. japonica* Sieb. & Zucc.,[156] *D. morrisiana* Hance[156] and probably in the roots of *D. lyciodes* Desf. subsp. *sericea* (Bernh. ex Kraus) de Winter[498] (Ebenaceae).

Physical properties: orange-red prisms, m.p. 226–228° (dec.), $[\alpha]_D^{22}$ −150° (CHCl$_3$), $\lambda_{max.}$ (EtOH) 221, 254, 436 nm (log ϵ 4·57, 4·39, 3·98), $\nu_{max.}$ (KBr) 1662, 1640, 1585 cm^{-1}.

Both diospyrin and its isomer, isodiospyrin, occur in the bark of *D. mespiliformis* and *D. chloroxylon*, and it was immediately clear from the ultraviolet-visible and infrared spectra that the two pigments were very similar.[160, 208] Isodiospyrin affords a dimethyl ether and a leuco-hexa-acetate, and the n.m.r. spectrum includes signals from two different aromatic methyl groups and two different *peri*-hydroxyls. Evidently isodiospyrin is another unsymmetrical methyljuglone dimer. By analogy with similar compounds the singlet (1H) at τ 2·33 could be assigned to a *peri*-hydrogen and the singlet (1H) at τ 2·67 to a β-aromatic proton in the other half of the molecule, and since the n.m.r. spectrum also provided evidence for four vinylic protons structure (100) could be assigned[160, 208] to isodiospyrin. The *meta*-relationship of the hydroxyl and methyl groups was at first assumed on biogenetic and phytochemical grounds, and was later confirmed by observation of the nuclear Over-hauser effects produced by irradiation of the methyl and methoxyl protons.[156] This structure accounts for the optical activity of this pigment, rotation about the central C-C bond being very restricted, and for the shielding of both methyl groups, but it is not clear why the vinylic protons in one quinone moiety exhibit an AB quartet whereas the other pair shows a two-proton singlet. However, in the dimethyl ether, this pair also appears as a pair of doublets.[156] The mass spectrum fully supports structure (100).

Biramentacéone (101), $C_{22}H_{14}O_6$

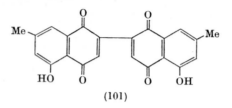

(101)

Occurrence: whole plants of *Drosera ramentacea* Burch.[138b] (Droseraceae).
Physical properties: orange-yellow rods, m.p. 235° (dec.), $\lambda_{max.}$ (EtOH) 215, 270, 430 nm (log ϵ 4·54, 4·32, 3·85), $\nu_{max.}$ (KBr) 1662, 1640 cm^{-1}.

Another dimer, $C_{22}H_{14}O_6$, has been found recently in *D. ramentacea* together with both 2-methyl- and 7-methyljuglone. Characteristic ultraviolet-visible and infrared absorption established the general structure and, from the simplicity of the n.m.r. spectrum, it was clearly a symmetrical compound. The spectrum consists of singlets at τ 7·58 (Ar–CH_3), 3·03 (Q-H) and −1·78 (*peri*-O*H*) and doublets at τ 2·55 and 2·92 (J, ca. 2 Hz) attributable to *meta*-coupled aromatic protons. The pigment must therefore be (101) or (102) which is fully consistent with its mass spectrum. The biquinone (102), obtained[159] by oxidation of diospyrol (94) with Fremy's salt, is almost identical spectroscopically with biramentacéone but there are slight differences in R_F values. Structure (101) was therefore preferred.[138b] By analogy with diospyrol structure (102) would have been expected for biramentacéone [this was later found *in vivo* (see below)], and the formation of (101) suggests that the dimerization step takes place *in vivo* at the naphthalene triol stage (1,4,5-trihydroxy-7-methylnaphthalene), possibly by oxidation of a glycoside similar to α-hydrojuglone 4-β-D-glucoside (p. 221). The tautomeric form of the triol is already known in *Diospyros ebenum* (p. 227).

Mamegakinone (102), $C_{22}H_{14}O_6$

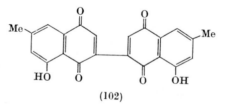

(102)

Occurrence: roots of *Diospyros lotus*† L.[156] (Ebenaceae).
Physical properties: orange crystals, m.p. 253°, $\lambda_{max.}$ (dioxan) 253, 275, 434 nm (log ϵ 4·34, 4·28, 3·96), $\nu_{max.}$ (KBr) 1652, 1625 cm^{-1}.

† Japanese name, mamegaki.

This was the fourth dimer of 7-methyljuglone to be isolated from *Diospyros* spp. and was consequently easily recognised. Its spectral properties, especially n.m.r. data, accord with a symmetrical dimer of this type in which the quinone rings are directly linked. As it is different from biramentacéone (101) the structure must be (102)[156] and the compound is identical with the biquinone obtained[159] by oxidation of diospyrol (94) with Fremy's salt.

Elliptinone (103), $C_{22}H_{14}O_6$

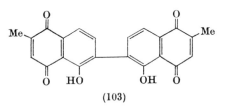

(103)

Occurrence: bark of *Diospyros siamang* Bakh[208] (= *D. elliptifolia* Merr.), bark and dried fruit of *D. ehretioides* Wall.[552] (= *D. mollis*) (Ebenaceae).

Physical properties: orange needles, subl. >290°, dec. >310°, λ_{max} (MeOH) 230, 263, 443 nm (log ϵ 4·45, 4·29, 3·96), ν_{max} (KBr) 1667, 1645, 1607 cm^{-1}.

This is yet another *Diospyros* dimer, $C_{22}H_{14}O_6$, and the ultraviolet-visible and infrared spectra confirm that it belongs to the same group as those above. The n.m.r. spectra of the diacetate and dimethyl ether show that the dimer is symmetrical, each half possessing a methyl group attached to the quinone ring which is coupled to a vinylic proton (quartet at $\tau \sim 3·3$), and two *ortho*-coupled aromatic protons (doublets at τ 1·93 and 2·42). Comparison with the n.m.r. spectrum of juglone acetate showed that signals from the C-6 and C-6' protons were absent from the spectrum of the dimer. Accordingly, structure (103) was assigned[208] to elliptinone which was confirmed by the spectral identity of its dimethyl ether with the biquinone previously obtained[159] by oxidation of diospyrol (94) tetramethyl ether. Hence elliptinone is 6,6'-biplumbagin. The monomer plumbagin is a co-metabolite in the bark of *D. siamang*.[208]

Bis-isodiospyrin (104), $C_{44}H_{26}O_{12}$

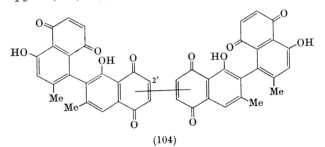

(104)

Occurrence: root of *Diospyros lotus* L.,[156] *D. japonica* Sieb. & Zucc.,[156] and *D. morrisiana* Hance[156] (Ebenaceae).

Physical properties: orange prisms, m.p. >320°, $[\alpha]_D^{21}$ −678° (CHCl$_3$), λ_{max}. (CHCl$_3$) 257, 444 nm (log ϵ 4·69, 4·25), ν_{max}. (KBr) 1658, 1639, 1601 cm^{-1}.

Diospyros spp. provide several examples of phenolic coupling and in this case two consecutive dimerisations have produced a unique tetra-quinone.† Once again 7-methyljuglone is clearly the monomeric unit as the ultraviolet-visible and infrared spectra of the pigment and its tetramethyl ether are virtually the same as those of 7-methyljuglone and isodiospyrin and their methyl ethers, respectively. The molecular formula establishes the tetrameric nature of the compound, and comparison of the n.m.r. spectrum with those of related compounds in the series shows unequivocally that the new member is a symmetrically coupled dimer of isodiospyrin. The n.m.r. spectra of (104) and (100) show that of the two quinonoid protons in isodiospyrin (100) which give a singlet, one has disappeared and the signal from the other has shifted downfield [exactly the same effect is observed when the spectra of 7-methyljuglone and mamegakinone (102) are compared]. Accordingly Natori and his co-workers[156] assign structure (104) to this tetraquinone, and prefer symmetrical coupling at the 2′-position by analogy with related *Diospyros* extractives. As the o.r.d. curve of (104) has the same sign as that of (100), the tetramer is assumed to be formed by dimerisation of an isodiospyrin precursor.

Droserone (105; R = H), C$_{11}$H$_8$O$_4$
Droserone-5-methyl ether (105; R = Me), C$_{12}$H$_{10}$O$_4$

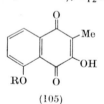

(105)

† Added in proof: Sidhu and Prasad have obtained a triquinone, xylospyrin, from *D. chloroxylon* believed[553] to have the following structure (or the irregular isomer, compare trianellinone p. 242).

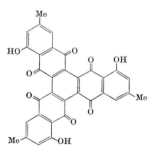

Occurrence: (105; R = H) rhizomatous roots of *Drosera whittakeri* Planch.[163] and *D. peltata* Sm.[137] (Droseraceae).

(105; R = Me) heartwood of *Diospyros melanoxylon* Roxb.[497] (Ebenaceae).

Physical properties: (100; R = H) yellow needles, m.p. 181°, $\lambda_{max.}$ (EtOH) 227, 245(sh), 285, 409 nm (log ϵ 4·19, 4·12, 4·08, 3·66), $\nu_{max.}$ (KBr) 3308, 1643(sh), 1625, 1580 cm^{-1}, $\nu_{max.}$ (CHCl$_3$) 1629 cm^{-1}.

(100; R = Me) yellow needles, m.p. 173–174° (171°), $\lambda_{max.}$ (95% EtOH) 228, 277, 376 nm (log ϵ 4·35, 4·34, 3·82), $\nu_{max.}$ (CHCl$_3$) 3380, 1653(sh), 1645, 1582 cm^{-1}.

The Australian sundew *D. whittakeri* contains two acidic pigments, droserone and hydroxydroserone, which were first examined by Rennie[163] in 1877. They showed redox properties, gave acetic acid on oxidation with chromic acid, and formed di- and tri-acetates, respectively. They were therefore considered to be di- and trihydroxymethyl-naphthaquinones. It was found later[164] that droserone formed a mono-pyridine salt and a monoboroacetate which indicated that one hydroxyl group was in the quinone ring and the other occupied a *peri-* position, and when hydroxydroserone was shown to be 3,5,8-trihydroxy-2-methyl-1,4-naphthaquinone it appeared that droserone was either (105) or (106). The former was favoured as the ultraviolet-visible absorption[166] was very similar to that of 3-hydroxyjuglone[167] and when that compound was C-methylated[168] with acetyl peroxide it gave 3,5-dihydroxy-2-methyl-1,4-naphthaquinone identical with droserone. Earlier, by a synthesis of (106), Asano and Hase[169a] had shown that droserone must

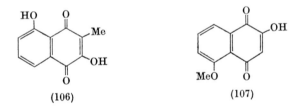

(106)   (107)

have the alternative structure (105) which they proceeded to confirm by converting plumbagin into droserone both directly, by careful oxidation with hydrogen peroxide in aqueous barium hydroxide,[169b] and indirectly, by Thiele acetylation and subsequent hydrolysis and oxidation.[170] In another approach, Cooke and Segal[171] treated the quinone (107) with propionyl peroxide and the 3-ethyl derivative so obtained was converted into droserone by a Hooker oxidation and final demethylation.

α-*Caryopterone* (108), $C_{15}H_{12}O_4$

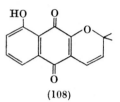

(108)

Occurrence: roots of *Caryopteris clandonensis* Bunge[548] (Verbenaceae).
Physical properties: red needles, m.p. 143·5–144·5°, $\lambda_{max.}$ (iso-octane) 240, 280, 426 nm (log ε 4·19, 4·07, 3·62), $\nu_{max.}$ (KBr) 1675(w), 1645, 1623 cm⁻¹.

This is the first prenylated juglone to be found in Nature. Evidence for the juglone structure comes from the spectroscopic data given above, the violet colour in aqueous sodium hydroxide, and the *peri*-hydroxyl proton and aromatic ABX system revealed by the n.m.r. spectrum. This also includes signals for a *gem*-dimethyl group and an AB system which are in close agreement with those from the pyran ring of dehydro-α-lapachone (25). The pigment should therefore be (108) or the isomer with the hydroxyl group in the alternative *peri*-position. The problem was resolved by reduction to a yellow dihydro derivative which was rearranged in concentrated sulphuric acid to an orange-red *o*-quinone showing $\nu_{max.}$ (CHCl₃) 1660(sh), 1639 cm⁻¹ and τ −1·9. On methylation (methyl iodide-silver oxide-chloroform) it gave a methyl ether showing $\nu_{max.}$ (CHCl₃) 1689, 1647 cm⁻¹, very similar to the spectrum of β-lapachone (19). Clearly the *o*-quinone obtained by rearrangement of

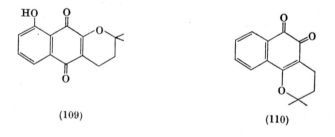

(109)   (110)

dihydro-α-caryopterone is chelated and must therefore be (110) (dihydro-β-caryopterone), whence the natural compound must be (108),[548] and its dihydro derivative (109).

6-Hydroxy-5-methoxy-2-methyl-1,4-naphthaquinone (diomelquinone A)
(111; R = H), $C_{12}H_{10}O_4$.
5,6-Dimethoxy-2-methyl-1,4-naphthaquinone (111; R = Me), $C_{13}H_{12}O_4$.

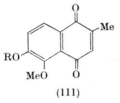

(111)

Occurrence: heartwood of *Diospyros melanoxylon* Roxb.,[175, 497] and (111; R = H) in the root bark of *D. discolor* Willd.[556]

Physical properties: (111; R = H) orange needles, m.p. 152–153°, λ_{max}. (95% EtOH) 216, 259, 308, 543·5 nm (log ε 4·33, 4·16, 3·86, 3·55), ν_{max}. (CCl$_4$) 3505, 1660, 1630 cm^{-1}.

(111; R = Me) yellow needles, m.p. 184–185°, λ_{max}. (95% EtOH) 214, 260, 395 nm (log ε 4·65, 4·47, 3·82), ν_{max}. (CHCl$_3$) 1655, 1625 cm^{-1}.

These quinones are the only natural *o*-naphthazarin (5,6-dihydroxy-1,4-naphthaquinone) derivatives known. The phenolic quinone contains one C–Me and one O–Me group, and forms a mono-acetate and a monomethyl ether (with diazomethane). The n.m.r. spectrum established that a methyl group is the sole substituent in the quinone ring and doublets at τ 2·00 and 2·75 (J, 8 Hz) could be assigned to *ortho*-coupled aromatic protons, one of which (at lower field) is *peri* to a quinone carbonyl group. It follows that the hydroxyl and methoxyl groups must be adjacent and in this case the hydroxyl is *not* in a *peri* position (ν_{HO} 3505, τ_{HO} 3·20). Structure (111; R = H) was preferred[175] to the 3-methyl isomer on phytochemical grounds and this was confirmed when the monomethyl ether was found to be identical to 2-methyl-5,6-dimethoxy-1,4-naphthaquinone previously obtained by oxidation of (112) with persulphate.[125b] The same quinone (111; R = Me) has also

(112)   (113)

been prepared by a similar oxidation of the isomer (113).[497] It should be noted that (111; R = H) is soluble in aqueous sodium bicarbonate, and on addition of sodium hydroxide to an ethanolic solution the long-wave maximum at 543·5 nm intensifies but does not shift.

Stypandrone (114), $C_{13}H_{10}O_4$

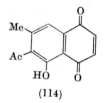

(114)

Occurrence: roots of *Stypandra grandis* White[176] and *Dianella intermedia* Endl.[1] (Liliaceae).

Physical properties: orange needles, m.p. 135·5–136°, $\lambda_{max.}$ (EtOH) 212, 249, 425 nm (log ε 4·40, 4·19, 3·64), $\nu_{max.}$ (KCl) 1692, 1667, 1623 cm⁻¹.

The possibility of finding naphthaquinones in *Dianella* arose from an observation that crude extracts gave a transient violet colour with alkali, and the identification[178] of dianellin and its aglycone, dianellidin (115)† in *Dianella* spp. In the expectation that related quinones might be found in Nature Cooke and Sparrow[176] oxidised dianellidin with

(115)   (116)

Fremy's salt to give the isomers (114) and (116). The former (stypandrone) was subsequently discovered in *D. intermedia*[177] but was first found in another Liliaceous plant, *Stypandra grandis*,[176] together with its dimer, dianellinone.

The structure of stypandrone follows from the *in vitro* synthesis, its light absorption and the n.m.r. spectrum which distinguishes it from the isomer (116).

Dianellinone (117), $C_{26}H_{18}O_8$

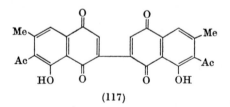

(117)

† Dianellidin = musizin = nepodin also occurs in *Maesopsis eminii* Engl.[183] (Rhamnaceae), *Rumex nepalensis* Wall.,[184] *R. japonicus* Houtt.,[184] and *R. obtusifolius* L.[185] (Polygonaceae).

Occurrence: roots of *Dianella revoluta* R. Br.,[176] *D. intermedia* Endl.[177] and *Stypandra grandis* White[176] (Liliaceae).

Physical properties: orange needles, m.p. 285–295° (dec.), λ_{max}. (dioxan) 223, 429 nm (log ϵ 4·72, 3·89), ν_{max}. (KCl) 1695, 1667, 1645, 1623 cm^{-1}.

All the spectroscopic properties of this compound show a marked similarity to those of its co-pigment, stypandrone, and its dimeric nature was apparent from the high melting point, low solubility in most solvents and, of course, the molecular formula, $C_{26}H_{18}O_8$ (mass spectrum). The n.m.r. spectrum (in CF_3CO_2H) showed that dianellinone was a symmetrical bi-stypandrone possessing two vinylic protons, and could be represented by (118). The remaining problem is therefore the position of the carbon–carbon bond linking the two halves of the molecule.

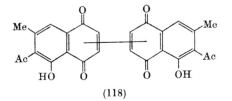

(118)

Dianellinone dimethyl ether is a yellow, light-sensitive, compound and on exposure to daylight in ethanol/chloroform it undergoes a

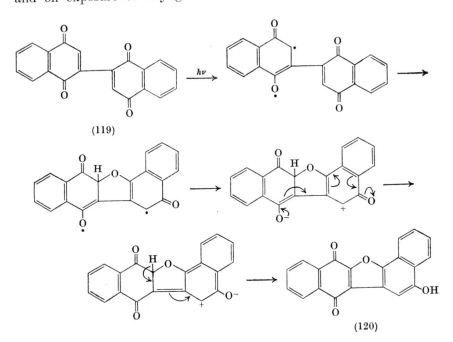

(119)

(120)

photochemical rearrangement forming a deep purple isomer, dimethyl-isodianellinone. This, in fact, occurs during methylation of dianellinone with methyl iodide-silver oxide in chloroform unless precautions are taken to remove the ethanol normally present. A similar rearrangement occurs during normal acetylation to give isodianellinone triacetate, but this reaction may be acid catalysed. Dimethylisodianellinone dissolves in aqueous ethanolic alkali with a bright blue colour, and forms a red monomethyl ether, and a leuco*triacetate* on reductive acetylation. These properties showed that isomerisation to dimethylisodianellinone results in the formation of a phenolic group and the disappearance of one of the quinone rings, and the reaction is thus another example of a well-known biquinone rearrangement which leads to the formation of dibenzo(dinaphtho)-furanquinones (e.g. 120).[179, 180] The transformation can be effected thermally, photochemically (119 → 120) or by acid catalysis. As dianellinone itself is not photolabile the rearrangement seems to be inhibited by the intramolecular hydrogen bonding which probably raises the energy required for the initial n → π* transition.[181] Apparently the rearrangement of the dimethyl ether involves only the carbonyl groups which, in the natural pigment, are hydrogen-bonded,

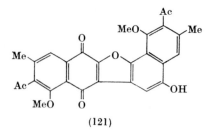

(121)

and this is accounted for if dianellinone has structure (117).[176] It follows that dimethylisodianellinone must be (121).

Trianellinone (122), $C_{39}H_{24}O_{12}$

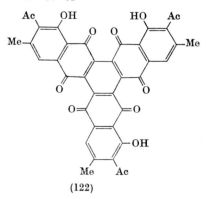

(122)

Occurrence: roots of *Dianella revoluta* R. Br.[549] (Liliaceae).
Physical properties: orange-yellow solid, dec. >300°, λ_{max}. (EtOH) 228, 280, 410 nm, ν_{max}. (KCl) 1695, 1680 cm^{-1}.

This minor pigment from *D. revoluta* is very similar, spectroscopically, to its co-pigment, dianellinone (117). The n.m.r. spectrum (CF$_3$CO$_2$H) >τ 0·0 comprises three singlets at τ 2·25, 7·10 and 7·43 of relative intensity 1:3:3 (cf. dianellinone τ 2·24, 2·64, 7·16 and 7·49; 1:1:3:3). The presence of the structural unit (123) is clearly indicated and from the molecular formula, C$_{39}$H$_{24}$O$_{12}$, there must be three of those for which two arrangements (122) and (124) are possible. The irregular structure (122) must

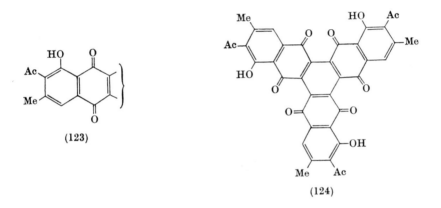

(123)

(124)

be correct[549] as the n.m.r. spectrum (CDCl$_3$) shows *two* singlets at low field (τ −1·74 and −1·64, relative intensity 1:2), and similarly, the trianellinone trimethyl ether shows *two* methoxyl signals at τ 5·76 and 5·78 in the ratio 1:2. The unusually high carbonyl frequencies which trianellinone exhibits must be associated, in part, with the close proximity of adjacent quinone carbonyl groups. The trimeric structure was confirmed synthetically[549] by treating equimolecular amounts of dianellinone and stypandrone with a mixture of nitrobenzene, pyridine, and acetic acid, in the absence of air. This gave trianellinone but it could not be obtained from stypandrone alone.

It is of interest that stypandrone and dianellinone, but not trianellinone, were found in *Stypandra grandis*, while *Dianella revoluta* yielded dianellinone and trianellinone but no stypandrone.

Mollisin (125), $C_{14}H_{10}Cl_2O_4$

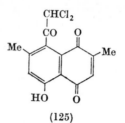

(125)

Occurrence: cultures of *Mollisia fallens* (Karst.) Karst.[187, 188] and *M. caesia* Sacc. sensu Sydow.[187, 188]

Physical properties: orange-yellow needles, m.p. 202–203° (dec.), $\lambda_{max.}$ (EtOH) 259, 280(sh), 420 nm (log ϵ 4·26, 3·90, 3·52), $\nu_{max.}$ (KBr) 1720, 1649, 1618 cm⁻¹.

In 1956 Gremmen[187] described some new strains of two *Mollisia* spp. which elaborated a characteristic yellow crystalline pigment when cultivated on malt agar. The new compound, mollisin, showed strong antifungal activity, and the subsequent chemical investigations of Van der Kerk and Overeem[188, 189, 190, 192] revealed a number of unusual features of which the high chlorine content (>20%) is the most obvious.

Mollisin shows a general resemblance to juglone in its ultraviolet-visible absorption and colour reactions although it failed to yield either an acetate or a leucotriacetate. However treatment with methyl iodide-silver oxide afforded a methyl ether, and this on degradation with alkaline hydrogen peroxide gave the acid (126), the structure of which was established by synthesis.[190] Mollisin can therefore be partly

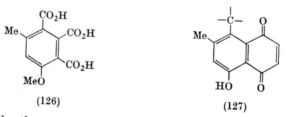

(126)   (127)

represented by the structure (127).

The halogen atoms were originally[188] assigned to the quinone ring as they were obviously reactive and could be eliminated easily by reduction with acid stannous chloride or hydriodic acid,[194] giving dechloro-mollisin and another compound, leucoanhydrodechloromollisin. However, Kuhn-Roth oxidations on mollisin and dechloromollisin indicated the presence of two and three C-methyl groups, respectively, leading to the conclusion that replacement of the two chlorine atoms by hydrogen had created a new terminal methyl group. Rather better Kuhn-Roth results on leucoanhydrodechloromollisin confirmed the formation of a third C-methyl group. These facts, together with the infrared data, can only be explained if mollisin has structure (128). A carbonyl band at

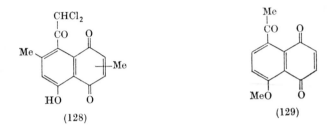

(128)   (129)

1720 cm^{-1}, which shifts to 1696 cm^{-1} in dechloromollisin, can be assigned to the α-dichloroketo function at C-8, while the two quinone carbonyl groups absorb at relatively low wave-numbers, 1649 and 1618 cm^{-1} (cf. naphthazarin, ν_{CO} 1620 cm^{-1}). Dechloromollisin methyl ether shows ν_{CO} 1695 and 1656 cm^{-1} which may be compared with 1685 and 1653 cm^{-1} for the acetylquinone (129). The n.m.r. spectrum of mollisin, determined later,[207] is in complete accord with structure (128) and comprises singlets at τ −2·05 (OH), 2·81 (ArH), 3·67 (COCHCl$_2$) and 7·58 (ArCH$_3$), and a doublet at τ 7·84 (Q-CH$_3$) coupled to a quartet at τ 3·16 (Q-H). No definite evidence for the location of the quinone methyl group was obtained but its position at C-2 was preferred[189] on biogenetic grounds and was confirmed by a synthesis[192] of dechloromollisin methyl ether.

The formation of the naphthalene derivative (130; R = Ac) by self-condensation of α,α′-diacetylacetone in the presence of base was one of Collie's[191] early demonstrations of the formation of aromatic compounds

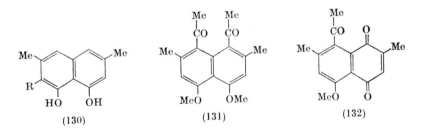

(130)   (131)   (132)

from β-diketones. When heated in a mixture of acetic acid and hydro-bromic acid it behaved as an enolic β-diketone and was hydrolysed to the naphthalenediol (130; R = H). Methylation and Friedel–Crafts acetylation then gave the diacetylnaphthalene (131) which could be oxidised to 8-acetyl-5-methoxy-2,7-dimethyl-1,4-naphthaquinone (132) which was identical with dechloromollisin methyl ether. Attempts to synthesise mollisin methyl ether via dichloroacetylation of the dimethyl ether of (130; R = H) were unsuccessful.

Reference was made earlier to the reduction of mollisin which gives a mixture of dechloromollisin $(C_{14}H_{12}O_4)$ and leucoanhydrodechloromollisin $(C_{14}H_{12}O_3)$. The latter can be converted into dechloromollisin by careful oxidation with hydrogen peroxide or chromic acid, and this reaction can be reversed by reduction with sodium dithionite. Since reduction of dechloromollisin methyl ether yields only the corresponding quinol, it appears that a hydroxyl group at C-5 is required for the formation of the anhydro compound. As the acetyl group must obviously take

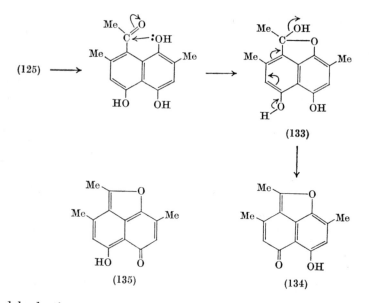

(125) ⟶

(133)

(135)   (134)

part dehydration presumably involves the hemi-acetal (133), and leucoanhydrodechloromollisin must be the quinone methide (134). This structure is consistent with the n.m.r. spectrum,[207] the formation of a golden-green fluorescent complex with boroacetic anhydride, and a positive Craven test. Infrared bands at 1647 and 1680 cm^{-1} were assigned to the chelated carbonyl group and the enol ether double bond, respectively. The experimental evidence does not completely eliminate the tautomeric structure (135) but it appears to be energetically less favourable.

The formation of anhydroleucodechloromollisin provides an explanation for the mixture of products obtained when reductive acetylation of mollisin was attempted. Reductive acetylation of dechloromollisin likewise failed but its methyl ether gave the expected leucodiacetate.

However the same treatment of mollisin methyl ether gave a new compound, $C_{17}H_{16}O_5$, to which structure (136) (ν_{CO} 1757, 1675 cm^{-1}) was ascribed.[192] This accords with the n.m.r. spectrum,[207] in particular the methylene group resonates as a singlet (2H) at τ 5·27. It is presumably

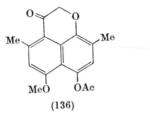

(136)

formed by initial reduction to leucomollisin methyl ether, followed by intramolecular elimination of hydrochloric acid and removal of the remaining chlorine atom by reduction.

Methylnaphthazarin (137), $C_{11}H_8O_4$

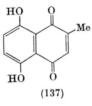

(137)

Occurrence: whole plants of *Drosera intermedia* Hayne,[495] *D. anglica* Huds.[495] and *D. rotundifolia* L.[496] (Droseraceae).

Physical properties: bronze-green plates, m.p. 174–175°, λ_{max} (EtOH) 214, 278, 480, 512, 550 nm (log ϵ 4·53, 3·95, 3·75, 3·78, 3·55), ν_{max} (KBr) 1612 cm^{-1}.

Methylnaphthazarin has recently been detected in trace amounts in several *Drosera* spp. Only the mass spectrum and a qualitative ultra-violet-visible spectrum could be recorded; identity was confirmed by co-chromatography.[495]

2,7-Dimethylnaphthazarin (138), $C_{12}H_{10}O_4$

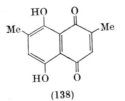

(138)

Occurrence: cultures of *Mollisia caesia* Sacc. sensu Sydow.[186]
Physical properties: red needles, m.p. 125–126°, $\lambda_{max.}$ (EtOH) 217, 280, 480, 510, 550 nm (log ϵ 4·55, 3·90, 3·77, 3·79, 3·58), $\nu_{max.}$ (KBr) 1611 cm⁻¹.

This compound is one of the minor metabolites of *M. caesia*; it was obtained only as a red oil, which partly crystallised, during the sublimation of the major product, mollisin. The quinone was not obtained pure but its visible absorption and various colour reactions suggested the presence of a naphthazarin component, most probably the 2,7-dimethyl derivative in view of its co-occurrence with mollisin (125). Comparison of synthetic and natural materials showed that they had identical visible spectra and were indistinguishable by paper chromatography (in one system). Gas–liquid chromatography of both the red oil and the yellow product obtained by trimethylsilylation revealed the presence of up to twenty compounds including one, in each case, with a retention time identical with that of 2,7-dimethylnaphthazarin and its di-O-trimethyl-silyl derivative, respectively. There seems little doubt that 2,7-dimethyl-naphthazarin (138) is one of the products of *M. caesia* metabolism.

Condensation of toluquinol with citraconic anhydride gives a mixture of 2,6- and 2,7-dimethylnaphthazarin;[117, 186] the product of lower melting point is the 2,7-isomer, the other having been prepared by an alternative route from 2,6-dimethylnaphthalene.[118]

Alkannin (139), $C_{16}H_{16}O_5$
Shikonin (140), $C_{16}H_{16}O_5$

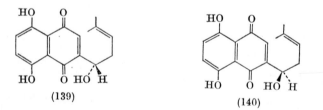

(139)   (140)

Occurrence: (139) in the roots of *Alkanna tinctoria* Tausch.,[213, 214] *Onosma echiodes* L.,[220] *Arnebia nobilis* Rech.,[499]† both free and/or esterified.

(140) in the roots of *Arnebia tibetana* Kurz,[532] *Lithospermum erythrorhizon* Sieb. & Zucc.,[215, 216, 222, 223, 343]† *L. officinale* L.,[478] *L. officinale* (L.) var. *erythrorhizon*,[534] *L. euchromum* Royle,[222, 223] *Macrotomia ugamensis* M. Pop.,[219a]† *Onosma caucasicum* (Levin),[219b]† *O. zerizaminum* Lipsky[531] and *Echium rubrum* (Forsk.)[219b]† both free and/or esterified, and the (±) form (shikalkin) in the roots of *Arnebia hispidissima* DC.[217] (all Boraginaceae); esters of (140) also in trunk and branch of *Jatropha glandulifera* (Roxb.)[505] (Euphorbiaceae).

Physical properties: (139) brown-red needles, m.p. 148°, $[\alpha]_{500}$ −280° (MeOH), $\lambda_{max.}$ (MeOH) 225, 280, 480, 510, 546 nm (log ϵ 3·91, 3·84, 3·74, 3·78, 3·60), $\nu_{max.}$ (KBr) 3320(br), 1610 cm⁻¹.

† Optical rotation not reported, could be (139) or (140).

(140) red-violet needles, m.p. 147–149°, $[\alpha]_{600}^{20}$ +272° (EtOH), λ_{max}. (EtOH) 275, 485, 516, 555 nm (log ϵ 3·83, 3·74, 3·78, 3·58), ν_{max}. (KBr) 3440, 1603 cm^{-1}.

The following side-chain esters of alkannin have been reported: acetate,[499] dark red crystals, m.p. 104–105° and β,β-dimethylacrylate,[499] dark red crystals, m.p. 116–117° (*Arnebia nobilis*), angelate,[221] deep red oil (*Alkanna tinctoria*), tetracrylate,[223] reddish violet, amorphous solid, $[\alpha]_{600}^{20}$ −169° (EtOH) (*L. erythrorhizon* and *L. euchromum*); and of shikonin, the acetate,[216, 223, 505] red prisms, m.p. 85–86° (reddish violet needles, m.p. 106–107°), $[\alpha]_D^{20}$ +26° (CHCl$_3$) ($[\alpha]_{600}^{17}$ +76·9 (EtOH)), β,β-dimethylacrylate,[222, 505] red prisms (red-violet needles), m.p. 113–114°, $[\alpha]_{600}^{20}$ +222° (EtOH), isobutyrate,[222] red-violet needles, m.p. 89–90°, $[\alpha]_{600}^{23}$ +125° (EtOH) and β-hydroxyisovalerate,[223] reddish violet needles, m.p. 90–92°, $[\alpha]_{600}^{20}$ +128° (EtOH) (*L. erythrorhizon* and *L. euchromum*); the acetate and β,β-dimethylacrylate also in *J. glandulifera*[505].

The use of Boraginaceous roots for dyeing is of considerable antiquity both in Europe and the Far East; *Alkanna tinctoria* was known to the Romans while the Japanese equivalent was shikon (= violet root), the root of *L. erythrorhizon*. Both the (+) (shikonin) and the (−) alkannin forms occur, usually esterified on the alcoholic group, but in some species the quinone seems to be reduced as the colouring matter does not appear until the plant material is dried. A similar but unidentified pigment of approximately twice the molecular weight has been reported in *A. tinctoria*.[224, 225] Although no longer of interest as dyes these pigments may still be used for the artificial colouring of wines, cosmetics and food.[217] The indicator properties of alkannin have received some attention[226] and it has been used as a reagent for the detection of magnesium, aluminium and beryllium.[227]

Alkannin is difficult to purify by classical methods and this, coupled with the reactivity of the side chain, misled many of the early investigators and a variety of molecular formulae were proposed. The formation of 2-methylanthracene on zinc dust distillation naturally led Liebermann[228] to conclude that his material was a derivative of anthraquinone whilst other erroneous interpretations arose from unsuitable isolation procedures which effected the elimination[229] or methylation[230] of the alcoholic function. The structure of alkannin was ultimately established by Brockmann;[214] the earlier proposal of Majima and Kuroda[216] for the structure of shikonin was incorrect with respect to the position of one hydroxyl group, the Japanese authors being unaware that their compound was optically active. Brockmann[214] showed later that both alkannin and shikonin could be converted into an identical optically inactive monomethyl ether and, in combination, yielded the racemic compound shikalkin, m.p. 148°.

The naphthazarin structure was recognised from the ultraviolet-visible absorption and the deep blue colour obtained in alkaline solution, and was confirmed by ozonolysis of the yellow triacetate which gave 3,6-dihydroxyphthalic acid and also acetone.[214] As the pigment is only slightly soluble in aqueous sodium carbonate the third hydroxyl group cannot be attached directly to the naphthazarin system and must therefore be located in a side chain, the nature of which was determined as follows. On catalytic hydrogenation alkannin absorbs three mols. of hydrogen giving, after reoxidation in air, a new quinone, alkannan, $C_{16}H_{18}O_4$, (151), which contains only two hydroxyl groups. Oxidation of alkannan with permanganate in acetone then gave 5-methylhexoic

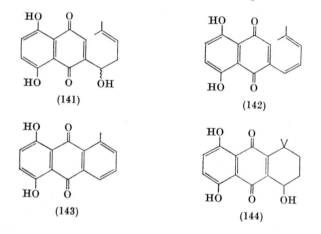

(141)   (142)

(143)   (144)

acid. When the racemic monomethyl ether (obtained by treating alkannin with cold 2% methanolic hydrochloric acid) was similarly hydrogenated it was possible to interrupt the reaction after absorption of two mols. of hydrogen, and degradation of the product then gave 2-methoxy-5-methylhexoic acid. It follows that alkannin has the α-hydroxyisohexenyl side chain shown in (141) which accounts for the optical activity and the "benzylic" reactivity of the alcoholic group, for example, formation of anhydro-alkannin (142) under acid conditions. The formation of 8-methylquinizarin (143) on dry distillation of alkannin, and its conversion into cyclo-alkannin (144) on treatment with stannic chloride in cold

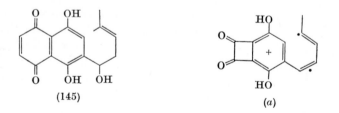

(145)   (a)

benzene, are acceptable transformations based on structure (141). Alkannin/shikonin are of course tautomeric and undergo addition reactions[214] with diazomethane and 2,3-dimethylbutadiene in the form (145). A significant peak in the mass spectrum of alkannin at $M—H_2O—Me—C_2H_2$ has been attributed[499] to the ion (a) derived from (145) but this ion is not apparent in the spectrum of shikonin.[505]

The remaining configurational uncertainty was resolved many years later by Arakawa and Nakazaki.[231] Ozonolysis of (+)shikonin in acetic acid, followed by treatment with hydrogen peroxide, gave malic acid (146) which was converted by ammonolysis of the dimethyl ester in liquid ammonia into D(+)-malamide (147). It follows that shikonin has the (R)- and alkannin the (S)-configuration.

(146)   (147)

The natural esters are labile and since their hydrolysis has no effect on the visible absorption only the side chain hydroxyl is esterified. Two structural variations recently found[499] in *Arnebia nobilis* are the esters (148)† in which the terminal double bond is hydrated. On hydrolysis they give, respectively, acetic and $\beta\beta$-dimethylacrylic acids, together

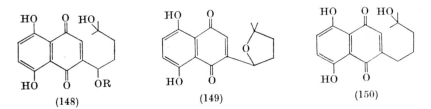

(148)   (149)   (150)

with (148; R = H) and (149). The position of the ester groups in (148) was established by dehydration to the corresponding alkannin esters.

A related compound from the same source has the structure (150).‡[499] In contrast to other members of the group there are two protons on the α-carbon atom next the quinone ring, revealed by a multiplet at τ 7·35 (2H), and the signal from the quinonoid proton is thus a triplet (τ 3·06). The mass spectrum and other features of the n.m.r. spectrum are consistent with structure (150).

† Arnebin-2 (R = Me_2C=CHCO), arnebin-6 (R = Ac).
‡ Arnebin-5.

Alkannan (151), $C_{16}H_{18}O_4$

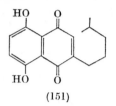

(151)

Occurrence: roots of *Alkanna tinctoria* Tausch.[214] (Boraginaceae).
Physical properties: red leaflets, m.p. 99°, λ_{max}. (CHCl$_3$) 485, 520, 558 nm.

Evidence for the structure of alkannan was given above. Simple naphthazarins can be obtained by condensing maleic anhydrides with quinols in an aluminium chloride-sodium chloride melt. The method apparently succeeded in the hands of Kuroda and Wada[232] (no yield stated) using isohexylquinol but Brockmann and Müller[233] obtained only a trace of alkannan in this way and showed it to be unstable under these conditions. More success attended their condensation of naphthazarin with isocaproaldehyde in hydrochloric acid-acetic acid which

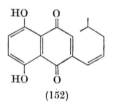

(152)

afforded dehydro-alkannan (152), leading to alkannan after hydrogenation. The direct formation of some alkannan (amount unstated) in this condensation was unexpected as was the appearance of another compound, regarded[233] as 6,7-di-isobutylquinizarin; formation of the latter implies that naphthazarin was doubly alkenylated in the same ring, followed by cyclisation.

Flaviolin (153), $C_{10}H_6O_5$
Flaviolin-2,7-dimethyl ether, $C_{12}H_{10}O_5$

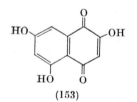

(153)

Occurrence: cultures of *Aspergillus citricus* (Wehmer) Mosseray.[195] (flaviolin) and an unidentified *Streptomyces* sp.[201] (the dimethyl ether).

Physical properties: (153) garnet-red rhombs, dec. ca. 250°, $\lambda_{max.}$ (EtOH) 215, 262, 308, 403, 452 nm (log ε 4·43, 4·12, 3·93, 3·30, 3·35), $\nu_{max.}$ (KBr) 3230, 1675, 1630, 1590 cm^{-1}.

2,7-Dimethyl ether, orange crystals, m.p. 266–268°, $\lambda_{max.}$ (CHCl$_3$) 261, 302, 435 nm (log ε 4·18, 4·07, 3·63), $\nu_{max.}$ (Nujol) 1667, 1613 cm^{-1}.

Cultures of *A. citricus* in an aqueous medium become gradually yellow but change, within a day, to a dark port-wine red as the solution becomes alkaline. This is characteristic of flaviolin which exhibits a striking series of colour changes as the pH is lowered from < 2·8 (yellow) to 7·0 (red) to >10 (deep violet). Such changes immediately suggest a hydroxy-quinone structure which was confirmed by its redox properties. The pigment formed a trimethyl ether, with difficulty, and a triacetate which could be oxidised with chromic acid to 3,5-dihydroxyphthalic acid. Flaviolin is therefore a 5,7-dihydroxy-1,4-naphthaquinone with another hydroxyl group (pK_a 4·7) in the quinone ring. The correct structure (153) was predicted[199a] on the basis of acetate biogenesis and established[198, 199] by synthesis of the trimethyl ethers of both isomers.

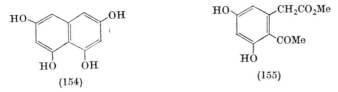

(154)   (155)

Flaviolin itself has been synthesised by Thiele acetylation of 7-hydroxyjuglone,[554] and by aeration,[475] in alkaline solution, of 1,3,6,8-tetrahydroxynaphthalene (154) which was prepared by Claisen cyclisation of (155) (cf. ref. 196). Other syntheses of the 5,7-dimethyl and the trimethyl ether of flaviolin have also been reported.[197, 200, 364] The natural dimethyl ether was identified as a dimethoxyjuglone by its chemical and spectroscopic properties, and most probably as the 2,7-isomer since its co-metabolite is 2,7-dimethoxynaphthazarin. This was confirmed[201] when vigorous methylation converted the pigment into flaviolin trimethyl ether.

6-Ethyl-2,7-dimethoxyjuglone (156), C$_{14}$H$_{14}$O$_5$

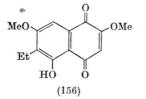

(156)

Occurrence: cultures of *Hendersonula toruloidea* Nattrass.[547]
Physical properties: yellow needles, m.p. 187°, λ_{max}. (MeOH) 221, 256(sh), 262, 305, 422 nm (log ϵ 4·48, 4·21, 4·22, 3·98), ν_{max}. (KBr) 1672, 1628, 1590 cm^{-1}.

The n.m.r. spectrum of this new pigment provided evidence for the presence of a *peri*-hydroxyl (τ −2·50) and two methoxyl groups, an ethyl group attached to an aromatic ring, and two isolated protons (τ 2·70 (ArH) and 3·96 (QH)). From this, the molecular formula and the visible absorption, structure (156), or the 3-OMe isomer, seemed likely. The quinone (156) was already known and direct comparison confirmed their identity.[547] It was previously obtained[372] by methylation of 2,7-dihydroxy-6-ethyljuglone, one of the many compounds formed by reduction of 3-acetyl-2,7-dihydroxynaphthazarin (p. 268) with sodium borohydride.

Hydroxydroserone (157), $C_{11}H_8O_5$

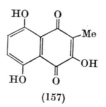

(157)

Occurrence: tuberous roots of *Drosera whittakeri* Planch.,[163] *D. erythrorhiza* Lindl.,[535] *D. stolonifera* Endl.,[535] *D. rosulata* Lehm.,[535] *D. zonaria* Planch.[535] and *D. gigantea* Lindl.[535] (tubers, stems, leaves, and particularly flowers and seed capsules[536]) (Droseraceae).
Physical properties: red plates, m.p. 192–193°, λ_{max}. (EtOH) 239(sh), 255(sh), 298, 488, 518(sh) nm (log ϵ 4·27, 3·95, 3·92, 3·83, 3·71), ν_{max}. (KCl) 3400, 1603 cm^{-1}.

Although the early investigations[163] indicated that hydroxydroserone was a trihydroxymethylnaphthaquinone there was little positive evidence, but later studies[172] established that it was a naphthazarin derivative by spectroscopic (u.v.) comparison with model compounds and the formation of a bis-boroacetate.[164] The third hydroxyl group was strongly acidic and was thought to be attached to the same ring as the methyl group since a nitrogeneous adduct was easily formed with diazomethane by reaction with the tautomeric 5,8-quinone structure. Substituted quinone rings can, in fact, add diazomethane but nevertheless structure (157) is correct. This was confirmed[165] synthetically by condensing maleic anhydride with 3-methoxy-2-methylquinol in fused aluminium chloride-sodium chloride. Subsequently, hydroxydroserone was obtained from 2-methylnaphthazarin by aeration in alkaline solution,[173]

(158)   (159)

and also by Diels–Alder addition of butadiene to 3-hydroxy-2-methyl-benzoquinone.[174] The adduct was aromatised in acetic anhydride followed by oxidation of the triacetate (158) to (159) with chromium trioxide.

Lomandrone (160), $C_{15}H_{16}O_6$

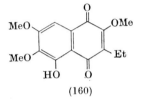

(160)

Occurrence: roots of *Lomandra hastilis* R. Br.[500] (Liliaceae).
Physical properties: orange prisms, m.p. 118°, λ_{max}. (EtOH) 260(sh), 266, 312, 427 nm (log ϵ 3·85, 3·90, 3·60, 3·18), ν_{max}. (KCl) 1667, 1635 cm^{-1}.

Lomandrone forms a methyl ether, an acetate and a leucotriacetate, and from its spectroscopic properties (ultraviolet-visible and infrared) and colour reactions it is evidently a juglone derivative. The n.m.r. spectrum revealed, in addition to the *peri*-hydroxyl group, an ethyl group and three methoxyls, and a signal (1H) at τ 2·8 attributable to an aromatic proton at a *peri*-position between a carbonyl and a methoxyl function. The ArH singlet was slightly broadened but irradiation of the methoxyl protons at τ 5·89 caused considerable sharpening, and vice versa. It may be concluded that one of the methoxyl groups is adjacent to the un-substituted *peri*-position which is consistent with the relatively large upfield shift (0·63 ppm) of that particular methoxyl resonance when the solvent was changed from chloroform to benzene (the signals from the more hindered methoxyl groups shifted only 0·13–0·14 ppm). Accordingly three structures can be considered for lomandrone, (160), (161) and (162);

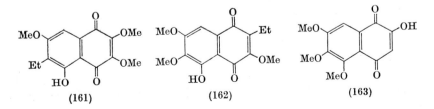

(161)   (162)   (163)

structure (161) was eliminated by synthesis, and compound (160) was already known[379] and seemed to resemble lomandrone closely in spectroscopic properties although there was a large discrepancy in melting point. That structure (160) is, in fact, correct was established by a synthesis of lomandrone methyl ether. This was accomplished by autoxidation of 5,6,7-trimethoxy-1-tetralone in basic solution to give the quinone (163) which was converted into the methyl ether of (160) by methylation, followed by C-ethylation with propionyl peroxide.[500]

Mompain (164), $C_{10}H_6O_6$
Mompain-2,7-dimethyl ether, $C_{12}H_{10}O_6$

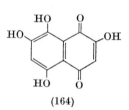

(164)

Occurrence: (164) mycelium of *Helicobasidium mompa* Tanaka,[202] and spines of *Echinothrix diadema* L.,[203] *E. calamaris* Pallis,[203] *Echinometra oblonga* Blain.[203] and *Tripneustes gratilla* L.; [203] the 2,7-dimethyl ether in cultures of an unidentified *Streptomyces* sp.[201]

Physical properties: (164) red needles, subl. 265–275°[203] (dec. >300°),[204] $\lambda_{max.}$ (EtOH) 228, 272, 318, 486, 517, 554 nm (log ε 4·43, 4·06, 3·93, 3·75, 3·80, 3·63), $\nu_{max.}$ (KBr) 3240, 1660(sh), 1625(w), 1602 cm^{-1}.

2,7-Dimethyl ether, dark red needles, m.p. 275–276°, $\lambda_{max.}$ (CHCl$_3$) 285, 308, 480, 512, 550 nm (log ε 3·94, 3·97, 3·85, 3·93, 3·74), $\nu_{max.}$ (Nujol) 1587 cm^{-1}.

Helicobasidium mompa is unusual in elaborating quinones of two completely different types, the sesquiterpenoid helicobasidins (p. 130) and the polyhydroxynaphthaquinone mompain, also found in several echinoids. The formation of a tetra-acetate ($\lambda_{max.}$ 248, 266(sh), 352 nm) and a leucohexa-acetate ($\lambda_{max.}$ 231, 295, 328(sh) nm) clearly indicates a tetrahydroxynaphthaquinone structure, and conversion into a dimethyl ether with diazomethane implies that two of the hydroxyl groups occupy β-positions and two are *peri*-substituted.[204] As mompain is different from 2,3-dihydroxynaphthazarin it appeared to be the 2,6- or 2,7-isomer, and the latter was favoured by the non-equivalence of the *peri*-hydroxyl protons in the n.m.r. spectrum of the dimethyl ether (τ −3·16 and −2·73)[204] (in 2,6-dimethoxynaphthazarin both hydroxyl protons resonate at τ −3·07).[328] Confirmation of this was forthcoming when Natori *et al.*[204] obtained mompain from 3-acetyl-2,7-dihydroxy-naphthazarin (p. 268) by de-acetylation in hot sulphuric acid. Subsequently it was synthesised by autoxidation of 5,7,8-trimethoxy-1-

tetralone to give 2-hydroxy-5,7,8-trimethoxy-1,4-naphthaquinone, which was demethylated.[205]

The 2,7-dimethyl ether, a co-metabolite of 2,7-dimethoxyjuglone, was identified[201] by its spectroscopic properties and colour reactions, and has been synthesised[475] by oxidising 1,8-dihydroxy-3,6-dimethoxynaphthalene with hydrogen peroxide and trifluoroacetic anhydride (another product is flaviolin).

SPINOCHROMES

These pigments were first discovered in the calcareous parts of sea urchins, that is, the spines and test (shell), and were so-named[331] to distinguish them from the "echinochrome" pigments found in the perivisceral fluid, eggs and internal organs. Historically, the name echinochrome has precedence, being first used in 1883 by MacMunn[314] for the red pigment in the perivisceral fluid of *Echinus esculentus* and other species, but echinochrome A† is actually a common spine pigment and since, in structure, it belongs to the spinochrome group, the name is now redundant. The spinochromes occur as mixtures of closely related compounds, and the inadequate methods used for their separation and purification together with the tendency for these pigments to form solvates, led to erroneous analytical data,‡ and molecular and structural formulae, which resulted in considerable confusion.[315] Happily this situation has been clarified by recent work; Gough and Sutherland[316] have shown that the quinones previously designated as spinochromes B, B_1, M_2, N and P_1 are, in fact, all identical (165), and Scheuer and his co-workers[317] have established that spinochromes A and M are identical (168), and likewise spinochromes C and F are the same (170). This considerably reduces the number of sea urchin quinones described in the older literature.[315, 320]§ Those most frequently encountered are spinochromes A–E and echinochrome A (165–170) whose structures are now firmly established. Gough and Sutherland[316] suggested that spinochrome B is the most appropriate trivial name for (165), "if one is needed". As the number of authentic spinochrome pigments is now more than twenty, and additions may be expected, the alphabetical

† Echinochromes B and C, found with echinochrome A in the ovaries of *Arbacia lixula*, have not been identified, and have not appeared in the literature since 1940.[319] Echinochrome B, a red compound, m.p. 173–175°, appears to be a naphthazarin, possibly identical with one of the known spinochromes (or a mixture). Little is known of the violet pigment, echinochrome C, which may not be a quinone.
‡ Elementary analysis has been virtually abandoned by one group.[318]
§ Spinochrome G[322] is (174).[504] Spinochrome P[350] was found to be essentially spinochrome A (168).[503]

9

system of nomenclature is clearly obsolete and has now been abandoned in favour of a semi-trivial system whereby the pigments are named as substituted juglones or naphthazarins.[318, 326]†

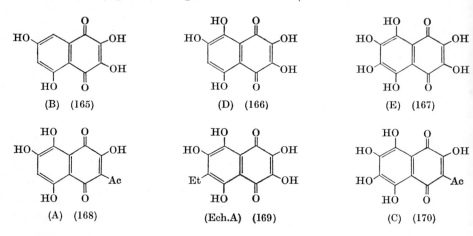

(B) (165)	(D) (166)	(E) (167)
(A) (168)	(Ech.A) (169)	(C) (170)

The pigments occur principally in spines and test as calcium (and magnesium) salts, and are isolated by treatment with hydrochloric acid and ether.[318, 326] Most species yield up to six pigments (see Table I) but additional trace components may be present, and the proportions may vary considerably from animal to animal, resulting in wide variations in colour within a population. By working on the kilogram scale Scheuer and his group[318] were able to isolate pigments present only to the extent of $<10^{-5}$ % in the spines. Startling exceptions to the general picture given by Table I are provided by the Hawaiian echinoids *Echinothrix diadema* and *E. calamaris* which elaborate[318] some thirty pigments (20 kg of *E. diadema* spines were processed). About half of these have been identified as spinochromes and one is a benzoquinone (p. 116). The remainder, which have not yet been characterised, are not necessarily quinones and one, at least, is nitrogeneous.[321] An abundance of pigments has also been found recently[321] in ophiuroids, *Ophiocoma erinaceus*, for example, yielding eight spinochromes as well as the nitrogeneous compound just mentioned.

The function of these pigments remains something of a mystery. Early suggestions that they have a respiratory function could not be con-

---

† Some account of the use of alphabetical suffixes is given elsewhere.[315, 349] The following list of synonyms may be helpful when referring to the literature.

Spinochrome A[317, 349] = B₃[354] = M[337] = M₁[352] = Aka₂[353] = P[326, 350]
     B[317, 322] = B[340] = B₁[338] = M₂[353] = N[315] = P₁[350]
     C[317, 322] = B₂[358] = F[337] = F₁[358] = M₃[359] = spinone A[357]
     = isoechinochrome[356]
     D[349] = Aka[337] = Aka₁[353]

firmed,[383, 385] and the view that the pigmented protein complex secreted by the eggs of *A. lixula* was a powerful sperm stimulator[386] could not be sustained.[387] The photosensitivity of sea urchins has been studied extensively[389, 390] and the correlation initially observed[391] between the action spectrum of *Diadema antillarum* and the visible absorption of the spinochromes in the skin appeared to be significant, but after further investigations[346] the participation of spinochromes in photoreception remains an open question. In any case some echinoids have little or no spinochrome pigmentation. The most recent idea is that spinochromes may function as algistats.[392] This is based on the use[393] of 2,3-dichloro-1,4-naphthaquinone to control the growth of *Cyanophyceae* and the observation that the growth of an algal culture could be inhibited by pieces of echinoid test. A blue-green alga has frequently been observed as a parasite on *E. esculentus*.[394]

In addition to spinochromes (165) to (170), the juglones (171) to (176) and the naphthazarins (177) to (191) have been found in Echinodermata. Some physical properties are given in Table II.† As spectroscopic data are now available[362, 363, 367, 529] for a large number of juglones and naphthazarins, it is possible to identify spinochromes largely by spectroscopic methods with little resort to chemical manipulation. Earlier structural investigations relied heavily on unreliable analytical data and the formation of derivatives, and very little degradative work was

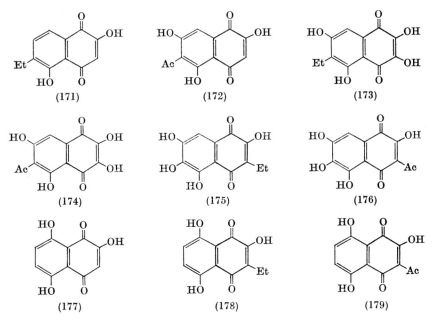

(171) (172) (173)

(174) (175) (176)

(177) (178) (179)

† Infrared spectra are not very informative in this series and are not included.

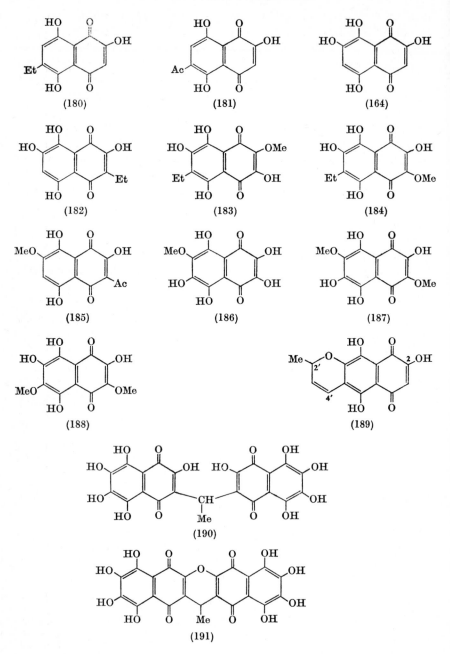

(180)   (181)   (164)

(182)   (183)   (184)

(185)   (186)   (187)

(188)   (189)

(190)

(191)

carried out. Consequently many incorrect structures were advanced, and in the outline of evidence given below only the essential facts in the early work are presented. Virtually all uncertainties have been removed by recent spectral studies and by synthesis, but unfortunately a new element of doubt has now appeared. For convenience in handling and separating complex mixtures it has been the practice in one laboratory[318] to methylate partially purified fractions and subsequently hydrolyse the individual methyl ethers after further purification. This was based on the assumption that spinochrome methyl ethers did not occur naturally but several have since been discovered. Consequently there is the possibility that some of the pigments reported in *Echinothrix* spp.[318] (see Table I) exist *in vivo* as their methyl ethers.

2,3,7-Trihydroxyjuglone (165). The formation[340] of a tetra-acetate and leucohexa-acetate, together with the molecular formula, $C_{10}H_6O_6$, show this to be a tetrahydroxynaphthaquinone, and the location of the substituents was deduced from the formation of 3,5-dimethoxyphthalic acid on degradation of the tetramethyl ether with hydrogen peroxide.[360] Synthetic confirmation was obtained[200] by treating the 2,3-epoxide of 5,7-dimethoxy-1,4-naphthaquinone with cold acetic anhydride to give the diacetate (192), followed by aeration in alkaline solution and demethylation of the resulting 2,3-dihydroxy-5,7-dimethoxy-1,4-naph-thaquinone.

2,3,6-Trihydroxynaphthazarin (166). The general nature of this spino-chrome is evident from the formation of a penta-acetate,[337, 338] a leuco-hepta-acetate[340] and a trimethyl ether (with diazomethane),[337, 338] but

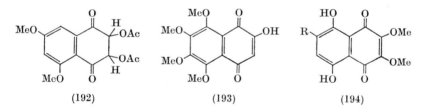

(192)   (193)   (194)

faulty analyses indicated a C_{11} molecular formula. However the n.m.r. spectrum of the penta-acetate revealed[365] the presence of one aromatic and fifteen acetate protons, and hence the parent compound is a tri-hydroxynaphthazarin. This has been established by two syntheses; in one procedure[365] the tetramethyl ether (193) was prepared by autoxida-tion of 5,6,7,8-tetramethoxy-1-tetralone in basic solution, and in the other,[366, 471] chloromaleic anhydride was condensed with 1,2-dihydroxy-3,4-dimethoxybenzene in an aluminium chloride-sodium chloride melt to give a mixture of naphthazarins, which, after methylation and chromatographic separation, afforded, principally (194; R = Cl), and

TABLE I.

Occurrence of spinochromes in *Echinodermata*

Class/Order	Species	Pigments						Others	Reference
		(165)	(166)	(167)	(168)	(169)	(170)		
CRINOIDEA (sea lilies, feather stars)									
ARTICULATA	*Antedon*[a]?				+			(164)	321
HOLOTHUROIDEA (sea cucumbers)[c]									
ASPIDOCHIROTA	*Polycheira rufescens* (Brandt)							(186)	324
ECHINOIDEA (sea urchins, sand dollars)									
CIDAROIDA (sea urchins)	*Chondrocidaris gigantea* (Ag.)		+		+				321
	Cidaris cidaris (Linn.)	+				+	+		326
	Eucidaris metularia (Lamk.)	+			+		+	(164)	321
	Eucidaris thouarsii (Val.)	+		+	+	+	+		326
	Goniocidaris tubaria (Lamk.)			+	+				333
	Goniocidaris tubaria var. *impressa* (Koehl.)				+				333
	Phyllacanthus irregularis (Mort.)			+	+	+			333
	Phyllacanthus parvispinus (Woods)			+	+				504
	Prionocidaris bispinosa (Lamk.)[a]	+			+	+	+		326
	Prionocidaris hawaiiensis (Ag. & H. L. Clark)		+		+		+		321

AULODONTA (sea urchins)

Species							References	Numbers
Centrostephanus nitidus (Koehl.)[d]								326
Centrostephanus rodgersii (Ag.)			+					504
Diadema antillarum (Philippi)			+	+			(183), (184)	326, 327, 401
Diadema paucispinus (Ag.)			+	+				328
Diadema setosum (Leske)			+	+	+			326,[d] 329
Echinothrix calamaris (Pallas)	+		+	+	+		(164), (171), (172), (173), (174), (177), (179), (180), (181), (182), (189)	318, 502
Echinothrix diadema (Linn.)	+		+	+	+		(164), (172), (173), (174), (177), (179), (181), (182), (189)	318

STIRODONTA (sea urchins)

Species							References	Numbers
Arbacia incisa (H. L. Clark)				+				326
Arbacia lixula (Linn.) (= A. pustulosa)		+	+	+				319, 322, 331, 332
Stomopneustes variolaris (Lamk.)			+	+				326,[d] 330
Tetraphygus niger (Molina)			+	+		+		555

CAMARODONTA (sea urchins)

Species							References	Numbers
Amblypneustes oveum (Lamk.)			+	+				333
Anthocidaris crassispina (Ag.)	+		+	+	+		b	326, 337, 338, 339
Colobocentratus atratus (Linn.)			+	+	+			328
Echinometra lucunter (Linn.)	+		+	+		+		326
Echinometra mathaei (Blain.)	+		+	+	+			326,[d] 465

TABLE I—continued

Class/Order	Species	Pigments							Reference
		(165)	(166)	(167)	(168)	(169)	(170)	Others	
CAMARODONTA—continued									
	Echinometra oblonga (Blain.)				+		+	(164), (174)?	317, 318, 328
	Echinostrephus aciculatus (Ag.)	+		+	+		+		336
	Echinus acutus (Lamk.)	+		+	+		+		326
	Echinus elegans (Düben & Koren)	+		+	+	+	+		326
	Echinus esculentus (Linn.)	+	+	+	+	+	+		326, 335
	Heliocidaris erythrogramma (Val.)	+	+						333
	Hemicentrotus pulcherrimus (Ag.) (= Strongylocentrotus pulcherrimus)	+			+		+		326, 340
	Heterocentrotus mammillatus (Linn.)	+			+		+		326, 337
	Loxechinus albus (Molina)					+			555
	Paracentrotus lividus (Lamk.)	+		+	+	+	+	(174)	317, 322, 326, 335, 341, 504
	Psammechinus miliaris (Gmel.)	+		+	+		+		326, 342
	Pseudocentrotus depressus (Ag.)	+	+						334
	Salmacis sphaeroides (Linn.)	+		+	+		+	(176)	316, 347
	Sphaerechinus granularis (Lamk.)			+	+	+	+		326
	Sterechinus neumayeri (Meiss.)[d]			+	+		+		326
	Strongylocentrotus drobachiensis (O. F. Müll.)	+		+	+		+	(191)	326, 401, 448
	Strongylocentrotus franciscanus (Ag.)	+		+	+		+		326
	Strongylocentrotus purpuratus (Stim.)	+		+	+		+		317, 326, 385
	Temnopleurus toreumaticus (Leske)	+		+	+		+	(173), (176)	326, 347, 401
	Tripneustes gratilla (Linn.)	+		+	+		+	(164)	318, 328
	Trioneustes ventricosus (Lamk.)	+		+			+		326

	CLYPEASTEROIDA (sand-dollars, cake urchins)				
Dendraster excentricus (Esch.)		+	+		326, 344
Echinarachnius parma (Lamk.)			+		448
Echinocyamus pusillus (O. F. Müll.)			+?		326
Encope emarginata (Leske)			+		326
Mellita sexiesperforata (Leske)	+		+		326
Scaphechinus mirabilis (Ag.) (= Echinarachnius mirabilis)			+		345
SPATANGOIDA (heart urchins)					
Echinocardium cordatum (Penn.)			+		326
Meoma ventricosa (Lamk.)		+	+		326
Spatangus purpureus (O. F. Müll.)		+	+	(190), (191)	326, 401
Spatangus raschi (Lovén)			+?	(191)	326, 401
ASTEROIDEA (starfishes)					
SPINULOSA					
Acanthaster planci (Linn.)				(187), (188)	346
OPHIUROIDEA (brittle stars)					
OPHIURAE					
Ophiocoma erinaceus (O. F. Müll. & Tros.)		+	+	(175), (178), (179), (180), (182), (185)	321
Ophiocoma insularia Lyman			+	(175), (178), (179), (180), (185)	312

[a] Species not yet identified. Spinochromes are possibly present in *Antedon bifida* (Pennant).[323]

[b] "Spinochrome M$_3$" is a mixture of (165), (168) and (170).[348] Testachromes M$_1$ and M$_2$[506] are probably similar mixtures.

[c] Unidentified naphthaquinones have been detected in the sea cucumber *Stichopus japonicus* Selenka.[351]

[d] Museum specimen only.

TABLE II.

Physical properties of spinochrome pigments

Pigment	Mol. formula	Physical form	m.p.	$\lambda_{max.}$ (EtOH or MeOH) nm	log ϵ
165	$C_{10}H_6O_6$	Red needles	>260° (dec.)	272, 323, 385, 480	4·38, 4·10, 4·04, 3·69
166	$C_{10}H_6O_7$	Red needles	285–290° (subl.)	266, 333, 463, 490(sh), 530	4·10, 3·44, 3·41, 3·40, 3·26
167	$C_{10}H_6O_8$	Red needles	>300°	270, 359, 450(sh), 477, 508(sh)	4·18, 3·70, 3·59, 3·64, 3·50
168	$C_{12}H_8O_7$	Purple needles	192–193°	251, 270(sh), 317, 490(sh), 520	3·98, 3·95, 3·94, 3·49, 3·52
169	$C_{12}H_8O_7$	Red needles	222–223°	260, 343, 470, 490(sh), 530	4·32, 3·99, 3·88, 3·87, 3·66
170	$C_{12}H_8O_8$	Red-orange needles	246–248°	240, 285, 330(sh), 463, 504	4·12, 4·21, 3·95, 3·77, 3·66
171	$C_{12}H_{10}O_4$	Orange prisms	219–220°	237, 291, 418(sh), 432, 455(sh)	3·91, 4·01, 3·54, 3·59, 3·49
172	$C_{12}H_8O_6$	Orange needles	215° (dec.)	241, 307, 374, 469	4·45, 4·16, 3·45, 3·47
173	$C_{12}H_{10}O_6$	Brick-red needles	265–269° (dec. and subl.)	270, 330, 417, 485(sh)	4·34, 4·02, 3·57, 3·18
174	$C_{12}H_8O_7$	Brick-red crystals	245–255° (subl.)	305, 388, 490	
175	$C_{12}H_{10}O_6$		220–226°		
176	$C_{12}H_8O_7$	Dark brown needles	280–285° (dec.)	273, 314, 424[b]	4·30, 4·12, 3·59
177	$C_{10}H_6O_5$	Orange brown crystals	200–210° (subl.)	290, 390(sh), 475, 488, 498(sh), 519(sh), 534(sh)	3·93, 3·23, 3·75, 3·77, 3·77, 3·63, 3·52
178	$C_{12}H_{10}O_5$		190·5–191·5°	296, 385(sh), 476(sh), 490(sh), 500, 524(sh), 538(sh)	3·88, 3·11, 3·74, 3·78, 3·80, 3·64, 3·57

179	$C_{12}H_8O_6$		163–164°	250(sh), 296, 490, 525(sh), 568(sh)	3·96, 4·02, 3·65, 3·60, 3·29
180	$C_{12}H_{10}O_5$	Brown fern-like crystals	204°	302, 479(sh), 505, 530, 544	3·93, 3·71, 3·80, 3·62, 3·63
181	$C_{12}H_8O_6$	Black needles	179–180° (dec.)	236, 297, 507	4·22, 3·90, 3·81
182	$C_{12}H_{10}O_6$	Dark red needles	190–192°	237, 265, 321, 420(sh), 450(sh), 489(sh), 513, 549(sh)	4·18, 3·88, 3·92, 3·43, 3·54, 3·69, 3·74, 3·55
183	$C_{13}H_{12}O_7$	Deep red prisms	202–204°	239, 256(sh), 337, 465, 493, 530	4·43, 4·40, 4·11, 3·40, 4·01, 3·88
184	$C_{13}H_{12}O_7$	Deep red prisms	179–184°	239, 263(sh), 337, 469, 499, 536(sh)	4·28, 4·12, 3·92, 3·78, 3·77, 3·58
185	$C_{13}H_{10}O_7$	Dark brown needles	246–248°	271, 313, 508	
186	$C_{11}H_8O_8$	Crimson prisms	218°	272, 361, 480	
187	$C_{12}H_{10}O_8$		252–254°	338, 459, 482, 520	
188	$C_{12}H_{10}O_8$		218–219°	342, 463(sh), 480, 520(sh)	
189	$C_{14}H_{10}O_6$		165–172° (dec.)	292, 340, 488, 564(sh)[a]	
190	$C_{22}H_{14}O_{14}$	Dark red needles	155–157°	262, 344, 470, 487, 526	4·55, 4·14, 4·07, 4·13, 4·03
191	$C_{22}H_{12}O_{13}$	Dark red needles	253–256°	268, 318, 468, 491, 523	4·50, 4·00, 4·12, 4·10, 3·95

[a] In $CHCl_3$.
[b] Before crystallisation; after crystallisation 249, 302, 424 nm.

after prolonged reaction with sodium methoxide in methanol, this gave the trimethyl ether (194; R = OMe). Demethylation of (193) and (194; R = OMe) gave (166). In the condensation with chloromaleic anhydride, just mentioned, the expected product is accompanied by appreciable amounts of 2,3-dihydroxynaphthazarin and its 6,7-dichloro derivative. There appears to be no precedent for this chlorine migration.

2,3,6,7-Tetrahydroxynaphthazarin (167). This compound contains ca. 50% of oxygen by weight and has the highest oxidation level of all the spinochromes. The structure is evident[368] from the molecular formula, $C_{10}H_6O_8$, and the formation of a hexa-acetate and a leuco-octa-acetate. The high fusion point and insolubility in ether (unusual for spinochromes) are not surprising. Methylation with diazomethane gives the 2,3,6,7-tetramethyl ether (195),[200] which was obtained synthetically[366, 471] from 6,7-dichloro-2,3-dimethoxynaphthazarin prepared as

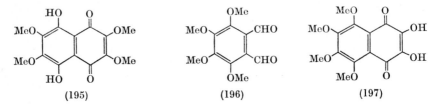

(195)   (196)   (197)

above, or in better yield using dichloromaleic anhydride. Prolonged refluxing with sodium methoxide gave, *inter alia*, 2,3,6-trimethoxy-7-hydroxynaphthazarin which was converted into the tetramethyl ether with diazomethane. In another synthesis,[369] using Weygand's method[108] for preparing isonaphthazarins (2,3-dihydroxy-1,4-naphthaquinones), the tetramethyl ether (197) was obtained by a crossed benzoin condensation of glyoxal with the phthalaldehyde (196). Both (195) and (197) gave tetrahydroxynaphthazarin on demethylation with hydrobromic acid.

3-Acetyl-2,7-dihydroxynaphthazarin (168). The correct structure for this pigment was deduced by Scheuer and his co-workers[328] mainly from spectral data. A naphthazarin structure is indicated by the ultraviolet-visible absorption, and the n.m.r. spectrum provided evidence for one aromatic proton and an acetyl group adjacent to hydroxyl on a quinone ring.[363] Together with the molecular formula, $C_{12}H_8O_7$, this suggests a 3-acetyl-2,6- or 2,7-dihydroxynaphthazarin structure, and the second possibility was found to be correct when reaction with methanolic hydrochloric acid led to the elimination of the acetyl group (hydrolysis of a β-diketo system) and methylation of the two β-hydroxyls resulting in the formation of 2,7-dimethoxynaphthazarin. In effect, this degradation was reversed in the subsequent synthesis[366, 471] of spinochrome

(168); methoxynaphthazarin was oxidized with lead tetra-acetate to the diquinone (198) which, without purification, was subjected to Thiele acetylation to give, after mild acid hydrolysis and reoxidation, a separable mixture of (199) and the 2,6-isomer. The leuco-acetate of (199) was

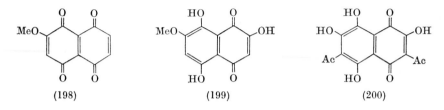

(198)   (199)   (200)

then acylated with boron trifluoride in cold acetic acid, and the product was gently hydrolysed to the diacetyldihydroxynaphthazarin (200). Partial deacylation was then effected by boiling in methanol solution with dilute hydrochloric acid to give, mainly, (168) identical with the natural material.

Methylation [379] of (168) with diazomethane proceeds as expected but the reaction with methyl sulphate is poor and gives a mixture of mono-, di- and tri-methyl ethers, the hydroxyl adjacent to the acetyl group remaining unattacked. With methanol and hydrogen chloride, methylation is accompanied by deacetylation. When (168) was reduced with a large excess of sodium borohydride a surprising mixture of eleven compounds resulted.[372] The majority were formed in <1% yield and would normally be neglected but detailed investigation [372] showed that they comprised six juglone and five naphthazarin derivatives. The major product was 2,7-dihydroxy-3-ethylnaphthazarin (11%) followed by 2,7-dihydroxy-6-ethyljuglone (4%), but in three of the minor products the acetyl group was retained intact while one, or even two, hydroxyls were removed. These reductions are not fully understood but can be of value in structural elucidations,[318, 379] despite the poor yields. (Phloroglucinol has been reduced to resorcinol in good yield with sodium borohydride.[381]) Elimination of hydroxyl groups from the quinonoid ring of juglones and some naphthazarins is readily effected with acid stannous chloride [370] and probably proceeds by tautomerisation of the initial quinol to the β-tautomer [for example (201) from 3-hydroxyjuglone], followed by elimination of water. Under alkaline conditions, using sodium stannite [371] as reducing agent, it is possible to remove one of the peri-hydroxyl groups from a naphthopurpurin system, for example, (202) → (204), a reaction which proceeds very smoothly in the anthraquinone series, under milder conditions.[544] In this case the resorcinol moiety of the leuco compound of (202) may react in the tautomeric form (203) and so lose its benzylic hydroxyl group.

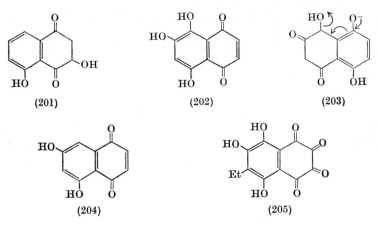

(201)   (202)   (203)

(204)   (205)

6-Ethyl-2,3,7-trihydroxynaphthazarin (169). This was the first sea urchin quinone to be isolated[373] and identified, the compound being obtained chiefly from the eggs and ovaries of *Arbacia lixula*.[331, 332, 374] The structure (169) was advanced by Kuhn and Wallenfels[332] on the basis of the molecular formula, $C_{12}H_{10}O_7$, the formation of a leucohepta-acetate, and of propionic acid (0·83 mol.) on chromic acid oxidation. This was confirmed synthetically[375] by condensing 2-ethyl-1-3,4-trimethoxybenzene with dibenzoyloxymaleic anhydride in an aluminium chloride-sodium chloride melt. The yield is poor (1·5%) owing to the instability of the anhydride component, and in a similar condensation[376] with quinol the product, obtained in good yield, was 2,5-dihydroxybenzo-phenone and no naphthazarin derivative could be detected. More recently, the pigment (169) has been obtained[365] by peroxide ethylation of trihydroxynaphthazarin (166) with propionyl peroxide. Like other 2,3-dihydroxy-1,4-naphthaquinones, (169) can be oxidised with silver oxide and other reagents, including hydrogen peroxide in the presence of a dehydrogenase,[377] to a yellow dehydro compound (205), known as dehydro-echinochrome.[378] As with other vicinal polyketones it is obtained initially as a monohydrate but can be converted into the anhydrous form on vacuum drying, whereas on exposure to the atmos-phere it rapidly forms a dihydrate. It is very easily reduced to (169) in an alkaline medium but is remarkably resistant to acid reduction, with the exception of hydrogen sulphide.

3-Acetyl-2,6,7-trihydroxynaphthazarin (170). From the molecular formula of this compound, $C_{12}H_8O_8$, it can be seen to possess one oxygen atom more, and two hydrogen atoms less, than the preceding pigment (169). This was attributed to the presence of a C-acetyl group as the compound gave acetic acid (0·93 mol.) on Kuhn–Roth oxidation, and formed a leuco-octa-acetate, and chiefly on this evidence, structure (170) was advanced.[380] This has been fully confirmed by recent spectroscopic observations,[317] the n.m.r. spectrum in dimethyl sulphoxide-d_6 showing only a singlet at τ 7·42 attributable to the methyl protons of an acetyl group adjacent to hydroxyl.† In the infrared spectrum the chelated ketonic carbonyl absorption is masked by the quinonoid carbonyl absorption but in the trimethyl ether spectrum there is a peak at 1704 cm^{-1}. The pigment has been synthesised,[366, 471] in low yield, by C-acetylation of hepta-acetoxynaphthalene [the leuco-acetate of (166)] using boron trifluoride in acetic acid, followed by hydrolysis with hot ethanolic aqueous hydrochloric acid. (The use of methanolic acid leads to removal of the C-acetyl group!)

6-Ethyl-2-hydroxyjuglone (171), a trace pigment in *Echinothrix calamaris*,[318] shows the ultraviolet-visible absorption of a juglone, probably monohydroxylated in the quinone ring, which accounts for its solubility in aqueous sodium bicarbonate and the formation of a monomethyl ether with diazomethane. The n.m.r. spectrum[372] of the latter reveals the presence of one vinylic proton, two *ortho*-coupled aromatic protons and an ethyl group, and the relative orientation of the substituents was revealed when the same ethyl-hydroxy-juglone was obtained as one of the products from the reduction of (168) with borohydride[372] (see above).

6-Acetyl-2,7-dihydroxyjuglone (172) was presumably identified[318] by comparison with one of the compounds obtained[372] by borohydride reduction of (168). The n.m.r. spectrum[372] of the latter confirms the presence of an aromatic proton at C-8 adjacent to a hydroxyl group (singlet at τ 2·92), an acetyl group chelated to hydroxyl (τ 7·18), and a vinylic proton (τ 3·75). It was later synthesised[554] by a Fries reaction on 1,2,4,5,7-penta-acetoxynaphthalene.

6-Ethyl-2,3,7-trihydroxyjuglone (173) is the major pigment in *Echinothrix diadema*.[318] The ultraviolet-visible absorption would fit a juglone structure having three β-hydroxyl groups, which would be consistent with the rapid formation of a dimethyl ether (two quinonoid hydroxyls) and the slower formation of a trimethyl ether, on treatment with diazomethane,[379] and the chemical shifts of the three methoxyl groups.[363] The n.m.r. spectrum of the natural pigment shows a singlet at τ 2·87 compatible with a proton at C-8 adjacent to hydroxyl, and signals for an ethyl group, the location of which (at C-6) is confirmed by the fragmentation pattern seen in the mass spectrum.[318]

6-Acetyl-2,3,7-trihydroxyjuglone (174). The structure was deduced[318] mainly from the n.m.r. spectrum of the dimethyl ether obtained with

---

† Intramolecular hydrogen bonding between acetyl and hydroxyl appears to be disrupted in this solvent.[363]

diazomethane. This showed methoxyl proton signals at τ 5·85 and 5·90 indicating that both groups are attached to the quinone ring, and the signal at τ 2·84 could be assigned to an aromatic proton at C-8 adjacent to a hydroxyl group at C-7. Two hydroxyl signals at τ −3·95 and −4·20 confirm that intramolecular hydrogen bonding involves both a quinone carbonyl group and the acetyl function (τ 7·17). The *Echinothrix* pigment was apparently unstable and "rather difficult to handle" whereas a compound of structure (174) has been synthesised[388] by a Fries reaction of 1,2,3,4,5,7-hexa-acetoxynaphthalene and was quite stable. (The synthetic compound has been found[504] to be identical with "spinochrome G" isolated from *P. lividus*.[322]) Methylation of the Fries reaction product with diazomethane gave a compound identical[554] with the dimethyl ether of (174).

3-Ethyl-2,6,7-trihydroxyjuglone (175). This ophiuroid spinochrome was identified[321] by conversion to 3-ethyl-2,6,7-trimethoxyjuglone[379] and by comparison with the borohydride reduction product of (176).

3-Acetyl-2,6,7-trihydroxyjuglone (176).† A peak in the infrared spectrum of this compound at 1684 cm^{-1}, attributed to an unchelated carbonyl group, and a singlet in the n.m.r. spectrum at τ 2·80 assigned to an aromatic proton at C-8 adjacent to a hydroxyl group, suggest that this spinochrome belongs to the juglone group. A three proton singlet at τ 7·38 is again indicative of an acetyl group adjacent to hydroxyl, and agreed with the Kuhn–Roth C-methyl estimation. From the molecular formula, $C_{12}H_8O_7$, the pigment must be an acetyltrihydroxyjuglone and the location of the side chain was ascertained by deacetylation in hot ethanolic hydrochloric acid to give 2,6,7-trihydroxyjuglone. The pigment is therefore (176)[347] and this was confirmed when the same quinone was obtained, albeit in very poor yield, by a Fries reaction on 1,2,4,5,6,7-hexa-acetoxynaphthalene.[458] When this quinone is crystallised it becomes very much less soluble and at the same time there is a change in the ultraviolet absorption (but not the visible). The change in solubility is presumably due to increased intermolecular hydrogen bonding, and some degree of aggregation evidently persists in solution which affects the ultraviolet absorption although not the chromatographic behaviour.

2-Hydroxynaphthazarin (*naphthopurpurin*) (177) is the simplest spinochrome pigment, long known as a synthetic compound. It may be prepared by condensing 1,2,4-trimethoxybenzene with maleic anhydride in fused aluminium chloride-sodium chloride,[166] and in other ways.

† "Spinochrome S".

3-Ethyl-2-hydroxynaphthazarin (178). The structure of this compound[321] was deduced both from its electronic and n.m.r. spectra, and from the fact that it can be obtained by reduction of (168) with borohydride.[372] It can be synthesised,[372] along with isomers, by Thiele acetylation of the diquinone obtained by oxidizing ethylnaphthazarin with lead tetra-acetate. Hydrolysis, in the presence of air, gives a mixture of 3-ethyl-2-hydroxy-, 6-ethyl-2-hydroxy- and 6-ethyl-3-hydroxynaphthazarins, separable by chromatography.

3-Acetyl-2-hydroxynaphthazarin (179). This is one of the pigments found in *Echinothrix diadema*[318] and also another of those obtained on reduction of (168) with borohydride.[372] It shows the electronic spectrum of a naphthazarin, the absorption being essentially identical with that of the 7-acetate of (168). This alone, indicates structure (179) which was supported by the mass and n.m.r. spectra,[372] the latter confirming the presence of an acetyl group adjacent to hydroxyl, and two *ortho*-coupled aromatic protons.

6-Ethyl-2-hydroxynaphthazarin (180). The n.m.r. spectrum of this pigment shows signals from an unhindered ethyl group attached to an aromatic ring.[363] As the compound is one of the naphthazarins derived from (168) by borohydride reduction, it follows that the hydroxyl group next to the acetyl group has been removed leaving structure (180).[372] It has been synthesised from ethylnaphthazarin (see above).[372]

6-Acetyl-2-hydroxynaphthazarin (181). The electronic spectrum of this quinone is compatible[529] with a naphthazarin having one β-hydroxyl group, while the n.m.r. spectrum shows the presence of an acetyl group (τ 7·27) and signals from one quinonoid (τ 3·57) and one benzenoid (τ 2·37) proton.[318] The acetyl and hydroxyl substituents are therefore in different rings, and their orientation was established when reduction of the pigment (as its methyl ether) with borohydride gave (180). It was synthesised[318] by acetylation of naphthopurpurin leucopenta-acetate with acetic anhydride-boron trifluoride, followed by mild basic hydrolysis in the presence of air. A little (2%) of the isomer (179) was also formed.

3-Ethyl-2,7-dihydroxynaphthazarin (182) was isolated from the spines of *E. diadema*, and was purified by methylation and subsequent acid hydrolysis.[318] It proved to be identical with the major borohydride reduction product of (168).[372] The formation of a dimethyl ether with diazomethane, the existence of an ethyl group (n.m.r.) and all the other spectroscopic evidence is consistent with structure (182). It was synthesised by Thiele acetylation of the diquinone arising from oxidation of 3-ethyl-7-methoxynaphthazarin, followed by acid hydrolysis.[372]

6-Ethyl-3,7-dihydroxy-2-methoxynaphthazarin (183)

6-Ethyl-2,7-dihydroxy-3-methoxynaphthazarin (184). These isomers occur together in *Diadema antillarum* and are difficult to separate. The n.m.r. and mass spectra showed that both compounds were monomethyl ethers of (169), both gave (169) on demethylation, and they could be identified[401] by comparison with known compounds. The isomer (184) has been obtained[379] by partial methylation of (169) with diazomethane, and the isomer (183) can be derived[379] from the tri-β-methyl ether of (169) by partial demethylation with ethanolic hydrochloric acid.

3-Acetyl-2-hydroxy-7-methoxynaphthazarin (185) is an ophiuroid pigment[321] and was found to be identical with the product of partial methylation of (168) with diazomethane.[379] As the n.m.r. signal from the acetyl protons of the latter appears at τ 7·12 (τ 7·15 in the parent compound) the acetyl group is still flanked by hydroxyl, and the structure is therefore (185).[363]

2,3,6-Trihydroxy-7-methoxynaphthazarin (namakochrome) (186), the first methoxylated spinochrome to be found, has already been mentioned. The original material was isolated from 28,000 muddy sea cucumbers (*Polycheira rufescens*) where it occurs as a protein complex.[324] Since demethylation[325b] gave the known (167) and further methylation[200] gave (195) the structure is not in doubt.

2,6-Dihydroxy-3,7-dimethoxynaphthazarin (187)

2,7-Dihydroxy-3,6-dimethoxynaphthazarin (188). These spinochromes occur in the starfish *Acanthaster planci*.[321] Partial methylation[379] of tetrahydroxynaphthazarin gives (186), both dimethyl ethers (187) and (188), and finally the tri- and tetramethyl ether, which can all be separated chromatographically.[379] The structures of the two isomers (187) and (188) were deduced from a comparison of the chemical shifts of the methoxyl protons with calculated values.[363, 379]

2-Hydroxy-2'-methyl-2'H-pyrano[2,3-b]naphthazarin (189). This is the only spinochrome known to have a C_4 unit attached to the naphthaquinone nucleus.[502] The presence of a methylpyran ring, frequently encountered in other groups of natural products, was deduced from the n.m.r. spectrum which shows a doublet of doublets for each vinylic proton (4'-H, J, 10 and 1·8 Hz; 3'-H, J, 10 and 4 Hz), a multiplet centred at τ 4·75 for the H-2 proton and a doublet (3H) for the methyl protons. Irradiation at τ 4·75 collapsed the methyl resonance to a singlet and eliminated the smaller couplings from the H-3' and H-4' signals. The pigment shows a normal naphthazarin spectrum in the visible region, and must contain a hydroxyl group attached to a quinone ring since it is soluble in aqueous sodium bicarbonate, forms a monomethyl ether with diazomethane, and the n.m.r. spectrum includes a singlet at τ 3·64 indicative of a quinonoid proton adjacent to a hydroxyl group. The hydroxyl group was assigned to C-2 (189) as the product obtained on catalytic hydrogenation, followed by reoxidation, has an ultraviolet-visible spectrum closely resembling that of 3-ethyl-2,7-dihydroxynaphthazarin (182). The reduction product is regarded as (206).[502]

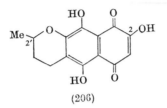

(206)

Ethylidene-3,3'-bis(2,6,7-trihydroxynaphthazarin) (190). Two novel bi-naphthaquinones have recently been isolated from *Spatangus purpureus*.[401] This one was clearly a naphthazarin, spectroscopically, and from its chromatographic behaviour it appeared to contain at least four β-hydroxyl groups. The n.m.r. spectrum showed that nuclear protons were absent, and a doublet (3H) at τ 8·23 coupled to a quartet

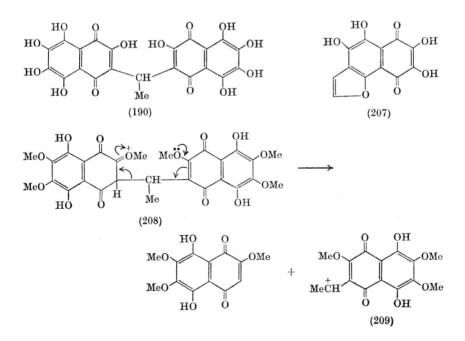

(190)

(207)

(208)

(209)

(1H) at τ 5·21 could be assigned to an ethylidene structure, $CH_3CH<$, linking two quinonoid rings. This suggested structure (190) which was confirmed synthetically by condensing trihydroxynaphthazarin with acetaldehyde. The base peak in the mass spectrum falls at m/e 238 and is due to the trihydroxynaphthazarin radical ion formed by fission of one of the central carbon-carbon bonds. This cleavage also occurs thermally, on chromatography on silica gel/oxalic acid plates, and to some extent by reduction with cold dithionite. Trihydroxynaphthazarin is the principal product in all cases but on heating in diphenyl ether the naphthazarin (207) (identified by mass spectrometry) can also be isolated. The hexamethyl ether of (190) (prepared with diazomethane) is similarly very prone to decompose on an acidic surface, and during chromatography on a column of acid-washed silica gel the breakdown products included 2,3,7-trimethoxynaphthazarin, 6-ethyl-2,3,7-trimeth-oxynaphthazarin, and two other compounds, one of which may have been 6-acetyl-2,3,7-trimethoxynaphthazarin. This acid-catalysed reac-tion is probably initiated by protonation of a ring carbon atom followed by cleavage as indicated (208) to give trimethoxynaphthazarin and the cation (209) from which the other products may be derived.[401]

Anhydro-ethylidene-3,3'-bis(2,6,7-trihydroxynaphthazarin) (191). On heat-ing the preceding binaphthazarin (190) in sulphuric acid it forms an anhydro compound identical with the accompanying pigment in *S. purpureus*. The simpler example (210) gives the β,β'-biquinone (211)

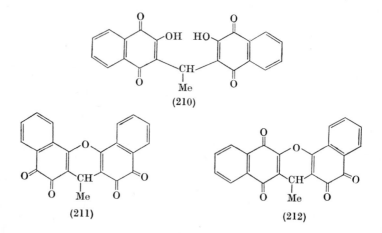

(210)

(211)   (212)

under the same conditions and the α,β-isomer (212) in the cold. Similar structures can be considered for the natural pigment but (191) is preferred[401] as there is no carbonyl absorption above 1621 cm^{-1}.

Cordeauxione† (213), $C_{14}H_{12}O_7$

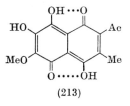

(213)

Occurrence: in the red glandular hairs under the leaves of *Cordeauxia edulis* (Hemsl.)[210] (Leguminosae).

Physical properties: garnet red prisms, m.p. 194°, $\lambda_{max.}$ (EtOH) 248, 305, 498 nm (log ϵ 4·26, 3·91, 3·91), $\nu_{max.}$ (CHCl₃) 1706, 1603 cm⁻¹.

This quinone was recognised[210] as a naphthazarin derivative by its ultraviolet-visible absorption in neutral and basic solution, and by various colour reactions. Chemical analyses indicated the presence of one O-methyl and two C-methyl groups, and *two* active hydrogen atoms but the formation of a yellow trimethyl ether revealed a third hydroxyl group, one of which must account for the solubility in sodium bicarbonate. The remaining substituent is an acetyl group which was detected by a positive iodoform test and conversion of the trimethyl ether into a 2,4-dinitrophenylhydrazone derivative. The arrangement of the substituents in one ring was established by oxidative degradation of the trimethyl ether with hydrogen peroxide in acetic acid which afforded

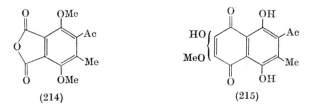

(214)   (215)

the anhydride (214), (ν_{CO} 1840, 1778, 1715 cm⁻¹) containing both the C-methyl and the acetyl group, the latter (ν_{CO} 1715 cm⁻¹) evidently not coplanar with the benzene ring. It was then possible for Karrer and his co-workers[210] to formulate cordeauxione as (215) (or a tautomer).

The orientation of the remaining substituents was finally determined by X-ray crystallography[211] (Fig. 1) when it was discovered that the most probable tautomeric form is the 1,5-naphthaquinone structure (213) (within the limits of error a small contribution from the 1,4-quinone structure (215) is not excluded), with intermolecular hydrogen bonds between the acetyl and β-hydroxyl groups. All the substituents deviate slightly from the plane of the ring system, being most marked in the case

† This shorter version of cordeauxiaquinone is now preferred.[450]

of the acetyl and methoxyl groups. This may have some bearing on the failure of this naphthazarin derivative to give a boroacetate test. All three crystalline forms of naphthazarin itself have a centrosymmetric 4,8-dihydroxy-1,5-naphthaquinone structure.[212]

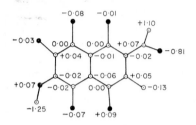

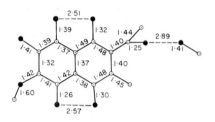

Deviations from the mean
plane through the ring atoms (Å)

Bond lengths (Å)

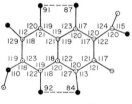

Bond angles (°)

FIG. 1. Structure of cordeauxione. From Fehlmann and Niggli *Helv. Chim. Acta* (1965). **48**, 305.

Lomastilone (216), $C_{15}H_{14}O_8$

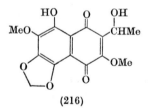

(216)

Occurrence: roots of *Lomandra hastilis* R. Br.[500] (Liliaceae).
Physical properties: red needles, m.p. 171–173°, $\lambda_{max.}$ (EtOH) 233, 253(sh), 305, 448(sh), 470, 500(sh) (log ε 4·70, 3·86, 3·79, 3·72, 3·75, 3·60), $\nu_{max.}$ (KCl) 3450, 1625, 1590 cm⁻¹.

The three quinones lomastilone, lomazarin (217) and lomandrone (p. 255) in *Lomandra hastilis* bear a striking resemblance to the spinochrome pigments found in sea urchins. Lomazarin corresponds to the trimethyl ether of the structure once postulated for spinochrome A, although in fact the C_2 side chains in the echinoid quinones are (as yet) invariably acetyl or ethyl groups.

Lomastilone shows one or two unusual features. Although it has only one *peri*-hydroxyl group (n.m.r.) and is therefore a juglone, it is unusually acidic (slightly soluble even in aqueous sodium bicarbonate), does not absorb in the usual quinone carbonyl region (~ 1660 cm^{-1}) of the infrared (solid state spectrum), and the visible absorption is appreciably displaced to wavelengths longer than normal. The n.m.r. spectrum shows that two methoxyl groups are present, a methylenedioxy group (2H singlet at τ 3·77, ν_{max} 931 cm^{-1}) which is extremely rare in natural quinones and a hydroxyethyl side chain. The latter is more clearly discerned in the spectra of the diacetate (—CH(OAc)CH$_3$, quartet at τ 3·99; —CH(OAc)CH_3, doublet at τ 8·48) and the dimethyl ether. The diacetate shows normal quinonoid absorption at 1669 cm^{-1} as well as bands at 1769 and 1739 cm^{-1} arising from the carbonyl stretching frequencies of the alcoholic and phenolic acetates, respectively. The dimethyl ether was obtained, with some difficulty, using methyl iodide-silver oxide, a monomethyl ether being formed simultaneously. The former did not possess a *peri*-hydroxyl group showing that the phenolic hydroxyl had been methylated, and the spectral properties of the compound (ν_{max} 3380(br), 1655, 1635 cm^{-1}, τ_{OH} 2·8) indicate that the alcoholic function must be hydrogen-bonded to a carbonyl group. Since hydrogen bonding of the side chain hydroxyl is not apparent in the natural pigment (ν_{HO} 3450 cm^{-1}, τ_{HO} 5·86) it was argued that both hydroxyls must be adjacent to the same carbonyl group, bonding of the alcoholic group being significant only after elimination of the hydrogen bond to the *peri*-hydroxyl by methylation, i.e., lomastilone is an 8-hydroxy-2-(α-hydroxyethyl)-1,4-naphthaquinone.

As the two methoxyl groups give rise to a degenerate n.m.r. signal they are unlikely to be in the same ring, and accommodation of the remaining substituents now demands that one must be located at C-3 and the other could be either at C-5 or C-7. Comparison with model compounds indicated that the chemical shift of the methoxyl signal (τ 5·81) was rather low for a *peri*-methoxyl group, usually $\sim \tau$ 6 ·1, whereas both methylated derivatives show methoxyl signals in this region. Accordingly the second methoxyl group was assigned to C-7 leading to structure (216) for lomastilone.[500]

Lomazarin (217), C$_{15}$H$_{16}$O$_8$

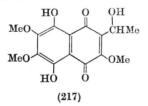

(217)

Occurrence: roots of *Lomandra hastilis* R. Br.[500] (Liliaceae).

Physical properties: dark red needles, m.p. 100–101°, λ_{max}. (EtOH) 273, 310, 460(sh), 484, 511 nm (log ϵ 4·38, 3·87, 3·82, 3·86, 3·72), ν_{max}. (KCl) 3450, 1598 cm⁻¹.

The spectroscopic data given above suggest that lomazarin, a minor constituent of *L. hastilis*, is a naphthazarin derivative. The n.m.r. spectrum is very like that of lomastilone except that the signal from the methylenedioxy protons is replaced by one arising from a third methoxyl group, and there are two singlets at low field from two different *peri*-hydroxyl protons. The hydroxyethyl group is still present, and as the naphthazarin system is fully substituted lomazarin must be (217).[500]

Lambertellin (218), $C_{14}H_8O_5$

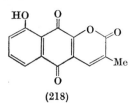

(218)

Occurrence: *Lambertella hicoriae* Whetzel,[234] *L. corni-maris* Hohn.[235] and other *Lambertella* spp.[236]

Physical properties: orange-red needles, m.p. 253–254°, λ_{max}. (EtOH) 284(sh), 290, 430 nm (log ϵ 4·08, 4·10, 3·68), ν_{max}. (Nujol) 1746, 1665, 1657, 1625 cm⁻¹.

Inoperculate discomycetes have a propensity to elaborate unusual quinones. Mollisin and mycochrysone are previous examples, and lambertellin is another, the uncommon feature in this case being the branched C_4 unit attached to the quinone ring. Inspection of the structure (218) suggests that it does not arise solely by acetate-malonate biogenesis. The only other quinone having the same carbon skeleton is the pyranonaphthaquinone (48) (p. 220) found in *Paratecoma peroba* along with lapachol and various other pigments containing a C_5 unit derived from mevalonate. This suggests that the C_4 unit in (48) may be a degraded C_5 entity and possibly the C_4 unit in lambertellin has a similar origin.

The presence of an antibiotic substance in a *Lambertella* culture[187,237] led to further chemical investigations.[234,235] Lambertellin shows the ultraviolet-visible absorption of a juglone and forms a monomethyl ether, an acetate, and a leucotriacetate. Degradation of the acetate with chromium trioxide gave 3-hydroxyphthalic acid. The n.m.r. spectrum of the sodium salt in deuterium oxide confirmed the presence of three aromatic protons, and showed a one-proton quartet at τ 3·23 coupled to a three-proton doublet at τ 7·29 (J, 1·5 Hz) characteristic of the group —CH=C—CH₃. As all the protons are now accounted for, lambertellin must be a 2,3-disubstituted juglone. That the remaining functional group is a lactone was indicated by titration, and by a carbonyl absorption band at 1746 cm^{-1} which is absent from the spectrum of the barium salt. This band remains when lambertellin is treated with sodium borohydride whereas the quinone carbonyl bands disappear. It is now clear that the lactone function must be fused to the quinone ring and it must include the allyl group; of the four possible arrangements, that shown in (218) was preferred by Armstrong and Turner[235] in order to explain the formation of propionaldehyde on alkaline hydrolysis of lambertellin. This reaction, they suggest, proceeds via ring opening and hydration of the side chain double bond, followed by a retro-aldal reaction (see 219) and decarboxylation of the β-aldehydo-acid split off. Sproston[234] reported the formation of acetaldehyde when lambertellin was degraded with alkaline permanganate. Like other coumarins,

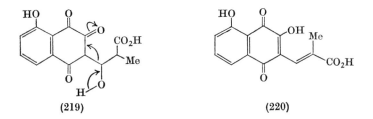

(219)   (220)

lambertellin can be recovered unchanged after brief treatment in alkaline solution but on prolonged reaction (one week in 0·2 N sodium hydroxide at room temperature exposed to daylight) the initial coumarinic acid rearranges to the more stable coumaric acid (220) which can be isolated. This reverts to lambertellin on long standing, in light, in acidified methanol.

Thus lambertellin should be either (218) or (221) but no decisive evidence to support either structure could be obtained. Armstrong and Turner[235] argued that alkaline hydrolysis of a compound of structure (218) should lead inter alia to 3-hydroxyjuglone whereas the isomer (221) should yield 2-hydroxyjuglone. From a comparison (t.l.c.) of the (unidentified) products obtained by alkaline treatment of all three compounds they concluded that lambertellin was probably (221).

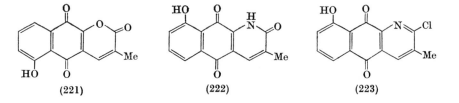

(221)   (222)   (223)

Sproston[234,238] ultimately favoured the other isomer (218) on the basis of some rather curious interpretations of the properties and spectra of the lactam (222) obtained by treating lambertellin with ammonia. The lactam reacts with phosphorus oxychloride to give a chloro compound which must be the aza-anthraquinone (223) (cf. ref. 238).

The actual structure of lambertellin is (218) which was established by synthesis of the methyl ethers of (218) and (221), and of lambertellin itself.[501] Condensation of the quinone (224) with formylpropionic ester (225) in pyridine containing a drop of piperidine gave the lactone (226)

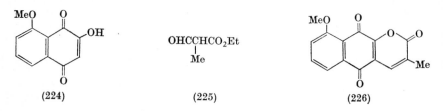

(224)   (225)   (226)

which afforded lambertellin† on demethylation.

(*9R,11S*)-*Eleutherin* (227)
(*9R,11R*)-*isoEleutherin* (228)

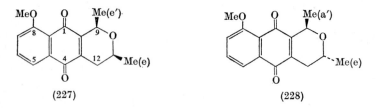

(227)   (228)

Occurrence: tubers of *Eleutherine bulbosa* (Mill.) Urb.[239,240] (Iridaceae).
Physical properties: eleutherin, yellow rods, m.p. 175°, $[\alpha]_D^{15}$ +346° (CHCl$_3$); isoeleutherin, yellow-orange needles, m.p. 176–177°, $[\alpha]_D^{20}$ −46° (CHCl$_3$), $\lambda_{max.}$ (EtOH) 246, 269(sh), 393 nm (log ϵ 4·13, 4·07, 3·58), $\nu_{max.}$ (KBr) 1662, 1635(w), 1589 cm^{-1}.

These two stereoisomeric quinones occur in *E. bulbosa* along with a related hydroxynaphthapyrone (eleutherinol) and a hydroxynaphthalide (eleutherol). Eleutherin was recognised as a juglone methyl ether by its ultraviolet-visible absorption and by the formation of 3-methoxyphthalic acid (and acetaldehyde) on ozonolysis. Reductive acetylation, however, gave only a leuco-*mono*-acetate and similar reductive alkylations gave mono-ethers, the products being crypto-phenolic 8-methoxy-1-naphthols. A more unusual reaction occurred

† The suggestion[234] that lambertellin is dimeric is not supported in other laboratories.[235,501]

during prolonged hydrogenation over platinum when three mols. of hydrogen were absorbed with the elimination of the methoxyl group. Similar reductions of leuco-eleutherin mono-acetate and mono-ethyl ether gave corresponding tetrahydrodesmethoxy derivatives (tetralins). When the product from the leuco-acetate was hydrolysed and methylated, and then oxidised with permanganate it yielded dimethoxy-

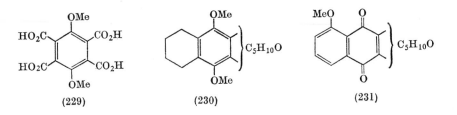

(229)   (230)   (231)

pyromellitic acid (229), thus demonstrating that eleutherin is a 2,3-disubstituted juglone methyl ether (231). The reduction product giving rise to (229) can be represented by (230). From the analytical data, the fragment $C_5H_{10}O$ (231) must contain an ether linkage, two C-methyl groups and two asymmetric carbon atoms (see later), and the only structures which meet these requirements, and account for the formation

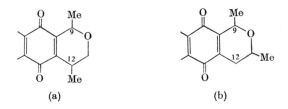

(a)   (b)

of acetaldehyde on oxidation, are represented by (a) and (b). These could be distinguished by examination of the optically active compound $C_{16}H_{18}O_3$, obtained by a Clemmensen reduction of eleutherin. This product, (232), was cryptophenolic, afforded 3-methoxyphthalic acid (but no acetaldehyde) when subjected to ozonolysis and formed a tetrahydrodesmethoxy derivative on hydrogenation. When treated with perphthalic acid it was slowly converted into the optically active hydroxyquinone (233) which passed into the optically inactive keto-quinone (234) (ψ-eleutherin) on further oxidation with chromic acid. From this absence of optical activity it follows that C-12 is not asymmetric in eleutherin which must therefore have structure (b), that is, (235) or the isomer with OMe at C-5.[239] The formation of (232) probably proceeds as indicated, acid-catalysed ring opening (236) being followed

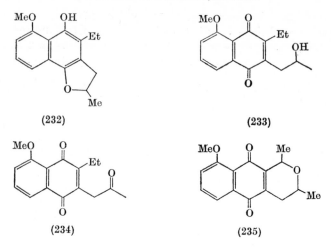

(232)   (233)

(234)   (235)

by cyclisation of the carbonium ion onto the hydroxyl group at C-4 (237), and reduction of the benzylic alcohol function. A by-product formed in the Clemmensen reduction (232; H in place of Et) may arise by a retro-aldol elimination of acetaldehyde from the benzyl alcohol.

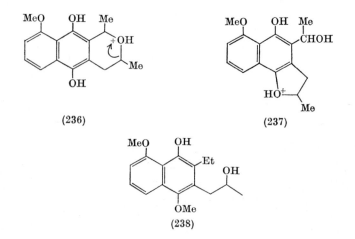

(236)   (237)

(238)

Finally, the location of the methoxyl group with respect to the pyran ring was deduced as follows. Clemmensen reduction of eleutherin affords the dihydrofuran (232), but the same treatment of leuco-eleutherin monomethyl ether gave the alcohol (238) [identified by oxidation to the quinone (233) with lead tetra-acetate]. As this compound does not cyclise to a dihydrofuran the side chain must be *ortho* to a methoxyl group, as shown, whereas in the reduction of eleutherin cyclisation occurs readily onto the free hydroxyl group at C-4. This is consistent with the cryptophenolic properties of (232). The Swiss workers[239] therefore concluded that eleutherin had structure (235) which was confirmed later by synthesis.

Eleutherin has two asymmetric centres; one of its stereoisomers, isoeleutherin, is also present in *E. bulbosa*, and the other two optically active forms were obtained *in vitro*.[240] Eleutherin epimerises in syrupy phosphoric acid giving an equilibrium mixture containing 83% of alloeleutherin, the enantiomer of isoeleutherin, and under similar conditions isoeleutherin partially racemises, the resulting mixture containing 18% of the fourth isomer, alloisoeleutherin, the enantiomer of eleutherin with which it forms a racemate. That eleutherin epimerises at C-9 was established by the observation that both it and alloeleutherin give the same optically pure dihydrofuran (232) on Clemmensen reduction, and the configuration at C-11, the stable asymmetric centre, was determined[241] by ozonolysis of both (232) and the derived quinone (233) which yielded (+)β-hydroxybutyric acid. As it proved impracticable to isolate degradation products containing the asymmetric centre at C-9, its configuration was deduced by reference to (+)eleutherol (239) which

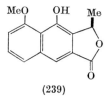

(239)

was degraded to D(—)lactic acid. Correlation of the molecular rotations of leucoeleutherin, leuco-iso-eleutherin and eleutherol, and of corresponding derivatives, clearly indicated that eleutherol and leuco-eleutherin had the same configuration at the benzylic position.[241] Accordingly eleutherin has the (9R,11S) configuration (227) and iso-eleutherin is the (9R,11R) isomer (228). The dihydropyran ring has a half-chair conformation which is inverted in (227) relative to (228) to relieve the steric strain that would otherwise arise from the two axial C–Me groups at C-9 and C-11. It is of interest that the methyl group at C-9 in both eleutherin and isoeleutherin is more stable in the pseudo-axial than the pseudo-equatorial configuration (see the equilibria mentioned above), and models show that the methyl group is further removed from the neighbouring carbonyl group when it is pseudo-axial.[507] The n.m.r. spectra[242] of eleutherin and isoeleutherin, which became available much later, display considerable fine structure arising from spin-spin interaction of the methylene protons with the adjacent axial proton at C-11 and also from homo-allylic coupling (CH—C=C—CH) with the pseudo-axial proton at C-9. By analysis[242] of these spectra the long-range and vicinal coupling constants were determined which fully supported the previously established stereochemistry, and in turn were used to elucidate the stereochemistry of protoaphin derivatives (Chapter 7).

The racemic eleutherins were synthesised[244] by converting 3-allyl-8-methoxy-1,4-naphthaquinone into the dihydrofuran (240) by reduction, followed by cyclisation under acid conditions. Reoxidation with ferric chloride led to the hydroxypropylquinone (241) which was condensed

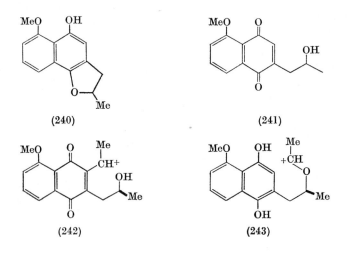

(240)   (241)

(242)   (243)

with acetaldehyde in phosphoric acid to give a separable mixture of (±)eleutherin and (±)isoeleutherin. In the final step it is unlikely that acetaldehyde condenses directly with the quinone, and additional experiments supported the view that an initial redox reaction generates a little of the quinol which undergoes electrophilic substitution and subsequent reoxidation. Eisenhuth and Schmid[244] also consider that direct condensation of the quinol nucleus with acetaldehyde is unlikely as this would lead to the formation of the cation (242), and hence both eleutherin and its thermodynamically more stable epimer should be formed. The absence of the latter shows that the reaction is under kinetic control, and probably proceeds via the cation (243) derived from the hemi-acetal resulting from the condensation of acetaldehyde with the side chain hydroxyl group.

Javanicin (solanione) (244), $C_{15}H_{14}O_6$

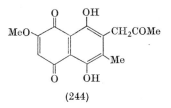

(244)

Occurrence: cultures of *Fusarium javanicum* Koorders,[245] *F. solani* (Mart.) App & Wr. (*var. rosa*) D_2 purple,[246, 247] *F. martii* App. & Wr. *var. pisi* F. R. Jones[249] and possibly *F. oxysporum* Schlecht ex Fr.,[255] and *Selenophoma donacis* (Pass.) Sprague & A. G. Johnson.[248]

Physical properties: red needles, m.p. 208°, $\lambda_{max.}$ (EtOH) 229, 309, 478, 512, 547 nm (log ϵ 4·60, 3·92, 3·78, 3·86, 3·70), $\nu_{max.}$ (Nujol) 1715, 1600 cm^{-1}.

Fusarium spp. are frequently brightly coloured and some, at least, of the pigments are quinones. One of these is a red antibiotic substance, $C_{15}H_{14}O_6$, which was studied, independently, by Arnstein and Cook,[245]† and by Weiss and Nord.[246]† It was found to contain one O–Me and two C–Me groups, three active hydrogens and a ketonic function, and from its ultraviolet-visible and infrared spectra and various colour reactions it was evidently a naphthazarin derivative. The compound formed only a diacetate but showed a marked tendency to eliminate the

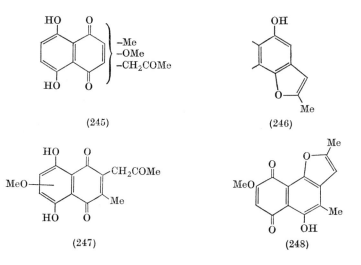

(245)   (246)

(247)   (248)

elements of water under acidic conditions so that acid-catalysed acetylation afforded anhydrojavanicin monoacetate. This information is summarized in (245), the formation of anhydrojavanicin, which is

† Javanicin[245] and solanione[246] were shown[250] to be identical by direct comparison.

accompanied by the disappearance of two active hydrogen atoms and the carbonyl band at 1715 cm^{-1}, being regarded as the condensation of the acetonyl side chain with an adjacent hydroxyl group to form a furan ring (246). The orientation of the three substituents in the naphthazarin system (245) remained uncertain until degradative work on fusarubin (see p. 291) established that the acetonyl and methyl side chains were attached to the same ring, i.e., (247 or a tautomer), the methoxyl group being placed in the other ring to explain its ready hydrolysis in alkaline solution to a bicarbonate-soluble hydroxyquinone. That javanicin is, in fact, the 7-methoxy isomer (244), as predicted[199a, 251] on biogenetical grounds, was finally established by synthesis. Anhydro-javanicin is therefore (248), its formation and reconversion to javanicin in alkaline solution being reminiscent of the chemistry of the lapachol series.

Hardegger and his co-workers have recently devised two synthetic routes to javanicin. In the most direct procedure,[252] the acetonyl side chain was introduced to the quinone (249) (prepared by conventional steps from 1,2,4-trimethoxybenzene) by Michael addition of benzyl acetoacetate, the final steps being selective demethylation of the

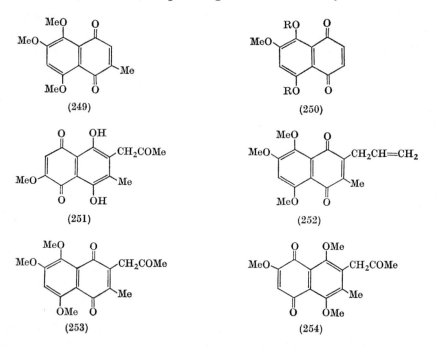

(249)   (250)

(251)   (252)

(253)   (254)

peri-methoxyl groups with aluminium chloride in nitrobenzene at room temperature (250; R = Me → 250; R = H) and hydrogenolysis of the benzyl ester (250; R = H) over palladium, accompanied by spontaneous decarboxylation. A similar sequence, starting from 2-methyl-5,7,8-trimethoxy-1,4-naphthaquinone (isomeric with 249) led to isojavanicin (251) which is scarcely distinguishable from javanicin (244) spectroscopically, but in admixture they give a large melting point depression.

In the other approach, developed by the Swiss group,[253,254] the allylquinone (252) was obtained by oxidation of 2-allyl-5,7,8-trimethoxy-3-methyl-1-naphthol with N-bromosuccinimide in dimethylformamide/acetic acid, and then converted into javanicin-5,8-dimethyl ether by the following sequence

$$\text{—CH}_2\text{CH=CH}_2 \;\rightarrow\; \overset{\overset{\displaystyle \text{OH}}{\displaystyle |}}{\text{—CH}_2\text{CHCH}_2\text{I}} \;\rightarrow\; \text{—CH}_2\text{COCH}_2\text{I} \;\rightarrow\; \text{—CH}_2\text{COMe.}$$

Selective demethylation then gave the natural product. In the n.m.r. spectrum[252] of javanicin the lone ring proton appears at τ 3·81 which favours the tautomer (244) but the compound may react of course in both tautomeric forms. This is illustrated by the formation of both dimethyl ethers (253) and (254) on heating with dimethyl sulphate-potassium carbonate in acetone.[253]

Solaniol (255), $C_{15}H_{14}O_6$

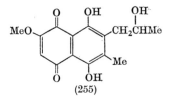

(255)

Occurrence: cultures of *Fusarium solani* (Mart.) App. & Wr. D$_2$ purple.[470]

Physical properties: dark red needles, m.p. 190–194° (dec.), $[\alpha]_{589}^{25}$ +122° (methanol), $\lambda_{max.}$ (dioxan) 227, 304, 472(sh), 500, 536(sh) nm (log ϵ 4·53, 3·97, 3·84, 3·91, 3·72), $\nu_{max.}$ (KBr) 3350, 3280, 1602 cm^{-1}.

The ultraviolet-visible curve of solaniol is very similar to that of javanicin whereas in the infrared the ketonic carbonyl band of the latter, near 1720 cm^{-1}, is absent and there is hydroxyl absorption in the 3 μ region. This suggests the presence of a 2-hydroxypropyl side chain in this pigment which would account for its optical activity, and is consistent with the n.m.r. and mass spectra (base peak at m/e M-44 by McLafferty rearrangement, and major peak at m/e 45). Structure (255) was confirmed when javanicin was reduced with sodium borohydride to give (after aerial reoxidation) racemic solaniol.[470]

10

Norjavanicin (256), $C_{14}H_{12}O_6$

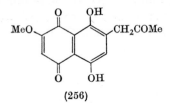

(256)

Occurrence: cultures of *Fusarium martii* App. & Wr. *var. pisi* F. R. Jones[508] and an unidentified *Fusarium* sp.[509]

Physical properties: red needles, m.p. 200–204°, $\nu_{max.}$ (CHCl$_3$) 1725, 1610, 1580 cm^{-1}.

This pigment shows a close resemblance to javanicin.† The n.m.r. spectrum provides evidence for a methoxyl group, two nuclear protons (τ 2·90 and 3·85), a C-methyl (τ 7·72) and a methylene group (τ 6·22), the last two functions being part of an acetonyl side chain as the compound, like javanicin, forms an anhydro-mono-acetate. This derivative no longer shows ketonic carbonyl absorption at 1725 cm^{-1} but the quinone carbonyl absorption has shifted to 1685 cm^{-1} (CHCl$_3$) and the C-methyl group, now attached to a furan ring, appears as a doublet (τ 7·56; J, 1 Hz) in the n.m.r. spectrum. All the principal peaks in the mass spectrum of norjavanicin are present also in the mass spectrum of javanicin but shifted by 14 m.u. with the exception of a common peak at *m/e* 43. Clearly norjavanicin only differs from javanicin by the absence of the ring methyl group and the structure is accordingly

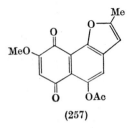

(257)

† *F. martii* also elaborates the hydroxymethyl derivative, novarubin.[530] Details are awaited.

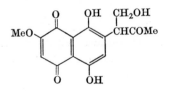

(256),[509] the position of the methoxyl group being assigned by analogy. The existence of the tautomeric form (256) (in solution) is supported by the fact that the methylene resonance is a singlet peak whereas a doublet would be expected if the acetonyl side chain was attached to a quinonoid ring. The anhydro-mono-acetate is (257).

Fusarubin (hydroxyjavanicin) (258), $C_{15}H_{14}O_7$
Anhydrofusarubin (259), $C_{15}H_{12}O_6$

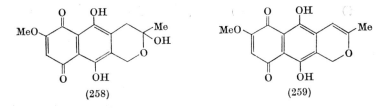

(258)   (259)

Occurrence:† cultures of *Fusarium javanicum* Koorders,[245] *F. solani* (Mart.) App. & Wr.,[256] *F. solani* D$_2$ purple,[247] *F. martii* App. & Wr. *var. pisi* F. R. Jones,[510] an unidentified *Fusarium* sp.,[509] and possibly *F. oxysporum* Schlecht ex Fr.[255]

Physical properties: (258) red prisms, m.p. 218°,‡ λ_{max}. (EtOH) 227, 307, 475(sh), 505, 538(sh) nm (log ε 4·32, 3·74, 3·57, 3·64, 3·46), ν_{max}. (KBr) 3358, 1600, 1575 cm^{-1}.

(259) violet-black needles, m.p. 193–201° (dec. >200°), λ_{max}. (EtOH) 237·5, 291, 540§ nm (log ε 4·32, 4·29, 4·00), ν_{max}. (Nujol) 1606 cm^{-1}.

The identity of the pigments, hydroxyjavanicin and fusarubin, obtained originally from *F. javanicum* and *F. solani*, was established[257] recently by direct comparison. The colour reactions and light absorption indicate that fusarubin is a naphthazarin derivative; it contains a C-methyl and an O-methyl group, and a hydroxyl which is not attached to the nucleus as the pigment does not dissolve in aqueous sodium bicarbonate. On hydrogenolysis over palladium in acetic acid it absorbs two mols. of hydrogen to give, after reoxidation, deoxyfusarubin, $C_{15}H_{14}O_6$ (loss of a hydroxyl group) but in ethanol or ethyl acetate the hydrogenolysis takes a different course and the product, which contains a ketonic function and two C–Me groups, was identified with javanicin. Evidently fusarubin undergoes hydrogenolysis in two forms and the experimental results suggest that the pigment contains a hemi-ketal structure (iii) which can give rise either to a deoxy compound (iii → iv) or a ketone (iii → ii → i). As the substituents (ii) must be adjacent,

† The O-demethyl derivative of (259) has now been found in cultures of *Gibberella fuji-kuroi*.[576]
‡ On a preheated block.
§ λ_{max}. (ether) 583 nm.

this leads to structure (258) for fusarubin.[256] The n.m.r. spectrum includes a singlet (1H) at τ 3·64[509] which does not particularly favour the tautomeric form shown but in the methyl ketal it shifts to τ 3·86[510]

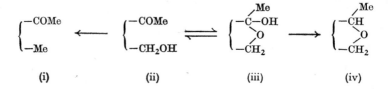

(i)   (ii)   (iii)   (iv)

which is more consistent with structure (260; R = Me). It is now obvious why fusarubin forms monoalkyl ethers, i.e., ketals (260) so readily when dissolved in alcohols containing hydrochloric acid, and its tendency to form the violet quinone, anhydrofusarubin (259), under acid contions, or on heating, is not surprising. The dehydration reaction, which is reversible, is characteristic of vinyl ethers, including pyran derivatives,

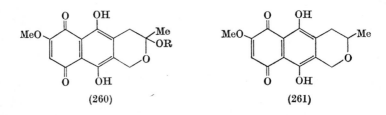

(260)   (261)

and an equilibrium can be established in aqueous acetic acid. Careful addition of water to a hot solution of anhydrofusarin in acetic acid changes the colour from violet to red (259 → 258); on boiling it becomes violet-red and reverts to red again on cooling. Similarly, anhydrofusarubin gives fusarubin monomethyl ether (260; R = Me) in hot methanolic hydrochloric acid, and the reaction can be regarded as an addition to a conjugated dienone system. Both fusarubin and anhydrofusarubin appear to be formed in *Fusarium* cultures, the compound isolated depending upon the isolation procedure, unless precautions are taken. Hydrogenation of anhydrofusarubin affords deoxyfusarubin (261) ($\lambda_{max.}$ 535 nm) the colour changing from violet to red as in all such reactions in which the double bond conjugated with the naphthazarin system is saturated.[256]

Fusarubinogen. After removal of fusarubin from the culture filtrate of *F. solani* by ether extraction, the aqueous phase remains yellow and contains the quinone in a reduced water-soluble modification; the strain *var. martii* (App. & Wr.) Wr. produces only this reduced material and no fusarubin at all.[258] When exposed to air at ca. pH 9 the yellow solution slowly becomes bluish-red as it autoxidises to the water-soluble pigment fusarubinogen which Ruelius and Gauhe[258] isolated as the ammonium

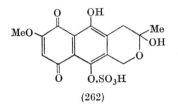

(262)

salt. On acid hydrolysis it liberates fusarubin and sulphuric acid, and since its ultraviolet-visible absorption resembles that of juglone it is evidently an O-sulphate of fusarubin in which one of the *peri*-hydroxyl groups is esterified (for example, 262). The colour of the reduced form present in *F. solani* cultures and its resistance to oxidation in neutral solution suggest that it may be a derivative of β-hydronaphthazarin (cf. proto-actinorhodin, p. 306).

Marticin, Isomarticin (263), $C_{18}H_{16}O_9$

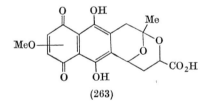

(263)

Occurrence: cultures of *Fusarium martii* App. & Wr. *var. pisi* F. R. Jones.[510]

Physical properties: marticin, red needles, m.p. 200–201°, $[\alpha]_{589}^{25}$ +132° (CHCl₃), $\lambda_{max.}$ 227, 305, 497 nm (log ϵ 4·52, 3·96, 3·89), $\nu_{max.}$ (Nujol) 3380, 3150, 1765, 1737, 1595 cm⁻¹.

Isomarticin, red crystals, m.p. 187–188°,† $[\alpha]_{589}^{25}$ +26° (CHCl₃), $\lambda_{max.}$ (EtOH) 227, 306, 497 nm (log ϵ 4·49, 3·94, 3·88), $\nu_{max.}$ (Nujol) 3230, 1735, 1600 cm⁻¹.

The plant pathogen *Fusarium martii* elaborates a group of quinones which act as wilting agents, particularly marticin and isomarticin. These new compounds show a general resemblance to their co-metabolites javanicin and fusarubin, particularly with respect to their ultraviolet-visible absorption. The pigments are thus naphthazarins, they each possess a methoxyl, a carboxyl and an aliphatic tert. methyl group, and so

† From benzene, or m.p. 168–169° from ethanol.

partial structure (264) can be written, a singlet (1H) at τ 3·92 indicating that the methoxyl group is attached to the quinonoid ring. The infrared spectrum (Nujol) of marticin suggests that the carboxyl group is present in both monomeric and dimeric forms in the solid state, since in dilute chloroform solution it shows only ν_{HO} 3430 and ν_{CO} 1775 cm^{-1}, and on methylation (MeOH HCl) the spectrum given above changes to

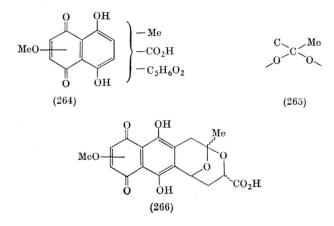

(264)   (265)

(266)

$\nu_{max.}$ 1727, 1605 cm^{-1} for the methyl ester. On the other hand, similar treatment of isomarticin gives three products, the main one being again marticin methyl ester, together with isomarticin methyl ester and a dark violet hydroxy-ester with a visible absorption spectrum closely resembling anhydrofusarubin (259). This bathochromic shift suggests that the hydroxy-ester contains a C=C double bond in conjugation with the quinone chromophore. The n.m.r. spectra of both marticin and isomarticin methyl esters show a singlet (3H) having the same chemical shift ($\sim\tau$ 8·40) as that of the tert. methyl group in fusarubin methyl ether. This implies that all three compounds contain the fragment (265) which leads to the postulated structure (266) for marticin and isomarticin, which fits the molecular formula. The ketal structure would account for the shift of the methyl signal to τ 7·90 in trifluoroacetic acid solution corresponding to the formation of a methyl ketone. Fission of the ketal structure also occurs to a limited extent when isomarticin is methylated with dimethyl sulphate–potassium carbonate in acetone. Whereas marticin gives two isomeric dimethyl ether–methyl esters (from both tautomeric 1,4-naphthaquinone forms) isomarticin affords, in moderate yield, a hydroxy-methyl ester dimethyl ether whose electronic spectrum is almost identical with that of anhydrofusarubin dimethyl ether. This compound may therefore be regarded as (267) and the n.m.r. spectrum includes signals appropriate to the –CH=C–Me system of the pyran ring. (It is noteworthy that both marticin and isomarticin give mono-methyl ether-methyl esters with diazomethane, as a result of partial methylation of one of the *peri*-hydroxyl groups.) By analysis of the n.m.r. spectrum of isomarticin methyl ester the 12-doublet resonances of an ABXY system (>CH–CH$_2$–CH<) was identified, in addition to

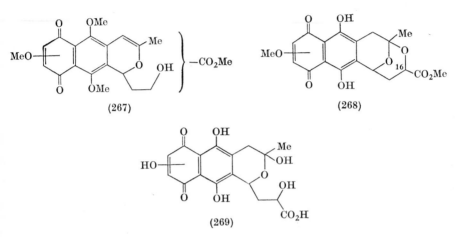

(267)   (268)

(269)

an A′B′ spectrum for the methylene group adjacent to the quinone ring, so that the ester group may be assigned to C-16 (268). Isomarticin and marticin can thus be represented as (263),[510] the methoxyl group being located most probably at C-7 by analogy with javanicin and fusarubin.

Marticin and isomarticin differ in the conformation of their ketal systems, the former having the 1,3-dioxo ring in the chair form (271)

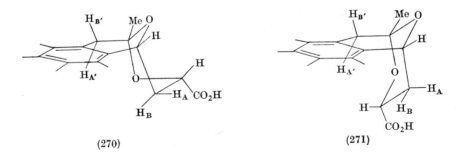

(270)   (271)

and the latter in the boat form (270). The coupling constants are in good agreement with those calculated from the dihedral angles measured on Dreiding models. In both conformations the carboxyl group remains equatorial, and in marticin the pyran ring is flattened so that the $H_{A'}$ and $H_{B'}$ protons make the same angle with the aromatic ring. Consequently the signal from these protons in marticin and its derivatives is a singlet and not an AB quartet.

These pigments undergo one or two interesting transformations centred on the cyclic ketal structure.[510] The formation of the hemiacetal (269) on treating isomarticin with sulphuric acid is not surprising (marticin forms a corresponding product merely on crystallisation from acetic acid) but the production of the anthraquinone (273) on heating in hydrobromic acid–acetic acid is intriguing. The structure is consistent with its chemical properties and the n.m.r. spectrum of the tetramethyl ether–methyl ester, and a possible reaction path is outlined below.

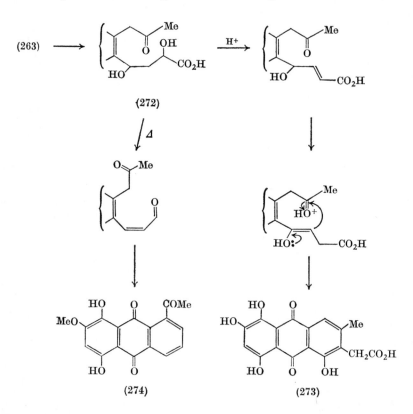

Another anthraquinone (274) arises on heating isomarticin in boiling tetralin; this reaction may involve the oxidative decarboxylation of the intermediate (272) followed by an aldol condensation. These rearrangements obviously have some bearing on the formation of anthracene on zinc dust distillation of isomarticin.

Erythrostominone (275), $C_{17}H_{16}O_8$
Deoxyerythrostominone (276), $C_{17}H_{16}O_7$
Deoxyerythrostominol (277), $C_{17}H_{18}O_7$

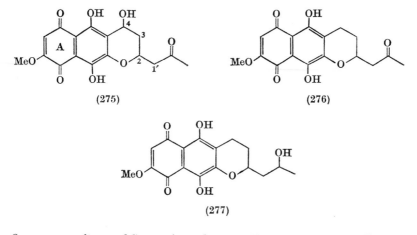

(275)　　　　　　　(276)

(277)

Occurrence: cultures of *Gnomonia erythrostoma* (Pers. ex Fr.) Auersw.[570]
Physical properties: (275) red needles, m.p. 184–186°, $[\alpha]_D^{25}$ +231° (acetone), $\lambda_{max.}$ (EtOH) 231·5, 280, 315, 480(sh), 509, 546 nm (logϵ 4·54, 3·89, 3·90, 3·86, 3·93, 3·72), $\nu_{max.}$ (CHBr$_3$) 3580, 1720, 1604 cm^{-1}; (276) red rods, m.p. 148–150°, $[\alpha]_D^{25}$ +277° (acetone), $\lambda_{max.}$ (EtOH) 234, 275, 317, 477(sh), 505, 542 nm (logϵ 4·49, 3·83, 3·88, 3·84, 3·91, 3·73), $\nu_{max.}$ (CHBr$_3$) 1713, 1601 cm^{-1}; (277) red needles, m.p. 139–141°, $[\alpha]_D^{25}$ +271° (acetone), $\lambda_{max.}$ (EtOH) 234, 275, 317, 475(sh), 505, 541 nm (logϵ 4·46, 3·85, 3·89, 3·84, 3·92, 3·73), $\nu_{max.}$ (Nujol) 3580, 1603 cm^{-1}.

Gnomonia erythrostoma produces a red broth in deep culture from which an antibacterial mixture of three red quinones can be isolated. From the spectral data above, the principal pigment, erythrosto-minone, is a naphthazarin which contains free hydroxyl and ketonic groups. Two singlets at low field (τ −2·6 and −3·19) in the n.m.r. spectrum confirm the naphthazarin structure, while singlets at τ 6·16 (3H) and 3·73 (1H) were attributed to the system in ring A of (275). The remainder of the n.m.r. spectrum is in complete agreement with the rest of structure (275), and further support was obtained by irradiation of the H-2 multiplet at τ 5·32 which reduced the H-1' multiplet to an AB system and simplified the H-3 multiplet at τ 8·25.[570] Hydrogenolysis over palladium-charcoal removed the benzylic hydroxyl group, the multiplet (1H) at τ 5·15 being replaced by another (2H) at τ 7·4, and further reduction with borohydride converted the methyl ketone into an alcohol with appropriate spectroscopic changes. The position of the methoxyl group in (275) was adopted on biogenetic grounds assuming an acetate-malonate origin.

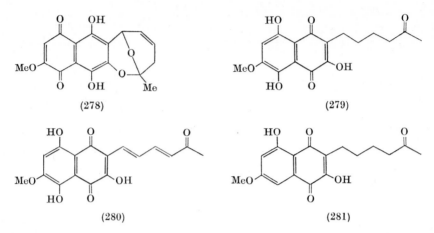

(278)   (279)

(280)   (281)

Several interesting chemical transformations support the structure deduced from spectroscopic evidence. On heating erythrostominone with toluene-*p*-sulphonic acid in benzene a 1,2-elimination leads to ring opening which is followed by cyclisation to the ketal (278) (and/or the double bond isomer). Hydrogenation of this in acetic acid gives several compounds including the ketone (279) formed by saturation of the double bond and hydrolysis of the ketal. Refluxing (278) in ethanolic hydrochloric acid effects hydrolysis and also elimination of water to give the unsaturated ketone (280) which, on hydrogenation, afforded (279) and also (281). Surprisingly a *peri*-hydroxyl group was eliminated from (275) while both β-oxygen functions remained intact. Oxidation of (281) with ruthenium dioxide-periodate afforded 6-oxoheptanoic acid which accounts for all the carbon atoms in the non-aromatic portion of erythrostominone.[570]

As already mentioned, hydrogenation of erythrostominone gave, *inter alia*, deoxyerythrostominone (276), and further reduction of that with sodium borohydride yielded, *inter alia*, deoxyerythrostominol (277), the trace pigment from *G. erythrostoma*.[570]

Granaticin (282), $C_{22}H_{20}O_{10}$

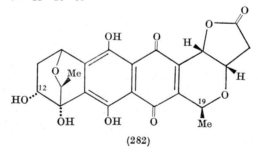

(282)

Occurrence: cultures of *Streptomyces olivaceus* (Waksman) Waksman & Henrici[482] and *S. violaceoruber* (Waksman & Curtis) Waksman *sensu* Waksman & Kutzner (together with a glycoside, granaticin B).[483]

Physical properties: garnet-like red crystals, m.p. 204–206° (dec.), $\lambda_{max.}$ (EtOH) 233, 286, 532, 576 nm (log ϵ 4·58, 3·76, 3·87, 3·75), $\nu_{max.}$ (KBr) 3420, 1780(sh), 1770, 1610 cm^{-1}.

The structure of granaticin was largely determined by chemical and spectroscopic methods but certain features, including the complete stereochemistry, were finally resolved by crystallographic analysis. It was identified[484] as a naphthazarin from its ultraviolet-visible absorption and blue colour in alkaline solution. However, as the molecular formula (C_{22}) suggests, it is unusually complicated and the absence of aromatic proton resonance in the n.m.r. spectrum shows that the naphthazarin system must be fully substituted. The infrared spectrum, in addition to hydroxyl and chelated carbonyl absorption, includes a band at ~1775 cm^{-1}† attributable to a γ-lactone, while that of the optically active tetra-acetate ([α]$_D^{20}$ −100° (CHCl$_3$)) has peaks at 1768 and 1740 cm^{-1} arising from phenolic and alcoholic O-acetyl groups, the quinone carbonyl band being shifted to 1665 cm^{-1}. Thus of the ten oxygen atoms in the granaticin molecule, two are present in the quinone system, two in a γ-lactone ring, there are two phenolic and two aliphatic hydroxyls, and the remaining two are evidently in ether linkages.

The presence of two CH$_3$–CH< groups was revealed by the n.m.r. spectrum as two doublets coupled to two quartets. Since one quartet, centred at τ 5·73, is shifted downfield to τ 4·9 in the tetra-acetate, this suggests the presence of a secondary alcohol function (CH$_3$–$\overset{|}{C}$HOH–), or, alternatively, the shift could derive from a structure CH$_3$–$\overset{|}{C}$H–OR having a nearby hydroxyl group (as 283) situated so that the carbonyl group of its acetate is in a pseudo-cyclic conformation with respect to the methine proton (as 284). Subsequently, the X-ray analysis showed

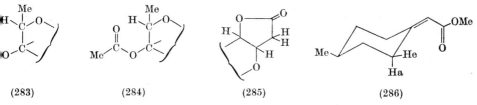

(283)   (284)   (285)   (286)

that structure (283) was correct. An analogous situation is found in the ester (286) where the signal for the equatorial proton at C-2 falls at τ 6·2 whereas the other methine and methylene protons all resonate

† The peak is split showing a maximum at 1770 with a shoulder at 1780 cm^{-1}, which is ascribed to Fermi resonance.

between τ 7·2 and 8·5.[486] The second quartet arising from a $CH_3–CH<$ function is centred at τ 4·58, and as it is not affected by acetylation it appears to be adjacent to an ether oxygen and (probably) to a benzene ring. Two other methine protons resonate at τ 4·3 and 4·9. The former is both benzylic and α to the lactone ring oxygen atom, and is coupled to the latter which is the X part of an ABX system, the AB signals (τ 6·5–7·2) being ascribed to a methylene group α to a carbonyl function. Together, these signals indicate the presence of a γ-lactone ring of structure (285).

As the naphthazarin system in granaticin is fully substituted, the central part of the molecule must be flanked by side chains or rings, and to obtain further information about these the compound was cleaved in the centre by ozonolysis. After methylation of the product with diazomethane, a crystalline diester, $C_{12}H_{14}O_7$, was obtained to which structure (287) was assigned. Spectroscopic data confirmed the existence of

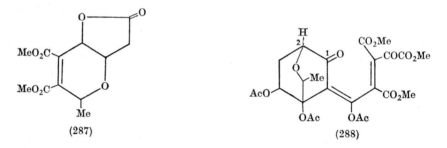

(287)   (288)

a γ-lactone ring, a tetrasubstituted double bond and an isolated $CH_3–CH<$ group. All the n.m.r. signals from this compound (except the OMe singlets) are present in the spectrum of granaticin itself showing that this fragment is present entire in the parent molecule, the carbon atoms of the double bond and the ester carbonyl groups being derived, presumably, from the naphthazarin system.

Ozonolysis of granaticin tetra-acetate conveniently gave, as the main product, a fragment derived from the other side of the molecule. The product was first methylated with diazomethane but unfortunately the resulting oil was difficult to purify. From previous findings it would be expected that the saturated part of this degradation product would contain an ether linkage, two aliphatic acetoxyl groups and a

$CH_3–\overset{|}{C}H–O–$ moiety. However, the n.m.r. spectrum revealed that three acetoxyl groups were present, one of which (τ 7·73) was attached to an unsaturated carbon atom and must arise from some part of the

naphthazarin system. The $CH_3–\overset{|}{C}H–O–$ fragment appeared to be linked to a quaternary carbon (289) as the quartet at τ 4·86 was not further split. Further analysis of the n.m.r. spectrum disclosed a complex ABXY system which was attributed to a part structure (290). Fragments (289) and (290) can be assembled in two ways to form the saturated part of this degradation product, both of which, (291) and (292), are consistent with the spectral data, and a decision in favour of (291) was

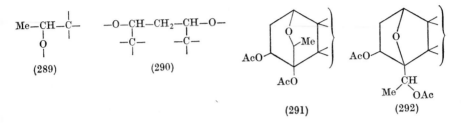

reached later following the X-ray crystallographic analysis. The cyclic structure (291) must be fused onto the naphthazarin system in granaticin, part of which is retained in this degradation product. Evidently ozonolysis of the tetra-acetate was incomplete and the structure (288) assigned to this compound is based on its mode of formation, the presence of three methoxycarbonyl groups and a vinylic acetoxy group, and the fact (n.m.r.) that all double bonds are tetrasubstituted. A carbonyl group was placed at C-1 to account for the chemical shift of the H-2 proton which appears at τ 3·8. The mass spectrum of the ozonolysis product was consistent with structure (288). The highest peak appears at m/e 494 and if this is assumed to be an M-60 fragment (loss of acetic acid from the molecular ion) the molecular weight is then 554 in agreement with (288).

Keller-Schierlein et al.[484] therefore concluded that the granaticin molecule comprised a central naphthazarin system flanked on one side by (293) and on the other by (294), two structures being possible in

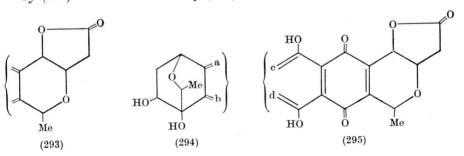

which the part structure (295) is linked to (294) either at a–c and b–d, or at b–c and a–d. The solution to this final problem was resolved by

X-ray crystallographic analysis of triacetylmono-iodoacetylgranaticin prepared from granaticin B (see below). Brufani and Dobler[485] thus established that granaticin has structure (282) with the stereochemistry as indicated.

Granaticin B was isolated[483] as an amorphous red powder from cultures of *S. violaceoruber*. On acid hydrolysis it affords granaticin and L-rhodinose, isolated as methyl L-rhodinoside. Solvolysis of granaticin B tetra-acetate in methanolic hydrochloric acid selectively removes the sugar leaving granaticin triacetate. This was converted into the tri-acetate-monoiodoacetate used in the X-ray analysis which established that the iodoacetoxy group was attached at C-12. Hence granaticin B is the 12-L-rhodinoside, the glycoside most probably having the α-configuration. This final problem could not be unambiguously resolved as the n.m.r. signal from the C-1′ proton in the sugar moiety overlaps with the quartet from the C-19 proton.

Actinorhodin (296), $C_{32}H_{26}O_{14}$

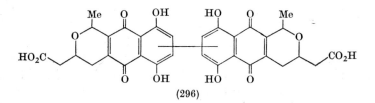

(296)

Occurrence: cultures of *Streptomyces coelicolor* (Müller) Waksman & Henrici.[259]
Physical properties: red needles, dec. ca. 270°, $\lambda_{max.}$ (dioxan) 285, 523, 531(sh) nm (log ε 4·25, 4·18, 4·05), $\nu_{max.}$ (KBr)† 1723, 1606 cm⁻¹.

A number of Actinomycetes elaborate "litmus-like" pigments. Under suitable conditions, cultures of *Streptomyces coelicolor* become bright blue and, when acidified, liberate a red pigment, actinorhodin, which has been studied extensively by Brockmann and his co-workers.[259-265, 466]

The indicator properties gave the first clue to the structure but the visible absorption, although similar to naphthazarin, is more than twice as intense. From this and other data, it gradually became clear that actinorhodin was a dimer which was confirmed by comparison with binaphthazarin (298).[261] The latter is not easily accessible and is best prepared[262] by Friedel-Crafts condensation of 2,5:2′,5′-tetramethoxy-biphenyl with succinic anhydride followed by cyclisation of the resulting di-acid (297) under nitrogen in fused aluminium chloride-sodium

† Diethyl ester.

chloride, and aeration of the product in alkaline solution. Other preliminary experiments[260, 261] disclosed the presence of two C–Me and

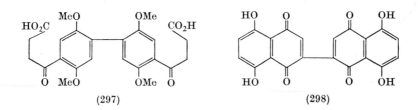

(297)   (298)

two carboxyl groups, and the formation of a dimethyl ester, a dimethyl ester tetra-acetate, a leuco-octa-acetate, and similar derivatives accounted for twelve of the fourteen oxygen atoms. The other two, on negative evidence, are in ether linkages. At this stage the structure of actinorhodin can be provisionally represented as (299).

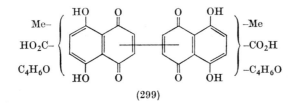

(299)

A key step in the subsequent elucidation of the structure was the discovery[261] of a new reaction by which actinorhodin could be cleaved on treatment with diazomethane. If 1,4-naphthaquinone is allowed to react in the normal way with etherial diazomethane (1·1 mol.) it undergoes 1,3-dipolar cyclo-addition and the rather unstable adduct (300) rapidly precipitates,[267] but if the reaction is conducted in dioxan-ether the adduct, kept in solution, is rapidly oxidised by air to the benzindazolequinone (301; R = H) which is slowly converted into the N-methyl derivatives (301 and 302; R = Me) in the presence of excess reagent.[263, 265] Similarly, 2,2'-binaphthaquinone yields, with a slight excess of diazomethane, the bis-adduct (303) which is also transformed, when excess diazomethane is used in dioxan–ether solution, into the benzindazolequinones (301 and 302; R = Me). The combined yield of the benzindazolequinones is the same (>80%) from both mono- and binaphthaquinones. The adduct (303) is probably a mixture of three stereoisomers (meso and racemic) resulting from *cis*-addition to each quinone ring, and the orientation was established[466] by oxidation to the bipyrazyl derivative (308) with ferricyanide. This, on alkali fusion, gave benzoic acid and the dimer (309; R = CO$_2$H) which was easily transformed into the parent

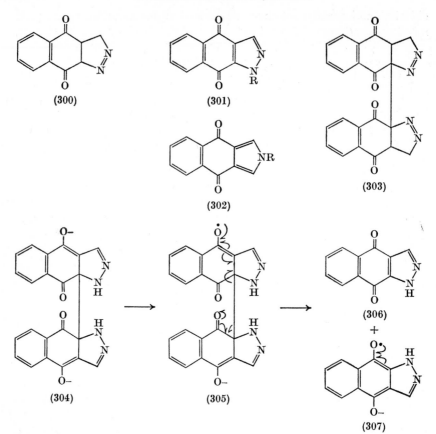

(300)   (301)   (303)

(302)

(304)   (305)   (306)
+
(307)

bipyrazyl (309; R = H). Fission of the bis-adduct (303) into monomeric benzindazolequinone can also be effected by shaking with aqueous sodium hydroxide but the yield is poorer. Brockmann *et al.*[263] consider

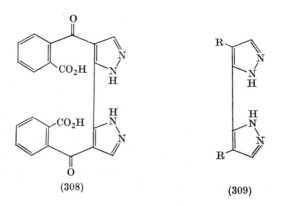

(308)   (309)

this to be an oxidative cleavage (oxygen is absorbed) in which a radical-anion (305), formed by removal of an electron from the initial di-anion (304), decomposes into the benzindazolequinone (306) and its semi-quinone (307) which then disproportionates into the quinone (306) and its quinol. The latter is rapidly oxidised to the former but both can be identified by acetylation (in the absence of air) to give the N-acetyl derivative of (306) and its leucotriacetate.[266] The driving force comes from the formation of two mesomerically stabilised products which liberates more than enough energy to break the central carbon–carbon bond.[264] As the adduct (300) is easily oxidised in air to the quinone (301) it is reasonable to assume that the dimer (303) can be similarly de-hydrogenated, cleavage occurring essentially by the mechanism already outlined.[263]

Returning now to the natural biquinone. Degradation [264] of actinorhodin dimethyl ester tetra-acetate with diazomethane gave (after hydrolysis and re-esterification with diazomethane) a single, homogeneous, optically active, benzindazolequinone (310).† This not only throws light on the C_4H_6O fragments but also shows (as assumed in (299)) that the

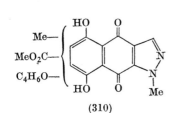

(310)

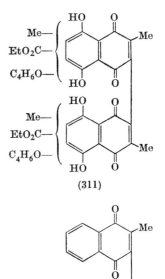

(311)

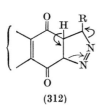

(312)

(313)

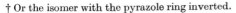

† Or the isomer with the pyrazole ring inverted.

two halves of the molecule have the same structure and configuration, and are linked symmetrically. The same cleavage product (310)† was also obtained by treatment of actinorhodin diethyl ester with diazomethane (after the same work-up). In this case the cleavage reaction is accompanied by another degradation which results in the loss of nitrogen and formation of the red biquinone (311). This is not unexpected as quinone-diazoalkane adducts are known[267, 268] to decompose, usually on warming or in base, to alkylated quinones (see 312). Another example is the formation of (313) on heating the bis-adduct (303) in toluene.[266]

Finally, the structural arrangement of the Me, CO_2H and C_4H_6O fragments attached to the flanks of the biquinone molecule was elucidated by oxidation of proto-actinorhodin (a reduced form of actinorhodin, see below) with alkaline hydrogen peroxide. This yielded lactic acid, β-hydroxyglutaric acid, and a viscous tricarboxylic acid which was further degraded, with hydriodic acid, to propionic acid and glutaric acid. Significantly, the major fragment contained eight carbon atoms which must therefore include two from the naphthazarin ring system; the structure (314) accords with the chemical data and is consistent with the n.m.r. spectrum of actinorhodin (in alkaline D_2O). Accordingly Brock-

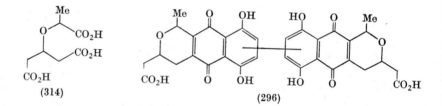

mann and his co-workers[264] assigned structure (296) to actinorhodin, the tautomeric form shown being preferred[513] as the two nuclear protons‡ resonate at τ 2·67. Further support for the structure of the dihydropyran rings was provided by the mass spectrum[264] of the cleavage product (310), the remaining uncertainties in the structure of the pigment being the stereochemistry of the heterocyclic rings and the location of the central carbon–carbon bond (2,2′ or 3,3′).

Proto-actinorhodin. Actinorhodin is normally isolated from cultures of *S. coelicolor* by extracting the mycelium with aqueous sodium hydroxide and acidification of the deep blue solution. Provided that the nutrient medium does not become alkaline most of the pigment remains in a reduced form, proto-actinorhodin, which can be isolated. It shows a blue fluorescence in solution and is rapidly oxidised to actinorhodin on

† Or the isomer with the pyrazole ring inverted.
‡ Measurements made on the more soluble dibutyl ester.

aeration in basic solution. From its relative stability it is unlikely to be

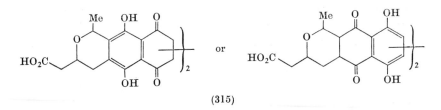

(315)

a 1,4,5,8-tetrahydroxynaphthalene derivative and must be assigned the β-hydronaphthazarin structure (315).

Frenolicin (316), $C_{18}H_{18}O_7$

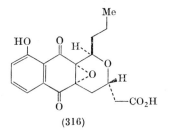

(316)

Occurrence: cultures of *Streptomyces fradiae* (Waksman & Curtis) Waksman & Henrici.[269]

Physical properties: pale yellow crystals, m.p. 161–162°, $[\alpha]_D^{25}$ −37·7° (MeOH), $\lambda_{max.}$ (EtOH) 234, 284(sh), 362 nm (log ε 4·26, 3·54, 3·72), $\nu_{max.}$ (KBr) 1710, 1650 cm^{-1}.

The structure of this novel antibiotic quinone was deduced[270] mainly from physical data, but chemically frenolicin behaves as a phenolic carboxylic acid (pK_a' = 10·0 and 5·2), and forms an acetate, a methyl ester (with diazomethane) and a methyl ester–methyl ether (with methyl iodide). The ultraviolet absorption curve resembles that of β-hydrojuglone (p. 222), and on oxidation of the methyl ester–methyl ether with alkaline permanganate it afforded 3-methoxyphthalonic acid, identical with that obtained by a similar oxidation of 1,5-dimethoxy-naphthalene. When the metabolite was treated with potassium iodide (or hydrogen bromide!) in refluxing acetic acid, or hydrogenated over palladised charcoal (two mols. of hydrogen were absorbed) and re-oxidised in air, it gave a yellow-orange quinone, deoxyfrenolicin, $C_{18}H_{18}O_6$, ($\nu_{max.}$ 1725, 1665, 1650, 1625 cm^{-1}, $\lambda_{max.}$ 246, 274, 420 nm) which was clearly a juglone derivative.

The remarkable ease of reduction suggested a juglone-2,3-epoxide structure with the remaining $C_7H_{13}O(CO_2H)$ portion linked in some way to the quinone ring (317); this part of the molecule must contain an ether linkage (not alkoxy) and the similarity of the ultraviolet-visible absorption of deoxyfrenolicin to that of the eleutherins (p. 282) suggested a useful analogy. The structure was deduced from the n.m.r. and mass spectra, and can be most readily understood by reference to the part structure (318). A three-proton triplet at τ 9·0 clearly arises from a primary C–Me group, and this, together with the observation that the major fragment ion in the mass spectra of frenolicin and its derivatives appears at (M-43), indicates the presence of an n-propyl group in the molecule. The grouping >CHCH$_2$CO$_2$H was deduced from a two-proton doublet at τ 7·40 attributed to signals from a methylene group deshielded by the carboxyl group and coupled to a methine proton, and a one-proton symmetrical triplet downfield at τ 5·42 could be ascribed to a

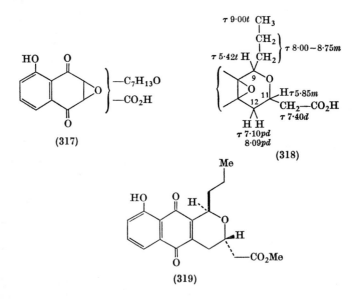

(317)

(318)

(319)

–CH$_2$CHO– group. By spin decoupling two pairs of doublets centred at τ 7·10 and 8·09 were diagnosed as part of an ABX system coupled to a one-proton multiplet at τ 5·85. This latter proton was also coupled to the doublet at τ 7·40, and hence the grouping

$$\begin{array}{c} \quad\;\; O- \\ \quad\;\; | \\ -CH_2CHCH_2CO_2H \end{array}$$

is probably present. Combination of these fragments then leads to the arrangement shown in (318) for which the mass spectrum provides additional support.

Combining the part structure (318) with the juglone epoxide system suggests that frenolicin may be represented by (316), or the isomer with the hydroxyl group at C-5, and the former was preferred by the Lederle group[270] on the grounds that the n.m.r. signal from the C-9 proton in frenolicin changes from a symmetrical triplet to a pair of doublets when the phenolic group is esterified or methylated, the signal from the C-12 protons being unchanged. The change in the environment at C-9 can reasonably be attributed to a difference between the magnetic anisotropy of the chelated carbonyl group in frenolicin and the free carbonyl group in the derivatives, or alternatively, there may be a change in the steric interaction between the carbonyl and the propyl group. Comparison of the coupling constants of the C-9, C-11 and C-12 protons in deoxy-frenolicin methyl ester (319) with those of the corresponding protons in isoeleutherin (p. 282) indicated that the C-11 proton was axial and the C-9 proton pseudo-equatorial (316). The probable stereochemistry of the epoxide group was deduced from the observation that frenolicin readily forms a chlorhydrin on reaction with hydrochloric acid, and as the ring opening can be reversed on treatment with base it is evidently *trans*-diaxial. From n.m.r. and other evidence the chlorhydrin appears to be

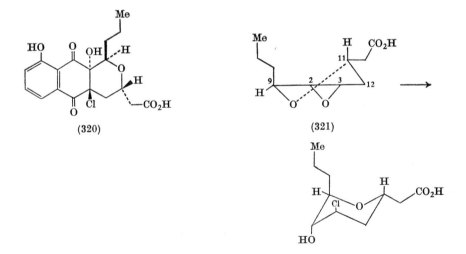

(320)

(321)

(320) with the proton at C-11 still axial. Inspection of Dreiding models suggests that this can only be so if the C-11 proton and the oxiran ring are *trans* to each other in frenolicin (as 321).

Ventilagone (322), $C_{16}H_{16}O_5$

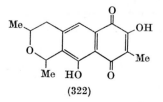

(322)

Occurrence: root bark of *Ventilago viminalis* Hook.[271] (Rhamnaceae).

Physical properties: red-orange needles, m.p. 205°, $[\alpha]_D^{18}$ +430° (CHCl₃), $\lambda_{max.}$ (EtOH) 227, 255, 294, 409 nm (log ε 4·22, 4·23, 4·11, 3·68), $\nu_{max.}$ (KCl) 3445, 1661, 1649, 1609 cm⁻¹; $\nu_{max.}$ (CHCl₃) 1661, 1610 cm⁻¹.

From the ultraviolet and infrared absorption ventilagone appeared to be a 2-hydroxyjuglone derivative but although it formed a diacetate, prolonged treatment with methyl iodide–silver oxide gave only a mono-methyl ether, the *peri*-hydroxyl group being unaffected. The rest of the structure was elucidated from n.m.r. data.[272] A peak at τ 7·91 was assigned to a methyl group attached to the quinone ring, and a signal (1H) at τ 2·57 showed that the molecule contained only one aromatic proton which was located at C-8 adjacent to a carbonyl group. The remaining $C_5H_{10}O$ fragment, which contains an ether link, must be attached to positions 6 and 7, and the arrangement shown in (323) was deduced as follows. At low field a quartet (1H) at τ 4·92, coupled to a

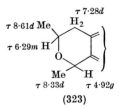

(323)

doublet (3H) at τ 8·33, was assigned to a benzylic proton adjacent to

oxygen. As both signals were shifted upfield on acetylation, the MeCHO– grouping was placed next to the *peri*-hydroxyl group, at the same time providing a steric explanation for its diminished reactivity. A one-proton multiplet at τ 6·29 also arises from a proton adjacent to oxygen which is coupled to a methyl group (doublet at τ 8·61) and two benzylic protons which appear as a doublet at τ 7·28. These observations are consistent with structure (323), and Cooke and Johnson[272] therefore regard ventilagone as (322). The stereochemistry is not yet known.

In *Ventilago viminalis* we have the only known example of the co-existence in the same plant of a naphthaquinone with anthraquinones of the chrysophanol–emodin type. Structurally, ventilagone appears to be "acetate-derived", like the anthraquinones, whereas naphthaquinone–anthraquinone combinations elsewhere, for example in Bignoniaceae, have a different origin.

Xanthomegnin (324), $C_{30}H_{22}O_{12}$

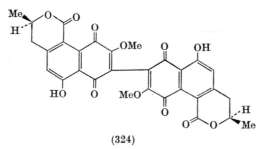

(324)

Occurrence: mycelium of *Trichophyton rubrum* (Cast.) Ota,[273,274] *T. megnini* Blanch.,[275] and *T. violaceum* Sab. apud Bodin.[281]

Physical properties: orange plates, dec. ca. 340°†, $[\alpha]_D^{22}$ −156° ($CHCl_3$), $\lambda_{max.}$ (dioxan) 228, 288(sh), 395 nm (log ε 4·73, 4·23, 4·00), $\nu_{max.}$ (KBr) 1718, 1678, 1616 cm^{-1}.

In the last decade the mixtures of pigments elaborated by *Trichophyton* spp. have been examined in several laboratories[273–279] and xanthomegnin was the first to be identified.[275] The spectroscopic properties, and the changes which result on acetylation, suggest a *peri*-hydroxy-1,4-naphthaquinone structure, the molecular weight[274b] indicates that the compound is dimeric, and from the n.m.r. spectrum it appeared to be symmetrical with one hydroxyl, one methoxyl and one C-methyl group in each half of the molecule. On hydrogenation of the diacetate two mols. of hydrogen were absorbed with concurrent disappearance of the quinone carbonyl bands in the infrared, and a shift of the 1718 cm^{-1} carbonyl band to 1658 cm^{-1}. As this band returned to 1718 cm^{-1} when the compound was acetylated (to leucoxanthomegnin hexa-acetate) there is clearly intramolecular hydrogen bonding between a quinol hydroxyl group and the carbonyl group responsible for the peak at 1718 cm^{-1}. This group is therefore adjacent to a quinone carbonyl in the natural pigment and was shown to be part of a lactone system by dissolution in aqueous alkali and conversion to the barium salt; the infrared spectrum of the latter showed carboxylate bands at 1550 and 1410 cm^{-1} but the 1718 cm^{-1} was absent. Further examination of the n.m.r. spectrum revealed only one aromatic proton per monomer unit, and evidence

† On preheated metal block.

for the presence, in each half of the molecule, of the fragment

$$Ar—CH_2—\overset{\displaystyle O^-}{\underset{\displaystyle |}{C}H}—Me$$

in the form of a multiplet at τ 5·53 coupled to signals from a methyl group (doublet at τ 8·51) and a benzylic methylene group (doublet at τ 7·08).

If this moiety is now associated with the lactone carbonyl group, structure (325) or (326) emerges for xanthomegnin. The location of the δ-lactone ring was supported by alkaline degradation of the pigment

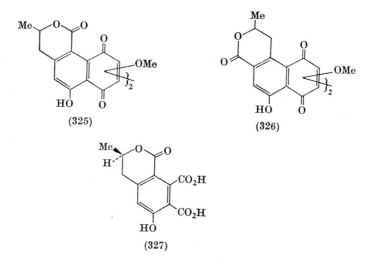

(325)   (326)

(327)

which gave an acid which was considered, on the basis of its infrared spectrum (ν_{CO} 1725, 1680 cm^{-1}) to have structure (327). Dimerisation at a quinonoid carbon atom accounts for the absence of a vinylic proton signal in the n.m.r. spectrum of the pigment, and a 3,3′ junction (324) follows from the assignment of the methoxyl groups to C-2 and C-2′ in accordance with the probable polyketide derivation of the compound *in vivo*. Facile hydrolysis of the methoxyl groups in alkali shows that they are indeed attached to the quinone rings.[274b] Finally the asymmetric centres were shown to have the R-configuration by reduction of the lactone (327) with lithium aluminium hydride, followed by ozonolysis and oxidation with silver oxide, which gave (R)-β-hydroxybutyric acid.[280] Structure (324) has therefore been adopted for xanthomegnin. The lactone carbonyl group must be very close to the quinone carbonyl at C-1 as the infrared absorption suggests that it is not significantly twisted out of the plane of the aromatic ring.

Fuscofusarin (328), $C_{30}H_{20}O_{11}$

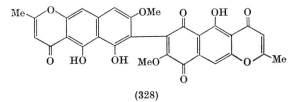

(328)

Occurrence: cultures of *Fusarium culmorum* (W. G. Smith) Sacc.[477]
Physical properties: brown powder, m.p. >300°, λ_{max}. (EtOH) 225, 281, 346, 405 nm (log ϵ 4·50, 4·63, 3·89, 4·01), ν_{max}. (CHCl₃) 1665, 1650, 1620, 1603 cm⁻¹.

This is a minor pigment produced by *F. culmorum* along with rubrofusarin (329)[196, 288] and aurofusarin (331). It shows a general resemblance in spectroscopic properties to its co-metabolites which extends to the di- and triacetates formed on acetylation in cold pyridine. The n.m.r. spectra of the acetates correspond closely to those of the acetates of

(329)   (330)

(329) and (331), and since the compound is dimeric it appeared to comprise a rubrofusarin moiety linked to a "half-molecule" of aurofusarin. The absence from the n.m.r. spectra of a signal corresponding to that from the C-7 proton of (330) (the monomer of aurofusarin obtained by oxidation of rubrofusarin) implies that the two halves of fuscofusarin are linked by a bond from C-7 of the quinone moiety to C-7 or C-9 in the rubrofusarin component (the presence of the C-10 proton was indicated by a resonance at τ 2·93 in the diacetate which shifted downfield to τ 2·38 in the triacetate). Accordingly fuscofusarin appears to be (328) and this was confirmed[477] by oxidation with Fremy's salt which gave aurofusarin (331).

Aurofusarin (331), $C_{30}H_{18}O_{12}$

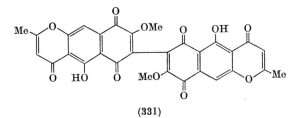

(331)

Occurrence: mycelium of *Fusarium culmorum* (W. G. Smith) Sacc.,[282] *F. graminearum* Schwabe,[282] *F. decemcellulare* Brick.,[477] *Hypomyces rosellus* (Alb. & Schio) Tul.,[283,473] and from the imperfect stage, *Dactylium dendroides* (Bull.) Fr.[283]

Physical properties: yellow needles, m.p. >330° (400°), $\lambda_{max.}$ (CHCl$_3$) 248, 269, 381, 422(sh) nm (log ϵ 4·69, 4·52, 3·99, 3·93), $\nu_{max.}$ (KBr) 1678, 1664, 1610, 1598 cm^{-1}.

Aurofusarin was first isolated in Raistrick's laboratory[282] in 1937. Structural investigations did not proceed far and the subject remained dormant for many years until the problem was taken up independently in four laboratories whose results,[283–285,473] leading to the complete structure, were reported in 1966–7. The following is a selection from the mass of experimental evidence provided.

Aurofusarin is a biquinone and affords a dimethyl ether, a diacetate and a leucohexa-acetate, the ultraviolet spectrum of the latter being very similar (but twice the intensity) to that of rubrofusarin diacetate. (Rubrofusarin is a co-metabolite in *F. culmorum*.) The ultraviolet-visible

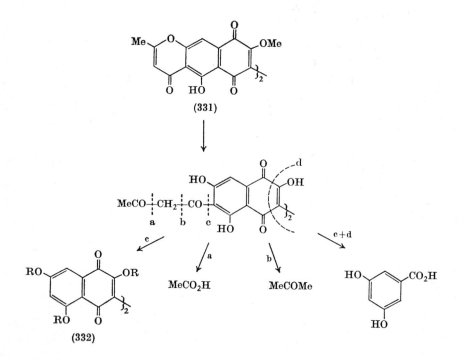

(331)

(332)

spectrum of the pigment is also similar to that of the biquinone xantho-megnin (324), and it too is a symmetrical dimer, while the n.m.r. spectrum (>τ 0·0) consists of four singlets only, attributable to an aromatic proton, a methoxyl group,† and the protons of an α-methyl-γ-pyrone system‡ (as in rubrofusarin) in each half of the molecule. The two methoxyl groups are evidently attached to the quinone rings as they can be readily hydrolysed with acid. Alkaline hydrolysis afforded methanol (from the methoxyl groups), acetone and acetic acid (from an α-methyl-γ-pyrone system), and a more significant fraction which yielded, after methylation, the biquinone (332; R = Me) (cf. flaviolin, p. 252) identical with a synthetic sample. Alkali fission of aurofusarin afforded (after methylation) 3,5-dimethoxybenzoic acid. This evidence leads logically to structure (331), the hydrolytic degradations proceeding as indicated.

Further confirmation comes from the extensive degradative work of Gray et al.[473] who obtained a series of binaphthyl derivatives by reductive methylation with zinc and dimethyl sulphate in alkaline solution, followed by treatment with diazomethane. This gave, inter alia, tetra

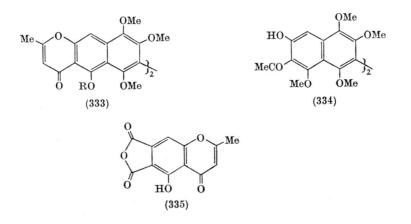

(333)

(334)

(335)

(333; R = H)- and hexa (333; R = Me)-O-methyltetrahydro-aurofusarin and the latter was hydrolysed in alkaline solution to the bis-acetonaph-thone (334). On the other hand, permanganate oxidation of aurofusarin dibenzoate destroyed the quinonoid rings and the chromone-anhydride (335) could be isolated. Like xanthomegnin and many natural biquinones, aurofusarin is apparently optically inactive although models suggest that there should be appreciable restriction of rotation about the central bond. Although aurofusarin and fuscofusarin are the only natural

† τ 5·58 shifting to τ 6·30 in the leucohexa-acetate.

‡ The signals from the CH₃—C=CH— moiety are frequently broad singlets in such compounds.

dimeric quinones of this type several related dimers are known (aura-sperones,[449] ustilaginoidins,[451] cephalochromin[452]) in which two nor-rubrofusarin moieties are coupled at a *peri*-position.

The biquinone (332; R = Me) was synthesised[284] by fusing the naph-thol (336) with cupric acetate[286] followed by oxidation of the resulting dimer (337) with Fremy's salt, and more recently, aurofusarin dimethyl ether has been derived from rubrofusarin. Oxidation of the appropriate

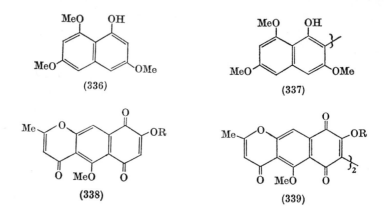

(336)   (337)

(338)   (339)

monomethyl ether[512] gave the quinone (338; R = Me). After acid hydrolysis, the hydroxyquinone (338; R = H) was converted into the dimer (339; R = H) by oxidation with persulphate, and then methylated to give (339; R = Me) identical with aurofusarin dimethyl ether.[511]

Rhodocladonic acid, $C_{15}H_{10}O_8$

Occurrence: numerous lichens of the genus *Cladonia*.[565-567]

Physical properties: red needles, dec. >300°, λ_{max}. (EtOH) 294, 318(sh), 328(sh), 449 nm (log ϵ 4·37, 4·18, 3·99, 3·65), ν_{max}. (KBr) 3385, 3300, 1660, 1630(sh) cm^{-1}.

Although this pigment has been known for more than sixty years the structure is still uncertain. For many years it was thought to be an anthraquinone[566] but recent work[568, 569] has shown this to be incorrect. The molecular formula is now found to be $C_{15}H_{10}O_8$. It contains a methoxyl group, and forms a triacetate and trimethyl ether, and a leucopenta-acetate and pentamethyl ether. The n.m.r. spectrum of the trimethyl ether includes a signal (3H) for a C-methyl group at τ 7·42 and a singlet (1H) at τ 1·98 attributed to a *peri*-proton in a naphtha-quinone system. The infrared spectrum of the leucopenta-acetate shows ketonic carbonyl absorption (in $CHCl_3$) at 1680[569] (1690[568]) cm^{-1}.

Degradation of rhodocladonic acid in hot potassium hydroxide solution gave[568] a mixture of products, one of which was assigned structure (340) since, *inter alia*, it gave on methylation a trimethyl ether whose mass spectrum was identical with that of 2,3,5,7-tetramethoxy-1,4-naphthaquinone. A second degradation product, $C_{13}H_8O_7$, was

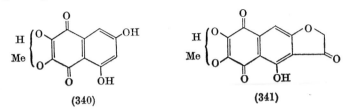

(340)   (341)

regarded as a de-acetylrhodocladonic acid as its trimethyl ether showed two aromatic proton signals at τ 2·06 and 2·63 but no C-methyl resonance, and was otherwise similar to the natural pigment. Hence it seemed possible to construct rhodocladonic acid by addition of a C_4 unit to position 6 of (340) in such a way as to form another ring as required by the molecular formula, and Baker and Bullock[568] regard structure (342)

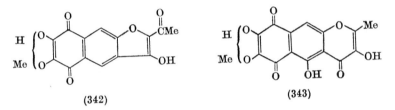

(342)   (343)

as the most reasonable formulation for rhodocladonic acid, keeping in mind a likely polyketide origin. The de-acetyl derivative is then (341). However another structure (343)[569] is equally consistent with the evidence although neither seems to account for Koller and Hamburg's observation[567] that rhodocladonic acid forms a mono-acetate simply on crystallisation from acetic acid.

Rubromycins†

Occurrence: cultures of *Streptomyces collinus* Lindenbein[289] and other Actinomycetes.[488]

Physical properties: α-rubromycin (collinomycin), $C_{27}H_{20}O_{12}$, orange needles, m.p. 278–281°, $\lambda_{max.}$ (CHCl₃) 319, 352, 365, 415, 484 nm, $\nu_{max.}$ (KBr) 1727, 1689, 1634 cm⁻¹.

β-rubromycin, $C_{27}H_{20}O_{12}$, red needles, m.p. 225–227°, $\lambda_{max.}$ (CHCl₃) 316, 350, 364, 504 nm, $\nu_{max.}$ (KBr) 1733, 1686, 1618 cm⁻¹.

γ-rubromycin, $C_{26}H_{18}O_{12}$, red needles, dec. ~235°, $\lambda_{max.}$ (CHCl₃) 317, 349, 364, 484, 513, 551 nm, $\nu_{max.}$ (KBr) 1733, 1686, 1600 cm⁻¹.

This group of closely related antibiotic pigments is being studied by Brockmann and his co-workers.[290] Abundant evidence has been obtained

† Added in proof: see Appendix ref. 4.

for the structure of the quinonoid portions of these molecules but information about the remainder (approximately half) is still fragmentary. The relationship of these compounds is set out below.

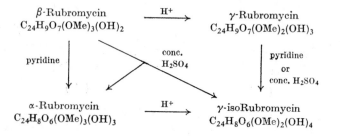

β-Rubromycin is quantitatively isomerised to α-rubromycin in boiling pyridine but on heating with mineral acids (and some organic acids) it is transformed into γ-rubromycin. The latter rearranges quantitatively in boiling pyridine into a fourth quinone γ-isorubromycin ($\lambda_{max.}$ 324, 362, 438, 469, 495, 532 nm, $\nu_{max.}$ 1727, 1686, 1603 cm^{-1}) which can also be derived from α- and β-rubromycins by acid treatment. It is possible that γ-rubromycin, isolated from cultures of S. collinus, is, in part, an artefact.

γ-Rubromycin is clearly a naphthazarin derivative from its visible absorption spectrum, and the bathochromic shifts with sodium hydroxide and with boroacetate. Furthermore the infrared band at 1600 cm^{-1} moves to 1647 cm^{-1} in the triacetate. The visible absorption of γ-rubromycin is very similar to that of 2,7-dimethoxynaphthazarin, and on pyrolysis in vacuo it gave a small yield of 2-hydroxy-7-methoxynaph-

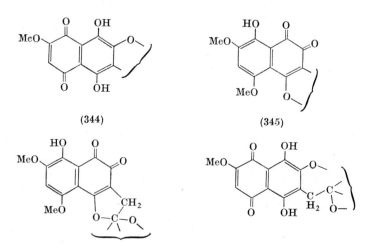

(344)   (345)

(346)   (347)

thazarin. Thus γ-rubromycin has the partial structure (344) which is consistent with a vinylic proton signal at τ 3·85 and the absence of a β-hydroxyl group.

β-Rubromycin forms a quinoxaline derivative, a violet anion and a boroacetate, and on acetylation the chelated carbonyl band at 1618 cm⁻¹ shifts to 1658 cm⁻¹. On this basis it is considered to be a derivative of 8-hydroxy-1,2-naphthaquinone, and since it is converted quantitatively into γ-rubromycin (344) with simultaneous demethylation of one methoxyl group, β-rubromycin appears to be (345). This is consistent with all the evidence and presumably the second quinone carbonyl band overlaps with that at 1686 cm⁻¹ (see below). The rearrangement is reminiscent of the chemistry of dunnione and β-lapachone with the important difference that the transformation is not reversible. As the *p*-quinone in this case is a naphthazarin, this is not surprising. The n.m.r. spectra of both β- and γ-rubromycin include an AB system (τ ~6·20, 6·48; J, 18·4 Hz) attributed to a methylene group attached to the naphthaquinone nucleus at C-3 and to a tert. carbon atom linked to oxygen (there is also supporting mass spectral evidence on this point) so that the structures of β- and γ-rubromycins can now be expanded to (346) and (347), respectively.

α-Rubromycin is a *p*-quinone isomeric with β-rubromycin. The additional hydroxyl group is located in the non-quinonoid part of the molecule (not yet considered) as the chromophore was shown by the usual tests to possess only one *peri*-hydroxyl group, as in β-rubromycin. The visible absorption has shifted to shorter wavelengths, relative to β-rubromycin, as expected, but surprisingly the ultraviolet absorption of α-rubromycin leucopenta-acetate shows a *bathochromic* shift relative to that of β- and γ-rubromycin leuco-acetates. This suggests that there is now a C=C double bond in conjugation with the main chromophore, and since the n.m.r. spectrum shows that the α-methylene group in (346) has been replaced by a vinylic proton, a partial structure for α-rubromycin could be (348). Conversion of (346) into (348) evidently

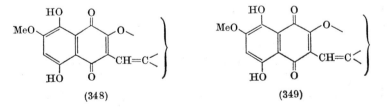

(348) (349)

entails, *inter alia*, abstraction of a proton from the methylene group by pyridine and elimination of the oxygen function on the adjacent carbon (thus providing the additional hydroxyl group in α-rubromycin).

γ-Isorubromycin is the naphthazarin (349) corresponding to α-rubromycin from which it can be obtained, for example, by treatment with hydrochloric acid in cold chloroform.

The relationship of these four quinones is shown again below. The

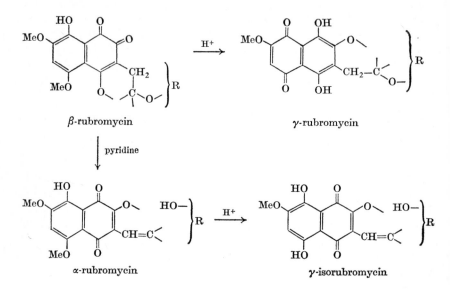

β-rubromycin   γ-rubromycin

pyridine

α-rubromycin   γ-isorubromycin

unidentified part R, which is responsible for the ultraviolet absorption ~$\lambda_{max.}$ 350, 365 nm, includes a methoxycarbonyl ($\nu_{max.}$ ~1730 cm^{-1}) and a chelated carbonyl ($\nu_{max.}$ ≃ 1686 cm^{-1}) group, two aromatic or vinylic protons, four aliphatic protons, an etherial oxygen and ten carbon atoms.[290]

Griseorhodin A, $C_{25}H_{16}O_{12}$[487]

Occurrence: cultures of an unidentified streptomycete.[291]

Physical properties: dark red, amorphous, material, m.p. 280–282° (dec.), $\lambda_{max.}$ (MeOH) 229, 259, 290, 431 nm (log ε 4·56, 4·47, 4·01, 4·13), $\nu_{max.}$ (KBr) 3400(br), 1700, 1660, 1620 cm^{-1}.

Griseorhodin A is one of a group of antibiotic pigments possibly related to the rubromycins. The partial structure (351) has been deduced by Eckardt[292] mainly from an examination of a red, crystalline degradation product (350) obtained by heating with methanolic hydrochloric acid. This compound, $C_{17}H_{14}O_9$ (? $C_{19}H_{16}O_{10}$), $\lambda_{max.}$ (EtOH) 232, 255, 315, 364, 508 nm, $\nu_{max.}$ 1700, 1660, 1620 cm^{-1}, retains the chromophore of griseorhodin A. The naphthopurpurin structure is suggested by the strong carbonyl peak at 1620 cm^{-1}, the positive pyroboroacetate test (comparison of the visible absorption with models), and the very close resemblance between the light absorption of the degradation product and naphthopurpurin, and their respective acetates. The acetyl and leuco-acetyl derivatives were found to be optically active. The expression

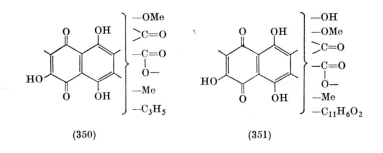

(350)   (351)

(351) represents the present state of knowledge;[292, 487] the unidentified C_{10} fragment is obviously aromatic and the structure includes an asymmetric carbon atom.

The prosthetic group of the antibiotic prunacetin A [489] (from *S. griseus* var. *purpureus*) may be related to griseorhodin A.

Maturinone (352), $C_{14}H_{10}O_3$
Maturone (353), $C_{14}H_{10}O_4$

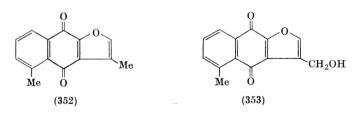

(352)   (353)

Occurrence: roots of *Cacalia decomposita* A. Gray [295] (Compositae).
Physical properties: (352) yellow needles, m.p. 168–169°, $\lambda_{max.}$ (EtOH) 251, 266, 287(sh), 355 nm (log ϵ 4·01, 4·39, 4·00, 3·74, 3·58), $\nu_{max.}$ (KBr) 1662, 1593, 1580 cm^{-1}.
(353) yellow needles, m.p. 169–170°, $\lambda_{max.}$ (95% EtOH) 250, 298, 354 nm (log ϵ 4·46, 3·81, 3·51), $\nu_{max.}$ (CHCl$_3$) 3500, 1690(sh), 1670, 1600 cm^{-1}.

These two quinones belong to a group of sesquiterpenes in the roots of the Mexican plant, *C. decomposita*, which have a furanonaphthalene or furanotetralin ring system. One of these is cacalol (354),[293, 294, 545] and when this was dehydrogenated with 2,3-dichloro-5,6-dicyanobenzoquinone (D.D.Q.) it gave a remarkably stable quinone methide (355) (cf. 10-methylene-anthrone), which could be oxidised with chromium trioxide in acetic acid to a quinone identical with maturinone, a minor constituent of *C. decomposita*. A similar oxidation of maturinin (356),

11

another constituent of the same plant, also afforded maturinone.[295] At the time when cacalol was regarded as the 4,8-dimethyl isomer of (354) Correa and Romo [295, cf. 546] proposed structure (357) for maturinone but it was shown later, by synthesis of both (357) and (352) that it is, in fact, the latter. This was accomplished [514] by alkylation of 2-hydroxy-5-methyl-1,4-naphthaquinone with β-chloro-α-methylpropionyl peroxide which gave the o-quinone (358). By ring opening in cold hydrochloric acid, followed by recyclisation in cold pyridine, it was partially rearranged

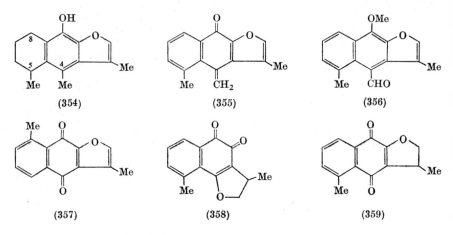

(354)   (355)   (356)

(357)   (358)   (359)

to the p-isomer (359) from which maturinone was derived by dehydrogenation with D.D.Q. Subsequently maturinone was obtained by condensing 2-acetoxy-5-methyl-1,4-naphthaquinone with propionaldehyde morpholine enamine.[545]

Biflorin (360), $C_{20}H_{20}O_3$

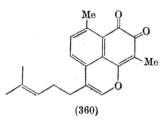

(360)

Occurrence: leaves and flowers of *Capraria biflora* L.[296] (Scrophulariaceae).
Physical properties: dark, brown-red, crystals, m.p. 159–160°, $\lambda_{max.}$ (EtOH) 234, 340, 555 nm (log ε 4·5, 3·75, 3·76), $\nu_{max.}$ (KBr) 1684, 1634, 1608, 1592 cm⁻¹.

This antibiotic quinone was the first oxaphenalene derivative to be found [296] in Nature and the presence of four C–Me groups in a C_{20} molecule implied that it was diterpenoid in character.[297] Biflorin forms

a leucodiacetate and a quinoxaline derivative, and is therefore an
o-quinone. From negative evidence, the third oxygen atom appeared
to be present in an ether bridge, probably attached to the quinone ring
to account for the relatively high dipole moment (7·19D). On catalytic
hydrogenation over platinum, the pigment readily absorbs one, two
and, more slowly, three mols. of hydrogen to give, after reoxidation in
air, biflorin and di- and tetrahydrobiflorin, respectively. The light
absorption of dihydrobiflorin is the same as that of the parent compound
showing that an isolated double bond has been saturated. This is
evidently an isopropylidene group as biflorin yields acetone on ozonolysis. However, uptake of a further mol. of hydrogen has a marked effect
on the visible absorption, the longwave band at 555 nm in dihydrobiflorin shifting to 450 nm in the tetrahydro compound, and the spectrum
of the latter showed a marked resemblance to that of the pyranonaphtha-

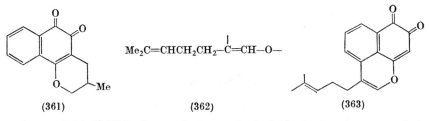

$$Me_2C=CHCH_2CH_2-\overset{|}{C}=CH-O-$$

(361)   (362)   (363)

quinone (361).[63] This shows that tetrahydrobiflorin (ν_{CO} 1687 cm^{-1}) is
an o-naphthaquinone and implies that biflorin and dihydrobiflorin
contain a double bond in conjugation with the quinone system. Vigorous
oxidation of biflorin with dilute nitric acid yielded benzene-1,2,3,4-
tetracarboxylic acid showing that the benzenoid ring carries carbon
substituents at positions 5,6 or 7,8 or 5,8.

The n.m.r. spectrum of biflorin includes, in addition to the expected
signals from the isopropylidene group, two singlets (3H) at τ 7·30 and
8·06 assigned to methyl groups attached to the benzenoid and the
quinonoid ring, respectively, and a symmetrical multiplet (4H) centred
at τ 7·6 attributed to the methylene protons of a fragment $>C=CCH_2$-
$CH_2C=C<$. This decreases in intensity during hydrogenation, accompanied by the appearance of methylene and methine peaks, of appropriate intensity, in the region τ 8–9. A singlet (1H) at τ 2·95, attributed to a
proton in a vinyl ether system, $>C=CH-O-$, changes, in tetrahydrobiflorin to a multiplet (2H) at τ 5·6 in agreement with a part structure
$>CH-CH_2-O-$. These fragments can now be linked as in (362),
and the whole assembly must be attached to an o-naphthaquinone
nucleus at positions 4 and 5 (see 363) to accord with the data cited earlier.
It follows that the quinone methyl group lies at C-3 and the aromatic
methyl was placed at C-8 because the signals from the two ortho-coupled

protons in the n.m.r. spectrum of biflorin collapse to a singlet (2H) at τ 2·75 in the tetrahydro derivative. Accordingly, Prelog and his co-workers[297] regard biflorin as (360) which obeys the isoprene rule. Dihydrobiflorin is therefore (364), and the high dipole moment and intense long wave absorption of these oxaphenalene systems can probably be attributed to contributions from the mesomeric structures (365).

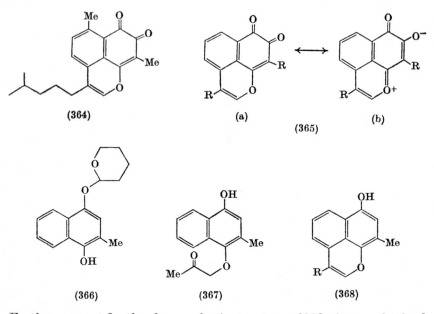

(364)   (a)   (b)
                          (365)

(366)   (367)   (368)

Further support for the chromophoric structure of biflorin was obtained by comparison with the model compound (365; R = Me).[298] This was prepared by condensing the dihydropyranyl ether (366) with chloro-acetone, followed by hydrolysis of the protecting group and cyclisation of (367) with polyphosphoric acid to the oxaphenalene (368). Oxidation with Fremy's salt then gave the quinone (365; R = Me) whose spectral properties ($\lambda_{max.}$ ca. 560 nm, $\nu_{max.}$ 1690 cm^{-1}) were in good agreement with those of biflorin.

Mansonones (369–377)

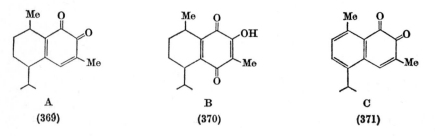

A   B   C
(369)   (370)   (371)

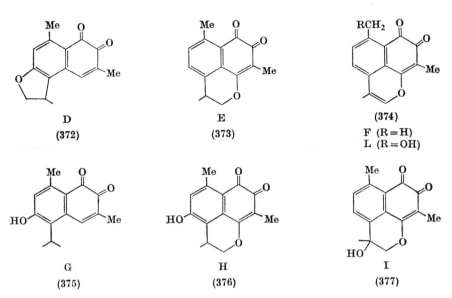

D   E   (374)

(372)   (373)   F (R=H)
                              L (R=OH)

G   H   I

(375)   (376)   (377)

Occurrence: heartwood of *Mansonia altissima* Chev.[299-301, 305, 542] (Sterculiaceea) Mansonone C also occurs in *Ulmus glabra* Mill.,[515, 516] *U. americana* L.,[516] *U. minor* (*plotii*) Mill.,[516] *U. campestris* L. (= *U. procera*)[516] and other *Ulmus* spp.[516] (Ulmaceae). See also p. 326.

The heartwood of *Mansonia altissima*, from tropical West Africa, is used for the manufacture of furniture but the sawdust frequently causes violent sneezing, vertigo and eczema.[299] This led to chemical investigations in three different laboratories[299-301] and the active principle, found in a chloroform extract, proved to be mixture of closely related naphthaquinones. So far ten have been found, mansonones A–L, but it is not known which are physiologically active. Biogenetically they are highly oxidized sesquiterpenes, and it is interesting that mansonone C (371) occurs in elm wood (*Ulmus* spp.) along with the naphthaldehydes (378; R = H and OMe) and other compounds having a cadalene

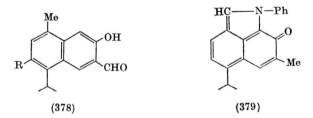

(378)   (379)

skeleton.[516, 517] The majority are *o*-quinones, recognisable by an absorption maximum in the visible region (except mansonone H) and a carbonyl peak in the infrared at relatively high frequency, but some anomalous values have been reported (see Table III). Several mansonones possess asymmetric carbon atoms although optical activity is recorded only for A, E and H. Most of the structural detail has been deduced from n.m.r. spectra.

Mansonone A, probably identical with the "Mansonia-quinone", m.p. 116–117°, first isolated[299] by steam distillation, has the spectral properties of an *o*-quinone (but fails to form a quinoxaline derivative with *o*-phenylenediamine) and its leucodiacetate shows benzenoid absorption ($\lambda_{max.}$ 274 nm). The n.m.r. spectrum is consistent with structure (369) and on dehydrogenation with chloranil it is converted into mansonone C (371).[301, 303] Mansonone A tends to disproportionate[541] into mansonone C and mansonone A quinol during chromatography (Al_2O_3) and on heating, so that the quinol obtained[541] from *M. altissima* could be an artefact. On heating in nitrobenzene mansonone A affords not only mansonone C and mansonone A quinol but also, surprisingly, mansonone F and a compound, $C_{21}H_{19}NO$, for which structure (379) has been proposed.[541]

Mansonone B is a hydroxyquinone which forms an oily acetate, and gives a marked bathochromic shift in alkaline solution ($\lambda_{max.}$ 525 nm). The n.m.r. spectrum revealed an isopropyl group, a secondary methyl group, and a methyl group attached to the quinone ring, but aromatic protons and benzylic methylene groups were absent. Structure (370) was therefore proposed[300] assuming, on biogenetic grounds, that the pigment had the same carbon skeleton as the other mansonones.

Mansonone C forms a quinoxaline with *o*-phenylenediamine, and a leucodiacetate which shows naphthalenoid ultraviolet absorption. The n.m.r. spectrum disclosed an isopropyl group, an aromatic methyl group (τ 7·38) assigned to C-8, two *ortho*-coupled aromatic protons, and a quinonoid methyl group coupled to a proton on the same ring. Assuming a common biogenesis with the associated pigments, mansonone C appears to be cadalene-7,8-quinone (371).[306] The two compounds have not been directly compared† but they agree in melting point as do their quinoxalines.[300, 301, 303] As already mentioned, related naphthols, including 7-hydroxycadalene, have been found[517] in elm woods (*Ulmus* spp.) which also elaborate mansonone C.[515, 516] *U. hollandica* cl. 390, which is resistant to Dutch elm disease, was found to produce two fungitoxic compounds after inoculation with *C. ulmi*.‡ They were identified[543] as mansonones E and F.

† Added in proof: they have now[515] and are identical
‡ Before inoculation only traces of C and E could be detected, in June, and none in September.

TABLE III.

Mansonones

Pigment	Formula	Appearance	m.p.	$\lambda_{max.}$ (MeOH or EtOH) nm	$\log \epsilon$	$\nu_{max.}$ (KBr) cm^{-1}
Mansonone A[a]	$C_{15}H_{20}O_2$	Red needles	119–120°	260(sh), 280(sh), 432	3·5, 3·1, 3·2	1685, 1670
Mansonone B	$C_{15}H_{20}O_3$	Golden-yellow prisms	68–69°	226, 272, 408	3·86, 3·69, 2·40	
Mansonone C	$C_{15}H_{16}O_2$	Orange needles	134–135°	206, 258, 380(sh), 430	4·14, 4·24, —, 3·39	1686, 1666
Mansonone D	$C_{15}H_{14}O_3$	Orange prisms	173–175°	219, 243, 278, 405	4·30, 4·10, 4·11, 3·88	1670, 1660
Mansonone E[b]	$C_{15}H_{14}O_3$	Orange-yellow needles	148–149°	220, 264, 375, 445	4·25, 4·31, 3·20, 3·38	1690, 1640
Mansonone F	$C_{15}H_{12}O_3$	Violet needles	214–215°	233, 336, 550	4·50, 3·70, 3·75	1685, 1630
Mansonone G	$C_{15}H_{16}O_3$	Orange needles	210–213°[d]	244, 274, 407	4·25, 4·40, 3·95	3300, 1660, 1645
Mansonone H[c]	$C_{15}H_{14}O_4$	Red needles	>320°	220, 265(sh), 274, 300, 387	4·25, —, 4·30, 3·95, 3·75	3160, 1660, 1620
Mansonone I	$C_{15}H_{14}O_4$	Orange-red needles	213–214°	265, 370, 440		3400, 1685, 1630, 1613
Mansonone L[e]	$C_{15}H_{12}O_4$	Violet crystals	165–167°	234, 255(sh), 555		

[a] $[\alpha]_D^{20}$ −600° (95% EtOH).
[b] $[\alpha]_D^{21}$ +92° (EtOH).
[c] Acetate, $[\alpha]_D^{21}$ +424° (CHCl$_3$).
[d] Or 201–203°.
[e] J and K are not in the Italian alphabet.

Mansonone D forms a quinoxaline and structure (372) is tentatively proposed[300] on the basis of its n.m.r. spectrum and biogenetic considerations. The quinone methyl signal is a doublet at τ 7·94 coupled to a quartet (1H) at τ 2·78, and there is one aromatic proton which resonates at τ 3·40. A peak at τ 7·35 (3H) is assigned to the methyl group at C-8, and the dihydrofuran ring is recognised from the methylene proton signals at τ 5·48 which form part of an ABX system where X is the benzylic proton (multiplet at τ 6·42) coupled also to the secondary methyl group (doublet at τ 8·62).

Mansonone E (373) is again a typical *o*-naphthaquinone and shows a marked resemblance to tetrahydrobiflorin in its ultraviolet-visible and infrared absorption. It forms a quinoxaline derivative, and the leucodiacetate has a naphthalene-like ultraviolet spectrum. As the n.m.r. spectrum shows the same features as that of tetrahydrobiflorin (except for the two *ortho*-protons which appear as an AB quartet and not a singlet) structure (373) can be assigned[300] to this pigment. Tanaka *et al.*[301, 304] deduce from the coupling constants of the ABX system that the methyl group attached to the asymmetric carbon atom is quasi-axial.

Mansonone F is a violet quinone which closely resembles dihydrobiflorin (364) in its spectroscopic properties, these being consistent with structure (374; R = H).[300] On catalytic hydrogenation it gives[304, 541] racemic mansonone E (373) and is probably identical with "Mansoniaazulene".[299]

Mansonone G is a hydroxyquinone which affords an oily acetate and methyl ether, and the latter formed a quinoxaline derivative with *o*-phenylenediamine. From its ultraviolet absorption the leucodiacetate is a naphthalene derivative and the ultraviolet-visible spectrum of the pigment, in both neutral and alkaline solution, is very similar to that of 6-hydroxy-1,2-naphthaquinone. The n.m.r. spectrum of the leucodiacetate reveals the presence of two aromatic methyl groups, an isopropyl group and two aromatic protons (singlets at τ 2·02 and 3·14). On this basis, and assuming the usual cadalene skeleton, mansonone G is

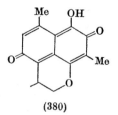

(380)

considered to have structure (375)[301,303,541] but it is not clear why methylation with diazomethane is "difficult". It has been suggested[301] that the green colour given with nickel acetate may be that of a chelate complex derived from the tautomeric structure (380).

Mansonone H. The formation of the usual derivatives shows that this pigment is another hydroxy-*o*-naphthaquinone in which the fourth oxygen atom must be in an ether bridge. It is noteworthy that there is no absorption maximum in the visible region which is also true for 2,6-dihydroxy-1,4-naphthaquinone whose ultraviolet absorption is somewhat similar to that of mansonone H. The latter, however, shows a much larger bathochromic shift in alkaline solution ($\lambda_{max.}^{HO^-}$ 412, 550 nm), approaching that of mansonone G. This suggests that the hydroxyl group is located at C-6 as does the green colouration with nickel acetate (cf. mansonone G). The n.m.r. spectrum of the acetate reveals a quinone methyl group, another at C-8, and one aromatic proton; in addition there are signals from a secondary methyl group attached to an ABX system of structure

$$\begin{array}{c} CH_3 \\ | \\ -CH-CH_2-O- \end{array}$$

almost identical with those in the spectrum of mansonone E. The coupling constants indicate that the methyl group attached to the asymmetric centre is quasi-axial. Assembling these components in a manner consistent with the isoprene rule leads to structure (376) for mansonone H. This is supported by oxidative degradation of the methyl ether with hydrogen peroxide which yielded, in addition to acetic acid,

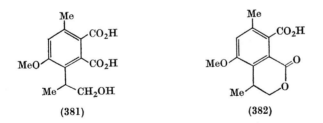

(381)   (382)

a hydroxyphthalic acid derivative, $C_{13}H_{16}O_6$ (381), which, on warming to 100°, formed a lactone regarded as (382) on the basis of its n.m.r. spectrum.[301,304]

Mansonone I shows ultraviolet-visible and infrared absorption very similar to the spectra of mansonone E and tetrahydrobiflorin with the addition of a hydroxyl band at 3400 cm^{-1}. The hydroxyl group is not attached to the chromophoric system (no shift of λ_{max}. in alkaline solution), and it disappears on treatment with acetic anhydride to give a violet quinone identical with mansonone F. The same dehydration takes place when mansonone I is condensed with o-phenylenediamine giving the quinoxaline derivative of mansonone F. The hydroxyl group is therefore attached to the heterocyclic ring, and must be located as in (377) to account for the observation that the aliphatic methyl signal is a singlet at τ 8·30 and there is a peak (2H) at τ 5·60 attributable to a methylene group attached to oxygen.[305]

Mansonone L shows a maximum in the visible region at 555 nm, like mansonone F, but contains an additional atom of oxygen. This is present as a hydroxyl group [M-17 (16%), mono-acetate] which must be attached to a methyl group, as in (374; R = OH),[542] since the n.m.r. spectrum of mansonone L closely resembles that of mansonone F except for the absence of the signal from the C-8 methyl group.

Coleon A (383), $C_{20}H_{22}O_6$

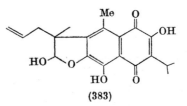

(383)

Occurrence: in the red glandular hairs under the leaves of *Coleus ignarius* (Schweinf.)[307] (Labiatae).

Physical properties: red prisms, m.p. 136°, $[\alpha]_D^{22}$ ca. +100° (EtOH), +80° (CHCl$_3$), λ_{max}. (iso-octane) 230, 252·5, 272·5, 315, 415(sh), 435, 460, 485(sh) nm (log ϵ 4·31, 4·24, 4·02, 4·03, 3·71, 3·80, 3·79, 3·58), ν_{max}. (CCl$_4$) 3521, 3322, 1669, 1626 cm^{-1}.

Coleon A occurs in some abundance (<1·9% dry wt.) on the leaves of the African shrub *C. ignarius*. The rather unusual structure has been determined by the extensive investigations of Eugster and his co-workers.[307, 308] The general character was apparent from its redox properties, colour reactions, and ultraviolet-visible and infrared spectra, all of which indicated a juglone type of molecule and, in particular, the ultraviolet-visible curve was almost parallel to that of droserone (p. 236) but shifted 25–30 nm to longer wavelengths. The n.m.r. spectrum shows that benzenoid and quinonoid protons are absent so that the naphthaquinone system is fully substituted, and there are signals arising from a quaternary methyl group, an aromatic methyl group, an isopropyl group attached to either a benzenoid or a quinonoid ring, and a terminal vinyl group. The latter can be reduced catalytically, giving dihydrocoleon

A, and must be the origin of the formaldehyde obtained on ozonolysis; moreover, as Kuhn–Roth oxidation of coleon A yields acetic and isobutyric acids and dihydrocoleon A gives, in addition, propionic acid, the vinyl group appears to be part of an allyl group. Eighteen carbon atoms have now been identified. Coleon A and its dihydro derivative are remarkably similar and have the same ultraviolet and infrared spectra, melting point, mixed melting point and R_F values; however, they differ in their n.m.r. spectra, optical rotation and chemical properties.

Coleon A forms a triacetate in cold pyridine but only a monomethyl ether with diazomethane; of the three hydroxyl groups, one is fairly acidic ($pK_{MCS} = 6.6$) and hence attached to a quinone ring, one is in a *peri*-position and its proton resonates at τ −3.7, and the nature of the third [$\nu_{max.}$ (CCl$_4$) 3597 cm^{-1}] was revealed, *inter alia*, by oxidation with manganese dioxide when its disappearance was accompanied by formation of a γ-enol-lactone (ν_{CO} 1820 cm^{-1}). As reduction of the lactone with potassium borohydride regenerated coleon A, the latter must contain a hemi-acetal structure. In solution, the lactone probably exists in equilibrium with the hydroxyaldehyde form (384 $\rightleftharpoons$ 385) since the

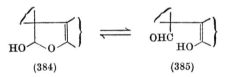

(384)　　　　　　(385)

hydroxyl proton and the adjacent quaternary methyl (see later) give weak doublet signals in the ratio 1:0.3. This doublet character disappears on oxidation to the lactone which may explain why the lactone but not coleon A, readily forms *crystalline* derivatives, for example, a leucotetra-acetate. All six oxygen atoms are thus accounted for, and with the inclusion of the cyclic hemi-acetal all twenty carbon atoms are now identified but there remains the very considerable task of determining the relationship of the various groups to each other.

When coleon A was oxidised with peracetic acid a new, stable, orange compound, $C_{20}H_{22}O_7$, was formed which had the same chromophore as the starting material. Dihydrocoleon A does not undergo this reaction. Spectroscopic examination† showed that the new product had lost both the allyl group and the hemi-acetal hydroxyl group, but the acetal proton was still present. This can be understood if the allyl and hydroxyl groups are adjacent, the reaction (see 386 → 387) leading, via an epoxide, to the formation of a furanofuran system. The n.m.r. spectrum is

† Doubling of several peaks in the n.m.r. spectrum suggests that the substance is a mixture of diastereoisomers.

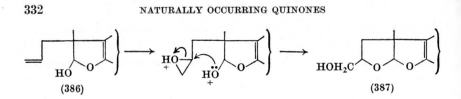

(386)   (387)

consistent with that of other compounds containing the same bicyclic system (for example, aversin, p. 491), and since the acetal proton signal at τ 4·0 is a singlet the quaternary methyl group must be attached to the adjacent carbon atom. The relationship of the fragment (384) to the main part of the coleon A molecule came to light in several ways. The γ-lactone, obtained by oxidation with manganese dioxide, dissolves in aqueous sodium hydroxide giving a red solution, but if the quinonoid hydroxyl group is first methylated with diazomethane it then gives a deep cornflower blue solution in alkali. This is characteristic of a naphthaquinone having two hydroxyl groups in 5,6- or 5,8-positions, and such a structure would arise if the lactone ring lies adjacent to the *peri-*

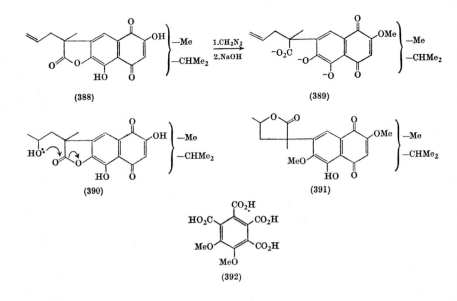

hydroxyl group as in (388); the blue anion would then be (389). This arrangement was confirmed by a sequence of reactions starting from the alcohol (390) which was obtained by hydration of the allyl group in the lactone (388) with cold sulphuric acid. On treatment with diazomethane another lactone ring was formed by an unusual intramolecular reaction (see 390) and the newly generated phenolic group was methylated to give (391). Further methylation of the *peri*-hydroxyl group with dimethyl sulphate, followed by degradation with alkaline permanganate, yielded the tetracarboxylic acid (392) which established the *ortho* orientation of the two oxygen functions in the benzenoid ring, and also confirmed the location of a carbon substituent at C-8. From the n.m.r. spectrum the latter must be methyl whence it follows that the remaining isopropyl group must be in the quinone ring at C-3 (383).[308]

Several other interesting rearrangements of coleon A support the evidence already cited for structure (383) and confirm the orientation of the quinonoid substituents with respect to those in the benzenoid ring. In sulphuric acid, the allyl group undergoes a Wagner–Meerwein migration to form the furan (393) (anhydro-isocoleon A) which, under appropriate conditions, is hydrated giving isocoleon A (394) isomeric with the natural quinone. Hydrogenation of (393) gives dihydro-anhydro-isocoleon A (395) also available by Wagner-Meerwein re-arrangement of dihydrocoleon A. Confirmation that the allyl (and

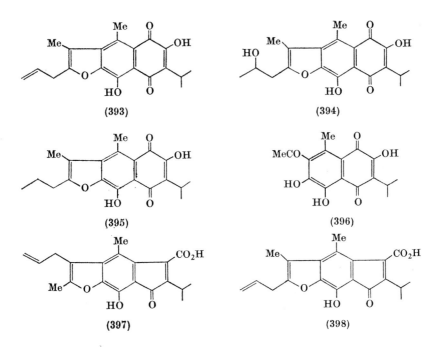

(393)

(394)

(395)

(396)

(397)

(398)

propyl) group, rather than the methyl group, had migrated, was obtained by ozonolysis of the diacetate of (395) which afforded butyric acid and a C-acetylquinone (396). The latter, incidentally, forms a cyclic carbonate on reaction with phosgene which further confirms the *ortho* relationship of the two oxygen functions in the benzenoid ring of coleon A.

Like other 2-hydroxy-3-alkylnaphthaquinones, coleon A rearranges in hot alkaline solution to give an indenone-carboxylic acid (397) (see p. 217). As its ketonic absorption band at 1667 cm^{-1} shifts, on acetylation, to 1698 cm^{-1}, it is clearly chelated, and this determines the orientation of the cyclopentadienone ring and hence that of the substituents in the original quinonoid ring. The initial alkaline rearrangement product was purified by vacuum distillation, at which stage a Wagner–Meerwein migration of the *methyl* group occurred as can be seen from inspection of (397). The structure of (397) follows from its spectroscopic properties, and from the fact that the compound is isomeric with the indenone-acid (398) obtained by alkaline rearrangement of anhydro-isocoleon A (393).

The biogenetic origin of coleon A is a matter for speculation. The presence of an isopropyl group in a C_{20} molecule inevitably suggests a diterpene, and Karanatsios and Eugster[308] have put foward the idea that coleon A may be essentially a highly oxidised 1,10-*seco*-diterpene of the abietic acid series, bearing in mind that oxidised diterpenes are

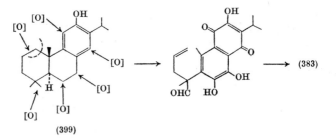

(399)

by no means uncommon in the Labiatae (e.g. tanshinones, p. 640 coleon B[577]). The hypothetical modification of ferruginol (399) is illustrative.

Streptovaricins (400)

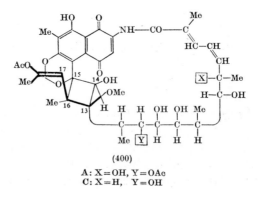

(400)

A: X=OH, Y=OAc
C: X=H, Y=OH

Occurrence: cultures of *Streptomyces spectabilis* n.sp.[309]
Physical properties:[311] λ_{max}. (95% EtOH) 245, 264(sh), 320(sh), 430 nm.

A, $C_{42}H_{53}NO_{16}$, m.p. 194–196°, $[\alpha]_D$ +610° ($CHCl_3$)
B, $C_{42}H_{53}NO_{15}$, m.p. 185–187°, $[\alpha]_D$ +576° ($CHCl_3$)
C, $C_{40}H_{53}NO_{14}$, m.p. 189–191°, $[\alpha]_D$ +602° ($CHCl_3$)
D, $C_{42}H_{55}NO_{15}$, m.p. 167–170°, $[\alpha]_D$ +436° ($CHCl_3$)
E, $C_{40}H_{55}NO_{13}$
G, $C_{40}H_{51}NO_{15}$, m.p. 190–192°, $[\alpha]_D$ +473° ($CHCl_3$)

Streptovaricin is a collective name for a group of orange antibiotics which show marked *in vivo* activity against *Mycobact. tuberculosis*.[310] Rinehart and his co-workers[311, 312, 518, 519] have determined the structure of two of these remarkable *ansa* macrolides and a full paper is awaited. Degradation of streptovaricins A, B, C and G with periodate–osmium tetroxide gives another orange compound, streptovarone, $C_{24}H_{23}NO_9$, having the same visible chromophore as the parent compounds, and this, on mild acid hydrolysis, yields formaldehyde, pyruvic acid and a purple quinone, dapmavarone, $C_{18}H_{16}O_7$. Oxidation of the latter with alkaline hydrogen peroxide gave the keto-acid (402) and the naphthaquinone (401) which was recognised by comparison of the light absorption of the

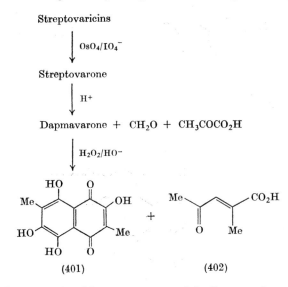

Streptovaricins

$\downarrow$ OsO_4/IO_4^-

Streptovarone

$\downarrow$ H+

Dapmavarone + CH_2O + CH_3COCO_2H

$\downarrow$ H_2O_2/HO^-

(401)   (402)

dimethyl ether (obtained by treatment with diazomethane) with that of 2,6-dimethoxynaphthazarin. In the n.m.r. spectrum of the dimethyl ether a signal (6H) at τ 7·87 is consistent with two C-methyl groups at β-positions (cf. the mansonones) and since both *peri*-hydroxyl protons resonate at τ −3·10 this excludes the 3,6-dimethyl isomer. The alternative 2,3-dimethyl-5,6,7,8-tetrahydroxy-1,4-naphthaquinone structure was eliminated by examination of the mass spectrum. Dapmavarone is regarded as (403) since its n.m.r. spectrum shows singlet peaks at τ 6·68 (2H) and 8·51 (3H) attributed to a moiety

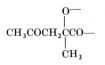

indicating that C-2 of the pentenoic acid (402) is linked via oxygen to the aromatic system. The formation of (401) and (402) can thus be seen as a Dakin oxidation in which one of the additional hydroxyl groups arises by replacement of a carbonyl group. This was assigned to C-5 since it showed ν_{CO} "below 1635 cm^{-1}" attributed to hydrogen bonding with a *peri*-hydroxyl group. As this must involve a 7-membered chelate ring it is surprising that the hydrogen bond is strong enough to hold the compound in the tautomeric *o*-quinonoid form and shift the ketonic carbonyl absorption to "below 1635 cm^{-1}".

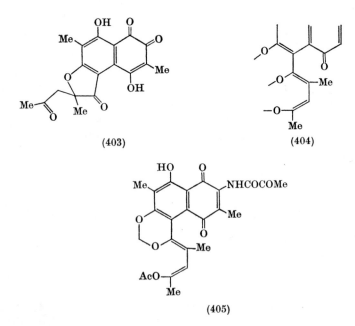

(403)   (404)

(405)

Streptovarone is also a naphthaquinone with one *peri*-hydroxyl group. As the nitrogenous group is readily eliminated as pyruvamide under both acid and alkaline conditions it is most probably attached to the quinone ring, and by analogy with the behaviour of the rifamycins,[521] this side chain was located at C-2. In the hydrolysis of streptovarone to dapmavarone, spectroscopic evidence also shows that there is loss of an enol acetate group and a methylenedioxy group [τ 4·5 (2H)] which is the origin of the formaldehyde cited earlier. These groups must be attached to the oxygen atoms of the part structure (404), the presence of which is deduced from the n.m.r. spectrum (C=CH at τ 3·45, C=C–CH$_3$ at τ 7·85 and 8·06) and the chemical evidence, and the acetate unit was assigned to the terminal oxygen atom to accord with the mass spectrum fragmentation pattern. Streptovarone was therefore regarded as (405).[311]

On irradiation[312] streptovarone undergoes an interesting change forming a yellow compound photostreptovarone, $C_{20}H_{17}NO_6$, which is the hydroxyphenalone (407). The reaction proceeds with elimination of

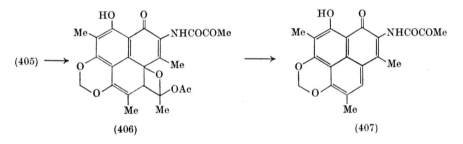

acetic anhydride, probably via the spiro-oxetan (406) for which there are many intermolecular analogies among simple quinones and alkenes.[313]

Streptovaricin A may be cleaved by oxidation with periodate into prestreptovarone, $C_{29}H_{29}NO_9$, and varicinal A. Further treatment of prestreptovarone with periodate–osmium tetroxide converts it into streptovarone by fission of a double bond in the side chain attached to nitrogen. From a study of the n.m.r. and mass spectra structures (408) and (409) were deduced for prestreptovarone[311] and varicinal A,[518] respectively. If streptovaricin A is oxidised with one mol. of periodate

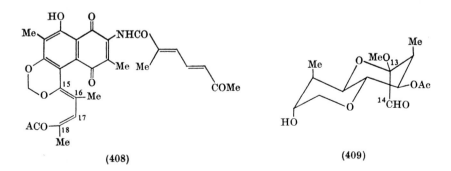

an intermediate compound, $C_{42}H_{51}NO_{16}$, is formed which breaks down to (408) and (409) on treatment with a further mol. of periodate. It follows that (408) and (409) must be linked at two places and it will be noted that varicinal A contains two aldehyde groups (one masked as a hemi-acetal). The structural unit (410) in streptovaricin A could be identified by n.m.r. and spin decoupling experiments, and a related structure (410) was found in streptovaricin C. This compound lacks one

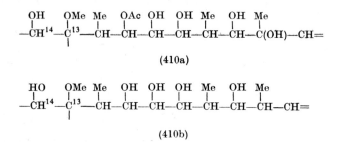

(410a)

(410b)

of the α-glycol units present in A, and both streptovaricin C and its triacetate consume only one mol. of periodate to give products containing structures analogous to varicinal A. As all the bonds in (410a) and (410b) could be accounted for except two, at C-13 and C-14, this must be the site where these units are attached to prestreptovarone in the construction of the streptovaricins. In prestreptovarone (408) the corresponding positions are considered to be C-15 and C-16 because, *inter alia*, on oxidative cleavage the signal from the protons of the C-16 methyl group move downfield to the olefinic region, and there is a corresponding shift of the C-17 vinylic proton signal. Thus combination of (409) and (408) at carbon atoms 13, 14, 15 and 16 leads to structure (400) for streptovaricins A and C which contains a cyclobutanol ring.[519] Cleavage of the four-membered ring would account for the uptake of a second mol. of periodate by A, and the single mol. by C and its triacetate, for which the following mechanism was suggested.

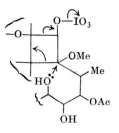

Tolypomycin Y. Another macrolide antibiotic which includes a naphthaquinone unit is tolypomycin Y, $C_{43}H_{54}N_2O_{14}$, (415),[526] a metabolite of *Streptomyces tolypophorus*. This compound is actually a quinone-imine formed by condensation with an amino-sugar. On mild acid hydrolysis it yielded tolyposamine, $C_6H_{13}NO_2$, (411) and a quinone, tolypomycinone, $C_{37}H_{43}NO_{13}$, which, under slightly more vigorous conditions, was degraded further to (412; R = OH), previously obtained by acid

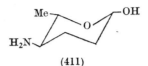

(411)

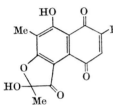

(412)

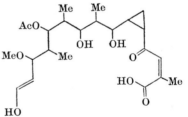

(413)

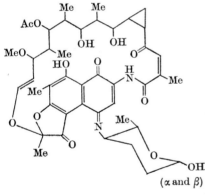

(415)

(414)

hydrolysis of rifamycin S,[520, 521] or to (412; R = NH$_2$). A second compound, tolypolide F, $C_{22}H_{30}O_7$, presumably derived† from tolypomycinone along with (411), possessed terminal lactone and aldehyde functions, for which structure (413) was deduced from chemical and n.m.r. studies. Related lactones were also obtained by ozonolysis of tolypomycinone and all the evidence indicated that this quinone contained the part structure (414). This was corroborated by spin decoupling experiments on tolypomycinone which led to the conclusion that it must have structure (415; = O in place of the amino-sugar

† Not explicitly stated in ref. 526.

residue). This was confirmed[537] subsequently by X-ray analysis of the tri-*m*-bromobenzoate.

THE VITAMINS K

Phylloquinone (K)† (416)
Menaquinones (MK-n)† (417; R = Me)
II-*Dihydromenaquinones* (MK-n(II-H$_2$))† (418)
Demethylmenaquinones (DMK-n)† (417; R = H)

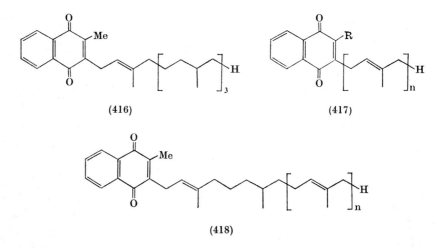

(416)   (417)

(418)

Occurrence: phylloquinone is generally associated with the green parts of plants[399, 400] and has been detected in various algae,[287, 399b, 400, 461, 479] in petals and fruits,[400] and in other chlorophyll-free tissues.[403] The menaquinones are all bacterial metabolites; the list below is not exhaustive.

Organism	Quinone	Reference
Aerobacter aerogenes	MK-8	416
Bacillus cereus	MK-7	404
B. licheniformis	MK-7	474
B. megaterium	MK-7	405
B. sphaericus	MK-7	414
Chromatium vinosum	MK-8	415
Corynebact. diphtheriae	MK-8(II-H$_2$)	409, 438
C. rubrum	MK-8(II-H$_2$)	419
Desulfovibrio vulgaris	MK-6	575

† Abbreviations proposed by IUPAC-IUB Commission on Biochemical Nomenclature.[395]

Organism	Quinone	Reference
Escherichia coli	MK-6, 7, 8, 9, DMK-7, 8	405, 528
Haemophilus parainfluenzae	DMK-5, 6, 7	411
Micrococcus luteus (= *Sarcina* *lutea*)	MK-7, 8, 9, MK-5(2-H), MK-6(2-H), MK-7(2-H), MK-8(2-H), MK-9(2-H)	405, 472, 558
M. lysodeikticus	MK-9	417
Mycobact. phlei	MK-8(II-H$_2$), MK-9(II-H$_2$), and (?) MK-10(II-H$_2$)	408, 437, 522
M. tuberculosis	MK-9 or MK-9(II-H$_2$)	436, 438, 439
Staph. albus	MK-7	405
S. aureus	MK-6, 7, 8, 9	472, 558
Strep. faecalis	DMK-7, 8, 9	410
Strep. unidentified	MK-9, MK-9(2-H), MK-9(4-H), MK-9(6-H), MK-9(8-H)	560
Unidentified (in putrefied fishmeal)	MK-6 and 7	413
Unidentified (in bovine liver)†	MK-10, 11, 12	559

Physical properties:[396, 397] yellow oils or crystals with m.p. ranging from 35° (MK-4) to 57° (MK-9); (416), $[\alpha]_D^{21}$ −0·28° (dioxan); (416), (417; R = Me) and (418), $\lambda_{max.}$ (EtOH) 239(sh), 242, 248, 260, 269, 325 nm, $\nu_{max.}$† (Nujol) 1660, 1618, 1595, 1299, 714 cm^{-1}; (417; R = H), $\lambda_{max.}$ (EtOH) 243, 248, 254, 263, 326 nm, $\nu_{max.}$‡ (CCl$_4$) 1672, 1585, 1300 cm^{-1}.

Numerous accounts of the discovery, chemistry and biological activity of the K vitamins can be found elsewhere, for example see refs. 431, 467, 468, 523. Martius[523] finds that MK-4 is the principal vitamin in the animal body and this is derived from exogeneous naphthaquinones by intestinal micro-organisms which replace the existing side chains by a geranylgeranyl group. It is generally accepted that its function is to maintain an adequate level of prothrombin and other clotting factors but the mechanism by which this antihaemorrhagic activity is effected remains obscure. The wide distribution of phylloquinone in the plant kingdom suggests a more fundamental role in cellular metabolism but its status in photosynthesis and in oxidative phosphorylation is also uncertain. It is interesting that phylloquinone also contains a C$_{20}$ side chain and perhaps surprising that higher plants, unlike bacteria, do not produce a family of phylloquinones corresponding to the menaquinones. The latter are involved in bacterial electron-transport systems.[361, 384, 572]

† Added in proof: see also Appendix ref. 9.
‡ Bands due to the isoprenoid side chain are found near 2940, 1450, 1390 and 900–750 cm^{-1}. By analysis of the variations in wavelength and shape Noll[398] was able to distinguish between K and MK compounds, and to determine chain length by measurement of relative intensities.

Phylloquinone. That phylloquinone is a 2,3-dialkylated 1,4-naphtha-quinone is apparent from its light absorption, a negative Craven test, and the formation of a leucodiacetate. Doisy and his associates [402, 420d] established the structure by oxidation with chromic acid which gave (419) in addition to phthalic acid, while a similar degradation of the leucodiacetate gave the acid (420) and a liquid ketone. The latter was more conveniently obtained by ozonolysis and proved to be 6,10,14-trimethylpentadecan-2-one (421) identical with the ketone prepared by oxidation of phytol. Clearly phylloquinone is 2-methyl-3-phytyl-1,4-naphthaquinone (416) which recent work has refined to (422), i.e., *trans*-(7′R,11′R)-phylloquinone. The configuration at C-7′ and C-11′ was determined [421] by comparison of the o.r.d. curves of the natural vitamin with those of the ketone (421) obtained both from natural phylloquinone and natural (7R,11R)-phytol, and the *trans* geometry at the 2′-3′ double bond was deduced [422] from the n.m.r. spectrum, *cis* and

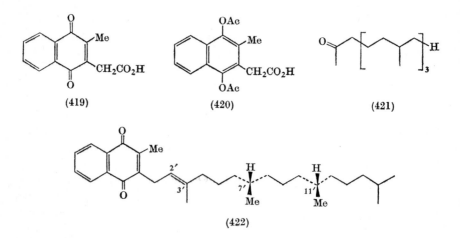

(419)   (420)   (421)

(422)

trans isomers being distinguishable by the chemical shift of the methyl group attached to C-3′.

    Phylloquinone can be synthesised by phytylation of 2-methylnaph-thaquinol followed by oxidation with silver oxide, and a number of variations on this theme have been developed. Synthetic work followed immediately after the structural elucidation, the first three syntheses appearing [420a, 423, 424] simultaneously in the same issue of the Journal of the American Chemical Society in 1939. In one of these,[424] 2-methyl-naphthaquinol was condensed with phytol in the presence of oxalic acid but the initial leucophylloquinone (424; R = H) was accompanied by some of the diketone (425),[425] resulting from electrophilic attack at

C-2, with consequent loss in yield. In later syntheses this was avoided by partial esterification[426, 427] of the quinol for example, (423; R = Ac, Bz), and improved yields were also obtained by using boron trifluoride as catalyst, and the tertiary allylic alcohol, isophytol (426)[428] as side

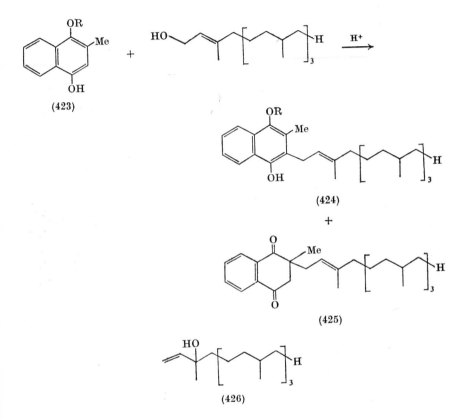

(423)

(424)

+

(425)

(426)

chain intermediate. The latter, however, condenses to give a *cis-trans* mixture (mainly *trans*), the resulting quinone having diminished biological activity. Isler and co-workers[422] synthesised phylloquinone identical in all respects, including optical rotatory dispersion, with the natural compound, using Lindlar's method.[427] The monobenzoate (424; R = Bz), prepared from natural phytol and the quinol ester (423; R = Bz) in the presence of boron trifluoride, was obtained as the pure *trans* form by crystallisation, and this, after hydrolysis, was converted into the quinone by aerial oxidation. Numerous analogues[429] and isotopically labelled compounds[430] have been prepared in similar fashion.

Hydroxyphylloquinone. Recent investigations[445] of the quinones in the blue-green alga *Anacystis nidulans* have shown that it contains a plastoquinone, phylloquinone, and a third quinone whose spectral features indicated a phylloquinone-type structure but chromatographically it appeared to be more polar. That this is due to a hydroxyl group is evident from its infrared spectrum ($\nu_{max.}$ 3630 cm^{-1})[445a] and mass spectrum[445b] which includes an M-18 peak more intense than the parent peak, and agrees with the molecular formula $C_{31}H_{46}O_3$. In the n.m.r. spectrum[445b] there is no absorption between τ 5·0 and 6·5 which indicates the absence of a structure >CH–O–. The hydroxyl group is therefore probably attached to one of the tertiary carbon atoms in the phytyl side chain, and the compound may be analogous to the hydroxylated plastoquinones.

Menaquinones. The second K vitamin (vitamin K_2) was isolated from putrefied fishmeal and was obviously similar to the plant product (vitamin K_1) in its light absorption and general properties, and the key to its structure was found by ozonolysis of its leucodiacetate. This reaction afforded the aldehyde (427), acetone and levulinic aldehyde clearly demonstrating that the vitamin was a 2-methyl-3-polyprenyl-1,4-naphthaquinone, and from the amount of levulinic aldehyde obtained, and previous estimates of the number of side chain double bonds

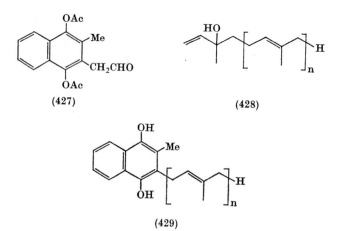

(427)   (428)

(429)

(by hydrogenation and bromination), and other analytical data, Doisy and his group[420c,d, 432] concluded that the new compound was the farnesylfarnesyl derivative (417; R = Me, n = 6). Some years later, however, in continuation of their work on the K vitamins the Hoffmann–

La Roche group synthesised several members of the all-*trans* mena-quinone series (417; R = Me, n = 2–7) for biological study. Direct comparison[433] then showed that Doisy's vitamin, m.p. 54°, was actually identical with the all-*trans* farnesylgeranylgeranyl isoprenologue, MK-7, (417; R = Me, n = 7), and furthermore, re-examination of putrefied fish-meal extracts revealed a second quinone, m.p. 50°, which did have structure (417; R = Me, n = 6), i.e., MK-6. The syntheses[433] followed the standard pattern, the greater part of the work being involved in the elaboration of the all-*trans* polyprenyl alcohols required as side chain components. Condensations were effected in the presence of boron trifluoride and zinc chloride, avoiding strongly acidic conditions which tend to promote cyclisation of the unsaturated side chains, the resulting quinol (429) being finally oxidised with silver oxide. In contrast to previous experience in the phylloquinone series, the all-*trans* tertiary alcohols (428) gave essentially all-*trans* products, and crystallisation of the menaquinones presented no difficulty. The higher homologues, MK-8 and MK-9, found later, in other bacteria, have also been synthesised[434–436] and their identity confirmed by direct comparison.

Dihydromenaquinones. Mycobacterium phlei[437] and several other *Mycobacteria*[438] elaborate a naphthaquinone which could not at first be identified, but obviously belonged to the vitamin K series. A close relationship to MK-9 was evident from its R_F value, on reversed phase chromatograms, which is intermediate[438] between that of MK-9 and MK-10 (or = MK-10[408]), and catalytic hydrogenation gave a compound of the phylloquinone series identical with that derived from MK-9 in the same way.[408] However, only nine mols. of hydrogen were absorbed[438] (that is, eight for the side chain double bonds and one for the quinone system) indicating that one of the nine isoprenoid units is saturated, and by analysis of the n.m.r. spectrum it could be shown that the saturated unit was not located at either end of the $(C_5)_9$ chain.[408, 438]

The position of the "missing" double bond in this MK-9(H$_2$) was neatly disclosed by the mass spectrum[438, 440] which, in addition to confirming the molecular weight, showed a fragmentation pattern consistent with successive loss of seven isoprene units. This means that the second C$_5$ unit from the quinone ring is saturated, as in (435), which was confirmed by chemical degradation as follows.[442]

The synthetic II†-dihydromenaquinone-3 (430) (obtained in the usual way from 2-methyl-1,4-naphthaquinol and 6,7-dihydrofarnesol[441]) was cyclised to the chromenol (431) by treatment with sodium hydride, acetylated immediately and then reduced to the chroman (432) with sodium in ethanol. The terminal double bond was then cleaved by

† Isoprene units are designated by Roman numerals starting from the quinone ring.

ozonolysis and the ozonide, on reduction with dimethylsulphide,[443] gave the aldehyde (433). The identical aldehyde was obtained when the same sequence of reactions was applied to the natural MK-9(H$_2$) whereas MK-9 itself afforded the lower homologue (434). It follows that the quinone from *M. phlei* is II-dihydromenaquinone-9 = (417; n = 7) (435). This was later confirmed by photolytic cleavage which yielded the ketone (436).[563] Other dihydromenaquinones are now known (see p. 340).

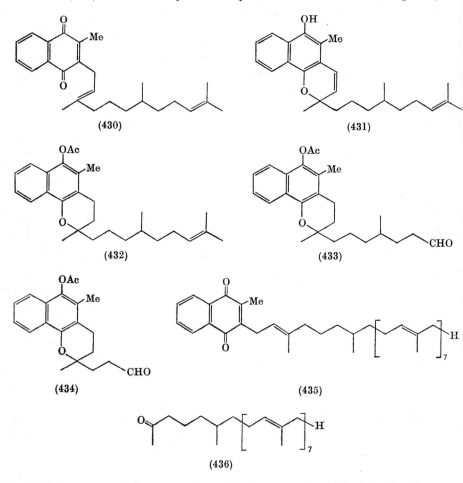

(430)   (431)

(432)   (433)

(434)   (435)

(436)

The presence of a *cis* isomer of MK-9(II-H$_2$) in *M. phlei* has been claimed[525] recently but this may be an artefact as other workers[522] could only detect this if ultraviolet light was used to detect the quinones on t.l.c. plates, or if the isolation procedure was not conducted in virtual darkness. The isolation of these bacterial quinones in a highly pure state can be very difficult.[522]

A new group of quinones belonging to the MK-9 series has been found[560] in an unidentified *Streptomycete*; in these compounds from one to four prenyl units are saturated but the sites of saturation have not yet been determined.

Demethylmenaquinones. The compounds in this group are distinguished from the rest of the menaquinone family by the absence of the methyl group at C-2. This is readily confirmed by n.m.r. spectroscopy which reveals the presence of a quinonoid proton by a triplet signal at τ 3·41 (J, 1·4 Hz). Further analysis led to the conclusion[411] that the principal component from *H. parainfluenzae* had a C_{30} polyprenyl side chain,

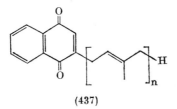

(437)

i.e., DMK-6 (437; n = 6). From comparative R_F values on reversed phase paper chromatograms, the minor components from this bacterium appear to be DMK-5 and DMK-7.[411] This was the first report of a terpenoid quinone having a C_{25} side chain and further confirmation is desirable.

Crane and his co-workers[444] have detected a new quinone in spinach chloroplasts which may be a demethylphylloquinone. The peak usually found at 269 nm is absent, a positive Craven test also indicates a free quinonoid position, and the infrared spectrum suggests a phylloquinone rather than a menaquinone.

Chlorobiumquinone (438), $C_{46}H_{62}O_3$

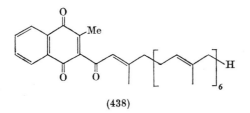

(438)

Occurrence: cultures of *Chlorobium thiosulphatophilum* Larsen[446] and "*Chloro-pseudomonas ethylicum* Shaposnikov *et al.*"[476]†

Physical properties: yellow crystals, m.p. 50–51°, $\lambda_{max.}$ (iso-octane) 249 nm (log ϵ 4·21), $\nu_{max.}$ (KBr) 1660, 1610 cm^{-1}.

† Possibly belongs to the *Chlorochromatium* genus.[453]

Both the green photosynthetic bacteria given above produce MK-7 and the modification, chlorobiumquinone. The new compound was thought at first[446] to have the molecular formula $C_{45}H_{62}O_2$ but after accurate mass measurement this was revised[524] to $C_{46}H_{62}O_3$. The mass spectra of menaquinones normally include a major peak at m/e 225 (b)

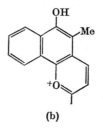

(b)

but this is very weak in the spectrum of chlorobiumquinone whereas there are significant peaks at m/e 241 ($C_{15}H_{13}O_3$), 201 ($C_{12}H_9O_3$) and 200 ($C_{12}H_8O_3$) which establish that the third oxygen atom in chloro-biumquinone must be on the first carbon atom of the side chain. As there is no hydroxyl group the structure must be (438)[524] which is consistent with the n.m.r. spectrum and the ultraviolet absorption which is very similar to that of 2-acetyl-1,4-naphthaquinone. Evidently both the ketonic and the quinonoid carbonyl absorptions overlap at 1660 cm⁻¹. A second, more polar, quinone present in both organisms, contains a hydroxyl group (infrared) and since it partially changes into chlorobiumquinone on storage it may be 1'-hydroxymenaquinone-7.[561]

Chlorobiumquinone has recently been synthesised.[562] All-trans-farnesylfarnesylacetone (439), elaborated from geraniol, was extended

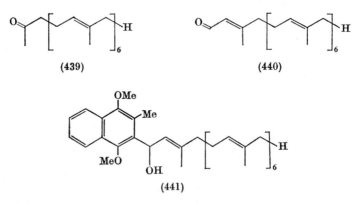

(439)   (440)

(441)

by a Wittig reaction with triethyl phosphonoacetate, followed by reduction of the ester with lithium aluminium hydride–aluminium chloride, and reoxidation with manganese dioxide to give the aldehyde (440). This was then condensed with 3-lithio-1,4-dimethoxy-2-methyl-naphthalene to form the alcohol (441) which was oxidised to the corresponding ketone with manganese dioxide. Finally, oxidation of the *all-trans*-isomer with argentic oxide in dioxan-phosphoric acid yielded the ketoquinone identical with chlorobiumquinone.

Like the analogous ubiquinones and plastoquinones, the terpenoid naphthaquinones readily cyclise under both acidic and basic conditions, to chroman derivatives, and early examples of this have already been

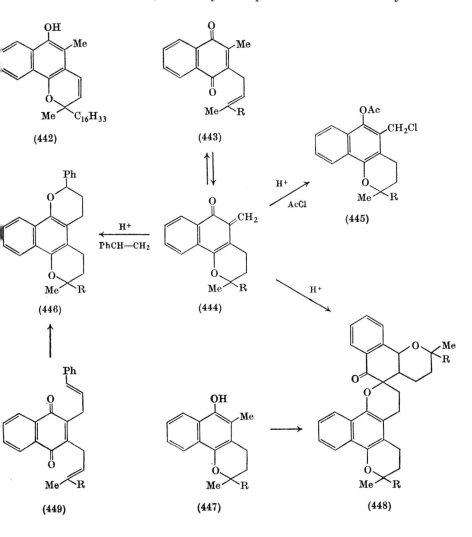

quoted. Phylloquinone also undergoes photochemical cyclisation to (442).[407] As the isomerisation (443) → (444) is energetically undemanding ($\Delta H_{(443)-(444)} = 0{\cdot}3$ kcal./mole),[454] reports[455] of the natural occurrence of K vitamins in a cyclised form are not surprising but there is the possibility that these compounds are artefacts. Recent interest[456] in the possible role of the K vitamins in oxidative phosphorylation, particularly in a scheme[454] involving the isomerisation (443) → (444), stimulated considerable chemical activity, notably by Lederer and his associates. Although, in the absence of supporting evidence *in vivo*, these speculations have been abandoned,[457] there is no doubt, from chemical evidence, that 2-methyl-3-allyl-1,4-naphthaquinones do react in the tautomeric quinone methide form (444). The characteristic dimerisation of *o*-quinone methides was effected by acid treatment[459, 460] of phylloquinone (443; R $= C_{16}H_{33}$) which gave the dimer (448); this may also be prepared by ferricyanide oxidation of the chromanol (447; R $= C_{16}H_{33}$), easily obtained by reductive cyclisation of phylloquinone.[460] The quinone methide (444; R $= C_{16}H_{33}$) can also be trapped with acetyl chloride in perchloric acid giving the chloromethylchroman (445),[459] while reaction with styrene under acid conditions affords the pyranochroman (446).[462] The structure of this compound was confirmed synthetically by reductive cyclisation of the bis-allylnaphthaquinone (449; R $= C_{16}H_{33}$) with acid stannous chloride.[464] Deuterium incorpora-

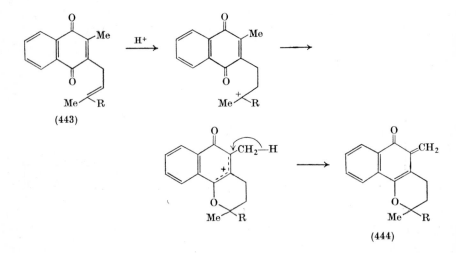

(443)

(444)

tion studies[463] leave no doubt that the first step in the acid-catalysed cyclisation of (443) is the protonation of the side-chain double bond at the β-position, followed by ring closure and loss of a proton from the C-2 methyl group.

REFERENCES

1. E. Melin, T. Wikén and K. Öblom, *Nature (London)* (1947), **159**, 840.
2. G. Bendz, *Arkiv Kemi* (1960), **15**, 131.
3. G. Bendz, *Acta Chem. Scand.* (1948), **2**, 192; (1951), **5**, 489; *Arkiv Kemi* (1952), **3**, 495.
4. A. R. Burnett, (1967), Ph.D. Thesis, University of Aberdeen.
5. S. Fairbank, *Amer. J. Pharm.* (1860), **32**, 254.
6. (a) H. Inouye, *J. Pharm. Soc. Japan* (1956), **76**, 976; (b) H. Inouye, T. Arai, Y. Yaoi and M. Ogawa, *Chem. Pharm. Bull. (Tokyo)* (1964), **12**, 255.
7. J. C. Peacock, *Amer. J. Pharm.* (1892), **64**, 295.
8. M. Proner, *Wiad. farm.* (1937), **64**, 623; R. Fischer and E. Linser, *Arch. Pharm.* (1930), **268**, 185.
9. E. S. Beshore, *Amer. J. Pharm.* (1887), **59**, 125; W. E. Ridenour, *Amer. J. Pharm.* (1895), **67**, 236.
10. G. Di Modica, S. Tira and E. Borello, *Gazz. Chim. Ital.*, 1953, **83**, 393; G. Di Modica and S. Tira, *Gazz. Chim. Ital.* (1956), **86**, 234.
11. R. Weissgerber and O. Kruber, *Ber.* (1919), **52**, 368.
12. H. Inouye and T. Arai, *Chem. Pharm. Bull. (Tokyo)* (1964), **12**, 533.
13. A. R. Burnett and R. H. Thomson, *J. Chem. Soc. (C)* (1968), 857.
14. K. H. Bolkart, M. Knobloch and M. H. Zenk, *Naturwissenschaften* (1968), **55**, 445.
15. G. Tommasi, *Gazz. Chim. Ital.* (1920), **50** I, 263.
16. B. A. Bohm and G. H. N. Towers, *Canad. J. Bot.* (1962), **40**, 677.
17. C. W. Glennie and B. A. Bohm, *Canad. J. Biochem.* (1965), **43**, 293.
18. S. Clevenger, *Arch. Biochem. Biophys.* (1958), **76**, 131.
19. A. Latif, *Indian J. Agric. Sci.* (1959), **29**, 147.
20. Song of Solomon, I, 14; IV, 13.
21. B.P., 236,557, U.S.P., 1,668,603.
22. J. B. Lal and S. Dutt, *J. Indian Chem. Soc.* (1933), **10**, 577.
23. L. F. Fieser, *J. Amer. Chem. Soc.* (1948), **70**, 3165.
24. K. Sheth, P. Catalforno and L. A. Sciuchetti, *Lloydia* (1967), **30**, 78.
25. J. E. Little, T. J. Sproston and M. W. Foote, *J. Biol. Chem.* (1948), **174**, 335.
26. B. L. Gilbert, J. E. Baker and D. M. Norris, *J. Insect Physiol.* (1967), **13**, 1453.
27. B. A. Bohm. Unpublished results quoted in D. Chen and B. A. Bohm, *Canad. J. Biochem.* (1966), **44**, 1389.
28. A. R. Burnett and R. H. Thomson, *J. Chem. Soc. (C)* (1967), 2100.
29. W. Sandermann and M. H. Simatupang, *Chem. Ber.* (1963), **96**, 2182.
30. H. H. Dietrichs, *Naturwissenschaften* (1964), **51**, 408.
31. W. Sandermann and M. H. Simatupang, *Holz Roh Werkst.* (1966), **24**, 190.
32. R. Craven, *J. Chem. Soc.* (1931), 1605.
33. A. R. Burnett and R. H. Thomson, *J. Chem. Soc. (C)* (1968), 850.
34. O. Isler, R. Rüegg, A. Studer and R. Jürgens, *Z. Physiol. Chem.* (1953), **295**, 290.
35. E. von Rudloff, *Canad. J. Chem.* (1955), **33**, 1714.
36. H. Orth, *Holzforschung* (1960), **14**, 89.
37. L. E. Wise, R. C. Rittenhouse and C. Garcia, *Tappi* (1951), **34**, 185.
38. W. Sandermann and H.-H. Dietrichs, *Holz Roh Werkst.* (1957), **15**, 281.
39. R. Livingstone and M. C. Whiting, *J. Chem. Soc.* (1955), 3631.
40. H. Matthes and E. Schreiber, *Ber. Deut. Pharm. Ges.* (1914), **24**, 385.

41. R. Hegnauer, "Chemotaxonomie der Pflanzen" (1964), Vol. 3, p. 271. Birkhauser, Basel.
42. T. Ikekawa, E. Lin Wang, M. Hamada, T. Takeuchi and H. Umezawa, *Chem. Pharm. Bull. (Tokyo)* (1967), **15**, 242.
43. C. R. Metcalfe and L. Chalk. "Anatomy of the Monocotyledons" (1950), Vol. 2, p. 1007. Oxford University Press, London.
44. T. H. Lee, *J. Chem. Soc.* (1901), **79**, 284.
45. S. R. Gupta, K. K. Malik and T. R. Seshadri, *Indian J. Chem.* (1969), **7**, 457.
46. "The Constitution and Properties of Lapachol, Lomatiol and Other Hydroxy-naphthaquinone Derivatives" (1937). Memorial Volume to Samuel C. Hooker 1864–1935 (ed. L. F. Fieser).
47. E. Paternò, *Gazz. Chim. Ital.* (1882), **12**, 337, 622.
48. S. C. Hooker, *J. Chem. Soc.* (1892), **61**, 611.
49. S. C. Hooker, *J. Chem. Soc.* (1896), **69**, 1355.
50. J. R. Price and R. Robinson, *J. Chem. Soc.* (1939), 1522.
51. L. F. Fieser, *J. Amer. Chem. Soc.* (1926), **48**, 3201.
52. L. F. Fieser, *J. Amer. Chem. Soc.* (1927), **49**, 857.
53. R. G. Cooke, *Austral. J. Sci. Res.* (1950), **3A**, 480.
54. M. Gates and D. L. Moesta, *J. Amer. Chem. Soc.* (1948), **70**, 614 (cf. L. F. Fieser and M. D. Gates, *J. Amer. Chem. Soc.* (1941), **63**, 2948).
55. G. R. Pettit and L. E. Houghton, *Canad. J. Chem.* (1968), **46**, 2471.
56. S. C. Hooker, *J. Amer. Chem. Soc.* (1936), **58**, 1163.
57. M. H. Zenk, M. Fürbringer and W. Steglich, *Phytochemistry* (1969), **8**, 2199.
58. R. G. Cooke and T. C. Somers, *Aust. J. Sci. Res.* (1950), **3A**, 466.
59. S. C. Hooker, *J. Amer. Chem. Soc.* (1936), **58**, 1168.
60. M. G. Ettlinger, *J. Amer. Chem. Soc.* (1950), **72**, 3472.
61. S. C. Hooker, *J. Amer. Chem. Soc.* (1936), **58**, 1190.
62. C. G. Casinovi, G. B. Marini-Bettòlo, O. Gonçalves Da Lima, M. H. Dalia Maia and I. L. D'Albuquerque, *Ann. Chim. (Rome)* (1962), **52**, 1184.
63. M. G. Ettlinger, *J. Amer. Chem. Soc.* (1950), **72**, 3090.
64. G. Arnaudon, *Compt. Rend.* (1858), **46**, 1152.
65. W. Stein, *J. prakt. Chem.* (1866), [1] **99**, 1.
66. A. R. Burnett and R. H. Thomson, *J. Chem. Soc. (C)* (1967), 1261.
67. E. Paternò and G. Minunni, *Gazz. Chim. Ital.* (1889), **19**, 601.
68. L. F. Fieser, *J. Amer. Chem. Soc.* (1928), **50**, 439.
69. R. G. Cooke, A. K. Macbeth and F. L. Winzor, *J. Chem. Soc.* (1939), 878.
70. L. F. Fieser, *J. Amer. Chem. Soc.* (1926), **48**, 3201.
71. B. O. Linn, C. H. Shunk, E. L. Wong and K. Folkers, *J. Amer. Chem. Soc.* (1963), **85**, 239.
72. A. F. Wagner, P. E. Wittreich, B. Arison, N. R. Trenner and K. Folkers, *J. Amer. Chem. Soc.* (1963), **85**, 1178.
73. E. H. Rennie, *J. Chem. Soc.* (1895), **67**, 784.
74. W. Sandermann. Personal communication (1966).
75. S. C. Hooker, *J. Amer. Chem. Soc.* (1936), **58**, 1181.
76. S. C. Hooker and A. Steyermark, *J. Amer. Chem. Soc.* (1936), **58**, 1198.
77. M. Gates, *J. Amer. Chem. Soc.* (1948), **70**, 617.
78. S. C. Hooker and A. Steyermark, *J. Amer. Chem. Soc.* (1936), **58**, 1207.
79. J. R. Price and R. Robinson, *Nature, (London)* (1938), **142**, 147.
80. J. B. Harborne, *Phytochemistry* (1966), **5**, 598.
81. S. C. Hooker, *J. Chem. Soc.* (1896), **69**, 1381.
82. J. R. Price and R. Robinson, *J. Chem. Soc.* (1940), 1493.

83. R. G. Cooke, *Nature (London)* (1948), **162**, 178; *Aust. J. Sci. Res.* (1950), **3A**, 481.
84. R. G. Cooke and T. C. Somers, *Aust. J. Sci. Res.* (1950), **3A**, 487.
85. D. H. R. Barton, P. de Mayo, G. A. Morrison and H. Raistrick, *Tetrahedron* (1959), **6**, 48.
86. L. F. Fieser, *J. Amer. Chem. Soc.* (1948), **70**, 3237.
87. L. A. Shchukina, A. P. Kondrat'eva and M. M. Shemyakin, *J. Gen. Chem.* (1948), **18**, 2121.
88. R. G. Cooke and T. C. Somers, *Nature (London)* (1950), **165**, 314.
89. L. F. Fieser and A. R. Bader, *J. Amer. Chem. Soc.* (1951), **73**, 681.
90. M. A. Oxman, M. G. Ettlinger and A. R. Bader, *J. Org. Chem.* (1965), **30**, 2051.
91. This is still of interest, U.S.P., 3,041,244, B.P., 889,813.
92. S. C. Hooker, *J. Amer. Chem. Soc.* (1936), **58**, 1174.
93. S. C. Hooker and A. Steyermark, *J. Amer. Chem. Soc.* (1936), **58**, 1179.
94. L. F. Fieser, J. L. Hartwell and A. M. Seligman, *J. Amer. Chem. Soc.* (1936), **58**, 1223.
95. (a) L. F. Fieser and M. Fieser, *J. Amer. Chem. Soc.* (1948), **70**, 3215; (b) L. A. Shchukina, Y. B. Shvetsov and M. M. Shemyakin, *J. Gen. Chem.* (1951), **21**, 346.
96. A. Vogel and C. Reischauer, *Büchner's Neues Rep. für Pharm.* (1856), **5**, 106; *Jahresber.* (1856), **9**, 693; (1858), **11**, 533; *Compt. Rend.* (1869), **69**, 1372.
97. A. Brissemoret and R. Combes, *Compt. Rend.* (1905), **141**, 838.
98. H. Horitsu and S. Sakamura, *J. Agr. Chem. Soc. Japan* (1956), **30**, 330.
99. D. R. Han, *Theses Collection Chungang Univ.* (1961), **5**, 263.
100. R. G. Cooke, H. Dowd and L. J. Webb, *Nature (London)* (1952), **169**, 974.
101. C. Daglish, *Biochem. J.* (1950), **47**, 458; P. Zuman, *Coll. Czech. Chem. Comm.* (1954), **19**, 1140.
102. C. Daglish, *Biochem. J.* (1950), **47**, 452.
103. W. Ruelius and A. Gauhe, *Annalen* (1951), **571**, 69.
104. A. Bernthsen, *Ber.* (1884), **17**, 1945; A. Bernthsen and A. Semper, *Ber.* (1887), **20**, 934.
105. A. J. Birch and K. A. M. Walker, *Tetrahedron Lett.* (1967), 3457.
106. H.-J. Teuber and N. Götz, *Chem. Ber.* (1954), **87**, 1236.
107. R. Ciusa, *Ann. Chim. Appl. (Rome)*, (1926), **16**, 127.
108. F. Weygand, *Ber.* (1942), **75**, 625; F. Weygand, K. Vogelbach and K. Zimmermann, *Chem. Ber.* (1947), **80**, 391.
109. N. F. Hayes and R. H. Thomson, *J. Chem. Soc.* (1955), 904; see also D. W. Cameron and J. C. A. Craik., *J. Chem. Soc. (C)* (1968), 3068.
110. O. Kubota and A. G. Perkin, *J. Chem. Soc.* (1925), **127**, 1889; A. G. Perkin and R. C. Storey, *J. Chem. Soc.* (1928), 229; A. G. Perkin and C. W. H. Story, *J. Chem. Soc.* (1929), 1399.
111. F. Mylius, *Ber.* (1884), **17**, 2411; (1885), **18**, 2567.
112. R. H. Thomson, *J. Chem. Soc.* (1950), 1737.
113. D. B. Bruce and R. H. Thomson, *J. Chem. Soc.* (1952), 2759.
114. R. H. Thomson, *J. Org. Chem.* (1951), **16**, 1082; J. W. McLeod and R. H. Thomson, *J. Org. Chem.* (1959), **25**, 36; F. G. Rothman, *J. Org. Chem.* (1958), **23**, 1049.
115. Y. P. Volkov, M. N. Kolosov, V. G. Kovobko and M. M. Shemyakin, *Izvest. Akad. Nauk SSSR, Ser. Khim.* (1964), 492.
116. D. C. Morrison and D. W. Heinritz, *J. Org. Chem.* (1962), **27**, 2229; V. V. Kozlov and N. N. Vorozhtzov, *Ber.* (1936), **69**, 416.
117. T. Kuroda, *Proc. Imp. Acad. Tokyo* (1939), **15**, 226.

118. D. B. Bruce and R. H. Thomson, *J. Chem. Soc.* (1955), 1089.
119. H. R. Bode, *Planta* (1958), **51**, 440.
120. H. K. Krogh, *Brit. J. Industr. Med.* (1964), **21**, 65.
121. M. M. Shemyakin and L. A. Shchukina, *Quart. Rev.* (*London*) (1956), **10**, 261.
122. R. G. Cooke, H. Dowd and L. J. Webb, *Nature* (*London*) (1952), **169**, 974; R. G. Cooke and H. Dowd, *Aust. J. Sci. Res.* (1952), **5A**, 760.
123. G. S. Sidhu and M. Pardhasaradhi, *Tetrahedron Lett.* (1967), 1313.
124. J. W. Loder, Stang Mongolsuk, A. Robertson and W. B. Whalley, *J. Chem. Soc.* (1957), 2233; Stang Mongolsuk and Ch. Sdarwonvivat, *J. Chem. Soc.* (1965), 1533.
125. (a) A. G. Brown, J. C. Lovie and R. H. Thomson, *J. Chem. Soc.* (1965), 2355; (b) A. G. Brown and R. H. Thomson, *J. Chem. Soc.* (1965), 4292.
126. R. H. Thomson. *In* "Chemistry of Natural and Synthetic Colouring Matters and Related Fields" (eds. T. S. Gore *et al.*) (1962), p. 99. Academic Press, New York.
127. R. G. Cooke and H. Dowd, *Aust. J. Chem.* (1953), **6**, 53.
128. Ref. 41, Vol. 4, p. 46.
129. Dulong d'Astafort, *Journal de Pharmacie* (1828), **14**, 441; Berzelius, *Jahresber.* (1830), **9**, 232.
130. J. B. Harborne, *Phytochemistry* (1967), **6**, 1415; personal communication, 1967.
131. F. Altamirano, *Materia Medica Mexicana* (1894), i, 79; F. A. Kincl and J. Rosenkranz, *Ciencia* (*Mex.*) (1956), **16**, 10.
132. W. Bettinck, *Nieuw Tijdschr. Pharm. Nederland* (1888) [2] **21**, 1; *Rec. Trav. Chim.* (1889), **8**, 319; M. Greshoff, *Ber.* (1890), **23**, 3543.
133. Flückiger, *Neues Handwörterbuch der Chemie* (1890), **5**, 723.
134. W. R. Witanowski, *Wiad. farm.* (1934), **61**, 420; (1935), **61**, 1.
135. (a) H. Dieterle, *Arch. Pharm.* (1935), **273**, 235; (b) H. Dieterle and E. Kruta, *Arch. Pharm.* (1936), **274**, 457.
136. A. Denoël, *J. Pharm. Belg.* (1949), **4**, 3.
137. M. Asano and J. Hase, *J. Pharm. Soc. Japan* (1943), **63**, 410; *Chem. Abs.* (1952), **46**, 93.
138. (a) R. Paris and P. Delaveau, *Ann. Pharm. Franç.* (1959), **17**, 585; (b) V. Krishnamoorthy and R. H. Thomson, *Phytochemistry* (1969), **8**, 1951.
139. H. Dieterle, *Arch. Pharm.* (1922), **260**, 45; *Apoth. Z.* (1927), **42**, 396.
140. R. G. Cooke. Personal communication, 1956.
141. R. Paris and H. Moyse-Mignon, *Compt. Rend.* (1949), **228**, 2063.
142. T. M. Meijer, *Rec. Trav. Chim.* (1947), **66**, 193.
143. A. C. Roy and S. Dutt, *J. Indian Chem. Soc.* (1928), **5**, 419.
144. A. Madinaveitia and M. Gallego, *Anal. Soc. Españ. Fís. Quím.* (1928), **26**, 263.
145. J. Sáenz de Buruaga and F. Verdú, *Anal. Soc. Españ. Fís. Quím.* (1934), **32**, 830.
146. L. F. Fieser and J. T. Dunn, *J. Amer. Chem. Soc.* (1936), **58**, 572.
147. D. B. Bruce and R. H. Thomson, *J. Chem. Soc.* (1952), 2759.
148. K. Heyne. "Nuttige Planten van Ned. Indie" (1927). 2nd Edit., p. 1252.
149. Extra Pharmacopeia: Martindale, (1967), 25th Edit.
150. U.S. Dispensatory (1947), 24th edit.
151. I. H. Burkill, "Dictionary of the Economic Products of the Malay Peninsula" (1935). Crown Agents, London.
152. R. S. Hisar, *Bull. Soc. Chim. France* (1954), 33; R. S. Hisar and R. E. Wolff, *Bull. Soc. Chim. France* (1955), 507.

153. U. C. Dutt. "The Materia Medica of the Hindus". (1877). Thacker, Spink and Co., Calcutta.
154. R. S. Kapil and M. M. Dhar, *J. Sci. Ind. Res.* (*India*) (1961), **20B**, 498.
155. G. S. Sidhu and M. Pardhasaradhi, *Tetrahedron Lett.* (1967), 1313.
156. K. Yoshihira, M. Tezuka and S. Natori, *Tetrahedron Lett.* (1970), 7.
157. S. Ito, *Engei Shikensho Hokoku* (1962), *Ser. B*, 1; *Chem. Abs.* (1963), **59**, 9083.
158. A. K. Ganguly and T. R. Govindachari, *Tetrahedron Lett.* (1966), 3373.
159. K. Yoshihira, S. Natori and Panida Kanchanapee, *Tetrahedron Lett.* (1967), 4857.
160. G. S. Sidhu and K. K. Prasad, *Tetrahedron Lett.* (1967), 2905.
161. R. Krahl, *Arzneimittel-Forsch.* (1956), **6**, 342.
162. R. H. Thomson and K. C. B. Wilkie, *Chem. Ind.* (*London*) (1961), 1712.
163. E. H. Rennie, *J. Chem. Soc.* (1877), **51**, 371; (1893), **63**, 1083.
164. A. K. Macbeth and F. L. Winzor, *J. Chem. Soc.* (1935), 334.
165. F. L. Winzor, *J. Chem. Soc.* (1935), 336.
166. J. W. H. Lugg, A. K. Macbeth and F. L. Winzor, *J. Chem. Soc.* (1937), 1597.
167. R. H. Thomson, *J. Org. Chem.* (1948), **13**, 870.
168. R. H. Thomson, *J. Chem. Soc.* (1949), 1277.
169. (a) M. Asano and J. Hase, *J. Pharm. Soc. Japan* (1943), **63**, 83; *Chem. Abs.* (1952), **46**, 92; (b) M. Asano, Y. Miyashita and J. Hase, *J. Pharm. Soc. Japan* (1943), **63**, 109; *Chem. Abs.* (1950), **44**, 7297.
170. M. Asano and J. Hase, *J. Pharm. Soc. Japan* (1943), **63**, 90; *Chem. Abs.* (1952), **46**, 92.
171. R. G. Cooke and W. Segal, *Aust. J. Sci. Res.* (1950), **3A**, 628.
172. A. K. Macbeth, J. R. Price and F. L. Winzor, *J. Chem. Soc.* (1935), 325.
173. C. Kuroda, *J. Sci. Res. Inst. Tokyo* (1951), **45**, 166.
174. F. Fariña, M. Lora-Tamayo and C. Suarez, *Anales Real. Soc. Españ. Fís. Quím.* (*Madrid*) (1963), *Ser. B*, **59**, 167.
175. G. S. Sidhu and A. V. B. Sankaram, *Annalen* (1966), **691**, 172.
176. R. G. Cooke and L. G. Sparrow, *Aust. J. Chem.* (1965), **18**, 218.
177. L. H. Briggs. Personal communication, 1966.
178. T. Batterham, R. G. Cooke, H. Duewell and L. G. Sparrow, *Aust. J. Chem.* (1961), **14**, 637.
179. A. J. Shand and R. H. Thomson, *Tetrahedron* (1962), **19**, 1919 and refs. therein.
180. D. Schulte-Frohlinde and V. Werner, *Chem. Ber.* (1961), **94**, 2726.
181. J. W. Sidman, *Chem. Rev.* (1958), **58**, 689.
182. K. Bournot, *Arch. Pharm.* (1913), **251**, 351.
183. C. J. Covell, F. E. King and J. W. W. Morgan, *J. Chem. Soc.* (1961), 702.
184. T. Murakami and A. Matsushima, *Chem. Pharm. Bull.* (*Tokyo*) (1961), **9**, 654.
185. R. E. Bowman, C. P. Falshaw, C. S. Franklin, A. W. Johnson and T. J. King, *J. Chem. Soc.* (1963), 1340.
186. R. Bentley and S. Gatenbeck, *Biochemistry* (1965), **4**, 1150.
187. J. Gremmen, *Antonie van Leeuwenhoek* (1956), **22**, 58.
188. G. J. M. van der Kerk and J. C. Overeem, *Rec. Trav. Chim.* (1957), **76**, 425.
189. J. C. Overeem and G. J. M. van der Kerk, *Rec. Trav. Chim.* (1964), **83**, 995.
190. J. C. Overeem and G. J. M. van der Kerk, *Rec. Trav. Chim.* (1964), **83**, 1023.
191. J. N. Collie and W. S. Myers, *J. Chem. Soc.* (1893), **63**, 122; J. N. Collie, *J. Chem. Soc.* (1893), **63**, 329. See also R. Kaushal, *J. Indian Chem. Soc.* (1946), **23**, 16 and J. R. Bethel and P. Maitland, *J. Chem. Soc.* (1962), 3751.
192. J. C. Overeem and G. J. M. van der Kerk, *Rec. Trav. Chim.* (1964), **83**, 1005.
193. A. R. Burnett and R. H. Thomson, *Chem. Ind.* (*London*) (1968), 1771.

194. D. B. Bruce and R. H. Thomson, *J. Chem. Soc.* (1954), 1428.

195. B. D. Astill and J. C. Roberts, *J. Chem. Soc.* (1953), 3303.

196. H. Tanaka and T. Tamura, *Tetrahedron Lett.* (1961), 151; H. Tanaka, Y. Ohne, N. Ogawa and T. Tamura, *Agr. Biol. Chem. (Tokyo)* (1963), **27**, 48.

197. B. W. Bycroft and J. C. Roberts, *J. Chem. Soc.* (1962), 2063.

198. J. E. Davies, F. E. King and J. C. Roberts, *Chem. Ind. (London)* (1954), 1110; *J. Chem. Soc.* (1955), 2782.

199. (a) A. J. Birch and F. W. Donovan, *Chem. Ind. (London)* (1954), 1047; (b) *Aust. J. Chem.* (1955), **8**, 529.

200. J. Smith and R. H. Thomson, *J. Chem. Soc.* (1961), 1008.

201. N. N. Gerber and B. Wieclawek, *J. Org. Chem.* (1966), **31**, 1496.

202. H. Nishikawa, *Agr. Biol. Chem. (Tokyo)* (1962), **26**, 696.

203. R. E. Moore, H. Singh and P. J. Scheuer, *J. Org. Chem.* (1966), **31**, 3645.

204. (a) S. Natori, Y. Kumada and H. Nishikawa, *Chem. Pharm. Bull. (Tokyo)* (1965), **13**, 633; (b) S. Natori, Y. Inouyé (née Kumada) and H. Nishikawa, *Chem. Pharm. Bull. (Tokyo)* (1967), **15**, 380.

205. A. C. Baillie and R. H. Thomson, *J. Chem. Soc.* (1966), 2184.

206. (a) O. Gonçalves de Lima, I. L. d'Albuquerque, M. P. Machado, G. P. Pinto and M. G. Maciel, *An. Soc. Biol. Pernambuco* (1956), **14**, 136; (b) O. Gonçalves de Lima, I. L. d'Albuquerque, C. Gonçalves de Lima and M. H. Dalia Maia, *Rev. Inst. Antibiót. Univ. Recife* (1962), **4**, 3; (c) O. Gonçalves de Lima, I. L. d'Albuquerque, M. A. Pereira and J. Francisco de Mello, *Rev. Inst. Antibiót. Univ. Recife* (1966), **6**, 23.

207. J. C. Overeem, Personal communication (1967).

208. A. L. Fallas and R. H. Thomson, *J. Chem. Soc. (C)* (1968), 2279.

209. R. G. Cooke. Personal communication, 1967.

210. J. H. Lister, C. H. Eugster and P. Karrer, *Helv. Chim. Acta* (1955), **38**, 215.

211. M. Fehlmann and A. Niggli, *Helv. Chim. Acta* (1965), **48**, 305.

212. C. Pascard-Billy, *Bull. Soc. Chim. France* (1962), 2282, 2293, 2299.

213. J. Pelletier, *Ann. Chim. (Phys.)* (1832) [2], **51**, 191; *Annalen*, 1833, **6**, 27; P. Bolley and R. Wydler, *Annalen* (1847), **62**, 141.

214. H. Brockmann, *Annalen* (1936), **521**, 1.

215. M. Kuhara, *Chem. News* (1878), **38**, 238; *Ber.* (1878), **11**, 2146.

216. R. Majima and C. Kuroda, *J. Chem. Soc. Japan*, 1918, **39**, 1051; *Acta Phytochim. (Tokyo)* (1922), **1**, 43.

217. A. C. Jain and S. K. Mathur, *Bull. Nat. Inst. Sci. India* (1965), No. 28, 52.

218. Ref. 41, p. 294.

219. (a) A. S. Romanova and A. I. Ban'kovskii, *Khim. Prirod. Soedin. Akad. Nauk. Uz. SSR* (1965), 226; (b) A. S. Romanova, N. V. Tareeva and A. I. Ban'kovskii, *Khim. Prirod. Soedin.* (1967), 71.

220. N. N. Boldyrev, *Farmatsiya* (1939), No. 10, 24; *Khim. ref. Zh.* (1940), No. 5, 108.

221. K. Brand and A. Lohmann, *Ber.* (1935), **68**, 1487.

222. I. Morimoto, T. Kishi, S. Ikegami and Y. Hirata, *Tetrahedron Lett.* (1965), 4737.

223. I. Morimoto and Y. Hirata, *Tetrahedron Lett.* (1966), 3677; I. Morimoto. Personal communication.

224. T. Y. Toribara and A. L. Underwood, *Anal. Chem.* (1949), **21**, 1352.

225. D. N. Majumdar and G. C. Chakravarti, *J. Indian Chem. Soc.* (1940), **17**, 272.

226. F. Burriel and J. Ramírez-Munoz, *Anal. Soc. Españ. Fís. Quim.* (1948), **44B**, 1149.

227. J. V. Dubský and E. Wagner, *Mikrochemie* (1935), **17**, 186; J. V. Dubský and E. Krametz, *Mikrochemie* (1936), **20**, 57; A. L. Underwood and W. F. Neumann, *Anal. Chem.* (1949), **21**, 1348.

228. C. Liebermann and M. Römer, *Ber.* (1887), **20**, 2428.

229. H. Raudnitz, L. Redlich and F. Fiedler, *Ber.* (1931), **64**, 1835.

230. H. Raudnitz and E. Stein, *Ber.* (1934), **67**, 1955.

231. H. Arakawa and M. Nakazaki, *Chem. Ind.* (*London*) (1961), 947.

232. C. Kuroda and M. Wada, *Proc. Imp. Acad. Japan* (1936), **12**, 239; *Sci. Papers Inst. Phys. Chem. Res. Tokyo* (1938), **34**, 1740.

233. H. Brockmann and K. Müller, *Annalen* (1939), **540**, 51.

234. T. Sproston, *Proc. Third Ann. Symp., Plant Phenolics Group of N. America, Toronto* (1963), p. 69; *Chem. Abs.* (1964), **61**, 8560.

235. J. J. Armstrong and W. B. Turner, *J. Chem. Soc.* (1965), 5927.

236. T. Sproston. Personal communication (1966).

237. R. K. S. Wood, *Trans. Brit. Mycol. Soc.* (1953), **36**, 109.

238. T. B. Tjio, T. Sproston and H. Tomlinson, *Lloydia* (1965), **28**, 359.

239. H. Schmid, A. Ebnöther and Th. M. Meijer, *Helv. Chim. Acta* (1950), **33**, 1751.

240. H. Schmid and A. Ebnöther, *Helv. Chim. Acta* (1951), **34**, 561.

241. H. Schmid and A. Ebnöther, *Helv. Chim. Acta* (1951), **34**, 1041.

242. D. W. Cameron, D. G. I. Kingston, N. Sheppard and Lord Todd, *J. Chem. Soc.* (1964), 98.

243. F. M. Dean, "Naturally Occurring Oxygen Ring Compounds" (1963), p. 467. Butterworths, London.

244. W. Eisenhuth and H. Schmid, *Experientia* (1957), **13**, 311; *Helv. Chim. Acta* (1958), **41**, 2021.

245. H. R. V. Arnstein, A. H. Cook and M. S. Lacey, *Nature* (*London*) (1946), **157**, 334; *J. Chem. Soc.* (1947), 1021.

246. S. Weiss, J. V. Fiore and F. F. Nord, *Arch. Biochem.* (1947), **15**, 326; S. Weiss and F. F. Nord, *Arch. Biochem.* (1949), **22**, 288.

247. G. P. Arsenault, *Tetrahedron Lett.* (1965), 4033.

248. K. Schofield and D. E. Wright, *J. Chem. Soc.* (1965), 6642.

249. Unpublished work mentioned by E. Hardegger, K. Steiner, E. Widmer, H. Corrodi, Th. Schmidt, H. P. Knoepfel, W. Rieder, H. J. Meyer, F. Kugler and H. Gempeler, *Helv. Chim. Acta* (1964), **47**, 1996.

250. A. Gauhe. Personal communication, 1955.

251. S. Gatenbeck and R. Bentley, *Biochem. J.* (1964), **92**, 21P.

252. E. Widmer, J. W. Meyer, A. Walser and E. Hardegger, *Helv. Chim. Acta* (1965), **48**, 538.

253. E. Hardegger, K. Steiner, E. Widmer and A. Pfiffner, *Helv. Chim. Acta* (1964), **47**, 2027.

254. E. Hardegger, E. Widmer, K. Steiner and A. Pfiffner, *Helv. Chim. Acta* (1964), **47**, 2031.

255. M. J. Carlile, *J. Gen. Microbiol.* (1956), **14**, 643.

256. H. W. Ruelius and A. Gauhe, *Annalen* (1950), **569**, 38.

257. G. P. Arsenault, *Canad. J. Chem.* (1965), **43**, 2423.

258. H. W. Ruelius and A. Gauhe, *Annalen* (1951), **570**, 121.

259. H. Brockmann, H. Pini and O. v. Plotho, *Chem. Ber.* (1950), **83**, 161.

260. H. Brockmann and V. Loeschcke, *Chem. Ber.* (1955), **88**, 778.

261. H. Brockmann and E. Hieronymus, *Chem. Ber.* (1955), **88**, 1379.

262. H. Brockmann and H. Vorbrüggen, *Chem. Ber.* (1962), **95**, 810.

263. H. Brockmann, K. van der Merve and A. Zeeck, *Chem. Ber.* (1964), **97**, 2555.

264. H. Brockmann, A. Zeeck, K. van der Merve and W. Müller, *Annalen* (1966), **698**, 209.

265. H. Brockmann and T. Reschke, *Tetrahedron Lett.* (1965), 4593.

266. A. Zeeck, *Angew. Chem.* (1967), **6**, 470.

267. F. M. Dean and P. G. Jones, *J. Chem. Soc.* (1963), 5342.

268. F. M. Dean, P. G. Jones, R. B. Morton and P. Sidisunthorn, *J. Chem. Soc.* (1963), 5336.

269. J. C. Van Meter, M. Dann and N. Bohonos, "Antibacterial Agents Annual— 1960" (1961), p. 77. Plenum Press, New York.

270. G. A. Ellestad, H. A. Whaley and E. L. Patterson, *J. Amer. Chem. Soc.* (1966), **88**, 4109; G. A. Ellestad, M. P. Kunstmann, H. A. Whaley and E. L. Patterson, *J. Amer. Chem. Soc.* (1968), **90**, 1325.

271. R. G. Cooke and B. L. Johnson, *Aust. J. Chem.* (1963), **16**, 695.

272. R. G. Cooke and B. L. Johnson, *Bull. Nat. Inst. Sci. India* (1965), No. 28, 66.

273. P. D. Mier, *Nature (London)* (1957), **179**, 1084.

274. (a) J. C. Wirth, P. J. O'Brien, L. F. Schmitt and A. Sohler, *J. Invest. Dermatol.* (1957), **29**, 47; (b) J. C. Wirth, T. E. Beesley and S. R. Anand, *Phytochemistry* (1965), **4**, 505.

275. G. Just, W. C. Day and F. Blank, *Canad. J. Chem.* (1963), **41**, 74.

276. M. G. McCabe and P. D. Mier, *J. Gen. Microbiol.* (1960), **23**, 1.

277. M. R. Baichwal and G. C. Walker, *Canad. J. Microbiol.* (1960), **6**, 383.

278. R. A. Zussman, I. Lyon and E. E. Vicher, *J. Bacteriol.* (1960), **80**, 708.

279. T. N. Ganjoo, R. Pohloudek-Fabini and H. Wollmann, *Arch. Pharm.* (1967), **300**, 414; R. Pohloudek-Fabini, H. Wollmann and T. N. Ganjoo, *Arch. Pharm.* (1967), **300**, 492; H. Wollmann, T. N. Ganjoo and R. Pohloudek-Fabini, *Arch. Pharm.* (1967), **300**, 674.

280. A. S. Ng, G. Just and F. Blank, *Canad. J. Chem.* (1969), **47**, 1223.

281. F. Blank, A. S. Ng and G. Just, *Canad. J. Chem.* (1966), **44**, 2873.

282. J. N. Ashley, B. C. Hobbs and H. Raistrick, *Biochem. J.* (1937), **31**, 385.

283. G. R. Birchall, K. Bowden, U. Weiss and W. B. Whalley, *J. Chem. Soc. (C)* (1966), 2237.

284. P. M. Baker and J. C. Roberts, *J. Chem. Soc. (C)* (1966), 2234.

285. S. Shibata, E. Morishita, T. Takeda and K. Sakata, *Tetrahedron Lett.* (1966), 4855; *Chem. Pharm. Bull. (Tokyo)*, 1968, **16**, 405.

286. W. W. Kaeding, *J. Org. Chem.* (1963), **28**, 1063.

287. K. Takamiya, M. Nishimura and A. Takamiya, *Plant and Cell Physiol.* (1967), **8**, 79.

288. G. H. Stout, D. L. Dreyer and L. H. Jensen, *Chem. Ind. (London)* (1961), 289.

289. H. Brockmann and K.-H. Renneberg, *Naturwissenschaften* (1953), **40**, 59, 166.

290. H. Brockmann, W. Lenk, G. Schwantje and A. Zeeck, *Tetrahedron Lett.* (1966), 3525; *Chem. Ber.* (1969), **102**, 126.

291. W. Treibs and K. Eckardt, *Naturwissenschaften* (1961), **48**, 430.

292. K. Eckardt, *Chem. Ber.* (1965), **98**, 24.

293. J. Romo and P. Joseph-Nathan, *Tetrahedron* (1964), **20**, 2331.

294. P. Joseph-Nathan, J. J. Morales and J. Romo, *Tetrahedron* (1966), **22**, 301.

295. J. Correa and J. Romo, *Tetrahedron* (1966), **22**, 685.

296. O. Gonçalves de Lima, W. Keller-Schierlein and V. Prelog, *Helv. Chim. Acta* (1958), **41**, 1386.

297. J. Comin, O. Gonçalves de Lima, H. N. Grant, L. M. Jackman, W. Keller-Schierlein and V. Prelog, *Helv. Chim. Acta* (1963), **46**, 409.

298. H. N. Grant, V. Prelog and R. P. A. Sneeden, *Helv. Chim. Acta* (1963), **46**, 415.

299. W. Sandermann and H.-H. Dietrichs, *Holz Roh Werkst.* (1959), **17**, 88.

300. G. B. Marini-Bettòlo, C. G. Casinovi and C. Galeffi, *Tetrahedron Lett.* (1965), 4857; G. B. Marini-Bettòlo, C. G. Casinovi, C. Galeffi and F. D. Monache, *Ann. Ist. Super. Sanità* (1966), **2**, 327.

301. N. Tanaka, M. Yasue and H. Imamura, *Tetrahedron Lett.* (1966), 2767.

302. N. Tanaka, M. Yasue and H. Imamura, *J. Japan Wood Res. Soc.* (1966), **12**, 289.

303. N. Tanaka, M. Yasue and H. Imamura, *J. Japan Wood Res. Soc.* (1967), **13**, 12.

304. N. Tanaka, M. Yasue and H. Imamura, *J. Japan Wood Res. Soc.* (1967), **13**, 16.

305. K. Shimada, M. Yasue and H. Imamura, *J. Japan Wood Res. Soc.* (1967), **13**, 126.

306. R. G. Lindahl, *Ann. Acad. Sci. Fennicae* (1953), *Ser. A*, II, No. 48, 7; *Chem. Abs.* (1955), **49**, 8223.

307. C. H. Eugster, H.-P. Küng, H. Kühnis and P. Karrer, *Helv. Chim. Acta* (1963), **46**, 530.

308. D. Karanatsios and C. H. Eugster, *Helv. Chim. Acta* (1965), **48**, 471.

309. P. Siminoff, R. M. Smith, W. T. Sokolski and G. M. Savage, *Ann. Rev. Tuberc.* (1957), **75**, 576; G. B. Whitfield, E. C. Olson, R. R. Herr, J. A. Fox, M. E. Bergy and G. A. Boyack, *Ann. Rev. Tuberc.* (1957), **75**, 584.

310. L. E. Rhuland, K. F. Stern and H. R. Reames, *Ann. Rev. Tuberc.* (1957), **75**, 588.

311. K. L. Rinehart, P. K. Martin and C. E. Coverdale, *J. Amer. Chem. Soc.* (1966), **88**, 3149, 3150.

312. R. J. Schacht and K. L. Rinehart, *J. Amer. Chem. Soc.* (1967), **89**, 2239.

313. D. Bryce-Smith and A. Gilbert, *Proc. Chem. Soc.* (1964), 87.

314. C. A. MacMunn, *Proc. Birm. Philos. Soc.* (1883), **3**, 380; *Quart. J. Microscop. Sci.* (1885), **25**, 469.

315. R. H. Thomson. "Naturally Occurring Quinones" (1957). Butterworths, London.

316. J. Gough and M. D. Sutherland, *Tetrahedron Lett.* (1964), 269.

317. C. W. J. Chang, R. E. Moore and P. J. Scheuer, *Tetrahedron Lett.* (1964), 3557.

318. R. E. Moore, H. Singh and P. J. Scheuer, *J. Org. Chem.* (1966), **31**, 3645.

319. R. Kuhn and K. Wallenfels, *Ber.* (1940), **73**, 458.

320. R. H. Thomson, *In* "Comparative Biochemistry" (eds. M. Florkin and H. S. Mason) (1962), Vol. IIIA, p. 631. Academic Press, New York.

321. H. Singh, R. E. Moore and P. J. Scheuer, *Experientia* (1967), **23**, 624.

322. E. Lederer, *Biochim. Biophys. Acta* (1952), **9**, 92.

323. E. J. Dimelow, *Nature* (1958), **182**, 812. R. I. T. Cromartie, personal communication, 1959.

324. T. Mukai, *Mem. Fac. Sci. Kyushu Univ.* (1958), *Ser. C*, **3**, 29.

325. (a) T. Mukai, *Bull. Chem. Soc. Japan* (1960), **33**, 453, 1234; (b) M. Yamaguchi, T. Mukai and T. Tsumaki, *Mem. Fac. Sci. Kyushu Univ.* (1961), *Ser. C*, **4**, 193.

326. H. A. Anderson, J. W. Mathieson and R. H. Thomson, *Comp. Physiol. Biochem.* (1969), **28**, 333.

327. N. Millott, *Proc. Zool. Soc. Lond.* (1957), **129**, 263.

328. C. W. J. Chang, R. E. Moore and P. J. Scheuer, *J. Amer. Chem. Soc.* (1964), **86**, 2959.

329. K. Nishibori, *Nature (London)* (1959), **184**, 1234.

330. K. Nishibori, *Nature (London)* (1961), **192**, 1293.

331. E. Lederer and R. Glaser, *Compt. Rend.* (1938), **207**, 454.

332. R. Kuhn and K. Wallenfels, *Ber.* (1939), **72**, 1407.
333. W. H. F. Sasse. Personal communications, 1964, 1967.
334. C. Kuroda and M. Okajima, *Proc. Japan Acad.* (1967), **43**, 41.
335. T. W. Goodwin and S. Srisukh, *Biochem. J.* (1950), **47**, 69.
336. K. Nishibori. Personal communication, 1967.
337. C. Kuroda and H. Ohshima, *Proc. Imp. Acad.* (*Tokyo*) (1940), **16**, 214.
338. C. Kuroda and K. Koyasu, *Proc. Imp. Acad.* (*Tokyo*) (1944), **20**, 23.
339. M. Okajima, *Sci. Papers Inst. Phys. Chem. Res. Tokyo* (1959), **53**, 356.
340. C. Kuroda and H. Iwakura, *Proc. Imp. Acad.* (*Tokyo*) (1942), **18**, 74.
341. L. Musajo and M. Minchilli, *Boll. Sci. Fac. Chim. Bologna* (1942), **3**, 113.
342. M. Yoshida, *J. Mar. Biol. Ass. U.K.* (1959), **38**, 455.
343. M. Krolikowska and L. Swiatek, *Diss. Pharm. Pharmacol.* (1966), **18**, 157.
344. T. W. Goodwin and D. L. Fox, *Experientia* (1955), **11**, 270.
345. K. Nishibori, *Bull. Jap. Soc. Sci. Fish.* (1957), **22**, 708.
346. N. Millott and H. Okumura, *Nature* (*London*) (1968), **217**, 92.
347. J. H. Gough and M. D. Sutherland, *Aust. J. Chem.* (1967), **20**, 1693.
348. C. Kuroda and M. Okajima, *Proc. Japan Acad.* (1965), **40**, 836. M. Okajima. Personal communication, 1965.
349. T. W. Goodwin, E. Lederer and L. Musajo, *Experientia* (1951), **7**, 375.
350. L. Musajo and M. Minchilli, *Gazz. Chim. Ital.* (1940), **70**, 287.
351. M. Yamaguchi, *Mem. Fac. Sci. Kyushu Univ.* (1961), *Ser. C*, **4**, 189.
352. C. Kuroda and M. Okajima, *Proc. Japan Acad.* (1950), **26**, 33.
353. C. Kuroda and M. Okajima, *Proc. Japan Acad.* (1953), **29**, 27.
354. C. Kuroda and M. Okajima, *Proc. Japan Acad.* (1962), **38**, 353.
355. M. Moir and R. H. Thomson, unpublished work.
356. R. Glaser and E. Lederer, *Compt. Rend.* (1939), **208**, 1939.
357. R. Kuhn and K. Wallenfels, *Ber.* (1941), **74**, 1594.
358. C. Kuroda and M. Okajima, *Proc. Japan Acad.* (1960), **36**, 424.
359. C. Kuroda and M. Okajima, *Proc. Japan Acad.* (1964), **40**, 836.
360. C. Kuroda and M. Okajima, *Proc. Japan Acad.* (1958), **34**, 616.
361. A. Asano and A. F. Brodie, *J. Biol. Chem.* (1964), **239**, 4280.
362. D. Becher, C. Djerassi, R. E. Moore, H. Singh and P. J. Scheuer, *J. Org. Chem.* (1966), **31**, 3650.
363. R. E. Moore and P. J. Scheuer, *J. Org. Chem.* (1966), **31**, 3272.
364. A. J. Birch, D. N. Butler and J. R. Siddall, *J. Chem. Soc.* (1964), 2941.
365. H. A. Anderson, J. Smith and R. H. Thomson, *J. Chem. Soc.* (1965), 2141; cf. ref. 205.
366. I. Singh, R. E. Moore, C. W. J. Chang and P. J. Scheuer, *J. Amer. Chem. Soc.* (1965), **87**, 4023.
367. G. P. Rabold, R. T. Ogata, M. Okamura, L. H. Piette, R. E. Moore and P. J. Scheuer, *J. Chem. Phys.* (1967), **46**, 1161.
368. J. Smith and R. H. Thomson, *Tetrahedron Lett.* (1960), 10.
369. H. A. Anderson and R. H. Thomson, *J. Chem. Soc.* (*C*) (1966), 426.
370. D. B. Bruce and R. H. Thomson, *J. Chem. Soc.* (1952), 2759.
371. J. F. Garden and R. H. Thomson, *J. Chem. Soc.* (1957), 2483.
372. R. E. Moore, H. Singh, C. W. J. Chang and P. J. Scheuer, *J. Org. Chem.* (1966), **31**, 3638.
373. J. F. McClendon, *J. Biol. Chem.* (1912), **11**, 435.
374. E. G. Ball, *J. Biol. Chem.* (1936), *114*, *Sci. Proc.*, XXV, vi; *Biol. Bull., Wood's Hole* (1934), **67**, 327.
375. K. Wallenfels and A. Gauhe, *Ber.* (1943), **76**, 325.
376. D. B. Bruce and R. H. Thomson, *J. Chem. Soc.* (1955), 1089.

377. K. Wallenfels and A. Gauhe, *Ber.* (1942), **75**, 413.
378. R. Kuhn and K. Wallenfels, *Ber.* (1942), **75**, 407.
379. R. E. Moore, H. Singh, C. W. J. Chang and P. J. Scheuer, *Tetrahedron* (1967) **23**, 3271.
380. R. Kuhn and K. Wallenfels, *Ber.* (1941), **74**, 1594.
381. G. I. Fray, *Tetrahedron* (1958), **3**, 316.
382. N. Millott, *Proc. Zool. Soc. Lond.* (1957), **129**, 263.
383. R. K. Cannon, *Biochem. J.* (1927), **21**, 184.
384. D. White, *J. Biol. Chem.* (1965), **240**, 1387.
385. A. Tyler, *Proc. Nat. Acad. Sci. U.S.* (1939), **25**, 523.
386. M. Hartmann, O. Schartau, R. Kuhn and K. Wallenfels, *Naturwissenschaften*, (1939), **27**, 433; M. Hartmann and O. Schartau, *Biol. Zentr.* (1939), **59**, 571; M. Hartmann, O. Schartau and K. Wallenfels, *Biol. Zentr.* (1940), **60**, 398.
387. I. Cornman, *Biol. Bull., Wood's Hole* (1941), **80**, 202; H. J. Bielig and P. Dohrn, *Z. Naturforsch.* (1950), **5b**, 316.
388. J. Gough, Ph.D. Thesis, University of Melbourne, 1967.
389. N. Millott. *In* "Comparative Biochemistry of Photoreactive Systems" (ed. M. B. Allen) (1960), p. 279. Academic Press, New York.
390. N. Millott. *In* "Light as an Ecological Factor" (eds. R. Bainbridge, G. C. Evans and O. Rackham) (1966). Brit. Ecol. Soc. Symposium, No. 6. Blackwell, Oxford.
391. N. Millott and M. Yoshida, *J. Exptl. Biol.* (1957), **34**, 394; M. Yoshida and N. Millott, *J. Exptl. Biol.* (1960), **37**, 390.
392. H. G. Vevers, *Proc. XVI Int. Congr. Zool. Washington*, D.C. (1963), **3**, 120.
393. G. P. Fitzgerald, G. C. Gerloff and F. Skoog, *Sewage Industr. Wastes* (1952), **24**, 888; G. P. Fitzgerald and F. Skoog, *Sewage Industr. Wastes* (1954), **26**, 1136.
394. T. Mortensen and L. K. Rosenvinge, *K. danske Vidensk. Selsk. Biol. Medd.* (1933), **10**, 1.
395. *Biochem. J.* (1967), **102**, 15.
396. J. F. Pennock, *Vitamins and Hormones* (1966), **24**, 307.
397. P. Sommer and M. Kofler, *Vitamins and Hormones* (1966), **24**, 349.
398. H. Noll, *J. Biol. Chem.* (1960), **235**, 2207.
399. (a) H. K. Lichtenthaler, *Planta* (1962), **57**, 731; (b) H. K. Lichtenthaler, *Planta* (1968), **81**, 140.
400. K. Egger, *Planta* (1965), **64**, 41.
401. J. W. Mathieson and R. H. Thomson, *J. Chem. Soc. (C)* (1970), in the Press.
402. S. B. Binkley, D. W. MacCorquodale, S. A. Thayer and E. A. Doisy, *J. Biol. Chem.* (1939), **130**, 219.
403. H. Dam and J. Glavind, *Biochem. J.* (1938), **32**, 485; H. K. Lichtenthaler, *Z. Pflanzenphysiol.* (1968), **59**, 195.
404. B. K. Jacobsen and H. Dam, *Biochim. Biophys. Acta* (1960), **40**, 211.
405. D. H. L. Bishop, K. P. Pandya and H. K. King, *Biochem. J.* (1962), **83**, 606.
406. R. J. Gibbons and L. P. Engle, *Science* (1964), **146**, 1307.
407. S. Fujisawa, S. Kawabata and R. Yamamoto, *J. Pharm. Soc. Japan* (1967), **87**, 1451.
408. P. H. Gale, B. H. Arison, N. R. Trenner, A. C. Page, K. Folkers and A. F. Brodie, *Biochemistry* (1963), **2**, 200.
409. P. B. Scholes and H. K. King, *Biochem. J.* (1965), **97**, 766.
410. R. H. Baum and M. I. Dolin, *J. Biol. Chem.* (1963), **238**, PC4109; (1965), **240**, 3425.

411. R. L. Lester, D. C. White and S. L. Smith, *Biochemistry* (1964), **3**, 949.
412. B. Frydman and H. Rapoport, *J. Amer. Chem. Soc.* (1963), **85**, 823; R. Powls and E. R. Redfearn, *Biochem. J.* (1967), **102**, 36.
413. O. Isler, R. Rüegg, L. H. Chopard-dit-Jean, A. Winterstein and O. Wiss, *Helv. Chim. Acta* (1958), **41**, 786.
414. P. H. Gale, A. C. Page, T. H. Stoudt and K. Folkers, *Biochemistry* (1962), **1**, 788.
415. L. K. Osnitskaya, D. R. Threlfall and T. W. Goodwin, *Nature* (*London*) (1964), **204**, 80.
416. R. J. Gibbons and L. P. Engle, *J. Bacteriol.* (1965), **90**, 561.
417. D. H. L. Bishop and H. K. King, *Biochem. J.* (1962), **85**, 550.
418. K. V. Rao, T. J. McBride and J. J. Oleson, *Cancer Research* (1968), **28**, 1952.
419. D. Misiti, H. W. Moore and K. Folkers, *Biochemistry* (1965), **4**, 1156.
420. (a) S. B. Binkley, L. C. Cheney, W. F. Holcomb, R. W. McKee, S. A. Thayer, D. W. MacCorquodale and E. A. Doisy, *J. Amer. Chem. Soc.* (1939), **61**, 2558; (b) *J. Biol. Chem.* (1939), **131**, 357; (c) S. B. Binkley, D. W. MacCorquodale, L. C. Cheney, R. W. McKee and E. A. Doisy, *J. Amer. Chem. Soc.* (1939), **61**, 1612; (d) R. W. McKee, S. B. Binkley, D. W. MacCorquodale, S. A. Thayer and E. A. Doisy, *J. Amer. Chem. Soc.* (1939), **61**, 1295.
421. H. Mayer, U. Gloor, O. Isler, R. Rüegg and O. Wiss, *Helv. Chim. Acta* (1964), **47**, 221.
422. L. M. Jackman, R. Rüegg, G. Ryser, C. von Planta, U. Gloor, H. Mayer, P. Schudel, M. Kofler and O. Isler, *Helv. Chim. Acta* (1965), **48**, 1332.
423. H. J. Almquist and A. A. Klose, *J. Amer. Chem. Soc.* (1939), **61**, 2557.
424. (a) L. F. Fieser, *J. Amer. Chem. Soc.* (1939), **61**, 2559, 2561; (b) L. F. Fieser, *J. Amer. Chem. Soc.* (1939), **61**, 3467.
425. M. Tishler, L. F. Fieser and N. L. Wendler, *J. Amer. Chem. Soc.* (1940), **62**, 2866.
426. R. Hirschmann, R. Miller and N. L. Wendler, *J. Amer. Chem. Soc.* (1954), **76**, 4592.
427. H. Lindlar, Sw.P., 320,582; B.P., 752,420, *Chem. Abs.* (1957), **51**, 9699.
428. O. Isler and K. Doebel, *Helv. Chim. Acta* (1954), **37**, 225.
429. J. Weichet, V. Kvita, L. Bláha and V. Trčka, *Coll. Czech. Chem. Comm.* (1959), **24**, 2754; J. Weichet, J. Hodrová and L. Bláha, *Coll. Czech. Chem. Comm.* (1964), **29**, 197; J. Weichet, L. Bláha, J. Hodrová, B. Kakáč and V. Trčka, *Coll. Czech. Chem. Comm.* (1966), **31**, 3607.
430. C. C. Lee, F. C. G. Hoskin, L. W. Trevoy, L. B. Jaques and J. W. T. Spinks, *Canad. J. Chem.* (1953), **31**, 769; R. J. Woods and J. D. Taylor, *Canad. J. Chem.* (1957), **35**, 941; U. Gloor, J. Würsch, H. Mayer, O. Isler and O. Wiss, *Helv. Chim. Acta* (1966), **49**, 2582.
431. R. A. Morton, "Biochemistry of Quinones" (1965). Academic Press, New York.
432. S. B. Binkley, R. W. McKee, S. A. Thayer and E. A. Doisy, *J. Biol. Chem.* (1940), **133**, 721.
433. O. Isler, R. Rüegg, L. H. Chopard-dit-Jean, A. Winterstein and O. Wiss, *Chimia* (1958), **12**, 69; *Helv. Chim. Acta* (1958), **41**, 786.
434. R. Rüegg, U. Gloor, A. Langemann, M. Kofler, C. von Planta, G. Ryser and O. Isler, *Helv. Chim. Acta* (1960), **43**, 1745.
435. C. H. Shunk, R. E. Erickson, E. L. Wong and K. Folkers, *J. Amer. Chem. Soc.* (1959), **81**, 5000.
436. H. Noll, R. Rüegg, U. Gloor, G. Ryser and O. Isler, *Helv. Chim. Acta* (1960), **43**, 433.

437. A. F. Brodie, B. R. Davis and L. F. Fieser, *J. Amer. Chem. Soc.* (1958), **80**, 6454.
438. S. Beau, R. Azerad and E. Lederer, *Bull. Soc. Chim. Biol.* (1966), **48**, 569.
439. H. Noll, *J. Biol. Chem.* (1958), **232**, 919.
440. E. Lederer, *I.U.B. Symposium Series* (1964), **33**, 63.
441. R. Azerad and M.-O. Cyrot, *Bull. Soc. Chim. France* (1965), 3740.
442. R. Azerad, M.-O. Cyrot and E. Lederer, *Biochem. Biophys. Res. Comm.* (1967), **27**, 249.
443. J. J. Pappas, W. P. Keaveney, E. Gancher and M. Berger, *Tetrahedron Lett.* (1966), 4273.
444. M. McKenna, M. D. Henninger and F. L. Crane, *Nature (London)* (1964), **203**, 524.
445. (a) M. D. Henninger, H. N. Bhagavan and F. L. Crane, *Arch. Biochem. Biophys.* (1965), **110**, 69; (b) C. F. Allen, H. Franke and O. Hirayama, *Biochem. Biophys. Res. Comm.* (1967), **26**, 562.
446. B. Frydman and H. Rapoport, *J. Amer. Chem. Soc.* (1963), **85**, 823.
447. R. J. Anderson and M. S. Newman, *J. Biol. Chem.* (1933), **101**, 773.
448. E. Bullock. Personal communication, 1969.
449. H. Tanaka, P. L. Wang, O. Yamada and T. Tamura, *Agr. Biol. Chem. (Tokyo)* (1966), **30**. 107; P. L. Wang and H. Tanaka, *Agr. Biol. Chem. (Tokyo)* (1966), **30**, 683.
450. C. H. Eugster. Personal communication, 1967.
451. S. Shibata and Y. Ogihara, *Chem. Pharm. Bull. (Tokyo)* (1963), **11**, 1576.
452. G. Tertzakian, R. H. Haskins, G. P. Slater and L. R. Nesbitt, *Proc. Chem. Soc.* (1964), 195.
453. J. M. Shewan. Personal communication, 1969.
454. M. Vilkas and E. Lederer, *Experientia* (1962), **18**, 546.
455. P. J. Russell and A. F. Brodie, *Biochim. Biophys. Acta* (1961), **50**, 76; R. Powls and E. R. Redfearn, *Biochem. J.* (1967), **102**, 3c.
456. E. Lederer and M. Vilkas, *Vitamins and Hormones* (1966), **24**, 409.
457. C. D. Snyder, S. J. Di Mari and H. Rapoport, *J. Amer. Chem. Soc.* (1966), **88**, 3868; F. Scherrer, R. Azerad and M. Vilkas, *Experientia* (1967), **23**, 360; J. Di Mari, C. D. Snyder and H. Rapoport, *Biochemistry* (1968), **7**, 2301; C. D. Snyder and H. Rapoport, *Biochemistry* (1968), **7**, 2318.
458. J. W. Mathieson, Ph.D. Thesis, University of Aberdeen (1969).
459. R. E. Erickson, A. F. Wagner and K. Folkers, *J. Amer. Chem. Soc.* (1963), **85**, 1535.
460. P. Mamont, P. Cohen, R. Azerad and M. Vilkas, *Bull. Soc. Chim. France* (1965), 2513.
461. E. Sun, R. Barr and F. L. Crane, *Plant Physiol.* (1968), **43**, 1935.
462. P. Mamont, P. Cohen, R. Azerad and M. Vilkas, *Bull. Soc. Chim. France* (1965), 2824.
463. P. Cohen and P. Mamont, *Bull. Soc. Chim. France* (1967), 1164.
464. R. Azerad and M.-O. Cyrot, *Bull. Soc. Chim. France* (1966), 1828.
465. K. Nishibori. Personal communication (1967).
466. H. Brockmann and A. Zeeck, *Chem. Ber.* (1967), **100**, 2885.
467. "The Vitamins" (eds. W. H. Sebrell and R. S. Harris) (1954). Vol. III. Academic Press, New York; 2nd edit. in the Press.
468. *Vitamins and Hormones* (1966), **24**.
469. G. S. Sidhu and M. Pardhasaradhi, *Tetrahedron Lett.* (1967), 4263.
470. G. P. Arsenault, *Tetrahedron* (1968), **24**, 4745.
471. I. Singh, R. E. Moore, C. W. J. Chang, R. T. Ogata and P. J. Scheuer, *Tetrahedron* (1968), **24**, 2969.

472. L. Jeffries, M. A. Cawthorne, M. Harris, A. T. Diplock, J. Green and S. A. Price, *Nature (London)* (1967), **215**, 257; L. Jeffries, M. Harris and S. A. Price, *Nature (London)* (1967), **216**, 808.

473. J. S. Gray, G. C. J. Martin and W. Rigby, *J. Chem. Soc. (C)* (1967), 2580.

474. M. R. J. Salton and M. D. Schmitt, *Biochim. Biophys. Acta* (1967), **135**, 196.

475. P. M. Baker and B. W. Bycroft, *Chem. Comm.* (1968), 71.

476. E. R. Redfearn and R. Powls, *Biochem. J.* (1968), **106**, 50P.

477. T. Takeda, E. Morishita and S. Shibata, *Chem. Pharm. Bull. (Tokyo)* (1968), **16**, 2213.

478. N. V. Tareeva, A. S. Romanova, A. I. Ban'kovskii and P. N. Kibal'chich, *Khim. Prirod. Soedin.* (1966), **2**, 359.

479. N. G. Carr, G. Exell, V. Flynn, M. Hallaway and S. Talukdar, *Arch. Biochem. Biophys.* (1967), **120**, 503.

480. G. Read, P. Shu, L. C. Vining and R. H. Haskins, *Canad. J. Chem.* (1959), **37**, 731; G. Read and L. C. Vining, *Canad. J. Chem.* (1959), **37**, 1881.

481. (a) G. Read and L. C. Vining, *Chem. Ind. (London)* (1963), 1239; (b) G. Read, A. Rashid and L. C. Vining, *J. Chem. Soc. (C)* (1969), 2059.

482. R. Corbaz, L. Ettlinger, E. Gaumann, J. Kalvoda, W. Keller-Schierlein, F. Kradolfer, B. K. Manukian, L. Neipp, V. Prelog, P. Reusser and H. Zähner, *Helv. Chim. Acta* (1957), **40**, 205.

483. S. Barcza, M. Brufani, W. Keller-Schierlein and H. Zähner, *Helv. Chim. Acta* (1966), **49**, 1736.

484. W. Keller-Schierlein, M. Brufani and S. Barcza, *Helv. Chim. Acta* (1968), **51**, 1257.

485. M. Brufani and M. Dobler, *Helv. Chim. Acta* (1968), 1269.

486. H. Gerlach, *Helv. Chim. Acta* (1966), **49**, 1291.

487. K. Eckardt, Personal communication, 1968.

488. G. Z. Yakubov, Yu. M. Khokhlova, O. I. Artamonova, L. N. Sergeeva and G. Ya. Kalmykova, *Microbiology* (1966), **35**, 875.

489. T. Arai, S. Kushikata, K. Takamiya, F. Yanagisawa and T. Koyama, *J. Antibiotics, Ser. A* (1967), **20**, 334.

490. W. Sandermann, M. H. Simatupang and W. Wendeborn, *Naturwissenschaften* (1968), **55**, 38.

491. M. H. Simatupang. Personal communication, 1968.

492. R. Paris and L. Prista, *Ann. Pharm. Franç.* (1954), **12**, 375.

493. L. Nogeira Prista, *Ann. Fac. Farm. Porto* (1954), **14**, 67; 1955, **15**. 73; L. Nogeira Prista and H. Romeira Prista, *Ann. Fac. Farm. Porto* (1954), **14**, 91.

494. E. W. Sheiffele and D. A. Shirley, *J. Org. Chem.* (1964), **29**, 3617; R. H. Thomson, unpublished.

495. G. Bendz and G. Lindberg, *Acta Chem. Scand.* (1968), **22**, 2722.

496. G. Bendz and G. Lindberg, *Acta. Chem. Scand.* (1970), **24**, 1082.

497. G. S. Sidhu, A. V. B. Sankaram and S. M. Ali, *Indian J. Chem.* (1968), **6**, 681.

498. A. Figueira de Sousa, *Bol. Inst. Invest. cient. Ang. (Luanda)* (1967), **4** (2), 7/28.

499. Y. N. Shukla, J. S. Tandon, D. S. Bhakuni and M. M. Dhar, *Experientia* (1969), **25**, 357; Y. N. Shukla, Ph.D. Thesis, University of Lucknow (1970).

500. R. G. Cooke, J. B. Robinson, J. R. Cannon and R. W. Retallack, *Aust. J. Chem.* (1970), **23**, 1029; R. G. Cooke and J. B. Robinson, *Aust. J. Chem.* (1970), **23**, 1695.

501. P. M. Brown, V. Krishnamoorthy, J. W. Mathieson and R. H. Thomson, *J. Chem. Soc. (C)* (1970), 109.

502. R. E. Moore, H. Singh and P. J. Scheuer, *Tetrahedron Lett.* (1968), 4581.

503. H. A. Anderson, Ph.D. Thesis, University of Aberdeen (1966).
504. M. D. Sutherland. Personal communication, 1968.
505. J. A. Ballantine, *Phytochemistry* (1969), **8**, 1587.
506. C. Kuroda and M. Okajima, *Proc. Japan Acad.* (1954), **30**, 982.
507. J. H. Bowie, D. W. Cameron, P. E. Schütz and D. H. Williams, *Tetrahedron* (1966), **22**, 1771.
508. E. Hardegger. Personal communication, 1967.
509. W. S. Chilton, *J. Org. Chem.* (1968), **33**, 4299.
510. A. Pfiffner, Thesis, Eidgenössischen Technischen Hochschule, Zürich (1963).
511. E. Morishita, T. Takeda and S. Shibata, *Chem. Pharm. Bull.* (*Tokyo*) (1968), **16**, 411.
512. S. Shibata, E. Morishita and Y. Arima, *Chem. Pharm. Bull.* (*Tokyo*) (1963), **11**, 821; (1967), **15**, 1757.
513. H. Brockmann and A. Zeeck, *Chem. Ber.* (1968), **101**, 4221.
514. P. M. Brown and R. H. Thomson, *J. Chem. Soc.* (*C*) (1969), 1184.
515. V. Krishnamoorthy and R. H. Thomson, unpublished work, 1968.
516. J. W. Rowe. Personal communication, 1968.
517. B. O. Lindgren and C. M. Svahn, *Phytochemistry* (1968), **7**, 1407; M. Fracheboud, J. W. Rowe, S. M. Fanega, A. J. Buhl and J. K. Toda, *Forest Prod. J.* (1968), **18**, 37.
518. K. L. Rinehart and H. H. Mathur, *J. Amer. Chem. Soc.* (1968), **90**, 6241.
519. K. L. Rinehart, H. H. Mathur, K. Sasaki, P. K. Martin and C. E. Coverdale, *J. Amer. Chem. Soc.* (1968), **90**, 6241.
520. W. Oppolzer, V. Prelog and P. Sensi, *Experientia* (1964), **20**, 336.
521. V. Prelog, *Pure Appl. Chem.* (1963), **7**, 551.
522. I. M. Campbell and R. Bentley, *Biochemistry* (1968), **7**, 3323.
523. C. Martius. *In* "Blood Clotting Enzymology" (ed. W. H. Seegers) (1967). Academic Press, London.
524. R. Powls, E. Redfearn and S. Trippett, *Biochem. Biophys. Res. Comm.* (1968), **33**, 408.
525. P. J. Dunphy, D. L. Gutnick, P. G. Phillips and A. F. Brodie, *J. Biol. Chem.* (1968), **243**, 398.
526. T. Kishi, M. Asai, M. Muroi, S. Harada, E. Mizuta, S. Terao, T. Niki and K. Mizuno, *Tetrahedron Lett.* (1969), 91; T. Kishi, S. Harada, M. Asai, M. Muroi and K. Mizuno, *Tetrahedron Lett.* (1969), 97.
527. W. J. Chamberlain and R. L. Stedman, *Phytochemistry* (1968), **7**, 1201.
528. I. M. Campbell and R. Bentley, *Biochemistry* (1969), **8**, 4651.
529. I. Singh, R. T. Ogata, R. E. Moore, C. W. J. Chang and P. J. Scheuer, *Tetrahedron* (1968), **24**, 6053.
530. H. Kern and S. Naef-Roth, *Phytopathol. Z.* (1967), **60**, 316; E. Hardegger. Personal communication, 1968.
531. U.S.S.R.P., 200,735; *Chem. Abs.* (1968), **68**, 62694s.
532. U.S.S.R.P., 200,737; *Chem. Abs.* (1968), **68**, 62695t.
533. M. Ruwet, *Bull. Soc. Roy. Sci. Liège* (1966), **35**, 142.
534. S. Miura, *Syoyakugaku Zasshi* (1963), **17**, 45.
535. M. C. Russell, *West. Austral. Naturalist* (1958), **6**, 11.
536. M. C. Russell, *West. Austral. Naturalist* (1959), **7**, 30.
537. K. Kamiya, T. Sugino, Y. Wada, M. Nishikawa and T. Kishi, *Experientia* (1969), **25**, 901.
538. B. L. Gilbert, J. E. Baker and D. M. Norris, *J. Insect Physiol.* (1967), **13**, 1453.
539. F. J. Schneiderhan, *Phytopath.* (1927), **17**, 529; E. F. Davis, *Amer. J. Bot.* (1928), **15**, 620; M. C. Strong, *Michigan Quart. Bull.* (1944), **26**, 194.

540. J. Vinkenborg, N. Sampara-Rumantir and O. F. Uffelie, *Pharm. Weekbl.* (1969), **104**, 45.

541. C. Galeffi, C. G. Casinovi, E. M. delle Monache and G. B. Marini-Bettòlo, *Ann. Ist. Super. Sanità* (1968), **4**, 305.

542. C. Galeffi, E. M. delle Monache, C. G. Casinovi and G. B. Marini-Bettòlo, *Tetrahedron Lett.* (1969), 3583.

543. J. C. Overeem and D. M. Elgersma, *Phytochemistry* (1970), **9**, 1949.

544. N. R. Ayyangar, A. V. Rama Rao and K. Venkataraman, *Indian J. Chem.* (1969), **7**, 533.

545. H. Kakisawa, Y. Inouye and J. Romo, *Tetrahedron Lett.* (1969), 1929.

546. R. M. Ruiz, J. Correa and L. A. Maldonado, *Bull. Soc. Chim. Fr.* (1969), 3612.

547. R. Howe and R. H. Moore, *Experientia* (1969), **25**, 474.

548. T. Matsumoto, C. Mayer and C. H. Eugster, *Helv. Chim. Acta* (1969), **52**, 808.

549. R. G. Cooke and J. G. Down, *Tetrahedron Lett.* (1970), 583.

550. R. F. Moore and W. A. Waters, *J. Chem. Soc.* (1953), 3405.

551. G. S. Sidhu and M. Pardhasaradhi, *Indian J. Chem.* (1970), **8**, 569.

552. S. Natori. Personal communication, 1969.

553. G. S. Sidhu and K. K. Prasad, *Tetrahedron Lett.* (1970), 1739.

554. H. Singh, T. L. Folk and P. J. Scheuer, *Tetrahedron* (1969), **25**, 5301.

555. F. Fariña and W. Heimlich, *Anal. Quim.* (1969), **65**, 713.

556. G. S. Sidhu and S. K. Pavanaram. Unpublished information.

557. K. H. Dudley and R. W. Chiang, *J. Org. Chem.* (1969), **34**, 120.

558. L. Jeffries, M. A. Cawthorne, M. Harris, B. Cook and A. T. Diplock, *J. Gen. Microbiol.* (1969), **54**, 365.

559. J. T. Matschiner, W. V. Taggart and J. M. Amelotti, *Biochemistry* (1967), **6**, 1243; J. T. Matschiner and J. M. Amelotti, *J. Lipid Res.* (1968), **9**, 176.

560. P. G. Phillips, P. J. Dunphy, K. L. Servis and A. F. Brodie, *Biochemistry* (1969), **8**, 2856.

561. R. Powls and E. R. Redfearn, *Biochim. Biophys. Acta* (1969), **172**, 429.

562. W. E. Bondinell, C. D. Snyder and H. Rapoport, *J. Amer. Chem. Soc.* (1969), **91**, 6889.

563. C. D. Snyder and H. Rapoport, *J. Amer. Chem. Soc.* (1969), **91**, 731.

564. K. Joshi, Ph.D. Thesis, University of Western Australia (1969).

565. W. Zopf, *Ber. Deut. Bot. Ges.* (1908), **26**, 51; O. Hesse, *J. Prakt. Chem.* (1911), **83**, 58; (1915), **92**, 448.

566. S. Shibata, *J. Pharm. Soc. Japan* (1941), **61**, 320.

567. G. Koller and H. Hamburg, *Monatsh.* (1936), **68**, 202.

568. P. M. Baker and E. Bullock, *Canad. J. Chem.* (1969), **47**, 2733.

569. S. McLean and C. J. Webster. Personal communication, 1969.

570. B. E. Cross, M. N. Edinberry and W. B. Turner, *Chem. Comm.* (1970), 209.

571. A. Rashid and G. Read, *J. Chem. Soc.* (*C*) (1969), 2053.

572. F. L. Crane and H. Low, *Physiol. Revs.* (1966), **46**, 662; F. L. Crane. In "Biological Oxidations" (ed. T. P. Singer) (1968). Interscience, New York.

573. S. H. Harper, A. D. Kemp and J. Tannock, *J. Chem. Soc.* (*C*) (1970), 626.

574. O. Gonçalves de Lima, I. L. d'Albuquerque, Dardano de Andrade Lima, G. Medeiros Maciel, *Rev. Inst. Antibiót., Univ. Recife* (1967), **7**, 3.

575. M. M. Weber, J. T. Matschiner and H. D. Peck, *Biochem. Biophys. Res. Comm.* (1970), **38**, 197.

576. B. E. Cross, P. L. Myers and G. R. B. Webster, *J. Chem. Soc.* (*C*) (1970), 930.

577. M. Ribi, A. Chang Sin-Ren, H. P. Kung and C. H. Engster, *Helv. Chim. Acta* (1969), **52**, 1685.

CHAPTER 5

Anthraquinones†

This is the largest group of natural quinones and, historically, the most important. Many Rubiaceae contain useful anthraquinone mordant dyes and these plants have been used for dyeing textiles in many parts of the world since ancient times. Another group, present particularly in *Cassia* (Leguminosae), *Rhamnus* (Rhamnaceae) and *Rheum* (Polygonaceae) spp., are associated with the well-known purgative extracts obtained from these plants (senna, cascara, rhubarb).

Undoubtedly many more anthraquinones await discovery. Isolation work frequently reveals the presence of minute amounts of unidentified anthraquinones, some of which are probably dimers. In suitable cases the actual isolation can be avoided by direct introduction of natural tissue into a mass spectrometer, making it possible to identify trace compounds with a minimum of labour and material. In the quinone field, only lichens[556] have been studied in this way but other plant and animal material might be examined with useful results.

Anthraquinone itself has been obtained from several natural sources but was probably an artefact in all cases. It was isolated[558] from the essential oil of tobacco leaf by chromatography of a fraction which had been distilled at 132–168°/4 mm., and has also been detected[559] in tannin extracts‡ although it does not appear to be present as such in the original wood and bark. The nature of the precursors is not known. Anthraquinone was also found at a burning coal seam on Mount Pyramide by a Norwegian expedition in 1921. This mineral, hoelite,[560] "crystallises in fine yellow needles, sitting directly on the coal, partly covered by a crust of sal-ammoniac", and was presumably formed by oxidation of anthracene. Polycyclic hydrocarbons have been found[561] in other mineral deposits. Graebeite,[562] a red material found in shale in a coal mine at Ölsnitz in Saxony, appears to be a mixture of polyhydroxyanthraquinones (cf. the fringelites, p. 594), and positive identification of 1,2,7-trihydroxyanthraquinone in a Posidonian shale has been claimed.[614] The anthraquinone obtained after extensive chemical treatment of humic acid,[563] extracted from volcanic ash soils in Japan, is of doubtful

† Protetrone is mentioned in Chapter 6.
‡ Quebracho extract from the heartwood of *Quebrachia lorentzii* Griseb. (Anacardiaceae) and black wattle extract from the dark of *Acacia decurrens* Willd. (Leguminosae).[559]

relevance to natural products but, on the other hand, solvent extraction of many samples of Irish soils[163, 620] has yielded a dimer of 1,8-dihydroxy-3-methylanthraquinone (p. 392).

Tectoquinone (1), $C_{15}H_{10}O_2$

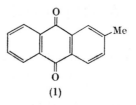

(1)

Occurrence: heartwood[1, 2] and sapwood[4] of *Tectona grandis* L. fil. (Verbenaceae), heartwood of *Tabebuia avellanedae* Lor. ex Griseb.[3] (Bignoniaceae), stems[5] and roots[25] of *Morinda umbellata* L. (Rubiaceae).

Physical properties: pale yellow needles, m.p. 178–179°, $\lambda_{max.}$ (EtOH) 255, 265, 274, 324 nm (log ϵ 4·65, 4·32, 4·24, 3·66), $\nu_{max.}$ (KBr) 1672, 1592 cm^{-1}.

Tectoquinone is one of the numerous quinones present in teak wood[6] (*Tectona grandis*) and in lapacho wood[3] (*Tabebuia avellanedae*), and its recent discovery in the Rubiaceae is of biogenetic significance. As tectoquinone is repellent to termites, it no doubt contributes to the durability of teak but it is ineffective against fungal attack.[7] Dry distillation of teak yields a tar containing 2-methylanthracene[2] from which the quinone can be obtained by re-oxidation with chromic acid.

2-Hydroxymethylanthraquinone (2; R = H), $C_{15}H_{10}O_3$
2-Acetoxymethylanthraquinone (2; R = Ac), $C_{17}H_{12}O_4$
Anthraquinone-2-aldehyde (3), $C_{15}H_8O_3$
Anthraquinone-2-carboxylic acid (4), $C_{15}H_8O_4$

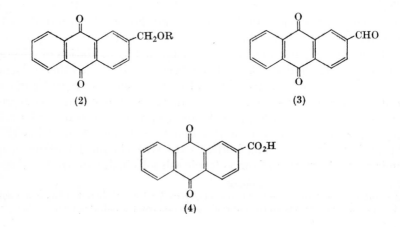

(2)   (3)

(4)

Occurrence: heartwood of *Tectona grandis* L. fil.[7] (Verbenaceae) and *Tabebuia avellanedae* Lor. ex Griseb.[3] (Bignoniaceae).

Physical properties: (2; R = H) pale yellow needles, m.p. 192–193°, $\lambda_{max.}$ (cyclohexane) 255, 268, 275, 324 nm (log ϵ 4·60, 4·30, 4·19, 3·61), $\nu_{max.}$ (KBr) 3250, 1681, 1595 cm^{-1}.

(2; R = Ac) pale yellow needles, m.p. 151–152°, $\lambda_{max.}$ (EtOH) 254, 263, 274, 323 nm (log ϵ 4·60, 4·28, 4·19, 3·61), $\nu_{max.}$ (KBr) 1740, 1680, 1591 cm^{-1}.

(3) pale yellow plates, m.p. 189°, $\lambda_{max.}$ (cyclohexane) 256, 266, 278, 328 nm (log ϵ 4·51, 4·27, 4·21, 3·61), $\nu_{max.}$ (KBr) 1714, 1672, 1589 cm^{-1}.

(4) fine yellow needles, m.p. 291°, $\lambda_{max.}$ (EtOH) 254·5, 325·5 nm (log ϵ 4·72, 4·03), $\nu_{max.}$ (KBr) 1696, 1680 cm^{-1}.

The hydroxymethyl derivative (2; R = H) is light sensitive,[7] and the aldehyde (3), although quite stable during chromatography, darkens on storage and is partly converted into the acid (4).[3] Both (3) and (4) may be artefacts. The acid (4) is normally prepared by oxidising 2-methyl-anthraquinone with chromic acid,[3,4,7] and the aldehyde (3) is most conveniently obtained[348] by side-chain bromination of the same compound followed by treatment with hexamine (Sommelet reaction). 2-Bromomethylanthraquinone is converted, by reaction with sodium acetate,[348] into the ester (2; R = Ac) which gives (2; R = HO) on hydrolysis.[349]

1-Hydroxyanthraquinone (5), $C_{14}H_8O_3$
1-Methoxyanthraquinone, $C_{15}H_{10}O_3$

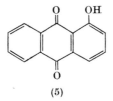

(5)

Occurrence: heartwood of *Tabebuia avellanedae* Lor. ex Griseb.[3] (Bignoniaceae).
Physical properties: yellow needles, m.p. 194–195°, $\lambda_{max.}$ (EtOH) 252, 266(sh), 277(sh), 327, 402 nm (log ϵ 4·46, 4·18, 4·44, 3·52, 3·74), $\nu_{max.}$ (KBr) 1670, 1635, 1591 cm^{-1}; methyl ether, pale yellow needles, m.p. 169–170°, $\lambda_{max.}$ (EtOH) 254, 262, 270(sh), 328, 378 nm (log ϵ 4·52, 4·36, 4·18, 3·46, 3·72), $\nu_{max.}$ (KBr) 1675, 1588 cm^{-1}.

These are the only anthraquinones in *T. avellanedae* which have no carbon side-chain. Synthetic material is most conveniently obtained from 1-aminoanthraquinone via the diazonium salt,[16] but a more interesting synthesis is that of Birch *et al.*[220] which proceeds via Diels-Alder addition of 1-methoxycyclohexa-1,3-diene to 1,4-naphthaquinone.

2-Hydroxyanthraquinone (6), $C_{14}H_8O_3$
2-Methoxyanthraquinone, $C_{16}H_{10}O_3$

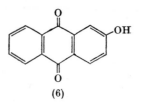

(6)

Occurrence: (6) root of *Oldenlandia umbellata* L.,[8] *Galium saxatile* L.,[13] *G. verum* L.,[13] *G. mollugo* L.,[13] *G. pumilum* L.,[13] *Asperula odorata* L.[13] and *Morinda umbellata* L.[25] The methyl ether occurs in the roots of four *Galium* spp.,[13] *Asperula odorata* L.,[13] *Rubia tinctorum* L.[565] and root and stems of *Morinda umbellata* L.[25] (all Rubiaceae).

Physical properties: yellow needles, m.p. 306°, $\lambda_{max.}$ (EtOH) 241, 271, 283, 330, 378 nm (log ϵ 4·31, 4·55, 4·46, 3·55, 3·55), $\nu_{max.}$ (KBr) 1671, 1590 cm^{-1}; methyl ether, pale yellow needles, m.p. 195–196°, $\lambda_{max.}$ (EtOH) 246, 267, 280, 329, 363 nm (log ϵ 4·21, 4·51, 4·38, 3·56, 3·60), $\nu_{max.}$ (KBr) 1673, 1590 cm^{-1}.

This is the simplest quinone found in chay root, or Indian madder, (*O. umbellata*) and in *Galium* spp. Chay root was formerly cultivated in India for use in dyeing, and both *Galium* spp. (bedstraws) and *A. odorata* (woodruff) have been used for the same purpose.[14]

Like the previous compound, 2-hydroxyanthraquinone is easily recognised from spectral data and colour reactions. It is best prepared from 2-aminoanthraquinone by way of the diazonium salt.[16]

1-Hydroxy-2-methylanthraquinone (7), $C_{15}H_{10}O_3$
1-Methoxy-2-methylanthraquinone, $C_{16}H_{12}O_3$

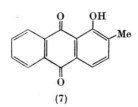

(7)

Occurrence: (7) heartwood of *Tabebuia chrysantha* Nichols[15] (Bignoniaceae), roots of *Rubia tinctorum* L.[565] and several *Galium* spp.,[13] roots and stems of *Morinda umbellata* L.[25] The methyl ether is found in the root of *G. aparine* L.,[13] *G. saxatile* L.,[13] *G. verum* L.,[13] *Asperula odorata* L.,[13] *Rubia tinctorum* L.[565] and *M. umbellata* L.[25] (all Rubiaceae), and the leaves of *Digitalis purpurea* L.[566] and *D. lanata* Ehrh.[566] (Scrophulariaceae)

Physical properties: (7) yellow needles, m.p. 185–186°, $\lambda_{max.}$ (EtOH) 255, 266(sh), 270(sh), 325, 402 nm (log ϵ 4·51, 4·33, 3·51, 3·74), $\nu_{max.}$ (KBr) 1670, 1637, 1593 cm^{-1}; methyl ether, pale yellow needles, m.p. 154–156°, $\lambda_{max.}$ (EtOH) 254, 275(sh), 346 nm (log ϵ 4·59, 4·13, 3·68), $\nu_{max.}$ (KBr) 1682, 1596 cm^{-1}.

The structure can be deduced from spectroscopic and analytical data. Synthetic material may be obtained via diazotization of 1-amino-2-methylanthraquinone[21] or from 1-hydroxyanthraquinone by methylation with formaldehyde in alkaline dithionite (the Marschalk reaction).[72]

3-Hydroxy-2-methylanthraquinone (8), $C_{15}H_{10}O_3$
3-Methoxy-2-methylanthraquinone, $C_{16}H_{12}O_3$

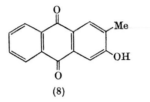

(8)

Occurrence: (8) heartwood of *Tectona grandis* L. fil.[10] (Verbenaceae), *Tabebuia avellanedae* Lor. ex Griseb.[3] and *T. chrysantha* Nichols[15] (Bignoniaceae), bark of *Coprosma lucida* Forst.[11] and probably *C. acerosa* A. Cunn.[12] (Rubiaceae). The methyl ether is found in the leaves of *Digitalis purpurea* L.,[566] *D. lanata* Ehrh.[566] and *D. canariensis* L. var. *isabelliana* (Webb) Lindinger[22] (Scrophulariaceae).

Physical properties: (8) yellow needles, m.p. 302°, $\lambda_{max.}$ (EtOH) 240, 244·5, 272·5, 383 nm (log ε 4·40, 4·40, 4·50, 3·67), $\nu_{max.}$ (KBr) 3400, 1667 cm^{-1}; methyl ether, yellow needles, m.p. 195–196°, $\lambda_{max.}$ (EtOH) 267·5, 363 nm (log ε 4·42, 3·30), $\nu_{max.}$ (KBr) 1667 cm^{-1}.

The structure can be deduced from spectroscopic and analytical data. It can be prepared in poor yield by direct condensation of *o*-cresol with phthalic anhydride in hot sulphuric acid, or better, in two steps, by the benzoylbenzoic acid method using the same components.[16]

Tabebuin (9), $C_{31}H_{26}O_5$

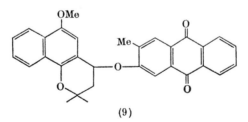

(9)

Occurrence: heartwood of *Tabebuia avellanedae* Lor. ex Griseb.[3] (Bignoniaceae).
Physical properties: pale orange rosettes, m.p. 211–212°, $[\alpha]_D^{18}$ +94·3° (CHCl₃), $\lambda_{max.}$ (EtOH) 250, 270(sh), 323, 333, 384 nm (log ε 4·59, 4·25, 3·81, 3·79, 3·16), $\nu_{max.}$ (KBr) 1662, 1628, 1590 cm^{-1}.

T. avellanedae elaborates two series of C_{15} compounds having either a naphthalene or an anthracene nucleus; tabebuin belongs to both. Some type of dimeric structure was immediately apparent from the molecular weight (478), and since the compound was a quinone (ν_{CO} 1662 cm^{-1}) and its ultraviolet absorption was very similar to that of dihydrolapachenole (10; R = H) (present in *T. chrysantha*[15]), it was no surprise when degradation with hydrobromic acid gave 3-hydroxy-2-methylanthraquinone and the dimer (12) which arises when lapachenole (11) is

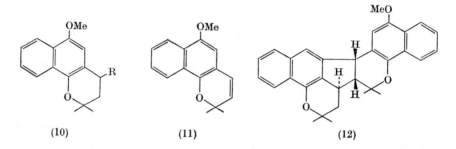

(10)   (11)   (12)

subjected to acid conditions.[17] As tabebuin contains no hydroxyl group it must have structure (9), and the racemic compound was readily obtained by alkylation of 3-hydroxy-2-methylanthraquinone with bromodihydrolapachenole (10; R = Br).[3] The structure is fully consistent with the n.m.r. and mass spectra; like the acid hydrolysis, cleavage under electron impact gives two major fragments *m/e* 240 and 238, corresponding to lapachenole and 3-hydroxy-2-methylanthraquinone, respectively. Lapachenole is isomeric with menaquinone-1 (p. 201) and its mass spectrum is very similar except for the absence of a peak at M-28.

Pachybasin (13), $C_{15}H_{10}O_3$
Pachybasin methyl ether, $C_{16}H_{12}O_3$

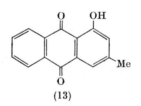

(13)

Occurrence: (13) cultures of *Pachybasium hamatum* (Bon.) Sacc. var. *candidum* (Sacc.)[18] (= *P. candidum* Sacc.), *Phoma foveata* Foister,[19] *Trichoderma viride* Pers. ex Fr.,[20] *Aspergillus crystallinus* Kwon & Fennell,[159] and two unidentified sterile fungi isolated from the heartwood of yellow cedar (*Chamaecyparis nootkatensis*);[601] in the heartwood of *Tectona grandis* L. fil.[9] (Verbenaceae). The methyl ether occurs in the leaves of *Digitalis canariensis* L. var. *isabelliana* (Webb) Lindinger[22] and *D. purpurea* L.[22] (Scrophulariaceae).

Physical properties: yellow needles, m.p. 178°, λ_{max}. (EtOH) 224, 252, 281, 403 nm (log ϵ 3·74, 4·01, 3·65, 3·02), ν_{max}. (Nujol) 1677, 1642, 1593 cm^{-1}; methyl ether, yellow needles, m.p. 190–191°, λ_{max}. (EtOH) 220, 255, 276(sh), 325, 383 nm (log ϵ 4·36, 4·45, 4·20, 3·45, 3·72), λ_{max}. (KBr) 1681, 1663(sh), 1610 cm^{-1}.

This is the simplest anthraquinone found in fungi and in the four species listed it occurs with the typical polyketide quinones, chrysophanol and emodin. In *Digitalis* (foxglove) leaves these latter pigments are absent but traces of other quinones can be detected. Pachybasin was recognised[18] after conversion to 1-hydroxyanthraquinone by chromic acid oxidation of the acetate, followed by hydrolysis and decarboxylation. It is most conveniently synthesised by direct condensation of *m*-cresol with phthalic anhydride in fused aluminium chloride–sodium chloride.[23]

Alizarin (14), $C_{14}H_8O_4$

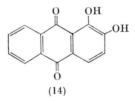

(14)

Occurrence: roots of *Rubia tinctorum* L.,[24,565] *Oldenlandia umbellata* L.,[8,26] *Galium* spp.,[13,27,230] *Asperula odorata* L.,[13] *A. ciliata* Rochel[27] and *Crucianella maritima* L.,[28] stems and roots of *Hedyotis auricularia* L.[31] and *Morinda umbellata* L.,[25] and the heartwood of *Morinda citrifolia* L.[78] In all these Rubiaceae it is probably present both free and as a glycoside. Alizarin has also been reported in rhubarb (*Rheum palmatum*?)[29] (Polygonaceae) and as a glycoside in *Libertia coerulescens* Kunth[30] (Iridaceae).

Physical properties: orange-red needles, m.p. 289–290°, λ_{max}. (EtOH) 247, 278, 330, 434 nm (log ϵ 4·45, 4·13, 3·46, 3·70), ν_{max}. (KBr) 3340, 1660, 1627 cm^{-1}.

Alizarin appears to be characteristic of several tribes of the Rubiaceae, and its presence in other families requires verification, particularly as it is reported[29,30] to occur with chrysophanol and emodin which, on biogenetical grounds, would not be expected. Alizarin is the principal colouring matter of madder, the ground root of *Rubia tinctorum*, one of the most ancient natural dyestuffs and in its heyday, probably the most important. The art of dyeing madder appears to have originated in the East, passed to the ancient Persians and Egyptians, and thence to the Greeks and Romans.[32] The famous Turkey Red, dyed on mordanted cotton, is the calcium salt of its aluminium chelate,[33] and was much favoured in the past for military uniforms, while the polygenetic character of alizarin was an advantage to the calico printer. Although alizarin is virtually obsolete as a dye it is still used for the study of bone growth and has not been superseded by radiocalcium.[343] *Rubia tinctorum* was widely cultivated in Western Europe from the Middle Ages until the advent of synthetic alizarin.[34] In young plants, the pigment is present as its primveroside, ruberythric acid, together with only a few related compounds, but in mature plants there is much of the aglycone and as many as nineteen anthraquinones have been isolated from the root.[565]

The chemical investigation† of madder dates back to 1826 when impure material was first isolated by Colin and Robiquet,[24] and the later work of Graebe and Liebermann[35] on its constitution and synthesis is one of the classics of organic chemistry. By the new procedure of zinc dust distillation they showed that alizarin could be reduced to anthracene, and from a comparison with other quinonoid compounds studied by Graebe[36] they deduced that it was a dihydroxy derivative of anthraquinone, at that time a new compound. They immediately proceeded to verify this hypothesis by synthesis. In the knowledge that two of the halogen atoms in chloranil could be replaced by hydroxyl groups on treatment with alkali, they prepared a dibromoanthraquinone assuming that it would behave in the same way, being also a halogenoquinone, and on fusion with sodium hydroxide it yielded alizarin, identical with the natural material obtained from madder. This was achieved in 1868. The astonishing success of this reaction could not be appreciated until the chemistry was more fully understood but its significance as the first laboratory synthesis of a natural pigment was immediately recognised, not least by the authors themselves. The position of the two hydroxyl groups was demonstrated later by the condensation of phthalic anhydride and catechol which gave a mixture of alizarin and hystazarin, and by the fact that both alizarin and quinizarin can be oxidised to purpurin (1,2,4-trihydroxyanthraquinone), whence it follows that alizarin must be 1,2-dihydroxyanthraquinone (14).

Graebe and Liebermann patented their synthesis but owing to the cost of bromine it was too expensive for commercial exploitation. An alternative process was soon developed based on the alkali fusion of anthraquinonesulphonic acid and this was also patented by Caro, Graebe and Liebermann in 1869, one day ahead of W. H. Perkin who had discovered the same process! However an amicable arrangement was reached and the first synthetic alizarin was sold in England in 1869 within a year of the original synthesis, and thereafter the use of madder rapidly declined. Contrary to the initial supposition, Perkin showed later that the compound which gave alizarin on alkali fusion was a

† For a full account see "The Centenary of the Discovery of Alizarin", A. Wahl, *Bull. Soc. Chim. France* (1927), **41**, 1417; "The Discovery of Synthetic Alizarin", L. F. Fieser, *J. Chem. Ed.* (1930), **7**, 2609.

*mono*sulphonic acid so that oxidation takes place during the course of the reaction. Nucleophilic substitution by hydroxyl ion displaces a hydride ion (see scheme below), and to prevent reduction of the quinone

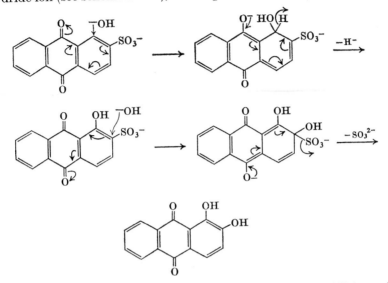

by the latter the process was later modified by the addition of an oxidising agent, such as sodium nitrate or chlorate, to the melt.[37] The brominated quinone which Graebe and Liebermann used in their original synthesis was actually 2,3-dibromoanthraquinone which suggests that one of the bromine atoms is displaced by hydride ion liberated in the hydroxylation step. Numerous other syntheses of alizarin have since been devised.[38]

Ruberythric acid, $C_{25}H_{26}O_{13}$, yellow needles, m.p. 258–260°, isolated by Rochleder[24] in 1851, was one of the first glycosides obtained in a crystalline state. Many years later the sugar was identified[39] as primverose, and, since methylation of ruberythric acid with methyl iodide and silver oxide, followed by hydrolysis, gave alizarin 1-methyl ether the sugar residue must be attached to the hydroxyl group at C-2. As hydrolysis can be effected by β-glucosidases,[39, 40] the linkage to the aglycone has the usual β-configuration, and the 2-β-primveroside structure was finally confirmed by synthesis[41] from alizarin and aceto-bromoprimverose.

Alizarin 1-methyl ether, $C_{15}H_{10}O_4$
Alizarin 2-methyl ether, $C_{15}H_{10}O_4$

Occurrence: both ethers are present in the root of *Galium* spp.,[13] *Asperula odorata* L.,[13] *Rubia tinctorum* L.[565] and the root and stem of *Morinda umbellata* L.,[25] and the 1-methyl ether also in the root of *Oldenlandia umbellata* L.,[8] *Morinda longiflora* G. Don,[42] *M. citrifolia* L.,[43] *M. tinctoria* Roxb.,[44] and possibly *Rubia tetragona* Schum.[45] and *Galium atherodes* Spreng.[46] (= *Relbunium atherodes*) (all Rubiaceae).

Physical properties: alizarin 1-methyl ether, pale yellow needles, m.p. 182–184°, $\lambda_{max.}$ (EtOH) 242(sh), 246·5, 270, 284, 330, 385 nm (log ϵ 4·37, 4·38, 4·37, 4·24, 3·47, 3·64), $\nu_{max.}$ (KBr) 3420, 1673 cm^{-1}; alizarin 2-methyl ether, yellow needles, m.p. 234°, $\lambda_{max.}$ (EtOH) 247, 275(sh), 424 nm (log ϵ 3·53, 3·20, 2·94), $\nu_{max.}$ (KBr) 1661, 1633, 1590 cm^{-1}.

These ethers are easily recognised as they give alizarin on demethylation, and can be distinguished by their alkali solubilities and infrared spectra. The 2-methyl ether is readily obtained from alizarin under normal methylation conditions as the strongly chelated *peri*-hydroxyl group is relatively unreactive. The 1-methyl ether is prepared via methylation of the 2-acetate with methyl sulphate–potassium carbonate–acetone, but treatment of the same ester with diazomethane gives mainly the acetate of alizarin 2-methyl ether.[47] Other 2-esters of alizarin undergo similar rearrangements.[48]

Hystazarin monomethyl ether (15), $C_{15}H_{10}O_4$

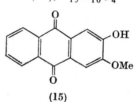

(15)

Occurrence: root of *Oldenlandia umbellata* L.[26,49] (Rubiaceae).
Physical properties: orange-yellow leaflets, m.p. 232°.

Perkin and Hummel[26,49] isolated a small amount of this minor constituent of chay root from 100 kg of plant material. It contained one methoxyl group and gave hystazarin on demethylation. Synthetically it can be obtained by partial demethylation of hystazarin dimethyl ether which may be prepared by cyclisation of 2-(3,4-dimethoxy-benzoyl)benzoic acid.[50] Despite many subsequent investigations of Rubiaceous plants, this is the only report of a naturally occurring hystazarin derivative.

3-Methylalizarin (16), $C_{15}H_{10}O_4$
Digitolutein (17), $C_{16}H_{12}O_4$

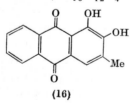

(16)

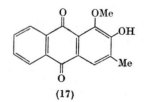

(17)

Occurrence: (16) leaves of *Digitalis purpurea* L.[25]; (17) leaves of *D. lutea* L.[51] *D. purpurea* L.,[52] *D. lanata* Ehrh.[25] and *D. viridiflora* Lindl.[571] (Scrophulariaceae). *Physical properties*: (16) orange needles, m.p. 250–251°, λ_{max}. (EtOH) 249, 268, 283(sh), 331, 436 nm (log ϵ 4·74, 4·68, 4·35, 3·58, 3·81), ν_{max}. (KBr) 3378, 1664, 1631, 1593 cm^{-1}; (17) yellow needles, m.p. 222°, λ_{max}. (EtOH) 242·5, 274, 386 nm (log ϵ 4·27, 4·40, 3·52), ν_{max}. (KBr) 3270, 1666, 1589 cm^{-1}.

Anthraquinones are uncommon leaf pigments but there are several in the leaves of the foxglove, *D. purpurea*. Digitolutein, first isolated from *D. lutea*, contains one C–Me and one O–Me function, and a hydroxyl group which forms an acetate and a methyl ether, and must be assigned to a β-position as the quinone is soluble in aqueous sodium carbonate. The C-methyl group also occupies a β-position as zinc dust distillation affords 2-methylanthracene, and the formation of phthalic acid on permanganate oxidation shows that all three substituents are in the same ring. As the pigment is different from rubiadin 1-methyl ether structure (17) is the only alternative.[53] Demethylation gives 3-methylalizarin which was discovered later in *D. purpurea*. 3-Methylalizarin can be prepared by direct[351] or indirect[53] hydroxylation of 2-hydroxy-3-methylanthraquinone. Acetylation in the presence of boroacetic anhydride gives the 2-acetate which, on methylation (methyl iodidesilver oxide) and hydrolysis gives the 1-methyl ether (17) identical[54] with digitolutein. Alternatively the 2-hydroxyl group can be protected by tosylation.[55]

6-Methylalizarin (18), $C_{15}H_{10}O_4$

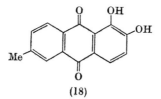

(18)

Occurrence: root of *Hymenodictyon excelsum* Wall.[567] and possibly *Coprosma parviflora* Hook.[56] (Rubiaceae).

Physical properties: yellow crystals, m.p. 222–224°, λ_{max}. (EtOH) 232·5, 262·5, 278·5 330, 429 nm (log ϵ 4·25, 4·53, 4·32, 3·55, 3·75), λ_{max}. (KBr) 3480, 1670, 1657, 1608 cm^{-1}.

This pigment (molecular weight 254) is very similar to alizarin in its colour reactions and spectroscopic properties which show that hydroxyl groups are present in the 1,2-positions. Its co-existence with morindone (p. 410) in *H. excelsum*, which has very similar chromatographic properties, suggested structure (18) which was confirmed[567] by direct comparison with a sample synthesised[21] by a two-step condensation of 3,4-dimethoxyphthalic anhydride with toluene. Although the method of synthesis is ambiguous, the natural pigment is almost certainly (18).

Xanthopurpurin (purpuroxanthin) (19), $C_{14}H_8O_4$
Xanthopurpurin 1-methyl ether, $C_{15}H_{10}O_4$
Xanthopurpurin 3-methyl ether, $C_{15}H_{10}O_4$
Xanthopurpurin dimethyl ether, $C_{16}H_{12}O_4$

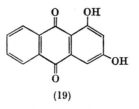

(19)

Occurrence: (19) roots of *Rubia tinctorum* L.,[58] *R. cordifolia* L.,[57] *R. sikkimensis* Kurz.,[57] several *Galium* spp.,[13, 46] *Asperula odorata* L.,[13] stems and roots of *Morinda umbellata* L.[25] Both monomethyl ethers occur in the root of *Rubia tetragona* Schum.[45] and *Relbunium hypocarpium* Hemsl.,[61] and the 3-methyl ether also in the root of *Rubia tinctorum* L.[565] The dimethyl ether is found in the roots of *R. tinctorum* L.,[565] *Asperula odorata* L.[13] and several *Galium* spp.[13] (all Rubiaceae).

Physical properties: (19) yellow needles, m.p. 269–270°, $\lambda_{max.}$ (EtOH) 246, 284, 415 nm (log ϵ 4·43, 4·36, 3·69), $\nu_{max.}$ (Nujol) 3413, 1675, 1637 cm^{-1}.

1-methyl ether, yellow leaflets, m.p. 311–313°.

3-methyl ether, light yellow needles, m.p. 193–194°.

Dimethyl ether, yellow needles, m.p. 154–155°, $\lambda_{max.}$ (EtOH) 227, 237, 277, 330, 390 nm (log ϵ 4·43, 4·30, 4·49, 3·56, 3·77), $\nu_{max.}$ (KBr) 1664, 1598 cm^{-1}.

Schutzenberger[58] first isolated xanthopurpurin from madder in 1865 and showed that it could be prepared from purpurin by reduction with sodium stannite. The location of the hydroxyl groups was established later, synthetically,[59] by condensing benzoic acid with 3,5-dihydroxy-benzoic acid in hot sulphuric acid. It is most conveniently prepared from purpurin by reduction with sodium dithionite in ammoniacal solution.[60] The 3-methyl ether can be obtained by direct partial methylation of xanthopurpurin, and the 1-isomer by methylation of its 3-acetate, and hydrolysis.[16]

Rubiadin (20), $C_{15}H_{10}O_4$
Rubiadin 1-methyl ether (21), $C_{16}H_{12}O_4$

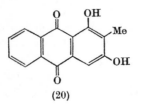

(20)

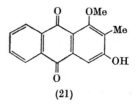

(21)

Occurrence: (20) root of *Rubia tinctorum* L.,[62] *Prismatomeris malayana* Ridley,[554] *Hymenodictyon excelsum* Wall.,[567] *Asperula odorata* L.,[13] and several *Galium* spp.,[13,27,63,230] the bark of *Coprosma lucida* Forst.,[11] *C. australis* Robinson,[64] *C. parviflora* Hook.[56] and probably *C. acerosa* A. Cunn.,[12] the root *Morinda citrifolia* L.,[65,610] root and stems of *M. umbellata* L.,[25] and the root bark of *M. jasminoides* A. Cunn.,[65] *Coelospermum reticulatum* Benth.,[66] *C. paniculatum* F. Muell.,[66] (it probably occurs in most of these Rubiaceae both free and as a glycoside), and in the heartwood of *Tectona grandis* L.[426] (Verbenaceae).

(21) Bark of several *Coprosma* spp.,[11,56,75-77] root bark of *Hymenodictyon excelsum* Wall.,[567] *Galium dasypodum* Klokov,[230] roots and leaves of *Morinda longiflora* G. Don,[42] roots and stems of *M. umbellata* L.,[25] root,[65,611] root bark[43,65] and heartwood[78] of *M. citrifolia* L., root bark of *M. jasminoides* A. Cunn.[65] and *Coelospermum paniculatum* F. Muell.[66] (all Rubiaceae). In some species it occurs as a glycoside, identified in *M. longiflora* as a primveroside, yellow needles, m.p. 216–218° ("longifloroside").[79]

Physical properties: (20) yellow plates, m.p. 302°, $\lambda_{max.}$ (EtOH) 241(sh), 245, 279, 414 nm (log ε 4·43, 4·48, 4·55, 3·89), $\nu_{max.}$ (KBr) 3390, 1660, 1620, 1589 cm^{-1}.

(21) yellow needles, m.p. 291°, $\lambda_{max.}$ (EtOH) 238·5, 243, 279, 372 nm (log ε 3·98, 3·97, 4·23, 3·27), $\nu_{max.}$ (KBr) 3300, 1672, 1649(w) cm^{-1}.

Schunk and Marchlewski[67,352] first realised that rubiadin was very similar to xanthopurpurin and showed, by oxidation to phthalic acid, that the methyl and hydroxyl groups were in the same ring. Rubiadin was therefore regarded as 4- or 2-methylxanthopurpurin and the latter was synthesised by condensing benzoic acid with 3,5-dihydroxy-4-methylbenzoic acid. However, owing to a discrepancy in the melting points of the "natural" and synthetic diacetates (in 1894 the routine determination of mixed melting points had not become normal practice) these two compounds were thought to be isomers, and it was initially concluded that rubiadin was 4-methylxanthopurpurin. That this was not so was demonstrated later by unambiguous syntheses[68,69] of both the 2- and the 4-methyl isomers. The natural pigment has also been obtained via its 3-methyl ether which may be prepared by a two-step condensation of phthalic anhydride and 2-methylresorcinol monomethyl ether,[70] or by methylation of xanthopurpurin 3-methyl ether with formaldehyde and dithionite.[71] Other syntheses of rubiadin have been reported.[74,291]

In *Galium* spp. rubiadin occurs[27,63,230] as the 3-β-primveroside, pale yellow plates, m.p. 248–250°, which, when hydrolysed with dilute sulphuric acid, gives D(+)-xylose and rubiadin 3-β-D-glucoside which was isolated from madder in an early investigation. The 3-primveroside has also been found in madder,[73] and it is possible[63] that Schunk and Marchlewski's glucoside[352] was an artefact of their isolation procedure. The 3-β-D-glucoside, yellow needles, m.p. 270–271°, was synthesised[70] in the usual way from rubiadin and acetobromoglucose, the location of the sugar moiety being established by methylation, followed by hydrolysis, to give rubiadin 1-methyl ether.

The 1-methyl ether can be prepared[70] by treatment of rubiadin with acetic anhydride-boroacetic anhydride to give the acetate-boroacetate (22) which decomposes in water leaving the 3-acetate. Methylation with

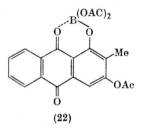

(22)

methyl iodide–silver oxide, and hydrolysis, then gives rubiadin 1-methyl ether.[70]

Lucidin (23), $C_{15}H_{10}O_5$

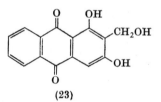

(23)

Occurrence: bark of *Coprosma lucida* Forst.,[11] *C. acerosa* A. Cunn.,[12] root bark of *Coelospermum reticulatum* F. Muell.,[66] root of *Asperula odorata* L.,[13] several *Galium* spp.,[13,230] *Rubia tinctorum* L.[82,83,565] (= *R. iberica*) (free and as its 3-primveroside), *Hymenodictyon excelsum* Wall.[567] and *Morinda umbellata* L.[25] (all Rubiaceae).

Physical properties: yellow needles, m.p. >330°, $\lambda_{\text{max.}}$ (EtOH) 242, 246, 280, 330, 415 nm (logϵ 4·39, 4·40, 4·38, 3·44, 3·26), $\nu_{\text{max.}}$ (KBr) 3410, 1663, 1620, 1592 cm^{-1}.

Lucidin is poorly soluble in organic solvents and hence easily separated from accompanying pigments. It forms a triacetate and tribenzoate, but methyl sulphate affords only mono(β)- and dimethyl ethers. Although a C-methyl estimation (Kuhn-Roth) was negative, a zinc dust distillation afforded 2-methylanthracene suggesting[80] the 2-hydroxymethyl structure (23) in view of the very similar light absorption of lucidin and rubiadin. As expected, oxidation with manganese dioxide[81] leads to the corresponding aldehyde, nordamnacanthal (27), while treatment with alkaline silver oxide[81,96] produces the acid, munjistin (33). The benzylic alcohol function is very reactive and by refluxing in acetic acid or ethanol (no catalyst) the ω-acetate and ω-ethyl ether are easily obtained,[100] respectively. It is partly converted[565] into the ω-ethyl ether merely on boiling in chloroform containing the normal 2% ethanol which probably accounts for the isolation of this compound from *R. tinctorum*[82] and *G. dasypodum*.[230]

Lucidin has been synthesised from xanthopurpurin[81] by hydroxy-methylation with formaldehyde, and from rubiadin diacetate[71] by side chain bromination, followed by treatment with sodium acetate and hydrolysis of the triacetate.

Damnacanthol (lucidin 1-methyl ether) (24), $C_{16}H_{12}O_5$

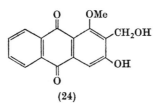

(24)

Occurrence: root of *Damnacanthus major* Sieb. & Zucc.,[84] *Morinda citrifolia* L.[65] and *Coelospermum paniculatum* F. Muell.[66] (all Rubiaceae).

Physical properties: yellow needles, m.p. 288°, $\lambda_{max.}$ (EtOH) 238·5, 243·5(sh), 280, 350–375 nm (log ε 4·35, 4·33, 4·54, 3·61), $\nu_{max.}$ (KBr) 3280, 1675, 1650 cm⁻¹.

Damnacanthol was recognised, by analysis, as the alcohol corresponding to the aldehyde, damnacanthal (28), which coexists in *D. major*. Methylation with diazomethane gave a product which appeared to be lucidin dimethyl ether.[90a] When isolated from *M. citrifolia*[65] and *C. paniculatum*[66] by extraction with acetone it was obtained as the dimethyl ketal (25) which can be prepared by refluxing damnacanthol in acetone in the presence of anhydrous copper(II) sulphate. Mild acid hydrolysis regenerates damnacanthol, more vigorous conditions giving lucidin. Damnacanthol has been synthesised from rubiadin 1-methyl ether 3-acetate[85] via side chain bromination and hydrolysis, and by

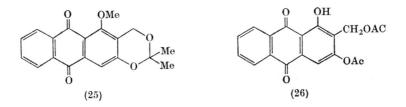

(25)   (26)

selective acetylation of lucidin[86] followed by methylation of the diacetate (26) with diazomethane in tetrachlorethane (47% yield), and final hydrolysis.

Nordamnacanthal (27), $C_{15}H_8O_5$

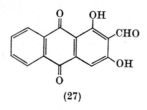

(27)

Occurrence: heartwood of *Morinda tinctoria* Roxb.,[44] roots of *M. citrifolia* L.,[65, 610] and *Hymenodictyon excelsum* Wall.,[567] root of *Damnacanthus major* Sieb. & Zucc.[567] and root bark of *Coelospermum paniculatum* F. Muell.[66] (all Rubiaceae).

Physical properties: yellow needles, m.p. 220–221°, $\lambda_{max.}$ (EtOH) 219, 262, 293, 421 nm (log ϵ 4·56, 4·73, 4·57, 4·09), $\nu_{max.}$ (KBr) 1683, 1651, 1634 cm^{-1}.

The synthetic compound, obtained by demethylation[84] of damnacanthal (28) or by oxidation[81] of lucidin with manganese dioxide, was known before nordamnacanthal was found in Nature. It may be recognised from its spectral properties and colour reactions, the aldehyde function being revealed by the chelated carbonyl absorption at 1651 cm^{-1} and the formation of a phenylhydrazone. It can be reduced to lucidin with sodium borohydride.[612] The pigment is soluble in aqueous sodium bicarbonate,[44] and the enhanced acidity of the β-hydroxyl group must be attributed to the effect of the neighbouring carbonyl function.

Damnacanthal (28), $C_{16}H_{10}O_5$

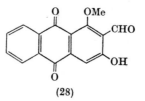

(28)

Occurrence: roots of *Damnacanthus major* Sieb. and Zucc.,[84] *D. major* Sieb. & Zucc. var. *parvifolius* Koidz.,[84] *D. indicus* Gaert. fil. var. *microphyllus* Makino,[84] *Hymenodictyon excelsum* Wall.,[567] heartwood of *Morinda tinctoria* Roxb.[44] and *M. citrifolia* L.[78] (also roots[610]), and possibly in the root bark of *M. umbellata* L.,[84, 97] stems of *Lasianthus chinensis* Benth.[88] (all Rubiaceae).

Physical properties: orange-yellow needles, m.p. 212°, $\lambda_{max.}$ (MeOH) 250, 281, 380 nm (log ϵ 4·41, 4·36, 4·38), $\nu_{max.}$ (KBr) 1677, 1661, 1650 cm^{-1}.

This quinone was first isolated from the Japanese genus *Damnacanthus* and examined in detail by Nonomura.[84, 90, 91] It contains one methoxyl and one hydroxyl group, and their relative positions were established by boiling with 50% aqueous potassium hydroxide which gave nordamnacanthal (27) and some xanthopurpurin. A formyl group (eliminated in the previous reaction) was characterised by the preparation of several derivatives and its position, adjacent to the hydroxyl function, was confirmed when a Perkin reaction yielded the coumarin (29) in addition to the triacetate (30). Oxidation of damnacanthal with alkaline hydrogen peroxide afforded damnacanthic acid which yielded munjistin (33) on demethylation. It follows that the formyl group must be located at C-2, and therefore structure (28) can be assigned[90] to damnacanthal, the methoxyl group being placed in an α-position as the infrared spectrum does not show chelated quinone carbonyl absorption (cf. nordamnacanthal, ν_{CO} 1634 cm^{-1}). Among other chemical reactions damnacanthal undergoes a Schmidt reaction to form the corresponding nitrile which gives munjistin on heating with phosphoric acid, and with diazomethane it appears to form the ketone (31) whereas with methyl iodide–silver oxide[91] the product was considered to be the dimethyl acetal (32). However acetal formation may have occurred during the final crystallisation from methanol. Oxidation[44] of damnacanthal with alkaline hydrogen peroxide under Dakin conditions gives anthragallol 1-methyl ether, while reduction with sodium borohydride affords lucidin 1-methyl ether.[612]

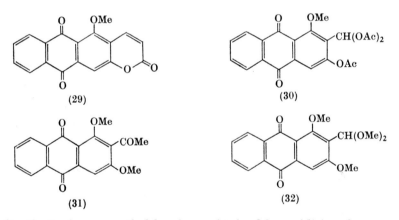

(29)   (30)

(31)   (32)

Synthetic damnacanthal has been obtained by oxidising damnacanthol with manganese dioxide.[85, 86]

Munjistin (33), $C_{15}H_8O_6$

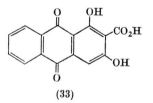

(33)

Occurrence: root of *Rubia cordifolia* L.,[92] *R. cordifolia* var. *munjista* Miq.,[95] *R. tinctorum* L.,[87] *R. sikkimensis* Kurz.,[93] *R. chinensis* Regel & Maack var. *glabrescens* Kitagawa,[557] *Relbunium hypocarpium* Hemsl.[604] *Morinda umbellata* L.[25] (Rubiaceae), and in the heartwood of *Tectona grandis* L.[426] (Verbenaceae).

Physical properties: orange leaflets, m.p. 232–233°, λ_{max}. (EtOH) 248·5, 288, 420 nm (log ε 4·52, 4·31, 3·71), ν_{max}. (KBr) 3250, 1711, 1672, 1640, 1591 cm^{-1}.

The original source of this pigment was munjeet, the root of *R. cordifolia*.[92] On heating above its melting point, munjistin loses carbon dioxide to form xanthopurpurin, and decarboxylation also occurs on bromination which gives 2,4-dibromoxanthopurpurin. Munjistin is thus xanthopurpurin-2- or 4-carboxylic acid and it was shown to be the former (33) by synthesis. The original methods[94] were very unsatisfactory but it may be conveniently prepared from lucidin by oxidation with alkaline silver oxide,[81, 96] or by reduction of pseudopurpurin (p. 408) with alkaline dithionite.[90b]

2-Benzylxanthopurpurin (34), $C_{21}H_{14}O_4$

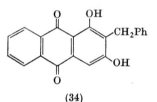

(34)

Occurrence: root of *Hymenodictyon excelsum* Wall.[567] and *Damnacanthus major* Sieb, and Zuce.[567] (Rubiaceae).

Physical properties: yellow crystals, m.p. 298–299°, λ_{max}. (EtOH) 241, 246, 280, 413 nm (log ε 4·40, 4·45, 4·51, 3·88). ν_{max}. (KBr) 3390, 1665, 1630, 1588 cm^{-1}.

A general resemblance to rubiadin in colour reactions and spectroscopic properties suggested an alkylated xanthopurpurin structure for this quinone (molecular weight 330). However, an unusual feature is the presence of a strong peak at m/e 91 (80%) in the mass spectrum which can be attributed to a tropylium cation. On this basis the benzylxanthopurpurin structure (34) seemed likely and this was confirmed by synthesis. Oxidation of 2-benzylquinizarin with lead tetra-acetate gave a diquinone which was subjected to Thiele acetylation. A mixture of

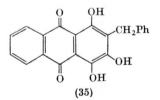

(35)

products resulted from which the quinone (35) could be obtained by hydrolysis, and this was converted to the xanthopurpurin (34) in the usual way with ammonia and sodium dithionite.[567]

6-Methylxanthopurpurin (36), $C_{15}H_{10}O_4$
6-Methylxanthopurpurin 3-methyl ether, $C_{16}H_{12}O_4$

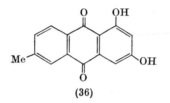

(36)

Occurrence: (36) root bark of *Morinda umbellata* L.,[97] (Rubiaceae); 3-methyl ether, cultures of *Alternaria porri* (Ell.) Saw.[294] (= *A. solani*).

Physical properties: (36) yellow needles, m.p. 269°, $\lambda_{max.}$ (96% EtOH) 242, 265(sh), 281, 320(sh), 411 nm (log ϵ 4·29, 4·34, 4·15, 4·22, 3·38), $\nu_{max.}$ (KBr) 3418, 1675(sh), 1640, 1625, 1595 cm^{-1}; methyl ether, orange needles, m.p. 184–185°, $\lambda_{max.}$ (96% EtOH) 226, 252, 261, 279, 409 nm (log ϵ 4·12, 4·25, 4·23, 4·23, 3·71), $\nu_{max.}$ (CHCl$_3$) 1676, 1635, 1610 cm^{-1}.

The status of 6-methylxanthopurpurin as a natural product is uncertain. It was reported by Perkin and Hummel[97] in *M. umbellata* but has not been found in any subsequent investigation of this or other *Morinda* sp. It formed a diacetate, gave 2(?)-methylanthracene on zinc dust distillation, and was thought to be identical with a compound of structure (36) obtained[352] by condensation of *p*-toluic and 3,5-dihydroxybenzoic acids. However there appears to have been no direct comparison of the synthetic and natural materials.

The 3-methyl ether was isolated recently, with related compounds, from cultures of *Alternaria porri*. The structure was deduced from the n.m.r. and other spectroscopic data, and confirmed by comparison with a synthetic specimen obtained by methylating 6-methylxanthopurpurin with diazomethane.[432]

2-Methylquinizarin (37), $C_{15}H_{10}O_3$

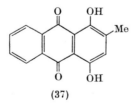

(37)

13

Occurrence: heartwood of *Tectona grandis* L. fil.[98] (Verbenaceae).

Physical properties: orange-red needles, m.p. 175–176°, $\lambda_{max.}$ (EtOH) 250, 286, 322, 482 nm (log ϵ 4·59, 4·06, 3·48, 4·29), $\nu_{max.}$ (CCl$_4$) 1628 cm^{-1}.

This is the only natural quinizarin so far identified with certainty.† It is recognisable[98] from its colour reactions and spectroscopic proper-ties, and may be synthesised by condensation of phthalic anhydride with toluquinol[99] or 2-methyl-4-chlorophenol.[382]

Soranjidiol (38), $C_{15}H_{10}O_4$

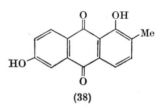

(38)

Occurrence: root bark of *Morinda citrifolia* L.[101] (as glycoside[65]) (and possibly *M. umbellata* L.[97, 102]), root of *Hymenodictyon excelsum* Wall.,[567] *Coprosma acerosa* A. Cunn.,[12] *C. australis* Robinson[64] and probably *C. lucida* Forst.[11] (all Rubiaceae).

Physical properties: orange-yellow needles, m.p. 275°,‡ $\lambda_{max.}$ (EtOH) 220, 245, 271·5, 280(sh), 292(sh), 337·5, 395(sh), 411, 425(sh) nm (log ϵ 4·45, 4·04, 4·46, 4·38, 4·18, 3·17, 3·71, 3·80, 3·71), $\nu_{max.}$ (KBr) 3420, 1680, 1637, 1600 cm^{-1}.

Morinda root, that is, the roots of *Morinda citrifolia* and *M. tinctoria*, was used at one time in India under the name Suranji. This is the origin of the name soranjidiol, the compound being first extracted from *M. citrifolia*. Analytical and recent spectroscopic[65] data provide evidence for one α- and one β-hydroxyl group. Assuming that the methyl group occupies a β-position, (38) is a likely structure (after eliminating known isomers) and the properties of soranjidiol are similar to those reported for the synthetic compound with this constitution. Despite some uncertainty, structure (38) was tacitly accepted for many years before it was confirmed[104] by direct comparison of natural and synthetic specimens.

1,6-Dihydroxy-2-methylanthraquinone was originally synthesised by stepwise condensation of anisole[103] and 3-methoxy-4-methylphthalic anhydride, and was obtained[104] recently by Marschalk methylation of 1-hydroxy-6-methoxyanthraquinone followed by demethylation with aluminium chloride.

† The presence of quinizarin and alizarin, together with chrysophanol, in *Libertia coeru-lescens* Kunth[30] (Iridaceae) requires verification.

‡ Kofler block; the melting points recorded range from 176–295°.

Phomarin (39), $C_{15}H_{10}O_4$
Phomarin 6-methyl ether, $C_{16}H_{12}O_4$

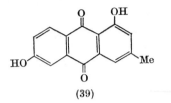

(39)

Occurrence: cultures of *Phoma foveata* Foister[19] and the leaves of *Digitalis purpurea* L.[567] and *D. viridiflora* Lindl.[569]; the 6-methyl ether occurs in the leaves of *D. purpurea*[567] (Scrophulariaceae).

Physical properties: (39) orange needles, m.p. 254–255°, λ_{max}. (EtOH) 220, 245, 271·5, 280(sh), 292(sh), 337·5, 395(sh), 411, 425(sh) nm (log ε 4·45, 4·04, 4·46, 4·38, 4·18, 3·17, 3·71, 3·80, 3·71), ν_{max}. (KBr) 3445, 1665, 1640, 1603 cm^{-1}; 6-methyl ether, yellow needles, m.p. 193°, λ_{max}. (EtOH) 218, 266·5, 291, 332·5, 392(sh), 411, 428(sh) nm (log ε 4·51, 4·55, 4·18, 3·29, 3·79, 3·87, 3·80), ν_{max}. (KBr) 1682(w), 1672, 1644, 1635, 1595 cm^{-1}.

During a study of *Phoma foveata*, a pathogen responsible for gangrene in potatoes, it was observed that ageing cultures frequently deposited crystalline yellow pigments. These were recently identified by Bick and Rhee[19] as a mixture of pachybasin (13), chrysophanol (40) and a new compound, phomarin, since found in *Digitalis* leaves and more fully characterised. From the molecular formula, $C_{15}H_{10}O_4$ (mass spectrum), and the presence of five aromatic protons (n.m.r.), there must be three substituents and these are a methyl and two hydroxyl groups. The quinone is thus isomeric with chrysophanol which was supported by comparison of their fragmentation patterns, but in this case the infrared and ultraviolet evidence clearly shows that one hydroxyl group occupies an α- and the other a β-position. In the n.m.r. spectrum, a singlet (3H) at τ 7·48 corresponds to a β-methyl group, and a pair of doublets at low field can be associated with two *meta* aromatic protons. From this, phomarin appears to be a 1-hydroxy-3-methylanthraquinone having a second hydroxyl group at C-6 or C-7. On biogenetical grounds the former (39) would be preferred and the pigment proved to be identical[567] with a compound of this structure previously made by Mühlemann.[653] This was prepared by the method employed to synthesise emodin (see p. 420) using a diketone analogous to (114) derived from acetone and ethyl *m*-methoxyphenylacetate.

Hallachrome, $C_{16}H_{12}O_4$

Occurrence: in the epidermal epithelial cells of the polychaete worms *Halla parthenopeia* (Della Chiaje)[647] and *Lumbriconereis impatiens* (Claparède).[648]

Physical properties: red prisms, m.p. 224–226° (dec.), λ_{max}. (MeOH) 250, 312, 500 nm (log ε 4·51, 4·52, 3·75), ν_{max}. 3330, 1695, 1652, 1610 cm^{-1}.

The red pigment in the skin of *Halla parthenopeia*, a rather rare marine worm found chiefly in the Bay of Naples, was first examined forty years ago. It was thought then to be an indolequinone derivative [647, cf. 649] and later, a nitrogeneous quinone sulphonic acid,[650] but it now seems likely that the compounds studied were artefacts as recent work [648] has shown that the molecular formula is $C_{16}H_{12}O_4$. Hallachrome shows redox properties, gives a green colour in aqueous sodium hydroxide, and forms a quinoxaline derivative, an acetate, and a leucotriacetate which has an anthracene-type absorption spectrum (λ_{max}. 253, 335, 351, 369, 389 nm). The n.m.r. and mass spectra provide evidence for the presence of one C–Me and one O–Me group, and hence the pigment is a hydroxy-methoxy-methyl-1,2-anthraquinone. The 1,2-quinone structure is confirmed by an AB quartet in the n.m.r. spectrum (DMSO-d_6) centred at τ 2·32 and 3·69 (J, 10 Hz), while in the aromatic region there are three 1H singlets, those at τ 1·58 and 2·20 showing good agreement with the singlets at τ 1·37 and 2·32 (CDCl$_3$) arising from the *meso* protons in the synthetic compound (40).[651] Thus hallachrome appears to be a most unusual anthraquinone which is unsubstituted at positions 9 and 10. Prota and his co-workers [648] favour the arrangement shown in (41);

(40)   (41)

although an M-18 peak in the mass spectrum suggests that the methoxyl and hydroxyl groups are adjacent, the latter being placed next to the ring proton (τ 2·50) which undergoes a significant acylation shift.

Hallachrome is the only anthraquinone known in the phylum Annelida. Arenicochrome, a pigment obtained from the lugworm, *Arenicola marina*, is a benzpyrenequinone but this is an artefact and the structure of the actual pigment in the skin of the animal is not yet known.[652]

Chrysophanol (chrysophanic acid, rumicin, archinin) (42), $C_{15}H_{10}O_4$

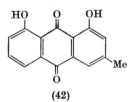

(42)

Occurrence:[299]† mainly in the root of numerous *Rheum*[105-115,174,360] and *Rumex*[116-129,356,358,555] spp. (Polygonaceae)[635] bark, leaves and fruit of *Rhamnus*[124,130-136,151] spp. (Rhamnaceae), wood, leaves and seeds of *Cassia*[137-144,183,548] spp. (Leguminosae). Aerial parts of *Cluytia similis* Muell. Arg.[145] (Euphorbiaceae), root of *Polygonum cuspidatum* Sieb. & Zucc.,[146] *P. multiflorum* All.[154] and *Muehlenbeckia hastulata* (J.S.N.) Standl. ex Macbr.[30] (Polygonaceae), wood of *Sonnerartia acida* L.[149] (Lythraceae) and *Maesopsis eminii* Engl.[522] (Rhamnaceae), heartwood of *Vatairea guianensis* Aubl.,[150] *V. lundellii* (Standley) Kill.,[150] and *Vataireopsis araroba* (Aguiar) Ducke[150] (= *Andira araroba* Aguiar) (Leguminosae), bark of *Harungana madagascariensis* Poir[156] (Guttiferae), roots, stems and leaves of *Libertia coerulescens* Kunth,[30] leaves of *L. ixioides* Spreng.[605] (Iridaceae), leaves and roots of *Bulbine annua* Willd.[152] and *B. asphodeloides* Spreng.[152] (Liliaceae), root of *Eremurus robustus* Regel,[153] *E. bungei* Bak.,[153] and *E. himalaicus* Bak.[153] (Liliaceae), *Asphodelus albus* Willd.[153] and *Asphodeline lutea* Reichb.,[153] leaves of *Aloe saponaria* (Ait.) Haw.[153] and other *Aloe* spp.[161,354] (Liliaceae)‡ and leaves of *Digitalis purpurea* L.§ (Scrophulariaceae). In many of these plants the pigment occurs both free and as a glycoside, and sometimes in a reduced (anthrone) form. Cultures of *Chaetomium elatum* Kunze ex Fr. (= *C. affine* Cda.),[157] *Penicillium islandicum* Sopp.,[536] *Sepedonium ampullosporum* Damon,[158] *Aspergillus crystallinus* Kwon & Fennell,[159] *Phoma foveata* Foister[19] and *Trichoderma viride* Pers. ex Fr.[20]

Physical properties: yellow leaflets, m.p. 196°, λ_{max}. (EtOH) 228, 257, 277, 287, 429 nm (log ϵ 4·30, 4·01, 4·03, 4·03), ν_{max}. (KBr) 1680, 1632, 1611 cm^{-1}.

As the list above shows, chrysophanol occurs fairly widely in both higher and lower plants, and it is of some interest that a chrysophanol dimer has been obtained from soil samples by solvent extraction (p. 392). It is normally associated with emodin and physcion, and related quinones, and the apparent exceptions (for example, *Cluytia similis*[145]) would probably prove false on re-examination. Early interest in this compound was stimulated by its presence in rhubarb, and Goa or Araroba powder, which were formerly important as purgative drugs. Medicinal rhubarb consists of the dried rhizome of *Rheum officinale* and is imported from China, while Araroba or Goa powder (exported from Goa) was obtained from a resin occurring in cavities in the wood of *Vataireopsis araroba* (= *Andira araroba*). Solvent extraction gave a product, chrysarobin, which contained a complex mixture of anthracene derivatives.[178,342,353]

As chrysophanol gave a diacetate, and yielded 2-methylanthracene when distilled over zinc, it appeared to be a dihydroxymethylanthraquinone,[35,164,165] the ultraviolet-visible absorption spectrum suggesting a chrysazin (1,8-dihydroxyanthraquinone) structure—an early example of the use of spectra in structure elucidation (Liebermann and Kostanecki, 1886).[166] This was confirmed when chrysophanol was converted

† Chrysophanol is reported in *Raputia magnifica*[155] (Rutaceae) but the original reference has not been traced.
‡ Chrysophanol has also been isolated (as an artefact) from bushfire-damaged *Xanthorrhea* (Liliaceae) resins.[162]
§ Incorrect, now known to be isochrysophanol, 1,8-dihydroxy-2-methylanthraquinone.[567]

into chrysazin, the side chain being removed by oxidation[165] (of the diacetate) and decarboxylation.[167] The location of the methyl group at C-3 was indicated *inter alia* by the formation of 5-hydroxyisophthalic acid on fusion[168] with potash, and later confirmed by synthesis. This has been achieved by stepwise condensation of *m*-cresol with 3-nitro-,[169] 3-acetylamino-,[170] or more directly, 3-methoxyphthalic anhydride,[171] and by a more lengthy, but unambiguous, method[172] in which the

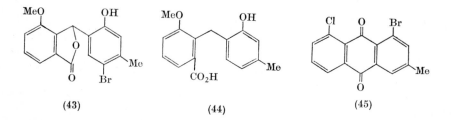

(43)   (44)   (45)

lactone (43), derived by interaction of 2-formyl-3-methoxybenzoic acid and 4-bromo-3-methylphenol, was reduced to the acid (44), followed by cyclisation, oxidation and demethylation. In another somewhat lengthy approach, the halogenated quinone (45) was prepared and converted into chrysophanol by heating under pressure with calcium hydroxide in the presence of copper bronze.[173]

In Chinese rhubarb,[174] *Rheum palmatum* var. *tangut.*,[175] *Rhamnus purshiana*[176] and *Rumex rechingerianus*,[555] the quinone occurs as a β-D-glucoside (chrysophanein), yellow needles, m.p. 245–246°,† [α]$_D^{20}$ −80·7° (acetone). It has been prepared[177] from chrysophanol by the standard method, and the location of the sugar moiety at position 8 was established[181] by methylation and hydrolysis which gave chrysophanol 1-methyl ether of known constitution.[182] This shows that the natural compound is also the 8-β-D-glucoside (however, see ref. 180).

In several Leguminosae (*Cassia siamea*,[140] *Vatairea guianensis*[150] and in chrysarobin[178]), in *Rumex crispus*,[128] and in *Rhamnus purshiana*,[135] chrysophanol occurs partly as its 9-anthrone (46). It has also been

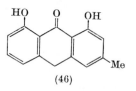

(46)

isolated from the heartwood of *Ferreira spectabilis* Allem.[179] The 9-anthrone is easily obtained by reduction with stannous chloride. Two

† M.p.s up to 256° are recorded.

glycosides[217] in the bark of *Rhamnus purshiana* may be chrysophanol-anthrone derivatives analogous to barbaloin (p. 401). (See also the rheidins (p. 404) and the palmidins (p. 421).) Chrysophanol-10,10′-bianthrone has been found (as a glycoside) in the bark of *Rhamnus purshiana.*[607]

Cassiamin C (47), $C_{30}H_{18}O_8$

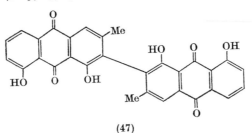

(47)

Occurrence: bark of *Cassia siamea* Lam.[564] (Leguminosae).
Physical properties: yellow crystals, m.p. >350°, $\lambda_{max.}$ (dioxan) 229, 262, 288, 430 nm (log ϵ 4·85, 4·78, 4·36, 4·43), $\nu_{max.}$ (Nujol) 1675, 1620 cm^{-1}.

This bichrysophanol is one of several bianthraquinones recently found in the bark of *C. siamea, Cassia* spp. being a well-known source of anthraquinones of the chrysophanol-emodin type. It yields a tetra-acetate and a tetramethyl ether, has a 1,8-dihydroxyanthraquinone chromophore, and is dimeric. The n.m.r. spectrum of the tetramethyl ether indicates a symmetrical structure and shows singlets at τ 7·85 (β-Me), 6·25 and 5·97 (OMe), an ABC pattern for three aromatic protons, and a singlet at τ 1·97 which can be attributed to the α-protons adjacent to the methyl groups by comparison with related spectra. On this basis, structure (47) is suggested[564] for cassiamin C; it may well be optically active as is cassiamin A (p. 393).

(+)-*Dianhydrorugulosin* (48), $C_{30}H_{18}O_8$

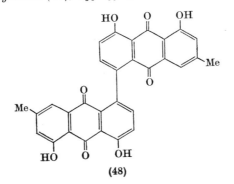

(48)

Occurrence: cultures of *Penicillium islandicum* Sopp.[568]
Physical properties: orange-red prisms, m.p. 321°, $\lambda_{max.}$ (dioxan) 282(sh), 439 nm (log ε 4·28, 4·33), $\nu_{max.}$ (KBr) 1680, 1631 cm^{-1}, optically active.

Shibata and his co-workers[508] have briefly reported the isolation of a further series of bianthraquinones from cultures of *P. islandicum*, all of which are new except dianhydrorugulosin. The structure of this pigment was already established by degradation and synthesis. On reductive cleavage it affords chrysophanol while oxidation of the tetra-acetate with chromic acid gave a di-acid converted, by decarboxylation, into the dimer (49).[519] This has been synthesised[521] by way of an Ullmann

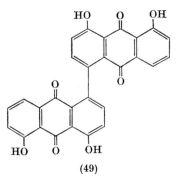

(49)

reaction on 5-iodo-1,8-dimethoxyanthraquinone. Dianhydrorugulosin (48) from *P. islandicum* is optically active and shows a positive Cotton effect; the (−) form is obtained on dehydration of (+)-rugulosin (p. 428).

Chrysotalunin (50), $C_{30}H_{18}O_8$

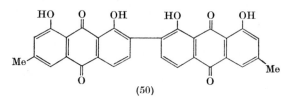

(50)

Occurrence: in soil.[163, 620]
Physical properties: orange crystals, m.p. >360°, $\lambda_{max.}$ (CHCl$_3$) 264, 289(sh), 440–460 nm (log ε 4·70, 4·39, 4·47), $\nu_{max.}$ (Nujol) 1672, 1624 cm^{-1}.

Although polycyclic aromatic hydrocarbons have been found in soil this is the first quinone from this source to be positively identified. It was found by McGrath[163, 620] as one of a mixture of pigments occurring in soils in the south-east of Ireland. About 15 mg/kg soil could be isolated by chloroform extraction.

Chrysotalunin is clearly a 1,8-dihydroxyanthraquinone from the above spectroscopic data. From the molecular formula and the simplicity of the n.m.r. spectrum it appeared to be a symmetrical dimer, and on reductive cleavage with dithionite it gave chrysophanol (42). However it differs from (47) and (48), and comparison of the n.m.r. spectrum of the tetra-acetate with that of chrysophanol diacetate showed that the structure must be (50).[621] The signal from H-7 in the monomer is missing while H-5 and H-6 form an AB system (J, 8 Hz).

Cassiamin A† (51), $C_{30}H_{18}O_9$

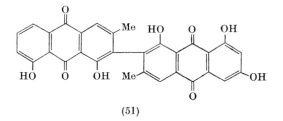

(51)

Occurrence: root bark of *Cassia siamea* Lam.[516, 564] (Leguminosae).
Physical properties: orange-yellow prisms, m.p. 356–357° (dec.), $\lambda_{max.}$ (EtOH) 228, 259, 288, 445 nm (log ϵ 4·92, 4·80, 4·69, 4·61), $\nu_{max.}$ (Nujol) 3320, 1672, 1623 cm^{-1}; penta-acetate, $[\alpha]_{589}^{30}$ −169·5° (CHCl₃).

In its light absorption and colour reactions cassiamin A shows a general resemblance to quinones of the chrysophanol-emodin group.[516] The molecular formula (mass spectrum) confirmed the dimeric nature of the compound and the formation of a pentamethyl ether and a penta-acetate identified five of the oxygen atoms. Four of the five hydroxyl groups occupy *peri*-positions, as revealed by four one-proton signals in the n.m.r. spectrum in the region τ −2·29 to −1·83, an unchelated (β) hydroxyl proton signal appearing upfield at τ 0·75. This accounts for the formation of a monomethyl ether on treatment with diazomethane. Analysis of the visible and infrared spectra leads to the conclusion that each half of the molecule contains a pair of 1,8-hydroxyl groups, and the n.m.r. spectra of cassiamin A and its derivatives establish the presence of two C-methyl groups and seven aromatic protons. Reductive fission with alkaline dithionite afforded only chrysophanol, which accounts for half of the molecule, and the other portion, which contains a β- and two

† Besides cassiamins A, B, and C, three other bianthraquinones, cassianin, siameanin and siameadin, have been isolated[515, 535] from *C. siamea*. It is not yet clear whether any of these pigments are identical with members of the cassiamin group.

α-hydroxyl groups, and a methyl group, has an emodin substitution pattern. The n.m.r. spectrum of the pentamethyl ether confirmed that one ring has a resorcinol structure. Comparison of the n.m.r. spectrum of chrysophanol dimethyl ether with that of cassiamin A pentamethyl ether shows that the proton at C-2 in the former is absent in the latter. It follows that the chrysophanol moiety is coupled at C-2 to the other half of the molecule and, as there is only one signal from the C-methyl protons, the two β-methyl groups clearly have the same environment. Accordingly, cassiamin A must be assigned structure (51),[516] its optical activity being due to restricted rotation.

THE JULICHROMES

Julimycin B-II (52), $C_{38}H_{34}O_{14}$
Julichrome $Q_{1.3}$[†] (53), $C_{38}H_{36}O_{15}$
Julichrome $Q_{2.3}$ (54), $C_{38}H_{34}O_{14}$
Julichrome $Q_{1.4}$ (55), $C_{38}H_{34}O_{15}$
Julichrome $Q_{3.4}$ (56), $C_{38}H_{36}O_{16}$

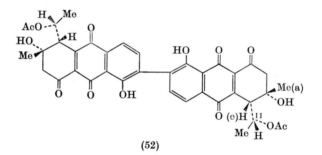

(52)

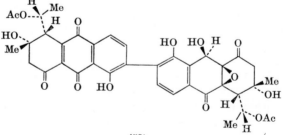

(53)

† Previously julimycin SV.[636]

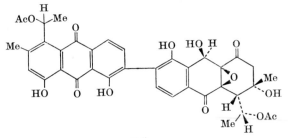

(54)

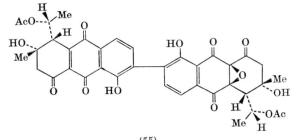

(55)

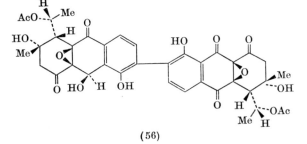

(56)

Occurrence: cultures of *Streptomyces shiodaensis* nov. sp.[636,640]

Physical properties: (52) reddish-orange plates, dec. 215–220°, optically active, $\lambda_{max.}$ (MeOH) 234, 456 nm (log ϵ 4·70, 4·07), $\nu_{max.}$ (Nujol) 3445, 1740, 1704, 1665, 1634 cm^{-1}.

(53) red prisms, m.p. 190–210°, optically active, $\lambda_{max.}$ (MeOH) 221, 260(sh), 273(sh), 450 nm (log ϵ 4·57, 4·33, 4·32, 3·81), $\nu_{max.}$ (Nujol) 3615, 3470, 1700–1725, 1675, 1633 cm^{-1}.

(54) yellow powder, m.p. 165–190° (dec.), optically active, $\lambda_{max.}$ (dioxan) 230, 268, 442 nm (log ϵ 4·54, 4·54, 4·15), $\nu_{max.}$ (Nujol) 3480, 1741, 1714, 1670, 1628 cm^{-1}.

(55) red prisms, m.p. 200–220°, optically active, $\lambda_{max.}$ (MeOH) 230, 247(sh), 390, 433(sh), nm (log ϵ 4·53, 4·51, 3·96, 3·89), $\nu_{max.}$ (Nujol) 3600, 3545, 3412, 3293, 1735–1720, 1711, 1669, 1645, 1633 cm^{-1}.

(56) yellow powder, m.p. >280°, optically active, $\lambda_{max.}$ (MeOH) 267, 370 nm (log ϵ 4·25, 3·94), $\nu_{max.}$ (CHCl$_3$) 3584, 3424, 1736, 1698, 1650 cm^{-1}.

Strep. shiodaensis elaborates a complex mixture of about twenty pigments of unusual structure, and so far Tsuji and his colleagues[637, 638, 640-642] have identified five of these. Julimycin B-II has antiviral and antibiotic properties.

Julimycin B-II. The n.m.r. spectrum of this compound, together with the molecular weight determination, indicates that it is a symmetrical dimer $(C_{19}H_{17}O_7)_2$, and as a biquinone it absorbs two mols. of hydrogen to form a yellow leuco compound, easily reoxidised in air. Chelated quinone carbonyl absorption at 1634 cm^{-1} disappears on methylation as does the low field singlet at τ −2·57. Other spectroscopic information (infrared and n.m.r.) provides evidence for the presence in each half of the molecule of a secondary and a tertiary methyl group, a tertiary hydroxyl and an acetoxyl group (ν_{CO} 1740 cm^{-1}; τ 8·23). The two phenolic groups can be preferentially acetylated in pyridine, whereas the tertiary hydroxyls are selectively acetylated in the presence of p-toluenesulphonic acid. Both diacetates give a tetra-acetate on further treatment. In each half of the molecule six atoms of oxygen are now accounted for and the seventh must be part of the carbonyl group absorbing at 1704 cm^{-1}.

On heating in pyridine julimycin B-II loses first one and then two molecules of water to form an optically active bisanhydro derivative which no longer absorbs at 3445 and 1704 cm^{-1}. The ultraviolet and the infrared spectra show that the bisanhydro compound is a 1,8-dihydroxyanthraquinone derivative, and the n.m.r. spectrum confirms that each half molecule contains two *peri*-hydroxyl groups and an isolated aromatic proton not present in the parent compound. The same spectrum also shows that the other substituents attached to the new benzenoid ring are methyl and –CH(OAc)Me. The orientation of these groups was partly established by degradation. The bis-p-nitrobenzoate of julimycin B-II was dehydrated in pyridine and, after conversion to the dimethyl ether, it was oxidised with permanganate to give the acid (57) which was isolated as the tetramethyl ester. By a similar oxidation of julimycin

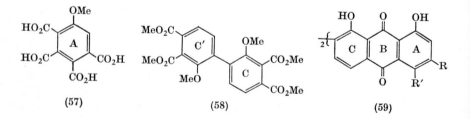

(57)   (58)   (59)

B-II dimethyl ether it was possible to isolate the biphenyl derivative (58) (after methylation) which establishes the position of the internuclear linkage. The bisanhydro compound is therefore (59), R and R′ being Me and CH(OAc)Me (see below).

Returning now to the parent compound, the structure of the reduced ring, which becomes ring A of (59) was shown to be (60) by analysis of the n.m.r. spectrum of julimycin B-II tetra-acetate. The carbonyl group is obviously located at C-1 as this provides the new *peri*-hydroxyl group

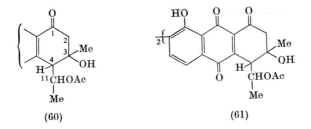

(60)   (61)

in (59), and since the methylene group exhibits a simple AB quartet there are no protons on the adjacent carbon atoms. Thus (60) is the only possible arrangement and hence the gross structure of julimycin B-II is (61).[637]

The stereochemistry shown in (52) was deduced from n.m.r. studies[638] and X-ray analysis.[639] Examination of the n.m.r. spectra of the julimycin B-II acetates showed that the signals from the methylene protons at C-2, the C-3 methyl group and the C-4 proton are broad, indicative of long range coupling, and this was confirmed by spin decoupling experiments.

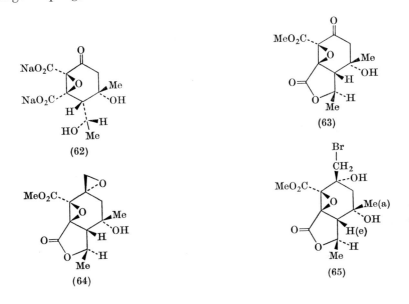

From this it was argued[638] that the C-4 proton is equatorial and the C-3 methyl group is axial (W-letter rule[645]). The conformation shown at C-11 (see 52) seemed most probable from spectroscopic evidence[638] and was confirmed by X-ray analysis of the degradation product (65).[639] This was obtained[638] by oxidation of julimycin B-II with alkaline hydrogen peroxide to give the product (62), which after acidification and methylation with methyl iodide afforded (63). Treatment with diazomethane then gave (64) which was converted to the bromohydrin (65) with hydrogen bromide. Julimycin is therefore correctly represented by (52).

The other members of this group, known as julichromes, are dimers derived from the units shown below. Thus, using this nomenclature, julimycin B-II is Q_1–Q_1 ($Q_{1.1}$) and its bisanhydro derivative is Q_2–Q_2 ($Q_{2.2}$).

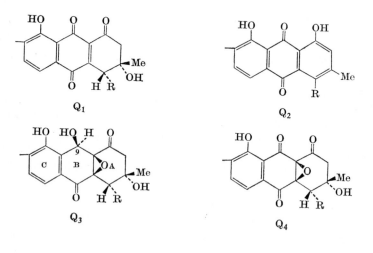

$$R = Me \overset{\cdots OAc}{\underset{H}{\bigwedge}}$$

The n.m.r. spectrum of julichrome $Q_{1.3}$ shows that this molecule has an asymmetrical structure and, by comparison with the spectrum of julimycin B-II, it includes the Q_1 unit. On heating in pyridine it forms a monoanhydro derivative, identical with julichrome $Q_{2.3}$, and by comparison of its n.m.r. spectrum with that of (59), it clearly contains the Q_2 unit. As the nomenclature implies the "other half" of these two pigments is a Q_3 unit, and its structure was deduced as follows. The n.m.r. spectra suggest that ring A has the same substitution pattern as ring A in Q_1, and as there is one phenolic group and two *ortho* aromatic protons as usual, ring C is also the same as in Q_1. The phenolic hydroxyl in ring C is not chelated however, as the carbonyl group at C_9 is replaced by >CHOH (n.m.r.). This leaves one atom of oxygen to be accounted for and this can only be located at the A/B ring juncture as an epoxide since both these carbon atoms are quaternary (n.m.r.) (cf. frenolicin, p. 307). Oxidation of $Q_{1.3}$ with chromic acid gave a product identical with julichrome $Q_{1.4}$ in which the >CHOH group had disappeared. Thus the structures of julichromes $Q_{1.3}$, $Q_{2.3}$ and $Q_{1.4}$ are Q_1-Q_3, Q_2-Q_3 and Q_1-Q_4, respectively.

Confirmation of the epoxide structure was obtained by reaction of $Q_{1.4}$ with potassium iodide in acetic acid which afforded Q_1-Q_1 (julimycin B-II), and the process could be reversed by epoxidation of julimycin B-II with hydrogen peroxide which gave $Q_{1.4}$ (and the di-epoxide). Dehydration of $Q_{1.4}$ with pyridine gave Q_2-Q_4, and when this was degraded with alkaline hydrogen peroxide it gave a product which could be isolated as the lactone (63) previously obtained from julimycin B-II. As (63) must have come from the Q_4 unit and not from Q_2, the configuration of the natural epoxide must be β.

This establishes[641] the stereochemistry of (53), (54) and (55) apart from that at C-9 in the Q_3 units, and the benzylic hydroxyl group was assigned a β configuration after a detailed study[642] of the spectroscopic properties of $Q_{1.3}$ and its epimer. The latter was obtained by solvolysis of the 9-O-acetate which is the main product when $Q_{1.3}$ is acetylated using p-toluenesulphonic acid as catalyst.

The structure of the remaining member of this group, julichrome $Q_{3.4}$ (56) is not in doubt since on treatment with potassium iodide in acetic acid it gave $Q_{1.3}$ (53) (reversed by epoxidation of $Q_{1.3}$), and oxidation with chromic acid converted it into the di-epoxide of julimycin B-II (52).[641]

Aloe-emodin (rhabarberone, iso-emodin) (66), $C_{15}H_{10}O_5$

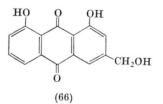

(66)

Occurrence:[299] leaves of *Aloe* spp.[152, 184−190, 354]† (see aloin below) *Asphodelus albus* Willd.[153] and possibly *A. lutea* Reichb.,[153] leaves and roots of *Bulbine annua* Willd.[153] and *B. asphodeloides* Spreng.[153] (Liliaceae), roots of *Rheum* spp.[187, 191, 192, 197] (Polygonaceae), bark of *Rhamnus* spp.[133, 193, 194] (Rhamnaceae), leaves and seeds of *Cassia* spp.[195, 196, 198, 199, 632] (Leguminosae). In many of these plants the quinone occurs both free and as a glycoside, or in a reduced (anthrone) form.
† Cf. ref. 608.

Physical properties: orange-yellow needles, m.p. 223–224°, $\lambda_{max.}$ (EtOH) 225, 254, 276·5, 287, 430, 457(sh) nm (log ϵ 4·59, 4·34, 4·01, 4·03, 4·04, 3·89), $\nu_{max.}$ (KBr) 3400, 1673, 1629 cm^{-1}.

As the name indicates, aloe-emodin is present in aloes where it occurs predominantly as aloin. It is highly probable that the compounds rhabarberone[191] and iso-emodin,[200] isolated from rhubarb, were impure samples of aloe-emodin.[192, 197] The structure was established by Oesterle[201] who showed that aloe-emodin could be oxidised by chromic acid to rhein (69), and reduced with hydriodic acid and red phosphorus to chrysophanolanthrone whence aerial oxidation in alkaline solution gave chrysophanol. As three hydroxyl groups were found to be present (tribenzoate, trimethyl ether), one must be attached to a side chain and the structure (66) was confirmed when the pigment was first synthesised[202] by stepwise reduction of the carboxyl group of rhein diacetate. More recently this has been effected in one step using diborane.[203] An alternative synthesis[204] starts from chrysophanol and proceeds by way of side-chain bromination with N-bromosuccinimide, followed by reaction with silver acetate in acetic anhydride and final hydrolysis. No explanation has been advanced to account for the unusual solubility of aloe-emodin (and the 1- or 8-methyl ether obtained by reaction with diazomethane) in aqueous sodium carbonate.[191a, 197]

The only glycoside of aloe-emodin which has been identified is the 8-mono-β-D-glucoside, yellow needles, m.p. 237–238°, isolated from *Rheum palmatum* var. *tangut.*,[175] *Rhamnus purshiana*[176] and *Cassia* spp.[205, 603] It has been synthesised[206] from aloe-emodin and aceto-bromoglucose, the location of the sugar residue being established by methylation with methyl sulphate to give the 1-methyl ether and the 1,ω-dimethyl ether, both of which could be oxidised with chromic acid to 1-methoxy-8-hydroxyanthraquinone-3-carboxylic acid. If the gluco-side is hydrolysed with acetic acid the aloe-emodin is partly converted into the ω-acetate.[630] On the other hand, several glycosides of aloe-emodin-9-anthrone are known, especially in *Aloe* spp.[153, 161, 214, 634] [The free anthrone has been reported[135] in the cathartic drug cascara sagrada (dried bark of *Rhamnus purshiana*).] Two commercial varieties of aloes, an exudation from the leaves of certain *Aloe* spp., are officially recognised for medical use,[207] Curaçao (Barbados) aloes obtained from *A. vera* cultivated in the West Indies, and Cape aloes derived from *A. ferox* and *A. perryi* grown in South and East Africa. Partial purifica-tion via the calcium salts yields a yellow powder known commercially as aloin, which has been used as a purgative for centuries, from which the major constituent barbaloin† (67), lemon-yellow needles, m.p. 148–148·5°, $[\alpha]_D^{20}$ +21° (water), may be obtained by crystallisation.

† Some authors use the names aloin (originally aloine[208]) and barbaloin synonymously.

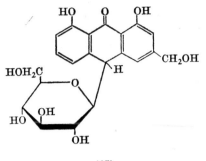

(67)

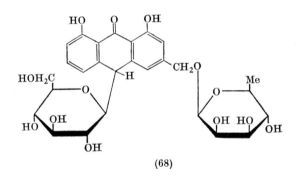

(68)

Considerable amounts are present, <25% in Curaçao aloes. Barbaloin was the first C-glycoside to be isolated[208] and over a hundred years elapsed before an acceptable structure (for a review see ref. 210) was put forward following Mühlemann's synthesis.[209] This was achieved by condensing acetobromoglucose with aloe-emodin-9-anthrone in acetone and aqueous sodium hydroxide, followed by deacetylation, reaction taking place at C-10. Oxidation with ferric chloride[211] gives aloe-emodin in good yield and, curiously, D-arabinose,[212] and both products are formed slowly on keeping barbaloin in contact with acid and air over several months,[213] although barbaloin is stable to acid under normal hydrolytic conditions. A similar oxidation may account for the small amounts of free quinone found in aloes. Isobarbaloin, $[\alpha]_D^{19}$ −19° (ethyl acetate), also present in Curaçao aloes, appears to be a stereoisomer of barbaloin.[210]

Several O-glycosides of barbaloin have been recognised. Aloinoside B, greenish needles, m.p. 233° (dec.), $[\alpha]_D^{20}$ −45·3° (aqueous dioxan), present in Socotra aloes (probably from *A. perryi*)[215] and Cape aloes,[216] and also isolated from *A. africana*[161] and *A. ferox*,[161] is the α-L-rhamnoside

(68);[215] aloinoside A, also in Socotra and Cape aloes, appears to be a stereoisomer of B.[215] According to Fairbairn and Simic,[217] the cascarosides A and B, occurring in the bark of *Rhamnus purshiana*, are O-glycosides of (+)- and (−)-barbaloin, and they are accompanied by two other compounds which may be O-glycosides of chrysaloin, the chrysophanol analogue of barbaloin.

Aloe-emodin-10,10′-bianthrone occurs in senna leaves,[244] in the ripe fruit of *Rhamnus frangula*,[240] and in the bark of *Rhamnus purshiana*,[607] and a glycoside of palmidin B, aloe-emodinchrysophanol-bianthrone (see p. 421) has been found in fresh rhubarb root.[323]

Rhein (69), $C_{15}H_8O_6$

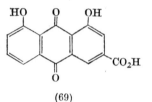

(69)

Occurrence:[299] roots of *Rheum*[174, 191a, 218, 219, 225] and *Rumex* spp.[127, 219] (Polygonaceae), the aerial parts of *Cassia* spp.[143, 196, 221, 222, 224, 226, 632] (Leguminosae), seeds of *Kniphofia aloides* Moen.[223] (Liliaceae), leaves and stems of *Haplopappus baylahuen* Remy[30, 605] (Compositae) and roots of *Muehlenbeckia hastulata* (J.S.N.) Stand. ex Macbr.[30, 605] (Polygonaceae). In many of these plants rhein occurs both free and as a glycoside.

Physical properties: orange needles, m.p. 321°, $\lambda_{max.}$ (MeOH) 228, 258, 431 nm (log ε 4·58, 4·37, 4·06), $\nu_{max.}$ (KBr) 3400, 1700, 1670, 1628 cm⁻¹.

The formation of a series of derivatives, for example, an ethyl ester-diacetate, demonstrated[227] the presence of a carboxyl and two phenolic groups, and since rhein can be obtained from chrysophanol by chromic acid oxidation of the diacetate, their relationship, as in (69), is established.[16] In a more recent synthesis,[173] the halogenated quinone (45) was oxidised to the corresponding 3-carboxylic acid, and then heated under pressure with calcium hydroxide in the presence of copper powder. Suggestions[228] that rhein has a dimeric structure have been refuted.[229]

The 8-mono-β-D-glucoside, m.p. 266–270°, has been isolated from *Rheum palmatum*[175] and *Cassia acutifolia*.[231, 232] The latter also contains another monoglucoside,[231] the 1,8-diglucoside and rhein-9-anthrone-8-glucoside.[232] Undoubtedly other simple glycosides await identification but most interest attaches to the more complex sennosides which are the most active constituents of senna,[207] the dried fruit and leaves of *Cassia senna*, *C. angustifolia* and *C. acutifolia*. Sennoside A is also present in *Rheum coreanum*,[237] and in *Terminalia chebula* (Combretaceae).[606] As in the other natural purgatives based on anthraquinone derivatives (cascara, rhubarb), cathartic action increases in the order anthraquinone < anthrone < bianthrone, glycosides are more effective than aglycones, and the mixture of glycosides found in *Cassia* leaves and pods is the most effective of all.[233, 234]

There are at least four dimeric glycosides, known as sennosides, in *Cassia* spp. Sennosides A and B are 8,8′-diglucosides of birhein-9-anthrone (70)[235] and sennosides C and D appear to be the corresponding diglucosides of aloe-emodin–rhein-bianthrone (71).[234, 236, 265] Sennosides A and C have the same stereochemistry and yield optically active aglycones (sennidines) on hydrolysis, which have restricted rotation about the 10,10′ bond; the bianthrone, sennidine B, is the optically inactive *meso* form. Hence in sennoside B (and D?) the optical activity is due solely to the sugar moieties. The three sennidines are also present in senna (*C. angustifolia*) leaves, and an amorphous glycoside, sennoside III, closely related to sennosides A and B, has been obtained[238] from both *C. acutifolia* and *C. angustifolia*. Oxidation of these dimeric com-

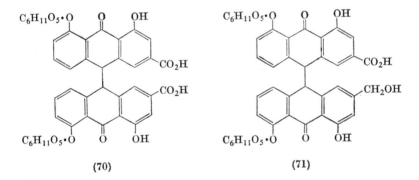

(70)   (71)

Sennoside A (70), $C_{42}H_{38}O_{20}$, m.p. 200–240° (dec.), $[\alpha]_D^{20}$ −147° (70% aqueous acetone)
Sennoside B (70), $C_{42}H_{38}O_{20}$, m.p. 180–186° (dec.), $[\alpha]_D^{20}$ −100° (70% aqueous acetone)
Sennoside C (71), $C_{42}H_{40}O_{19}$, m.p. 197–205° (dec.), $[\alpha]_D^{20}$ −20·6° (water)
Sennoside D (71), $C_{42}H_{40}O_{19}$, m.p. 210–220° (dec.), $[\alpha]_D^{20}$ +3·1° (water)

pounds with ferric chloride yields the corresponding anthraquinone(s), and reductive fission to form the anthrones can be effected with stannous chloride, with dithionite in aqueous bicarbonate solution, or by catalytic hydrogenolysis over palladium. This latter reaction can be reversed by catalytic autoxidation of the anthrone in alkaline solution, in fact merely passing air through an alkaline solution of the monomer (72) gives mixture of sennosides A and B, a very facile example of phenolic coupling.[235]

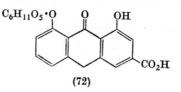

(72)

A group of closely related bianthrones has been found in rhubarb root (*Rheum palmatum*)[239, 243] and in frangula bark (*Rhamnus frangula*).[240–242] These include sennidine C, the rheidins[243]† and the palmidins (see p. 421). Rheidin A and aloe-emodin-bianthrone have also been detected in senna leaves.[244] Rheidin A is the rhein-emodin-bianthrone (73), rheidin B is the rhein-chrysophanol-bianthrone (74) and rheidin C is the rhein-physcion-bianthrone (75). Rheidin A has been obtained in

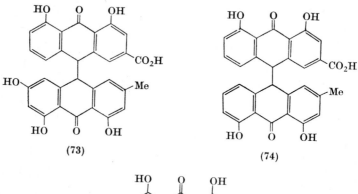

(73)   (74)

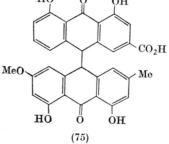

(75)

small yield by oxidising an equimolecular mixture of emodin- and rhein-9-anthrones with ferric chloride; the two symmetrical bianthrones are, of course, formed as well.[243]

† Alternative spelling, reidin.

Anthragallol (76), $C_{14}H_8O_5$

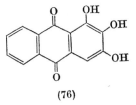

(76)

Occurrence: bark of *Coprosma lucida* Forst.,[11] roots of *Rubia tinctorum* L.[565] and heartwood of *Hymenodictyon excelsum* Wall.[567] (Rubiaceae).

Physical properties: orange needles, m.p. 312–313°, $\lambda_{max.}$ (EtOH) 245, 283, 335(sh), 413 nm (log ϵ 4·21, 4·40, 3·21, 3·72), $\nu_{max.}$ (KBr) 3450, 3360, 1655, 1625 cm^{-1}.

Anthragallol 2-methyl ether, $C_{15}H_{10}O_5$

Occurrence: bark of *Coprosma lucida* Forst.,[11] *C. acerosa* A. Cunn.[12] and *C. linariifolia* Hook[247] (as a primveroside).

Physical properties: yellow needles, m.p. 218°, $\lambda_{max.}$ (EtOH) 240, 281, 407 nm (log ϵ 4·51, 4·24, 3·57), $\nu_{max.}$ (Nujol) 3378, 1664, 1634 cm^{-1}.

Anthragallol 3-methyl ether, $C_{15}H_{10}O_5$

Occurrence: roots of *Rubia tinctorum* L.[565]

Physical properties: orange-red needles, m.p. 242–243°.

Anthragallol 1,2-dimethyl ether, $C_{16}H_{12}O_5$

Occurrence: roots of *Oldenlandia umbellata* L.,[49] bark of *Coprosma lucida* Forst.,[11] *C. acerosa* A. Cunn.,[12] root bark of *C. rhamnoides* A. Cunn.,[75] and roots of *Morinda citrifolia* L.[611] (= *M. geminata*) (all Rubiaceae).

Physical properties: yellow plates, m.p. 238°, $\lambda_{max.}$ (EtOH) 241, 281, 362 nm (log ϵ 3·92, 4·30, 3·36).

Anthragallol 1,3-dimethyl ether, $C_{16}H_{12}O_5$

Occurrence: root of *Oldenlandia umbellata* L.[49] and bark of *Coprosma linariifolia* Hook.,[247] whole plants of *Oldenlandia umbellata* L.[622] (free and as a glycoside).

Physical properties: yellow needles, m.p. 212–213°.

Anthragallol 2,3-dimethyl ether, $C_{16}H_{12}O_5$

Occurrence: heartwood of *Morinda citrifolia* L.[78] and root of *Rubia tinctorum* L.[565]

Physical properties: yellow needles, m.p. 166–168°, $\lambda_{max.}$ (EtOH) 244·5, 277·5, 335·5, 402 nm (log ϵ 4·32, 4·51, 3·42, 3·80), $\nu_{max.}$ (KBr) 1672, 1641 cm^{-1}.

Anthragallol trimethyl ether, $C_{17}H_{14}O_5$

Occurrence: whole plants of *Oldenlandia umbellata* L.[622]

Physical properties: yellow needles, m.p. 164–166°, $\lambda_{max.}$ (EtOH) 240, 276, 356 nm (log ϵ 4·6, 4·4, 3·5), $\nu_{max.}$ (Nujol) 1664 cm^{-1}.

By comparison with the other di- and trihydroxyanthraquinones found in the Rubiaceae, anthragallol is a comparatively recent addition. Two of the ethers were isolated by Perkin, and the parent compound may have escaped earlier detection on account of its ease of oxidation, especially under basic conditions. The characteristic green colour in aqueous sodium hydroxide quickly turns brown on exposure to air, autoxidation resulting in cleavage of the pyrogallol ring and formation of the naphthaquinone (77).[245] As more drastic oxidation with nitric

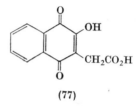

(77)

acid affords phthalic acid the three hydroxyl groups must be located in the same ring and their orientation (76), evident from the infrared spectrum, was established by synthesis. Anthragallol can be prepared[246] by heating a mixture of gallic and benzoic acids in sulphuric acid, or by a similar condensation of phthalic anhydride with pyrogallol.

The various methyl ethers found in Nature were synthesised by Perkin and co-workers taking advantage of rearrangements effected by diazo-

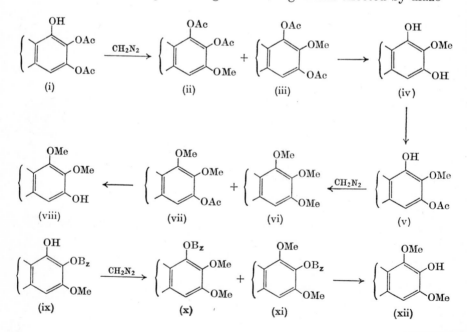

methane. Partial acetylation with cold acetic anhydride containing
potassium acetate afforded the 2,3-diacetate (i) which reacted with
diazomethane to give the diacetate of the 2-methyl ether (iii) and,
more remarkably, of the 3-methyl ether (ii). Hydrolysis of (iii) then gave
the 2-methyl ether (iv).[47] Partial acetylation of the latter led to the
3-acetate (v) which, with diazomethane, yielded a mixture of the
trimethyl ether (vi) and the 1,2-dimethyl ether-3-acetate (vii), whence
(viii) was derived.[60b] The 2,3-dimethyl ether is accessible by normal
methylation[78] with methyl sulphate (2·2 mol.)–potassium carbonate,
and the 1,3-isomer (xii) by treatment of the methyl ether–benzoate
(ix) with diazomethane which gives both the normal (xi) and the
rearranged (x) dimethyl ether; (xi) was hydrolysed to give (xii).[60b]

Purpurin (78), $C_{14}H_8O_5$

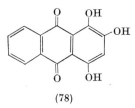

(78)

Occurrence: roots of *Rubia tinctorum* L.,[24a, 248, 565] *R. cordifolia* L.,[8, 92] *R.
cordifolia* var. *munjista* Miq.,[95] *R. sikkimensis* Kurz.[8] and *R. tetragona* Schum.,[45]
numerous *Galium* spp.,[13, 46] *Asperula odorata* L.,[13] *Relbunium hypocarpium*
Hemsl.[46] (all Rubiaceae).

Physical properties: red needles, m.p. 263°, $\lambda_{max.}$ (MeOH) 255, 290(sh), 457(sh),
485, 518, 542(sh) nm (log ϵ 4·41, 4·09, 3·79, 3·92, 3·86, 3·58), $\nu_{max.}$ (KBr) 3225,
1630, 1592 cm^{-1}.

Purpurin is probably the second most important pigment in madder
and the related *Rubia* spp. formerly used as dyestuffs. There is little free
pigment in young madder roots, and the purpurin in mature plants and
in dried roots is probably derived from pseudopurpurin (82) which is
present as a glycoside.[63] It is readily oxidised, and on blowing air
through an alkaline solution the red colour soon fades and phthalic acid
is obtained on acidification.[268] This demonstrates that all three hydroxyl
groups are in the same ring and their orientation (78) follows from the
infrared spectrum and from numerous syntheses and degradations. For
example, purpurin can be prepared by oxidation of either alizarin,[249]
xanthopurpurin[250] or quinizarin,[251] with manganese dioxide and
sulphuric acid; conversely purpurin can be reduced by various reagents
(conveniently dithionite and ammonia[48, 60]) to xanthopurpurin, and
with zinc and acetic acid, followed by heating with hydrochloric acid,
to quinizarin.[252] In these reductions the quinol formed initially isomerises

in acid solution to the diketo form (78a) which undergoes a 1,2-elimination. Under basic conditions however, the tautomer (78b) predominates

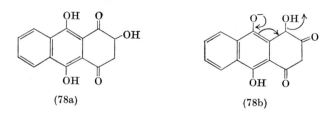

(78a)   (78b)

and elimination of the benzylic hydroxyl group occurs as indicated.[633] Catalytic hydrogenation of purpurin over platinum gives, after reoxidation in air, 2,9-dihydroxy-1,4-anthraquinone, an oxygen atom being removed from a *meso* position.[458b] The same product results when purpurin triacetate is reduced with dithionite, followed by reoxidation with ferric chloride.[458b] Oxidation with lead(IV) compounds results in diquinone formation, for example (79) from purpurin 2-acetate,[253] but most oxidations lead to subsequent ring cleavage. Dimroth and Schultze[254] obtained 3-acetyl-2-hydroxy-1,4-naphthaquinone (80) by using alkaline hydrogen peroxide and a cobalt catalyst, and Scholl and Dahll[255] isolated 2,5-dihydroxybenzoquinone from a ferricyanide oxidation. They considered (81) to be a likely intermediate in this reaction.

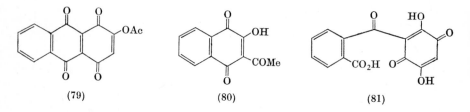

(79)   (80)   (81)

Pseudopurpurin† (82), $C_{15}H_8O_7$

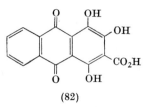

(82)

---

† Boletol, the isomeric 1,2,4-trihydroxyanthraquinone-5 or 8-carboxylic acid, which was alleged[260] to occur in various *Boletus* toadstools, and to be responsible for the blueing reaction which follows bruising, does not exist.[261-263]

Occurrence: roots of many Rubiaceae, mainly as a glycoside: *Rubia tinctorum* L.,[58, 256] *R. cordifolia* L.,[257] *R. peregrina* L.,[27] numerous *Galium* spp.,[13, 27] *Asperula* spp.,[27, 28] *Sherardia arvensis* L.[27] and *Relbunium* spp.[45, 46]

Physical properties: red-brown plates, m.p. 229·5–230·5° (dec.), $\lambda_{max.}$ (EtOH) 256·5, 285(sh), 487, 520(sh) nm (log ϵ 4·52, 4·10, 3·96, 3·85) $\nu_{max.}$ (KBr) 3140, 1705, 1580 cm^{-1}.

Two reactions establish the structure of this quinone; decarboxylation readily affords purpurin and reduction with alkaline dithionite removes an α-hydroxyl group to give munjistin (33).[259] It can be prepared[258] by oxidation of either 1,2- or 1,4-dihydroxyanthraquinone-3-carboxylic acid with manganese dioxide and sulphuric acid, or by condensing purpurin with formaldehyde in sulphuric acid, and oxidation of the product.[63] This however is not the 3-hydroxymethyl derivative, as was thought, but rather the cyclic acetal (83) formed by condensation with a

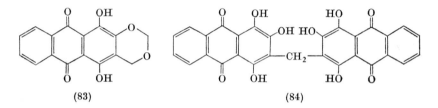

(83)   (84)

further molecule of formaldehyde.[266] Condensation of formaldehyde with purpurin under alkaline conditions may give either the hydroxy-methyl derivative[260] or the dimer (84).[266]

Pseudopurpurin occurs in *Rubia tinctorum*, *Relbunium hypocarpium* and various *Galium* spp. as the somewhat labile 1-β-D-primveroside, yellow needles, dec. >100°, known as galiosin.[27] The position of the sugar moiety was deduced[63] from its colour reactions, ease of hydrolysis, and the formation of munjistin by hydrogenolysis over palladium in cold aqueous solution.

2-Hydroxy-1,6-dimethoxy-3-methylanthraquinone (85)

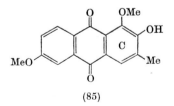

(85)

Occurrence: leaves of *Digitalis viridiflora* Lindl.[571] (Scrophulariaceae).

Physical properties: yellow needles, m.p. 227–229°, $\lambda_{max.}$ (MeOH) 276, 304, 366 nm (log ϵ 4·37, 4·11, 3·69), $\nu_{max.}$ (KBr) 3330, 1655 cm^{-1}.

This *Digitalis* leaf pigment is a methoxylated digitolutein. The infrared spectrum provides evidence for a β-hydroxyl group, and the n.m.r. spectrum accounts for a β-methyl group, two methoxyls and four aromatic protons, one of which gives a broad singlet at τ 2·03. It follows that one ring is trisubstituted and the arrangement of ring C shown in (85) fits the data. The remaining methoxyl group must occupy a β-position to accord with the rest of the n.m.r. spectrum at low field, and structure (85) is most likely from biogenetic considerations.[571]

Morindone (86), $C_{15}H_{10}O_5$

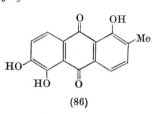

(86)

Occurrence: free and as glycosides in the root, root bark and heartwood of *Morinda citrifolia* L.,[65, 78, 270, 271, 272, 610] root bark and heartwood of *M. tinctoria* Roxb.,[44] root bark of *M. umbellata* L.,[97] roots and stems of *M. persicaefolia* Buch.-Ham.,[281] bark of *Coprosma australis* Robinson[76] and root of *Hymeno-dictyon excelsum* Wall.[567] (all Rubiaceae).

Physical properties: orange-red needles, m.p. 275° (285°), $\lambda_{max.}$ (EtOH) 232, 259·5, 292, 301, 448 nm (log ε 4·50, 4·53, 4·16, 4·18, 4·07), $\nu_{max.}$ (KBr) 3470, 1634, 1609 cm^{-1}.

All the above plant materials contain anthraquinones in abundance. Morinda root (root of *M. citrifolia* and *M. tinctoria*) was formerly an important natural dyestuff, particularly in India, *M. umbellata* was used for calico printing in Java, and the inner bark of *C. australis* was employed by the Maoris for dyeing flax. The quinones in the bark of *C. australis* may constitute 17% of the dry weight.[76]

Some early confusion[273] of this pigment with alizarin was resolved when Thorpe[274] established the molecular formula, $C_{15}H_{10}O_5$, and suggested a trihydroxy-methyl-anthraquinone structure in agreement with the formation of appropriate derivatives, and of 2-methylanthracene on zinc dust distillation. The orientation of the substituents was deduced by Simonsen[275] from the following data: (a) there appeared to be two α-hydroxyl groups as methylation with methyl iodide and sodium methoxide gave only a monomethyl ether; (b) morindone is a mordant dye[276] resembling alizarin so presumably it is a 1,2-dihydroxy-anthraquinone; (c) the monomethyl ether is readily attacked by alkaline permanganate, but as no aromatic products could be isolated it seemed likely that a hydroxyl group was present in the methoxyl-substituted ring; (d) morindone is quite stable in alkaline solution and is therefore not a derivative of anthragallol or purpurin, and (e) as it was not possible to oxidise the triacetate or trimethyl ether to an anthraquinone-carboxylic acid (Me → CO_2H) it seemed, by analogy with model compounds, that the methyl group was adjacent to a hydroxyl group.† Four formulae satisfy these data of which (86) and (87) are the more likely, and on the basis of comparative colour reactions Simonsen [275] regarded (86) as the most probable structure for morindone.

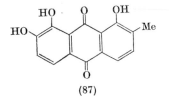

(87)

He subsequently showed this to be correct by a synthesis, first of 1,7,8-trimethoxy-2-methylanthraquinone [278] which differed from morindone trimethyl ether, and later, of morindone [279] itself.

Synthetic morindone was first obtained, however, by Jacobson and Adams [280] who condensed opianic acid with 4-bromo-2-methylphenol in cold sulphuric acid to give the phthalide (88) which was reduced to the benzylbenzoic acid (89) with zinc and aqueous sodium hydroxide; the

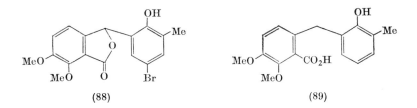

(88)   (89)

remaining steps were conventional. In the Simonsen synthesis [279] soranjidiol (38) was oxidised to morindone by fusion with potash containing sodium arsenate. Recently morindone has been obtained [104] by Marschalk methylation of 1,5,6-trihydroxyanthraquinone.

Morindone occurs to a large extent in glycosidic forms of which the following have been identified: the 6-β-rutinoside, m.p. 264·5° (dec.), $[\alpha]_D^{25}$ −90·9° (dioxan), "morindin" in *C. australis*,[282] the 6-β-primveroside, m.p. 264°, also called "morindin", in *M. persicaefolia*,[281] and morindonin, the 6-β-gentianoside, m.p. 255–256°, in *M. tinctoria*.[283] The sugar moiety was assigned to the C-6 hydroxyl group either because the glycoside was resistant to mild methylation,[281,282] or because vigorous

† This seems to be quite general.

methylation, followed by hydrolysis, gave morindone-1,5-dimethyl ether.[283] Whether the original glycosides ("morindins") obtained from *M. citrifolia*[270–272, 274, 275] and *M. umbellata*[97, 284] are identical with any of the above compounds is not known as the sugar components were not identified.

Obtusifolin† (90), $C_{16}H_{12}O_5$

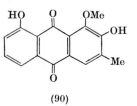

(90)

Occurrence: seeds of *Cassia obtusifolia* L.[141] (Leguminosae).
Physical properties: yellow needles, m.p. 237–238°, $\lambda_{max.}$ (EtOH) 225·5, 275·5, 402 nm (log ε 4·29, 4·33, 3·88), $\nu_{max.}$ (Nujol) 3302, 1656, 1631 cm⁻¹.

The seeds of *Cassia obtusifolia*, which contain a group of hydroxy-anthraquinones, have been used as a yellow dyestuff in Japan. Several new quinones have been identified by Takido[141, 505] of which obtusifolin is the simplest. It contains a methoxyl group and two hydroxyls, one being in an α-position (ν_{CO} 1631 cm⁻¹) and the other a β-position (ν_{OH} 3302 cm⁻¹, soluble in aqueous sodium carbonate). The key to its structure was obtained by reduction with red phosphorus and hydriodic acid; reoxidation of the crude anthrone gave chrysophanol (∼10%) showing that demethylation had been effected and also an unusual elimination of the β-hydroxyl group. As the ultraviolet-visible absorption of nor-obtusifolin, obtained by demethylation with hydrogen bromide, is almost identical with that of 1,2,8-trihydroxyanthraquinone, and yet the colour reaction with magnesium acetate indicates that obtusifolin is not a 1,2-diol, there appear to be two possible structures for this pigment, (90) or (91). However the dimethyl ether of obtusifolin is different from the known dimethyl ether of (91), which leaves structure (90) and this was confirmed by chromic acid oxidation of the di-ethyl ether of obtusifolin which gave 3-ethoxy-phthalic anhydride.

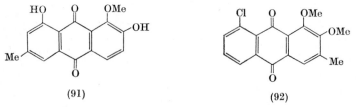

(91)   (92)

† Not to be confused with obtusifolin, a flavanone from *Guaphalium obtusifolium*.[570]

Obtusifolin has not been synthesised but nor-obtusifolin has been derived from the chloroquinone (92) (prepared by condensation of 3-chlorophthalic anhydride with 6-methoxy-2-methylphenol, and methylation) by reaction with sodium methoxide in methanol to give 1,2,8-trimethoxy-3-methylanthraquinone, which was demethylated.[89]

Homonataloin.† Nataloe-emodin 8-methyl ether (93), which is isomeric with the preceding quinone, occurs‡ as the anthrone C-glycoside, homonataloin (94), m.p. 202–204°, $[\alpha]_D^{28}$ −111·5° (EtOH), in numerous *Aloe* spp.[153, 161, 210b, 214, 285,]§[634] When oxidised with ferric chloride it gives D-arabinose and nataloe-emodin 8-methyl ether, whereas ozonolysis gives both D-arabinose and D-glucose.[285] The quinone contains one

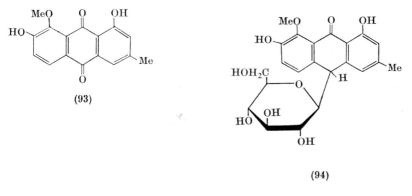

(93)

(94)

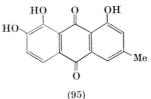

(95)

O–Me and one C–Me group, and two hydroxyls which, from spectroscopic considerations, appear to occupy 1,8-positions. Demethylation gives nataloe-emodin whose colour reaction with magnesium acetate indicates the presence of an alizarin structure, and this was confirmed by a permanganate oxidation of homonataloin dimethyl ether (obtained by methylation with diazomethane in ether-methanol) which afforded

† Léger's original[286] nataloin, $C_{22}H_{22}O_{10}$, could not be found in a recent re-investigation[285] and may have been a methanolate of homonataloin.[210a]

‡ The free quinone has been detected in *Aloe speciosa* Bak.[153]

§ The original botanical source of Natal aloes is obscure.[210a, 285]

3,4-dimethoxyphthalic anhydride. Nataloe-emodin is therefore a 1,2,8-trihydroxyanthraquinone and the C-methyl group was placed at C-6 assuming normal polyketide biogenesis.[†] This structure (95) was confirmed by synthesis: hemipinic anhydride was condensed under Friedel-Crafts conditions with *m*-cresol, and the resulting benzoyl-benzoic acid was cyclised in sulphuric acid with simultaneous demethylation to give (95) identical with nataloe-emodin. As nataloe-emodin methyl ether is stable to sodium metaperiodate at 0°, whereas nataloe-emodin is very easily oxidised (consuming 1·8 mol. of oxidant), the former clearly does not possess a catechol moiety, and accordingly Haynes and co-workers[285] assign it structure (93).

Juzunol (96), $C_{16}H_{12}O_6$

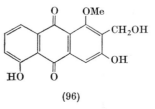

(96)

Occurrence: roots of *Damnacanthus major* Sieb. & Zucc.[85b] and *D. major* Sieb. & Zucc. var. *parvifolius* Koidz.[85b] (Rubiaceae).

Physical properties: orange-brown needles, m.p. >300°$\lambda_{max.}$ (Nujol) 3181, 1658, 1642, 1618 cm^{-1}.

Juzunol and the corresponding aldehyde, juzunal, are 5-hydroxy derivatives of the similarly related pair, damnacanthol and damnacanthal, all of which occur together in certain Japanese *Damnacanthus* spp. Juzunol and juzunal were obviously closely related and since the former can be oxidised to the latter (98) with manganese dioxide, juzunol is evidently (96).[85b] Nor-juzunol, obtained by demethylation with hydrogen bromide, has been synthesised by hydroxymethylation of 1,3,5-trihydroxyanthraquinone with formaldehyde in alkaline solution.[85b] This method is ambiguous but the position of the side chain was confirmed later by condensing 3-hydroxybenzoic acid with 3,5-dihydroxy-4-methylbenzoic acid in sulphuric acid–boric acid, to give, *inter alia*, 2-methyl-1,3,5-trihydroxyanthraquinone, which was converted to nor-juzunol by side chain bromination of the triacetate, followed by reaction with sodium acetate in acetic anhydride, and final hydrolysis.[288]

[†] The structure, 1,2,8-trihydroxy-7-methylanthraquinone, suggested[267] for cladofulvin, a pigment found in cultures of *Cladosporium fulvum*, has been shown to be incorrect by synthesis.[268]

Norjuzunal (97), $C_{15}H_8O_6$

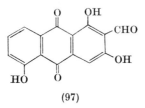

(97)

Occurrence: roots of *Damnacanthus major* Sieb. & Zucc.[567]
Physical properties: orange needles, m.p. 249–251° (266–267°), $\lambda_{max.}$ (hexane) 232·5, 245, 252, 260(sh), 293(sh), 302, 432, 474(sh) nm (log ϵ 4·50, 4·49, 4·51, 4·48, 4·30, 4·33, 4·13, 3·42), $\nu_{max.}$ (KBr) 1662, 1631, 1598 cm^{-1}.

Structure (97) is consistent with the above spectroscopic data and the compound proved to be identical with the product obtained by de-methylation of the co-pigment, juzunal (98).[567]

Juzunal (98), $C_{16}H_{10}O_6$

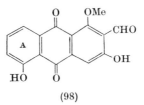

(98)

Occurrence: root bark of *Damnacanthus major* Sieb. & Zucc.,[84] *D. major* Sieb. & Zucc. var. *parvifolius* Koidz.[84] and *D. indicus* Gaertner fil. var. *microphyllus* Makino[84] (Rubiaceae).
Physical properties: orange needles, m.p. 248°, $\lambda_{max.}$ (?) 252, 305, 380 nm (log ϵ 4·05, 4·08, 3·92), $\nu_{max.}$ (KBr) 1676, 1661, 1646, 1639, 1596 cm^{-1}.

Juzunal is a hydroxy derivative of damnacanthal; it yields an oxime[289] and a tetra-acetate in which the formyl group is converted into a gem-diacetate. The colour reactions of juzunal and nor-juzunal (obtained by demethylation with 80% sulphuric acid) are somewhat similar to those of damnacanthal and nor-damnacanthal, respectively, and suggest the absence of vicinal hydroxyl groups. This places the extra hydroxyl group in the A-ring, and also in a *peri*-position, as juzunal shows no free hydroxyl absorption in the infrared, the hydroxyl groups being chelated to the formyl and a quinone carbonyl group. The free carbonyl absorption at 1676 cm^{-1} shifts to a lower frequency in nor-juzunal which means that both quinone carbonyl groups are now chelated. Nonomura[289] therefore concluded that nor-juzunal is 2-formyl-1,3,5-trihydroxyanthraquinone and so the natural pigment is probably (98) by analogy with the co-pigment, damnacanthal. Hirose[290] excluded the alternative 1,3,8-trihydroxy structure for nor-juzunal by synthesis, and then confirmed the 1,3,5- orientation by converting synthetic nor-juzunol into nor-juzunal by oxidation with manganese dioxide.[85b]

Macrosporin (99), $C_{16}H_{12}O_5$

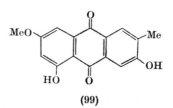

(99)

Occurrence: cultures of *Alternaria* (*Macrosporium*) *porri* (Ell.) Saw. (= *A. solani*)[292,293,294] and *A. cucumerina* (Ell. & Everh.) Ell.[294]

Physical properties: orange-yellow plates, m.p. 308–316° (dec.) (300–302°), $\lambda_{max.}$ (EtOH) 225, 284, 305(sh), 381 nm (log ε 3·60, 4·80, 4·40, 4·15), $\nu_{max.}$ (KBr) 3299, 1659, 1632 cm^{-1}.

The pathogen *Macrosporium porri* elaborates two pigments, one of which, macrosporin, was identified by Suemitsu and co-workers.[292,295–297] It contains one C–Me and one O–Me group, and two hydroxyls of which one is located at a β- and the other at an α-position. Selective ethylation of the β-hydroxyl group, followed by chromic acid oxidation, gave the phthalic acid (100) which determined

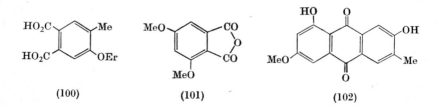

(100)   (101)   (102)

the relative positions of two of the substituents, and the *meta*-relationship of the others was shown by selective acetylation of the β-hydroxyl group in the presence of boro-acetic anhydride, followed by methylation with methyl iodide–silver oxide. The resulting methyl ether-acetate was then hydrolysed and oxidised with alkaline permanganate to give 3,5-dimethoxyphthalic anhydride (101). It follows[296] that macrosporin is either (99) or (102). Structure (99) was confirmed[297,298] by condensing the acids (103) and (104) in sulphuric acid-boric acid which afforded the

quinone (105) in poor yield. Methylation of this, first with diazomethane
and then with methyl sulphate, yielded the mono-[297] and dimethyl[298]

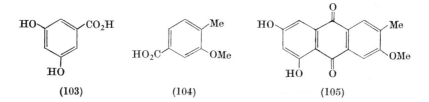

(103)   (104)   (105)

ethers, respectively, of macrosporin. The natural pigment is therefore
(99).

Coelulatin (106), $C_{15}H_{10}O_6$

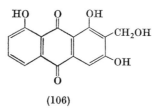

(106)

Occurrence: bark of *Coelospermum reticulatum* Benth.[66] (Rubiaceae).

Physical properties: orange needles, dec. >250°, $\lambda_{max.}$ (EtOH) 247·5, 270, 285,
320(sh), 437·5 nm (log ε 4·23, 4·28, 4·30, 3·54, 4·02), $\nu_{max.}$ (KCl) 3410, 1664, 1610
cm⁻¹.

Like the related compound, lucidin, this pigment is sparingly soluble
in most solvents. The presence of a hydroxymethyl group was deduced
from the ester carbonyl absorption of the tetra-acetate (1736 and 1764
cm⁻¹), and the formation of a monomethyl ether when that compound
was hydrolysed with methanolic hydrochloric acid (cf. lucidin, p. 308).
The ether formed a tri-acetate showing only phenolic ester carbonyl
absorption (1770 cm⁻¹). Similarly, direct methylation with dimethyl
sulphate and potassium carbonate in acetone gave a trimethyl ether
whose mono-acetate revealed only saturated ester carbonyl absorption
at 1733 cm⁻¹. As the colour reactions and ultraviolet-visible absorption
of coelulatin indicated a 1,3,8-trihydroxyanthraquinone structure, and
as rubiadin and lucidin are also present in *C. reticulatum*, it appeared to
Cooke *et al.*[66] that the new quinone had structure (106). This was confirmed by direct comparison with authentic 1,3,8-trihydroxy-2-hydroxy-
methylanthraquinone obtained[86] by hydroxymethylation of 1,3,8
trihydroxyanthraquinone, and from 1,3,8-trihydroxy-2-methylanthra-
quinone via side chain bromination.[290]

14

A compound, $C_{17}H_{14}O_6$, orange-yellow needles, m.p. 216–218°, $\lambda_{max.}$ (KCl) 1670, 1645 cm^{-1}, isolated[66] in very small amount from the bark of *C. paniculatum* is thought to be coelulatin dimethyl ether. On methylation it gave a product which appeared to be coelulatin trimethyl ether.

Deoxyerythrolaccin (107), $C_{15}H_{10}O_5$

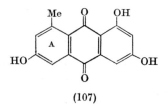

(107)

Occurrence: from the insect *Laccifer lacca* Kerr.[264] (Coccidae).
Physical properties: orange needles, dec. >300°, $\lambda_{max.}$ (EtOH) 220, 280, 430 nm (log ϵ 4·29, 4·39, 3·68), $\nu_{max.}$ (Nujol) 3250, 1662, 1628 cm^{-1}.

Deoxyerythrolaccin is the simplest anthraquinone obtained from stick lac (see p. 462). It affords a triacetate and a trimethyl ether (with diazomethane only a dimethyl ether), and the n.m.r. spectrum of the latter provides evidence for an α-methyl group (τ 7·23), a resorcinol system (doublets at τ 2·68 and 3·26) and two other *meta* coupled protons whose chemical shifts (τ 2·47 and 3·08) agree with the orientation shown in

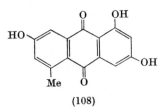

(108)

ring A (107) or the alternative 5,7-arrangement (108). Structure (107) is preferred because the anthrone obtained on reduction with acid stannous chloride still contains a chelated carbonyl group, and the signals from the two α-protons in the n.m.r. spectrum are shifted to higher field as a consequence of reducing the unchelated carbonyl group.[596] This structure is consistent with those of the other coccid anthraquinones and was confirmed by synthesis.[264] Condensation of 3,5-dimethoxyphthalic anhydride with *m*-methoxytoluene in an aluminium chloride-sodium chloride melt gave mainly emodin together with a small amount (5%) of (107).

Emodin (frangula-emodin, rheum-emodin, archin) (109), $C_{15}H_{10}O_5$

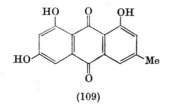

(109)

Occurrence:[299] numerous species of *Rheum*,[113, 115, 134, 174, 193, 300, 304] *Rumex*,[120] [126, 127, 128, 134, 219, 301, 305, 356, 555] *Polygonum*,[146, 219] *Rhamnus*[133, 136, 219, 302, 303] and *Cassia*,[139, 147, 219, 548] leaves of *Muehlenbeckia hastulata* (J.S.N.) Standl. ex Macbr.[605] (Polygonaceae), *Libertia coerulescens* Kunth.[30] and *L. ixioides* Spreng.[605] (Iridaceae), possibly in the root of *Simethis bicolor* Kunth.[313] (Liliaceae), in the wood of *Sonneratia acida* L.[149] (Lythraceae) and in chrysarobin[178] (see p. 389). In cultures of *Penicilliopsis clavariaeformis* Solms-Laubach,[306] *Penicillium islandicum* Sopp.,[307] *P. brunneum* Udagawa[357, 365] and *P. frequentans* Westl.,[552] *Aspergillus terreus* Thom.[545] *Talaromyces avellaneus* (Thom & Turreson) C. R. Benjamin (= *Penicillium avellaneum*),[308] *Trichoderma viride* Pers. ex Fr.,[20] *Phoma foveata* Foister,[19] *Cladosporium fulvum* Cooke,[309] *Chaetomium elatum* Kunze ex Fr. (= *C. affine* Cda.),[157] in *Valsaria rubricosa* (Fr.) Sacc.,[624] *Polystictus versicolor* (L.) Fr.,[160] *Cortinarius sanguineus* (Wulf. ex Fr.) Fr.[310] and other *Cortinarius* spp.,[311] in the lichens *Xanthoria elegans* (Link) Th. Fr.,[540] *X. parietina* (L.) Th. Fr.,[551, 556] *Fulgensia fulgida* (Nyl.) Szat.,[148] *Anaptychia obscurata* (Nyl.) Vain,[543] *Protoblastenia testacea* (Hoffm.) Clauz. & Rond.,[556] *P. rupestris* (Scop.) Stnr.,[556] and *Nephroma laevigatum* Ach. non auct. nonn.[598] (syn. *N. lusitanicum* Schaer.), *Byssoloma tricholomum* (Mont.) Zahlbr.,[646] and in several *Caloplaca* spp.,[623] and from the insect *Eriococcus confusus* Maskell[312] (Coccidae).

Physical properties: orange needles, m.p. 255°, $\lambda_{max.}$ (EtOH) 253, 266, 289, 436 nm (log ε 4·31, 4·29, 4·36, 4·14), $\nu_{max.}$ (KBr) 3390, 1675, 1631 cm^{-1}.

Emodin is perhaps the most ubiquitous natural anthraquinone, occurring in several higher plant families, both higher and lower fungi, in lichens, and in the animal kingdom. In higher plants it is present chiefly in glycosidic combinations, and in reduced and dimeric forms. Dimers are also found in fungi. Like the related quinones, aloe-emodin, chrysophanol and rhein, it occurs in several ancient carthartic drugs (still listed in modern Pharmacopoeia), particularly in rhubarb, the rhizome of *Rheum palmatum*, in the bark of the common buckthorn (*Rhamnus cathartica*), the alder buckthorn (*Rh. frangula*) and in cascara, the dried bark of *Rh. purshiana*. As already mentioned, the free quinones show little activity, the active principles being reduced derivatives present as glycosides.

Emodin was first isolated[304] in 1858, and the confused history of its structural investigation has been reviewed by Eder and Widmer.[314a] It was shown to contain three hydroxyl groups, of which one could be easily methylated and hence occupied a β-position, and a β-methyl group as zinc dust reduction gave β-methylanthracene. As emodin and chrysophanol frequently occurred together in plants, emodin was regarded as a β-hydroxy-chrysophanol and of the three possibilities,

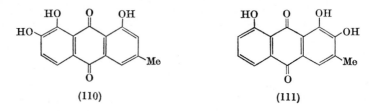

(110)   (111)

(110) and (111) seemed unlikely on the basis of comparative colour reactions. This left (109) as the most likely structure which was confirmed by synthesis. Recently, the structure was elucidated entirely from spectroscopic data.[422]

Several syntheses[173, 314–317] of emodin have been reported of which the most direct is that of Posternak *et al.*[316] who effected a Friedel-Crafts acylation of methyl 3,5-dimethoxybenzoate with 5-bromo-2-methoxy-4-methylbenzoyl chloride to give (112) which was cyclised in oleum, and then heated with hydriodic acid to eliminate the bromine and complete the demethylation. Reoxidation of the resulting anthrone gave emodin.

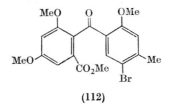

(112)

The cyclisation step is very difficult if the halogen is absent. A more interesting approach was that of Mühlemann[317] who devised a synthesis based on his early biogenetic scheme.[318] The diketone (114), obtained by Claisen condensation of ethyl 3,5-dimethoxyphenylacetate and acetone,

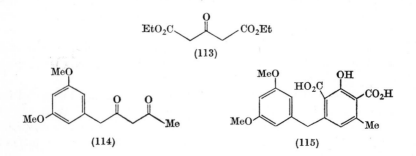

(113)

(114)   (115)

was allowed to react with the keto-ester (113), forming the acid (115), which was converted into emodin anthrone by heating with hydrogen bromide in acetic acid. The anthrone has been detected in the flowers of *Hypericum hirsutum* L.[325]

There is little or no free emodin in fresh plant material where it exists mainly in glycosidic combination, often in a reduced form. Emodin 8-β-D-glucoside, m.p. 190–191°, has been found in *Polygonum cuspidatum*,[579] *Rh. purshiana*,[176] *Rh. undulatum*[322] and *Rh. palmatum* var. *tangut*,[175] and a second monoglucoside in the latter is regarded[321] as the 6-isomer. In *Rhamnus* barks the glycosidic mixture may be complex[176, 240] which accounts for some confusion in earlier work. Recent studies,[176, 319, 320, 628] chiefly by the Hörhammers and Wagner, have established the presence of the following glycosides in the bark of *Rh. frangula*; others have still to be identified.

Frangulin A,† emodin-6-β-L-rhamnoside, m.p. 228°
Frangulin B,† emodin-6-α-L-rhamnoside, m.p. 196°
Glucofrangulin A, emodin-6-β-L-rhamno-8-β-D-glucoside, amorphous.
Glucofrangulin B, amorphous; like glucofrangulin A it gives emodin, rhamnose and glucose on hydrolysis.

Both glucofrangilins are converted into the corresponding frangulins by enzymic hydrolysis. The frangulins are probably derived from the dimeric compounds found in *fresh Rh. frangula* bark which are converted, on storage, into anthrone and anthraquinone glycosides, and emodin.[324] Structural details are not yet complete but they include[240, 241, 242] a mono- and bis-rhamnoside of emodin-10,10′-bianthrone, and a monorhamnoside of palmidin C. Mühlemann and Schmid[324] appear to have obtained another glycoside corresponding to a bi-glucofrangulin. A glycoside of palmidin C also occurs in the ripe fruit of *Rh. frangula*. The palmidins are unsymmetrical bianthrones, at least three of which exist as glycosides in fresh rhubarb root (*Rheum palmatum*).[323] Palmidin A is aloe-emodin-emodin-bianthrone (116), palmidin B is aloe-emodin-chrysophanol-bianthrone (117), and palmidin C is emodin-chrysophanol-bianthrone (118). They each give two anthraquinones on oxidation with ferric chloride, and two anthrones on reduction with stannous chloride. Kinget[607] has found palmidins A, B and C and emodin-, aloe-emodin- and chrysophanol-bianthrones in cascara bark (*Rh. purshiana*). A fourth dimer, palmidin D (chrysophanol-physcion-bianthrone), has been detected in *Cassia occidentalis* along with related bianthrones including glycosides of emodin- and physcion-bianthrones.[583]

† The mixed isomers were previously known as frangulin, franguloside or rhamnoxanthin.

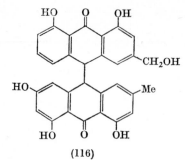

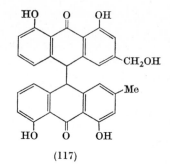

(116)   (117)

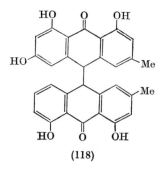

(118)

Although anthraquinones frequently occur in glycosidic combination in higher plants, this had not been found in the fungal quinones until it was observed recently that emodin (and dermocybin, p. 511) are present in the toadstool *Cortinarius sanguineus* mainly as glycosides.[542]

Cassiamin B (119), $C_{30}H_{18}O_{10}$

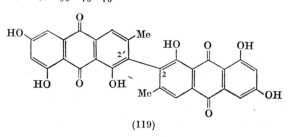

(119)

Occurrence: root bark of *Cassia siamea* Lam.[564] (Leguminosae).

Physical properties: red plates, m.p. >350°, $\lambda_{max.}$ (EtOH) 226, 256, 269, 289, 448 nm (log ϵ 4·61, 4·46, 4·49, 4·52, 4·36), $\nu_{max.}$ (Nujol) 3310, 1670, 1629 cm^{-1}.

This is the third bianthraquinone found in *C. siamea* and it is symmetrical. The n.m.r. spectrum of the hexamethyl ether (molecular weight 622) shows singlets (ratio 1:1:2) at τ 7·83, 6·25 and 5·98 attributable to β-methyl and methoxyl proton signals, and characteristic doublets at τ 2·55 and 3·15 (J, 2·5 Hz) ascribed to the protons of a resorcinol ring. The molecule is therefore a bi-emodin linked symmetrically either at the 2,2'- or 4,4'-positions, and the former is evidently correct as the aromatic protons at H-4 and H-4' resonate at τ 2·3.[564]

(+)–*Skyrin* (endothianin) (120), $C_{30}H_{18}O_{10}$

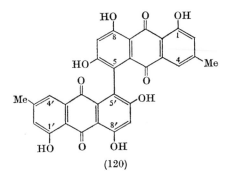

(120)

Occurrence: cultures of *Penicillium islandicum* Sopp.,[361] *P. rugulosum* Thom,[362] *P. wortmanni* Klöcker,[362] *P. tardum* Thom,[363] *P. piceum* Raper & Fennell,[364] *P. variabile* Sopp,[365] *P. brunneum* Udagawa,[365] *Penicilliopsis clavariaeformis* Solms-Laubach,[366] *Endothia parasitica* (Murr.) P. J. & H. W. And.,[367, 370] *E. fluens* (Sow.) Shear & Stevens,[367] *E. gyrosa* (Schw.) Fr.,[368] *E. longirostris* Earle,[368] *E. tropicalis* Shear & Stephens,[368] *Sepedonium ampullosporum* Damon[366] and *Preussia multispora* (Saito & Minoura) Cain.[369]

Physical properties: dark orange needles, m.p. >300°, λ_{max}. (dioxan) 258, 290 448 nm (log ϵ 4·60, 4·52, 4·40), ν_{max}. (Nujol) 3497, 3226, 1681, 1613 cm^{-1}; optically active.

This pigment was discovered independently in the laboratories of Raistrick[361, 362] and Shibata,[367] and proved to be the first, and the most widely distributed, of the bianthraquinone group. It forms a hexamethyl ether and a hexa-acetate, the molecular weight of the latter confirming the dimeric structure. The key to its structural elucidation lay in the observation[361] that reductive cleavage occurred smoothly on treatment with alkaline dithionite to give the monomer, emodin, in high yield. Fission of the central carbon-carbon bond can be effected in other ways,[371, 379] and with lithium aluminium hydride[379] the presence of oxygen functions *ortho* or *para* to the carbon–carbon linkage [as in (122)] is not a structural requirement.† Skyrin is thus a bi-emodin which was confirmed by two further cleavage reactions which gave emodin derivatives. Skyrin reacts with diazomethane forming the 6,6'-dimethyl ether (bi-physcion) which was reduced with alkaline dithionite to physcion (134), and similarly, mild oxidation of skyrin hexa-acetate with chromic acid afforded a dicarboxylic acid (skyric acid) from which emodic acid (157) was obtained by reductive cleavage.[371]

† 1,1'- and 2,2'-Bianthraquinones have now been cleaved with alkaline dithionite.[539]

Of the several possible bi-emodin structures, those which involve coupling at C-2 or C-4 were excluded by the formation of 3-methoxy-6-methylphthalic acid when the hexamethyl ether was vigorously oxidised with chromic acid. Of the others, the 5,5'-dimeric structure (120) was preferred,[371, 372] and this was confirmed synthetically. Bromination of emodin trimethyl ether gives the derivative (121), and when this was

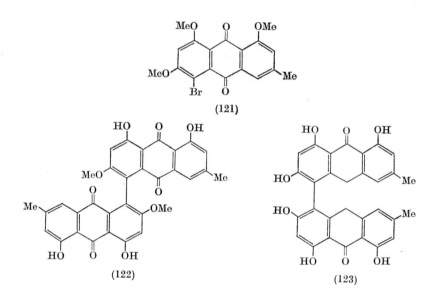

subjected to an Ullmann reaction in boiling naphthalene it afforded skyrin hexamethyl ether, and partial demethylation of this with hydrogen bromide then gave the β,β'-dimethyl ether (122) identical with that formed on treating skyrin with diazomethane.[373] Skyrin itself has also been obtained by oxidation of derivatives of penicilliopsin (123),[366, 375] an orange pigment which accompanies skyrin in cultures of *Penicilliopsis clavariaeformis*.[366, 374]

If skyrin is kept in cold methanolic solution, or better, heated in the presence of sulphuric acid, it is converted into a compound, $C_{30}H_{16}O_8(OMe)_2$,[361, 371] isomeric with the dimethyl ether obtained by methylation with diazomethane. It is insoluble in aqueous sodium carbonate and forms a tetra-acetate. Similar compounds are obtained with other alcohols, from all of which skyrin can be regenerated by treatment with cold alkali or by boiling in acetic acid. These compounds are derivatives of pseudoskyrin (124; R = H) and studies[376] with a series of model compounds showed that hemiketal formation only occurs with 1,1'-bianthraquinones having a free hydroxyl group at C-2 and C-2', for example (125) forms a bis-hemiketal but skyrin hexamethyl ether does not. Dimethyl pseudoskyrin (124; R = Me) shows no free hydroxyl or carbonyl absorption in the infrared which confirms that the

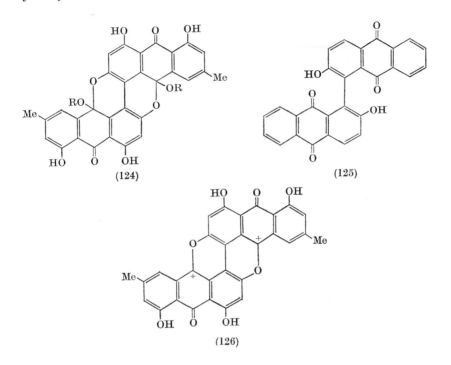

(124)   (125)

(126)

β-hydroxyl groups and the carbonyl groups at C-10 and C-10' are involved in hemiketal formation. Skyrin itself has been obtained in two crystalline forms, orange rods and yellow plates, and was considered by Howard and Raistrick[361] to be dimorphic. However, they appear to be chemically different, the orange form giving a yellow solution in concentrated sulphuric acid which turns rapidly green, the yellow crystals giving an immediate green solution. It is possible that the yellow form is actually pseudoskyrin (124; R = H), the green colour in concentrated sulphuric acid being that of the mesomeric dication (126). If skyrin is treated briefly in cold concentrated acid and then poured into water, a green precipitate is obtained from which a greenish-yellow substance, m.p. >360°, can be extracted whose light absorption and general behaviour are almost identical with pseudoskyrin dimethyl ether. This compound is probably pseudoskyrin (124; R = H)[376] but it has not been compared with Raistrick's "yellow skyrin".

(+)—*Oxyskyrin* (127), $C_{30}H_{18}O_{11}$

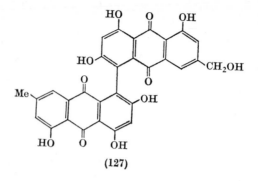

(127)

Occurrence: cultures of *Penicillium islandicum* Sopp.[377,378] and *Endothia parasitica* (Murr.) P. J. & H. W. And.[378]

Physical properties: orange-red needles, m.p. >360°, $\lambda_{max.}$ (EtOH) 257, 300, 462 nm (log ϵ 4·69, 4·50, 4·37); optically active.

This pigment is very closely related to skyrin, their ultraviolet-visible absorption curves being identical. The structure (127)[378] follows from the formation of a hepta-acetate which can be oxidized with chromic acid to skyric acid hexa-acetate, and the formation of both emodin and ω-hydroxyemodin (71) on reductive fission with alkaline dithionite.

(+)—*Skyrinol* (128), $C_{30}H_{18}O_{12}$

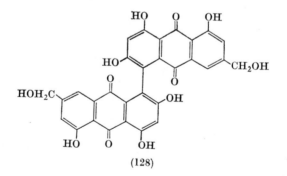

(128)

Occurrence: cultures of *Penicillium islandicum* Sopp.[377] and possibly *Endothia longirostris* Earle.[368]

Physical properties: red crystals, m.p. >360°, $\lambda_{max.}$ (dioxan) 258, 290, 448 nm, $\nu_{max.}$ (Nujol) 3436, 1691, 1631, 1608(sh) cm^{-1}; optically active.

Structure (128) is indicated by colour reactions and the spectroscopic evidence above, and by reductive fission, on a microscale,[377] which gave ω-hydroxyemodin.

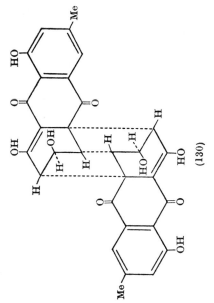

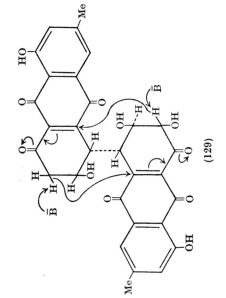

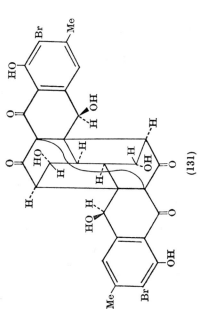

Rugulosin. Closely allied to the foregoing bianthraquinones is an optically active, non-quinonoid, dimeric, compound, (+)-rugulosin, $C_{30}H_{22}O_{10}$, which occurs with skyrin in a number of moulds.[362, 363, 366, 367c] The general structure of this yellow pigment is apparent from its facile dehydration in the presence of acid to give (−)-dianhydro-rugulosin[519] (p. 391), and the formation of chrysophanol on reductive cleavage, and both chrysophanol and emodin on pyrolysis.[362] This accounts for all the oxygen atoms, two of which are in alcoholic groups (removed on dehydration) since the hexa-acetate shows carbonyl absorption at both 1773 and 1751 cm⁻¹. As a result of recent n.m.r. studies the relatively straightforward dimeric structure first pro-posed[522] has now been revised[573] in favour of the remarkable structure (130) in which the two halves of the molecule are joined by *three* carbon-carbon bonds. Shibata and his co-workers suggest[573] that this may arise from a dimeric precursor (129) by a double intramolecular Michael reaction. It is surprising, however, that rugulosin forms an alcoholic diacetate merely by heating with acetic acid. An even more remarkable compound in which the two halves of the molecule are linked by *four* carbon–carbon bonds is the (+)-dibromo compound (131) obtained by treatment of (+)-tetrahydrorugulosin with "dioxan dibromide". The absolute stereochemistry of (131) has been established by X-ray crystal-lographic analysis.[574]

Another non-quinonoid, dimeric, pigment which belongs to the same biogenetic group is flavomannin (132),[499] a metabolite of *Penicillium*

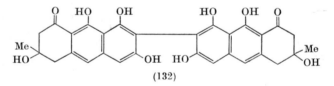

(132)

wortmanni which also elaborates skyrin and rugulosin.

(+)–*Auroskyrin* (133), $C_{30}H_{18}O_9$

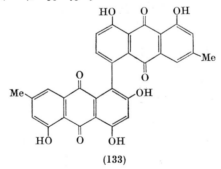

(133)

Occurrence: cultures of *Penicillium islandicum* Sopp.[568]
Physical properties: m.p. >300°, optically active.

On reductive cleavage with dithionite auroskyrin gives emodin and chrysophanol, and the position of the internuclear linkage was deduced from a comparison of the n.m.r. spectra of the dimer and the two monomer peracetates.[568]

Physcion (parietin, parmel yellow, lichen-chrysophanic acid, rheochrysidin) (134), $C_{16}H_{12}O_5$

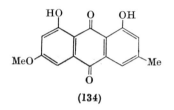

(134)

Occurrence:[299] in species of *Rheum*,[174, 191a, 192, 193, 219] *Rumex*,[120, 122, 125, 219] and *Polygonum*[219, 326, 327] (Polygonaceae), *Rhamnus*[148, 219] (Rhamnaceae) and *Cassia*[328-330, 548] (Leguminosae); root bark of *Ventilago maderaspatana* (Gärtn.)[331] (Rhamnaceae), bark of *Harungana madagascariensis* Poir.[156] (Guttiferae), heartwood of *Vatairea guianensis* Aubl.[150] and *Vataireopsis araroba* (Aguiar) Ducke[150] (Leguminosae), heartwood of *Maesopsis eminii* Engl.,[572] (Rhamnaceae) and in chrysarobin;[178] cultures of *Penicillium herquei* Bain. & Sart.,[332] several spp. of the *Aspergillus glaucus* series,[333, 335, 355] in *Valsaria rubricosa* (Fr.) Sacc.,[624] *Cortinarius sanguineus* (Wulf. ex Fr.) Fr.[334] and in the following lichens: *Xanthoria* (= *Physcia* = *Parmelia*) *parietina* (L.) Th. Fr.,[336, 551]† *X. fallax* (Hepp.) Arn.,[337] *X.* (= *Caloplaca*) *elegans* (Link) Th. Fr.,[340, 433] *Teloschistes flavicans* (Sw.) Norm.,[338] *T. exilis* (Michaux) Wainio,[339, 546] *Stereocaulon corticulatum* Nyl. var. *procerum* Lamb.,[359] *Protoblastenia testacea* (Hoffm.) Clauz. & Rond.,[556] *P. rupestris* (Scop.) Stnr.,[556] *Caloplaca cinnabarina* (Ach.) Zahlbr.[602] and other *Caloplaca* spp.,[623] *Fulgensia fulgida* (Nyl.) Szat.,[148] *Polycauliona regalis* (Wain.) Hue[467] and others.[644]

Physical properties: orange-yellow leaflets, m.p. 207°, $\lambda_{max.}$ (EtOH) 257(sh), 266, 288, 431 nm (log ε 4·35, 4·36, 4·35, 4·20), $\nu_{max.}$ (Nujol) 1678, 1623 cm⁻¹.

Physcion frequently occurs, sometimes as a glycoside, in association with emodin and chrysophanol. Physcion-9-anthrone is present in the heartwood of *Vatairea guianensis*[150] and *V. araroba*,[150] while both the 9- and 10-anthrones have been found in the root bark of *Ventilago maderaspatana*,[331] in rhubarb,[341] in chrysarobin,[178] and in cultures of several of the *Aspergillus glaucus* group.[333b] The 8-β-D-glucoside, m.p. 230–232°, has been identified in *Polygonum cuspidatum*,[579] *Rheum palmatum* (rheochrysin, physcionin)[174, 175] and *Rhamnus purshiana*.[176] It was synthesised from physcion by a standard method,[346] and the structure was established by acylation shift measurements.[616]

† The pigment from this source was originally named chrysophanic acid.

As physcion contains one methoxyl group and is insoluble in aqueous sodium carbonate but gives emodin on demethylation,[344] it must be the 6-monomethyl ether (134) of emodin. It can be prepared by partial methylation[345] of emodin, or by partial demethylation[173] of emodin trimethyl ether by heating with hydrobromic acid in acetic acid.

The lichen *Xanthoria parietina* elaborates emodin, physcion, fallacinol (154), fallacinal (156) and parietinic acid (158), but pure cultures of the fungal component, *Xanthoriomyces parietinae*, produced only physcion and emodin, and there was no sign of side chain oxidation.[592]

Questin (135), $C_{16}H_{12}O_5$

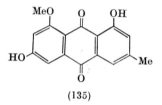

(135)

Occurrence: cultures of *Penicillium frequentans* Westl.[360] and *Aspergillus terreus* Thom.[552]

Physical properties: yellow or orange needles, m.p. 301–303°, $\lambda_{max.}$ (EtOH) 224, 248, 285, 425 nm (log ε 4·59, 4·12, 4·37, 3·97), $\nu_{max.}$ (Nujol): yellow form – 3250, 1670, 1630, 1580 cm⁻¹, orange form – 3250, 1670, 1650(w), 1630, 1580 cm⁻¹.

As questin contains one C-methyl, one O-methyl, and two hydroxyl groups, and gives emodin on demethylation, it is evidently another monomethyl ether of the latter. Reaction with diazomethane gave a monomethyl ether which could be degraded to 3,5-dimethoxyphthalic acid and is therefore emodin-6,8-dimethyl ether. It follows that questin is either the 6- or the 8-monomethyl ether, and it must be (135)[360] as it is different from physcion (134), and has a β-hydroxyl group (ν_{HO} 3250 cm⁻¹).

Madagascin (136), $C_{20}H_{18}O_5$

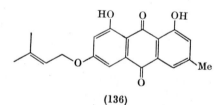

(136)

Occurrence: bark of *Harungana madagascariensis* Poir.[156] (Guttiferae).

Physical properties: orange crystals, m.p. 156–157°, $\lambda_{max.}$ (95% EtOH) 224, 254, 265, 286, 438 nm (log ε 4·55, 4·26, 4·28, 4·27, 4·07), $\nu_{max.}$ (Nujol) 1662, 1626 cm⁻¹.

One of the pigments in the bark of *H. madagascariensis* is physcion whose ultraviolet-visible absorption is identical with that of madagascin. The n.m.r. spectra of the two pigments are also similar except for the absence of a methoxyl signal from madagascin and additional signals easily recognisable as those arising from a γγ-dimethylallyl ether. Accordingly the new pigment is (136) the 3-(γγ-dimethylallyl) ether of emodin which Ritchie and Taylor[156] confirmed by the formation of emodin on treatment with hydrobromic acid, by its mass spectrum[460] (identical with that of emodin below the base peak at *m/e* 270), and by synthesis from emodin using γγ-dimethylallyl bromide and potassium carbonate in dimethylformamide. Madagascin is the only prenylated

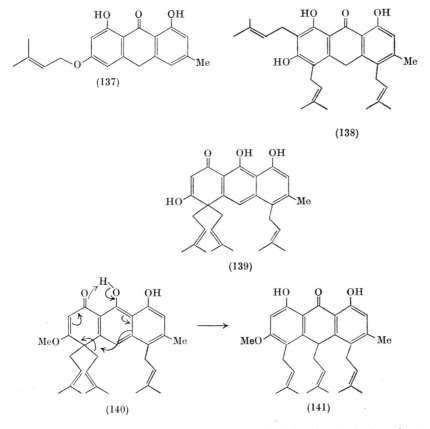

(137)

(138)

(139)

(140)

(141)

anthraquinone. The quinone is accompanied in the bark by the 9-anthrone (137) which was recognised by its ultraviolet absorption ($\lambda_{max.}$ 220, 254, 272, 306, 357 nm) and n.m.r. spectrum which is similar to that of madagascin with the addition of a two-proton singlet at τ 5·90 attributed to the methylene group at C-10.[156]

Another compound present in *H. madagascariensis* is harongin anthrone, probably (138);[156] the corresponding quinone has not been detected but the first pigment to be isolated from the bark, harunganin, proved to have the unique structure (139). Stout and co-workers[461] established this solely by X-ray crystallography without the use of heavy atom labelling. Chemical and spectroscopic studies came later,[156] the most interesting observation being the rearrangement of the mono-methyl ether (140) to the anthrone (141), possibly as indicated, which is effected by brief heating at 180°.

7-Chloro-emodin (142), $C_{15}H_9ClO_5$

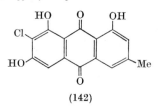

(142)

Occurrence: in the lichens *Nephroma laevigatum* Ach. non auct. nonn.[544,598] (= *N. lusitanicum* Schaer.), *Anaptychia obscurata* (Nyl.) Vain.,[543] *Lecidea quernea* (Dicks.) Arn.,[556] *Lasallia papulosa* (Ach.) Llano[584] and *L. papulosa* (Ach.) Llano var. *rubiginosa* (Pers.),[624] *Byssoloma tricholomum* (Mont.) Zahlbr.,[646] *Caloplaca arenaria* (Pers.) Müll. Arg[623] and *C. percrocata* (Arn.) Stein,[623] and in cultures of *Valsaria rubricosa* (Fr.) Sacc.,[381,624] and *Aspergillus fumigatus* Fr.[545]

Physical properties:† orange needles, m.p. 286–287° (271–273°), $\lambda_{max.}$ (EtOH) 220·5, 255, 273, 282, 308(sh), 437, 458, 515(sh) nm (log ϵ 4·45, 4·16, 4·32, 4·33, 4·07, 4·01, 3·96, 3·17) [$\lambda_{max.}$ (EtOH) 217·5, 258, 312, 324(sh), 437, 510 nm (log ϵ 4·49, 4·24, 4·22, —, 3·87, 3·72)], $\nu_{max.}$ (KBr) 3333, 1663, 1611 cm^{-1}.

A group of chloro-emodins has been found recently in lichens, and fungi, usually accompanied by emodin itself. The simplest of these, (142), is smoothly converted into emodin by reduction either with alkaline dithionite or with zinc and acetic acid, followed by aeration.[543] The position of the chlorine atom follows from the n.m.r. spectrum [543,544] which reveals the presence of three aromatic protons, the orientation of which was easily established by comparison with the spectra of other emodin derivatives. This showed that the chlorine atom must be in the resorcinol ring, and if located at C-7 (142)[543] the chemical shift of the H-5 singlet at τ 2·80 is in fair agreement with the calculated[538] value, and this also accounts for the marked downfield shift of one of the *peri*-hydroxyl protons (τ −3·03, cf. emodin −1·89 and −2·01).[545] Accordingly, the new chloroquinone was assigned structure (142) and it could be related to fragilin (see below) by mass spectral and methylation studies.[543,544]

† Data in parentheses from ref. 545, the other data from ref. 543.

7-Chloro-emodin has been synthesised[626] from physcion by chlorination to give the 4,5,7-trichloro derivative from which the α-chlorine atoms were removed by reduction with hydrazine hydrate and palladised charcoal; the product mixture was methylated to give (143) which

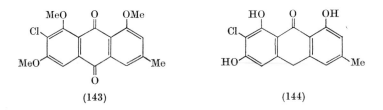

(143)   (144)

afforded 7-chloro-emodin on heating with pyridine hydrochloride.

The chlorinated pigments found in cultures of *A. fumigatus*[545] include the anthrone (144); when the chloride in the medium was replaced by bromide the corresponding bromo-anthrone was produced but apparently not the bromoquinone.

7-Chloro-emodin 6-methyl ether (fragilin) (145), $C_{16}H_{11}ClO_5$
7-Chloro-emodin 1-methyl ether (146), $C_{16}H_{11}ClO_5$
7-Chloro-emodin 1,6-dimethyl ether (147), $C_{17}H_{13}ClO_5$

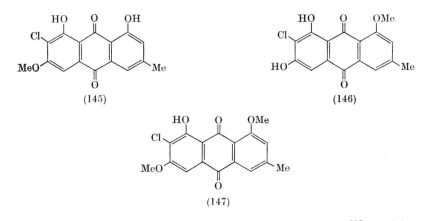

(145)   (146)

(147)

Occurrence: (145) in the lichens *Sphaerophorus fragilis* Pers.,[387] *S. globosus* Vain.[387] (= *S. corraloides* Pers.), *Nephroma laevigatum* Ach. non auct. nonn.[544, 598] (= *N. lusitanicum* Schaer.), *Byssoloma tricholomum* (Mont.) Zahlbr.,[646] *Caloplaca arenaria* (Pers.) Müll. Arg.,[623] and *C. percrocata* (Arn.) Stein[623]; (146) and (147) also in *Nephroma laevigatum*.[544, 598]

Physical properties: (145) orange-yellow needles, m.p. 267–268°, $\lambda_{max.}$ (CHCl$_3$) 271·5, 312·5, 434·5 nm (log ϵ 4·56, 4·15, 4·18), $\nu_{max.}$ (KBr) 1680, 1630 cm^{-1}; (146) m.p. 279–281°, $\lambda_{max.}$ (EtOH) 258, 288, 424 nm, $\nu_{max.}$ (?) 1660, 1620 cm^{-1}; (147) m.p. 254–255°, $\lambda_{max.}$ (EtOH) 289, 427 nm, $\nu_{max.}$ (?) 1670, 1630 cm^{-1}.

Zopf[386] first encountered fragilin in 1898 as a minor constituent of two lichens but the structure was determined only recently. It contains one atom of chlorine (mass spectrum), a β-methyl and an O-methyl group (n.m.r.), and two hydroxyl groups (diacetate, dimethyl ether) which occupy 1,8-positions (infrared). The n.m.r. spectrum also revealed two *meta* aromatic protons, and one other which gives a singlet signal at τ 2·53. Fragilin is thus a penta-substituted anthraquinone, with two *meta*-substituents in one ring. A 1-hydroxy-3-methylanthraquinone structure accords with the chemical shifts of the *meta* protons, and that of the methyl group (τ 7·58) suggested that it was not adjacent to hydroxyl, methoxyl or chlorine. As the second hydroxyl is at C-8 the methoxyl group was assigned to C-6, on biogenetic grounds, and thus fragilin was seen to be a chlorophyscion. Structure (145) was preferred[387] as the chemical shift of the isolated proton (τ 2·53) is closest to the estimated value if it is assigned to C-5. It has been synthesised[626] by partial demethylation of (143) with hydrobromic acid.

Emodin, 7-chloro-emodin, and the three ethers (145), (146) and (147) constitute the pigment mixture in *Nephroma laevigatum* known as nephromin[599] which was recently separated into its components.[544, 598] All the chloroquinones give the same product on complete methylation. Methylation of fragilin (145) with diazomethane gives a trace of the natural dimethyl ether, the major product being an isomer regarded as 7-chloro-emodin 6,8-dimethyl ether on the assumption that the acidity of the 8-hydroxyl group is enhanced by the chlorine atom at C-7. The natural dimethyl ether is consequently (147), and since the natural monomethyl ether, isomeric with fragilin, gives (147) on treatment with diazomethane, it must be the 8-methyl ether (146).[544, 598]

5,7-Dichloro-emodin (148), $C_{15}H_8Cl_2O_5$

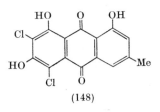

(148)

Occurrence: in the lichen *Anaptychia obscurata* (Nyl.) Vain.[543]
Physical properties: orange needles, m.p. 267–269°, $\lambda_{max.}$ (EtOH) 263, 320·5. 455, 525(sh) nm (log ϵ 4·34, 4·22, 3·98, 3·88), $\nu_{max.}$ (KBr) 3317, 1666, 1623 cm^{-1}.

The presence of two atoms of chlorine in this lichen pigment was immediately apparent from the isotopic peaks in the mass spectrum. On reduction with alkaline dithionite at room temperature, followed by aerial reoxidation, it yields the monochloro compound (142) and emodin, both of which are present in *A. obscurata*. The n.m.r. spectrum shows two broad singlets at τ 2·47 and 2·92 clearly attributable to the H-4 and H-2 protons, respectively, by comparison with other emodin derivatives. The structure is therefore (148).[543] The related optically-active bianthronyls, the flavo-obscurins A (149), B$_1$ and B$_2$, (150) occur in the same lichen.[575] B$_1$ and B$_2$ are rotational isomers. All the flavo-

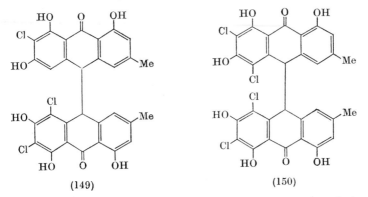

(149)   (150)

obscurins are sensitive to light and their photochemistry, possibly leading to compounds of the hypericin type (see Chapter 7), should be interesting. The presence of four chlorine atoms recalls the lichen depsidone diploicin.[594]

Citreorosein (ω-hydroxyemodin) (151), $C_{15}H_{10}O_6$

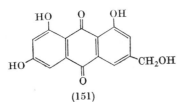

(151)

Occurrence: cultures of *Penicillium cyaneo-fulvum* Biourge[383] (= *P. citreoroseum* Dierckx), *P. cyclopium* Westl.,[384] *P. islandicum* Sopp.,[382] and *Talaromyces avellaneus* (Thom & Turreson) C. R. Benjamin[308] (= *P. avellaneum*), and in the lichen *Xanthoria parietina* (L.) Beltram var. *aureola* (Ach.).[551]

Physical properties: orange needles, m.p. 288°, $\lambda_{max.}$ (EtOH) 221, 252, 266(sh), 290, 438, 458(sh) nm (logϵ 4·41, 4·19, 4·16, 4·19, 3·95, 3·94), $\nu_{max.}$ (KBr) 3508, 3432, 1674, 1630, 1595 cm^{-1}.

Posternak[383] observed that citreorosein was very similar to emodin in its ultraviolet-visible absorption and colour reactions, yet formed a tetra-acetate and tetrabenzoate. Since 2-methylanthracene was obtained on zinc dust distillation, the fourth hydroxyl group appeared to be attached to the side chain, and this was established[384,385] by oxidation of the tetra-acetate with chromic acid which yielded the triacetate of emodic acid (157). Further reduction of the pigment with hydriodic acid and red phosphorus, and reoxidation of the resulting anthrone, gave emodin. Citreorosein is therefore ω-hydroxyemodin (151). The structure was confirmed by synthesis, initially[204] by side chain bromination of emodin triacetate with N-bromo-succinimide, followed by reaction with silver acetate in acetic anhydride and final hydrolysis, and more recently[203] via reduction of emodic acid triacetate with diborane.

7-Chlorocitreorosein (152), $C_{15}H_9ClO_6$

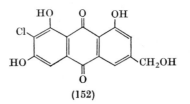

(152)

Occurrence: cultures of *Aspergillus fumigatus* Fr.[545]
Physical properties: orange-red powder, m.p. 297°, $\lambda_{max.}$ (EtOH) 220, 249(sh), 306, 332(sh), 435, 453(sh), 520(sh?) nm (logε 4·50, —, 4·16, —, 3·96, —, 3·62), $\nu_{max.}$ (KBr) 1660, 1620 cm⁻¹.

This quinone was found with 7-chloro-emodin in *A. fumigatus* and was obviously similar. The differences lie in the absence of a C-methyl group, and the formation of a tetra-acetate (ν_{CO} 1775 and 1735 cm⁻¹) whose n.m.r. spectrum showed a downfield shift of the *meta* coupled proton signals relative to those of 7-chloro-emodin triacetate, and a signal (2H) at τ 4·78 attributable to a benzylic methylene group of the type Ar–CH_2–OAc. Accordingly the pigment is considered to have structure (152).[545]

Questinol (153), $C_{16}H_{12}O_6$

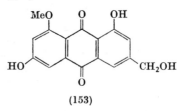

(153)

Occurrence: cultures of *Penicillium frequentans* Westl.[360]
Physical properties: orange needles, m.p. 280–282°, $\lambda_{max.}$ (EtOH) 224, 247, 286, 432 nm (logε 4·57, 4·15, 4·34, 3·95), $\nu_{max.}$ (Nujol) 3500, 3150, 1665, 1625, 1605 cm⁻¹.

This quinone is produced along with questin (p. 430) in cultures of *P. frequentans*. Both compounds contain one methoxyl group but whereas questin forms a diacetate and, on methylation, emodin trimethyl ether, questinol forms a triacetate and is converted into ω-hydroxyemodin tetramethyl ether when treated with dimethyl sulphate and alkali. The new compound is therefore a monomethyl ether of ω-hydroxyemodin. Since methylation (diazomethane), and then oxidation, yielded 3,5-dimethoxyphthalic acid (cf. questin) it is evidently the 6- or 8-monomethyl ether and as questinol is different from fallacinol (154) it must be the 8-methyl ether (153).

Fallacinol (teloschistin) (154), $C_{16}H_{12}O_6$

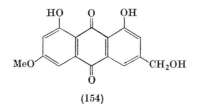

(154)

Occurrence: in the lichens *Teloschistes flavicans* (Sw.) Norm.,[388] *T. exilis* (Michaux) Wainio,[546] *Xanthoria fallax* (Hepp.) Arn.,[389] *X. elegans* (Link) Th. Fr.,[550] *X. parietina* (L.) Th. Fr.,[551] *Caloplaca ferruginea* (Huds.) Th. Fr.[623] and other *Caloplaca* spp.[623]

Physical properties: orange needles, m.p. 245–247°,† $\lambda_{max.}$ (EtOH) 224, 250, 265, 287, 433, 455(sh) nm (log ε 4·49, 4·21, 4·23, 4·21, 4·06, 4·03), $\nu_{max.}$ (KBr) 3520, 3450, 1670(w), 1631, 1624 cm⁻¹.

Fallacinol contains a methoxyl group and may be converted into emodin by heating with hydriodic acid and red phosphorus, followed by reoxidation. Since it formed a triacetate, but only a dimethyl ether on mild treatment with methyl sulphate, one hydroxyl group appeared to be alcoholic, indicating that fallacinol was a monomethyl ether of ω-hydroxyemodin.[388] The ω-acetate is formed merely by heating in acetic acid and was, in fact, obtained as an artefact when *X. parietina* was extracted with acetone containing 1% acetic acid.[551] The position of the methoxyl group was established[390] by a controlled oxidation of fallacinol triacetate with chromic acid in acetic anhydride which yielded the diacetate of emodic acid 6-methyl ether. A similar oxidation of physcion diacetate gave the same product. Fallacinol is thus ω-hydroxy-physcion (154) prepared originally[384, 391] by partial methylation of ω-hydroxyemodin, and later[389, 392] from physcion diacetate by side-chain bromination and hydrolysis.

† Also 229–230°, 234–235°, 238–239°.

Carviolin (roseopurpurin) (155), $C_{16}H_{12}O_6$

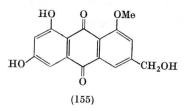

(155)

Occurrence: cultures of *Penicillium roseo-purpureum* Dierckx[383b, 385, 393] (= *P. carmino-violaceum* Dierckx).
Physical properties: yellow needles, m.p. 286°.

This mould metabolite was discovered and characterised independently by Posternak[385] and by Hind,[393] who also isolated a second pigment, carviolacin. Carviolin contains a methoxyl group and forms a triacetate, a leucopenta-acetate and, under vigorous conditions, a trimethyl ether identical with citreorosein tetramethyl ether. It is therefore another monomethyl ether of ω-hydroxyemodin, isomeric with fallacinol. As the compound is soluble in aqueous sodium carbonate and gives colour reactions reminiscent of xanthopurpurin, the hydroxyl groups at C-6 and C-8 appear to be free, and the location of the methoxyl group at C-1 (155) was confirmed by a two-stage oxidation with permanganate (first alkaline, then acid) which gave 6-methoxybenzene-1,2,4-tricarboxylic acid.

Carviolacin, $C_{20}H_{16}O_7$, light brown needles, dec. 243°, the second colouring matter isolated[393] from *P. roseo-purpureum* appears to be closely related to carviolin and gives the same colour reactions. It contains a methoxyl group, forms a trimethyl ether, and yields 2-methylanthracene on zinc dust distillation. The structure is not known.

Fallacinal (156), $C_{16}H_{10}O_6$

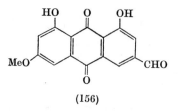

(156)

Occurrence: in the lichens *Xanthoria fallax* (Hepp.) Arn.,[389] *X. elegans* (Link) Th. Fr.,[550] *X. parietina* (L.) Th. Fr.,[551] *Teloschistes flavicans* (Sw.) Norm.,[391] *Caloplaca ferruginea* (Huds.) Th. Fr.[623] and other *Caloplaca* spp.[623]

Physical properties: orange-yellow needles, m.p. 251–252°, $\lambda_{max.}$ (EtOH) 244, 264, 280, 340, 425 nm, $\nu_{max.}$ (KBr) 1710, 1670, 1625 cm^{-1}.

The colouring matters of the lichen *Xanthoria fallax*, growing on mulberry trees, were first examined by Asano and Fuziwara[394] who isolated a pigment, fallacin. This was then shown[337] to contain physcion, and two other components, fallacinol and fallacinal, which were separated later by Murakami.[389] The related species *X. elegans* contains the same three pigments and other anthraquinones.[540, 550] Fallacinal, $C_{15}H_7O_5$ (OMe), contains an aldehyde group (ν_{CO} 1710 cm^{-1}), and forms a 2,4-dinitrophenylhydrazone and a tetra-acetate (cf. damnacanthal), and in view of its colour reactions and spectral properties, and the co-existence of physcion, structure (156) seemed very probable. This was confirmed by synthesis.[389] Physcion diacetate was oxidized with chromium trioxide to the corresponding emodic acid derivative which was converted, by Rosenmund reduction of its acid chloride, to the aldehyde. On final hydrolysis the quinone (156) was obtained identical with the natural pigment.

In alkaline solution fallacinal undergoes a Cannizzaro reaction giving fallacinol and parietinic acid (158).[551] The aldehyde function condenses readily with acetone in the presence of acetic acid to give a quinone with a C_4 (–CH=CHCOCH$_3$) side chain. This was obtained, as an artefact, when *X. parietina* was extracted with acetone containing 1% acetic acid.[551]

Emodic Acid (157), $C_{15}H_8O_7$

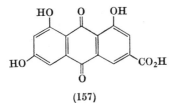

(157)

Occurrence: cultures of *Penicillium cyclopium* Westl.[384] and *Talaromyces avellaneus* (Thom & Turreson) C. R. Benjamin,[369] and in the lichen *Xanthoria parietina* (L.) Th. Fr. var. *aureola* (Ach.).[551]

Physical properties: orange needles, m.p. 363–365°, $\lambda_{max.}$ (EtOH) 222, 248·5, 271·5, 293(sh), 437 nm (log ϵ 4·40, 4·18, 4·18, 4·14, 4·02), $\nu_{max.}$ (KBr) 1700, 1670(sh), 1625 cm^{-1}.

This pigment is elaborated by *P. cyclopium* along with ω-hydroxy-emodin, and in *T. avellaneus* emodin is a co-pigment. It is soluble in aqueous sodium bicarbonate, forms a methyl ester and a methyl ester-triacetate, and was easily recognised[384] as the acid (157) prepared[576] earlier by oxidation of emodin triacetate with chromic acid.

Parietinic Acid (158), $C_{16}H_{10}O_7$

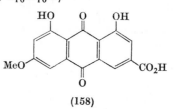

(158)

Occurrence: in the lichens *Xanthoria parietina* (L.) Th. Fr.,[551] *X. parietina* var. *aureola* (Ach.),[395, 551] *X. elegans* (Link) Th. Fr.,[540] *Caloplaca ferruginea* (Huds.) Th. Fr.[623] and other *Caloplaca* spp.[623]

Physical properties: dark yellow needles, subl. ca. 300°, λ_{max}. (acetone) 325, 435 nm (log ϵ 4·25, 4·30), ν_{max}. (KBr) 1700, 1629 cm^{-1}.

From the work of Murakami[389] it was known that *Xanthoria parietina* contains physcion, fallacinol and fallacinal, and when a fourth pigment was discovered, which could be separated by extraction into aqueous sodium carbonate, it was logical to consider (158) as a likely structure. The presence of a carboxyl group was confirmed by its infrared spectrum and the structure, emodic acid 6-methyl ether (158), was established[395] by direct comparison with a synthetic sample prepared[314b] by chromic acid oxidation of physcion diacetate.

Endocrocin (clavoxanthin) (159), $C_{16}H_{10}O_7$

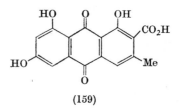

(159)

Occurrence: in the lichen *Nephromopsis endocrocea* Asahina,[396] ergot [*Claviceps purpurea* (Fr.) Tul.],[397] cultures of *Aspergillus amstelodami* (Mangin) Thom & Church[335] and a mutant of *Penicillium islandicum* Sopp.,[398] and in the toadstools *Dermocybe sanguinea* (Wulf. ex Fr.)[615] and *D. semisanguinea* (Fr.).[615]

Physical properties: orange leaflets, m.p. 340° (dec.), λ_{max}. (MeOH) 274, 287(sh), 311(sh), 442 nm (log ϵ 4·32, 4·18 3·92, 4·02), ν_{max}. (KBr) 3390, 1718, 1666, 1615 cm^{-1}.

Endocrin is distinctly acidic (pK_a 4·2),[399] forms a methyl ester and a methyl ester-trimethyl ether, and loses carbon dioxide on heating to give emodin.[396, 398] The position of the carboxyl group was ascertained[396] by parallel oxidations of emodin trimethyl ether and endocrocin trimethyl ether-methyl ester. The former gave the phthalic acid (160), while the latter yielded a methoxycarbonyl derivative which was regarded† as (161) since it differed from the isomer (162) previously derived from kermesic acid (p. 457). Endocrocin therefore appeared to

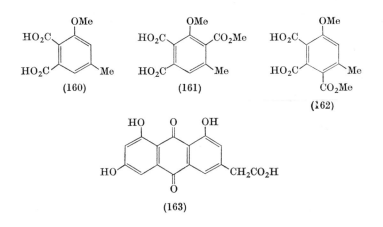

(160)   (161)   (162)

(163)

have structure (159) but this was doubted later when an examination[401] of its infrared spectrum showed ν_{CO} 1718 cm^{-1} which is a rather high carbonyl stretching frequency for an aromatic carboxylic acid. However, when a compound having the alternative structure (163) (ν_{CO} 1721 cm^{-1}) was synthesised,[401] it proved to be different from endocrocin, and in fact, the n.m.r. spectrum[399, 402] of the natural pigment is consistent with structure (159) and shows the presence of three aromatic protons. Other pigments of this type, for example, dermorubin (296) also show high frequency carboxyl carbonyl absorption although 1-hydroxy-3-methyl-anthraquinone-2-carboxylic acid shows only two carbonyl stretching frequencies at 1670 and 1633 cm^{-1}, the carboxyl carbonyl absorption overlapping that of the unchelated quinone carbonyl group.[401]

Venkataraman and his co-workers[401] have synthesised endocrocin by two methods starting from 2,3-dimethylanthraquinone. In one route, this was dinitrated, and the desired isomer partially reduced with boiling dimethylaniline to the aminoquinone (164) which was converted into the hydroxy-trinitro compound (165) with hot nitric acid. Reduction then gave a triamine from which a tetrahydroxyquinone was obtained by diazotization and hydrolysis, and further reduction with alkaline dithionite then afforded 2-methylemodin (166). Bromination of the latter with N-bromosuccinimide gave mainly the 2-bromomethyl derivative, from which (159) was derived by conversion to the 2-acetoxymethyl compound and oxidation with silver oxide in alkaline solution. In the other route[401] (164) was converted into (167) by bromination and

† Now established; unpublished information quoted in ref. 401.

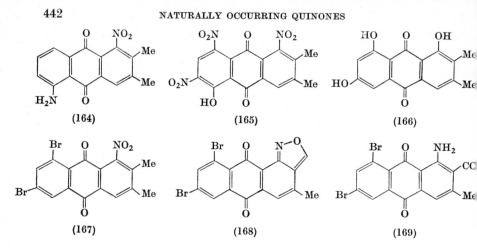

(164)   (165)   (166)

(167)   (168)   (169)

deamination, and treated with 20% oleum at 30°, or with aluminium chloride at 150°, to obtain the oxazole (168). This was then hydrolysed with methanolic barium hydroxide to give (169) which has a carboxyl group in the desired position. Finally bromo and amino substituents were replaced, in stages, by hydroxyl groups.

A much more convenient synthesis has been reported [627] very recently. Mühlemann's intermediate (115) was cyclised to the anthrone which gave endocrocin after oxidation and demethylation (or vice versa).

Dermolutein† (170), $C_{17}H_{12}O_7$

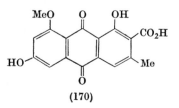

(170)

Occurrence: mainly as a glycoside in *Dermocybe sanguinea* (Wulf. ex Fr.)[615] and *D. semisanguinea* (Fr.).[615]

Physical properties: red crystals, dec. ~270°, λ_{max}. (EtOH) 247, 286, 432 nm (log ϵ 4·32, 3·36, 3·95), ν_{max}. (KBr) 1745(sh), 1724(sh), 1700, 1672, 1613 cm^{-1}.

Dermolutein forms a methyl ester, by which it was purified, and a diacetate, and its structure was established [615] by conversion into two known compounds. On thermal decarboxylation it gave questin (135) while demethylation in hot sulphuric acid afforded endocrocin (159), a co-pigment in *Dermocybe*.

† Cinnalutein, the isomeric 6-methyl ether, has been found in *Dermocybe cinnabarina*.[627] Details are awaited.

5-Chlorodermolutein (171), $C_{17}H_{11}ClO_7$

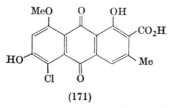

(171)

Occurrence: mainly as a glycoside in *Dermocybe sanguinea* (Wulf. ex Fr.)[615] and *D. semisanguinea* (Fr.).[615]

Physical properties: brown felted needles, dec. >280°, $\lambda_{max.}$ (EtOH) 230, 253, 290, 436 nm (log ϵ 4·52, 4·23, 4·28, 4·00), $\nu_{max.}$ (KBr) 1698, 1675, 1644 cm^{-1}.

This is a monochloro derivative of the preceding pigment which it closely resembles. The n.m.r. spectra of their pertrimethylsilyl derivatives are essentially the same except that the signal from the C-5 proton is absent from that of the chloro compound, and the resonance from H-7 is a singlet. The chlorine atom is therefore located at C-5 (171).[615]

Laccaic Acid D (xanthokermesic acid) (172), $C_{16}H_{10}O_7$

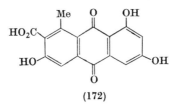

(172)

Occurrence: from the insect *Laccifer lacca* Kerr.[264] (Coccidae).

Physical properties: yellow needles, dec. >300°, $\lambda_{max.}$ (EtOH) 236, 286, 436 nm (log ϵ 3·92, 3·88, 3·75), $\nu_{max.}$ (Nujol) 3415, 1690, 1662, 1630 cm^{-1}.

Decarboxylation of laccaic acid D in boiling diethylaniline yields deoxyerythrolaccin (p. 418). The position of the carboxyl group is defined by the n.m.r. spectrum of the trimethyl ether-methyl ester which includes a pair of doublets at τ 2·69 and 3·23 characteristic for a 1,3-dimethoxyanthraquinone, and a singlet from a third aromatic proton at τ 2·37 corresponding to an α-proton adjacent to a methoxyl group. Laccaic acid D can therefore be represented as (172).[264]

(S)-Rhodoptilometrin (173), $C_{17}H_{14}O_6$

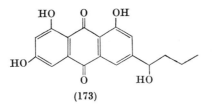

(173)

Occurrence: in the crinoid *Ptilometra australis* Wilton[403] (Ptilometridae).

Physical properties: orange-red needles, m.p. 217–218°, $\lambda_{max.}$ (EtOH) 253, 266, 291, 437 nm (log ϵ 4·28, 4·24, 4·27, 4·03), $\nu_{max.}$ (Nujol) 3500, 3410, 1670, 1630, 1604 cm^{-1}; trimethyl ether, $[\alpha]_D$ −41·6 (CHCl$_3$).

The Australian crinoid *Ptilometra australis* contains a mixture of anthraquinones readily extracted into cold acetone. From nearly four hundred animals Powell and Sutherland[403] obtained >4 g of crude black pigment which yielded 0·8 g of a red oil containing rhodoptilometrin and isorhodoptilometrin.

Rhodoptilometrin yields a tetra-acetate containing both aromatic ($\nu_{max.}$ 1780 cm^{-1}) and aliphatic ($\nu_{max.}$ 1740 cm^{-1}) ester groups, and with methyl sulphate, a trimethyl ether (showing ν_{OH} 3470 cm^{-1}) and also a tetra-methyl ether in small amount. The infrared and ultraviolet-visible spectra of rhodoptilometrin favour a 1,8-dihydroxyanthraquinone structure, a third hydroxyl group is clearly in a side chain, and the fourth must be a β-substituent to account for the solubility of the pigment in aqueous sodium carbonate and the formation of a monomethyl ether with diazomethane. As colour reactions and chromatographic behaviour indicate the absence of vicinal hydroxyl groups, rhodoptilometrin must be a 1,3,8-trihydroxyanthraquinone with three carbon atoms and a hydroxyl group in a saturated side chain(s). Rhodoptilometrin contains only one C-methyl group, and since both acetic and propionic acids were evolved during the Kuhn-Roth estimation, this defines the side chain as –CHOHCH$_2$CH$_3$ in complete accord with the n.m.r. spectrum of the trimethyl ether. This spectrum also confirms the *meta* relationship of two methoxyl groups, and a pair of doublets at τ 2·48 and 2·77 (J, 1·3 Hz), somewhat broadened by weak coupling with the benzylic protons of the side chain, can only arise from two *meta* protons in the ring containing the side chain. Structure (173) can therefore be assigned[403] to rhodoptilometrin. The natural pigment is optically active and by a slight extension of the empirical rules[520] relating the sign of rotation and configuration of aromatic compounds, rhodoptilometrin can be represented by (174),

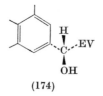

(174)

i.e. S(−)-1,6,8-trihydroxy-3-(1-hydroxypropyl)anthraquinone.[403]

*iso*Rhodoptilometrin (175), $C_{17}H_{14}O_6$

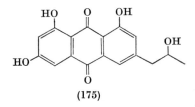

(175)

Occurrence: in the crinoid *Ptilometra australis* Wilton[403] (Ptilometridae).
Physical properties: orange-red needles, m.p. 275–277°, $\lambda_{max.}$ (EtOH) 252, 265, 290, 438 nm, $\nu_{max.}$ (Nujol) 3340, 1663, 1624, 1609 cm^{-1}; trimethyl ether, $[\alpha]_D$ +22° (?).

As the above spectral data show, this pigment is very similar to the previous isomer, and the pair were difficult to separate.[403] It affords a tetra-acetate and, with dimethyl sulphate, a trimethyl ether (ν_{HO} 3480 cm^{-1}) while treatment with excess diazomethane gives both a mono- and a dimethyl ether, the former being identical with nalgiovensin (176). Methylation of the mixture of the two isomers isolated from *P. australis*, and oxidation of the crude product with hypoiodite yielded emodic acid trimethyl ether. The side chain, in this case, has the 2-hydroxypropyl structure, $-CH_2CHOHMe$, as defined by the n.m.r. spectrum of the tetra-acetate which includes a doublet (3H) at τ 8·64 and another (2H) at τ 6·98, both coupled to a multiplet (1H) at τ 4·73. At low field, two distinct AB systems are seen which correspond closely to the parallel situation in rhodoptilometrin. Isorhodoptilometrin is therefore (175).[403]

Nalgiovensin (176), $C_{18}H_{16}O_6$

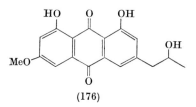

(176)

Occurrence: cultures of *Penicillium nalgiovensis* Laxa.[404]
Physical properties: orange needles, m.p. 199–200°, $[\alpha]_{5790}^{20}$ +39·7° (CHCl₃), $\lambda_{max.}$ (EtOH) 266, 287, 437 nm (log ϵ 4·27, 4·22, 4·09), $\nu_{max.}$ (Nujol) 3380, 1679, 1626, 1597 cm^{-1}.

Nalgiovensin and nalgiolaxin (p. 446) are metabolites of *P. nalgiovensis*, the only known source of which is the Ellischauer cheese produced in Czechoslovakia. Nalgiovensin has one O-Me and one C-Me group, and yields a triacetate, a dimethyl ether (methyl sulphate-alkali) and a trimethyl ether (methyl iodide-silver oxide). The orientation of the substituents was ascertained by chromic acid oxidation of the dimethyl ether and the triacetate which gave the corresponding derivatives of emodic acid, the pigment itself, under mild conditions, giving a little emodic acid 6-methyl ether and an optically inactive ketone, dehydro-nalgiovensin, $C_{18}H_{14}O_6$, showing carbonyl absorption at 1772 cm^{-1}.[405] As nalgiovensin gives a positive iodoform reaction these facts are accounted for if the side chain has the structure $-CH_2CHOHMe$ and is located at C-3. This is supported by the formation of a neutral compound, as one of the chromic acid oxidation products of nalgiovensin triacetate, which is another optically inactive ketone. This shows ν_{CO} 1772, 1740, 1705 and 1675 cm^{-1}, the ketonic group being adjacent to an aromatic ring, and as both aliphatic and aromatic ester groups are present, the side chain must be $-COCH(OAc)Me$.[405] As the asymmetric centre is now adjacent to a carbonyl group racemization is not surprising. Nalgiovensin is therefore (176) and this was confirmed by a recent examination of the n.m.r. spectrum of the dimethyl ether.[517] Reduction of nalgiovensin with hydriodic acid and red phosphorus gives an anthrone which can be reoxidised to 1,6,8-trihydroxy-3-n-propylanthraquinone. The structure of the latter has been confirmed by synthesis.[406]

Nalgiolaxin (177), $C_{18}H_{15}ClO_6$

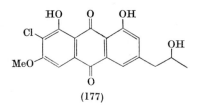

(177)

Occurrence: cultures of *Penicillium nalgiovensis* Laxa.[404]

Physical properties: yellow needles, m.p. 248°, $[\alpha]_{5790}^{22}$ +40·3° (CHCl$_3$), $\lambda_{max.}$ (EtOH) 226, 270, 310, 435, 452(sh) nm (log ϵ 4·49, 4·49, 4·03, 4·08, 4·04), $\nu_{max.}$ (KBr) 3304, 3230, 1670, 1630, 1600 cm^{-1}.

Nalgiolaxin differs from nalgiovensin solely in the replacement of a hydrogen atom by chlorine, and the yield can be increased ten-fold if 0·1% of sodium chloride is added to the culture medium. The skeletal structure is clearly the same as that of nalgiovensin as nalgiolaxin can also be converted into 1,6,8-trihydroxy-3-n-propylanthraquinone by reduction with hydriodic acid and reoxidation. Oxidation of nalgiolaxin triacetate gives an acid, $C_{15}H_4ClO_4(OMe)$, very similar to the emodic acid derivative obtained from nalgiovensin triacetate, so that the chlorine atom is again attached to an aromatic ring. The 2-hydroxypropyl side chain is confirmed by the n.m.r. spectrum[517] of the dimethyl ether which differs from the spectrum of nalgiovensin dimethyl ether only in the absence of a signal from the proton at C-7, and the presence of a signal arising from the proton at C-5 which is now a singlet shifted slightly downfield to τ 2·47. It follows that the chlorine atom is located at C-7 (177).[517]

Ptilometric Acid (178), $C_{18}H_{14}O_7$

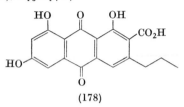

(178)

Occurrence: in the crinoids *Ptilometra australis* Wilton[403] (Ptilometridae) and *Tropiometra afra* Hartlaub[403] (Comasteridae).

Physical properties: red needles, m.p. 298–299° (dec.), $\lambda_{max.}$ (EtOH) 225, 278, 442 nm (log ϵ 4·35, 4·23, 3·94), $\nu_{max.}$ (Nujol) 3400, 1711, 1675, 1626, 1605 cm^{-1}.

Tropiometra afra is a large ten-armed comatulid, abundant in Queensland waters, which usually appears black in the sea but is actually deep purple, while some specimens are yellow. Ptilometric acid can be extracted from both varieties with cold acetone but most of the colouring matter remains behind. The spectroscopic data, given above, indicate that ptilometric acid is a 1,8-dihydroxyanthraquinone. It possesses an additional hydroxyl and also a carboxyl group, since it forms a triacetate, which is soluble in aqueous sodium bicarbonate, and a tetramethyl derivative showing ν_{CO} 1741 and 1663 cm^{-1}. On pyrolysis, the natural pigment afforded 1,6,8-trihydroxy-3-n-propylanthraquinone previously derived from nalgiovensin. The location of the carboxyl group in ptilometric acid was deduced from the n.m.r. spectrum of the tetramethyl derivative. This includes signals from the *meta* protons of a resorcinol system and a singlet (1H) at τ 2·07. The latter can be assigned to an α-proton in the other benzenoid ring by comparison with the C-4 proton in fully methylated endocrocin which resonates at τ 2·12, and the signal from the C-2 proton in 1,3,6,8-tetramethoxy-4-methoxycarbonyl-anthraquinone which appears at τ 3·30. Ptilometric acid is therefore (178).[403]

Islandicin (funiculosin, rhodomycelin) (179), $C_{15}H_{10}O_5$

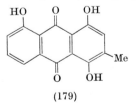

(179)

Occurrence: cultures of *Penicillium islandicum* Sopp.[407] and *P. funiculosum* Thom,[408] root bark of *Ventilago viminalis* Hook.[151] and heartwood of *Maesopsis eminii* Engl.[572] (Rhamnaceae).

Physical properties: dark-red plates, m.p. 218°, λ_{max}. (EtOH) 231·5, 251·5, 289, 460(sh), 479·5(sh), 491, 513, 527 nm (log ϵ 4·56, 4·34, 3·93, 4·01, 4·08, 4·14, 4·00, 3·99), ν_{max}. (KBr) 1603 cm^{-1}.

The mould *Penicillium islandicum* elaborates[269] a complex mixture of anthraquinone pigments amounting to as much as 20% of the dried mycelium, of which islandicin was the first to be identified. Recently, it has been found in higher plants. The structure includes one C-Me group, and three hydroxyl groups (triacetate, trimethyl ether) which must occupy α-positions as the compound is insoluble in sodium carbonate solution, and can be converted into cynodontin (p. 504) by oxidation with manganese dioxide in sulphuric acid. As reduction with hydriodic acid and red phosphorus leads to chrysophanol (p. 388), islandicin must have either structure (179) or (183). The latter, however, has been assigned to helminthosporin, and hence islandicin is (179).[407] The structure has been confirmed by several syntheses.[173, 409, 410] In one of these, 3-methoxy-phthalic anhydride was condensed with *m*-cresol to give (180), which was converted to the quinol (181) by an Elbs persulphate oxidation and then cyclised in hot sulphuric acid containing boric acid.[410]

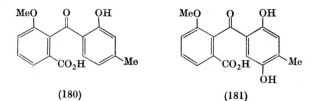

(180)   (181)

(+)-*Iridoskyrin* (182), $C_{30}H_{18}O_{10}$

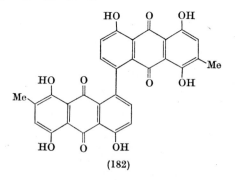

(182)

Occurrence: cultures of *Penicillium islandicum* Sopp.[518]

Physical properties: iridescent red plates, m.p. 358–360°, λ_{max}. (dioxan) 286, 502 nm (log ϵ 4·26, 4·39), ν_{max}. (CHCl$_3$) 1602 cm^{-1}; optically active.

That iridoskyrin is a dimer of islandicin is shown by the similarity of their ultraviolet-visible spectra, and by the formation of islandicin on reductive cleavage of iridoskyrin with alkaline dithionite. Reduction with hydriodic acid removes two α-hydroxyl groups and the product, after reoxidation, is dianhydrorugulosin (p. 391). Iridoskyrin is therefore (182).[519] It is optically active and shows a positive Cotton effect in the long wavelength region of the ORD curve.[568] The (−) form has been derived from (+)-rugulosin by dehydration to (−)-dianhydrorugulosin (48) followed by peracid oxidation.[573]

(+)−*Roseoskyrin* (183), $C_{30}H_{18}O_9$

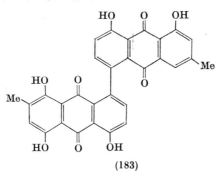

(183)

Occurrence: cultures of *Penicillium islandicum* Sopp.[568]
Physical properties: m.p. >300°, optically active.

On reductive cleavage with dithionite roseoskyrin gives islandicin and chrysophanol, and the position of the internuclear linkage was deduced from a comparison of the n.m.r. spectra of the dimer and the two monomer peracetates.[568] It was obtained synthetically from dianhydro-rugulosin (48) by peracid oxidation which confirms structure (183).

(+)−*Rhodoislandin A* (184), $C_{30}H_{18}O_{10}$
(+)−*Rhodoislandin B* (185), $C_{30}H_{18}O_{10}$

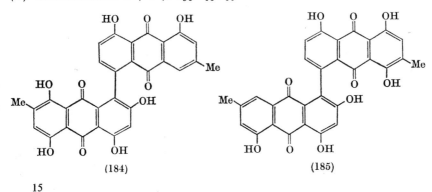

(184)   (185)

15

Occurrence: cultures of *Penicillium islandicum* Sopp.[568]
Physical properties: m.p. >300°, optically active.

This pair of isomers was examined as a mixture as chromatographic separation either of the pigments or their acetates was unsuccessful. They are similar to the other dimers obtained from. *P. islandicum* and their structures were deduced[568] from the products of dithionite cleavage. These are chrysophanol and catenarin (1:1), and emodin and islandicin (1:1), the amount of emodin + islandicin being about five times that of chrysophanol + catenarin.

Helminthosporin (186), $C_{15}H_{10}O_5$

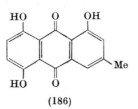

(186)

Occurrence: cultures of *Pyrenophora graminea* (Rabenh.) Ito & Kurib. (= *Helminthosporium gramineum* Rabenh.),[411] *H. tritici-vulgaris* Nishik.,[412] *H. catenarium* Drechsler,[412] *H. cynodontis* Marignoni,[413] *Deuterophoma tracheiphila* Petri,[414] *Phoma violacea* (Bertel) Eveleigh (= *P. pigmentivora*),[415] and possibly in the root bark of *Ventilago viminalis* Hook. (Rhamnaceae).[151]

Physical properties: dark maroon needles, m.p. 227°, $\lambda_{max.}$ (EtOH) 231, 255, 289, 480, 490, 510, 525 nm (log ϵ 4·53, 4·15, 3·79, 3·99, 4·01, 3·89, 3·81), $\nu_{max.}$ (KBr) 1603 cm^{-1}.

Helminthosporin was first isolated[411] from the plant pathogen *Pyrenophora graminea* which, in laboratory culture, produces a mixture of helminosporin and catenarin (p. 494) constituting 30% of the dry weight of the mycelium. Routine acetylation, and distillation with zinc dust, showed it to be a trihydroxymethylanthraquinone, isomeric with islandicin. Chromic acid oxidation of the triacetate yielded tri-acetyl-helminthosporic acid which was converted, by hydrolysis and decarboxylation, into 1,4,5-trihydroxyanthraquinone. More drastic oxidation of the natural pigment with nitric acid gave a mixture containing a considerable amount of 5-hydroxy-3-methyl-2,4,6-trinitro-benzoic acid, arising, evidently, from the monohydroxylated ring in which the methyl group is a *meta* substituent. Raistrick, Robinson and Todd[411] therefore concluded that helminthosporin had structure (186) which they verified,[416] synthetically, by a two-stage condensation of 3-methoxy-5-methylphthalic anhydride and quinol dimethyl ether. More recently, it has been prepared from (180; OH and OMe inter-changed) by an Elbs oxidation, followed by cyclisation.[417]

Digitopurpone (187), $C_{15}H_{10}O_5$

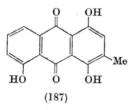

(187)

Occurrence: leaves of *Digitalis purpurea* L.[567] (Scrophulariaceae).

Physical properties: red needles, m.p. 209–211°, λ_{max}. 232, 251·5, 292, 465(sh), 413(sh), 494, 512(sh), 530(sh) nm (log ϵ 4·62, 4·35, 3·97, 4·06, 4·09, 4·15, 4·03, 3·99), ν_{max}. (KBr) 1603 cm^{-1}.

As can be seen from the above data this trace pigment from foxglove leaves is very similar to islandicin (179) and helminthosporin (186) and clearly possesses three α-hydroxyl groups. The n.m.r. spectrum includes a signal from a β-methyl group (τ 7·61) ,and as the compound is different from the isomers (179) and (186), it can only be (187) or (188). The former is correct[567] as the low field region of the spectrum consists of

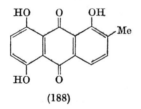

(188)

signals from an ABC system and a broad singlet (1H) at τ 2·85 which sharpens and intensifies on irradiation at τ 7·61.

Copareolatin (areolatin) (189), $C_{15}H_{10}O_6$

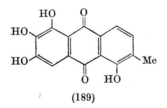

(189)

Occurrence: bark of *Coprosma areolata* Cheesem.[418] (Rubiaceae).

Physical properties: brownish-red needles, dec. >300°, λ_{max}. (EtOH) 258(sh), 288, 432 nm (log ϵ 3·96, 4·34, 3·92), ν_{max}. (Nujol) 3472, 1621, 1590 cm^{-1}.

Copareolatin occurs abundantly in the bark of *C. areolata* and, with rubiadin, accounts for 23% of the dry weight. It was characterised as a tetrahydroxy-2-methylanthraquinone by standard chemical methods. Degradation of the tetramethyl ether with chromic acid gave a little 3-methoxy-4-methylphthalic anhydride which defined the substituents in one ring, and the failure to oxidise the tetra-acetate to a carboxylic acid ($CH_3 \rightarrow CO_2H$) suggested (by analogy with such quinones as cynodontin, catenarin and morindone) that the methyl group was adjacent to an α-hydroxyl group. Colour reactions and dyeing properties favoured an anthragallol structure for the other benzenoid ring, and the relative orientation of structure (189) was preferred, by Briggs and his co-workers,[418] to that of (190) as no 1,8-dihydroxyanthraquinones were

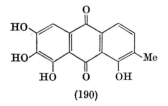

(190)

known, at that time, in the Rubiaceae. These deductions were confirmed by synthesis;[419] equimolecular amounts of gallic acid and 3-hydroxy-4-methylbenzoic acid were condensed in hot sulphuric acid to give a mixture of pigments from which 1,2,3,5-tetrahydroxy-6-methylanthraquinone, identical with copareolatin, could be separated.

Copareolatin dimethyl ether (191), $C_{17}H_{14}O_6$

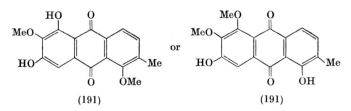

(191)   (191)

Occurrence: stem bark of *Coprosma australis* Robinson[420] (Rubiaceae).
Physical properties: orange plates, m.p. 271°, $\lambda_{max.}$ (EtOH) 281, 407 nm (log ε 4·29, 3·68), $\nu_{max.}$ (Nujol) 3300, 1653, 1629 cm^{-1}.

This quinone has two O-methyl groups and affords copareolatin tetramethyl ether on methylation. It is isomeric with copareolatin 6,7-dimethyl ether which was obtained as a by-product in the methylation of copareolatin with dimethyl sulphate and potassium carbonate in acetone. The pigment from *C. australis* is soluble in aqueous sodium carbonate (free β-hydroxyl group), and on heating in 80% sulphuric acid at 100° one methoxyl group is hydrolysed, presumably in an α-position. As both the natural material and the derived monomethyl ether can be displaced by acetic acid on a column of magnesia, vicinal hydroxyl groups are probably absent and the structure of the dimethyl ether is limited therefore to those shown above (191).[420]

7-Hydroxyemodin (192), $C_{15}H_{10}O_6$

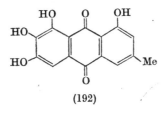

(192)

Occurrence: from the insect *Eriococcus confusus* Maskell.[422] (Coccidae).
Physical properties: red needles, m.p. 296–298° (dec.), $\lambda_{max.}$ (96% EtOH) 215, 289, 433 nm (log ε 4·44, 4·38, 4·19), $\nu_{max.}$ (Nujol) 3430, 3180, 1670, 1620 cm⁻¹.

The coccid insects from which this quinone was extracted were collected from eucalyptus trees growing in Canberra. After removal of wax the pigments were separated into water-soluble and water-insoluble fractions, the latter containing emodin and 7-hydroxyemodin.

The structure of the new pigment was deduced by Chan and Crow[422] almost entirely from spectroscopic information, the general similarity of the mass, ultraviolet-visible and infrared spectra to those of emodin providing a useful starting point. The spectral data provided clear evidence for a tetrahydroxymethylanthraquinone structure, there being two α- (1,8) and two β-hydroxyl groups. The n.m.r. spectrum confirmed the presence of three aromatic protons, and comparison with the spectrum of emodin revealed that the upfield doublet at τ 3·54, assigned to the proton at C-7, was absent but the chemical shift of the methyl group (τ 7·62) was identical in both compounds. On this basis, structure (192) was tentatively assigned[422] to the insect pigment, some confirmatory evidence being provided by its chromatographic behaviour on a magnesium carbonate column, and by a positive Bargellini test.[423] This has been confirmed by synthesis;[564] condensation of 3,4,5-trimethoxy-phthalic anhydride with 3-methoxytoluene in fused aluminium chloride-sodium chloride gave (192), identical (as its dimethyl ether) with the natural pigment, together with the isomer (196).

Dermoglaucin (193), $C_{16}H_{12}O_6$

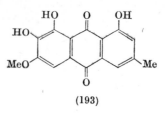

(193)

Occurrence: mainly as a glycoside in *Cortinarius sanguineus* (Wulf. ex Fr.) Fr.[334] (= *Dermocybe sanguinea*) and *Dermocybe semisanguinea* (Fr.).[615]

Physical properties: brick red crystals, m.p. 236°, λ_{max}. (EtOH) 248, 254, 282, 435, 580 nm (logϵ 3·94, 3·97, 4·30, 3·86, 3·11), ν_{max}. (KBr) 3400, 1661(sh), 1621 cm^{-1}.

The principal pigments, emodin and dermocybin, in *Cortinarius sanguineus* were originally isolated by Kögl and Postowsky.[310] In a recent investigation, Steglich and Austel[334, 615] isolated ten colouring matters from 17 kg of fresh fungus, dermoglaucin (0·004%) being one of minor components.

The visible spectrum of dermoglaucin suggests the presence of two α-hydroxyl groups, and monomethylation with diazomethane establishes the existence of a β-hydroxyl group, while signals from an O-methyl group and a C-methyl group are seen in the n.m.r. spectrum. This spectrum is very similar to that of physcion with respect to the methyl groups and the broad singlets emanating from the C-2 and C-4 protons, but the doublets from the protons at C-5 and C-7 in physcion are replaced by a sharp singlet at τ 2·90. This can be attributed to a proton at C-5 or C-8 in structure (194), and since dermoglaucin has both free and chelated

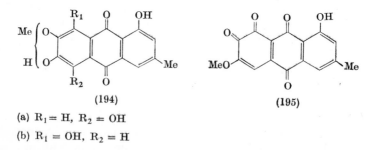

(194)   (195)

(a) $R_1 =$ H, $R_2 =$ OH

(b) $R_1 =$ OH, $R_2 =$ H

carbonyl groups (ν_{CO} 1661 and 1621 cm^{-1}), structure (194a) can be excluded. Finally, the position of the methoxyl group was determined by oxidation with lead tetra-acetate in acetic anhydride to give the red-violet diquinone (195) which was converted, on treatment with sulphuric acid, into a mixture of products containing dermocybin (p. 511). It follows that dermoglaucin has structure (193)[334] which is further confirmed by the identity of the 7-methyl ether with the 6,7-dimethyl ether of 7-hydroxyemodin isolated from *Eriococcus confusus*.[541]

Majoronal (196), $C_{16}H_{10}O_7$

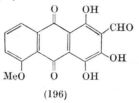

(196)

Occurrence: roots of *Damnacanthus major* Sieb. & Zucc.[567]

Physical properties: red needles, m.p. 267–269°, $\lambda_{max.}$ (hexane) 254, 262(sh), 297, 490, 540(sh) nm (log ϵ 4·42, 4·44, 4·44, 4·05, 3·13), $\nu_{max.}$ (KBr) 1652, 1634, 1590 cm^{-1}.

Majoronal is a co-pigment of nordamnacanthal (27) and norjuzunal (97). Its solubility in aqueous sodium carbonate suggests the presence of a β-hydroxyl group, probably chelated to an aldehyde function (as in 27 and 97) as the infrared spectrum shows a peak at 1652 cm^{-1} but no free hydroxyl absorption. The spectroscopic evidence, given above, points to a 1,4-dihydroxyanthraquinone structure so that, by analogy with the co-pigments, majoronal might be (196) or (197; R = Me). The former was

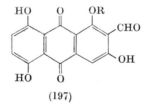

(197)

preferred when demethylation gave a product which was different from (197; R = H) obtained by synthesis.

isoErythrolaccin (198), $C_{15}H_{10}O_6$

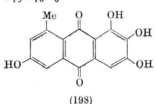

(198)

Occurrence: from the insect *Laccifer lacca* Kerr.[564] (Coccidae).

Physical properties: orange needles, dec. >320°, $\lambda_{max.}$ (EtOH) 220, 286, 412 nm (log ϵ 4·36, 4·63, 3·85), $\nu_{max.}$ (Nujol) 3250, 1665, 1635 cm^{-1}.

Isoerythrolaccin forms a tetramethyl ether whose n.m.r. spectrum includes a singlet (3H) at τ 7·22 assigned to an α-methyl group, two doublets arising from *meta* coupled aromatic protons, and a singlet (1H) at τ 2·42 which must be attributed to an α-proton in the other (trisubstituted) benzene ring. On this basis the two most probable structures are (198) and (199). Condensation of 2,3-dimethoxytoluene with 3,5-dimethoxyphthalic anhydride in fused aluminium chloride-sodium

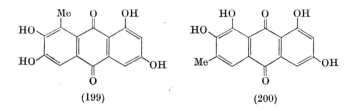

(199)   (200)

chloride gave (199) and (200), both of which were different from isoerythrolaccin. It was therefore concluded[564] that the new pigment must be (198) although it gave a negative Bargellini test. This was confirmed[564] by a similar condensation of 3,4,5-trimethoxyphthalic anhydride and 3-methoxytoluene which yielded (198), identical with isoerythrolaccin, and the isomeric insect pigment (192).

Ceroalbolinic Acid (201), $C_{16}H_{10}O_8$

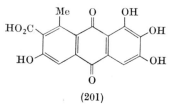

(201)

Occurrence: from the insect *Ceroplastes albolineatus* Cockerell[421] (Coccidae).
Physical properties: dark red prisms, dec. >290°, $\lambda_{max.}$ (95% EtOH) 294, 418 nm (logϵ 4·50, 3·76), $\nu_{max.}$ (CHCl$_3$) 3500, 1725, 1675, 1643, 1590 cm^{-1}.

In Mexico, the shrub *Senecio praecox* is infested during the rainy season with female coccid insects of the species *Ceroplastes albolineatus*. A thick covering of wax can be removed by treatment with chloroform, and thereafter extraction with boiling ethanol yields the red colouring matter, ceroalbolinic acid; 8 g were obtained from 120 kg of insect material.[421] The preparation of numerous derivatives established the presence of four hydroxyl groups and a carboxyl function. Thus acetylation with acetic anhydride and perchloric acid afforded a tetraacetate, soluble in aqueous sodium bicarbonate, and converted by diazomethane into a methyl ester (ν_{CO} 1790, 1740, 1680 cm^{-1}). Treatment of the insect pigment with diazomethane gives a trimethyl ether-methyl ester (ν_{CO} 1740, 1668, 1630 cm^{-1}) which forms a monoacetate (ν_{CO} 1768, 1730, 1668 cm^{-1}). Evidently one hydroxyl group is chelated. When the trimethyl ether-methyl ester was distilled over zinc "1-methylanthracene was obtained"† while decarboxylation over copper chromite in quinoline gave a tetrahydroxymethylanthraquinone, characterised as its tetraacetate and trimethyl ether (with diazomethane).

That one ring has the substitution pattern shown in (202) was established by permanganate oxidation of the trimethyl ether-methyl

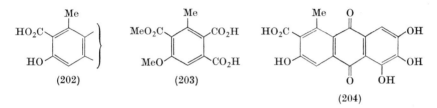

ester which gave the cochenillic acid[442] derivative (203). The other ring therefore contains three hydroxyl groups and must have an anthragallol structure as only one quinone carbonyl group is chelated. This accounts for the instability of the compound in alkaline solution. Two structures (201) and (204) can therefore be considered for ceroalbolinic acid but the chemical shifts of the aromatic protons (in two derivatives) do not permit a distinction to be made (cf. ref. 421). The problem has been resolved by a synthesis[577] of the "wrong" isomer by condensing the anhydride of cochenillic acid methyl ether with pyrogallol trimethyl ether. The product, which was different from ceroalbolinic acid, had structure (204) since decarboxylation in boiling diethylaniline and subsequent methylation with diazomethane gave a compound isomeric with isoerythrolaccin (198) trimethyl ether.[596] Ceroalbolinic acid is therefore regarded[577] as (201), a structure consistent with those of the other coccid anthraquinones. Like endocrocin, the carboxyl carbonyl stretching frequency of this compound is normal.

Kermesic Acid (205), $C_{16}H_{10}O_8$

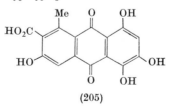

(205)

† This statement is not supported by experimental details.

Occurrence: from the insect *Kermococcus ilicis* L.[437] (Coccidae).

Physical properties: red needles, dec. ca. 250°, $\lambda_{max.}$ (EtOH) 292, 498, 538 nm (log ϵ 3·73, 3·04, 2·94), $\nu_{max.}$ (Nujol) 1670, 1623 cm^{-1}.

Kermes is perhaps the most ancient dyestuff on record.† In the Old Testament, "scarlet" is probably a translation of a Hebrew phrase denoting the red colour obtained by dyeing with kermes, and according to Bancroft[438] our words vermilion and crimson also originate from terms describing this colouring matter. The pigment was obtained from the insect *K. ilicis* which infests the kermes-oak (*Quercus coccifera*), kermes consisting of the wingless females which were collected, killed by exposure to vinegar and dried in the sun. The European supply came mainly from France, Spain, Portugal and Morocco, and was used extensively for dyeing shades of red on an aluminium mordant until supplanted by cochineal.

Despite its antiquity, kermes has attracted comparatively little chemical interest and for many years has been an exceedingly rare material. The colouring principle, kermesic acid, was first obtained pure by Heise,[437] most of the structural work being carried out later by O. Dimroth and his co-workers.[439-441] It was isolated from the insects, after removal of wax, by extraction with ethereal hydrochloric acid (5 kg of kermes yielded 50–55 g kermesic acid).

Kermesic acid forms a tetra-acetate, affords 1-methylanthracene on zinc dust distillation, and loses carbon dioxide on heating. It is therefore a tetrahydroxy-1-methylanthraquinone-carboxylic acid in agreement with the molecular formula, $C_{16}H_{10}O_8$, recently confirmed.[578] However, Dimroth's molecular formula was $C_{18}H_{12}O_9$ (the C and H values are almost the same), and the structural formula he finally adopted included a C-acetyl group, although there was no supporting chemical evidence. The position of four of the substituents was established by reduction with zinc and acetic acid, followed by heating in sulphuric acid until gas evolution ceased. This yielded 1,4,6-trihydroxy-8-methylanthraquinone, the structure of which was established by synthesis. The position of the carboxyl group can be deduced from the formation of the cochinellic acid derivative (203) when kermesic acid dimethyl ether-methyl ester was degraded with alkaline permanganate. It follows that kermesic acid should be either (205) or the 1,2,4,6-tetrahydroxy isomer, both of which are consistent with a recent n.m.r. study.[578] Structure (205) is correct as the pigment can be converted into xanthokermesic acid (= laccaic acid D, p. 443) by reduction with alkaline dithionite or hydrogenolysis over palladised charcoal.[578] An attempted synthesis[578] by condensing 2-methoxyquinol with the anhydride (206) in the presence of boron trifluoride etherate gave a good yield of the "wrong" isomer (isokermesic

† Cf. ref. 578.

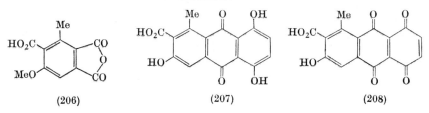

(206)   (207)   (208)

acid). However, by oxidising (207) to the diquinone (208), followed by Thiele acetylation and hydrolysis, a separable mixture of both kermesic and isokermesic acids was obtained.[564]

Carminic Acid (209), $C_{22}H_{20}O_{13}$

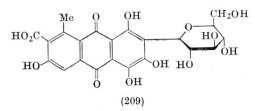

(209)

Occurrence: from the insect *Dactylopius coccus* Costa[443] (Coccidae).

Physical properties: red prisms, no m.p., $[\alpha]_{645}^{15}$ +51·6° (water), $\lambda_{max.}$ (MeOH) 278, 311(sh), 491, 540(sh) nm (log ε 4·50, 4·18, 3·89, 3·64), $\nu_{max.}$ (Nujol) 3250(br), 1700, 1600 cm^{-1}.

Cochineal is another ancient insect colouring matter. It was used in the Americas for many centuries before its introduction to Europe, and there is evidence[444] that cochineal dyeing (using *D. coccus* or other *Dactylopius* spp.) was known to the Incas in Peru. The dyestuff consists of the dried bodies of *D. coccus*, the insects being found on various cacti, especially *Nopalea coccinellifera* which was extensively grown in Mexico for the sake of the dye. After 1830 the cultivation of cochineal spread to Spain, Algeria, the Canary Islands, and to Java, but with the arrival of synthetic dyes its importance rapidly declined. The colouring matter was obtained from the female insects which were heat-killed and then ground to a powder containing about 10% of the pigment, carminic acid. It dyes wool and silk a magnificent scarlet on a tin mordant and was especially favoured for ceremonial uniforms. The British penny lilac postage stamp, first issued in 1881, was printed with the iron lake of carminic acid and not with Perkin's mauve as generally assumed.[553] Cochineal is no longer of commercial interest apart from such minor uses as the colouring of food and biological staining.

Carminic acid is easily extracted from cochineal with water and the first crystalline specimen was obtained in 1858.[443] It was the subject of numerous researches in the nineteenth century which led to the isolation of many benzenoid and naphthalenic degradation products, but carminic acid was not recognised as an anthraquinone prior to the investigations of Dimroth (1909–1920). The nature of the glycosidic moiety was elucidated by Ali and Haynes,[445] and the complete structure was finally settled in 1962 when Overeem[442] determined the correct position of the carboxyl group.

Oxidation of carminic acid with potassium persulphate[446] yields tribasic cochenillic acid, $C_{10}H_8O_7$, and dibasic α-coccinic acid, $C_9H_8O_5$. Liebermann and Voswinkel[446] considered the latter to be the same as the known[448] m-hydroxyuvitic acid (211), and since cochenillic acid could be degraded to (211), (212) and (213), they concluded that the tribasic acid had structure (210). However, as Overeem and Van der Kerk[442]

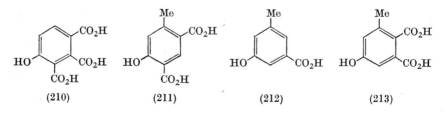

(210)   (211)   (212)   (213)

have recently pointed out, the selective decarboxylation, (210) → (211), occurring *meta* to the phenolic group is improbable; α-coccinic acid was not directly compared with m-hydroxyuvitic acid in the early work, and in fact cochenillic acid is *not* the same as mollisic acid of established[442] structure (210). [Mollisic acid methyl ether was obtained by oxidative degradation of mollisin (p. 244).] The problem is resolved if cochenillic acid is actually (214), decarboxylation *ortho* and *para* to the hydroxyl

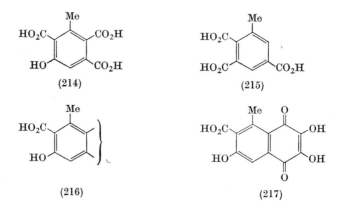

(214)   (215)

(216)   (217)

group leading to (212), (213) and (215) (i.e., α-coccinic acid). Two syntheses[442, 447] of cochenillic acid methyl ether have established that structure (214) is correct, and therefore one ring of carminic acid can be represented by (216). In agreement, the single aromatic proton in the natural pigment resonates at τ 2·33.[447] By cautious oxidation with cold acid permanganate, and other oxidants, Dimroth[449, 450] was able to isolate several intermediate compounds, including the red naphthaquinone, carminazarin (217). Like other isonaphthazarins it gives a blue colour in alkaline solution and can be oxidized to a tetraketone; further degradation leads to cochenillic acid.

In 1867, Hlasiwetz and Grabowski[588] found that fusion of carminic acid with potassium hydroxide afforded a pale yellow compound, coccinin, $C_{17}H_{14}O_6$, which gave a yellow solution in alkali, passing through green to purple on aerial oxidation to give coccinone, $C_{17}H_{12}O_7$. Dimroth[450] showed later that coccinone could be reduced to coccinin with zinc and ammonia, and recognised that these compounds had an anthrone-anthraquinone relationship. The anthraquinone (coccinone) yielded a triacetate and the anthrone (coccin) a tetra-acetate, and as coccinone could be degraded with alkaline peroxide to cochenillic acid, it was formulated as (218). There was no direct evidence for the presence of a second methyl group, and as already mentioned, ring A should have the substitution pattern represented by (216). It was then apparent that carminic acid was a polyhydroxyanthraquinone, and this was supported by the formation of a mixture of hydrocarbons on zinc dust distillation, from which anthraquinone and 1-methylanthraquinone were obtained after oxidation.

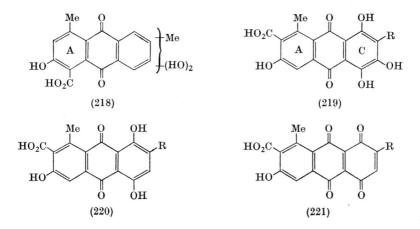

(218) (219) (220) (221)

Meanwhile Dimroth and Fick[441] had concluded that kermesic acid (205) was a substituted hydroxyanthrapurpurin(1,3,4,6-tetrahydroxyanthraquinone), and from the similarity of their visible spectra and dyeing properties it was deduced[451] that the two insect pigments had the same hydroxylation pattern, that is, carminic acid should be (219; $R = C_6H_{11}O_5$) adopting the revised structure for ring A. The purpurin structure of ring C was confirmed by reduction with zinc and acetic acid, and reoxidation in alkaline solution to give deoxycarminic acid (220; $R = C_6H_{11}O_5$). This was then converted back into carminic acid by oxidation with lead tetra-acetate to the diquinone (221; $R = C_6H_{11}O_5$) and Thiele acetylation.

As carminic acid forms an octa-acetate, the side-chain, $C_6H_{11}O_5$, must contain four hydroxyl groups, the fifth oxygen being in an ether linkage. As it is resistant to normal acid hydrolysis, the nature of the side-chain baffled early workers until Ali and Haynes [445] demonstrated that it was a C-glycoside, analogous to barbaloin (p. 401). Both glycosides yield D-arabinose and D-glucose on oxidation with ferric chloride, or better, by ozonolysis in the case of carminic acid. Treatment of the pigment with a large excess of diazomethane in ether/methanol methylates *all* the phenolic groups as well as the carboxyl function. As this compound rapidly consumes 2·1 mol. of periodate at 0° with the formation of formic acid, the substituent at C-2 must be D-glucopyranosyl. Carminic acid can therefore be represented as (209), the only example of a quinone C-glycoside.

LACCAIC ACIDS

Coccid insects are characterised by the secretion of a waxy or resinous protective covering containing a mixture of metabolic products, including anthraquinones.† This is most highly developed in the Lacciferidae or lac insects, of which the Indian species, *Laccifer lacca* Kerr, is the most important. The crude product, stick-lac, was used in ancient China [462] and was a valuable dye in Eastern countries for centuries before it was known in Europe. On a tin mordant it gave a scarlet, equivalent to cochineal. The Indian lac industry is now concerned only with shellac and the dye (perhaps 100,000 kg per annum [457]) is lost in the water washings. The female insects are collected from the host trees on which they thrive (*Schleichera oleosa*, *Zizyphus mauritania*, *Butea monosperma*, and others) by cutting down the twigs and scraping off the lac encrustation. This material (stick-lac) is then crushed, and the water-soluble colouring matter, "laccaic acid", is extracted with water or dilute sodium carbonate solution. The residual material, seed-lac,

† Little is known of the distribution of the quinones. In addition to those mentioned in the text they may occur in the European species, *Porphyrophora polonicus*, formerly used for dyeing,[438] and in the Australian species, *Austrotachardia acaciae*.[452] See also 7-hydroxyemodin (p. 453).

contains the water-insoluble quinones, erythrolaccin, deoxyerythro-laccin and isoerythrolaccin.

Until recently, relatively little was known of the chemistry of "laccaic acid". It was examined initially by Schmidt,[453] and later by Dimroth and Goldschmidt.[454] A general resemblance to carminic acid was recognised and it was partially formulated as a polyhydroxyanthra-quinone. However, recent work by Venkataraman[456, 580, 581, 582] and Schofield[458] and their co-workers, has shown that "laccaic acid" is a mixture of pigments which can be separated chromatographically, albeit with some difficulty. Some of these contain nitrogen which was over-looked by the earlier workers, who, in any case, were handling mixtures. The present account deals only with the recent investigations.† The compounds isolated and purified so far are designated by Schofield as laccaic acid A_1, A_2 and B while Venkataraman's pigments are named laccaic acid A, B, C, D and E, of which A, B and C are also present in lac larvae,[564] C being partly associated with protein. It is fairly certain that $A = A_1$, but all the others appear to be different.

Laccaic Acid A (A_1) (222), $C_{26}H_{19}NO_{12}$

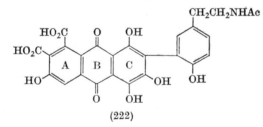

(222)

Occurrence: from the insect *Laccifer lacca* Kerr. (Coccidae).[456d]

Physical properties: red plates, dec. 230°, λ_{max}. (EtOH) 290, 340, 500, 530 nm (log ϵ 4·80, 4·31, 4·32, 4·12), ν_{max}. (Nujol) 1715, 1692, 1667, 1626 cm^{-1}.

Oxidation of laccaic acid A with alkaline hydrogen peroxide gave the acid (223), identified[456] by synthesis of its fully methylated deri-vative. This shows that one benzenoid ring has either structure (224) or (225), and the former was preferred after comparison of the pK_a values[458] with those estimated from the pK_a values of

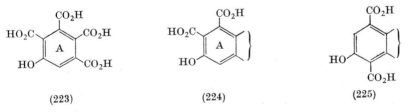

(223)   (224)   (225)

† For a summary of the early work see ref. 455.

anthraquinone-1,2-dicarboxylic acid and carminic acid, and is consistent with a one proton singlet at $\sim \tau$ 2·2 in the n.m.r. spectra of laccaic acid A and its derivatives.[456] Various colour reactions suggest that the other benzenoid ring has a 1,2,4-trihydroxylation pattern and it undergoes the characteristic purpurin-xanthopurpurin reduction on treatment with alkaline dithionite.[456, 458] The product, xantholaccaic acid A (ν_{CO} 1703, 1675, 1623 cm^{-1}), contains one atom of oxygen less, and one aromatic proton (singlet at τ 2·42) more than the parent compound, and forms an amorphous triacetate whereas laccaic acid A gives a tetra-acetate. At this stage the pigment appears to be a tetrahydroxyanthraquinone dicarboxylic acid but the ultraviolet-visible absorption of laccaic acid A and various derivatives are not wholly consistent with this concept, and of course a significant part of this C_{26} molecule has still to be considered. Having settled the structure of ring A the Indian workers[456] then proceeded to examine a series of methylation products of laccaic acid and xantholaccaic acid, and from a study of their n.m.r. and mass spectra were able to make the surprising announcement in 1963[456] that these pigments are derivatives of 2-phenylanthraquinone. The n.m.r. spectrum of fully methylated xantholaccaic acid (molecular weight 605), for example, shows signals derived from six methoxyl groups (two of which are esters) and five aromatic protons, two of which (at low field) could be assigned to α-positions in an anthraquinone nucleus. A multiplet (4H) centred near τ 7·25 was attributed to the methylene chain in an ArCH$_2$CH$_2$NHAc system, and, by comparison with a model compound, the broad triplet at $\sim\tau$ 4·2 appeared to originate from an –NH– group in a similar acetylamino function, which was consistent also with a C-methyl signal at τ 8·08. Thus most of the structure of xantholaccaic acid A, and hence laccaic acid A, could be deduced from the n.m.r. spectrum of this derivative. Xantholaccaic acid A has four, and laccaic acid A five phenolic groups, and as four of the lattter are located on the anthraquinone nucleus (one in ring A and three in the purpurin system of ring C), the fifth must be in the additional benzene ring. The formation of only a triacetate by xantholaccaic acid A and a tetra-acetate by laccaic acid A may possibly be explained[458] by the fact that the phenolic group adjacent to carboxyl in ring A is easily hydrolysed during work-up. Three of the aromatic protons, revealed by the n.m.r. spectra of laccaic acid A and its derivatives, are coupled and can only be assigned to a trisubstituted benzene ring. One substituent is methoxyl, as we have seen, and the other must be acetylaminoethyl as the coupling characteristics of these three protons change when this side chain is modified.

Chemical evidence for the –CH$_2$CH$_2$NHAc group was obtained[458] by alkaline hydrolysis of laccaic acid A which gave acetic acid (1 mol.) and deacetyl laccaic acid A as a red, amorphous, solid. When this was methylated with dimethyl sulphate and potassium carbonate in acetone it gave a mixture of products, one of which, a quaternary methosulphate, was subjected to a Hofmann degradation to eliminate the nitrogen (as trimethylamine). The product contained the expected carbon-carbon double bond and when submitted to Lemieux oxidation (periodate-osmium tetroxide) it yielded formaldehyde. This sequence of reactions (see below) indicated the presence of a side chain $-CH_2CH_2NHAc$ which is fully supported by the n.m.r. spectrum of the pigment,

$$ArCH_2CH_2NHAc \rightarrow ArCH_2CH_2NH_2 \rightarrow Ar'CH_2CH_2\overset{+}{N}Me_3(MeSO_4^-)$$

laccaic acid A   deacetyl laccaic
                            acid A

$$Ar'CHO + CH_2O \leftarrow Ar'CH{=}CH_2 + NMe_3$$

as already mentioned, and of the Hofmann product. The peak at 1677 cm^{-1} in the infrared spectrum of laccaic acid A can now be assigned to the amide carbonyl group.

One of the fully methylated derivatives of laccaic acid A contains only *five* O-methyl groups and one of the fully methylated xantholaccaic acid A derivatives has *six* O-methyl groups. As xantholaccaic acid A has one hydroxyl *fewer* than laccaic acid A this means that methylation of laccaic acid A is accompanied by a cyclisation in which the elements of water (or methanol) are lost. As the cyclisation has a marked effect on the chemical shifts of the three protons in the additional benzene ring, the hydroxyl group attached to this ring is probably involved in the reaction. The cyclisation is readily understood if an *o*-hydroxyphenyl group is placed at C-2 (222). After methylation of the relatively acidic hydroxyl at C-3, nucleophilic substitution by the hydroxyl group (or its anion) on the adjacent phenyl ring could then proceed very easily if the quinizarin derivative (226) reacted in the tautomeric form (b). This would lead to the formation of a furan ring (227).[458] (This structure was also deduced from n.m.r. and mass spectral evidence.[456c]) The reaction is illustrated by the conversion of the dihydroxyquinone (228) into the brazanquinone (229) by treatment with dimethyl sulphate and potassium carbonate in boiling acetone.[459] Support for the tautomerisation depicted in (226) comes from the observation that xantholaccaic acid, which has only one *peri*-hydroxyl group, does not undergo this cyclisation. Analysis of the n.m.r. spectra of laccaic acid A and its derivatives shows that the three aromatic protons in the phenyl ring form an ABX system, and their chemical shifts indicate that the acetylaminoethyl side chain is located *para* to the hydroxyl group.

There remains the final problem of the relative orientation of rings A and C. On catalytic hydrogenation laccaic acid A undergoes an unusual

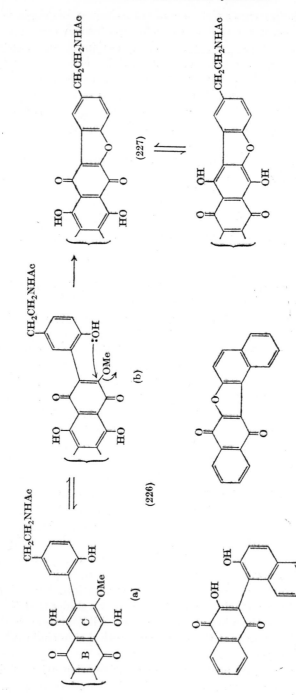

reduction in which an oxygen atom is removed from a *meso* position. The product, deoxylaccaic acid, which shows unchelated carbonyl absorption at 1667 cm^{-1} and forms a triacetate ($\lambda_{max.}$ 459 nm), is formulated as a derivative of 3,10-dihydroxy-1,4-anthraquinone by analogy with the behaviour of purpurin (see p. 407). Deoxylaccaic acid A could be either (230) or (231). In the n.m.r. spectrum of 3,10-dihydroxy-1,4-anthra-quinone signals arising from the protons at H-9 and H-5 appear at τ 1·63 and 1·42, shifted to τ 1·25 and 2·09, respectively, in the spectrum of deoxylaccaic acid A. Allowing for the influence of the hydroxyl group at

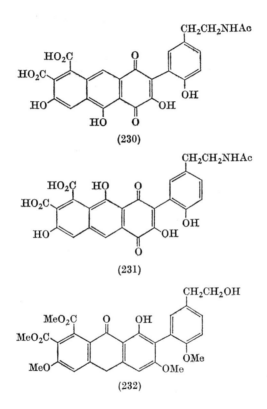

(230)

(231)

(232)

C-6 (on H-5) and the carboxyl group at C-8 (on H-9), these values are consistent with structure (230) but not (231). It follows that the structure of laccaic acid A is (222).[458] The relative orientation of rings A and C has also been determined conveniently by stannous chloride reduction of a xantholaccaic acid B derivative to give the anthrone (232). This results in an upfield shift of the signals from *both* α-protons in agreement with the structure assigned.[609]

Laccaic Acids B, C and E.† Several of the numerous methylation products derived from "laccaic acid" have been identified by Venkataraman and his co-workers,[456c,d, 582] chiefly by n.m.r. spectroscopy. One of these, a carbonate methyl ester (232b), obtained by treatment with methyl sulphate in the presence of potassium carbonate, was also derived from laccaic acid B, which can be separated from A by chromatography of a butanolic hydrochloric acid solution on a column of polycaprolactam. Accordingly, B was considered[456c] to be an intermolecular or intra-molecular carbonate of (232a; R = OH). However, further work,[582] with

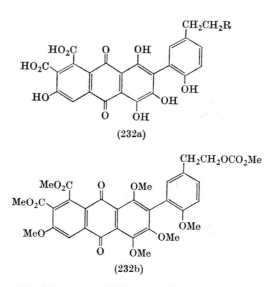

(232a)

(232b)

more highly purified laccaic acid B, showed that it is, in fact, the alcohol (232a; R = OH), the carbonate ester being formed by the action of dimethyl sulphate and potassium carbonate via dimethyl carbonate. Similarly, β-phenylethanol yields β-phenylethyl methyl carbonate under the same conditions.

When "laccaic acid" is chromatographed on a column of cellulose powder the major components A and B form a single band, and a minor fraction separates into laccaic acids C and E. Laccaic acid C, dark red needles, dec. >360°, contains nitrogen and closely resembles A and B in colour reactions and ultraviolet-visible spectrum, but lacks the N-acetyl group of A. Acetylation, followed by hydrolysis of the O-acetyl groups, gave an N-acetyl derivative which was different from laccaic acid A. The corresponding xantholaccaic acid was obtained by catalytic hydrogenation in alkaline solution, and this, on methylation (methyl iodide-silver oxide-dimethylformamide) gave a heptamethyl derivative to which

† Experimental details are not yet available.

structure (233) can be assigned on the basis of n.m.r. and mass spectral data. The chemical shifts of the five aromatic protons are fully consistent

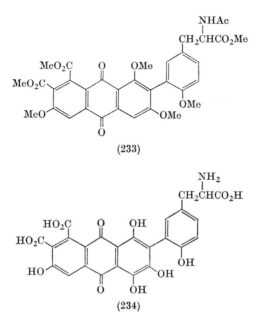

(233)

(234)

with those of related compounds, and the presence of an N-acetyl group is confirmed by a singlet (3H) at τ 8·03 and broad absorption at τ 3·9. Unlike other methylated xantholaccaic acids, the usual signals from two methylene groups are replaced by multiplets at τ 7·00 (2H) and 5·15 (1H) which correspond with a side chain of structure $-CH_2CH(NHAc)CO_2Me$ and are in good agreement with the spectra of model compounds. In the mass spectrum of (233) ($M^{+\cdot}$ 663) the base peak occurs at m/e 533 ($M-CH(NHAc)CO_2Me$), and the next most abundant ion appears at M-59 resulting from loss of CH_3CONH_2 as in other methylated laccaic acid A derivatives. It follows from the structure of this derivative (233) that laccaic acid C must be the amino-acid (234);[581] it gives a positive ninhydrin reaction. This is a unique example of a natural quinone bearing an amino-acid side chain. It is obviously derived from tyrosine and is presumably introduced *in vivo* by phenolic coupling with a purpurin derivative.

Laccaic acid E has not yet been obtained completely free from C, but is probably deacetyl laccaic acid A (232; $R = NH_2$). Acetylation, followed by hydrolysis of O-acetyl groups, gave a product chromatographically identical with laccaic acid A.[581]

Laccaic acid D was considered on p. 443.

Erythrolaccin (235), $C_{15}H_{10}O_6$

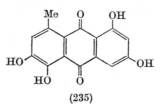

(235)

Occurrence: from the insect *Laccifer lacca* Kerr.[424, 425] (Coccidae).
Physical properties: orange needles, m.p. 314° (dec.), λ_{max}. (EtOH) 228, 266, 295, 466 nm (log ϵ 4·26, 4·18, 4·13, 3·74), ν_{max}. (KBr) 3465, 1629(sh), 1623, 1602(sh) cm^{-1}.

Erythrolaccin is a minor pigment in stick-lac, first isolated by Tschirch and Lüdy [424] who correctly considered it to be a tetrahydroxy-methylanthraquinone but provided little evidence. Later work by Venkataraman and his co-workers [425] established the presence of two β-hydroxyl groups (methylated with diazomethane) and two α-hydroxyls in the 1,4- or 1,5-positions (no free carbonyl absorption). Colour reactions and dyeing properties excluded a 1,2- ,1,4- or 1,2,3- arrangement of hydroxyls but a 1,3- orientation seemed likely from the ease of hydroxy-methylation with alkaline formaldehyde. This reaction occurs readily at C-2 in 1,3-dihydroxyanthraquinones and appears to be diagnostic for such structures. As zinc dust distillation provided 2-methylanthracene these results led logically to structure (236) for erythrolaccin.[425] However

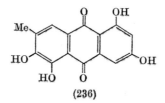

(236)

this is incorrect. Examination [425, 427] of the n.m.r. spectrum of the tetramethyl ether revealed the expected doublets arising from the *meta*-protons in the resorcinol ring but the remaining aromatic singlet, at τ 3·01, is too far upfield for an α-proton at C-8 in (236) [cf. alaternin (237) tetramethyl ether whose corresponding α-proton signal falls at τ 2·12 [428]]. Accordingly the structure of erythrolaccin was revised [427, 428] to (235) which Yates *et al*.[427] confirmed synthetically by condensing 3,5-di-methoxyphthalic anhydride with 4-methylcatechol in an aluminium chloride-sodium chloride melt. Another synthesis, starting from 3-chlorophthalic anhydride and 4-hydroxy-3-methoxytoluene, has been reported.[436] Evidently a methyl migration occurred during the zinc dust distillation of erythrolaccin. Conversely, distillation of 1-methyl-5,8-dihydroxyanthraquinone over zinc is reported[429] to give some 2-methylanthracene as well as the 1-isomer.

Coccid pigments. All the anthraquinones isolated from coccid insects are set out below with the exception of 7-hydroxyemodin. Largely owing to the work of Venkataraman and his colleagues in Poona, it can now be seen that they adhere to a common structural pattern characterised by adjacent carbon side chains in α,β-positions or an α-methyl group only

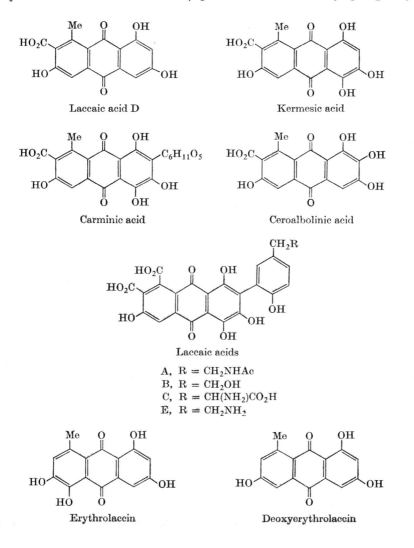

Laccaic acid D

Kermesic acid

Carminic acid

Ceroalbolinic acid

Laccaic acids

A, R = CH$_2$NHAc
B, R = CH$_2$OH
C, R = CH(NH$_2$)CO$_2$H
E, R = CH$_2$NH$_2$

Erythrolaccin

Deoxyerythrolaccin

where decarboxylation may have occurred. In this respect they differ from the anthraquinones found in plants yet they appear to be of polyketide origin; the exceptional 7-hydroxyemodin is, of course, of the type found in plants and it was isolated along with emodin from *Eriococcus confusus*. Although there is an obvious biosynthetic relationship[578] between these coccid quinones, it is by no means certain that they are all natural compounds. One or two, at least, are almost certainly artefacts either of the method of isolation or of post-mortem enzymic changes. The Coccoidea and Aphidoidae are related superfamilies, and analogy suggests that some of these pigments may arise from glycosides which occur *in vivo* and undergo changes after the death of the insect (cf. Chapter 7).

Alaternin (237), $C_{15}H_{10}O_6$

Alaternin 6-methyl ether, $C_{16}H_{12}O_6$

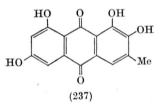

(237)

Occurrence: (237) bark of *Rhamnus alaternus* L.[303] (Rhamnaceae); the 6-methyl ether occurs in cultures of *Alternaria porri* (Ell.) Saw.[294] (= *A. solani*).

Physical properties: (237) brick-red plates, m.p. 310°, λ_{max}. (EtOH) 229, 252, 283·5, 317·5, 430 nm (log ε 4·32, 4·11, 4·40, 4·06, 4·01), ν_{max}. (Nujol) 3367, 3195, 1686, 1613 cm^{-1}.

6-methyl ether, orange crystals, m.p. 267–268°, λ_{max}. (96% EtOH) 229, 280, 312, 425 nm (log ε 4·36, 4·43, 4·00, 4·03), ν_{max}. (KBr) 3480, 1665, 1622 cm^{-1}.

The bark of *Rh. alaternus* contains an abundance of hydroxyanthraquinone glycosides (22% of the dry weight) of which emodin and alaternin are the principal aglycones. Analyses, and the formation of derivatives, showed that the new pigment was another tetrahydroxymethylanthraquinone. The spectroscopic data indicated that two hydroxyls are located at the 1,8-positions, while colour and dyeing tests were consistent with a 1,2- but not a 1,2,3- or 1,2,7,8- arrangement of phenolic groups. From this a 1,2,6,8- orientation seemed the most likely and, assuming that the methyl group occupies a β-position as in other Rhamnaceous anthraquinones, Briggs and co-workers[303] favoured the hydroxyemodin structure (237) for alaternin on phytochemical grounds. A positive reaction with alkaline formaldehyde provided some confirmation[425] and structure (237) was established[430] by acylating 3,5-dimethoxybenzoic ester with 2,3-dimethoxy-4-methylbenzoyl chloride to give (238) which was converted into alaternin on brief treatment in molten aluminium chloride-sodium chloride. Another synthesis was mentioned on p. 456 (200–237).

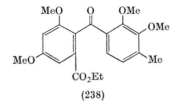

(238)

The monomethyl ether found in an *Alternaria* culture was considered to be the 6-methyl ether as this pigment alone, of the *Alternaria* anthraquinones, gave an olive-brown colour with ferric chloride which was attributed to the presence of a catechol structure.[294]

Altersolanol A (239), $C_{16}H_{16}O_8$

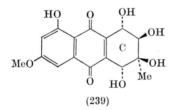

(239)

Occurrence: cultures of *Alternaria porri* (Ell.) Saw.[293, 294] (= *A. solani*).

Physical properties: m.p. 218° (dec.), $[\alpha]_D^{23}$ −292° (pyridine), $\lambda_{max.}$ (96% EtOH) 219, 240, 268, 285(sh), 422 nm (log ϵ 4·57, 3·96, 4·15, 3·84, 3·65), $\nu_{max.}$ (KBr) 3650–3000, 1671, 1635, 1603, 1595 cm^{-1}.

In addition to the toxin, alternaric acid, the plant pathogen *Alternaria porri* elaborates several pigments, two of which have modified anthraquinone structures. Altersolanol A, $C_{16}H_{16}O_8$, can be dehydrated by sublimation† to give, principally, alaternin-6-methyl ether, $C_{16}H_{12}O_6$, a

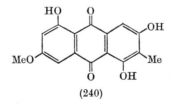

(240)

co-metabolite, and a minor pigment characterised spectroscopically as the quinone (240). Whereas these dehydration products contain three

† For a period of 17 days!

phenolic groups, the n.m.r. spectrum of altersolanol A reveals the presence of one chelated phenolic group, and four alcoholic hydroxyls. Other signals can be attributed to an aromatic methyl ether, a C-methyl group (singlet at τ 8·69), two *meta*-aromatic protons and three other protons [τ 5·46 (d), 5·62 (br. s), and 6·32 (d)] which shift downfield on acetylation. A tetra-acetate, prepared in pyridine, still retains an alcoholic hydroxyl which can be acetylated using perchloric acid as catalyst. Hydrolysis of the penta-acetate regenerates altersolanol A. This information shows that altersolanol A has one tertiary and three secondary hydroxyl groups, and the combined evidence is consistent with the gross structure of (239). Additional confirmation of the structural assignments at C-2 and C-3 was obtained by a periodate oxidation which afforded equimolecular amounts of acetic and formic acids.

Attempts to methylate altersolanol A were unsuccessful but the reaction with diazomethane provided unexpected support for the naphthaquinonoid nature of the molecule. It is well-known[431] that diazomethane will add to the 2,3-double bond of naphthaquinones and the product, in this case, was a mixture of benzindazolequinones of which (242) (or an isomer) was a major product. To account for this Stoessl[294] suggests that the initial adduct (241) collapses, as indicated, and then undergoes methylation on nitrogen, and also, at some stage, on the

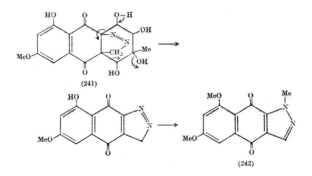

(241)

(242)

peri-hydroxyl group as well. In accordance with this postulate altersolanol A penta-acetate gave a single stable product with diazomethane, analogous to (241), which on treatment with alkali afforded a mixture of benzindazolequinones.

Regarding the stereochemistry [432] of altersolanol A, the protons at C-1
and C-2 give rise to a pair of doublets with J ~7 Hz which suggests, by
analogy with related compounds (for example the eleutherins, p. 282),
a *trans*-axial-pseudoaxial relationship. Further, the chemical shift of the
C-4 proton in altersolanol A tetra-acetate moves downfield when the
remaining, tertiary, hydroxyl at C-3 is acetylated, which implies [587] that
the deshielded proton is pseudoequatorial and the vicinal acetoxyl group
is axial. It was also established, by careful study, that the isopropylidene
derivative formed by altersolanol A involves the hydroxyls at C-2 and
C-3, and consequently the stereochemistry of ring C can be represented

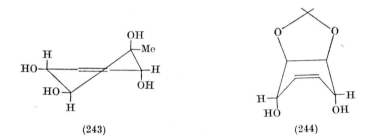

(243)   (244)

by (243) or its mirror image.[432] In the isopropylidene derivatives, the H_1,
H_2 coupling constants are reduced to ~2·6 Hz showing that a change of
conformation to the half-boat form (244) has occurred. On heating in
pyridine the acetonide triacetate aromatises to form (245), in which an

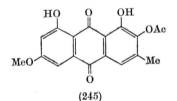

(245)

acetyl group has migrated from the oxygen at C-1 to that at C-2.[432]

Altersolanol B (246), $C_{16}H_{16}O_6$

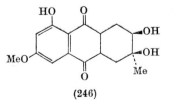

(246)

Occurrence: cultures of *Alternaria porri* (Ell.) Saw.[294] (= *A. solani*).

Physical properties: lustrous red-brown plates, m.p. 228–230°, $\lambda_{max.}$ (96% EtOH) 217, 265, 285, 421 nm (log ε 4·52, 4·14, 3·97, 3·58), $\nu_{max.}$ (KBr) 3490, 3420, 1679, 1635, 1619, 1599 cm^{-1}.

Altersolanol B contains two hydroxyl groups less than A but is otherwise very similar. The n.m.r. spectra of the parent compound and its triacetate reveal the presence of two benzylic methylene groups, and on aromatisation with thionyl chloride in pyridine it gave 6-methylxanthopurpurin 3-methyl ether. This implies that altersolanol B is (246), the *cis*-diol structure being supported by the ready formation of an isopropylidene derivative.[294]

Bostrycin (247), $C_{16}H_{16}O_8$

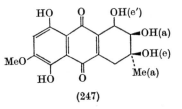

(247)

Occurrence: cultures of *Bostrychonema alpestre* Ces.[585]

Physical properties: red crystals, m.p. 222–224° (dec.), $\lambda_{max.}$ (EtOH) 228, 303, 472, 505, 542 nm (log ε 4·27, 3·69, 3·47, 3·60, 3·39), $\nu_{max.}$ (KBr) 3510, 3480, 3360, 1595 cm^{-1}.

This new antibiotic pigment is another tetrahydro-anthraquinone derivative. The infrared and ultraviolet-visible spectra suggest that bostrycin is a naphthazarin. This is supported by the n.m.r. spectrum (singlets at τ −3·12 and −2·42) which also includes signals attributable to three more hydroxyl groups, a methoxyl, an aliphatic methyl and a benzylic methylene group, one aromatic and three aliphatic protons. One of the hydroxyl groups is evidently benzylic as it is removed on hydrogenolysis giving, after reoxidation, deoxybostrycin, which now contains two benzylic methylene groups (n.m.r.).

On refluxing in formic acid bostrycin is converted into three hydroxyanthraquinones which were isolated as their acetates. The major product, arising by dehydration, is regarded as the triacetate (248), and the minor products whose formation must entail oxidation and reduction as well as dehydration, are considered to be (249) and (250). The n.m.r. spectra are consistent with these structures but give no information concerning the relative orientation of ring C with respect to ring A, that shown being adopted on the basis that a 3,6- arrangement of methyl and methoxyl groups is common to a number of natural anthraquinones. (However,

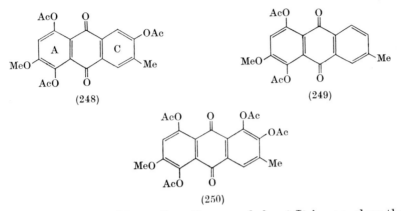

(248)   (249)

(250)

there is crystallographic confirmation—see below.) It is now clear that ring C of bostrycin contains three hydroxyl groups and they must occupy vicinal positions as shown in (247). Deoxybostrycin is consequently (251) and the unstable leuco compound formed on hydrogenolysis is (252).

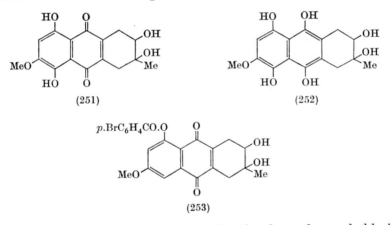

(251)   (252)

(253)

When this compound (252) was treated with *p*-bromobenzoyl chloride in pyridine it formed an amorphous 5,8-di-*p*-bromobenzoate of (251) (which gave a crystallisable monobromobenzoate after adsorption on silica gel) and also, quite remarkably, the mono-*p*-bromobenzoate (253) which appears to be a derivative of altersolanol B (or its mirror image).† The same compound was also obtained by hydrogenolysis of the amorphous di-bromobenzoate. The elimination of a *peri*-hydroxyl group by treatment with *p*-bromobenzoyl chloride in pyridine is difficult to explain but the structure of the product has been established by X-ray crystallographic analysis.[586] This showed that ring C has the half-chair

† On hydrolysis it gave, *inter alia*, a compound indistinguishable (u.v., t.l.c.) from altersolanol B.[595]

conformation (254) and as the chemical shift of the axial methyl group in bostrycin is not affected by acetylation of the C-1 hydroxyl group, the latter must have a pseudoequatorial configuration. Bostrycin is thus (247).[585]

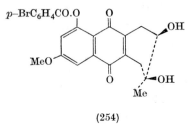

(254)

With acetic anhydride in pyridine bostrycin forms a triacetate, and it was considered that the tertiary hydroxyl and the more hindered phenolic group remained free (255). It shows a juglone-type electronic spectrum

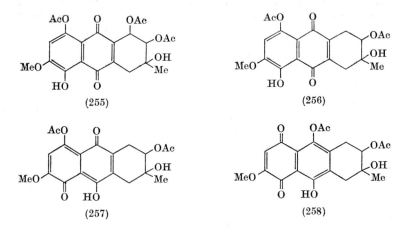

(255)   (256)

(257)   (258)

($\lambda_{max.}$ 430 nm), chelated carbonyl absorption at 1637 cm^{-1} but apparently no free carbonyl absorption ca. 1660 cm^{-1}.† Hydrogenolysis of (255) eliminates the benzylic acetoxyl group to give the diacetate (256). In the n.m.r. spectrum of this compound signals at τ 6·11 (OCH_3), 4·02 (ArH) and −2·59 (*peri*-OH) are accompanied by peaks of 10% intensity at τ 6·04, 3·32 and −2·96, and this was attributed to the existence of the tautomeric equilibrium (256 $\rightleftharpoons$ 257). It seems more likely, however, that the diacetate is a mixture of (256) and (258) formed by partial migration of an acetyl group to a neighbouring *peri*-hydroxyl during the hydrogenation (for another example see ref. 589). The major isomer which shows nuclear proton resonance at τ 4·02 must be (258).

† The authors[585] assign bands at 1637 and 1602 cm^{-1} to free and chelated carbonyl absorption, respectively.

Xanthorin (lauropurpone) (259), $C_{16}H_{12}O_6$

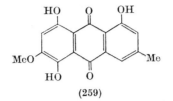

(259)

Occurrence: in the lichens *Xanthoria* (= *Caloplaca*) *elegans* (Link) Th. Fr.[433] and *Laurera purpurina* (Nyl.) Zahlbr.,[600] and the heartwood of *Maesopsis eminii* Engl.[572] (Rhamnaceae).

Physical properties: red needles, m.p. 253°, $\lambda_{max.}$ (ethanol:dioxan, 10:1) 234, 256, 302, 457, 487, 507, 520 nm (log ϵ 4·50, 4·56, 3·79, 4·06, 4·16, 4·01, 4·01), $\nu_{max.}$ (KBr) 1597 cm^{-1}.

In a recent re-examination of the lichen *X. elegans*, 2·6 kg of dried plant material yielded 27 g of physcion (p. 429) and 0·26 g of a new quinone, xanthorin. It is a major pigment in the dark red lichen *Laurera purpurina* which does not, however, contain physcion. The presence of one O–Me and one C–Me group was revealed by the n.m.r. spectrum, and the similarity of the ultraviolet-visible absorption to that of dermocybin, which was familiar to Steglich *et al.*,[334] indicated that three α-hydroxyl groups were included in the molecular structure. The relative positions of the five substituents were immediately apparent when the n.m.r. spectra of xanthorin and emodin (run as their tris-trimethylsilyl ethers in carbon tetrachloride) were compared. The signals from the C-methyl group and the aromatic protons at C-2 and C-4 are almost identical in position and shape in both spectra, but in the xanthorin spectrum a singlet at τ 3·52, from the C-7 proton, replaces a doublet and there is no signal from a proton at C-5. This defines the structure of xanthorin as (259).[433] A compound of this structure (m.p. 8° lower) was previously derived[373] from 5-bromo-emodin trimethyl ether but was obtained by Steglich *et al.*[433] from emodin by direct hydroxylation with potassium hydroxide, followed by partial methylation of the main product. The hydroxylation step is ambiguous but the product must have structure (259) as it is different from erythroglaucin (304). Xanthorin has also been derived from physcion, in low yield, by Elbs persulphate oxidation, the minor products including a trace of erythroglaucin.

The tetrahydroxyquinone obtained by demethylation of (259) has also been detected (t.l.c.) in extracts of *L. purpurina*.[600]

Valsarin I (papulosin?) (260), $C_{15}H_9ClO_6$

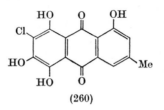

(260)

Occurrence: in the fungus *Valsaria rubricosa* (Fr.) Sacc.[381, 624] and in the lichens *Lasallia papulosa* (Ach.) Llano[584] and *L. papulosa* (Ach.) Llano var. *rubiginosa* (Pers.).[624]

Physical properties:† red needles, m.p. 273–274°, λ_{max}. (C_6H_{12}) 245, 257, 305, 485, 518 nm, ν_{max}. (KBr) 1605 cm^{-1}.

Valsarin I forms a tetra-acetate and a tetramethyl ether, and the above spectral data suggest that three of the hydroxyl groups occupy α-positions. This is consistent with the observation that the product obtained on oxidation with manganese dioxide and sulphuric acid shows visible absorption almost identical with that of 1,4,5,8-tetrahydroxy-anthraquinone. As this compound still contains chlorine it must be assigned a β-position and if this is *ortho* to the β-hydroxyl group it would account for the unusual acidity of valsarin (soluble in aqueous sodium bicarbonate).[381] The relative positions of the chlorine atom and β-hydroxyl group were assumed on phytochemical grounds and the co-existence of 7-chloro-emodin, and the characteristic substitution pattern of the other benzenoid ring was established by the n.m.r. spectrum of the tetra-trimethylsilyl ether.[584] Hence valsarin I is (260).[381, 584]

Bohman[624] found that valsarin from some specimens of *V. rubricosa* could be separated into two components, valsarins I and II, which are

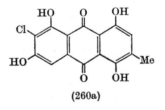

(260a)

also present in the lichen *L. papulosa* var. *rubiginosa*. It is suggested[624] that valsarin II may be the isomer (260a).

† Of valsarin;[381] papulosin,[584] reddish-brown plates, m.p. 268–269°, λ_{max}. (MeOH) 239, 264, 307, 351, 465, 491, 520, 572 nm, ν_{max}. (?) 3480, 1590 cm^{-1}, may be an isomer.

Clavorubin (261), $C_{16}H_{10}O_8$

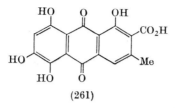

(261)

Occurrence: in ergot (*Claviceps purpurea* (Fr.) Tul.[397]).
Physical properties: dark red powder, dec. ca. 232°, $\lambda_{max.}$ (MeOH) 265, 340, 496, 528, 564 nm, $\nu_{max.}$ (KBr) 3390, 1720, 1639, 1623, 1600 cm⁻¹; methyl ester, red needles, m.p. 259°, $\lambda_{max.}$ (di-isopropyl ether) 260, 305, 380, 471, 500, 536 nm (log ε 4·37, 3·85, 3·28, 3·85, 3·99, 3·86), $\nu_{max.}$ (KBr) 3487, 1719, 1611 cm⁻¹.

The pigments of ergot (*C. purpurea*)[590] include minor amounts of two closely related anthraquinones, endocrocin and clavorubin. The latter (1·6 g crude material from 3·2 kg of the fungus)[435] is somewhat intractable and was more easily handled as its crystalline methyl ester. The new pigment contains a C-methyl and a carboxyl group (ν_{CO} 1720 cm⁻¹, cf. endocrocin), and the spectroscopic and analytical data suggest the presence of one β- and three α-hydroxyl groups.[434] The mass spectrum of the methyl ester, $C_{17}H_{12}O_8$, has a base peak at M-32 which can be attributed to loss of methanol from a methyl ester of structure (262),

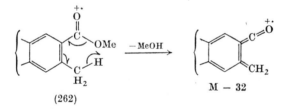

(262)   M − 32

showing that the C-methyl group is adjacent to carboxyl in clavorubin. Similarly, endocrocin shows an intense M-18 peak, and as both pigments show carboxyl carbonyl absorption at unusually high frequency it may be concluded that clavorubin is also a 1-hydroxy-3-methylanthraquinone-2-carboxylic acid. The close relationship to endocrocin was confirmed[435] when the latter was oxidised with manganese dioxide and sulphuric acid to give, *inter alia*, clavorubin, and it was shown that hydroxylation had taken place at C-5 (and not C-4) as the product obtained on decarboxylation of clavorubin is *not* catenarin (p. 494). Clavorubin is therefore the quinone (261).[435]

16

Averythrin (263; R = H), $C_{20}H_{18}O_6$

Averythrin 6-methyl ether (263; R = Me), $C_{21}H_{20}O_6$

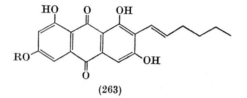

(263)

Occurrence: (263; R = H) cultures of *Aspergillus versicolor* (Vuill.) Tiraboschi;[470] (263; R = Me) in the lichen *Solorina crocea* (L.) Ach.[629]

Physical properties: (263; R = H) red prisms, m.p. 229–231° (dec.), $\lambda_{max.}$ (EtOH) 223, 255(sh), 266, 294, 324, 453 nm (log ϵ 4·47, 4·12, 4·18, 4·45, 4·02, 3·96), $\nu_{max.}$ (KBr) 3350, 1650, 1617, 1580, 981 cm^{-1}; (263; R = Me) orange-red crystals, m.p. 215°, $\lambda_{max.}$ (dioxan) 223, 255(sh), 267, 292, 320, 445 nm, $\nu_{max.}$ (KBr) 1670, 1610 cm^{-1}.

Averythrin is one of a small group of pigments, elaborated chiefly by strains of *A. versicolor*, which have a C_6 side chain at C-2. As the name implies *A. versicolor* is a multicoloured organism and up to fifteen pigments (not all quinones) have been detected in one strain.[481] Averythrin forms a tetra-acetate and a tetramethyl ether, and yields dihydroaverythrin on catalytic hydrogenation (and aerial reoxidation), the infrared band at 981 cm^{-1} disappearing. The dihydro derivative also gives a tetramethyl ether whose light absorption is almost identical with that of 1,3,6,8-tetramethoxy-2-methylanthraquinone. This substitution pattern was supported by the n.m.r. spectrum so that dihydroaverythrin is a 2-substituted 1,3,6,8-tetrahydroxyanthraquinone, the substituent being C_6H_{13} and most probably (n.m.r.) *n*-hexyl. The natural pigment is therefore the corresponding hexenyl derivative and the position of the double bond next to the ring was established, *inter alia*, by ozonolysis of the tetra-acetate which afforded valeraldehyde, while similar treatment of the tetramethyl ether led to the isolation of 1,3,6,8-tetramethoxy-anthraquinone-2-carboxylic acid. Averythrin is therefore (263).[470] This has been confirmed by a synthesis first of the tetramethyl ether[471] and then of averythrin[631] itself. The synthetic[472] trimethyl ether (264) was converted into the tetra-acetate (265; R = H) and so to the bromo compound (265; R = Br) with N-bromosuccinimide. Dehydrobromination and hydrolysis was then effected in one step using 1,5-diazabicyclo-[3.4.0]non-5-ene.

Spectroscopic evidence, including the n.m.r. and mass spectra, shows that the pigment from *Solorina crocea* is a β-methyl ether of averythrin, and it gives averythrin on demethylation. Structure (263; R = Me) is preferred[629] by analogy with the co-pigment, solorinic acid (267; R = Me).

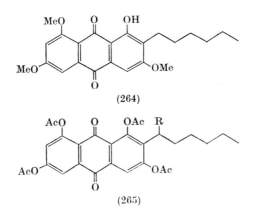

(264)

(265)

Averantin (266), $C_{20}H_{20}O_7$

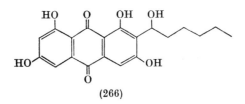

(266)

Occurrence: cultures of *Aspergillus versicolor* (Vuill.) Tiraboschi.[473, 474]

Physical properties: orange needles, m.p. 233–234°, $[\alpha]_{579}^{22}$ −178° (EtOH), $\lambda_{max.}$ (EtOH) 223, 258(sh), 266, 287(sh), 294, 325, 454 nm (log ε 4·53, 4·18, 4·24, 4·47, 4·53, 4·01, 4·03), $\nu_{max.}$ (KBr) 1650, 1620 cm⁻¹.

Like averythrin, this pigment is elaborated by *A. versicolor*; it differs in molecular formula by the elements of one molecule of water, is optically active, and lacks the peak at 981 cm⁻¹ (ethylenic C–H out-of-plane bending) in the otherwise similar infrared spectrum. As the ultraviolet absorption is very similar to that of dihydroaverythrin, structure (266) seemed very reasonable. This is consistent with the n.m.r. spectrum and that of the oily acetate, and with the (surprisingly difficult) conversion of averantin into averythrin which was effected by prolonged heating in benzene with phosphoric acid.[474] The racemic tetramethyl ether has been synthesised[471, 631] by reduction of solorinic acid (267) trimethyl ether leuco-acetate with lithium aluminium hydride, followed by aerial oxidation.

Solorinic Acid (267; R = Me), $C_{21}H_{20}O_7$

Norsolorinic Acid (267; R = H), $C_{20}H_{18}O_7$

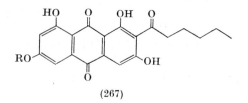

(267)

Occurrence: (267; R = Me and H) in the lichen *Solorina crocea* (L.) Ach.[475, 476]; (267; R = H) also in cultures of *Aspergillus versicolor* (Vuill.) Tiraboschi[537] and in the lichen *Lecidea piperis* (Spreng.) Nyl.[625]

Physical properties: (267; R = Me) red-orange needles, m.p. 201°, $\lambda_{max.}$ (EtOH) 269, 281, 311, 460 nm (log ε 4·43, 4·40, 4·45, 3·98), $\nu_{max.}$ (KBr) 1681, 1624 cm⁻¹; (267; R = H) red plates, m.p. 269–270°, $\lambda_{max.}$ (EtOH) 270, 286, 312, 453(sh), 466 nm (log ε 4·32, 4·35, 4·44, 3·95, 4·00), $\nu_{max.}$ (KBr) 3400, 1680(w), 1625 cm⁻¹.

Early investigations on this pigment by Zopf[478] and Hesse[479] were based on erroneous molecular formulae. Later, Koller and Russ[475] established the formula $C_{20}H_{17}O_6$.OMe, and prepared a triacetate, a trimethyl ether and an oxime. The latter was obtained under conditions in which polyhydroxyanthraquinones do not react with hydroxylamine. The nature of the side chain was suggested by the formation of n-hexanoic acid on oxidation with cold alkaline permanganate. The same acid also appeared when the quinone was treated with hydriodic acid in phenol at 150°, the other products being methyl iodide and an anthrone which was (a) reduced with zinc to anthracene, and (b) oxidised to 1,3,6,8-tetra-hydroxyanthraquinone. At this stage structure (267; R = Me) was proposed[475] although no experimental evidence was available to support the locations of the methoxyl group and the side chain. The reported isolation of 2-methylanthracene on zinc dust distillation is difficult to explain.

Recent work[476] has fully confirmed the Koller and Russ structure. Both the mass spectrum of the natural pigment and the n.m.r. spectrum of the trimethyl ether are in complete accord with (267; R = Me). The position of the side chain was also established by oxidation of the trimethyl ether with alkaline permanganate which gave 1,3,6,8-tetra-methoxyanthraquinone-2-carboxylic acid, while the position of the methoxyl group is evident from the absence of free hydroxyl absorption in the infrared spectrum, the insolubility of solorinic acid in aqueous sodium carbonate, and by the formation of 1,3,8-trihydroxy-6-methoxy-anthraquinone on degradation with hot alkaline dithionite. Structure (267; R = Me) was finally confirmed by synthesis.[476] The trimethyl ether was secured by reaction of 1,3,6,8-tetramethoxyanthraquinone-2-carbonyl chloride with di-n-pentylcadmium, from which solorinic acid was obtained by refluxing with hydrogen bromide in acetic acid. The *peri*-methoxyl groups are dealkylated first, followed by the methoxyl group at C-3. The 6-methoxyl group is best demethylated with pyridine hydrochloride which yields norsolorinic acid (267; R = H), identical with one of the minor pigments in *S. crocea*. The natural nor- compound was recognised[476] by the similarity of its ultraviolet-visible and infrared spectra to those of solorinic acid, the additional hydroxyl absorption at 3400 cm^{-1}, and by conversion into solorinic acid trimethyl ether.

Averufin (268), $C_{20}H_{16}O_7$

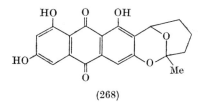

(268)

Occurrence: cultures of *Aspergillus versicolor* (Vuill.) Tiraboschi.[481]
Physical properties: orange-red laths, m.p. 280–282° (dec.), $[\alpha]_D^{22} \not> 1°$ (EtOH). [Triacetate, $[\alpha]_D^{22}$ −14·9° (CHCl$_3$)], $\lambda_{max.}$ (EtOH) 223, 256(sh), 266, 286(sh), 294, 324, 453 nm (logϵ 4·54, 4·21, 4·26, 4·43, 4·52, 3·98, 4·05), $\nu_{max.}$ (KBr) 3588, 3381, 1675, 1622, 1611, 1598 cm^{-1}.

Averufin forms a triacetate and, with difficulty, a trimethyl ether. The spectroscopic evidence suggests that two of the hydroxyl groups occupy 1,8-positions and the third a β-position which, from the n.m.r. data, is sited *meta* to one of the others. Colour reactions, and comparisons with

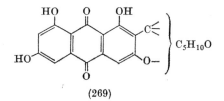

(269)

emodin, also indicate that averufin is a 1,6,8-trihydroxyanthraquinone, and the partial structure (269) seems likely in view of the fact that the ultraviolet absorption of the triacetate resembles that of 2-methoxy-anthraquinone. A singlet at τ 2·43 in the n.m.r. spectrum of the trimethyl ether can be assigned to the proton at C-4. Averufin contains no unsaturation additional to that shown in (269) so, assuming that two well-defined bands in the infrared spectrum, at 1099 and 1129 cm^{-1}, are cyclic ether

C–O–C stretching frequencies, valency considerations show that the $C_{20}H_{16}O_7$ molecule must be pentacyclic. The nature of the $C_5H_{10}O$ fragment was disclosed by the n.m.r. spectrum of the trimethyl ether which includes a C-methyl peak at τ 8·36 superimposed on a broad absorption "hump", representing six protons, and a deformed triplet (one proton) at τ 4·5. This information leads to the conclusion that the bicyclic side chain has either the cyclic ketal (270) or the cyclic acetal (271) structure but does not decisively distinguish between them.[481, 487] Subsequent comparison[488] with the n.m.r. spectra of model 1,3-benzo-dioxans appeared to support structure (271) but deuteriation experiments[488, 489] show that (270) is correct. On treatment with acid a ketal of

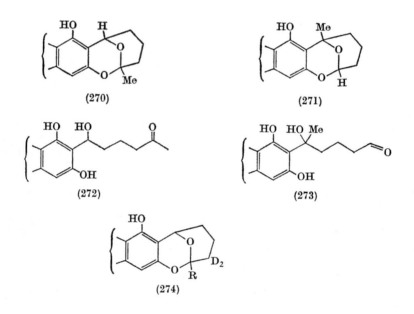

(270)   (271)

(272)   (273)

(274)

structure (270) should equilibrate with the open chain hydroxyketone (272), and the acetal (271) with the hydroxyaldehyde (273). Hence acid-catalysed deuteriation should lead to the exchange of five protons α to the carbonyl group in (272), but only two protons in (273). In fact, refluxing averufin trimethyl ether with phosphorus pentachloride in tetrahydrofuran and deuterium oxide[490] led to a mixture of two compounds containing seven and eight deuterium atoms, three of which had replaced the aromatic protons. The positions of the others were determined by mass spectrometry, comparison with the spectrum of the non-deuterated parent compound showing that $CD_3CO\cdot$, $CD_3COCD_2\cdot$, $CD_3COCD_2CH_2\cdot$, $CD_3COCD_2CH_2CH_2\cdot$, were significant losses from the octadeutero derivative, while the heptadeutero derivative lost $CHD_2CO\cdot$, $CHD_2COCD_2\cdot$, etc. This means that the deuteriated product was a mixture of (274; R = CHD_2 and CD_3), whence it follows that averufin has the cyclic ketal structure (270) = (268).[489]

A dimethyl ether of averufin has also been found[593] in a culture of *Aspergillus versicolor*. It is considered to be the 6,8-isomer as an α-hydroxyl group is present and the singlet signal from the H-4 proton (τ 2·85) moves downfield on acetylation (τ 2·42), and then upfield to τ 2·79 on exchange of the acetate with trifluoroacetic acid, the chemical shifts of the *meta* protons remaining almost constant.[593]

Avermutin (275), $C_{20}H_{18}O_7$

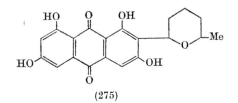

(275)

Occurrence: cultures of *Aspergillus flavus* Link ex Fries[597] and a u.v.-mutant of *A. versicolor* (Vuill.) Tiraboschi.[489]

Physical properties: orange crystals, m.p. 271° (260°), $[\alpha]_D^{22}$ +226° (dioxan), $\lambda_{max.}$ (MeOH) 223, 263, 292, 312, 452 nm (log ε 4·30, 4·16, 4·25, 4·00, 3·94), $\nu_{max.}$ (KBr) 3407, 1666, 1618 cm⁻¹.

This pigment forms a tetramethyl ether whose n.m.r. spectrum (*meta*-coupled aromatic protons at τ 2·66 and 3·21, singlet at τ 2·46) shows clearly, by comparison with other members of this group, that it is a 1,3,6,8-tetramethoxyanthraquinone substituted at C-2. The nature of the substituent was determined by ozonolysis of a dimethyl ether of avermutin (one of two ethers obtained from a u.v.-mutant of *A. versicolor*) which gave the acid (276).[489] Avermutin is therefore (275), and

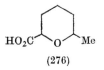

(276)

the two derivatives found in an u.v.-mutant of *A. versicolor* are the 8-mono- and the 6,8-dimethyl ethers.[489]

Versicolorin A (277), $C_{18}H_{10}O_7$

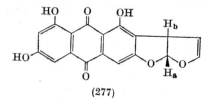

(277)

Occurrence: cultures of *Aspergillus versicolor* (Vuill.) Tiraboschi.[480]
Physical properties: orange-yellow needles, m.p. 289° (dec.), $[\alpha]_D^{18}$ −354° (dioxan), $\lambda_{max.}$ (EtOH) 255, 267, 290, 326, 450 nm (log ϵ 4·13, 4·26, 4·40, 3·83, 3·85), $\nu_{max.}$ (KBr) 3340, 1679, 1625, 1610 cm^{-1}.

The first anthraquinone isolated from *A. versicolor* was a pigment known as "versicolorin",[482] but later work, in the same laboratory,[480] showed this to be a mixture of three closely related quinones, versicolorins A, B and C. The electronic absorption of versicolorin A is very similar to that of 1,3,6,8-tetrahydroxy-2-methylanthraquinone. It forms a trimethyl ether which, on oxidation with permanganate, gave 3,5-dimethoxyphthalic acid, and a hydroxytrimethoxyanthraquinone carboxylic acid, convertible, by methylation, into the methyl ester of 1,3,6,8-tetramethoxyanthraquinone-2-carboxylic acid. As the larger degradation product absorbs in the 3 μ region only at 1690 and 1660 cm^{-1}, it was considered[480] to be the acid (278) although the evidence is not compelling. This implies that versicolorin A is a 1,6,8-trihydroxy-

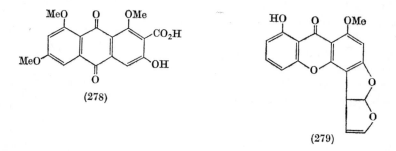

(278)   (279)

anthraquinone with a side chain at C-2 and an ether linkage at C-3 which is consistent with the n.m.r. spectrum of the trimethyl ether.

Catalytic hydrogenation of the trimethyl ether (and reoxidation in air) gives a dihydro derivative and, as the reaction is not accompanied by any appreciable change in the electronic absorption, the double bond is evidently not conjugated with the anthraquinone system. Infrared bands in the spectrum of versicolorin A, at 1610, 1032 and 726 cm^{-1}, were attributed to a vinyl ether system [cf. sterigmatocystin (279), a co-metabolite] for which chemical justification was obtained by the formation of a hemi-acetal acetate (ii) when the trimethyl ether was dissolved in cold acetic acid. There is precedent for this in the behaviour of 2,3-dihydrofuran systems,[484] and by hydrolysis and chromic acid oxidation the hemi-acetal (iii) was converted into a γ-lactone (iv) (ν_{CO} 1802 cm^{-1}). Hence the vinyl ether structure in versicolorin A is part of a

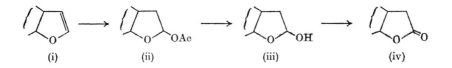

(i)   (ii)   (iii)   (iv)

dihydrofuran ring (i). Analysis of the n.m.r. spectrum of the trimethyl ether leads to the conclusion that versicolorin A has the same bicyclic side chain as sterigmatocystin (279),[483, 485] in particular the H$_a$ proton shows a doublet at τ 3·28 which is coupled to H$_b$ at τ 5·15. Hamasaki et al.[480] therefore regarded versicolorin A as the linear structure (277) but found no decisive evidence against the angular isomer (280). (However,

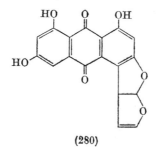

(280)

see aversin = versicolorin B dimethyl ether, below.)

Versicolorin B (C) (287), C$_{18}$H$_{12}$O$_7$

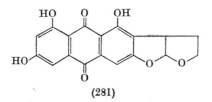

(281)

Occurrence: cultures of *Aspergillus versicolor* (Vuill.) Tiraboschi.[480]

Physical properties: orange-yellow needles, m.p. 298° (dec.), $[\alpha]_D^{25}$ −223° (dioxan), $\lambda_{max.}$ (EtOH) 255, 266, 291, 324, 450 nm (log ϵ 4·13, 4·29, 4·38, 4·11, 3·94), $\nu_{max.}$ (KBr) 3340, 1670, 1625 cm^{-1}.

Versicolorin B is very similar to versicolorin A and the structure (243) follows, *inter alia*, from the fact that its trimethyl ether is identical with dihydroversicolorin A trimethyl ether.[480]

Versicolorin C, orange-red needles, m.p. 350°, is a racemate of versicolorin B, found in cultures of *A. versicolor*[480] and *A. flavus*.[597]

Versiconol (282), $C_{18}H_{16}O_8$

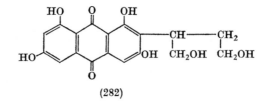

(282)

Occurrence: cultures of *Aspergillus versicolor* (Vuill.) Tiraboschi.[617]

Physical properties: orange-red needles, m.p. 265° (dec.), $[\alpha]_D^{25}$ −35·8° (dioxan), $\lambda_{max.}$ (EtOH) 224, 255, 265, 295, 322, 460 nm (log ϵ 4·65, 4·35, 4·35, 4·53, 4·14, 4·02), $\nu_{max.}$ (KBr) 3400, 1675, 1620 cm^{-1}.

This recent addition to the group of anthraquinones elaborated by *A. versicolor* could be recognised by comparison with its congeners and related compounds. Thus the above infrared and ultraviolet-visible data are in good agreement with a 2-alkyl-1,3,6,8-tetrahydroxyanthraquinone structure, and this is also supported by the n.m.r. spectra ($< \tau$ 3·5) of several derivatives. Versiconol forms a hexa-acetate containing both phenolic and alcoholic ester groups (ν_{CO} 1780, 1740, 1675 cm^{-1}; four 3H singlets between τ 7·52 and 7·67, and two at τ 8·03 and 8·07), and a tetramethyl ether which gives a diacetate ($\nu_{acetate}$ 1740 cm^{-1}). The nature of the side chain at C-2 was deduced from the n.m.r. spectrum (in DMSO-d$_6$) of the tetramethyl ether which included signals attributable to a 2-Ar-substituted-butan-1,4-diol system (see 283). The assignments

$$-\underset{\underset{\tau\ 6\cdot28(d)}{CH_2OH}}{CH}\!-\!\!\!-\!\!\!-\underset{\underset{\tau\ 6\cdot60(t)}{CH_2OH}}{CH_2}\ \tau\ 8\cdot03(m)$$

(283)

were confirmed by appropriate downfield shifts on acetylation and by spin decoupling experiments on the diacetate, when *inter alia* a signal at τ 5·92 from the benzylic proton was observable. The structure of versiconol is therefore (282) and its relationship to the versicolorins is obvious. In replacement cultures of *A. versicolor* supplemented with versiconol, there was increased production of versicolorin C (287) indicating that (282) is a precursor of (287).

Aversin (284), $C_{20}H_{16}O_7$

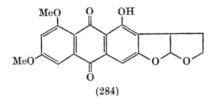

(284)

Occurrence: cultures of *Aspergillus versicolor* (Vuill.) Tiraboschi.[486]
Physical properties: golden needles, m.p. 217°, $[\alpha]_D^{20}$ −222° (CHCl$_3$), $\lambda_{max.}$ (EtOH) 224, 251, 285, 313, 363, 440 nm (log ϵ 4·56, 4·13, 4·53, 3·95, 3·70, 3·89), $\nu_{max.}$ (KBr) 1665, 1628, 1595 cm^{-1}.

This quinone, along with 6-methoxysterigmatocystin, was isolated from a variant strain of *A. versicolor* in small amount. Aversin has two O-methyl groups, and forms a monomethyl ether whose ultraviolet-vis. absorption closely resembles that of 1,3,6,8-tetramethoxy-2-methyl-anthraquinone. Chelated carbonyl absorption at 1628 cm^{-1} indicates that the hydroxyl group occupies an α-position, and the n.m.r. spectrum includes signals appropriate to a bicyclic side chain as in dihydro-sterigmatocystin and versicolorin B. This suggests a structure of the type represented by (284) which is also consistent with the pattern of aromatic proton signals, a singlet at τ 2·76 is assigned to H-4 as this shifts to τ 2·65 in the methyl ether close to the signal from the lower field *meta*-coupled proton at C-5 which resonates at τ 2·70. On acetylation of aversin the aromatic singlet shifts downfield to τ 2·48, and on exchange of the acetate with trifluoroacetic acid it moves upfield to τ 2·78, the *meta* proton signals remaining constant (cf. ref. 486). This shows that the hydroxyl group is in the same ring as the isolated aromatic proton and hence aversin must have structure (284).[593] It has been shown by direct comparison,[477] that aversin methyl ether is identical with versicolorin B trimethyl ether.

Rhodocomatulin-6-methyl ether† (285; R = H), $C_{19}H_{16}O_7$

Rhodocomatulin-6,8-dimethyl ether (285; R = Me), $C_{20}H_{18}O_7$

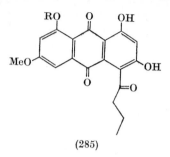

(285)

Occurrence: in the crinoids *Comatula pectinata* L.[491] and *C. cratera* A. H. Clark[491] (Comasteridae).

Physical properties: (285; R = H) red-orange prisms, m.p. 250–252° (dec.), $\lambda_{max.}$ (acid EtOH) 256, 263, 293, 317, 366, 456 nm (log ε 4·24, 4·25, 4·42, 3·96, 3·46, 4·05), $\nu_{max.}$ (Nujol) 1669, 1627, 1592 cm⁻¹, $\nu_{max.}$ (dioxan) 3515, 3450, 1702, 1663, 1625, 1596 cm⁻¹; (285; R = Me) orange-yellow needles, m.p. 208·5–209° or 229·5–230·5°, $\lambda_{max.}$ (acid EtOH) 256(sh), 268(sh), 287, 310(sh), 361, 448 nm (log ε 4·14, 4·22, 4·42, 3·95, 3·56, 3·93), $\nu_{max.}$ (Nujol) 1668, 1621, 1588, $\nu_{max.}$ (dioxan) 3519, 3450, 1700, 1666, 1627, 1596 cm⁻¹.

Crinoids (sea lilies) are highly coloured animals but although there have been many casual examinations of their pigments in the last hundred years (see ref. 491) no chemical investigation was undertaken until Sutherland and Wells began their work in 1957.[491] Both the bright or dark red 10-armed comatulid, *C. pectinata*, collected under coral boulders, and the less highly coloured *C. cratera*, taken from the nets of Queensland prawn trawlers, contain a similar mixture of anthraquinones easily extracted with cold acetone. The crude pigment from *C. pectinata* amounted to 4–5% of the dry weight and ca. 0·7% in the case of *C. cratera*.

The monomethyl ether has three hydroxyl groups and the dimethyl ether (the major pigment) has two, both giving the same tetramethoxy compound on complete methylation. The essential structure became apparent when the dimethyl ether was refluxed with hydrobromic acid–acetic acid which gave 1,3,6,8-tetrahydroxyanthraquinone and n-butyric acid; at an intermediate stage it was possible to isolate 1,3,8-trihydroxy-6-methoxyanthraquinone. The infrared spectra of the natural pigments in solution show aromatic carbonyl absorption near 1700 cm⁻¹ in addition to the quinone carbonyl peaks, and this together with the preparation of an oxime from the tetramethyl ether establishes the presence of an

---

† Rhodocomatulin is the name given to the parent compound (285; R = H, OH in place of OMe). It has not yet been found in Nature.

n-butyryl side chain. The butyryl group must be at C-4 as oxidation of the tetramethyl ether with alkaline permanganate gave 1,3,6,8-tetra-methoxyanthraquinone-4-carboxylic acid, which was also synthe-sised,[492] although initially analogy with solorinic acid (267) had suggested that the side chain would be at C-2. The tetramethyl ether is

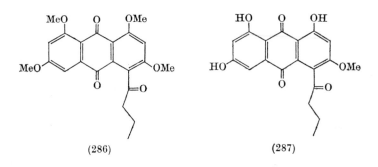

(286)   (287)

therefore (286) which is in complete accord with its n.m.r. spectrum, and the natural dimethyl ether must be (286) since it affords 3,5-dimethoxy-phthalic acid on oxidative degradation.

The monomethyl ether yields 1,3,8-trihydroxy-6-methoxyanthra-quinone on pyrolysis (and in other ways), and is therefore either (285) or (287). The former is preferred on n.m.r. evidence, in particular the AB system of the triacetate shows signals at τ 2·47 and 3·13 very similar to those of the C-2 and C-4 protons of 1-acetoxy-3,6,8-trimethoxyanthra-quinone which resonate at τ 2·36 and 3·12. The infrared spectrum of the monomethyl ether, in solution, shows bands due to unchelated ketone and free hydroxyl groups which can be explained by the crowded situation of the side chain which must cause the carbonyl group to twist out of the plane of the ring. This is also consistent with the solubility of the monomethyl ether in aqueous sodium carbonate, and the easy methylation of the β-hydroxyl group with diazomethane.

Several methods for the extrusion of the side chain have been men-tioned and this can also be achieved with Raney nickel alloy, or better, with dithionite in alkaline solution, provided there is one free phenolic group in the butyryl ring. In this way the natural dimethyl ether afforded 1,3-dihydroxy-6,8-dimethoxyanthraquinone, and the synthetic dimethyl ether (from diazomethane methylation) gave 1,8-dihydroxy-3,6-dimethoxyanthraquinone. The reaction is not without precedent but the rhodocomatulins are much more susceptible to cleavage than either solorinic acid or typical acylresorcinols; this may be associated with the relief of steric strain achieved by the elimination of a butyryl group from C-4.

Catenarin (288), $C_{15}H_{10}O_6$

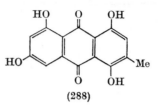

(288)

Occurrence: cultures of *Helminthosporium catenarium* Drechsler,[412] *H. velutinum* Link,[412] *H. tritici-vulgaris* Nishik.,[412] *Pyrenophora graminea* (Rabenh.) Ito & Kuribayashi[411]† (= *H. gramineum* Rabenh.), *Aspergillus amstelodami* (Mangin) Thom & Church,[494] and *Penicillium islandicum* Sopp.[307, 377]

Physical properties: red plates, m.p. 246°, λ_{max}. (EtOH) 231, 257, 282, 317(sh), 462(sh), 479(sh), 492, 513(sh), 525 nm (log ϵ 4·51, 4·20, 4·21, 4·02, 4·02, 4·10, 4·13, 4·05, 3·99), ν_{max}. (KBr) 3450(br), 1598 cm⁻¹.

Catenarin is elaborated by several *Helminthosporium* spp., and from *H. catenarium* yields exceeding 15% of the dry weight of the mycelium have been obtained.[493] It was recognised as a tetrahydroxymethylanthraquinone by the formation of a tetra-acetate and reduction to 2-methylanthracene, and two key experiments established the orientation pattern.[493] Reduction with hydriodic acid and red phosphorus gave emodin (after reoxidation), whilst degradation of the tetramethyl ether with chromic acid yielded a mixture of 3,5-dimethoxy- and 3,6-dimethoxy-4-methylphthalic anhydrides. It follows that catenarin has structure (288). It was synthesised by Anslow and Raistrick[493] from the intermediate (289; R = Br) which had been used in an earlier preparation

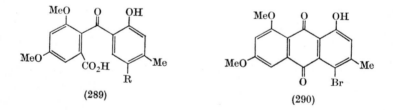

(289)   (290)

of emodin; on heating in sulphuric acid–boric acid at 100° the keto-acid cyclised to the quinone (290) which was converted into (288) by raising the temperature to 150–160°. More recently, Seshadri and co-workers[495] converted (289; R = H) into (289; R = HO), by an Elbs persulphate oxidation, from which catenarin was derived by cyclisation and demethylation.

† In this paper catenarin was called hydroxy*iso*helminthosporin.

5,6-Dihydrocatenarin (291), $C_{15}H_{12}O_6$

5,6,7,8-Tetrahydrocatenarin (292), $C_{15}H_{14}O_6$

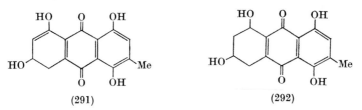

(291)   (292)

Occurrence: cultures of *Penicillium islandicum* Sopp.[382]
Physical properties: (251) red crystals, m.p. ca. 95–105° (and resolidifies), λ_{max}. (EtOH) 495, 530(sh), 570(sh) nm; (222) red crystals, m.p. ca. 130° (dec.), λ_{max}. (EtOH) 485, 514, 554 nm.

These two modified anthraquinones are recent additions to the list of pigments elaborated by *P. islandicum*. Both are relatively labile. The visible absorption of the compound, $C_{15}H_{14}O_6$, resembles naphthazarin and on dissolution in cold sulphuric acid it was rapidly dehydrated forming 2-methylquinizarin, $C_{15}H_{10}O_4$. Similarly, the mass spectrum showed major peaks at M-18 and M-36, the latter and the rest of the spectrum in the lower mass range corresponding to the spectrum of 2-methylquinizarin. By analogy with the structure of all the other *islandicum* metabolites it seems most likely that this new compound is (292).[382] It decomposed when acetylation was attempted and a suitable solution for n.m.r. study could not be obtained.

The compound, $C_{15}H_{12}O_6$, is similar to the above, but much less stable and it deteriorates on keeping. When heated to ca. 100° it melts transiently and then resolidifies as islandicin, and the same compound (a co-metabolite) is formed on treatment with cold sulphuric acid. This, together with biogenetical considerations, suggests that the compound, $C_{15}H_{12}O_6$, is probably (291).[382] Not surprisingly, the molecular ion could not be found in the mass spectrum of (291) but the expected molecular ion was recorded for the tetrakistrimethylsilyl derivative, and that spectrum included a peak at M-89 attributed to loss of a trimethylsilyloxy group from the secondary hydroxyl group. When (291) is heated at higher temperatures (150–200°) a more complex decomposition occurs, catenarin being formed along with islandicin and unidentified material, and similarly, in the mass spectrometer, the spectrum of islandicin appears first followed, after further heating, by the spectrum of catenarin and additional peaks at much higher m/e ratios. This suggests that at higher temperatures the simple dehydration of (291) to islandicin is accompanied by another aromatisation process in which it is dehydrogenated to catenarin and more complex products are formed.

$(+)-Dicatenarin$ (293), $C_{30}H_{18}O_{12}$

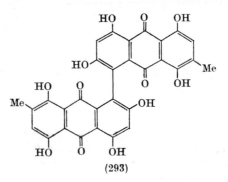

(293)

Occurrence: cultures of *Penicillium islandicum* Sopp.[523]
Physical properties: m.p. >300°, optically active.

Very little information is available on this pigment. Reductive cleavage with dithionite yielded catenarin,[523] and the position of the internuclear linkage was ascertained by comparing the n.m.r. spectra of the dimer and monomer peracetates.[568]

$(+)-Punicoskyrin$ (294), $C_{30}H_{18}O_{11}$

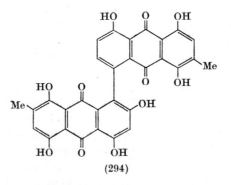

(294)

Occurrence: cultures of *Penicillium islandicum* Sopp.[568]
Physical properties: m.p. >300°, optically active.

On reductive cleavage with dithionite punicoskyrin gives islandicin and catenarin, and the position of the internuclear linkage was deduced from a comparison of the n.m.r. spectra of the dimer and the two monomer peracetates.[568]

Aurantioskyrin (295), $C_{30}H_{18}O_{11}$

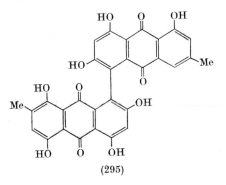

(295)

Occurrence: cultures of *Penicillium islandicum* Sopp.[568]
Physical properties: m.p. >300°, optically active.

On reductive cleavage with dithionite aurantioskyrin gives emodin and catenarin, and the position of the internuclear linkage was deduced from a comparison of the n.m.r. spectra of the dimer and the two monomer peracetates.[568] It was obtained synthetically from skyrin (120) by peracid oxidation which confirms structure (295).

Dermorubin (296), $C_{17}H_{12}O_8$

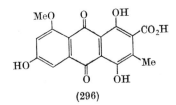

(296)

Occurrence: mainly as a glycoside in *Dermocybe sanguinea* (Wulf. ex Fr.)[615] and *D. semisanguinea* (Fr.).[615]
Physical properties: dark red needles, dec. >300°, λ_{max}. (EtOH) 233, 279, 291(sh), 481 nm (log ϵ 4·54, 4·30, 4·14, 4·04), ν_{max}. (KBr) 3500–2300, 1724, 1590 cm^{-1}.

This is another member of the group of anthraquinone carboxylic acids found in *Dermocybe*. It contains a methoxyl group, and forms a methyl ester and a triacetate. On heating in sulphuric acid it undergoes demethylation, and decarboxylation of the product with copper powder in boiling quinoline then gives catenarin (288). As dermorubin contains two *meta*-coupled protons (n.m.r.), the demethylation product must be catenarin-2-carboxylic acid, and since both quinone carbonyl groups are chelated (infrared) the substituent at C-4 must be hydroxyl. The methoxyl group is therefore located either at C-8 (296) or C-1 (297), and

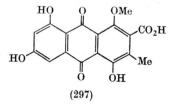

(297)

the former must be correct[615] as the H-5 and H-7 protons show acelation shifts[616] of −0·44 and −0·49 ppm, respectively, corresponding to a 1-methoxy-3-hydroxy substitution pattern.

5-Chlorodermorubin (298), $C_{17}H_{11}ClO_8$

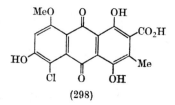

(298)

Occurrence: mainly as a glycoside in *Dermocybe sanguinea* (Wulf. ex Fr.)[615] and *D. semisanguinea* (Fr.).[615]

Physical properties: dark red needles, m.p. >300°, $\lambda_{max.}$ (EtOH) 255, 283, 486 nm (log ϵ 4·14, 4·25, 4·09), $\nu_{max.}$ (KBr) 3390, 1715, 1597 cm^{-1}.

This is the second chlorinated quinone in *Dermocybe*. It is closely related to dermorubin and comparison of the n.m.r. spectra of the pertrimethylsilyl derivatives shows that the signal from the H-5 proton is absent from that of the chloro compound and the H-7 resonance at τ 3·43 is a singlet. The pigment is therefore 5-chlorodermorubin (298).[615]

(−)-*Rubroskyrin* (299), $C_{30}H_{22}O_{12}$

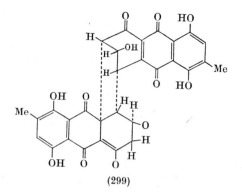

(299)

Occurrence: cultures of *Penicillium islandicum* Sopp.[518]

Physical properties: dark red plates, m.p. 281° (dec.), $\lambda_{max.}$ (CHCl$_3$) 275, 415, 435, 530, 540 nm (log ϵ 4·31, 4·10, 4·10, 3·97, 3·97), $\nu_{max.}$ (Nujol) 3559, 3341, 1706, 1623 cm^{-1}.

Howard and Raistrick[518] showed that this dimeric compound was readily dehydrated, for example, by dissolution in cold concentrated sulphuric acid, the product being (+)-iridoskyrin (182). Similarly, normal acetylation in the presence of sodium acetate affords iridoskyrin hexa-acetate. Iridoskyrin is also formed on pyrolysis of rubroskyrin, but the major product is islandicin (179) accompanied by a minor amount of

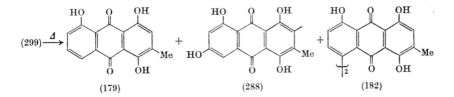

(179)   (288)   (182)

catenarin (288). The appearance of the latter suggests that the remaining oxygen functions are located at C-6 and C-6', and these are evidently alcoholic hydroxyl groups since a diacetate can be isolated after careful treatment with acetyl chloride in the cold.[525] The diacetate (ν_{CO} 1749, 1708, 1620 cm^{-1}) was converted into iridoskyrin on keeping, or, more rapidly, on heating, in ethanol.

To account for all these, and other facts, Shibata and his co-workers[524, 525] originally proposed structure (300) for rubroskyrin, the infrared band at 1706 cm^{-1} being assigned to the ketonic groups at C-8 and C-8'. They have recently revised[573] this to the remarkable structure (299) following a detailed examination of the n.m.r. spectrum of rubro-skyrin triacetate ($[\alpha]_D^{31}$ −333° (dioxan), $\nu_{max.}$ (CHCl$_3$) 1770(sh), 1745,

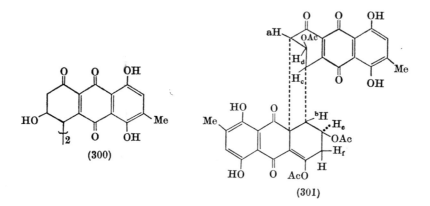

(300)

(301)

1705, 1655(sh), 1640, 1620 cm⁻¹) which is now considered to be (301). Only one methylene group was apparent (two doublets at ∼τ 8·6 (J, 17·5 Hz)), and a series of spin decoupling experiments established that the following protons were coupled: e–f, a–d–c, c–b. This, together with the respective chemical shifts, establishes the sequence,

$$-\overset{|}{\underset{|}{C}}-\overset{|}{CH}-CH(OH)-\overset{|}{CH}-\overset{|}{CH}- \quad \ldots \quad -CH(OH)-CH_2-\overset{|}{\underset{|}{C}}-$$
$$\quad\quad a\quad\quad d\quad\quad\quad\quad c\quad\quad b\quad\quad\quad\quad\quad\quad e\quad\quad\quad f$$

and bearing in mind the previous chemical evidence, it can only be fitted into the rubroskyrin structure if there is an additional bond between the two halves of the molecule as indicated (301). As rubroskyrin is easily recovered from the triacetate by alkaline hydrolysis it should be (299), and the formation of such a molecule can be envisaged as a single intramolecular Michael reaction analogous to that shown on p. 427. Only one half of the molecule is quinonoid. Examination of Dreiding models showed that the dihedral angle between H_b and H_c is about 90° which explains the absence of coupling between these protons.

Luteoskyrin. Closely related to the preceding compound is a yellow, optically active, pigment (–)-luteoskyrin, which is another metabolite of *P. islandicum.*[377, 526, 527, 528, 552] It was identified as the toxic factor in rice infected with this organism, and has been shown to cause damage, and sometimes cancer, in the liver of experimental animals.[530] Luteoskyrin is isomeric with rubroskyrin, and the latter is converted into the

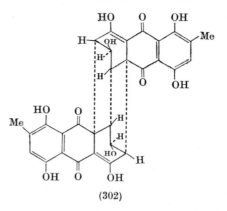

(302)

former in pyridine solution. Recent n.m.r. studies indicate that luteoskyrin has the structure (302) (cf. rugulosin) and is probably formed from rubroskyrin by intramolecular Michael addition (see p. 427). On exposure to light and air, a yellow solution of luteoskyrin in acetone becomes deep

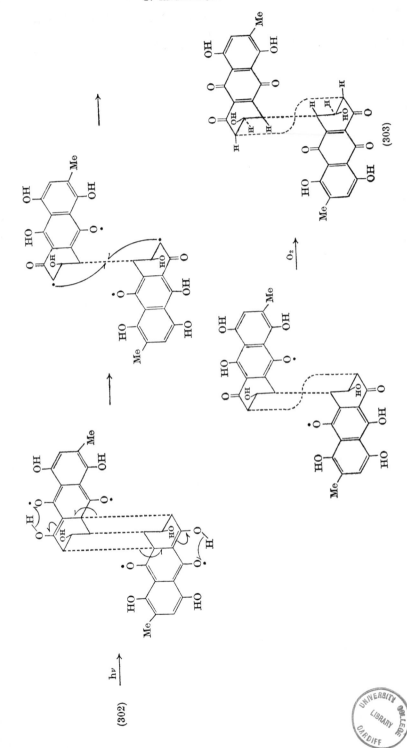

(302)

$h\nu$ →

(303)

red, and dark violet crystals of lumiluteoskyrin (303) are deposited.[591] This is an intriguing rearrangement in which two bonds linking the two halves of luteoskyrin are broken, and a new one is formed together with two double bonds which results in aromatisation of two rings. The final product, after aerial oxidation, is a biquinone. The mechanism on p. 501 has been suggested.[591]

Erythroglaucin (304), $C_{16}H_{12}O_6$

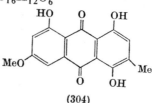

(304)

Occurrence: cultures of *Aspergillus ruber* (Bremer) Thom & Raper,[400] *A. umbrosus* Bain. & Sart.,[496] *A. echinulatus* (Delacr.) Thom & Church,[400] *A. niveo-glaucus* Thom & Raper,[400] *A. chevalieri* (Mangin) Thom & Church,[400] and in the lichen *Xanthoria elegans* (Link) Th. Fr.[540]

Physical properties: deep red plates, m.p. 205–206°, $\lambda_{max.}$ (EtOH) 231, 255, 275, 302·5, 460(sh), 475(sh), 489, 511(sh), 523 nm (log ϵ 4·44, 4·17, 4·13, 3·92, 3·95, 4·01, 4·06, 3·95, 3·90), $\nu_{max.}$ (KBr) 1596 cm^{-1}.

Gould and Raistrick[400] isolated a red pigment, rubroglaucin, in a survey of the *Aspergillus glaucus* series, which was later found[333b] to be a mixture of physcion and a new compound erythroglaucin. When grown under optimum conditions, the pigments of *A. umbrosus* may constitute 40–70% of the dry weight of the mycelium.[496] As erythroglaucin is insoluble in aqueous sodium carbonate, and its trimethyl ether is identical with the tetramethyl ether of catenarin, the structure must be (304).[493] It can be obtained by partial methylation of catenarin[493] or by partial demethylation of (290; HO in place of Br), an intermediate in the catenarin synthesis referred to above.[495]

Fusaroskyrin (305), $C_{32}H_{22}O_{12}$

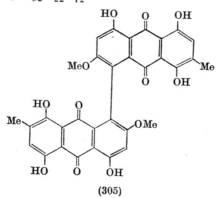

(305)

Occurrence: cultures of an unidentified *Fusarium* sp.[533,534]
Physical properties: dark red needles, m.p. >300°, λ_{max}. (CHCl$_3$) 256, 277, 305–310, 505 nm, ν_{max}. (?) 1615 cm^{-1}.

One of the pathogens responsible for the "purple speck" disease of soya beans is a *Fusarium* species which elaborates a mixture of pigments. The principal colouring matter, fusaroskyrin, was isolated and purified via its hexa-acetate, $C_{44}H_{34}O_{18}$.[534] It contains two O-methyl and two C-methyl groups, and on reduction with alkaline dithionite or lithium aluminium hydride it gives erythroglaucin. Similarly, demethylation with hydriodic acid, followed by reductive fission with dithionite, yielded catenarin. It follows that fusaroskyrin is a bi-erythroglaucin and structure (305) was preferred[534] on the basis of its behaviour in concentrated sulphuric acid. Fusaroskyrin gives a red-violet colour but the demethylated compound gave a green solution, suggesting the conversion of a 2,2'-dihydroxy-1,1'-bianthraquinone into a bishemiketal (cf. pseudoskyrin, p. 425) or the corresponding dication (306).

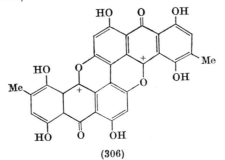

(306)

Tritisporin (307), $C_{15}H_{10}O_7$

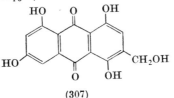

(307)

Occurrence: cultures of *Helminthosporium tritici-vulgaris* Nisik.[412]
Physical properties: brown needles, m.p. 260–262°, λ_{max}. (EtOH) 231, 254, 279, 299·5, 462(sh), 474(sh), 488, 508, 522 nm (log ϵ 4·41, 4·14, 4·15, 3·96, 3·86, 4·00, 4·05, 3·97, 3·92), ν_{max}. (KBr) 3400(br), 1642(w), 1590 cm^{-1}.

Both catenarin and tritisporin are produced by *H. tritici-vulgaris* and their ultraviolet-visible absorption spectra are virtually identical. Tritisporin, however, forms a penta-acetate, but since it does not contain a C-methyl group and yet forms 2-methylanthracene when distilled with zinc, the evidence points to the ω-hydroxycatenarin structure

(307).[412, 497] It was synthesised[497] from catenarin tetra-acetate via side-chain bromination, reaction with sodium acetate, and final hydrolysis of the penta-acetate.

Cynodontin (308), $C_{15}H_{10}O_6$

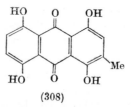

(308)

Occurrence: cultures of *Helminthosporium cynodontis* Marignoni,[413] *H. euchlaenae* Zimm.,[413] *H. victoriae* Meehan & Murphy,[498] and *H. oryzae* Breda de Haan,[643] *Deuterophoma tracheiphila* Petri,[414] *Phoma violacea* (Bertel) Eveleigh (= *P. pigmentivora*),[415] *Pyrenophora avenae* (Eidam) Ito & Kurib.[412] (= *H. avenae* Eidam), and *Pyrenochaeta terrestris* (Hansen) Gorez *et al.* (= *Phoma terrestris* Hansen);[500, 501] heartwood of *Maesopsis eminii* Engl.[572] (Rhamnaceae).

Physical properties: bronze needles, m.p. 260–263°, $\lambda_{max.}$ (EtOH) 241, 295, 471(sh), 483, 503(sh), 514, 539, 552 nm (log ε 4·56, 4·06, 4·06, 4·14, 4·31, 4·38, 4·37, 4·42), $\nu_{max.}$ (Nujol) 1572 cm^{-1}.

The coexistence of cynodontin and helminthosporin in the mycelium of *H. cynodontis* suggested a close relationship, and this was confirmed when helminthosporin (with three hydroxyl groups) was converted into cynodontin (which has four) by oxidation with manganese dioxide in sulphuric acid. As cynodontin does not dissolve in aqueous sodium carbonate it must have structure (308).[413] This was established synthetically[502] by a two-stage condensation of quinol dimethyl ether and 3,6-dimethoxy-4-methylphthalic anhydride. Alternative syntheses have since been reported.[410, 613]

1,4,5,8-Tetrahydroxy-2,6-dimethylanthraquinone (309), $C_{16}H_{12}O_6$

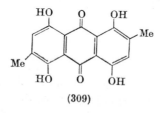

(309)

Occurrence: cultures of *Curvularia lunata* (Wakker) Boedijn,[503] and several other *Curvularia* spp.,[547] *Cochliobus spicifer* Nelson[547] and *C. sativus* (Ito & Kurib.) Drechsler ex Dastur[549] (= *Helminthosporium sativum*).

Physical properties: red crystals, m.p. 263·5°, $\lambda_{max.}$ (EtOH) 241, 295, 454(sh), 470(sh), 483, 504, 514, 540, 552 nm (log ε 4·58, 3·98, 3·56, 3·95, 4·09, 4·26, 4·34, 4·32, 4·37), $\nu_{max.}$ (KCl) 1576 cm^{-1}.

This somewhat unusual dimethylanthraquinone was isolated from culture fluids of *C. lunata*, an organism which is used for the microbiological oxidation of steroids. Its electronic absorption, and colour reactions, indicated a polyhydroxyanthraquinone structure, and since it gave a tetra-acetate it was clear from the infrared spectrum that all four hydroxyl groups were strongly chelated. The near identity of its visible absorption with that of 1,4,5,8-tetrahydroxyanthraquinone provided confirmatory evidence. This pigment contains one carbon atom more than cynodontin and reference to the literature revealed that 1,4,5,8-tetrahydroxy-2,6-dimethylanthraquinone (309) had properties akin to those of the *C. lunata* metabolite. Identity was established by direct comparison with synthetic material, secured by self-condensation of 2,5-dihydroxy-4-methylbenzoic acid in sulphuric acid.[504]

Aurantio-obtusin (310), $C_{17}H_{14}O_7$

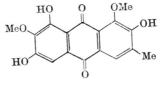

(310)

Occurrence: seeds of *Cassia obtusifolia* L.[505] (Leguminosae).

Physical properties: orange needles, m.p. 265–266°, $\lambda_{max.}$ (EtOH) 226, 286, 314, 388 nm (log ε 4·20, 4·58, 3·94, 3·82), $\nu_{max.}$ (Nujol) 3325, 1663, 1629 cm^{-1}.

In addition to chrysophanol, physcion and its isomer, obtusifolin, the seeds of *C. obtusifolia* contain three O-methylated pentahydroxyanthraquinones, aurantio-obtusin, obtusin and chryso-obtusin. All three give the same pentamethoxy derivative when fully methylated. Aurantio-obtusin has three hydroxyl and two methoxyl groups, and gives a pentahydroxyquinone, $C_{15}H_{10}O_7$, on demethylation. Useful information regarding the substitution pattern in these quinones was gained from a rather unusual reduction of obtusin, using hydriodic acid and red phosphorus in acetic acid, which gave emodin, in poor yield, after reoxidation of the intermediate anthrone. On this basis, the compound, $C_{15}H_{10}O_7$, is a dihydroxyemodin and from the spectroscopic data (see

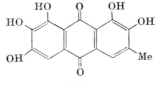

(311)

above) it must be (311) which has two α-hydroxyls and an unchelated carbonyl group. The colour reactions of aurantio-obtusin suggest the absence of vicinal hydroxyl groups while its solubility in aqueous sodium bicarbonate implies the presence of at least two β-hydroxyl groups. As the dimethyl ether, prepared with diazomethane, possesses an α-hydroxyl group (ν_{CO} 1670, 1623 cm^{-1}), aurantio-obtusin must have structure (310).

Obtusin (312), $C_{18}H_{16}O_7$

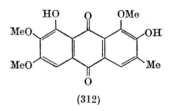

(312)

Occurrence: seeds of *Cassia obtusifolia* L.[505] (Leguminosae).

Physical properties: yellowish-brown needles, m.p. 242–243°, $\lambda_{max.}$ (EtOH) 227, 285, 317, 382 nm (log ε 4·15, 4·42, 3·85, 3·75), $\nu_{max.}$ (KBr) 3318, 1653, 1628 cm^{-1}.

Obtusin has two hydroxyl and three methoxyl groups, and is therefore a trimethyl ether of (311) (see above). From the infrared spectrum and the solubility in aqueous sodium carbonate, the hydroxyl groups occupy α- and β-positions, and colour reactions indicated that these were not vicinal. Treatment of obtusin with diazomethane gives a mono-O-methyl derivative identical with the dimethyl ether of aurantio-obtusin, prepared in the same way. When obtusin is heated with concentrated sulphuric acid it yields a mono-O-methyl derivative of (311), and under the same conditions anthragallol trimethyl ether forms the 3-methyl ether. Evidently the 3-methoxyl group is resistant to demethylation, and

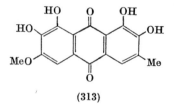

(313)

accordingly, Takido[505] concluded that obtusin has structure (312), and the mono-O-methyl derivative must be (313).

Chryso-obtusin (314), $C_{19}H_{18}O_7$

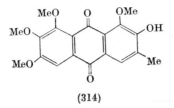

(314)

Occurrence: seeds of *Cassia obtusifolia* L.[505] and *C. tora* L.[277] (Leguminosae). *Physical properties*: yellow needles, m.p. 219–220°, $\lambda_{max.}$ (EtOH) 222, 284, 306(sh), 356 nm (log ε 4·14, 4·02, 3·94, 3·67), $\nu_{max.}$ (Nujol) 3325, 1676, 1582 cm⁻¹.

Takido[505] established, by routine operations, that this pigment was a tetramethyl ether of (311) having a free β-hydroxyl group. As partial demethylation with concentrated sulphuric acid again gave the mono-methyl ether (313), the free β-hydroxyl must be at C-2 or C-7. However chryso-obtusin ethyl ether is identical with obtusin monomethyl-monoethyl ether, prepared by partial alkylation of obtusin in two steps. As this compound (315) must have a methoxyl group at C-7, it follows

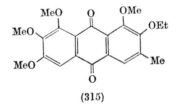

(315)

that the ethoxyl lies at C-2, and hence chryso-obtusin has structure (314).

Rubrocomatulin 6-methyl ether† (316), $C_{19}H_{16}O_8$

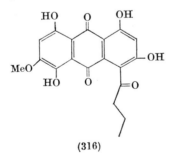

(316)

Occurrence: in the crinoids *Comatula pectinata* L.[491] and *C. cratera* A. H. Clark[491] (Comasteridae).

† Rubrocomatulin is the name given to the parent compound (316; OH in place of OMe). It has not yet been found in Nature.

Physical properties: scarlet prisms, m.p. 298–299° (dec.), $\lambda_{max.}$ (EtOH/1% HOAc) 237, 261, 275, 292, 317, 370, 475(sh), 489(sh), 500(sh), 523, 527 nm (log ϵ 4·31, 4·41, 4·22, 3·90, 4·12, 3·26, 4·04, 4·12, 4·21, 4·06, 4·10), $\nu_{max.}$ (Nujol) 1673, 1635, 1593 cm^{-1}.

This pigment is similar to the rhodocomatulin derivatives (p. 492) present in the same animals. It possesses four hydroxyl groups and one O-methyl group, and on prolonged heating with hydrobromic acid-acetic acid the side chain is eliminated leaving a pentahydroxyanthraquinone. Colour reactions and the n.m.r. spectrum of its pentamethyl ether, suggested a 1,2,4,5,7- or 1,2,4,6,8-orientation, and the former was established by synthesis.[492] After short acid hydrolysis, a tetrahydroxy-monomethoxyanthraquinone could be isolated which gave the same pentamethyl ether on methylation. This resistance to demethylation implies the presence of a β-methoxy group, and as colour reactions indicated the absence of vicinal groups, the product must be 1,4,5,7-tetrahydroxy-2-methoxyanthraquinone (317). In the formation of (317) a C_4H_6O fragment is lost, and both analogy with the rhodocomatulin pigments, and the odour of n-butyric acid which is detectable when the

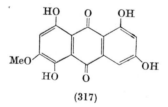

(317)

compound melts with decomposition, indicate the presence of an n-butyryl side chain. Further evidence is provided by a carbonyl band at 1692 cm^{-1} in the infrared spectrum of the tetra-acetate, and is confirmed by the n.m.r. spectrum of the tetramethyl ether. The signals at high field closely simulate those of rhodocomatulin tetramethyl ether, and at low field two one-proton peaks at τ 3·23 and 3·32 must be assigned to β-protons. The side chain therefore occupies the remaining α-position in (317) so that rubrocomatulin monomethyl ether is (316).

Ventimalin (318), $C_{15}H_{10}O_7$

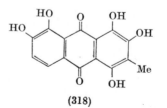

(318)

Occurrence: root bark of *Ventilago viminalis* Hook.[151] (Rhamnaceae).

Physical properties: dark red plates, m.p. 277–279° (dec.), $\lambda_{max.}$ (EtOH) 227, 272, 312, 418(sh), 468(sh), 495 nm (log ϵ 4·23, 4·39, 4·03, 3·62, 3·95, 4·06), $\nu_{max.}$ (KCl) 3401, 1603, 1587 cm^{-1}.

V. viminalis contains an interesting group of quinones including chrysophanol, islandicin, ventilagone (a naphthaquinone) and the pentahydroxyanthraquinone ventimalin. This new compound contains a C-methyl group, and forms a penta-acetate and a pentamethyl ether. Both the visible and the infrared absorption suggest that three α-hydroxyl groups are present, and various colour tests provided evidence for vicinal hydroxyl groups. The similarity of the electronic absorption to that of quinalizarin (1,2,5,8-tetrahydroxyanthraquinone) suggested the orientation of four of the hydroxyl groups, and the fifth must be assigned to a β-position as ventimalin readily undergoes the purpurin-xanthopurpurin reaction when treated with alkaline dithionite. The product is a tetrahydroxymethylanthraquinone which gives a positive zirconium nitrate test and shows only one carbonyl band (ν_{CO} 1597 cm^{-1}) in the infrared. This compound is therefore (319; R = H) or (320; R = H),

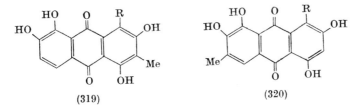

(319)　　　　　　　　(320)

assuming that the methyl group occupies a β-position as usual, and ventimalin must be represented by (319; R = OH) or (320; R = OH). Structures (319; R = H and OH) are, in fact, correct, and this was

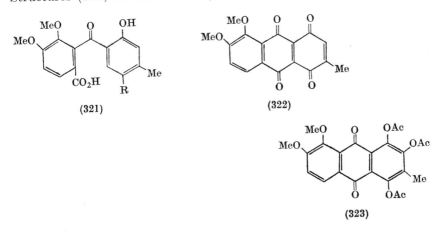

(321)　　　　　　　　(322)

(323)

established by a synthesis of (319; R = OH). The keto-acid (321; R = H) was oxidised to (321; R = OH) with persulphate, and then cyclised to 1,4-dihydroxy-7,8-dimethoxy-3-methylanthraquinone. Further oxidation with lead tetra-acetate then afforded the diquinone (322) which, without purification, was subjected to Thiele acetylation to give (323) whence ventimalin was obtained by hydrolysis and demethylation.[151]

Viminalin (324), $C_{17}H_{14}O_7$

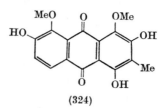

(324)

Occurrence: root bark of *Ventilago viminalis* Hook.[151] (Rhamnaceae).
Physical properties: orange-red needles, m.p. 278·5–279°, ν_{max}. (KCl) 3435, 1672, 1629, 1587 cm⁻¹.

Viminalin has two O-methyl groups and gives ventimalin on demethylation. As the infrared spectrum shows that both free and chelated carbonyl groups are present, and a test for vicinal hydroxyl groups was negative, viminalin must have structure (324).[151]

Asperthecin (325), $C_{15}H_{10}O_8$

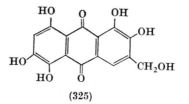

(325)

Occurrence: cultures of *Aspergillus quadrilineatus* Thom & Raper,[506] *A. nidulans* (Eidam) Wint.,[506] *A. nidulans* (Eidam) Wint. mut. *alba* Yuill,[506] and *A. rugulosus* Thom & Raper.[506]

Physical properties: brown needles, m.p. >370°, λ_{max}. (EtOH) 237·5, 262, 286·5, 318·5, 484, 511, 550(sh) nm (log ε 4·38, 4·44, 4·30, 4·05, 4·20, 4·22, 3·83) ν_{max}. (KBr) 3410, 1600 cm⁻¹.

Asperthecin dissolves in aqueous sodium bicarbonate and gives a neutral hexa-acetate. This shows the absence of a carboxyl group and the presence of six hydroxyls, of which at least two must occupy β-positions. The existence of a hydroxymethyl group was deduced from the formation of only a pentamethyl ether on prolonged treatment with dimethyl sulphate-potassium carbonate-acetone or with excess diazomethane in methanol, and from a negative Kuhn-Roth estimation, despite the formation of 2-methylanthracene on zinc dust distillation. As reduction with hydriodic acid and red phosphorus gave emodin† (after reoxidation), the partial structure (326) may be written,[506] and this was advanced to

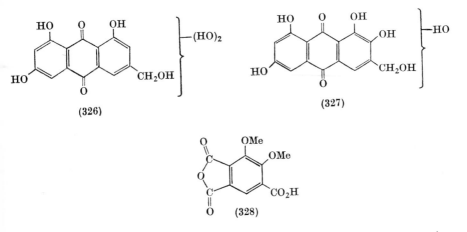

(326)

(327)

(328)

(327) following the isolation of the anhydride (328) from a permanganate oxidation of asperthecin pentamethyl ether.[507] One hydroxyl group remains to be located, and both the visible and infrared spectra strongly indicate that it lies at C-5, as in (325). This was confirmed by oxidising the side chain of asperthecin with nitrous acid in sulphuric acid to give asperthecic acid, decarboxylation of which yielded 1,2,5,6,8-penta-hydroxyanthraquinone, identified by synthesis. This was achieved by way of a two-stage condensation of hemipinic anhydride with 1,2,4-trimethoxybenzene, the structure of the intermediate o-benzoylbenzoic acid being established by an alternative synthesis.[508]

A closely related pigment, possibly a reduced or tautomeric form of asperthecin was observed by Howard and Raistrick[506] in a second strain of *A. quadrilineatus*. Attempts to purify this material led only to asperthecin, and acetylation of the crude product gave asperthecin hexa-acetate.

Dermocybin (329), $C_{16}H_{12}O_7$

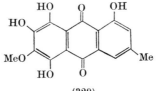

(329)

† Cf. aurantio-obtusin, p. 505.

Occurrence: mainly as a glycoside in *Cortinarius sanguineus* (Wulf. ex Fr.) Fr.[310] (= *Dermocybe sanguinea*) and *Dermocybe semisanguinea* (Fr.).[615]

Physical properties: red needles, m.p. 228–229°, $\lambda_{max.}$ (EtOH) 262·5, 279, 459, 486, 521 nm (log ϵ 4·38, 4·33, 4·07, 4·14, 4·00), $\nu_{max.}$ (KBr) 1597 cm^{-1}.

Studies on the pigments of the higher fungi are frequently hampered by the difficulty of collecting a large amount of material quickly, and by their irregular appearance in successive seasons. These problems beset the first two investigations[310, 509] on the pigments of *C. sanguineus* but in the most recent work Steglich and Austel[334, 615] were able to secure 17 kg of fungal material, the yield of dermocybin being 0·15%.

Dermocybin contains one O-methyl and one C-methyl group, and forms a tetra-acetate. The principal visible absorption maximum at 486 nm is strongly indicative of three α-hydroxyl groups, and the fourth must be in a β-position to allow for the formation of a monomethyl ether (with diazomethane) and a mono-acetate (acetic anhydride, 1 min). Under the same conditions, nordermocybin, obtained by demethylation of the natural pigment, gave a diacetate and dimethyl ether, the latter being identical with the dermocybin monomethyl ether just mentioned. Evidently dermocybin contains a β-methoxyl group. It was known at this stage that emodin was a co-pigment in *C. sanguineus*, and it was a reasonable assumption that nordermocybin might be a dihydroxyemodin, for which there are four possible structures. Two of these, (330) and (331), were synthesised by Birkinshaw and Gourlay,[509] and the latter was found to be identical with nordermocybin. It was obtained by condensation of

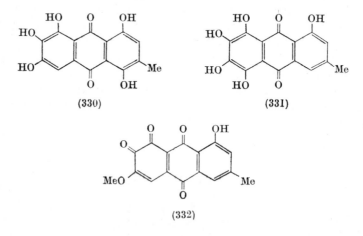

(330)   (331)

(332)

3-methoxy-5-methylphthalic anhydride with 1,2-dihydroxy-3,4-dimethoxybenzene in a melt of aluminium chloride-sodium chloride. Finally, the location of the β-methoxyl group was ascertained by conversion of dermoglaucin (p. 454), another co-pigment in *C. sanguineus*, into dermocybin via the diquinone (332).[334] This shows that the methoxyl group is at C-6 and dermocybin is therefore (329). It occurs in *C. sanguineus* mainly as its 1-glucoside.[616]

Like other purpurins, dermocybin reacts with ammonia at room temperature to give an amino compound (333) by replacement of a *peri*-hydroxyl group. Structure (333) was established by hydrolysis of

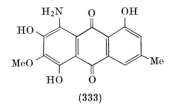

(333)

the diazonium salt to give (334) which does not contain vicinal hydroxyl groups and both quinone carbonyl groups are chelated.[615]

2,5,7-Trihydroxyemodin (334), $C_{15}H_{10}O_8$

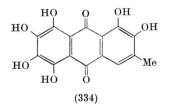

(334)

Occurrence: in the lichen *Mycoblastus sanguinarius* (L.) Norm.[619]
Physical properties: red crystals, m.p. >365°, $\lambda_{max.}$ (EtOH) 294, 313(sh), 350(sh), 497 nm (rel. % 100, 91, 63, 33), $\nu_{max.}$ (KBr) 3380, 1631, 1614 cm^{-1}.

This is the most highly hydroxylated anthraquinone yet found in Nature. It was first reported[619] in the lichen *M. sanguinarius* as an unextractable pigment but was recently isolated[618] after extraction with hot acetone for two weeks. A maximum in the visible region at 497 nm suggests the presence of three α-hydroxyl groups and there are also three β-hydroxyls as the pigment forms a trimethyl ether with diazomethane and is distinctly acidic (soluble in aqueous sodium bicarbonate) although lacking a carboxyl group (infrared). The n.m.r. spectrum (in $(CD_3)_2SO$) shows a singlet (3H) at τ 7·83 assigned to a β-methyl group and another (1H) at τ 2·58 which can be attributed to an α-proton adjacent to a methyl group. Accordingly, this quinone is 2,5,7-trihydroxyemodin (334).[619]

17

AZA-ANTHRAQUINONES

Bostrycoidin (335), $C_{15}H_{11}NO_5$

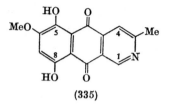

(335)

Occurrence: cultures of *Fusarium bostrycoides* Wollenweber[463,464] and *F. solani* (Mart.) App. & Wr. D_2 purple.[465]

Physical properties: red or brown laths, m.p. 243–244°, $\lambda_{max.}$ (dioxan) 250, 319, 488 nm (log ϵ 4·6, 3·9, 4·0), $\nu_{max.}$ (KBr) 1615 cm^{-1}.

In the original study[464] of this antibiotic pigment the molecular formula $C_{18}H_{14}O_7$ was adopted and the presence of nitrogen was overlooked. However a methoxyl group, two hydroxyls and possibly a methyl group, were detected, and from the spectroscopic evidence (above) a naphthazarin structure was suggested. Arsenault[465] has recently confirmed these findings and revised the molecular formula to $C_{15}H_{11}NO_5$ which means that a third aromatic ring must be present. The implication that bostrycoidin is an aza-anthraquinone was strongly supported by comparison with model compounds, the ultraviolet-visible absorption being very similar to that of 5,8-dihydroxy-2-aza-9,10-anthraquinone (336). In the n.m.r. spectrum broad singlets (each 1H) at τ 0·53 and 2·09 which shift upfield (0·17 and 0·15 ppm, respectively)

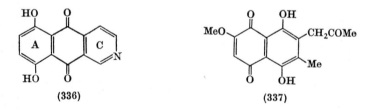

(336)   (337)

on acetylation, are attributed to *para*-protons in the heterocyclic ring. Comparison with (336) shows that the signals from the aromatic protons in ring A move downfield on acetylation while those in ring C shift upfield ($\alpha \geqslant 0.15$ ppm, $\beta < 0.10$ ppm). The methyl singlet at τ 7·22 (7·27 in the diacetate) suggests that this group occupies a β-position as the corresponding signal in the spectrum of 3-methyl-isoquinoline falls at τ 7·32. Direct evidence for the position of the methoxyl group was not obtained but, since javanicin (337), fusarubin and solaniol (see Chapter 4) are co-metabolites, it can safely be placed at C-6. Consequently bostry-coidin is (335).[465] In fusarubin the ring-methyl group of javanicin (337) is replaced by –CH_2OH, and it seems likely that the pyridine ring of bostrycoidin is formed *in vivo* by condensation of the corresponding aldehyde with ammonia.[466]

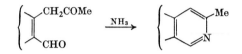

Phomazarin (338), $C_{19}H_{17}NO_8$

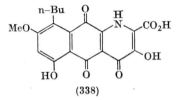

(338)

Occurrence: cultures of *Pyrenochaeta terrestris* (Hansen) Gorez *et al.*[510] (= *Phoma terrestris* Hansen).

Physical properties: orange needles, m.p. 197° (dec.), $\lambda_{max.}$ (EtOH) 235, 274, 325(sh), 390(sh), 439 nm (log ε 4·30, 4·48, 3·83, 3·78, 3·88), $\nu_{max.}$ (KBr) 3445, 3250, 1718(sh), 1627, 1596 cm^{-1}, $\nu_{max.}$ (CHCl$_3$) 1729(sh), 1701, 1635, 1600, 1583 cm^{-1}.

Pyrenochaeta terrestris, the fungus responsible for the "pinkroot" disease of onions, elaborates the aza-anthraquinone, phomazarin, in laboratory culture. Much of the structural information comes from the extensive degradative studies of Kögl and his co-workers[510–512] who established the molecular formula, and the existence of one O-methyl group, two hydroxyl groups and a carboxylic acid function, in addition to a *p*-quinone system. The presence of an n-butyl group was deduced from the formation of n-butyric and valeric acids when phomazarin was oxidized with hydrogen peroxide in sulphuric acid, and larger fragments containing this side chain were obtained by degradation of the triacetate with chromic acid. The main product proved to be the phthalic acid (339) which thus established the substitution pattern in one ring. The second compound isolated from this oxidation was an orange-yellow quinone, $C_{14}H_{13}O_4$.OMe, soluble in aqueous sodium carbonate, which was regarded as a naphthaquinone (340). The nitrogen function appeared to be tertiary but could not be quaternised, and as one derivative (the methyl ester triacetate) showed basic properties, it was concluded that the nitrogen atom must be located in an aromatic ring system. From the molecular formula, the pigment must be tricyclic, the parent compound being $C_{13}H_9N$, that is, an aza-anthracene or an azaphenanthrene. A phenanthraquinone structure was excluded by the remarkable stability of phomazarin under alkaline conditions, being unchanged after three

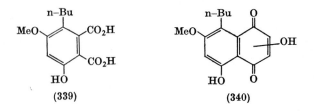

(339)   (340)

hours boiling in alcoholic potash. Phomazarin is therefore an aza-anthraquinone, and the pyridine ring must carry the remaining substituents.

By mild alkaline treatment of phomazarin trimethyl ether-methyl ester Kögl and co-workers obtained "dimethylphomazarin hydrate", $C_{21}H_{23}NO_9$, to which they assigned a ring-opened malonic acid structure. However, Birch et al.[513] have since found that this compound can be recrystallised from non-aqueous solvents to give a product, $C_{21}H_{21}NO_8$, and probably Kögl's compound was, in fact, a hydrate of di-O-methyl-phomazarin. On melting it loses carbon dioxide to give di-O-methyl-decarboxyphomazarin (341) which can be obtained directly from the trimethyl ether-methyl ester in one step by heating in sulphuric acid. In this reaction one methyl ether function is hydrolysed which must therefore be located either *peri* to a quinone carbonyl group, *ortho* or *para* to the nitrogen atom of the pyridine ring, or both. That it is located in the pyridine ring is demonstrated[513] by the replacement of the derived hydroxyl group by chlorine on treatment with phosphoryl chloride, a

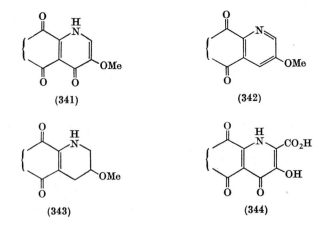

(341)   (342)

(343)   (344)

reaction which can be reversed with alkali. Hydrogenolysis of the chloro compound over palladised charcoal, gave, after reoxidation, di-O-methyldeoxydecarboxyphomazarin (342) and the corresponding tetra-hydro derivative (343).

The following evidence supports the partial structures (341), (342) and (343), and hence (344) for phomazarin.[513] The decarboxylated compound (341) contains an additional aromatic proton revealed by a singlet in the n.m.r. spectrum at τ 1·33, consistent with its location adjacent to aromatic nitrogen, and in the deoxydecarboxy derivative (342) this is *meta* coupled (J, 3 Hz) to a further proton signalled at τ 2·12. The *meta* relationship of the methoxyl group to the nitrogen atom is confirmed by its stability to acid hydrolysis and failure to rearrange to an N-methyl-pyridone. Turning to the infrared data, it was observed that whereas in (342) both quinonoid carbonyl groups absorbed at 1673 cm^{-1} the spectrum of the tetrahydro compound (343) shows ν_{CO} 1668 and 1620 cm^{-1} (all in CHCl$_3$). The latter is indicative of a vinylogous amide system which locates the nitrogen function adjacent to the quinonoid ring as shown, and in agreement, the tetrahydro compound is non-basic. This defines the situation in the pyridine ring and there remains the problem of its orientation relative to the benzenoid ring.

In the infrared spectrum of (341) (in CHCl$_3$) two carbonyl peaks at 1675 and 1640 cm^{-1} can be attributed to the stretching frequencies of a free quinone carbonyl group, and the two carbonyl groups γ to the nitrogen function (vinylogous amide structures), respectively. Decarboxyphomazarin shows carbonyl absorption (in CHCl$_3$) at 1665 and

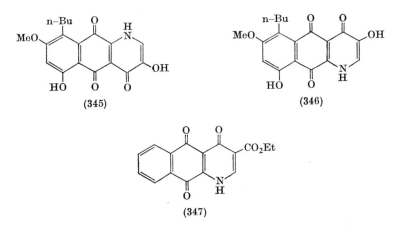

(345)   (346)

(347)

1630 cm^{-1} from which Birch and co-workers[513] concluded that a quinone carbonyl group is present which is neither involved in a vinylogous amide structure nor chelated to a *peri*-hydroxyl group. They argued that this is only possible if the compound has structure (345) and not (346), and therefore phomazarin should have structure (338) [cf. (347),[514] ν_{CO} 1720, 1680, 1630 cm^{-1}]. Most of the molecule is of polyketide origin, the nitrogenous portion being derived from an amino-acid, possibly glycine.[501]

REFERENCES

1. R. Romanis, *J. Chem. Soc.* (1887), **51**, 868; *Proc. Chem. Soc.* (1888), **4**, 116.
2. K. Kafuku and K. Sebe, *Bull. Chem. Soc. Japan* (1932), **7**, 114.
3. A. R. Burnett and R. H. Thomson, *J. Chem. Soc.* (C) (1967), 2100.
4. V. K. Ahluwalia and T. R. Seshadri, *J. Sci. Ind. Res. (India)* (1957), **16B**, 323.
5. W. H. Hui and C. W. Yee, *Phytochemistry* (1967), **6**, 441.
6. W. Sandermann and M. H. Simatupang, *Holz Roh Werkst.* (1966), **24**, 190.
7. P. Rudman, *Chem. Ind. (London)* (1960), 1356.
8. A. G. Perkin and J. J. Hummel, *J. Chem. Soc.* (1893), **63**, 1160.
9. M. H. Simatupang and G. P. Sastry. Personal communication (1968).
10. S. K. Panavaram and L. R. Row, *J. Sci. Ind. Res. (India)* (1957), **16B**, 409.
11. L. H. Briggs and G. A. Nicholls, *J. Chem. Soc.* (1949), 1241.
12. L. H. Briggs and B. R. Thomas, *J. Chem. Soc.* (1949), 1246.
13. A. R. Burnett and R. H. Thomson, *J. Chem. Soc.* (C) (1968), 854.
14. E. Bancroft, "Philosophy of Permanent Colours", (1813), Vol. II, p. 303. Cadell and Davies, London.
15. A. R. Burnett and R. H. Thomson, *J. Chem. Soc.* (C) (1968), 850.
16. "Elsevier's Encyclopaedia of Organic Chemistry", Ser. III, Vol. 13.
17. R. Livingstone and M. C. Whiting, *J. Chem. Soc.* (1955), 3631; W. D. Cotterill, R. Livingstone, K. D. Bartle and D. W. Jones, *Tetrahedron* (1968), **24**, 1981.
18. S. Shibata and M. Takido, *Pharm. Bull. (Tokyo)* (1955), **3**, 156.
19. I. R. C. Bick and C. Rhee, *Biochem. J.* (1966), **98**, 112.
20. G. P. Slater, R. H. Haskins, L. R. Hogge and L. R. Nesbitt, *Canad. J. Chem.* (1967), **45**, 92.
21. P. C. Mitter and H. Biswas, *J. Indian Chem. Soc.* (1928), **5**, 769.
22. S. K. Pavanaram, P. Hofer, H. Linde and K. Meyer, *Helv. Chim. Acta* (1963), **46**, 1377; K. Meyer. Personal communication (1968).
23. H. Waldmann and P. Sellner, *J. prakt. Chem.* (1938), **150**, 145.
24. (a) Colin and Robiquet, *Ann. Chim. Phys.* (1827), **34**, 225; (b) E. Schunk, *Annalen* (1848), **66**, 174; *Phil. Trans.* (1851), 433; (c) Rochleder, *Annalen* (1851), **80**, 321.
25. A. R. Burnett and R. H. Thomson, *Phytochemistry* (1968), **7**, 1421.
26. A. G. Perkin and J. J. Hummel, *J. Chem. Soc.* (1895), **67**, 817.
27. R. Hill and D. Richter, *Proc. Roy. Soc.* (1937), **121B**, 547.
28. A. Juillet, J. Suspluga and V. Massa, *Bull. Pharm. Sud-Est* (1937), **41**, 374.
29. H. Müller, *J. Chem. Soc.* (1911), **99**, 967.
30. Nora Henriquez Ulloa, *Anal. Fac. Quim. Farm. Univ. Chile* (1960), **12**, 113 (*Chem. Abs.*, 1962, **56**, 5121).
31. A. N. Ratnagiriswaran and K. Venkatachalam, *J. Indian Chem. Soc.* (1942), **19**, 389.
32. G. Schaefer, *Ciba Rev.* (1941), **4**, 1407, 1417.

33. E. G. Kiel and P. M. Heertjes, *J. Soc. Dyers and Colourists* (1963), **79**, 21.
34. G. Schaefer, *Ciba Rev.* (1941), **4**, 1398.
35. C. Graebe and C. Liebermann, *Annalen Suppl.* (1870), **7**, 257.
36. C. Graebe, *Annalen* (1868), **146**, 1.
37. B.A.S.F., G.P. 186,526.
38. Thorpe's Dictionary of Applied Chemistry (1937), Vol. I, p. 216. Longmans Green, London.
39. D. Richter, *J. Chem. Soc.* (1936), 1701.
40. E. T. Jones and A. Robertson, *J. Chem. Soc.* (1933), 1167.
41. G. Zemplén and R. Bognar, *Ber.* (1939), **72**, 913.
42. M. Barrowcliff and F. Tutin, *J. Chem. Soc.* (1907), **91**, 1907.
43. J. L. Simonsen, *J. Chem. Soc.* (1920), **117**, 561.
44. V. V. S. Murti, S. Neelakantan, T. R. Seshadri and B. Venkataramani, *J. Sci. Ind. Res. (India)* (1959), **18B**, 367.
45. G. A. Fester and S. G. Lexow, *Rev. Fac. Quím. Ind. Agr.* (1942–43), **11–12**, 84.
46. G. A. Fester, A. F. Suarez and M. A. Gargallo, *Rev. Fac. Quím. Ind. Agr.* (1946–47), **15–16**, 74; *An. Asoc. quím. argent.* (1948), **36**, 43.
47. O. Kubota and A. G. Perkin, *J. Chem. Soc.* (1925), **127**, 1889.
48. A. G. Perkin and R. C. Storey, *J. Chem. Soc.* (1928), 229.
49. A. G. Perkin and J. J. Hummel, *J. Chem. Soc.* (1907), **91**, 2066.
50. K. Lagodinski, *Annalen* (1905), **342**, 90.
51. Adrian and A. Trillat, *Compt. Rend.* (1899), **129**, 889.
52. R. Paris, *C.R. Soc. Biol.* (1940), **133**, 46; *Compt. Rend.* (1954), **238**, 932.
53. M.-M. Janot, J. Chabasse-Massonneau, P. de Graeve and R. Goutarel, *Bull. Soc. Chim. France* (1955), 108.
54. J. C. Lovie and R. H. Thomson, *J. Chem. Soc.* (1959), 4139.
55. K. Chandrasenan, S. Neelakantan and T. R. Seshadri, *J. Sci. Ind. Res. (India)* (1960), **19A**, 28.
56. L. H. Briggs, P. S. Rutledge and R. N. Seelye, unpublished.
57. A. G. Perkin and J. J. Hummel, *J. Chem. Soc.* (1893), **63**, 1157.
58. P. Schutzenberger, *Bull. Soc. Chim. France* (1865), **4**, 12.
59. E. Noah, *Ber.* (1886), **19**, 332.
60. (a) E. de B. Barnett and J. W. Cook, *J. Chem. Soc.* (1922), **121**, 1376; (b) A. G. Perkin and C. W. H. Story, *J. Chem. Soc.* (1929), 1399; G. P. 212,697.
61. G. A. Fester, A. F. Suarez and J. Huck, *Rev. Fac. Quím. Ind. Agr.* (1945), **14**, 163.
62. E. Schunk, *Trans. Roy. Soc.* (1853), **143**, 67.
63. R. Hill and D. Richter, *J. Chem. Soc.* (1936), 1714.
64. L. H. Briggs, G. A. Nicholls and R. M. L. Paterson, *J. Chem. Soc.* (1952), 1718.
65. J. H. Bowie and R. G. Cooke, *Aust. J. Chem.* (1962), **15**, 332.
66. J. H. Bowie, R. G. Cooke and P. E. Wilkin, *Aust. J. Chem.* (1962), **15**, 336.
67. E. Schunk and L. Marchlewski, *J. Chem. Soc.* (1893), **63**, 969; (1894), **65**, 182.
68. F. D. Stouder and R. Adams, *J. Amer. Chem. Soc.* (1927), **49**, 2043; P. C. Mitter, M. Sen and P. K. Paul, *J. Indian Chem. Soc.* (1927), **4**, 535.
69. P. C. Mitter and P. Gupta, *J. Indian Chem. Soc.* (1928), **5**, 25; cf. T. Kusaka, *J. Pharm. Soc. Japan* (1935), **55**, 682.
70. E. T. Jones and A. Robertson, *J. Chem. Soc.* (1930), 1699.
71. B. S. Joshi, N. Parkash and K. Venkataraman, *J. Sci. Ind. Res. (India)* (1955), **14B**, 87.

72. Ch. Marschalk, F. Koenig and N. Ouroussoff, *Bull. Soc. Chim. France* (1936), 1545; Ch. Marschalk, *Bull. Soc. Chim. France* (1939), 655.
73. E. Rybacki, *Dissert. pharm.* (1965), **17**, 335, 339; *Int. Abstr. Biol. Sci.* (1966), **42**, 668.
74. (a) P. C. Mitter and S. C. Pal, *J. Indian Chem. Soc.* (1930), **7**, 259; (b) P. C. Mitter and H. Biswas, *J. Indian Chem. Soc.* (1930), **7**, 839.
75. L. H. Briggs and A. R. Taylor, *J. Chem. Soc.* (1955), 3298.
76. L. H. Briggs and J. C. Dacre, *J. Chem. Soc.* (1948), 564.
77. L. H. Briggs, M. R. Craw and J. C. Dacre, *J. Chem. Soc.* (1948), 568.
78. S. Balakrishna, T. R. Seshadri and B. Venkataramani, *J. Sci. Ind. Res. (India)* (1961), **20B**, 331.
79. R. Paris and N. Abiusso, *Ann. Pharm. Franç.* (1958), **16**, 660.
80. L. H. Briggs and G. A. Nicholls, *J. Chem. Soc.* (1953), 3068.
81. N. R. Ayyangar and K. Venkataraman, *J. Sci. Ind. Res. (India)* (1956), **15B**, 359.
82. V. A. Stikhin, A. I. Bankovskii and M. E. Perel'son, *Khim. Prirod. Soedin.* (1966), 12; (1967), 276.
83. V. A. Stikhin, A. I. Bankovskii and M. E. Perel'son, *Khim. Prirod. Soedin.* (1968), 273.
84. S. Nonomura, *J. Pharm. Soc. Japan* (1955), **75**, 219.
85. (a) Y. Hirose, *J. Pharm. Soc. Japan* (1956), **76**, 1448; (b) *Chem. Pharm. Bull. (Tokyo)* (1960), **8**, 417.
86. N. R. Ayyangar, B. S. Joshi and K. Venkataraman, *Tetrahedron* (1959), **6**, 331.
87. E. Schunk and H. Roemer, *Ber.*, 1877, **10**, 172, 790; *J. Chem. Soc.* (1877), **31**, 666; (1878), **33**, 422; H. Plath, *Ber.* (1877), **10**, 616.
88. W. H. Hui, S. K. Szeto and C. W. Yee, *Phytochemistry* (1967), **6**, 1299.
89. A. V. Patwardhan, R. Rao and K. Venkataraman, unpublished work.
90. (a) S. Nonomura, *J. Pharm. Soc. Japan* (1955), **75**, 222, 225; (b) S. Nonomura and Y. Hirose, *J. Pharm. Soc. Japan* (1955), **75**, 1305; Y. Hirose, J. Kusuda, S. Nonomura and H. Fukui, *Chem. Pharm. Bull. (Tokyo)* (1968), **16**, 1377.
91. S. Nonomura, *Pharm. Bull. (Tokyo)* (1957), **5**, 365.
92. J. Stenhouse, *Annalen* (1864), **130**, 325.
93. A. G. Perkin and J. J. Hummel, *J. Chem. Soc.* (1893), **63**, 1157.
94. P. C. Mitter and H. Biswas, *Nature (London)* (1930), **126**, 761; (1931), **127**, 666; *Ber.* (1932), **65**, 622.
95. Y. Takagi, *J. Chem. Soc. Japan, Pure Chem. Sect.* (1961), 82, 1561.
96. Y. Hirose, *Chem. Pharm. Bull. (Tokyo)* (1962), **10**, 985.
97. A. G. Perkin and J. J. Hummel, *J. Chem. Soc.* (1894), **65**, 851.
98. W. Sandermann and M. H. Simatupang, *Naturwissenschaften* (1965), **52**, 262.
99. R. Nietzki, *Ber.* (1877), **10**, 2011.
100. Y. Hirose, H. Ueno, M. Iwashita and E. Kawagishi, *Chem. Pharm. Bull. (Tokyo)* (1963), **11**, 531.
101. O. A. Oesterle and E. Tisza, *Arch. Pharm.* (1908), **246**, 158.
102. P. C. Mitter and H. Biswas, *J. Indian Chem. Soc.* (1928), **5**, 769.
103. J. L. Simonsen and M. G. Rau, *J. Chem. Soc.* (1921), **119**, 1339.
104. N. S. Bhide, Ph.D. Thesis, University of Poona (1966).
105. R. Brandes, *Annalen* (1834), **9**, 85.
106. Ph. L. Geiger, *Annalen* (1834), **9**, 91.
107. J. Schlossberger and O. Döpping, *Annalen* (1844), **50**, 196.
108. F. Beilstein, *Ber.* (1882), **15**, 901.

109. A. Tschirch and J. Edner, *Arch. Pharm.* (1907), **245**, 139.
110. F. Gstirner and H. Holtzem, *Pharmazie* (1949), **4**, 333.
111. L. F. Bley and E. Diesel, *Arch. Pharm.* (1847), **99**, 121.
112. J. J. Holmström, *Schweiz. Apoth.-Ztg.* (1921), **59**, 183.
113. R. Segal, I. Milo-Goldzweig and D. V. Zaitschek, *Lloydia* (1964), **27**, 237.
114. K. V. Taraskina and T. K. Chumbalov, *Izvest. Akad. Nauk Kaz. S.S.R. Ser. Khim.* (1962), No. 2, 83.
115. T. K. Chumbalov and G. M. Nurgalieva, *Khim. Prirod. Soedin.* (1967), 144.
116. Ph. L. Geiger, *Annalen* (1834), **9**, 310.
117. O. Hesse, *Annalen* (1896), **291**, 305; (1899), **309**, 48.
118. Schroff, *Wochenbl. Ztschr. Ges. Ärzte. Wien* (1864), **20**, 27, 35.
119. Grothe, *Chem. Centralblatt.* (1862), 107.
120. G. D. Beal and R. E. Okey, *J. Amer. Chem. Soc.* (1919), **41**, 693.
121. P. Q. Keegan, *Chem. News* (1916), **114**, 74.
122. F. Tutin and H. W. B. Clewer, *J. Chem. Soc.* (1910), **97**, 1.
123. R. L. Khazanovitch, *Farmatsiya* (1941), No. 2, 30.
124. W. Poethke and H. Behrendt, *Pharm. Zentralhalle* (1965), **104**, 549.
125. W. E. Hillis, *Aust. J. Chem.* (1955), **8**, 290.
126. T. K. Chumbalov and K. V. Tarashina, *Izvest. Akad. Nauk Kaz. S.S.R. Ser. Khim.* (1956), No. 9, 61.
127. K. Tsukida, *J. Pharm. Soc. Japan* (1954), **74**, 386.
128. A. A. Khan, *Canad. J. Chem.* (1963), **41**, 1622.
129. T. K. Chumbalov and R. A. Muzychkina, *Khim. Prirod. Soedin. Akad. Nauk Uz. S.S.R.* (1965), 360.
130. A. Tschirch and J. F. A. Pool, *Arch. Pharm.* (1908), **246**, 320.
131. A. Tschirch and H. Bromberger, *Arch. Pharm.* (1911), **249**, 222.
132. A. Tschirch and L. Monikowski, *Arch. Pharm.* (1912), **250**, 92.
133. R. W. Liddell, C. G. King and G. D. Beal, *J. Amer. Pharm. Ass.* (1942), **31**, 161.
134. K. Tsukida, *J. Pharm. Soc. Japan* (1954), **74**, 398, 401.
135. K. Tsukida and N. Suzuki, *J. Pharm. Soc. Japan* (1955), **75**, 982.
136. B. Akačic and B. Poje, *Acta Pharm. Jugoslav.* (1960), **10**, 57.
137. M. Greshoff, *Ber.* (1890), **23**, 3537.
138. T. Peckolt, *Arch. Pharm.* (1868), **184**, 37.
139. H. W. Youngken and R. A. Walsh, *J. Am. Pharm. Ass.* (1954), **43**, 139.
140. K. Iwakawa, *Arch. exp. Path. Pharmak.* (1911), **65**, 315.
141. M. Takido, *Chem. Pharm. Bull. (Tokyo)* (1958), **6**, 397.
142. R. D. Tiwari and T. Joshi, *Proc. Nat. Acad. Sci. India* (1965), **35A**, 448.
143. S. P. Bhutani, S. S. Chibber and T. R. Seshadri, *Curr. Sci.* (1966), **35**, 363.
144. J. W. Fairbairn and A. B. Shrestha, *Phytochemistry* (1967), **6**, 1203.
145. F. Tutin and H. W. B. Clewer, *J. Chem. Soc.* (1912), **101**, 2221.
146. K. Tsukida and M. Yoneshige, *J. Pharm. Soc. Japan* (1954), **74**, 379.
147. T. T. Kariyone and K. Tsukida, *J. Pharm. Soc. Japan* (1954), **74**, 223, 225.
148. G. Hauschild, M. Steiner and K. W. Glombitza, *Naturwissenschaften* (1968), **55**, 346.
149. G. R. Chaudry and S. Siddiqui, *J. Sci. Ind. Res. (India)* (1950), **9B**, 118; G. R. Chaudry, V. N. Sharma and S. Siddiqui, *J. Sci. Ind. Res. (India)* (1950), **9B**, 142.
150. M. H. Simatupang, H. H. Dietrichs and H. Gottwald, *Holzforschung* (1967), **21**, 89.
151. R. G. Cooke and B. L. Johnson, *Aust. J. Chem.* (1963), **16**, 695.
152. M. C. B. van Rheede van Oudtshoorn, *Planta Med.* (1963), **11**, 332.

153. M. C. B. van Rheede van Oudtshoorn, *Phytochemistry* (1964), **3**, 383.
154. K. Tsukida and M. Yokota, *J. Pharm. Soc. Japan* (1954), **74**, 230.
155. J. C. Th. Uphof, "Dictionary of Economic Plants", (1959), 2nd edit. Steckert-Hafner, New York.
156. E. Ritchie and W. C. Taylor, *Tetrahedron Lett.* (1964), 1431.
157. V. Arkley, F. M. Dean, P. Jones, A. Robertson and J. Tetaz, *Croat. Chem. Acta* (1957), **29**, 141.
158. S. Shibata, J. Shoji, A. Ohta and M. Watanabe, *Pharm. Bull. (Tokyo)* (1957), **5**, 380.
159. T. M. Farley, *Diss. Abs.* (1965), **25**, 6193.
160. T. Shimano, K. Taki and K. Goto, *Ann. Proc. Gifu Coll. Pharm.* (1953), No. 3, 43.
161. M. C. B. van Rheede van Oudtshoorn and K. W. Gerritsma, *Pharm. Weekblad.* (1964), **99**, 1425.
162. H. Duewell, *J. Chem. Soc.* (1954), 2562.
163. D. McGrath, *Nature (London)* (1967), **215**, 1414.
164. C. Graebe and C. Liebermann, *Annalen Suppl.* (1870), **7**, 306; C. Liebermann and F. Giesel, *Annalen* (1876), **183**, 174.
165. O. Fischer, F. Falco and H. Gross, *J. prakt. Chem.* (1911), **83**, 208.
166. C. Liebermann and St. v. Kostanecki, *Ber.* (1886), **19**, 2327.
167. O. A. Oesterle, *Schweiz. Wschr. Chem. Pharm.* (1911), **49**, 661.
168. E. Léger, *Compt. Rend.*, 1912, **154**, 281; *J. Pharm. Chim.* (1912), **5**, 281.
169. R. Eder and C. Widmer, *Helv. Chim. Acta* (1922). **5**, 3.
170. R. Eder and C. Widmer, *Helv. Chim. Acta* (1923), **6**, 419.
171. A. C. Bellaart and C. Koningsberger, *Rec. Trav. Chim.* (1960), **79**, 1289.
172. C. A. Naylor and J. H. Gardner, *J. Amer. Chem. Soc.* (1931), **53**, 4109, 4114.
173. N. R. Ayyangar, D. S. Bapat and B. S. Joshi, *J. Sci. Ind. Res. (India)* (1961), **20**, 493.
174. E. Gilson, *Arch. intern. Pharmacodyn.* (1905), **14**, 455.
175. H. Wagner, L. Hörhammer and L. Farkas, *Z. Naturforsch.* (1963), **18b**, 89.
176. L. Hörhammer, G. Bittner and H. P. Hörhammer, *Naturwissenschaften* (1964), **51**, 310.
177. R. Takahashi, *J. Pharm. Soc. Japan* (1925), **45**, 969; H. Foster and J. H. Gardner, *J. Amer. Chem. Soc.* (1936), **58**, 597; H. Mühlemann, *Pharm. Acta Helv.* (1949), **24**, 315.
178. F. Tutin and H. W. B. Clewer, *J. Chem. Soc.* (1912), **101**, 290.
179. F. E. King, M. F. Grundon and K. G. Neill, *J. Chem. Soc.* (1952), 4580.
180. L. Hörhammer, H. Wagner and E. Müller, *Chem. Ber.* (1965), **98**, 2859; L. Hörhammer. Personal communication (1967).
181. A. C. Bellaart, *Chem. Ber.* (1966), **99**, 2471.
182. A. Stoll and B. Becker, *Rec. Trav. Chim.* (1950), **69**, 553.
183. M. O. Farooq, M. A. Aziz and M. S. Ahmad, *J. Am. Oil Chemists' Soc.* (1956), **33**, 21.
184. A. Tschirch, *Ber. Deut. Pharm. Ges.* (1898), **8**, 174; A. Tschirch and G. Pedersen, *Arch. Pharm.* (1898), **236**, 200.
185. J. Aschan, *Arch. Pharm.* (1903), **241**, 340.
186. A. Tschirch and J. Klaveness, *Arch. Pharm.* (1901), **239**, 241.
187. A. Tschirch and U. Cristofoletti, *Schweiz. Wschr. Chem. Pharm.* (1904), **42**, 456.
188. A. Tschirch and O. Hoffbauer, *Arch. Pharm.* (1905), **243**, 399.
189. E. Léger, *Bull. Soc. Chim. France* (1900) [3] **23**, 785.
190. G. Condò-Vissicchio, *Arch. Pharm.* (1909), **247**, 81.

191. (a) O. Hesse, *Annalen* (1899), **309**, 32; (b) *J. prakt. Chem.* (1908), [2] **77**, 321, 350.
192. F. Tutin and H. W. B. Clewer, *J. Chem. Soc.* (1911), **99**, 946.
193. S. Shibata and M. Takido, *J. Pharm. Soc. Japan* (1952), **72**, 1311.
194. P. F. Jorgensen, *Dansk. Tidsskr. Farm.* (1950), **24**, 111.
195. A. Tschirch and E. Hiepe, *Arch. Pharm.* (1900), **238**, 427.
196. F. Tutin, *J. Chem. Soc.* (1913), **103**, 2006.
197. M. Uchibayashi and T. Matsuoka, *Chem. Pharm. Bull. (Tokyo)* (1961), **9**, 234.
198. S. P. Bhutani, S. S. Chibber and T. R. Seshadri, *Curr. Sci.* (1966), **35**, 363.
199. A. S. Romanova and A. I. Ban'kovskii, *Khim. Prirod. Soedin.* (1965), 294.
200. P. A. A. F. Eijken, *Pharm. Weekblad.* (1904), **41**, 202.
201. O. A. Oesterle, *Arch. Pharm.* (1903), **241**, 604; (1911), **249**, 455.
202. P. C. Mitter and D. Banerjee, *J. Indian Chem. Soc.* (1932), **9**, 375
203. D. S. Bapat, B. C. Subba Rao, M. K. Unni and K. Venkataraman, *Tetrahedron Lett.* (1960), No. 5, 15; D. S. Bapat and B. C. Subba Rao, *J. Sci. Ind. Res. (India)* (1962), **21B**, 179.
204. T. R. Rajagopalan and T. R. Seshadri, *Proc. Indian Acad. Sci.* (1956), **44A**, 418.
205. J. W. Fairbairn and A. B. Shrestha, *J. Pharm. Pharmac.* (1966), **18**, 467.
206. L. Hörhammer, L. Farkas, H. Wagner and E. Müller, *Chem. Ber.* (1964), **97**, 1662.
207. British Pharmacopoeia (1963).
208. T. Smith and H. Smith, *Pharm. J.* (1851), **11**, 23.
209. H. Mühlemann, *Pharm. Acta Helv.* (1952), **27**, 17.
210. L. J. Haynes, *Adv. Carbohydrate Chem.*, (a) (1963), **18**, 227; (b) (1965), **20**, 357.
211. R. S. Cahn and J. L. Simonsen, *J. Chem. Soc.* (1932), 2573.
212. J. E. Hay and L. J. Haynes, *J. Chem. Soc.* (1956), 3141.
213. E. Léger, *Ann. Chim.* (1916), **6**, 318.
214. M. C. B. van Rheede van Oudtshoorn and K. W. Gerritsma, *Naturwissenschaften* (1965), **52**, 35, 186.
215. L. Hörhammer, H. Wagner and G. Bittner, *Z. Naturforsch.* (1964), **19b**, 222.
216. H. Boehme and L. Kreutzig, *Deut. Apotheker-Z.* (1963), **103**, 505.
217. J. W. Fairbairn and S. Simic, *J. Pharm. Pharmac. Suppl.* (1960), **12**, 45T; (1963), **15**, 292T; (1964), **16**, 450.
218. O. Hesse, *Pharm. J.* (1895 [4] **1**, 325; A. Tschirch and P. A. A. F. Eijken, *Schweiz. Wschr. Chem. Pharm.* (1904), **42**, 549.
219. K. Tsukida, N. Suzuki and M. Yokota, *J. Pharm. Soc. Japan* (1954), **74**, 224.
220. A. J. Birch, D. N. Butler and J. R. Siddall, *J. Chem. Soc.* (1964), 2941.
221. F. K. Modi and M. L. Khorana, *Indian J. Pharm.* (1952), **14**, 61; G. J. Kapadia and M. L. Khorana, *Lloydia* (1962), **25**, 55; A. Kumar, C. S. Pande and R. K. Kaul, *Indian J. Chem.* (1966), **4**, 460; V. K. Murty, T. V. P. Rao and V. Venkateswarlu, *Tetrahedron* (1966), **23**, 515.
222. N. N. Kaji and M. L. Khorana, *Indian J. Chem.* (1965), **27**, 338.
223. L. Boross, *Acta Chim. Hung.* (1963), **35**, 195.
224. M. Anchel, *J. Biol. Chem.* (1949), **177**, 169.
225. E. Schratz, *Planta Med.* (1960), **8**, 301.
226. H. Hauptmann and L. L. Nazario, *J. Amer. Chem. Soc.* (1950), **72**, 1492.
227. R. Robinson and J. L. Simonsen, *J. Chem. Soc.* (1909), **95**, 1085.
228. H. Wagner and I. Köhler, *Naturwissenschaften* (1957). **44**, 260; L. Hörhammer, H. Wagner and I. Köhler, *Naturwissenschaften* (1958), **45**, 389; L. Hörhammer, H. Wagner and I. Köhler, *Arch. Pharm.* (1959), **292**, 591.

229. H. Auterhoff and F. C. Scherff, *Arch. Pharm.* (1960), **293/65**, 918; H. Nawa, M. Uchibayashi and T. Matsuoka, *Chem. Pharm. Bull. (Tokyo)* (1960), **8**, 566; *J. Org. Chem.* (1961), **26**, 979.

230. N. S. Zhuravljev and M. I. Borisov, *Khim. Prirod. Soedin.* (1969), 118, 176.

231. A. S. Romanova and A. I. Ban'kovskii, *Khim. Prirod. Soedin.* (1966), 143.

232. J. K. Crellin, J. W. Fairbairn, C. A. Friedmann and H. A. Ryan, *J. Pharm. Pharmac.* (1961), **13**, 639.

233. J. W. Fairbairn, *Lloydia* (1964), **27**, 79; *Pharm. Weekblad.* (1965), **100**, 1493.

234. W. Schmid and E. Angliker, *Helv. Chim. Acta* (1965), **48**, 1911.

235. A. Stoll and B. Becker, *Fortschr. Chem. Org. Naturstoffe* (1950), **7**, 248.

236. J. Lemli, *Pharm. Tijdschr. Belg.* (1962), **39**, 67.

237. M. Miyamoto, S. Imai, M. Shinohara, S. Fujioka, M. Goto, T. Matsuoka and H. Fujimura, *J. Pharm. Soc. Japan* (1967), **87**, 1040.

238. P. K. Hietala and A. Penttilä, *Acta Chem. Scand.* (1966), **20**, 575.

239. J. Lemli, R. Dequeker and J. Cuveele, *Pharm. Weekblad.* (1963), **98**, 500.

240. J. Lemli, *Lloydia* (1965), **28**, 63.

241. J. Lemli and J. Cuveele, *Pharmazie* (1965), **20**, 396.

242. J. Lemli, *Verh. Konink. Vlaam. Acad. Geneeskunde Belg.* (1966), **28**, 49.

243. J. Lemli, R. Dequeker and J. Cuveele, *Pharm. Weekblad.* (1963), **98**, (a) 529, (b) 613, (c) 655; (d) (1964), **99**, 613.

244. J. Lemli, R. Dequeker and J. Cuveele, *Pharm. Weekblad.* (1964), **99**, 589.

245. M. Bamberger and A. Praetorius, *Monatsh.* (1901), **22**, 587; (1902), **23**, 688.

246. C. Seuberlich, *Ber.* (1877), **10**, 38.

247. L. H. Briggs and J. F. Beachen, unpublished.

248. F. F. Runge, *J. prakt. Chem.* (1835), **5**, 362; H. Debus, *Annalen* (1848), **66**, 351; (1853), **86**, 117; J. Wolff and A. Strecker, *Annalen* (1850), **75**, 1.

249. F. de Lalande, *Compt. Rend.* (1874), **79**, 669.

250. A. Rosenstiehl, *Compt. Rend.* (1874), **79**, 764.

251. A. Baeyer and H. Caro, *Ber.* (1875), **8**, 152.

252. D. B. Bruce and R. H. Thomson, *J. Chem. Soc.* (1952), 2759.

253. O. Dimroth, O. Friedmann and H. Kämmerer, *Ber.* (1920), **53**, 481.

254. O. Dimroth and E. Schultze, *Annalen* (1916), **411**, 339.

255. R. Scholl and P. Dahll, *Ber.* (1924), **57**, 80.

256. A. Rosenstiehl, *Ann. Chim.* (1878), [5] **13**, 248; C. Liebermann and H. Plath, *Ber.* (1877), **10**, 1618.

257. M. Wada, *Science (Japan)* (1941), **11**, 415.

258. Bayer and Co., G.P. 260,765; 272,301.

259. P. C. Mitter and H. Biswas, *Ber.* (1932), **65**, 622.

260. Y. Hirose, *Chem. Pharm. Bull. (Tokyo)* (1963), **11**, 533.

261. P. C. Beaumont, R. L. Edwards and G. C. Elsworthy, *J. Chem. Soc. (C)* (1968), 2968.

262. W. Steglich, W. Furtner and A. Prox, *Z. Naturforsch.* (1968), **23b**, 1044.

263. A. Bräm and C. H. Eugster, *Helv. Chim. Acta* (1969), **52**, 165.

264. A. R. Mehandale, A. V. Rama Rao, I. N. Shaikh and K. Venkataraman, *Tetrahedron Lett.* (1968), 2231.

265. J. Lemli and J. Cuveele, *Pharm. Acta Helv.* (1965), **40**, 667.

266. N. Y. Ayyangar, *Indian J. Chem.* (1967), **5**, 530.

267. G. Agosti, J. H. Birkinshaw and P. Chaplen, *Biochem. J.* (1962), **85**, 828.

268. V. M. Chari, S. Neelakantan and T. R. Seshadri, *Tetrahedron Lett.* (1967), 999.

269. S. Shibata, *Chem. Brit.* (1967), **3**, 110.

270. T. Anderson, *Trans. Roy. Soc. Edin.* (1849), **16**, 435; *Annalen* (1849), **71**, 216; *J. prakt. Chem.* (1849), **47**, 43.
271. O. A. Oesterle and E. Tisza, *Arch. Pharm.* (1907), **245**, 534; (1908), **246**, 150.
272. J. L. Simonsen, *J. Chem. Soc.* (1918), **113**, 766.
273. Rochleder, *Annalen* (1852), **82**, 205; J. Stenhouse, *J. Chem. Soc.*, 1864, **17**, 333.
274. T. E. Thorpe and T. H. Greenall, *J. Chem. Soc.* (1887), **51**, 52; T. E. Thorpe and W. J. Smith, *J. Chem. Soc.* (1888), **53**, 171.
275. J. L. Simonsen, *J. Chem. Soc.* (1918), **113**, 766.
276. C. D. Mell, *Textile Colorist* (1928), **50**, 531.
277. S. Shibata, E. Morishita, M. Kaneda, Y. Kimura, M. Takido and S. Takahashi, *Chem. Pharm. Bull.* (*Tokyo*) (1969), **17**, 454.
278. J. L. Simonsen, *J. Chem. Soc.* (1924), **125**, 721.
279. R. Bhattacharya and J. L. Simonsen, *J. Indian Inst. Sci.* (1927), **10A**, 6.
280. R. A. Jacobson and R. Adams, *J. Amer. Chem. Soc.* (1924), **46**, 2788; (1925), **47**, 283.
281. R. Paris and Ng. Ba Tuoc, *Ann. Pharm. Franç.* (1954), **12**, 794.
282. L. H. Briggs and P. W. Le Quesne, *J. Chem. Soc.* (1963), 3471.
283. S. Balakrishna, T. R. Seshadri and B. Venkataramani, *J. Sci. Ind. Res.* (*India*) (1960), **19B**, 433.
284. A. G. Perkin, *Proc. Chem. Soc.* (1908), **24**, 149.
285. L. J. Haynes and J. I. Henderson, *Chem. Ind.* (*London*) (1960), 50; L. J. Haynes, J. I. Henderson and J. M. Tyler, *J. Chem. Soc.* (1960), 4879.
286. E. Léger, *Ann. Chim.* (1917), **8**, 265.
287. L. H. Briggs, M. Kingsford and R. N. Seelye, unpublished.
288. Y. Hirose, N. Sasaki, E. Kawagishi and S. Nonomura, *Chem. Pharm. Bull.* (*Tokyo*) (1962), **10**, 634.
289. S. Nonomura, *Kumamoto Pharm. Bull.* (1959), No. 4, 239.
290. Y. Hirose, *J. Pharm. Soc. Japan* (1958), **78**, 947.
291. V. M. Chari, S. Neelakantan and T. R. Seshadri, *Indian J. Chem.* (1966), **4**, 330.
292. R. Suemitsu, Y. Matsui and M. Hiura, *Bull. Agric. Chem. Soc. Japan* (1957), **21**, 1.
293. A. Stoessl, *Chem. Comm.* (1967), 307.
294. A. Stoessl, *Canad. J. Chem.* (1969), **47**, 767.
295. R. Suemitsu and M. Hiura, *Bull. Agric. Chem. Soc. Japan* (1959), **23**, 337.
296. R. Suemitsu, M. Nakajima and H. Hiura, *Bull. Agric. Chem. Soc. Japan* (1959), **23**, 547.
297. R. Suemitsu, M. Nakajima and M. Hiura, *Agr. Biol. Chem. Japan* (1961), **25**, 100.
298. S. Nonomura and Y. Hirose, *Chem. Pharm. Bull.* (*Tokyo*) (1961), **9**, 510.
299. Additional references to old work may be found in Beilstein's "Handbuch der Organischen Chemie", Vol. VIII, and supplements, and Elsevier's "Encyclopaedia of Organic Chemistry", Vol. XIII.
300. G. K. Ray, R. C. Guha, A. B. Bose and B. Mukerji, *Indian J. Pharm.* (1944), **6**, 55; H. W. Youngken, *J. Am. Pharm. Ass.* (1946), **35**, 148.
301. Y. Murayama and T. Itagaki, *J. Pharm. Soc. Japan* (1921), 470, 327; G. Kurono and T. Ishida, *Ann. Rept. Fac. Pharm. Kanazawa Univ.* (1951), **1**, 1.
302. C. Vutyrakis, *Chim. Ind.* (*Paris*) (1937), **41**, 315.
303. L. H. Briggs, F. E. Jacombs and G. A. Nicholls, *J. Chem. Soc.* (1953), 3069.

304. W. de la Rue and H. Müller, *J. Chem. Soc.* (1858), **10**, 298; *J. prakt. Chem.* (1858), **73**, 443; Dragendorf, *Pharm. J.* (1877/78), [3] **8**, 826; O. Hesse, *Pharm. J.* (1895), [4] **1**, 325.
305. A. Tschirch and F. Weil, *Arch. Pharm.* (1912), **250**, 20.
306. S. Shibata, J. Shoji, A. Ohta and M. Watanabe, *Pharm. Bull. (Tokyo)* (1957), **5**, 380.
307. S. Gatenbeck, *Acta Chem. Scand.* (1958), **12**, 1985.
308. S. Natori, F. Sato and S. Udagawa, *Chem. Pharm. Bull. (Tokyo)* (1965), **13**, 385.
309. G. Agosti, J. H. Birkinshaw and P. Chaplen, *Biochem. J.* (1962), **85**, 828.
310. F. Kögl and J. J. Postowsky, *Annalen* (1925), **444**, 1.
311. M. Gabriel, *Ann. Univ. Lyons, Sci. Sect. C* (1959–60), No. 11–12, 67.
312. A. W. K. Chan and W. D. Crow, *Aust. J. Chem.* (1966), **19**, 1701.
313. Daniel Penas Goas, *Farmacognosia* (1950), **10**, 229.
314. (a) R. Eder and C. Widmer, *Helv. Chim. Acta* (1923), **6**, 966; (b) R. Eder and F. Hauser, *Helv. Chim. Acta* (1925), **8**, 126.
315. R. A. Jacobson and R. Adams, *J. Amer. Chem. Soc.* (1924), **46**, 1312.
316. T. Posternak, J.-P. Jacob and H. Ruelius, *Rev. farm. B. Aires* (1942), **84**, 264.
317. H. Mühlemann, *Pharm. Acta Helv.* (1951), **26**, 195.
318. H. Mühlemann, *Festschrift Paul Casparis* (1949), 159.
319. H. Wagner and H. P. Hörhammer, *Naturwissenschaften* (1966), **53**, 585.
320. H. Mühlemann and H. Wernli, *Pharm. Acta Helv.* (1965), **40**, 534.
321. A. S. Romanova, A. I. Ban'kovskii, M. E. Perel'son and A. A. Kir'yanov, *Khim. Prirod. Soedin., Akad. Nauk Uz. S.S.R.* (1966), **2**, 83.
322. J. H. Gardner, *J. Amer. Pharm. Ass.* (1939), **28**, 143.
323. J. Lemli, R. Dequeker and J. Cuveele, *Planta Med.* (1964), **12**, 107.
324. H. Mühlemann and H. Schmid, *Pharm. Acta Helv.* (1955), **30**, 363.
325. H. Brockmann and W. Sanne, *Naturwissenschaften* (1953), **40**, 509.
326. A. G. Perkin, *J. Chem. Soc.* (1895), **67**, 1084.
327. K. Tsukida and M. Yokota, *J. Pharm. Soc. Japan* (1954), **74**, 301.
328. M. Takido, *Chem. Pharm. Bull. (Tokyo)* (1958), **6**, 397.
329. M. Takido, T. Nakamura and K. Nitta, *Nippon Daigaku Yakugaku Hokuku* (1960), **4**, 18.
330. N. M. King, *J. Am. Pharm. Ass.* (1957), **46**, 271.
331. A. G. Perkin and J. J. Hummel, *J. Chem. Soc.* (1894), **65**, 923.
332. J. A. Galarraga, K. G. Neill and H. Raistrick, *Biochem. J.* (1955), **61**, 456; K. G. Neill and H. Raistrick, *Biochem. J.* (1957), **65**, 166.
333. (a) H. Raistrick, R. Robinson and A. R. Todd, *J. Chem. Soc.* (1937), 80; (b) J. N. Ashley, H. Raistrick and T. Richards, *Biochem. J.* (1939), **33**, 1291.
334. W. Steglich and V. Austel, *Tetrahedron Lett.* (1966), 3077.
335. S. Shibata and S. Natori, *Pharm. Bull. (Tokyo)* (1953), **1**, 160.
336. Herberger, *Rep. für Pharm.* (1834), **47**, 179; F. Rochleder and W. Heldt, *Annalen* (1843), **48**, 1.
337. M. Asano and Y. Arata, *J. Pharm. Soc. Japan* (1940), **60**, 521.
338. T. R. Seshadri and S. S. Subramanian, *Proc. Indian Acad. Sci.* (1949), **30A**, 67.
339. W. B. Mors, *Bol. inst. Quím. agric.* (1951), No. 23, 7.
340. S. Neelakantan and T. R. Seshadri, *J. Sci. Ind. Res. (India)* (1952), **11B**, 126.
341. D. Marković, *Mikrochemie* (1951), **38**, 419.

342. O. Hesse, *Annalen* (1912), **388**, 65; (1917), **413**, 350.

343. A. D. Dixon and D. A. N. Hoyte, *Anat. Record* (1963), **145**, 101.

344. O. Hesse, *Annalen* (1912), **388**, 97.

345. H. A. D. Jowett and C. E. Potter, *J. Chem. Soc.* (1903), **83**, 1330; R. Eder and F. Hauser, *Helv. Chim. Acta* (1925), **65**, 923.

346. L. Hörhammer, L. Farkas, H. Wagner and S. Imre, *Acta Chim. Acad. Sci. Hung.* (1964), **40**, 309.

347. C. Liebermann and G. Glock, *Ber.* (1884), **17**, 888.

348. M. D. Bhavsar, B. D. Tilak and K. Venkataraman, *J. Sci. Ind. Res. (India)* (1957), **16B**, 392.

349. G. Izoret, J. Camier, J.-J. Brun and G. Lonchambon, *Bull. Soc. Chim. France* (1964), 822.

350. K. Venkataraman, *J. Sci. Ind. Res. (India)* (1966), **25**, 97.

351. P. C. Mitter and M. Sen, *J. Indian Chem. Soc.* (1928), **5**, 631.

352. L. Marchlewski, *J. Chem. Soc.* (1893), **63**, 1137.

353. H. A. D. Jowett and C. E. Potter, *J. Chem. Soc.* (1902), **81**, 1575.

354. G. H. Mahran, M. D. Sayed and M. A. El-Keiy, *Egyptian Pharm. Bull.* (1958), **40**, 149.

355. J. Kitamura, U. Kurimoto and M. Yokoyama, *J. Pharm. Soc. Japan* (1956), **76**, 972.

356. P. B. Lukić, *Planta Med.* (1959), **7**, 400.

357. I. H. Tsunoda, *Shokuhin Eiseigaku Zasshi* (1961), **2**, No. 2, 33.

358. O. K. Bagrii, *Farmatsevt. Zh. (Kiev)* (1963), **18**, 25.

359. S. Huneck and G. Follmann, *Z. Naturforsch.* (1965), **20b**, 1012.

360. A. Mahmoodian and C. E. Stickings, *Chem. Ind. (London)* (1962), 1718; *Biochem. J.* (1964), **92**, 369.

361. B. H. Howard and H. Raistrick, *Biochem. J.* (1954), **56**, 56.

362. J. Breen, J. C. Dacre, H. Raistrick and G. Smith, *Biochem. J.* (1955), **60**, 618.

363. Y. Yamamoto, A. Hamaguchi, I. Yamamoto and S. Imai, *J. Pharm. Soc. Japan* (1956), **76**, 1428.

364. S. Shibata and H. Tokutake, unpublished.

365. S. Shibata and S. Udagawa, *Chem. Pharm. Bull. (Tokyo)* (1963), **11**, 402.

366. S. Shibata, J. Shoji, A. Ohta and M. Watanabe, *Pharm. Bull. (Tokyo)* (1957), **5**, 380.

367. (a) S. Shibata, O. Tanaka, G. Chihara and H. Mitsuhashi, *Pharm. Bull. (Tokyo)* (1953), **1**, 302; (b) S. Shibata, T. Murakami, O. Tanaka, G. Chihara, I. Kitagawa, M. Sumimoto and C. Kaneko, *Pharm. Bull. (Tokyo)* (1955), **3**, 160; (c) S. Shibata, T. Murakami, O. Tanaka, G. Chihara and M. Sumimoto, *Pharm. Bull. (Tokyo)* (1955), **3**, 274.

368. L. H. Briggs and P. W. Le Quesne, *J. Chem. Soc.* (1965), 2290.

369. S. Natori, F. Sato and S. Udagawa, *Chem. Pharm. Bull. (Tokyo)* (1965), **13**, 385.

370. E. Gäumann and S. Naef-Roth, *Pflanzenschutz-Ber.* (1957), **19**, 9.

371. S. Shibata, O. Tanaka and I. Kitagawa, *Pharm. Bull. (Tokyo)* (1955), **3**, 278.

372. S. Shibata, O. Tanaka and I. Kitagawa, *Pharm. Bull. (Tokyo)* (1956), **4**, 143.

373. O. Tanaka and C. Kaneko, *Pharm. Bull. (Tokyo)* (1955), **3**, 284.

374. A. E. Oxford and H. Raistrick, *Biochem. J.* (1940), **34**, 790.

375. H. Brockmann and H. Eggers, *Angew. Chem.* (1955), **67**, 706; *Chem. Ber.* (1958), **91**, 81.

376. O. Tanaka, *Chem. Pharm. Bull. (Tokyo)* (1958), **6**, 203.

377. S. Shibata, M. Takido and T. Nakajima, *Pharm. Bull.* (*Tokyo*) (1955), **3**, 286.

378. S. Shibata, M. Takido, A. Ohta and T. Kurosu, *Pharm. Bull.* (*Tokyo*) (1957), **5**, 573.

379. S. Matsueda, Y. Fujimatsu and S. Mitsui, *Chem. Ind.* (*London*) (1965), 88.

380. N. L. Dutta, A. C. Ghosh, P. M. Nair and K. Venkataraman, *Tetrahedron Lett.* (1964), 3023.

381. L. H. Briggs and D. R. Castaing, *Bull. Nat. Inst. Sci. India* (1965), No. 28, 71.

382. J. D. Bu'Lock and J. R. Smith, *J. Chem. Soc.* (*C*) (1968), 1941.

383. (a) T. Posternak, *C.R. Soc. Phys. Hist. Nat. Genève* (1939), **56**, 29; (b) T. Posternak and J.-P. Jacob, *Helv. Chim. Acta* (1940), **23**, 237.

384. W. K. Anslow, J. Breen and H. Raistrick, *Biochem. J.* (1940), **34**, 159.

385. T. Posternak, *Helv. Chim. Acta* (1940), **23**, 1046.

386. W. Zopf, *Annalen* (1898), **300**, 322; (1905), **340**, 276.

387. T. Bruun, D. P. Hollis and R. Ryhage, *Acta Chem. Scand.* (1965), **19**, 839.

388. T. R. Seshadri and S. S. Subramanian, *Proc. Indian Acad. Sci.* (1949), **30A**, 67.

389. T. Murakami, *Pharm. Bull.* (*Tokyo*) (1956), **4**, 298.

390. S. Neelakantan, S. Rangaswami, T. R. Seshadri and S. S. Subramanian, *Proc. Indian Acad. Sci.* (1951), **33A**, 142.

391. T. R. Rajagopalan and T. R. Seshadri, *Proc. Indian Acad. Sci.* (1959), **49A**, 1.

392. S. Neelakantan and T. R. Seshadri, *J. Sci. Ind. Res.* (*India*) (1954), **13B**, 884; S. Neelakantan, T. R. Seshadri and S. Subramanian, *Proc. Indian Acad. Sci.* (1956), **44A**, 42.

393. H. G. Hind, *Biochem. J.* (1940), **34**, 67, 577.

394. M. Asano and N. Fuziwara, *J. Pharm. Soc. Japan* (1936), **56**, 1007.

395. W. Eschrich, *Biochem. Z.* (1958), **330**, 73.

396. Y. Asahina and F. Fuzikawa, *Ber.* (1935), **68**, 1558.

397. B. Franck and T. Reschke, *Angew. Chem.* (1959), **71**, 407; *Chem. Ber.* (1960), **93**, 347.

398. S. Gatenbeck, *Acta Chem. Scand.* (1959), **13**, 386.

399. B. Franck, *Planta Med.* (1960), **8**, 420.

400. B. S. Gould and H. Raistrick, *Biochem. J.* (1934), **28**, 1640.

401. B. S. Joshi, S. Ramanathan and K. Venkataraman, *Tetrahedron Lett.* (1962), 951.

402. K. Venkataraman, *J. Sci. Ind. Res.* (*India*) (1966), **25**, 97.

403. V. H. Powell and M. D. Sutherland, *Aust. J. Chem.* (1967), **20**, 541.

404. H. Raistrick and J. Ziffer, *Biochem. J.* (1951), **49**, 563.

405. A. J. Birch and R. A. Massy-Westropp, *J. Chem. Soc.* (1957), 2215.

406. A. J. Birch and C. J. Moye, *J. Chem. Soc.* (1961), 4691.

407. B. H. Howard and H. Raistrick, *Biochem. J.* (1949), **44**, 227.

408. H. Agarashi, *J. Agric. Chem. Soc. Japan* (1939), **15**, 225; H. Nishikawa, *J. Fac. Agri. Tohoku Univ. Japan* (1955), **5**, 285.

409. B. S. Joshi, B. D. Tilak and K. Venkataraman, *Proc. Indian Acad. Sci.* (1950), **32A**, 348.

410. S. Neelakantan, T. R. Rajagopalan and T. R. Seshadri, *Proc. Indian Acad. Sci.* (1959), **49A**, 234.

411. J. H. V. Charles, H. Raistrick, R. Robinson and A. R. Todd, *Biochem. J.* (1933), **27**, 499.

412. H. Raistrick, R. Robinson and A. R. Todd, *Biochem. J.* (1934), **28**, 559.

413. H. Raistrick, R. Robinson and A. R. Todd, *Biochem. J.* (1933), **27**, 1170.
414. A. Quilico, C. Cardani, F. Piozzi and P. Scrivani, *Atti Accad. Naz. Lincei, Rend. Classe Sci. fis. mat. e nat.* (1952), **12**, 650.
415. K. Schofield and D. E. Wright, *J. Chem. Soc.* (1965), 6642.
416. H. Raistrick, R. Robinson and A. R. Todd, *J. Chem. Soc.* (1933), 488.
417. K. Chandrasenan, S. Neelakantan and T. R. Seshadri, *J. Indian Chem. Soc.* (1961), **38**, 907.
418. L. H. Briggs, M. R. Craw and J. C. Dacre, *J. Chem. Soc.* (1948), 568.
419. L. H. Briggs, J. C. Dacre and G. A. Nicholls, *J. Chem. Soc.* (1948), 990.
420. L. H. Briggs, G. A. Nicholls and R. M. L. Paterson, *J. Chem. Soc.* (1952), 1718.
421. Tirso Ríos, *Tetrahedron* (1966), **22**, 1507.
422. A. W. K. Chan and W. D. Crow, *Aust. J. Chem.* (1966), **19**, 1701.
423. G. Bargellini, *Gazz. Chim. Ital.* (1919), **49**, 47.
424. A. Tschirsch and F. Lüdy, *Helv. Chim. Acta* (1923), **6**, 994.
425. K. G. Dave, B. S. Joshi, A. V. Patwardhan and K. Venkataraman, *Tetrahedron Lett.* (1959), No. 6, 22.
426. W. Sandermann and M. H. Simatupang. Personal communication (1967).
427. P. Yates, A. C. Mackay, L. M. Pande and M. Amin, *Chem. Ind. (London)* (1964), 1991.
428. N. S. Bhide, A. V. Rao and K. Venkataraman, *Tetrahedron Lett.* (1965), 33.
429. R. Majima and C. Kuroda, *Acta Phytochim.* (1922), **1**, 43.
430. J. C. Lovie and R. H. Thomson, *J. Chem. Soc.* (1961), 485.
431. F. M. Dean, P. G. Jones and P. Sidisunthorn, *J. Chem. Soc.* (1962), 5186 and subsequent papers.
432. A. Stoessl, *Can. J. Chem.* (1969), **47**, 777.
433. W. Steglich, W. Lösel and W. Reininger, *Tetrahedron Lett.* (1967), 4719.
434. B. Franck and T. Reschke, *Chem. Ber.* (1960), **93**, 347.
435. B. Franck and I. Zimmer, *Chem. Ber.* (1965), **98**, 1514.
436. N. S. Bhide and A. V. Rama Rao, *Indian J. Chem.* (1969), **7**, 996.
437. R. Heise, *Arbeit Kaiserl. Gesund.* (1895), **11**, 513.
438. E. Bancroft, "Philosophy of Permanent Colours" (1813), Vol. I. Cadell and Davies, London.
439. O. Dimroth, *Ber.* (1910), **43**, 1387.
440. O. Dimroth and W. Scheurer, *Annalen* (1913), **399**, 43.
441. O. Dimroth and R. Fick, *Annalen* (1916), **411**, 315.
442. J. C. Overeem, *Indust. Chim. Belg.* (1962), **27**, 529; J. C. Overeem and G. J. M. van der Kerk, *Rec. Trav. Chim.* (1964), **83**, 1023.
443. Pelletier and Caventou, *Ann. Chim. Phys.* (1818), **8**, 250; P. Schützenberger, *Ann. Chim. Phys.* (1858), **54**, 52.
444. G. A. Fester, *Isis* (1953), **44**, 13.
445. M. A. Ali and L. J. Haynes, *J. Chem. Soc.* (1959), 1033.
446. C. Liebermann and H. Voswinkel, *Ber.* (1897), **30**, 688, 1731.
447. S. B. Bhatia and K. Venkataraman, *Indian J. Chem.* (1965), **3**, 92.
448. A. Oppenheim and S. Pfaff, *Ber.* (1874), **7**, 929.
449. O. Dimroth, *Ber.* (1909), **42**, 1611.
450. O. Dimroth, B. Kerkovius, G. Weuringh and L. Holch, *Annalen* (1913), **399**, 1.
451. O. Dimroth and H. Kämmerer, *Ber.* (1920), **53**, 471.
452. H. F. Lower, *Trans. Roy. Soc. S. Aust.* (1959), **82**, 175.
453. R. E. Schmidt, *Ber.* (1887), **20**, 1285.
454. O. Dimroth and S. Goldschmidt, *Annalen* (1913), **399**, 62; see also ref. 424.

455. R. H. Thomson. "Naturally Occurring Quinones" (1957), 1st edit., p. 229. Butterworths, London; "Thorpe's Dictionary of Applied Chemistry" (1946), 4th edit., Vol. VII, p. 158.

456. (a) N. S. Bhide, B. S. Joshi, A. V. Patwardhan, R. Srinivasan and K. Venkataraman, *Bull. Nat. Inst. Sci. India* (1965), No. 28, 114; (b) N. S. Bhide, R. Srinivasan and K. Venkataraman, *Abstr. IUPAC 4th Int. Sym. Chem. Nat. Prods., Kyoto* (1964), p. 205; (c) E. D. Pandhare, A. V. Rama Rao, R. Srinivasan and K. Venkataraman, *Tetrahedron* (1966), *Suppl.* **8**, 229; (d) E. D. Pandhare, A. V. Rama Rao and I. N. Shaikh, *Indian J. Chem.* (1969), **7**, 977.

457. K. Venkataraman, *J. Sci. Ind. Res. (India)* (1966), **25**, 97.

458. R. Burwood, G. Read, K. Schofield and D. E. Wright, *J. Chem. Soc.* (a) (1965), 6067; (b) *J. Chem. Soc. (C)* (1967), 842.

459. B. L. Kaul, unpublished, quoted in ref. 456c.

460. W. Ritchie and W. C. Taylor, *Tetrahedron Letters* (1964), 1437.

461. G. H. Stout, R. A. Alden, J. Kraut and D. F. High, *J. Amer. Chem. Soc.* (1962), **84**, 2653.

462. Y. Yonezawa, *Mem. Inst. Oriental Cultures, Tokyo Univ.* (1956), No. 11, 375.

463. M. A. Hamilton, M. S. Knorr and F. A. Cajori, *Antibiotics and Chemotherapy* (1953), **3**, 854.

464. F. A. Cajori, T. T. Otani and M. A. Hamilton, *J. Biol. Chem.* (1954), **208**, 107.

465. G. P. Arsenault, *Tetrahedron Lett.* (1965), 4033.

466. G. P. Arsenault, *Tetrahedron* (1968), **24**, 4745.

467. S. Huneck and G. Follmann, *Z. Naturforsch.* (1966), **21b**, 91.

468. H. El Khadem and Z. M. El-Shafei, *Tetrahedron Lett.* (1963), 1887.

469. J. A. Mills, *Aust. J. Chem.* (1964), **17**, 277.

470. J. C. Roberts and P. Roffey, *J. Chem. Soc.* (1965), 3666.

471. J. A. Elix, P. Roffey and M. V. Sargent, *Tetrahedron Lett.* (1967), 3161.

472. J. C. Roberts and P. Roffey, *J. Chem. Soc. (C)* (1966), 160.

473. J. H. Birkinshaw and I. M. M. Hammady, *Biochem. J.* (1957), **65**, 162.

474. J. H. Birkinshaw, J. C. Roberts and P. Roffey, *J. Chem. Soc. (C)* (1966), 855.

475. G. Koller and H. Russ, *Monatsh.* (1937), **70**, 54.

476. H. A. Anderson, R. H. Thomson and J. W. Wells, *J. Chem. Soc. (C)* (1966), 1727.

477. J. C. Roberts. Personal communication.

478. W. Zopf, *Annalen* (1895), **284**, 107; (1909), **364**, 273.

479. O. Hesse, *J. prakt. Chem.* (1915), **92**, 425.

480. T. Hamasaki, Y. Hatsuda, N. Terashima and M. Renbutsu, *Agr. Biol. Chem. (Tokyo)* (1965), **29**, 166, 696; (1967), **31**, 11.

481. D. F. G. Pusey and J. C. Roberts, *J. Chem. Soc.* (1963), 3542.

482. Y. Hatsuda and S. Kuyama, *J. Agr. Chem. Soc. Japan* (1954), **28**, 989; Y. Hatsuda, S. Kuyama and N. Terashima, *J. Agr. Chem. Soc. Japan* (1955), **29**, 11.

483. J. E. Davies, D. Kirkaldy and J. C. Roberts, *J. Chem. Soc.* (1960), 2169.

484. D. H. R. Barton, H. T. Cheung, A. D. Cross, L. M. Jackman and M. Martin-Smith, *J. Chem. Soc.* (1961), 5061.

485. E. Bullock, J. C. Roberts and J. G. Underwood, *J. Chem. Soc.* (1962), 4179.

486. E. Bullock, D. Kirkaldy, J. C. Roberts and J. G. Underwood, *J. Chem. Soc.* (1963), 829.

487. P. Roffey, Ph.D. Thesis, University of Nottingham (1965).

488. P. Roffey and M. V. Sargent, *Chem. Comm.* (1966), 913; P. Roffey, M. V. Sargent and J. A. Knight, *J. Chem. Soc. (C)* (1967), 2328.

489. J. S. E. Holker, S. A. Kagal, L. J. Mulheirn and P. M. White, *Chem. Comm.* (1966), 911.

490. J. Seibl and T. Gäumann, *Helv. Chim. Acta* (1963), **46**, 2857.

491. M. D. Sutherland and J. W. Wells, *Chem. Ind. (London)* (1959), 291; *Aust. J. Chem.* (1967), **20**, 515.

492. T. F. Low, R. J. Park, M. D. Sutherland and I. Vessey, *Aust. J. Chem.* (1965), **18**, 182.

493. W. K. Anslow and H. Raistrick, *Biochem. J.* (1940), **34**, 1124.

494. S. Shibata and S. Natori, *Pharm. Bull. (Tokyo)* (1953), **1**, 160.

495. K. Chandrasenan, S. Neelakantan and T. R. Seshadri, *Proc. Indian Acad. Sci.* (1960), **51A**, 296.

496. L. J. Carbone and G. T. Johnson, *Mycologia* (1964), **56**, 185.

497. S. Neelakantan, A. Pocker and H. Raistrick, *Biochem. J.* (1956), **64**, 464.

498. R. B. Pringle, *Biochim. Biophys. Acta* (1958), **28**, 198.

499. J. Atherton, B. W. Bycroft, J. C. Roberts, P. Roffey and M. E. Wilcox, *J. Chem. Soc. (C)* (1968), 2560.

500. D. E. Wright and K. Schofield, *Nature (London)* (1960), **188**, 233.

501. A. J. Birch, R. I. Fryer, P. J. Thomson and H. Smith, *Nature (London)* (1961), **190**, 441.

502. W. K. Anslow and H. Raistrick, *Biochem. J.* (1940), **34**, 1546.

503. F. Bohlmann, W. Lüders and W. Plettner, *Arch. Pharm.* (1961), **294**, 521.

504. G. Flumiani, *Monatsh.* (1924), 45, 44.

505. M. Takido, *Chem. Pharm. Bull. (Tokyo)* (1960), **8**, 246.

506. B. H. Howard and H. Raistrick, *Biochem. J.* (1955), **59**, 475.

507. S. Neelakantan, A. Pocker and H. Raistrick, *Biochem. J.* (1957), **66**, 234.

508. J. H. Birkinshaw and R. Gourlay, *Biochem. J.* (1961), **81**, 618.

509. J. H. Birkinshaw and R. Gourlay, *Biochem. J.* (1961), **80**, 387.

510. F. Kögl and J. Sparenburg, *Rec. Trav. Chim.* (1940), **59**, 1180.

511. F. Kögl and F. W. Quackenbush, *Rec. Trav. Chim.* (1944), **63**, 251.

512. F. Kögl, G. C. van Wessem and O. I. Elsbach, *Rec. Trav. Chim.* (1945), **64**, 23.

513. A. J. Birch, D. N. Butler and R. W. Rickards, *Tetrahedron Lett.* (1964), 1853.

514. K. H. Dudley and R. L. McKee, *J. Org. Chem.* (1967), **32**, 3210.

515. A. Chatterjee and S. R. Bhattacharjee, *J. Indian Chem. Soc.* (1964), **41**, 415.

516. N. L. Dutta, A. C. Ghosh, P. M. Nair and K. Venkataraman, *Tetrahedron Lett.* (1964), 3023.

517. A. J. Birch and K. S. J. Stapleford, *J. Chem. Soc. (C)* (1967), 2570.

518. B. H. Howard and H. Raistrick, *Biochem. J.* (1954), **57**, 212.

519. S. Shibata, T. Murakami, I. Kitagawa and T. Kishi, *Pharm. Bull. (Tokyo)* (1956), **4**, 111.

520. H. El Khadem and Z. M. El-Shafei, *Tetrahedron Lett.* (1963), 1887; J. A. Mills, *Austral. J. Chem.* (1964), **17**, 277.

521. H. Brockmann, R. Neeff and E. Mühlmann, *Chem. Ber.* (1950), **83**, 467.

522. S. Shibata, T. Murakami and M. Takido, *Pharm. Bull. (Tokyo)* (1956), **4**, 303.

523. M. Kikuchi, *Bot. Mag. Tokyo* (1962), **75**, 158.

524. S. Shibata and I. Kitagawa, *Pharm. Bull. (Tokyo)* (1956), **4**, 309.

525. S. Shibata and I. Kitagawa, *Chem. Pharm. Bull. (Tokyo)* (1960), **8**, 884.

526. Y. Yamamoto, T. Yamamoto, S. Kanamoto and K. Tanimichi, *J. Pharm. Soc. Japan* (1956), **76**, 192.

527. T. Tatsuno, M. Tsukioka, Y. Sakai, Y. Suzuki and Y. Asami, *Pharm. Bull. (Tokyo)* (1955), **3**, 476.

528. S. Shibata, I. Kitagawa and H. Nishikawa, *Pharm. Bull.* (*Tokyo*) (1957), 5, 383.
529. S. Shibata, T. Ikekawa and T. Kishi, *Chem. Pharm. Bull.* (*Tokyo*) (1960), 8, 889.
530. D. O. Schachtschabel, *Planta Med.* (1968), (Suppl.) 77 and refs. therein.
531. I. Kitagawa and S. Shibata, *Chem. Ind.* (*London*) (1960), 1054.
532. S. Shibata and I. Kitagawa, *Chem. Pharm. Bull.* (*Tokyo*) (1961), 9, 352.
533. M. Shibata and M. Masumura, *Tohoku Seibutsu Kenkyu* (1951), II, 16.
534. S. Fujise, S. Hishida, M. Shibata and S. Matsueda, *Chem. Ind.* (*London*) (1961), 1754; S. Matsueda, *J. Org. Chem.* (1965), 30, 744.
535. A. Chatterjee and S. R. Bhattacharjee, *Bull. Nat. Inst. Sci. India* (1965), No. 31, 141.
536. B. H. Howard and H. Raistrick, *Biochem. J.* (1950), 46, 49.
537. T. Hamasaki, M. Renbutsu and Y. Hatsuda, *Agric. Biol. Chem. Japan* (1967), 31, 1513.
538. J. A. Ballantine and C. T. Pillinger, *Tetrahedron* (1967), 23, 1691.
539. S. Matsueda. Personal communication.
540. W. Steglich and W. Reininger, *Z. Naturforsch.* (1969), 24b. 1196.
541. W. D. Crow, unpublished observation.
542. W. Steglich and W. Lösel, unpublished information.
543. I. Yosioka, H. Yamauchi, K. Morimoto and I. Kitagawa, *Tetrahedron Lett.* (1968), 1149.
544. G. Bendz, G. Bohman and J. Santesson, *Acta Chem. Scand.* (1968), 21, 2889.
545. Y. Yamamoto, N. Kiriyama and S. Arahata, *Chem. Pharm. Bull.* (*Tokyo*) (1968), 16, 304.
546. R. Juliani, M. N. Graziano and J. D. Coussio, *Phytochemistry* (1968), 7, 507.
547. R. G. Coombe, J. J. Jacobs and T. R. Watson, *Aust. J. Chem.* (1968), 21, 783.
548. R. Anton and P. Duquénois, *Compt. Rend.* (1968), 266, Ser. D, 1523.
549. R. G. Coombe. Personal communication (1968).
550. R. N. Knanna and T. R. Seshadri, *Curr. Sci.* (1968), 37, 72.
551. M. Piattelli and M. Giudici de Nicola, *Phytochemistry* (1968), 7, 1183.
552. A. Platel, Y. Ueno and P. Fromageot, *Bull. Soc. Chim. France* (1968), 50, 678.
553. G. W. Midglow, quoted by W. H. Cliffe, *J. Soc. Dyers Col.* (1959), 75, 282.
554. H. H. Lee, *Phytochemistry* (1969), 8, 501.
555. K. V. Taraskina, T. K. Chumbalov and L. K. Kuznetsova, *Khim. Prirod. Soedin.* (1968), 188.
556. J. Santesson, *Arkiv. Kemi* (1969), 30, 363.
557. M. Aritomi, *Mem. Fac. Educ. Kumamoto Univ.* (1964), 12 (I), 13.
558. I. Onishi, T. Fukuzumi, K. Yamamoto and H. Takahara, *Sci. Papers, Cent. Res. Inst. Japan Monop. Corp.* (1961), No. 103, 25.
559. K. S. Kirby and T. White, *Biochem. J.* (1955), 60, 583; A. R. Burnett, Ph.D. Thesis, University of Aberdeen (1967).
560. I. Oftedal, *Result. Norske Spitzbergenexsped.* (1922), I, Nr. 3, 9.
561. T. A. Geissman, K. Y. Sim and J. Murdoch, *Experientia* (1967), 23, 793.
562. A. Treibs and H. Steinmetz, *Annalen* (1933), 506, 171.
563. K. Kumada, A. Suzuki and K. Aizawa, *Nature London* (1961), 191, 415.
564. V. B. Patil, A. V. Rama Rao and K. Venkataraman, *Indian J. Chem.* (1970), 8, 109.
565. A. R. Burnett and R. H. Thomson, *J. Chem. Soc.* (*C*) (1968), 2437.
566. A. R. Burnett and R. H. Thomson, *Phytochemistry* (1968), 7, 1423.

567. E. J. C. Brew and R. H. Thomson, *J. Chem. Soc.* (c) (1971), in the press.
568. Y. Ogihara, N. Kobayashi and S. Shibata, *Tetrahedron Lett.* (1968), 1881.
569. S. Imre, *Phytochemistry* (1969), **8**, 315.
570. P. Narayanan, K. Zechmeister, M. Röhrl and W. Hoppe, *Tetrahedron Lett.* (1970), 3643.
571. S. Imre and H. Wagner, *Phytochemistry* (1969), **8**, 1601.
572. A. M. Cumming and R. H. Thomson, *Phytochemistry* (1970), **9**, in press.
573. U. Sankawa, S. Seo, N. Kobayashi, Y. Ogihara and S. Shibata, *Tetrahedron Lett.* (1968), 5557 (cf. S. Shibata, Y. Ogihara, S. Seo and I. Kitagawa, *Tetrahedron Lett.* (1968), 3179).
574. N. Kobayashi, Y. Iitaka, U. Sankawa, Y. Ogihara and S. Shibata, *Tetrahedron Lett.* (1968), 6135.
575. I. Yosioka, H. Yamauchi, K. Morimoto and I. Kitagawa, *Tetrahedron Lett.* (1968), 3749.
576. O. Fischer and H. Gross, *J. prakt. Chem.* (1911), **84**, 369.
577. D. D. Gadgil, A. V. Rama Rao and K. Venkataraman, *Tetrahedron Lett.* (1968), 2229.
578. D. D. Gadgil, A. V. Rama Rao and K. Venkataraman, *Tetrahedron Lett.* (1968), 2223.
579. T. Murakami, K. Ikeda and M. Takido, *Chem. Pharm. Bull. (Tokyo)* (1968), **16**, 2299.
580. E. D. Pandhare, A. V. Rama Rao, I. N. Shaikh and K. Venkataraman, *Tetrahedron Lett.* (1967), 2427.
581. A. V. Rama Rao, I. N. Shaikh and K. Venkataraman, *Indian J. Chem.* (1969), **7**, 188.
582. E. D. Pandhare, A. V. Rama Rao, I. N. Shaikh and K. Venkataraman, *Tetrahedron Lett.* (1967), 2437; N. S. Bhide, E. D. Pandhare, A. V. Rama Rao, I. N. Shaikh and R. Srinivasan, *Indian J. Chem.* (1969), **7**, 987.
583. R. Anton and P. Duquénois, *Ann. Pharm. Franç.* (1968), **26**, 673.
584. C. H. Fox, W. S. G. Maass and T. P. Forrest, *Tetrahedron Lett.* (1969), 919.
585. T. Noda, T. Take, M. Otani, K. Miyauchi, T. Watanabe and J. Abe, *Tetrahedron Lett.* (1968), 6087.
586. A. Takenaka, A. Furusaki, T. Watanabé, T. Noda, T. Take, T. Watanabe and J. Abe, *Tetrahedron Lett.* (1968), 6091.
587. K. Yamada, S. Takada, S. Nakamura and Y. Hirata, *Tetrahedron* (1968), **24**, 1267; C. R. Narayanan and M. R. Sarma, *Tetrahedron Lett.* (1968), 1553.
588. H. Hlasiwetz and A. Grabowski, *Annalen* (1867), **141**, 329.
589. N. F. Hayes and R. H. Thomson, *J. Chem. Soc.* (1955), 904.
590. B. Franck, *Planta Med.* (1960), **8**, 420.
591. S. Seo, U. Sankawa, Y. Ogihara and S. Shibata, *Tetrahedron Lett.* (1969), 767. For previous work see refs. 531 and 532.
592. M. Piattelli and M. Giudici de Nicola, *Ric. Sci.* (1968), **38**, 850; R. Tomaselli, *Arch. Bot. Biogeog. Ital.* (1963), **39** [4, VIII], 4.
593. J. S. E. Holker. Personal communication (1969).
594. T. J. Nolan, J. Algar, E. T. McCann, W. A. Manahan and N. Nolan, *Sci. Proc. Roy. Dublin Soc.* (1948), **24**, 319.
595. A. Stoessl. Personal communication (1969).
596. K. Venkataraman. Personal communication (1969).
597. J. G. Heathcote and M. F. Dutton, *Tetrahedron* (1969), **25**, 1497.
598. G. Bohman, *Arkiv Kemi* (1969), **30**, 217.

599. E. Bachmann, *Ber. Deut. Bot. Ges.* (1887), **5**, 192; O. Hesse, *J. prakt. Chem.* (1898), **57**, 443; (1903), **68**, 52.

600. K.-E. Stensiö and C. A. Wachtmeister, *Svensk Kem. Tidskr.* (1967), **79**, 579; *Acta Chem. Scand.* (1969), **23**, 144.

601. A. J. Cserjesi and R. S. Smith, *Mycopath. Mycol. Appl.* (1968), **35**, 91.

602. S. H. Harper and R. M. Letcher, *Proc. Trans. Rhodesia Sci. Ass.* (1966), **51**, 156.

603. A. S. Romanova, N. D. Semakina and A. I. Ban'kovskii, *Rast. Resur.* (1968), **4**, 342.

604. Fernando Perez Barre, *Soc. Venez. Cienc. Natur. Bol.* (1967), **27**, 314.

605. Ricardo Urtubia Marin, *Anal. Fac. Quim. Farm. Univ. Chile* (1966), **18**, 19.

606. K. N. Gaind and T. S. Saini, *Indian J. Pharm.* (1968), **30**, 233.

607. R. Kinget, *Planta Med.* (1967), **15**, 233.

608. T. J. McCarthy, *Planta Med.* (1968), **16**, 348.

609. K. Venkataraman. Personal communication, 1969.

610. L. Nogeira Prista, A. Spinola Roque, M. A. Ferreira and A. Correia Alves, *Garcia Orta* (1965), **13**, 19.

611. L. Nogeira Prista, A. Spinola Roque, M. A. Ferreira and A. Correia Alves, *Garcia Orta* (1965), **13**, 39.

612. L. Nogeira Prista, A. Spinola Roque, M. A. Ferreira and A. Correia Alves, *Garcia Orta* (1965), **13**, 45.

613. V. M. Chari, S. Neelakantan and T. R. Seshadri, *Indian J. Chem.* (1969), **7**, 40.

614. H. Kroepelin, *Adv. Org. Geochem.* (1964), **15**, 165.

615. W. Steglich, W. Lösel and V. Austel, *Chem. Ber.* (1969), **102**, 4104.

616. W. Steglich and W. Lösel, *Tetrahedron* (1969), **25**, 4391.

617. Y. Hatsuda, T. Hamasaki, M. Ishida and S. Yoshikawa, *Agr. Biol. Chem. (Tokyo)* (1969), **33**, 131.

618. G. Bohman, *Tetrahedron Lett.* (1970), 445.

619. W. Zopf, *Annalen* (1899), **306**, 305.

620. D. McGrath, *Trans. Internat. Sym.* "Humus et Planta IV" (1967), 236.

621. D. McGrath, *Chem. Ind. (London)* (1970), 1353.

622. K. K. Purushothaman, S. Saradambal and V. Narayanaswami, *Leather Sci. (India)* (1968), **15**, 49.

623. G. Bohman, *Phytochemistry* (1969), **8**, 1829.

624. G. Bohman, *Acta Chem. Scand.* (1969), **23**, 2241.

625. J. Santesson, *Acta Chem. Scand.* (1969), **23**, 3270.

626. M. V. Sargent, D. O'N. Smith and J. A. Elix, *J. Chem. Soc. (C)* (1970), 307.

627. W. Steglich and W. Reininger, *Chem. Comm.* (1970), 178.

628. H. Wagner and H. P. Hörhammer, *Z. Naturforsch.* (1969), **24b**, 1408.

629. Y. Ebizuka, U. Sankawa and S. Shibata, *Phytochemistry* (1970), **9**, 2061.

630. N. D. Gyanchandani and I. C. Nigam, *J. Pharm. Sci.* (1969), **58**, 833.

631. M. V. Sargent, D. O'N. Smith, J. A. Elix and P. Roffey, *J. Chem. Soc. (C)* (1969), 2763.

632. C. S. Shah and M. V. Shinde, *Indian J. Pharm.* (1969), **31**, 27.

633. N. R. Ayyangar, A. V. Rama Rao and K. Venkataraman, *Indian J. Chem.* (1969), **7**, 533.

634. T. J. McCarthy, *Planta Med.* (1969), **17**, 1.

635. R. Hegnauer, "Chemotaxonomie der Pflanzen" (1969), Vol. 5, p. 363. Birkhauser, Basel.

636. J. Shoji, Y. Kimura and K. Katagiri, *J. Antibiotics, Ser. A* (1964), **17**, 156.

637. N. Tsuji, *Tetrahedron* (1968), **24**, 1765.

638. N. Tsuji and K. Nagashima, *Tetrahedron* (1968), **24**, 4233.
639. H. Nakai, M. Shiro and H. Koyama, *J. Chem. Soc.* (*B*) (1969), 498.
640. N. Tsuji, K. Nagashima, T. Kimura and H. Kyotani, *Tetrahedron* (1969), **25**, 2999.
641. N. Tsuji and K. Nagashima, *Tetrahedron* (1969), **25**, 3007.
642. N. Tsuji and K. Nagashima, *Tetrahedron* (1969), **25**, 3017.
643. G. T. Johnson and J. P. White, *Mycologia* (1969), **61**, 661.
644. C. F. Culberson, "Chemical and Botanical Guide to Lichen Products" (1969). University N. Carolina Press, Chapel Hill.
645. A. Rassat, C. M. Jefford, J. M. Lehn and B. Waegell, *Tetrahedron Lett.* (1964), 233.
646. J. Santesson, *Acta Chem. Scand.* (1970), **24**, 371.
647. F. P. Mazza and G. Stolfi, *Boll. Soc. Ital. Biol. Sper.* (1930), **5**, 74; *Arch. Sci. Biol.* (*Bologna*) (1931), **16**, 183.
648. G. Prota, M. D'Agostino and G. Misuraca, *Experientia*, in the press.
649. J. D. Bu'Lock, J. Harley-Mason and H. S. Mason, *Biochem. J.* (1950), **47**, xxxii.
650. H.-J. Bielig and H. Möllinger, *Z. Physiol. Chem.* (1960), **321**, 276.
651. P. Boldt and K.-P. Paul, *Chem. Ber.* (1966), **99**, 2337.
652. I. Morimoto, M. I. N. Shaikh, R. H. Thomson and D. G. Williamson, *Chem. Comm.* (1970), 550.
653. H. Nühlemann, *Helv. Chim. Acta* (1951), **26**, 195.

CHAPTER 6

Anthracyclinones

In the search for new antibiotics many hundreds of *Streptomycetes* have been screened revealing, *inter alia*, the existence of numerous pigments, many of them being quinones. The anthracyclinones[1] form the major group, and are unique insofar as the naphthacene- or 7,8,9,10-tetrahydronaphthacenequinone system is elaborated only by these bacteria. However, the same tetracyclic carbon skeleton is found in the tetracycline antibiotics,[2] for example (1), which are also *Streptomycete*

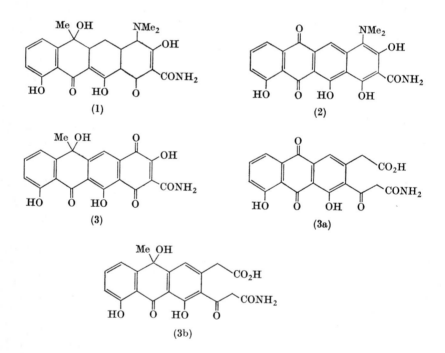

(1)

(2)

(3)

(3a)

(3b)

metabolites. Here too, quinones have been detected in studies with mutants;[3] "tetramid-green"† (3) appears to be responsible for the green

---

† The compound is actually red and the green colour has been attributed[3b] to a tautomer; it is firmly held in *S. aureofaciens* cultures in an insoluble form involving magnesium and/or calcium ion and an organic macromolecule.

colour of a non-antibiotic producing strain of *S. aureofaciens* and another mutant produces "tetramid-blue"† (2). Full details are not yet available but the isolation of (2) is described in a patent,[73] and (3) has been obtained by oxidation *in vitro* of the precursors 6-methylpretetramid and 4-hydroxy-6-methylpretetramid.[74] Recently,[79] the orange quinone, protetrone (3a), and the anthrone (3b) have been isolated from blocked mutants of the same organism, but these incompletely cyclised poly-ketides are not intermediates in the biosynthetic pathway leading to the tetracycline antibiotics.

The anthracyclinones have been studied extensively by Brockmann and his co-workers[1] at Göttingen. They occur both free and as glycosides (anthracyclines) in combination with various sugars including amino-sugars, not all of which have been identified, but rhodosamine[4] is frequently present. In consequence the anthracyclines are basic which

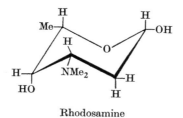

Rhodosamine

facilitates their isolation as hydrochlorides or perchlorates after separation by countercurrent extraction and chromatography. Most of the anthracyclinones have a polyhydroxyanthraquinone chromophore in which 2-4 hydroxyl groups are located solely in α-positions. In this respect they differ from the fungal anthraquinones, many of which possess β-hydroxyl groups. The trivial nomenclature used in association with the orientation of the phenolic hydroxyl groups is shown below; nogalamycin and the reticulomycins, recent additions to the group, appear to have a 1,4-dihydroxyanthraquinone chromophore but other-wise structural variations are chiefly confined to ring A. The stereo-chemistry of ring A, which is the same in nearly all cases, is discussed on p. 547. Brockmann[1] distinguishes individual members of the pyrromy-cinone, rhodomycinone and isorhodomycinone series, by Greek letters arranged alphabetically according to the order of increasing R_F values. Originally α-rhodomycinone was the rhodomycinone with the lowest R_F value but others with still lower values were found later, and were

† So-called on account of its blue colour in alkaline extracts.

designated α_1-, α_2- and α_3-rhodomycinones. The other groups will no doubt be subdivided in the same way as new pigments are discovered.

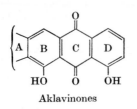

Aklavinones

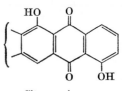

Citromycinones

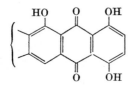

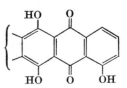

Rhodomycinones[a]

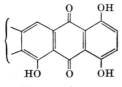

Pyrromycinones

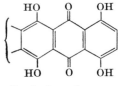

Iso-rhodomycinones

[a] Daunomycinone and adriamycinone also belong to this group.

7S,9R,10R-*Aklavinone* (4), $C_{22}H_{20}O_8$

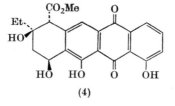

(4)

Occurrence: as the glycosides aklavin, in cultures of an unidentified *Streptomycete*,[5]† and of *Strep. galileus* Ettlinger et al., [80] and galirubin B[6] also from *Strep. galileus*.

Physical properties: orange needles, m.p. 171–172°, $[\alpha]_D^{19}$ +142° (CHCl₃), $[\alpha]_D^{28}$ +213° (dioxan), $\lambda_{max.}$ (MeOH) 229, 258, 278(sh), 288, 430 nm (log ϵ 4·64, 4·23, 4·08, 4·06, 4·12), $\nu_{max.}$ (Nujol) 1740, 1728, 1674, 1623, 1575 cm⁻¹, $\nu_{max.}$ (CCl₄) 3592, 3516, 1737 cm⁻¹.

† Isolated from soil collected at Aklavik, Northwest Territory, Canada.

Aklavinone was first obtained[7] by mild acid hydrolysis of aklavin. It was characterised as a triacetate which showed both phenolic and alcoholic acetate carbonyl absorption, and a peak at $3610\,cm^{-1}$ indicating the presence of a tertiary hydroxyl group. Spectral data, and alkaline hydrolysis, confirmed the existence of a methyl ester function, and an isolated ethyl group was revealed by the n.m.r. spectrum. The infrared, and ultraviolet-visible spectra of aklavinone (see above) and its triacetate are consistent with a 1,8-dihydroxyanthraquinone chromophore, and, like related anthracyclinones, it could be converted, with the aid of toluene-p-sulphonic acid, into a bisanhydro derivative having the spectral properties (see p. 541) of a 1,11-dihydroxytetracene-5,12-quinone. By analogy with bisanhydro-ϵ-pyrromycinone (p. 545) this degradation product was allotted structure (5), the positions of the ethyl and methoxycarbonyl groups being confirmed by the formation of benzene-1,2,3,4-tetracarboxylic acid on permanganate oxidation, and later, by

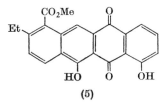

(5)

synthesis[9] (p. 541). It follows that ring A in aklavinone is saturated and contains two alcoholic groups of which the tertiary can be assigned to C-9 thus isolating the ethyl side chain, and the secondary hydroxyl must be at C-7 to accord with the n.m.r. spectrum. This includes a multiplet (1-H) at τ 4·67 attributed to the benzylic proton at C-7, and also signals for a methylene group and a singlet at τ 5·70 arising from the other benzylic proton at C-10. Accordingly, Ollis and co-workers[7] assigned structure (4) (without stereochemistry) to aklavinone.

The 9R,10R-configuration emerges from the similarity of the o.r.d. curves[11] of aklavinone, 7-deoxyaklavinone and ζ-pyrromycinone (p. 548), and the 7S configuration is deduced from the chemical shift of the proton at C-7 which is the same as that of H-7 in β-rhodomycinone (7S) (p. 554) whereas in α-rhodomycinone (7R) (p. 552) the corresponding resonance occurs at lower field.[55]

The glycoside aklavin[5,7] shows interesting antiphage activity and is probably a mixture of closely related substances. It has the same electronic absorption spectrum as aklavinone, forms a number of salts (hydrochloride, picrate, etc.) and appears to contain an aminosugar (possibly $C_8H_{17}NO_4$ isomeric with amosamine and mycaminose) linked to the secondary hydroxyl group of aklavinone. The other aklavinone glycoside, galirubin B, an amorphous orange-red substance [$\lambda_{max.}$ (MeOH) 230, 259, 291(sh), 431 nm], gives two sugars on hydrolysis, at least one of which must be basic.[6]

9R,10R-7-Deoxyaklavinone (galirubinone D) (6), $C_{22}H_{20}O_7$

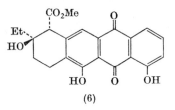

(6)

Occurrence: in cultures of *Strep. galileus* Ettlinger *et al.*[6] and an unidentified *Strep.* sp.[8].

Physical properties: orange needles, m.p. 224–225°, [α]$_D^{28}$ +81° (dioxan), $\lambda_{max.}$ (MeOH) 229, 259, 290, 431 nm (log ϵ 4·56, 4·47, 4·01, 4·13), $\nu_{max.}$ (KBr) 3580, ~3450, 1735, 1680, 1630, 1580 cm^{-1}.

This is the main pigment obtained by neutral extraction of the strain of *Strep. galileus* studied by Eckardt and Bradler.[6, 32] It contains one hydroxyl group less than the co-pigment, aklavinone, and yields a diacetate still containing an intact tertiary hydroxyl group (infrared). Spectroscopically it shows a close resemblance to aklavinone and since it gives bisanhydro-aklavinone (another co-pigment) on heating with palladium black/palladium charcoal the structure must be (6), that is, 7-deoxyaklavinone.[6] On heating with hydrogen bromide in acetic acid

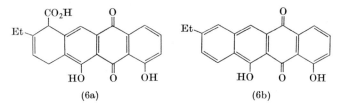

(6a)   (6b)

the dehydro-acid (6a) is formed and this, when heated above its melting point, is transformed into (6b) which has been synthesised.[24] The stereochemistry was deduced from the o.r.d. curve which is very similar to that of ζ-pyrromycinone.[11]

Bisanhydro-aklavinone (galirubinone B$_1$) (5), $C_{22}H_{16}O_6$

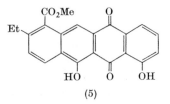

(5)

Occurrence: in cultures of *Strep. galileus* Ettlinger *et al.*[6]

Physical properties: orange needles, m.p. 236°, $\lambda_{max.}$ (hexane) 242, 262, 279, 290, 440, 462, 474 nm (log ϵ 4·65, 4·68, 4·28, 4·29, 4·28, 4·16, 4·19), $\nu_{max.}$ (Nujol) 1740, 1678, 1627, 1602, 1579 cm^{-1}, $\nu_{max.}$ (KBr) 1724, 1667, 1616, 1600, 1572 cm^{-1}.

As this pigment was already known as a degradation product of aklavinone (see p. 538) it was quickly recognised when isolated from *Strep. galileus*.[6b] It has been synthesised[9] by way of a Friedel-Crafts condensation of 3-methoxyphthalic anhydride with the naphthol (7) to give the keto-acid (8). After methylation and reduction of the ketonic

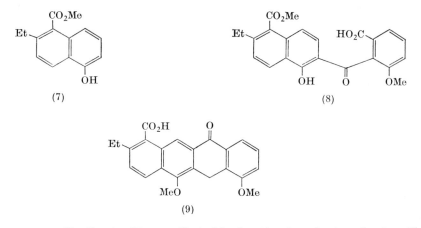

(7)   (8)

(9)

group, cyclisation to (9) was effected by heating in polyphosphoric acid. Chromic acid oxidation of the methyl ester then gave the corresponding naphthacenequinone which afforded bisanhydro-aklavinone on de-methylation with boron tribromide.

7S,9R,10R-α-*Citromycinone* (10), $C_{20}H_{18}O_7$

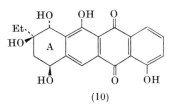

(10)

Occurrence: as a glycoside in cultures of *Strep. purpurascens* Lindenbein.[10]

Physical properties: yellow needles, m.p. 135–137°, $\lambda_{max.}$ (C_6H_{12}) 418, 436 nm (log ϵ 4·04, 4·04), $\nu_{max.}$ (KBr) 3370, 1622, 1598 cm^{-1}.

The yellow citromycinones are minor components of the mixture of pigments obtained on acid hydrolysis of the anthracycline fraction of *Strep. purpurascens*. They are distinguished from all other anthra-cyclinones by virtue of a 1,5-dihydroxyanthraquinone chromophore, their visible absorption and that of their red acetoboric esters being virtually identical with those of anthrarufin and its acetoborate, respectively. Unlike all the other anthracyclinones, there is a strong doublet in the 6 μ region which is also seen in the i.r. spectrum of 1,5-dihydroxyanthraquinone. α-Citromycinone contains one oxygen atom less than that of its co-pigment, α_2-rhodomycinone (p. 553), and its mass spectrum is very similar in the higher mass range except that m/e values are 16 mass units smaller, with prominent peaks *inter alia* at M-18, M-36, and M-72 (retro-Diels-Alder fragmentation). This suggests that ring A (11) is the same in both pigments which was confirmed by catalytic

(11)   (12)

hydrogenolysis of α_2-rhodomycinone (using palladised barium sulphate in triethanolamine-ethanol at room temperature) which gave, after reoxidation, a mixture of α- and γ-citromycinones. Removal of an α-hydroxyl group from an anthraquinone under such mild conditions is unusual, and possibly proceeds via the tautomeric form (12) of the leuco compound, by reduction of a carbonyl group and dehydration. α-Citromycinone is therefore 1-deoxy-α_2-rhodomycinone and comparison[11] of the o.r.d. curves showed that it had the same absolute configuration as α_2-rhodomycinone. Hence it can be formulated as (10).

Very dilute solutions of α-citromycinone (and α_2-rhodomycinone) in chloroform and benzene form gels at room temperature. Remarkably, no other anthracyclinones have this property.

γ-Citromycinone (13), $C_{20}H_{18}O_6$

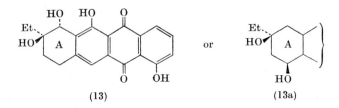

(13)   or   (13a)

Occurrence: as a glycoside in cultures of *Strep. purpurascens* Lindenbein.[10]
Physical properties: yellow needles, m.p. 207° (dec.), $\lambda_{max.}$ (C_6H_{12}) 421, 438 nm (log ϵ 4·08, 4·09).

As already indicated γ-citromycinone has a 1,5-dihydroxyanthra-quinone chromophore and is formed on hydrogenolysis of α_2-rhodo-mycinone. It has one hydroxyl group less than α-citromycinone which means that there are only two in ring A, and as the retro-Diels-Alder fragmentation leads to an ion m/e 282 (m/e 300 in the case of α-citro-mycinone) there can be only one hydroxyl group in a benzylic position.

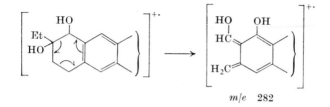

m/e 282

γ-Citromycinone is consequently (13) or (13a), the former being favoured by the observation[10, 12] that hydrogenolysis of a hydroxyl group at C-7 is often difficult in the absence of a neighbouring hydroxyl at C-6.

Isoquinocycline A (14), $C_{33}H_{32}N_2O_{10} \cdot HX$

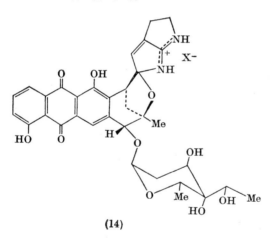

(14)

Occurrence: cultures of *Streptomyces aureofaciens* Duggar.[68]

Physical properties: the yellow, crystalline, hydrochloride shows $[\alpha]_D^{25} +12\cdot0°$ (HOAc), $\lambda_{max.}$ (0·1 N HCl) 231, 260, 292, 425–435 nm (log ε 4·74, 4·28, 4·03, 4·11, 4·12), $\nu_{max.}$ (KBr) 3311, 1718, 1626, 1605 cm^{-1}.

Under suitable conditions a strain of *Strep. aureofaciens* was found in the Lederle Laboratories[68a, 69] to produce chlortetracycline and two other antibiotics, quinocycline A and isoquinocycline A. From a related strain, cultured in the Pfizer Research Laboratories,[68b] a "quinocycline complex" was obtained, containing as major components two highly active, but relatively labile, antibiotics, quinocyclines A and B, which partly isomerised during the isolation procedure to the less active, but more stable, isoquinocyclines A and B. Preliminary chemical and spectroscopic studies [69] of isoquinocycline A, suggested the presence of a 1,5-dihydroxyanthraquinone unit (ultraviolet-visible spectrum and colour reactions) and from the isolation of a product closely resembling tetracene, by distillation with a highly active form of zinc dust, [70] a tetracyclic structure was considered. The structure of one ring was established by permanganate oxidation of a crude methylation product which afforded 3-methoxyphthalic anhydride. Acid hydrolysis gave an unusual C_8 sugar which was isolated as its crystalline anhydro derivative (15). [72] The parent compound is evidently (16) which is presumably the syrupy reducing sugar present with (15) in the hydrolysate. The aglycone still retains both

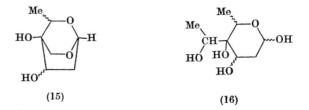

(15)　　　　　　　　　(16)

nitrogen atoms and was isolated as its crystalline hydrochloride, $[\alpha]_D^{25}$ +27·2° (HOAc). The visible spectrum is identical in λ_{max} to that of the parent compound showing that the sugar moiety is not attached to a phenolic group. One nitrogen atom appeared to be tertiary but both acid and alkaline hydrolysis liberated ammonia.

These data allowed Cosulich et al. [69] to put forward a partial structure for isoquinocycline A, and Tulinsky [71] later determined the complete structure (14) by X-ray crystallographic analysis of the hydrochloride and hydrobromide†, using isomorphous replacement procedures. While retaining the general form of an anthracycline (but with a side chain methyl group in place of the more usual ethyl), isoquinocycline A shows several unusual features. Ring A is highly modified by a fused tetrahydrofuran ring which is joined to a *spiro* pyrrolidino-pyrrole system. The nitrogenous bicyclic system is planar and hence the C–N double bond is delocalised as indicated in (14); both nitrogen-to-bridge-carbon distances are equal and shorter than a normal C–N single bond. The two secondary hydroxyl groups of the sugar are hydrogen bonded to dioxan molecules. Solvation is a troublesome feature of isoquinocycline A and

† Crystallised from aqueous dioxan.

its aglycone, and added considerably to the difficulties of the chemical investigation.

7S,9R,10R-ε-*Pyrromycinone* (rutilantinone) (17), $C_{22}H_{20}O_9$

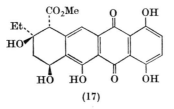

(17)

Occurrence: in cultures of various *Streptomyces* (some unidentified), chiefly as the glycosides pyrromycin,[13] rutilantins A, B and C,[14] cinerubins A[15,80] and B[15] [*Strep. antibioticus* (Waksmann and Woodruff) Waksmann and Henrici,[15] *Strep. galileus* Ettlinger *et al.*[15,80] and *Strep. niveoruber* Ettlinger *et al.*[15]], and galirubin A[6] (*Strep. galileus*).

Physical properties: fiery red needles, m.p. 213–214°,† $[\alpha]_{Cd}^{20}$ +143° ($CHCl_3$) $\lambda_{max.}$ (EtOH) 234, 257, 293, 494, 525 nm (log ε 4·63, 4·26, 3·86, 4·02, 3·92), $\nu_{max.}$ (KBr) 3410, 1740, 1620 cm^{-1}.

This pigment (either free or as a glycoside) was discovered independently in three laboratories. Structural investigations were reported in 1959[15–17] and the proposal of Ollis and co-workers[17] was accepted[18] the following year. That the pyrromycinones have a 1,4,5-trihydroxyanthraquinone chromophore is clear from the visible and infrared spectral data given above, and from the bathochromic shift in the visible region which results on addition of boroacetic anhydride. Reductive acetylation gave a product showing anthracene-like ultraviolet absorption. Acetylation afforded a yellow tetra-acetate which still retained a free (that is tertiary) hydroxyl function, and showed both phenolic and alcoholic acetate carbonyl absorption. The presence of an ethyl group was suggested by the formation of propionic and acetic acids on chromic acid oxidation, while the existence of an ester group, indicated by the carbonyl absorption at 1743 cm^{-1}, was confirmed by heating with hydrogen iodide to give a carboxylic acid and methyl iodide. All the oxygen atoms are thus accounted for as three phenolic hydroxyls, two alcoholic hydroxyls, a methoxycarbonyl group and the quinone carbonyls. On heating alone, or in acid solution, the elements of two molecules of water (including the alcoholic hydroxyl groups) are lost forming a bisanhydro compound, η-pyrromycinone, of structure (29) (see p. 550), which is a co-pigment in certain *Streptomycetes*. It is now evident that the hydroxyl groups in ε-pyrromycinone which are eliminated during aromatisation are located in ring A. Initially they were placed as shown in (18) on biogenetic

† Changes into η-pyrromycinone on heating.

18

grounds, assuming an acetate derivation for the molecule; this arrangement is also consistent with observations that one hydroxyl group can be cleaved by hydrogenolysis,[19] and the other is tertiary and probably β to

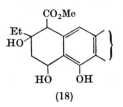

(18)

the ester function since elimination can also be effected easily under basic conditions. These orientations (17) were fully confirmed later by n.m.r. studies.[1]

The glycoside, pyrromycin, isolated as an orange, crystalline, hydrochloride, $C_{30}H_{35}NO_{11}.HCl$, m.p. 162–163° (dec.), $[\alpha]_D^{20}$ +132° (±27°) (MeOH),[13] gives 1 mol. of ϵ-pyrromycinone and 1 mol. of rhodosamine on mild acid hydrolysis. As the glycoside and aglycone show the same λ_{max}. values the sugar must be attached to one of the alcoholic hydroxyl groups, and since the glycoside forms only a tetra- and not a penta-acetate, the tertiary alcoholic group at C-9 must be intact. It follows that

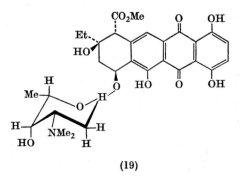

(19)

the sugar residue must be linked to the aglycone at C-7. As pyrromycin, in contrast to rhodosamine, shows no reducing properties the hydroxyl group at C-1 in the sugar moiety must be involved in the glycosidic link (19), the configuration of which is not known.[13]

Cinerubins A and B are isomeric crystalline glycosides, $C_{43-45}H_{57-61}NO_{18}$, containing one O–Me, two N–Me and at least four C–Me groups.[15] Spectroscopically they behave as 1,4,5-trihydroxy-anthraquinones showing that the sugar residues are linked to ring A. On acid hydrolysis they each liberate three sugars, two of which appear to be the same in each case.[15] Brockmann and Waehneldt[21] identified the aminosugar as rhodosamine, and one of the other nitrogen-free sugars as 2-deoxy-L-fucose; under very mild conditions pyrromycin could be detected during the hydrolysis of both cinerubins but the complete structure of these glycosides is not yet known.

The rutilantins have been obtained as hydrochlorides, picrates and citrates, and yield unidentified sugars, some of which are basic, on mild acid hydrolysis.[14] On hydrolysis of galirubin A, five sugars can be detected by paper chromatography, at least one of which is basic; only ring A is glycosidically bound.[6]

STEREOCHEMISTRY OF THE ANTHRACYCLINONES

The stereochemistry of ring A in ϵ-pyrromycinone and other anthracyclinones was first correlated by H. Brockmann Jr. and Legrand[30] by circular dichroism measurements. Comparison of the c.d. curves in the region 270–390 nm (where all these pigments show very similar absorption) led to the conclusion that all the natural compounds they examined have the same absolute configuration at their asymmetric centres. All the curves have the same characteristic S-shape; the dichroism is very large and is chiefly influenced by the asymmetry at C-10 and C-7 (where applicable), whereas the asymmetric centre at C-9 being further from the chromophore has relatively little effect. Comparison with derivatives possessing asymmetry only at C-10 shows that introduction of a further asymmetric centre at C-9 always increases the amplitude of the c.d. curve, indicating that the configuration at C-9 is always the same. When a third asymmetric centre is present at C-7 the amplitude of the curve is considerably increased, and by almost the same amount in each case, and since the $\Delta\epsilon_{max}$ value is almost doubled it can be concluded that the substituents at C-10 and C-7 have the same steric relationship to the chromophore and consequently are *trans* to each other. Since γ-rhodomycinone (p. 558) is reluctant to react with periodate the hydroxyl groups at C-9 and C-10 also have a *trans* relationship, whence it is possible to represent the relative stereochemistry of ring A in those pigments which are hydroxylated at C-9 and C-10 as in (20).[30] The corresponding

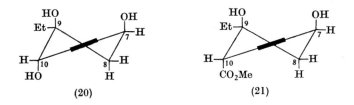

(20)   (21)

arrangement (21) for ϵ-pyrromycinone, and other pigments having a methoxycarbonyl substituent at C-10, could be assigned in the light of the following considerations.[30] The pyrolytic formation of anhydro derivatives presumably proceeds chiefly by a *cis*-elimination of water, and hence it was concluded that in, for example, ζ-pyrromycinone (below), the hydroxyl at C-9 is *cis* to the proton at C-10. Further, ϵ-rhodomycinone and ϵ-isorhodomycinone tetramethyl ethers (pp. 560 and 569) yield isopropylidene derivatives with acetone dimethyl ketal, and hence their hydroxyl groups at C-7 and C-9 must be *cis* to each other. Both conformations (20) and (21) gain stability by intramolecular hydrogen bonding between the *cis*-diaxial hydroxyl groups and by virtue of the fact that the ethyl group, (usually) the largest substituent, occupies an equatorial position.

Following the determination of the absolute stereochemistry of daunomycinone, which has the 7S,9S configuration, by Arcamone *et al.*[63] (see p. 564) it was possible to show[55] that ϵ-pyrromycinone and several other anthracyclinones also have the 7S configuration by correlation of their c.d. curves with those of daunomycinone and 7-deoxydaunomycinone.[51] This information, together with n.m.r. data and chemical relationships such as those mentioned above, leads to the absolute configurations given throughout this chapter.[55]

9R,10R-ζ-*Pyrromycinone* (galirubinone C) (22), $C_{22}H_{20}O_8$

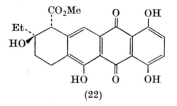

(22)

Occurrence: cultures of *Strep. galileus* Ettlinger *et al.*[6] and other *Streptomyces*.[22a]
Physical properties: fiery red needles, m.p. 216°, $[\alpha]_{Cd}^{20}$ +74° ($\pm6°$) (CHCl$_3$), $\lambda_{max.}$ (C$_6$H$_{12}$) 483, 494, 516, 528 nm, $\nu_{max.}$ (KBr) 3510, 1725, 1610(sh), 1590 cm^{-1}.

ζ-Pyrromycinone contains one oxygen atom less than ϵ-pyrromycinone and its structure was defined by two experiments. On heating with palladium it is transformed into η-pyrromycinone (29) (p. 550) with loss of $H_2 + H_2O$ which establishes the complete carbon skeleton,[16,20] and since it can be obtained via hydrogenolysis[23] of ϵ-pyrromycinone which removes the benzylic hydroxyl group, it follows that ζ-pyrromycinone has structure (22). It is of interest that η-pyrromycinone was a minor product of the reduction which was conducted in triethanolamine-ethanol solution. The more hindered ester group in η-pyrromycinone withstands the basic conditions, but in the starting material ring A is non-planar and the more accessible methoxycarbonyl group is hydrolysed, the main reaction product being ζ-pyrromycinone acid. Methylation with diazomethane then gave ζ-pyrromycinone.

Structure (22) is fully supported by the n.m.r. spectrum[1] and by several other reactions. Acetylation gives mainly a triacetate but the tetraacetate, involving esterification of the tertiary hydroxyl, can be obtained.[23] Facile elimination of the tertiary alcoholic group would be expected and it is intriguing to find that dehydration proceeds in three directions if suitable conditions are chosen.[1,16,20] Heating at 230° gives the anhydro compound (23) (*cis*-elimination) accompanied by a bathochromic shift indicating conjugation of the double bond with the aromatic system; further treatment with palladium converts it into

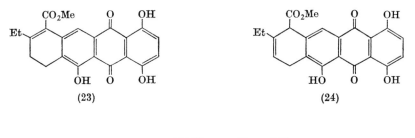

(23)   (24)

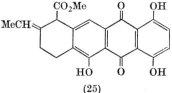

(25)

η-pyrromycinone. On the other hand, heating with phosphorus pentoxide (*trans*-elimination) gives the isomeric product (24) whose light absorption is practically identical with that of the starting material. The third anhydro compound, which can only be (25), is obtained by heating in toluene with toluene-p-sulphonic acid (*trans*-elimination); on warming in basic solution it rearranges to (23).

On heating alone, a second compound, isomeric with ζ-pyrromycinone, is formed as well as (23). This product[20] shows an additional carbonyl band at 1720 cm^{-1} and on warming in pyridine-triethylamine it is reconverted into optically inactive ζ-pyrromycinone. On this basis it is regarded[1] as (26), and the presence of abundant M-29 and M-57 peaks in the mass spectrum provides confirmatory evidence for the ketonic side chain.[56] Its formation by a "pyrolytic retro-aldol" reaction may be considered as a reversal of the *in vivo* reaction by which optically active

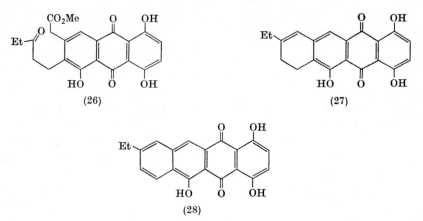

(26) (27)

(28)

ζ-pyrromycinone is formed. Heating ζ-pyrromycinone in hydrobromic acid-acetic acid results in hydrolysis, decarboxylation, and elimination of water to form (27), which can be dehydrogenated to demethoxy-carbonyl-η-pyrromycinone (28).[16, 20]

η-*Pyrromycinone* (galirubinone B$_2$, ciclacidine) (29), C$_{22}$H$_{16}$O$_7$.

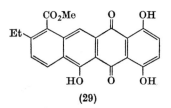

(29)

Occurrence: cultures of *Strep. galileus* Ettlinger *et al.*,[6] *Strep. capoamus* sp. nov.[81] and other *Streptomyces*.[22b]

Physical properties: red crystals, m.p. 236–237° (242°), λ$_{max.}$ (C$_6$H$_{12}$) 252, 275, 486, 496, 508, 519, 530 nm (log ε 4·80, 4·56, 4·35, 4·38, 4·32, 4·41, 4·34), ν$_{max.}$ (KBr) 1724, 1640, 1615, 1594, 1580 cm^{-1}.

η-Pyrromycinone is identical with bisanhydro-ε-pyrromycinone (= bisanhydrorutilantinone) and accordingly it contains an ethyl and a methoxycarbonyl group. The ultraviolet-visible absorption is virtually identical with that of 1,4,6-trihydroxytetracenequinone and the same is true of their triacetates;[17] the bathochromic shifts on treatment with boroacetic anhydride also confirm that all three hydroxyl groups are *peri* to the quinone carbonyl groups.[16] As oxidation of η-pyrromycinone with alkaline permanganate gave[17] benzene-1,2,3,4-tetracarboxylic acid it follows that the ethyl and methoxycarbonyl substituents occupy adjacent positions in ring A, and they were located as shown in (29) where the hindered position of the ester group (*o*-ethyl and *peri*-hydrogen) would account for its relative stability to alkaline hydrolysis. A hydroxyl group *peri* to the ester function would provide greater steric hindrance, but the hydroxyl group in ring B was assigned to C-6 rather than C-11 as η-pyrromycinone showed no tendency to form a lactone. (However, it is perhaps doubtful whether a strongly hydrogen bonded hydroxyl at C-11 would lactonise with an acid or ester group at C-10.) That the ethyl group occupies a β-position was also confirmed by hydrolysis and decarboxylation to give demethoxycarbonyl-η-pyrromycinone (28) which yielded benzene-1,2,4-tricarboxylic acid when degraded with nitric acid.[19] These views were supported by several syntheses, first of (28)[8] and (30)[25] (from (28) by oxidation with manganese dioxide and sulphuric acid), and later of η-pyrromycinone itself. The natural pigment was obtained via

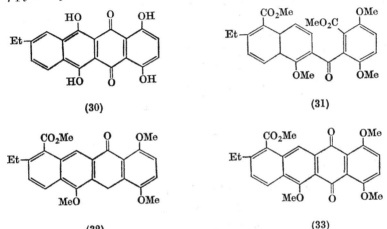

(30)   (31)

(32)   (33)

Friedel-Crafts condensation of the naphthol (7) with 3,6-dimethoxy-phthalic anhydride which gave, after methylation, the keto-diester (31). Reduction of the ketonic group, followed by cyclisation in polyphosphoric acid afforded the tetracyclic ketone (32) and this on oxidation with chromic acid afforded η-pyrromycinone trimethyl ether (33). Finally, selective demethylation with cold boron tribromide yielded η-pyrromycinone.[26]

η_1-*Pyrromycinone* (34), $C_{21}H_{14}O_7$

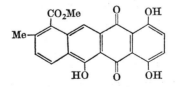

(34)

Occurrence: cultures of an unidentified *Streptomyces* sp.[27]

Physical properties: red crystals, m.p. 290–292°, $\lambda_{max.}$ (C_6H_{12}) 252, 275, 482, 492, 504, 515, 526 nm, $\nu_{max.}$ (Nujol) 1700, 1630, 1600(sh), 1575 cm^{-1}.

In this latest addition to the pyrromycinone series the usual β-ethyl group is replaced by methyl, and the compound does not show antibiotic properties. The ultraviolet-visible absorption is identical with that of η-pyrromycinone, and the zinc dust distillation product has the same electronic absorption as tetracene. The infrared spectrum provides evidence for an aromatic ester† and chelated quinone carbonyl groups. η_1-Pyrromycinone forms a triacetate. The n.m.r. spectrum of the parent compound includes 3-proton singlets at τ 7·45 and 5·92 arising from an aromatic methyl group and a methoxyl group, respectively, as well as three 1-proton singlets at low field attributable to three hydrogen bonded hydroxyl groups. Hydrolysis of the ester group, followed by decarboxylation, converts η_1-pyrromycinone into a demethoxycarbonyl derivative with the same visible absorption (in cyclohexane) as the parent compound. This implies[41] that there is no hydroxyl group at C-11 *peri* to the methoxycarbonyl group at C-10. On the other hand oxidation of the pigment with manganese dioxide in sulphuric acid introduced a fourth hydroxyl group and gave a product whose light absorption closely resembled that of 1,4,6,11-tetrahydroxy-8-ethyltetracenequinone. These observations lead to the conclusion[27] that η_1-pyrromycinone has structure (34).

7R,9R,10R-α-*Rhodomycinone* (35), $C_{20}H_{18}O_8$

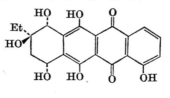

(35)

Occurrence: as a glycoside in cultures of *Strep. purpurascens* Lindenbein.[28, 29] and other *Actinomycetes*.[66]

Physical properties: red prisms, m.p. 217–220° (dec.), $\lambda_{max.}$ (C_6H_{12}) 483, 494, 517, 529 nm, $\nu_{max.}$ (KBr) 3420, 1603 cm^{-1}.

α-Rhodomycinone was found to be isomeric with the previously identified β-rhodomycinone (p. 554) and their ultraviolet-visible, n.m.r. and mass spectra are virtually identical. Clearly both pigments have the same chromophore, and the same gross structure in ring A, and the orientation of ring A with respect to the rest of the molecule was established by heating with hydrogen chloride in acetic acid which gave the bis-anhydro

† ν_{CO} (Nujol) 1700 cm^{-1}, cf. 1724 cm^{-1} (KBr) in η-pyrromycinone and 1724 cm^{-1} (KBr), 1740 cm^{-1} (Nujol) in bisanhydro-aklavinone.

compound (36) previously obtained from β-rhodomycinone using hydrogen iodide. However, treating α-rhodomycinone with hydrobromic acid in acetic acid gave a much lower yield of (36), the major product

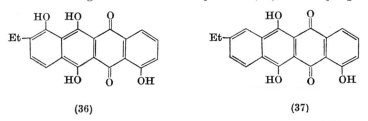

(36)   (37)

being the tetracenequinone (37) which was later synthesized.[33] This was unexpected but it is of interest that the mass spectrum of α-rhodomycinone includes a peak at m/e 334 attributed to a deoxybisanhydro fragment, possibly arising by an intermolecular redox reaction prior to electron bombardment.[37] As both α- and β-rhodomycinone yield γ-rhodomycinone (p. 558) on hydrogenolysis over palladium in triethanolamine-ethanol, they are clearly stereoisomers which differ only in their configuration at C-7. Selective esterification of the C-7 hydroxyl group can be effected by treatment with cold trifluoroacetic acid; mild alkaline hydrolysis of the 7-trifluoroacetate of both α- and β-rhodomycinone gives a mixture of both epimers.[31] As the hydroxyl groups at C-7 and C-10 in β-rhodomycinone are situated in a *trans* relationship (p. 556), α-rhodomycinone must have the corresponding *cis* arrangement expressed in (35).[31]

The α- and β-rhodomycinones show curious differences in acidity. On shaking a chloroform solution of α-rhodomycinone with 2N sodium carbonate the aqueous phase becomes blue-violet but there is no colouration with the β-isomer. In piperidine solution, the visible absorption curve of α-(and γ)-rhodomycinone shows longwave maxima at 560 and 601 nm, but in the β-isomer these are shifted to 570 and 612 nm. In neutral solvents both isomers show the same absorption.

α-Rhodomycinone is the aglycone of the antibiotic glycoside mycetin C.[45]

7S,9R,10R-α₂-*Rhodomycinone* (38),† $C_{20}H_{18}O_8$

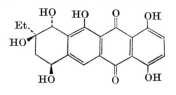

(38)

† Or (38) with the opposite configuration at all three asymmetric centres.

Occurrence: as a glycoside in cultures of *Strep. purpurascens* Lindenbein.[10]

Physical properties: red leaflets, m.p. 207–209 (dec.), $\lambda_{max.}$ (C_6H_{12}) 482, 494, 515, 529 nm, $\nu_{max.}$ (KBr) 3400, 1592 cm^{-1}.

The visible absorption of this recently discovered pigment, which is isomeric with α- and β-rhodomycinone, is again indicative of a 1,4,5-trihydroxyanthraquinone structure, and the lack of carbonyl absorption above 1600 cm^{-1} confirms the absence of an ester group. As was found in the case of α-rhodomycinone, heating with hydrobromic acid-acetic acid eliminates all the hydroxyl groups from ring A leaving the naphthacene-quinone (39). The structure of (39) was deduced from spectral evidence

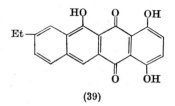

(39)

and its close similarity to the isomeric demethoxycarbonyl-η-pyrro-mycinone (28). The n.m.r. spectrum revealed that three aromatic protons were present, two of which, at C-2 and C-3, are equivalent and "appear" as a singlet at τ 2·71 (as in δ-rhodomycinone, p. 560), the third (τ 1·72) being assigned to C-6. Further analysis[10] of the n.m.r. spectrum established that ring A has the same structure, relative configuration and conformation as ring A in β-rhodomycinone (below), which was supported by comparison[11] of their o.r.d. spectra. Accordingly α$_2$-rhodo-mycinone can be written as (38) or its antipode.[10] α$_2$-Rhodomycin II is its 7,10-dirhodosaminide.[84]

The structures of α$_1$- and α$_3$-rhodomycinone are not yet known.[10]

7S,9R,10R-*β-Rhodomycinone* (40), $C_{20}H_{18}O_8$

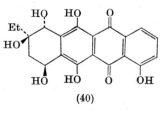

(40)

Occurrence: as the glycosides rhodomycin A and B in cultures of *Strep. purpurascens* Lindenbein[38] and unidentified *Streptomyces* species.[44, 66]

Physical properties: red needles, m.p. 224–225° (dec.), $\lambda_{max.}$ (C_6H_{12}) 483, 494, 517, 529 nm, $\nu_{max.}$ (KBr) 3460, 1608 cm^{-1}.

β-Rhodomycinone was the first member of the group to be studied[34] and was regarded initially as a monoalkylated anthraquinone, the formation of a tetracene-like hydrocarbon on zinc dust distillation being attributed to cyclisation of a side chain. β-Rhodomycinone forms an optically active penta-acetate, ($[\alpha]_D^{20} -34\cdot4^\circ (\pm 5^\circ)$ (MeOH)), still retaining a free hydroxyl group, and hence of the eight oxygen atoms five are involved in the 1,4,5-trihydroxyanthraquinone chromophore (see electronic spectrum above) and three are alcoholic hydroxyl groups of which one is tertiary. All the aliphatic hydroxyls are removed on heating with hydriodic acid, one being replaced by hydrogen, the others being eliminated as water with consequent aromatization of ring A, to give (37).[36] This compound was already known as demethoxycarbonyl-bisanhydro-ε-rhodomycinone, and was subsequently synthesised[33] (see also p. 561). On heating with hydrochloric acid however, the reaction is

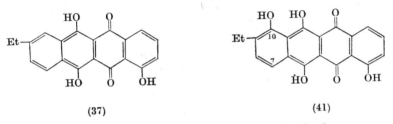

(37)　　　　　　　　(41)

limited to the elimination of water giving a bisanhydro compound containing four hydroxyl groups to which the tetracenequinone structures (41) or (41, HO at C-7 instead of C-10) could be assigned.[36] Relative to (37), the longwave absorption of (41) or its isomer is shifted ca. 20 nm into the red, a bathochromic displacement which can be attributed (by comparison with model compounds[25]) to the introduction of a hydroxyl group in ring A *peri* to another in ring B. The relative positions of the phenolic groups in bisanhydro-β-rhodomycinone (as 41) were settled later by comparison of its spectral properties, and that of its *peri*-carbonate, with those of the model quinones (42) and (43), and their *peri*-carbonates. These were prepared by condensation of 3-hydroxy-phthalic anhydride with β-hydrojuglone in the presence of boric acid, followed by treatment with phosgene.[35] Regarding the positions of the

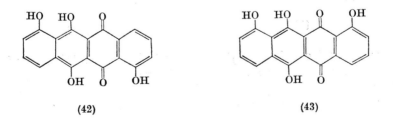

(42)　　　　　　　　(43)

remaining hydroxyl groups in ring A of β-rhodomycinone, the tertiary hydroxyl is clearly located at C-9 and the other was assigned to the benzylic position at C-7 as it could be removed by catalytic hydrogenolysis to give γ-rhodomycinone, accompanied by the expected small hypsochromic shift (observed in piperidine solution).[36] This defines the structure of ring A as (44) which is consistent with the n.m.r.[36] and mass[37] spectra. β-Rhodomycinone is therefore (45).

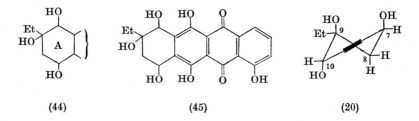

(44)   (45)   (20)

The stereochemistry of ring A was deduced from various considerations, which were discussed previously (p. 547). Circular dichroism studies showed[30] that the hydroxyl groups at C-7 and C-10 were *trans* and the relative resistance to periodate oxidation indicated[36] that the hydroxyls at C-9 and C-10 were also in a *trans* relationship. Analysis of the coupling constants of the ABX system of β-rhodomycinone, derived from the 100 MHz n.m.r. spectrum (in d_5-pyridine), led to the conclusion[31] that the half-chair conformation carried a quasi-equatorial proton at C-7, the C-7 hydroxyl being quasi-axial. Ring A may therefore be represented by (20) and β-rhodomycinone by (40).

β-Rhodomycinone is the aglycone of β-rhodomycins I and II (formerly rhodomycins B and A, respectively[39]), both of which have the same visible absorption as the aglycone. β-Rhodomycin I gives γ-rhodomycinone (50) and rhodosamine on hydrogenolysis and is thus (46a;

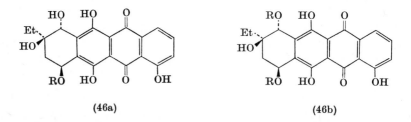

(46a)   (46b)

R = rhodosamine residue).[84] β-Rhodomycin II affords β-rhodomycin I and rhodosamine on partial hydrolysis, while hydrogenolysis leads ultimately to 10-deoxy-γ-rhodomycinone (51). The structure is therefore (46b; R = rhodosamine residue).[84] Hydrolysis of β-rhodomycin IV gives β-rhodomycinone, rhodosamine (2 mols.), 2-deoxy-L-fucose (1 mol.) and L-rhodinose. As catalytic hydrogenation yields the γ-rhodomycinone rhodosaminide (46c), which can also be derived in the same way from β-rhodomycin II, β-rhodomycin IV must have a deoxy-L-fucose-L-rhodinose disaccharide moiety at C-7 and a rhodosamine residue at C-10.[84]

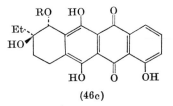

(46c)

β-Rhodomycinone is probably also the aglycone of the antibiotic mycetin B₁.[45]

9R,10R-β₁-*Rhodomycinone* (47),† $C_{19}H_{16}O_7$

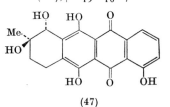

(47)

Occurrence: as a glycoside in cultures of *Strep. purpurascens* Lindenbein.[29]

Physical properties: red prisms, dec. ca. 260°, $\lambda_{max.}$ (C_6H_{12}) 495, 517, 531 nm, $\nu_{max.}$ (KBr) 3400, 1590 cm⁻¹.

This pigment is less soluble than the other rhodomycinones and as the amount available to Brockmann and his co-workers[29] was small, its structure has been deduced chiefly from spectral data. It was the first example in this series of a pigment containing a methyl side chain. It has a C_{19} molecular formula and the infrared spectrum shows no ester carbonyl band; the normal 1,4,5-trihydroxyanthraquinone chromophore is present and the visible absorption of its blue-violet solution in piperidine ($\lambda_{max.}$ 570, 601 nm) closely resembles that of γ-rhodomycinone (48; R = Et). Significant similarities are also seen in the mass spectra. Like all anthracyclinones which lack a methoxycarbonyl group at C-10, a retro-Diels-Alder fragmentation is of major importance, resulting, in the case of γ-rhodomycinone, in successive loss of fragments of 72 and 28 mass units giving peaks at *m/e* 298 and 270. The same peaks are present in the mass spectrum of β₁-rhodomycinone but the initial fragment lost (49) is only 58 mass units showing that R must be methyl. This means

† Or the isomer with HO at C-1 instead of C-4.

that ring A in β_1-rhodomycinone is as shown in (48; R = Me) which is

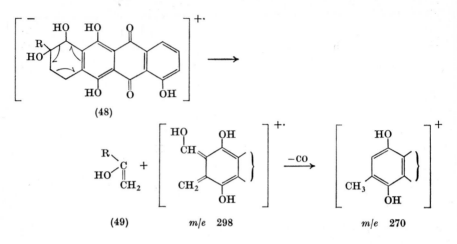

(48)   (49)   m/e 298   m/e 270

fully consistent with the n.m.r. spectrum. The latter also reveals the presence of three vicinal aromatic protons so that ring A must be fused to the dihydroxylated benzenoid ring, and hence β_1-rhodomycinone must be (47) or the isomer with HO at C-1 instead of C-4. Analogy with the other rhodomycinones and biogenetic considerations favour structure (47).[29]

9R,10R-γ-*Rhodomycinone* (50), $C_{20}H_{18}O_7$

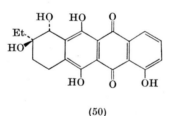

(50)

Occurrence: in several glycosidic combinations in cultures of *Strep. purpurascens* Lindenbein[28] and unidentified *Streptomycetes*.[44,66]

Physical properties: fiery red needles, dec. 230–240°, $\lambda_{max.}$ (C_6H_{12}) 483, 494, 517, 529 nm, $\nu_{max.}$ (KBr) 3450, 1590 cm⁻¹.

As γ-rhodomycinone can be derived from both α- and β-rhodomycinone by catalytic hydrogenolysis[29, 36] over palladium, so cleaving a benzylic hydroxyl group at C-7, it must have structure (50). On heating with hydrochloric acid in acetic acid it gives (with other products) some of the 10-epimer in which the hydroxyl groups at C-9 and C-10 have a *cis*-relationship. Accordingly, this epimer can be oxidized with periodate readily; in parallel experiments the glycol function of 10-*epi*-γ-rhodomycinone was completely oxidised in 15 min but the natural pigment remained unchanged.[31]

γ-Rhodomycinone is the aglycone of a group of basic glycosides, the γ-rhodomycins, which are separable on a column of cellulose, and on mild acid hydrolysis the following sugars are liberated.[28] It is clear from their

	Hydrolysis products (mol. ratio)			
Glycoside	γ-Rhodomycinone	Rhodosamine	2-Deoxy-L-L-fucose	Rhodinose[40]
γ-Rhodomycin I	1	1	—	—
γ-Rhodomycin II	1	2	—	—
γ-Rhodomycin III	1	2	1	—
γ-Rhodomycin IV	1	2	2	1

visible absorption curves that none of the phenolic groups is involved in glycosidic linkages so that all the sugar residues must be attached to ring A, presumably via the secondary hydroxyl group. γ-Rhodomycin IV is thus a tetrasaccharide, and since it can be partially hydrolysed to γ-rhodomycins III, II and I, the sequence of sugar residues is evidently γ-rhodomycinone–rhodosamine–rhodosamine–2-deoxy-L-fucose–rhodinose.

9R-10-*Deoxy-γ-rhodomycinone* (51), $C_{20}H_{18}O_6$

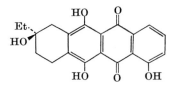

(51)

Occurrences: as a glycoside in cultures of *Strep. purpurascens* Lindenbein.[29]

Physical properties: red needles, m.p. 232°, λ_{max}. (C_6H_{12}) 467, 485, 496, 519, 532 nm, ν_{max}. (KBr) 3500, 1590 cm⁻¹.

The electronic absorption of this minor pigment is very similar to that of the preceding pigment (50) and its structure (51) was established[29] when it was obtained, in 20% yield, by catalytic hydrogenolysis of γ-rhodomycinone in triethanolamine-ethanol solution. This was confirmed by the n.m.r. spectrum which showed the complete absence of benzylic hydroxyl groups, and by the mass spectrum which includes a major peak at m/e 282 attributable to loss of $EtC(OH){=}CH_2$ by retro-Diels-Alder fragmentation.[29]

$7S,9R,10R\text{-}\delta\text{-}Rhodomycinone$ (52), $C_{22}H_{20}O_9$

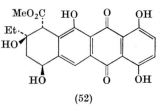

(52)

Occurrence: cultures of *Strep. purpurascens* Lindenbein.[41]

Physical properties: red prisms, m.p. 195–197°, λ_{max}. (C_6H_{12}) 483, 495, 517, 530 nm, ν_{max}. (KBr) 3480, 1725, 1710, 1615, 1592 cm^{-1}.

δ-Rhodomycinone is isomeric with ε-pyrromycinone, has the same functional groups and virtually the same spectral properties. They differ only in the location of the phenolic group in ring B which, by difference, must be at C-11 in δ-rhodomycinone. This assignment is supported by the visible absorption curve of the bisanhydro derivative (53), prepared by heating with hydrobromic acid, which shows maxima (in cyclohexane)

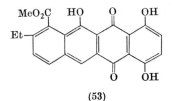

(53)

at slightly longer wavelengths (4–5 nm) than that of 1,4,6-trihydroxy-tetracenequinone. According to the Brockmanns[41] this is a characteristic of such quinones in which a methoxycarbonyl group in ring A is *peri* to a hydroxyl group in ring B. It will be seen that the arrangement of hydroxyl groups in this pigment is different from that in the other rhodomycinones and in the pyrromycinones.

$7S,9R,10R\text{-}\epsilon\text{-}Rhodomycinone$ (54), $C_{22}H_{20}O_9$

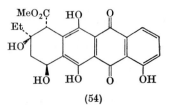

(54)

Occurrence: cultures of *Strep. purpurascens* Lindenbein[34] and as a glycoside from an unidentified *Streptomyces* sp.[44]

Physical properties: red needles, m.p. 210°,† λ_{max}. (C_6H_{12}) 483, 493, 515, 529 nm, ν_{max}. (KBr) 3450, 1724, 1600, 1580(sh) cm^{-1}.

ε-Rhodomycinone is isomeric with the preceding pigment and with ε-pyrromycinone, and they differ only in the orientation of the phenolic groups. Of the two alcoholic hydroxyls, one is tertiary and the other can be cleaved by hydrogenolysis (to give ζ-rhodomycinone), and both are eliminated by heating either alone, or with acid. The bisanhydro product is a tetracenequinone (55; R = H) which could be oxidised with manganese dioxide in sulphuric acid to the known η-isopyrromycinone (55; R = OH).[23, 42] This establishes the side chain positions in ring A,

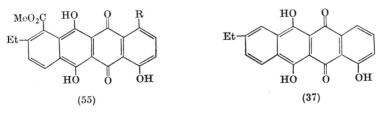

(55)　　　　　　　　　　　　　　(37)

and the location of the hydroxyl group in ring D was confirmed by a synthesis[33] of the demethoxycarbonyl-bisanhydro derivative (37). ε-Rhodomycinone is therefore (54) which is consistent with its mass[37, 56] and n.m.r.[1] spectra.

9R,10R-ζ-Rhodomycinone (56), $C_{22}H_{20}O_8$

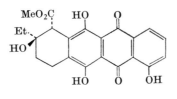

(56)

Occurrence: cultures of *Strep. purpurascens* Lindenbein[43] and as a glycoside from an unidentified *Streptomyces* sp.[44]

Physical properties: fiery red crystals, m.p. 274–275°, ν_{max}. (KBr) 3520, 3400, 1712, 1595 cm^{-1}.

The structure of this pigment is determined by the fact that it can be obtained by hydrogenolysis of ε-rhodomycinone (54) in triethanolamine-ethanol solution. The initial product is an acid, owing to basic hydrolysis of the methoxycarbonyl group, but on treatment with diazomethane it gives a pigment identical in all respects with ζ-pyrromycinone (56).[23]

† Converted to the bisanhydro compound on heating.

As this pigment forms only a triacetate evidently the hydrogenolysis removes the secondary hydroxyl groups from ε-rhodomycinone. Heating ζ-rhodomycinone with hydrobromic acid-acetic acid results in hydrolysis, decarboxylation, and loss of 1 mol. of water to give a dihydrotetracenequinone which can be converted, with the aid of palladised charcoal, into (37).

θ-Rhodomycinone (57), $C_{22}H_{20}O_9$

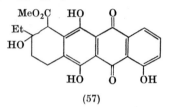

(57)

Occurrence: as a glycoside in cultures of an unidentified *Streptomyces* sp.[44]
Physical properties: red-brown needles, dec. 220°, $[\alpha]_D^{25}$ +191·5° (THF), $\lambda_{max.}$ (95% EtOH) 234, 258, 296, 480, 493, 513, 527 nm (log ε 4·60, 4·35, 3·82, 4·10, 4·13, 4·01, 3·95), $\nu_{max.}$ (KBr) 3478, 1712, 1642, 1622, 1581 cm^{-1}.

This pigment appears to be a stereoisomer of ε-rhodomycinone and contains the same array of functional groups. However, it differs from related pigments in that the ester carbonyl function absorbs at 1712 cm^{-1} (cf. the usual range, 1725–1740 cm^{-1}) possibly due to hydrogen bonding with a neighbouring phenolic group, and hydrolysis of the ester occurred relatively easily. On heating with hydrogen bromide in acetic acid (or with N-sodium hydroxide) the demethoxycarbonyl-bisanhydro compound (37; R = H) was the chief product. Some of the acid (37;

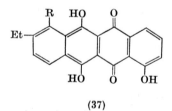

(37)

R = CO_2H) was also isolated and gave (37; R = H) on heating, but all attempts to obtain the ester (37; R = CO_2Me) failed. Bowie and Johnson[49] therefore suggest that θ-rhodomycinone has the gross structure (57), which is consistent with the n.m.r. spectrum of the tetra-acetate, but differs from the other rhodomycinones and pyrromycinones in having the ester group in an equatorial position (see p. 547). This would facilitate hydrolysis relative to the usual axial position in which the ester function is hindered by bulky neighbouring groups. The configuration of the rest of the ring has not been determined.

7S,9S-*Daunomycinone* (58), $C_{21}H_{18}O_8$

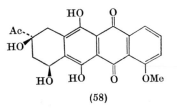

(58)

Occurrence: as a glycoside in cultures of *Strep. peucetius* Grein *et al.*[46] and *Strep. coeruleorubidus* Gause *et al.* [47, 52]

Physical properties: red crystals, m.p. 213–214°, $[\alpha]_D^{20}$ +193° (dioxan), λ_{max}. (MeOH) 234, 252, 290, 480, 495, 532 nm, ν_{max}. (KBr) 3470, 1718, 1615, 1580 cm⁻¹.

The antibiotic daunomycin (rubidomycin,[47] rubomycin C[52]) was isolated in the laboratories of Farmitalia[46] and Rhône-Poulenc,[47] and later in Russia,[52] and shows considerable antitumour activity.[48, 78] It is the first antibiotic to show a therapeutic effect in the treatment of acute leukaemia in man.[77] The structure of the aglycone, daunomycinone, was determined by Arcamone *et al.*[49, 62] It shows electronic absorption similar to that of 1,4,5-trihydroxyanthraquinone, yields tetracene (u.v.) on zinc dust distillation, and it contains a methoxyl group. There is also a carbonyl group absorbing at 1718 cm⁻¹, but this is not due to the presence of the usual methoxycarbonyl function but rather to a methyl ketone as daunomycinone readily forms a 2,4-dinitrophenyl-hydrazone, and the n.m.r. spectrum includes a three-proton singlet at τ 7·31. Heating with acetic anhydride in pyridine affords a tetra-acetate showing both phenolic and alcoholic acetate absorption, but reaction with methyl sulphate-acetone-potassium carbonate produced only a trimethyl ether still containing a free hydroxyl group. Treatment with either acid or alkali gives a bisanhydro derivative $C_{21}H_{14}O_6$; this exhibits conjugated ketone absorption at 1685 cm⁻¹ and retains two phenolic hydroxyl groups (diacetate, ν_{CO} 1765 cm⁻¹) and the methoxyl group. Reduction with sodium borohydride, followed by periodate oxidation, afforded acetaldehyde in good yield confirming the presence of an acetyl side chain clearly attached to a hydroxylated carbon atom. That this is a β ring carbon atom is evident from the formation of benzene-1,2,4-tricarboxylic acid in high yield when the bisanhydro compound is oxidised with permanganate. The simultaneous formation of 3-methoxyphthalic acid defines another ring in this molecule which can thus be represented as (59).

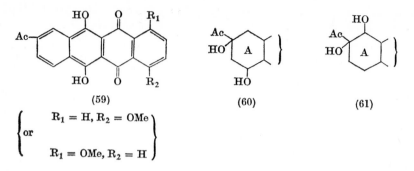

(59)   (60)   (61)

$$
\left\{ \begin{array}{c} R_1 = H,\ R_2 = OMe \\ \text{or} \\ R_1 = OMe,\ R_2 = H \end{array} \right\}
$$

It is now obvious that ring A of daunomycinone contains two hydroxyl groups, one attached to the carbon atom bearing the acetyl group and the other is evidently benzylic as it can be removed by hydrogenolysis over palladised barium sulphate (apparently without reduction of the ketonic group). The structure of ring A is therefore either (60) or (61), only the former being consistent with the n.m.r. spectra of daunomycinone and its derivatives. In particular, the spectrum of the trimethyl ether shows a four line signal at τ 5·08 arising from the proton at C-7 which is the X part of an ABX system, the AB part of which consists of two pairs of doublets at τ 7·58 and 8·13 (J_{AB}, 15 Hz) attributable to a methylene group β to an aromatic ring. (This is seen more clearly in the 220 MHz n.m.r. spectrum of daunomycin which indicates that the C-7 proton is pseudo-equatorial.[75]) The benzylic methylene group at C-10 is represented by two doublets at τ 6·78 and 6·98 (J, 18·5 Hz).

The stereochemistry of ring A was determined as follows.[63] Partial demethylation of daunomycinone trimethyl ether with aluminium chloride in benzene gave the 7-0-methyl derivative. Oxidation of this with lead tetra-acetate gave the diquinone (62) which, without isolation, was subjected to a Lemieux oxidation (permanganate-periodate)[64] to give, inter alia, S(−)-methoxysuccinic acid (63), isolated by preparative paper chromatography of the ammonium salt. The S-configuration can

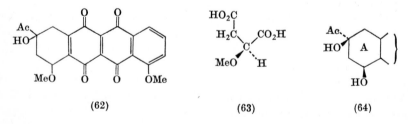

(62)   (63)   (64)

therefore be assigned to the asymmetric centre at C-7, and since daunomycinone forms an isopropylidene derivative the hydroxyl groups at C-7 and C-9 must have a *cis* relationship, and consequently the stereochemistry at C-9 is also of S-configuration. Ring A has therefore the usual stereochemistry (64)[63] with an acetyl group replacing the ethyl group found in most anthracyclinones.

The final structural problem concerns the position of the methoxyl group in daunomycinone, already assumed in (62). This was determined by chromic acid oxidation to form the diketone (65) which was converted into (66) on treatment with base. The trihydroxyquinone (66) showed the bathochromic shift relative to bisanhydrodaunomycinone (59) expected on the introduction of a *peri*-hydroxyl group. Demethylation,

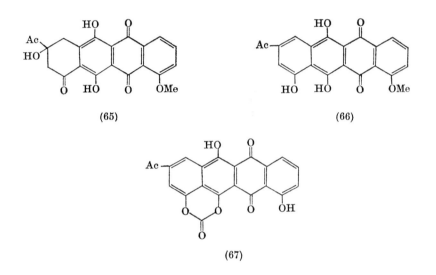

(65)   (66)

(67)

followed by treatment with phosgene in pyridine, then led to the formation of the carbonate (67) whose structure is confirmed by the absence of free carbonyl absorption near 1670 cm^{-1}. Similarly, the derivative (67; Ac replaced by PhCHMe) showed no absorption in the region 1620–1800 cm^{-1}. This confirms that daunomycinone has structure (58) with the methoxyl group at C-4. If the methoxyl group were at C-1, then the derivative corresponding to (67) would have had an unchelated carbonyl group.

The glycoside daunomycin (rubidomycin, rubomycin C) is basic and may be isolated as its crystalline hydrochloride, $C_{27}H_{29}NO_{10}\cdot HCl$, m.p. 188–190° (dec.), $[\alpha]_D$ +253° (MeOH). On mild acid hydrolysis it yields daunomycinone and the aminosugar daunosamine (68).[50, 65] As both the glycoside and the aglycone show very similar electronic absorption, the sugar moiety is obviously attached to ring A, and in fact to the hydroxyl group at C-7 as hydrogenolysis of daunomycin affords 7-deoxydaunomycinone and daunosamine.[62, 75] The stereochemistry of the glycosidic

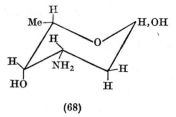

(68)

link was deduced by Arcamone *et al.*[63] from the 100 MHz n.m.r. spectrum of N-acetyldaunomycin (see also ref. 75). In particular, the sum of the

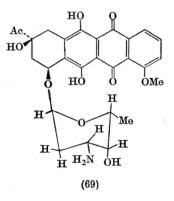

(69)

coupling constants for the 1'- and 2'-protons (5·0–5·5 Hz) excludes axial-axial interaction and so the anomeric proton must be equatorial. Daunomycin is therefore represented by (69).[63] A simple glucosidic analogue with the same configuration at C-7 has been synthesised.[51] The c.d. curves of daunomycin and daunomycinone are very similar in the 280 nm region, the main contribution arising from the substituent at C-7 as 7-deoxydaunomycinone shows a much weaker dichroic effect.[75]

Rubomycin C appears to be the same as daunomycin and is one of a mixture of related but unidentified pigments, rubomycins A_1, A_2, B_0 and B_1[52, 76] (for further references see ref. 77b).

7S,9S-*Adriamycinone* (70), $C_{21}H_{18}O_9$

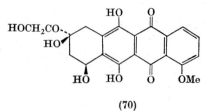

(70)

Occurrence: cultures of a *Streptomyces peucetius* mutant.[82]

Physical properties: red crystals, m.p. 223–224°, $[\alpha]_D^{23}$ +188° (dioxan), ultraviolet-visible spectrum superposable on that of (58), ν_{max} 3440, 1727, 1615 cm^{-1}.

Adriamycin has been compared favourably with daunomycin as an antitumour antibiotic.[83] On hydrolysis it gives daunosamine and a red aglycone, adriamycinone, which differs from daunomycinone only in the presence of an additional hydroxyl group. As dehydration of adriamycinone with hydrogen bromide in acetic acid gives a product whose visible absorption is identical with that of bisanhydrodaunomycinone, the additional hydroxyl group must be located on the acetyl side chain as shown in (70).

Hydrogenolysis of adriamycin over palladised barium sulphate gives daunosamine and 7-deoxyadriamycinone which forms a tetra-acetate. In the n.m.r. spectrum of the latter, the benzylic CHOAc signal observed in the spectrum of adriamycinone penta-acetate. is absent This establishes that the glycosidic linkage in adriamycin is at C-7, and comparison of its c.d. curve with that of daunomycin confirmed that the stereochemistry was the same at C-7 and C-9 [i.e., (S)]. As the optical

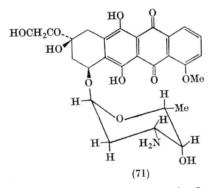

(71)

rotations of adriamycin and daunomycin, and of adriamycinone and daunomycinone, are virtually the same, the sugar moiety must make the same optical contribution in both glycosides. Adriamycin is therefore (71).[82]

7S,9R,10R-β-*Isorhodomycinone* (72), $C_{20}H_{18}O_9$

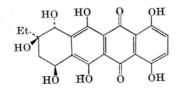

(72)

Occurrence: as a glycoside in cultures of *Strep. purpurascens* Lindenbein[34, 38d, 53] and other *Actinomycetes*.[66]

Physical properties: dark red needles, no m.p., $\lambda_{max.}$ ($CHCl_3$) 491, 514, 524, 552, 564 nm, $\nu_{max.}$ (KBr) 3400, 1585 cm^{-1}.

The visible spectrum of β-isorhodomycinone is almost identical with that of 1,4,5,8-tetrahydroxyanthraquinone, and like the other isorhodomycinones it gives a blue solution in aqueous sodium hydroxide and in concentrated sulphuric acid. On heating with hydrobromic acid-acetic acid it gave the known tetracenequinone (73) together with a smaller amount of a bisanhydro derivative (74). This is analogous to the

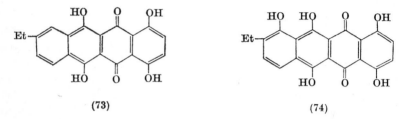

(73)   (74)

behaviour of β-rhodomycinone, two hydroxyl groups being eliminated from ring A as molecules of water and a third by reductive fission. If hydrobromic acid is replaced by hydrochloric acid only the bisanhydro-β-isorhodomycinone (74) is obtained. The positions of the substituents in ring A were deduced[53] from a comparison of visible absorption spectra. Relative to (73), the absorption maxima of (74) show a bathochromic shift of 22–27 nm which, as model compounds indicate, would only arise if the additional hydroxyl group occupied an α-position in ring A. Since the isolated ethyl group in (73) has practically no effect on the visible absorption of the parent tetrahydroxytetracenequinone but exerts a small bathochromic effect in (74) relative to its parent pentahydroxy-tetracenequinone, it was deduced that the ethyl group is adjacent to the hydroxyl in ring A of bisanhydro-β-isorhodomycinone. It follows that in ring A of β-isorhodomycinone one hydroxyl group is at C-10, and the others involved in dehydration are most probably located at C-7 and C-9 by analogy with related compounds. This was confirmed[53] by the n.m.r. spectrum (in d_5-pyridine) which was almost identical above τ 4 with that of β-rhodomycinone. It follows that ring A in β-isorhodomycinone has the same conformation as in β-rhodomycinone, and the o.r.d. curve[11] shows that it has the same absolute configuration. Accordingly its structure can be represented by (72).[31] Like β-rhodomycinone it can be epimerised at C-7 by mild hydrolysis of the 7-trifluoroacetate to give α-isorhodomycinone which has not yet been found in Nature.[31]

The glycoside β-isorhodomycin II (formerly isorhodomycin A[38]) yields β-isorhodomycinone on hydrolysis and contains two rhodosamine residues which, from n.m.r. comparisons, are placed at C-7 and C-10.[84]

9R,10R-γ-*Isorhodomycinone* (75), $C_{20}H_{18}O_8$

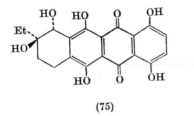

(75)

Occurrence: as a glycoside in cultures of *Strep. purpurascens* Lindenbein.[53]
Physical properties: amorphous red solid, $\lambda_{max.}$ (CHCl$_3$) 491, 514, 525, 552, 564 nm, $\nu_{max.}$ (KBr) 3530, 1565 cm^{-1}.

This minor pigment from *Strep. purpurascens* was only obtained in very small amount and was identified[53] as (75) by comparison (infrared, R_F) with the product formed on hydrogenolysis of β-isorhodomycinone in which, by analogy, the benzylic hydroxyl group at C-7 is removed.

7S,9R,10R-ε-*Isorhodomycinone* (76), $C_{22}H_{20}O_{10}$

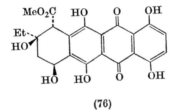

(76)

Occurrence: cultures of *Strep. purpurascens* Lindenbein.[34]
Physical properties: dark red crystals, m.p. 227–229° (dec.), $\lambda_{max.}$ (dioxan) 242, 298, 523, 535, 550, 560 nm, $\nu_{max.}$ (KBr) 3480, 1740, 1598 cm^{-1}.

This pigment, one of the first anthracyclinones to be studied, was regarded initially[34] as a monoalkylated anthraquinone. Subsequently the correct structure was elucidated by Brockmann and Boldt.[54] On

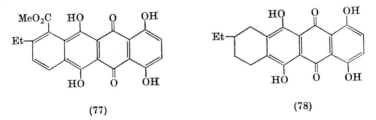

(77)   (78)

heating alone, or better with hydrobromic acid-acetic acid, a bisanhydro compound was obtained which proved to be identical with η-isopyrromycinone (77). The carbon skeleton is thus defined. ϵ-Isorhodomycinone forms a penta-acetate still containing a free hydroxyl group. Of the six hydroxyls, two are replaced by hydrogen on heating with phenol and hydriodic acid, the methoxycarbonyl group being simultaneously eliminated to give a product (78) which retains the 1,4,5,8-tetrahydroxy-anthraquinone chromophore of the original pigment, and yields β-ethyladipic acid (and other dibasic acids) when oxidised with permanganate in pyridine. Most of ring A is now defined. The tertiary hydroxyl can be assigned to C-9 as propionic acid is readily obtained by permanganate degradation of ϵ-isorhodomycinone, and the benzylic situation of the remaining hydroxyl group is shown by hydrogenolytic cleavage (and formation of ζ-isorhodomycinone) resulting in a small hypsochromic shift (6–7 nm) in piperidine solution. This leads to structure (76) for ϵ-isorhodomycinone which is fully consistent with its n.m.r. spectrum[54] and o.r.d. curve.[11]

$9R,10R$-ζ-*Isorhodomycinone* (79), $C_{22}H_{20}O_9$

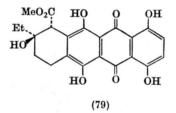

(79)

Occurrence: cultures of *Strep. purpurascens* Lindenbein.[54]

Physical properties: dark red needles, m.p. 258–260°, $\lambda_{max.}$ (dioxan) 244, 298–300, 521, 548, 560 nm, $\nu_{max.}$ (KBr) 3540, 1720, 1600 cm^{-1}.

ζ-Isorhodomycinone is identical with the deoxy compound formed on hydrogenolysis of ϵ-isorhodomycinone over palladised barium sulphate in triethanolamine-ethanol solution. Its structure is therefore (79).[54]

Other Anthracyclinones

Several other antibiotic pigments have been isolated from *Streptomycetes*,[66] some of which show anti-tumour properties, whose visible absorption spectra suggest that they may belong to this group. Minomycin,[57] aquayamycin[67] and the ayamycins[58] may be included here, and also the ruticulomycins,[60] and nogalomycin,[59] for which an incomplete structure has been advanced.

Nogalamycin

Nogalarol, $C_{29}H_{31}NO_{13}$

Occurrence: as the glycoside nogalamycin in cultures of *Strep. nogalater* var. *nogalater* Bhuyan & Dietz.[61]

Physical properties: red crystals, m.p. ca. 220° (dec.), λ_{max}. (MeOH) 234, 258, 280, 475 nm (log ε 4·73, 4·39, 3·99, 4·19), ν_{max}. (Nujol) 3440, 1725, 1655, 1615 cm^{-1}.

The antibiotic nogalamycin is the most recent addition to the anthracycline family.[59] Spectral data, especially the visible absorption, show that it differs from the others, as does the aglycone, nogalarol, in possessing a 1,4-dihydroxyanthraquinone chromophore. Further, the visible absorption of the bisanhydro compound, nogalarene, which is formed along with nogalarol on hot acid hydrolysis of nogalamycin, is similar to that of 6,11-dihydroxytetracene-5,12-quinone. A carbonyl band at 1725 cm^{-1} and an appropriate n.m.r. signal, indicate the presence of a methoxycarbonyl group. The hydroaromatic ring evidently contains two hydroxyl substituents, one of which must be attached to a carbon atom bearing a methyl group, as an aliphatic C-methyl singlet in nogalarol appears as an aromatic C-methyl in nogalarene. The other hydroxyl group is benzylic since it can be replaced by methoxyl on treatment with methanolic hydrochloric acid, and on heating with acid O-methylnogalarol is converted into nogalarene. O-methylnogalarol is also formed, *inter alia*, on methanolysis of nogalamycin suggesting that the glycosidic link involves a benzylic hydroxyl group. This is supported by the fact that the sugar, nogalose, can be detached from nogalamycin by catalytic hydrogenolysis. In the n.m.r. spectrum of nogalarol a doublet of doublets (1H) centred at τ 4·83 is attributed to a benzylic proton attached to a carbon atom carrying a hydroxyl group and adjacent to a methylene group, and a singlet (1H) at τ 6·24 is assigned to another benzylic proton attached to a carbon atom bearing the methoxycarbonyl substituent. Ring A in nogalarol is therefore formulated as (80) which agrees with the formation of benzene-1,2,3,4-tetracarboxylic acid

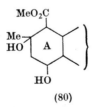

(80)

when nogalarene is oxidised with permanganate. As no other aromatic acid could be isolated from this reaction it seems that both benzenoid rings are oxygenated. The Upjohn group [59] therefore propose the partial structure (81) for nogalarol which allows for the observation that only two aromatic protons are present. However, these give rise to singlets at τ 2·81 and 3·40 (2·9 and 3·0 in nogalamycin) which seems to be inconsistent with the proposed structure.

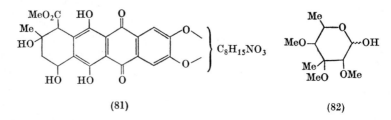

(81)   (82)

Nogalamycin, probably $C_{39}H_{49}NO_{17}$, m.p. 195–196° (dec.), $[\alpha]_D^{25°}$ +479° ($CHCl_3$), can be hydrolysed with hot dilute acid to give nogalarol and nogalarene, as already mentioned, and a sugar, nogalose, to which structure (82) has been assigned.[59] This is known (see above) to be attached to the aglycone at the C-7 hydroxyl group. The unidentified fragment, $C_8H_{15}NO_3$, in (81) appears to be an aminosugar residue, and must be the origin of the dimethylamine evolved when nogalamycin is treated with base. However, it is much more firmly bound than is the case with other anthracyclinones and as yet, has not been isolated (cf. isoquinocycline A). It is still attached to nogalarene which is optically active.

Direct comparison[59] has shown that nogalamycin is very closely related to (or identical with?) ruticulomycin A isolated in the Lederle laboratories.[60] Ruticulomycin A, dark red plates, m.p. 183–184° (dec.) and ruticulomycin B, red rosettes, m.p. 179–180° (dec.) are antibiotics elaborated by *Strep. rubrireticuli* Waksmann. They show very similar infrared and ultraviolet absorption and were independently recognised[60] as anthracyclines with a 1,4-dihydroxyanthraquinone chromophore and a firmly bound nitrogeneous residue.

REFERENCES

1. H. Brockmann, *Prog. Chem. Org. Nat. Prods.* (1963), **21**, 121.
2. H. Muxfeldt and R. Bangert, *Prog. Chem. Org. Nat. Prods.* (1963), **21**, 80; T. Money, *Prog. Org. Chem.* (1968), **7**, 1.
3. (a) J. R. D. McCormick. *In* "Biogenesis of Antibiotic Substances" (eds. Z. Vaněk and Z. Hošťálek) (1965), p. 73. Czech. Acad. Sci. Prague; (b) Personal communication, 1968.
4. H. Brockmann and E. Spohler, *Naturwissenschaften* (1955), **42**, 154; H. Brockmann and Th. Waehneldt, *Naturwissenschaften* (1963), **50**, 92; H. Brockmann, E. Spohler and Th. Waehneldt, *Chem. Ber.* (1963), **96**, 2925.
5. F. Strelitz, H. Flon, U. Weiss and I. N. Asheshov, *J. Bacteriol.* (1956), **72**, 90.

6. (a) K. Eckardt and G. Bradler, *Naturwissenschaften* (1965), **52**, 539; (b) K. Eckardt, *Chem. Ber.* (1967), **100**, 2561.

7. J. J. Gordon, L. M. Jackman, W. D. Ollis and I. O. Sutherland, *Tetrahedron Lett.* (1960), No. 8, 28.

8. W. D. Ollis, I. O. Sutherland and P. L. Veal, *Chem. Comm.* (1960), 349.

9. Z. Horii, H. Hakusui and T. Momose, *Chem. Pharm. Bull.* (*Tokyo*) (1966), **14**, 802; (1968), **16**, 1262; Z. Horii, H. Hakusui, T. Momose and E. Yoshino, *Chem. Pharm. Bull.* (*Tokyo*) (1968), **16**, 1251.

10. H. Brockmann and J. Niemeyer, *Chem. Ber.* (1968), **101**, 1341.

11. J. Niemeyer and G. Maass, unpublished work.

12. H. Brockmann and H. Brockmann Jr., *Chem. Ber.* (1963), **36**, 1771.

13. H. Brockmann and W. Lenk, *Chem. Ber.* (1959), **92**, 1904.

14. I. N. Asheshov and J. J. Gordon, *Biochem. J.* (1961), **81**, 101; W. D. Ollis and I. O. Sutherland. *In* "Chemistry of Natural Phenolic Compounds" (ed. W. D. Ollis) (1961), p. 212. Pergamon, London.

15. L. Ettlinger, E. Gäumann, R. Hütter, W. Keller-Schierlein, F. Kradolfer, L. Neipp, V. Prelog, P. Reusser and H. Zähner, *Chem. Ber.* (1959), **92**, 1867.

16. H. Brockmann and W. Lenk, *Chem. Ber.* (1959), **92**, 1880.

17. W. D. Ollis, I. O. Sutherland and J. J. Gordon, *Tetrahedron Lett.* (1959), No. 16, 17.

18. H. Brockmann, H. Brockmann Jr., J. J. Gordon, W. Keller-Schlierlein, W. Lenk, W. D. Ollis, V. Prelog and I. O. Sutherland, *Tetrahedron Lett.* (1960), No. 8, 25.

19. H. Brockmann and H. Brockmann Jr., *Naturwissenschaften* (1960), **47**, 135.

20. H. Brockmann and W. Lenk, *Naturwissenschaften* (1960), **47**, 135.

21. H. Brockmann and Th. Waehneldt, *Naturwissenschaften* (1961), **48**, 717.

22. (a) H. Brockmann and W. Lenk, *Angew. Chem.* (1957), **69**, 477; (b) H. Brockmann, L. Costa Plà and W. Lenk, *Angew. Chem.* (1957), **69**, 477.

23. H. Brockmann and H. Brockmann Jr., *Chem. Ber.* (1961), **94**, 2681.

24. H. Brockmann and J. Niemeyer, *Chem. Ber.* (1968), **101**, 2409.

25. H. Brockmann and E. Wimmer, *Chem. Ber.* (1963), **96**, 2399.

26. Z. Horii, T. Momose and Y. Tamura, *Chem. Pharm. Bull.* (*Tokyo*) (1964), **12**, 1262; (1965), **13**, 635, 797.

27. J. R. Hegyi and N. N. Gerber, *Tetrahedron Lett.* (1968), 1587.

28. H. Brockmann and Th. Waehneldt, *Naturwissenschaften* (1961), **48**, 717.

29. H. Brockmann, J. Niemeyer, H. Brockmann Jr. and H. Budzikiewicz, *Chem. Ber.* (1965), **98**, 3785.

30. H. Brockmann Jr. and M. Legrand, *Naturwissenschaften* (1962), **49**, 374; *Tetrahedron* (1963), **19**, 395.

31. H. Brockmann and J. Niemeyer, *Chem. Ber.* (1967), **100**, 3578.

32. G. Bradler, K. Eckardt and R. Fügner, *Zeit. Allg. Mikrobiol.* (1966), **6**, 361.

33. H. Brockmann, R. Zunker and H. Brockmann Jr., *Annalen* (1966), **696**, 145.

34. H. Brockmann and B. Franck, *Chem. Ber.* (1955), **88**, 1792.

35. H. Brockmann and E. Wimmer, *Chem. Ber.* (1965), **98**, 2797.

36. H. Brockmann and P. Boldt, *Naturwissenschaften* (1957), **44**, 616; H. Brockmann, P. Boldt and J. Niemeyer, *Chem. Ber.* (1963), **96**, 1356.

37. H. Brockmann Jr., H. Budzikiewicz, C. Djerassi, H. Brockmann and J. Niemeyer, *Chem. Ber.* (1965), **98**, 1260.

38. (a) H. Brockmann and K. Bauer, *Naturwissenschaften* (1950), **37**, 492; (b) H. Brockmann, K. Bauer and I. Borchers, *Chem. Ber.* (1951), **84**, 700; (c) H. Brockmann and I. Borchers, *Chem. Ber.* (1953), **86**, 261; (d) H. Brockmann and P. Patt, *Chem. Ber.* (1955), **88**, 1455.

39. H. Brockmann and E. Spohler, *Naturwissenschaften* (1961), **48**, 716.
40. H. Brockmann and Th. Waehneldt, *Naturwissenschaften* (1963), **50**, 43.
41. H. Brockmann and H. Brockmann Jr., *Naturwissenschaften* (1963), **50**, 20; *Chem. Ber.* (1963), **96**, 1771.
42. H. Brockmann and H. Brockmann Jr., *Naturwissenschaften* (1961), **48**. 161.
43. P. Boldt, Dissertation, Göttingen, 1958.
44. J. H. Bowie and A. W. Johnson, *J. Chem. Soc.* (1964), 3927.
45. G. Z. Yakubov, N. O. Blinov, L. N. Sergeeva, O. I. Artamonova and A. S. Khokhlov, *Antibiotiki* (1965), **10**, 771.
46. A. Grein, C. Spalla, A. Di Marco and G. Canevazzi, *Giorn. Microbiol.* (1963), **11** 109; G. Cassinelli and P. Orezzi, *Giorn. Microbiol.* (1963), **11**, 167.
47. M. Dubost, P. Ganter, R. Maral, L. Ninet, S. Pinnert, J. Preud'homme and G.-H. Werner, *Compt. Rend.* (1963), **257**, 1813.
48. A. Di Marco, M. Gaetani, P. Orezzi, B. M. Scarpinato, R. Silvestrini, M. Soldati, T. Dasdia and L. Valentini, *Nature, Lond.* (1964), **201**, 706; R. Despois, M. Dubost, D. Mancy, R. Maral, L. Ninet, S. Pinnert, J. Preud'homme, Y. Charpentie, A. Belloc, N. de Chezelles, J. Lunel and J. Renaut, *Arzneim.-Forsch.* (1967), **17**, 934.
49. F. Arcamone, G. Franceschi, P. Orezzi, G. Cassinelli, W. Barbieri and R. Mondelli, *J. Amer. Chem. Soc.* (1964), **86**, 5334.
50. F. Arcamone, G. Cassinelli, P. Orezzi, G. Francheschi and R. Mondelli, *J. Amer. Chem. Soc.* (1964), **86**, 5335.
51. J. P. Marsh, R. H. Iwamoto and L. Goodman, *Chem. Comm.* (1968), 589.
52. M. G. Brazhnikova, N. V. Konstantinova, V. A. Pomaskova and B. V. Zacharov, *Antibiotiki* (1966), **11**, 763.
53. H. Brockmann, J. Niemeyer and W. Rode, *Chem. Ber.* (1965), **98**, 3145.
54. H. Brockmann and P. Boldt, *Naturwissenschaften* (1960), **47**, 134; *Chem. Ber.*, 1961, **94**, 2174.
55. H. Brockmann, H. Brockmann Jr. and J. Niemeyer, *Tetrahedron Lett.* (1968), 4719.
56. R. I. Reed and W. K. Reid, *Tetrahedron* (1963), **19**, 1817.
57. H. Nishimura, K. Sasaki, M. Mayama, N. Shimaoka, K. Tawara, S. Okamoto and K. Nakajima, *J. Antibiotics, Ser. A* (1960), **13**, 327.
58. K. Sato, *J. Antibiotics, Ser. A* (1960), **13**, 321.
59. P. F. Wiley, F. A. MacKellar, E. L. Caron and R. B. Kelly, *Tetrahedron Lett.* (1968), 663.
60. L. A. Mitscher, W. McCrae, W. W. Andres, J. A. Lowery and N. Bohonos, *J. Pharm. Sci.* (1964), **53**, 1139.
61. B. K. Bhuyan, R. B. Kelly and R. M. Smith, U.S.P. 3,183,157; *Chem. Abs.* (1965), **63**, 3588.
62. F. Arcamone, G. Francheschi, P. Orezzi, S. Penco and R. Mondelli, *Tetrahedron Lett.* (1968), 3349.
63. F. Arcamone, G. Cassinelli, G. Francheschi, P. Orezzi and R. Mondelli, *Tetrahedron Lett.* (1968), 3353.
64. R. V. Lemieux and E. von Rudloff, *Canad. J. Chem.* (1955), **33**, 1711.
65. J. P. Marsh, C. W. Mosher, E. M. Acton and L. Goodman, *Chem. Comm.* (1967), 973.
66. G. Z. Yakubov, Y. M. Khokhlova, O. I. Artamonova, L. N. Sergeeva and G. Y. Kalmykova, *Microbiology* (1966), **35**, 875.
67. M. Sezaki, T. Hara, S. Ayukawa, T. Takeuchi, Y. Okami, M. Hamada, T. Nagatsu and H. Umezawa, *J. Antibiotics* (1968), **21**, 91.

68. (a) J. H. Martin, A. J. Shay, L. M. Pruess, J. N. Porter, J. H. Mowat and N. Bohonos, *Antibiotics Annual* (1954–1955), 1020; (b) W. D. Celmer, K. Murai, K. V. Rao, F. W. Tanner and W. S. Marsh, *Antibiotics Annual* (1957–1958), 484.

69. D. B. Cosulich, J. H. Mowat, R. W. Broschard, J. B. Patrick and W. E. Meyer, *Tetrahedron Lett.* (1963), 453; 1964, 750 and further corrections.

70. F. Kögl and W. B. Deijs, *Annalen* (1935), **515**, 10; J. S. Webb, R. W. Broschard, D. B. Cosulich, W. J. Stein and C. F. Wolf, *J. Amer. Chem. Soc.* (1957), **79**, 4563.

71. A. Tulinsky, *J. Amer. Chem. Soc.* (1964), **86**, 5368.

72. J. S. Webb, R. W. Broschard, D. B. Cosulich, J. H. Mowat and J. E. Lancaster, *J. Amer. Chem. Soc.* (1962), **84**, 3183.

73. U.S.P. 3,074,975.

74. J. R. D. McCormick and E. R. Jensen, *J. Amer. Chem. Soc.* (1965), **87**, 1794; C. H. Hassall and T. E. Winters, *J. Chem. Soc. (C)* (1968), 1558.

75. R. W. Iwamoto, P. Lim and N. S. Bhacca, *Tetrahedron Lett.* (1968), 3891.

76. M. G. Brazhnikova, N. V. Konstantinova, A. S. Mezentsev and V. A. Pomaskova, *Antibiotiki* (1968), **13**, 781.

77. (a) J. Bernard, Cl. Jacquillat, M. Boiron, Y. Najean, M. Seligmann, J. Tanzer, M. Weil and P. Lortholary, *Presse Méd.* (1967), **75**, 951; (b) "Rubidomycin" (eds. J. Bernard, R. Paul, M. Boiron, Cl. Jacquillat and R. Maral). Springer-Verlag, Berlin, 1969.

78. D. Gottlieb and P. D. Shaw, "Antibiotics, Mechanism of Action", Vol. 1. Springer-Verlag, Berlin, 1967.

79. J. R. D. McCormick and E. R. Jensen, *J. Amer. Chem. Soc.* (1968), **90**, 7126; J. R. D. McCormick, E. R. Jensen, N. H. Arnold, H. S. Corey, U. H. Joachim, S. Johnson, P. A. Miller and N. O. Sjolander, *J. Amer. Chem. Soc.* (1968), **90**, 7127.

80. M. G. Brazhikova, M. K. Kudinova, A. S. Mezentsev, G. B. Fedorova, R. S. Ukholina, G. V. Kochetkova, T. C. Maksimova, N. P. Nechaeva and O. K. Rossolimo, *Antibiotiki*, (1968) **13**, 963.

81. O. Gonçalves da Lima, F. Delle Monache, I. L. d'Albuquerque and G. B. Marini Bettòlo, *Tetrahedron Lett.* (1968), 471.

82. F. Arcamone, G. Francheschi, S. Penco and A. Selva, *Tetrahedron Lett.* (1969), 1007.

83. A. Di Marco, M. Gaetani and B. Scarpinato, *Cancer Chemother. Reps.* (1969), **53**, 33.

84. H. Brockmann, Th. Waehneldt and J. Niemeyer, *Tetrahedron Lett.* (1969), 415.

CHAPTER 7

Extended Quinones

The pigments collected together under this heading include some of the most highly condensed aromatic compounds found in Nature. There is little doubt that they arise from simpler precursors by phenolic coupling and in most cases the appropriate monomer, or a close relative, is also a natural product. The perylenequinone (1) occurs with methyl ethers of 1,8-dihydroxynaphthalene, and emodinanthrone has been found along with hypericin (14). The protoaphins, in aphids, are essentially binaphthyl derivatives, but for convenience they are discussed in this chapter as they are converted, post-mortem, into more highly condensed pigments by processes which evidently take place *in vivo* in the case of rhodoaphin (77). Although not very numerous, extended quinones have been discovered in fungi and higher plants, in fossil crinoids and living arthropods; undoubtedly more will be found.

4,9-Dihydroxyperylene-3,10-quinone (1), $C_{20}H_{10}O_4$

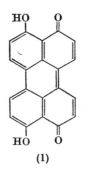

(1)

Occurrence: sporophores of *Daldinia concentrica* (Bolt) Ces. & de Not.[1,2]
Physical properties: dark red, almost black needles, dec. >350°, λ_{max}. ($CHCl_2CHCl_2$) 265, 340, 419, 444, 493, 526, 567 nm, ν_{max}. (Nujol) 1630 cm^{-1}.

The fungus *Daldinia concentrica* is one of the larger Ascomycetes, frequently encountered in Britain as a parasite on ash trees (*Fraxinus excelsior*). When mature it produces numerous fruit-bodies as black hemispherical lumps, part of the colour arising from the presence of the quinone (1).[1,2] Although extremely insoluble in most solvents, it can be extracted with acetone giving a reddish-brown solution which rapidly deposits the pigment as a black precipitate. It has been suggested that this relative solubility initially may be due to association[1] with other more polar constituents (mainly mannitol) in the fungus or to the formation of derivatives which are easily hydrolysed.[2] The principal product isolated is the binaphthyl (3)[2] but after exhaustive extraction the sporophores are still black.

Allport and Bu'Lock[3] found that *D. concentrica* was a variable species in culture, some strains being only slightly pigmented, and from a white mycelium they isolated the mono- and dimethyl ethers of 1,8-dihydroxy-naphthalene (2; R = H, R' = Me; R = R' = Me). The parent diol was not detected but there seems little doubt that it must be a metabolic intermediate. In some strains which lack an oxidase enzyme† it is

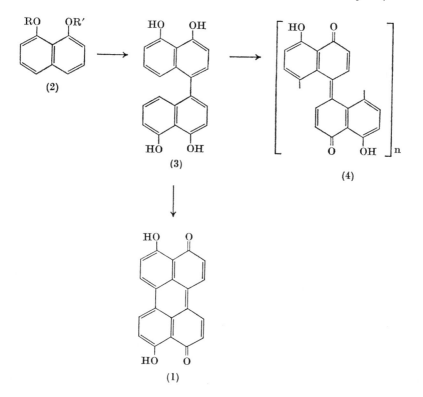

methylated, but in others and in the "wild-type", oxidative coupling leads to the binaphthyl (3) and hence to the quinone (1). The binaphthyl (3) can be oxidised enzymically with an aqueous extract of *D. concentrica* (wild-type) mycelium, and in other ways, to a black product containing

† Some strains remain relatively unpigmented as the formation of benzenoid phenolic compounds predominates over naphthalenediol biosynthesis.

19

the quinone (1) and polymeric material. Similarly, 1,8-dihydroxy-naphthalene blackens on exposure to air and old samples contain (1) and a black polymer. The conversion of (3) into (1) proceeds via intramolecular coupling but a similar intermolecular process could lead to polymers of type (4) and such compounds, although not coplanar, would appear black and be highly insoluble. The major part of the black material (containing 2% nitrogen) in *D. concentrica* sporophores probably consists of a "melanochitin" formed by reaction of polymers (4) with amino-sugar residues in the cell-wall polysaccharides. It can be reversibly bleached with reducing agents, and since it yields 1,8-dihydroxynaph-thalene and catechol on alkali fusion this "fungal melanin" may be a copolymer.[100] Plant melanins are usually based on catechol.

4,9-Dihydroxyperylene-3,10-quinone can be prepared by nitrating perylene, which gives mainly the 3,4,9,10-tetranitro derivative, followed by treatment with hot concentrated sulphuric acid.[4] It shows characteristic longwave absorption in the visible region, forms a

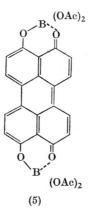

(5)

sparingly soluble greenish-blue sodium salt and a crystalline bis-boro-acetate (5),[4] and yields mellitic acid when oxidised with nitric acid. In most respects it behaves as an "extended naphthazarin".

ELSINOCHROMES†

Elsinochrome A (6), $C_{30}H_{24}O_{10}$
Elsinochrome B (7), $C_{30}H_{26}O_{10}$
Elsinochrome C (8), $C_{30}H_{28}O_{10}$
Elsinochrome D (9), $C_{30}H_{26}O_{10}$

† The pigments isolated[7] from "*Phyllosticta caryae* Peck" were named phycarones but the term was abandoned when this fungus was found[8] to be an imperfect stage of *Elsinoë randii* Bitanc. & Jenkins.

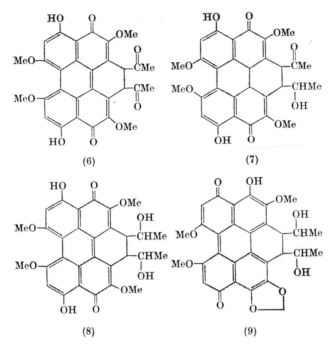

(6)

(7)

(8)

(9)

Occurrence: cultures of numerous species of *Elsinoë* (e.g. *E. anonae* Bitanc. and Jenkins, *E. phaseoli* Jenkins, *E. australis* Bitanc. & Jenkins) and of its conidial stage *Sphaceloma* (e.g. *S. randii* Bitanc. & Jenkins, *S. populi* (Sacc.) Jenkins).[5, 6]

Physical properties: (A) dark red needles, m.p. 248° $[\phi]_{600}$ +10,307°, $[\phi]_{480}$ −35,504°, $[\phi]_{380}$ +34,495°, $[\phi]_{293}$ −37,795° (CHCl₃), $\lambda_{max.}$ (EtOH) 219, 262, 280, 330, 430, 455, 525, 563 nm (log ϵ 4·76, 4·59, 4·55, 3·74, 4·40, 4·44, 4·11, 4·21), $\nu_{max.}$ (KBr) 1715, 1623, 1595 cm⁻¹; (B) red crystals, m.p. 208°, $\lambda_{max.}$ (EtOH) 262, 339, 455, 525, 563 nm (log ϵ 4·54, 3·70, 4·33, 4·12, 4·17), $\nu_{max.}$ (KBr) 3400, 1713, 1620, 1585 cm⁻¹; (C) red-brown crystals, m.p. 293°, $\lambda_{max.}$ (EtOH) 272, 339, 355, 455, 523, 563 nm (log ϵ 4·49, 3·67, 3·62, 4·26, 4·05, 4·13), $\nu_{max.}$ (KBr) 3400, 1617, 1585 cm⁻¹; (D) orange crystals, m.p. 159–161°, $\lambda_{max.}$ (EtOH) 224, 253, 267, 348, 463, 526, 560 nm (log ϵ 4·70, 4·50, 4·46, 3·66, 4·28, 4·05, 3·15), $\nu_{max.}$ (KBr) 3400, 1634, 1620, 925 cm⁻¹.

A number of moulds of the genus *Elsinoë* (and the conidial stage *Sphaceloma*) produce a mixture of red photodynamic[5, 9] pigments, four of which have been identified by joint Dutch-American investigations.[9] Elsinochrome A contains four methoxyl groups, and forms a diacetate, and a leucotetra-acetate having the characteristic ultraviolet-visible absorption of a perylene derivative. The corresponding spectrum of the parent pigment shows a striking resemblance to that of erythroaphin (see p. 603) and 4,9-dihydroxyperylene-3,10-quinone. The presence of this chromophore is confirmed by a band at 1623 cm⁻¹ in the infrared

(ν_{HO} absent), the green colour in alkaline solution (reversibly changed to pink with dithionite), and by the formation of purple complexes with boro-acetate and stannic chloride, and of mellitic acid on oxidation with nitric acid. Zinc dust fusion afforded[7] an alkylated 1,12-benzperylene (identified by u.v.) while a carbonyl peak at 1715 cm^{-1}, together with a positive iodoform reaction, suggested the presence of a non-conjugated acetyl group(s).

The n.m.r. spectrum of elsinochrome A is very simple and consists of six singlets at τ 7·95 (COCH_3), 5·92 (OCH_3), 5·68 (OCH_3), 4·80 ,3·37 (ArH), and −6·20 (peri-OH). It is evident that the molecule is symmetrical, and the integration curve showed that two hydroxyl groups, two pairs of methoxyls, two aromatic protons and two C-acetyl groups must be present. This accounts for six of the eight available nuclear positions in the dihydroxyperylenequinone chromophore (two "aromatic" protons and four methoxyl groups) leaving only two positions for the remaining four substituents, that is, the two non-conjugated acetyl groups and the two protons which resonate at τ 4·80. These can only be accommodated by the symmetrical structure (10), the signal at τ 4·80 being

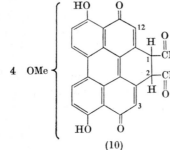

(10)

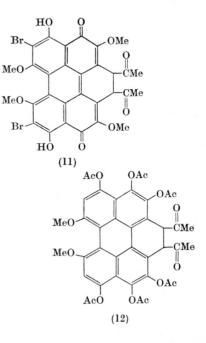

(11)

(12)

attributed to the benzylic protons. Structure (10) is consistent with all the foregoing data and the chemical shift of the aromatic protons at τ 3·37 is in accord with a tautomeric system of the naphthazarin type.

This is borne out by the formation of two dimethyl ethers of elsinochrome A in which the nuclear proton resonances are at τ 3·12 (aromatic) and 3·87 (quinonoid), respectively.

The four methoxyl groups in elsinochrome A can be symmetrically disposed in (10) in three different ways. After some initial uncertainty[7, 101] the orientation was settled by brief bromination of the pigment in acetic acid which gave a nuclear dibromo derivative. Since the hydroxyl and one of the methoxyl signals in the n.m.r. spectrum of A had shifted on bromination, it followed that reaction had occurred in the vicinity of these groups, and this is only possible if the natural quinone has structure (6),[9] the dibromo compound being (11). Additional evidence for the presence of methoxyl groups at C-3 and C-12 was obtained by partial demethylation with aluminium chloride which selectively attacked one pair of methoxyl groups. Reductive acetylation of the crude product gave the hexa-acetate (12), very similar to the leucoacetate of elsinochrome A except for a shift of the benzylic proton signal. This confirms that a pair of methoxyls are located close to the benzylic groups.[9]

Elsinochrome A (6) is optically active and shows a negative Cotton effect. The optical activity could arise from the non-planarity of the chromophore, from the asymmetric centres, or both. The first of these is unlikely since, although the dihydrobenzperylenequinone system itself is non-planar, substituents larger than methoxyl would be required at C-7 and C-8 in order to prevent racemisation (prolonged refluxing of A in chloroform has little effect on the o.r.d. curve). If the alicyclic ring is held in a fixed conformation the two acetyl groups could have either a *cis* or a *trans* relationship. The former would require one of the benzylic protons to be axial and the other equatorial but this is excluded by the n.m.r. spectrum (sharp benzylic singlet in both $CDCl_3$ and C_6D_6). If the acetyl groups are fixed rigidly *trans* the two benzylic protons could be either axial or equatorial. The latter is clearly unlikely as both acetyl groups would then be axial. The alternative is favoured by the observation that complete equilibration in aqueous pyridine results in a loss of only 15% of the optical activity, the original compound being largely unchanged. This agrees with model studies[102] of cyclohexyl methyl ketones which show that the *trans*-isomer with the acetyl group in the equatorial position predominates. However the signals from the ^{13}C satellites of the C-1 and C-2 carbon atoms, which are coupled to the benzylic protons (J_{CH}, 130 Hz), are split into a doublet of only 5 Hz, which means that the *trans*-configuration of the acetyl groups is not rigid. Hence the optical activity of elsinochrome A is attributed to the *trans*-orientation of the acetyl groups which are attached to a rapidly interconverting alicyclic system.[9]

The two dimethyl ethers of elsinochrome A, already mentioned, were assigned[111] structures (i) (yellow, $\lambda_{max.}$ 426 nm; QH τ 3·87) and (ii) (red, $\lambda_{max.}$ 456 nm; ArH τ 3·12) on the basis of solvent shifts ($\Delta_{C_6H_6}^{CDCl_3}$) and NOE studies. For the red isomer irradiation of the methoxyl resonances

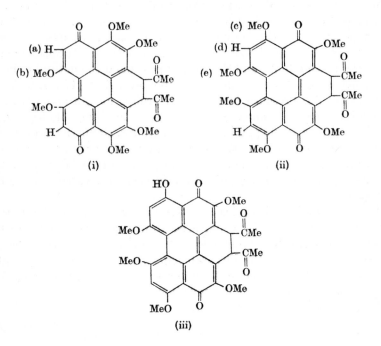

(c) and (e) (both of which undergo a solvent shift in benzene) increased the intensity (17%) of the signal from proton (d), while for the yellow isomer only irradiation at the methoxyl (b) resonance (the only one which undergoes a solvent shift in benzene) intensifies (21%) the signal from proton (a). Less prolonged methylation of elsinochrome A affords two orange monomethyl ethers, one of which was obtained pure and is regarded as (iii).[111] The n.m.r. spectrum showed signals for two aromatic protons at τ 3·10 and 3·20, the *peri*-hydroxyl proton resonance appearing far downfield at τ −5·3. On further methylation (iii) gave (ii).

Elsinochromes B and C are more polar than A, but have the same ultraviolet-visible spectrum. However the ketonic carbonyl band in the infrared spectrum of B is much weaker than that in A, and is absent from the spectrum of C. Similarly, the methyl ketone resonance is missing from the n.m.r. spectrum of C and is replaced by a doublet (6H) centred at τ 8·79. There is also a hydroxyl signal (2H) at τ 8·30, and multiplets at

τ 6·1 and 6·3 replace the sharp benzylic singlets of A. From this it seems highly probable that C has structure (8), and since the n.m.r. spectrum of elsinochrome B is a combination of those of A and C, B is evidently (7). The conversion of C into A and B by oxidation with chromic acid, confirmed these conclusions.[9]

Elsinochrome D[105] is similar to C and the leuco-acetates of both have the same ultraviolet spectrum. However somewhat different chromophoric systems are indicated by the broad carbonyl absorption of D with maxima at 1634 (free) and 1620 cm^{-1} (chelated), the presence of only one *peri*-hydroxyl group (τ −6·0), and by the brownish-green solution in alkali which contrasts with the typical emerald green displayed by its congeners. Further differences are the presence of three methoxyl groups and a methylenedioxy group [ν_{max}. 925 cm^{-1}, τ 3·70 (2H, slightly split)], and two nuclear protons which resonate at relatively high field (τ 3·77 and 3·90) showing that they are attached to a quinonoid system. In benzene solution two of the methoxyl signals are displaced revealing two multiplets (each 2H) centred at the same positions as those in the n.m.r. spectrum of elsinochrome C. This means that the central rings in C and D are the same, and the presence of a broad hydroxyl signal at τ 8·0 and doublets (3H) in the methyl region confirms that they have the same side chains. Lousberg, Salemink and Weiss[105] conclude that elsinochrome D is represented by structure (9). Methylenedioxy groups are extremely rare among naturally occurring quinones and this is only the third example (see pp. 156 and 278).

Cladochrome A (13), $C_{38}H_{38}O_{12} \cdot 2H_2O$

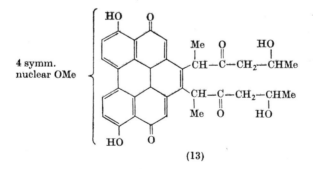

(13)

Occurrence: *Cladosporium cucumerinum* Ell. & Arth. parasitic on cucumber seedlings.[10]

Physical properties: red crystals, m.p. 197–199°, λ_{max}. (EtOH) 270, 335–340, 478, 540(sh), 585 nm, ν_{max}. (KBr) 3440, 1724, 1608 cm^{-1}.

When cucumber seedlings are attacked by *C. cucumerinum* greyish spots develop after a few days but when etiolated seedlings are infected, a wine-red discolouration results. From the investigations of Overeem *et al.*[10] it is fairly certain that the interaction of both organisms is required to produce this pigmentation, and it is most conveniently induced by spraying seedlings growing in the dark with a spore suspension of *C. cucumerinum*. The red material, containing at least five pigments, can then be extracted with solvent, cladochrome A being the main component.

The molecular formula, $C_{38}H_{42}O_{14}$, includes four methoxyl groups. The electronic absorption spectrum, together with the deep green colour in alkaline solution provide evidence for a chromophore of the 4,9-dihydroxy-3,10-perylenequinone type, and this accounts for the chelated carbonyl absorption at 1608 cm^{-1}. There is also infrared absorption due to aliphatic hydroxyl and carbonyl groups. Methylation with methyl iodide-silver oxide-chloroform gave a series of products (only one obtained crystalline). Two red compounds, evidently monomethyl ethers, were formed initially and then replaced by several others on continued methylation; of these two red (λ_{max}. ca. 470 nm) and two yellow (λ_{max}. ca. 445 nm) compounds were isolated. Both phenolic groups in all four compounds were methylated, and in the red and yellow derivative of highest R_F value the alcoholic groups were also methylated (infrared). A close resemblance to the elsinochromes was soon recognized since, by coincidence, the Utrecht laboratory where this work was carried out[10] is immediately adjacent to Salemink's laboratory where the elsinochromes were under investigation.

Assuming an elsinochrome type of polycyclic structure, the n.m.r. spectrum of cladochrome A showed, in addition to an "aromatic" proton singlet at τ 3·41 (2H), that the four methoxyl groups were symmetrically disposed (6H singlets at τ 5·75 and 5·95), and that two identical side chains were also present. Spin decoupling experiments established the existence in each chain of two secondary methyl groups, one lying between an aromatic ring and a carbonyl group (6H doublet at τ 9·35), the other (6H doublet at τ 7·7) being linked to a carbon atom bearing a hydroxyl group, and a tertiary proton (1H multiplet centred at τ 5·17) which is spin coupled to a methylene group adjacent to a carbonyl group. These assignments are consistent with side chains of structure –CH(Me)COCH$_2$CH(OH)Me. Overeem and his co-workers[10] tentatively suggest that these are located as shown in (13) but, as they point out, further work is necessary to establish the structure of this pigment.

Cercosporin, $C_{29}H_{26}O_{10}$

Occurrence: cultures of *Cercospora kikuchii* (M. & T.) Hara.[103]

Physical properties: deep red prisms, m.p. 237–241°, $[\alpha]^{20}_{700}$ +470° (CHCl$_3$), $\lambda_{max.}$ (MeOH) 223, 260, 271, 275, 325, 472, 564 nm, $\lambda_{max.}$ (N NaOH) 478, 610, 620, 645 nm, $\nu_{max.}$ (KBr) 3400, 1619, 1585 cm^{-1}.

This pigment, derived from the fungus responsible for the "purple speck" disease of soya beans, is clearly related to those already discussed although the complete structure is not yet known. The presence of a 4,9-dihydroxyperylene-3,10-quinone chromophore is indicated[103, 104] by the ultraviolet-visible absorption, a carbonyl peak at 1619 cm^{-1} (1638 cm^{-1} in the tetra-acetate), the green colour in alkaline solution and the formation of mellitic acid on oxidation with nitric acid. Whereas the ultraviolet-visible spectrum of the leucohexa-acetate is like that of perylene the hydrocarbon product of zinc dust fusion is evidently a benz(*ghi*)-perylene showing that the perylene system in cercosporin is either alkylated or bridged at the C-1, C-12 positions. Two methoxyl, two alcoholic hydroxyl, and at least two >CHMe groups[103, 110] are present but the disposition of these has yet to be determined. The molecular formula has recently been found[110] to be C$_{29}$H$_{26}$O$_{10}$ and not C$_{30}$H$_{28}$O$_{10}$ on which Kuyama's numerous derivatives and degradation products were based. The pigment is optically active, the rotation changing from +47° to −152° on heating. By chromatography[103] of the equilibrium mixture it was possible to isolate isocercosporin, $[\alpha]^{20}_{700}$ −826°.

Other Pigments

In connection with the foregoing perylenequinones of fungal origin it is of interest to mention a so-called green pigment [it is actually reddish-brown but gives a green colour ($\lambda_{max.}$ 430, 450, 570 and 615 nm) in alkaline solution] present in "P type" humic acid found in soil samples from various parts of the world.[46] The pigment seems to be associated with small black sclerotia, probably derived from *Cenococcum graniforme*, and is therefore regarded as a fungal metabolite. It is considered to be some type of perylenequinone (unpublished data) possibly related to a purplish-black crystalline compound isolated from olive-green patches of lateritic and podzolic soils in certain parts of Australia, apparently associated with decomposed roots of *Eucalyptus obliqua*.[47] (7 kg of soil yielded 24 mg of pigment.) This pigment, C$_{20}$H$_4$Cl$_6$O$_5$, [$\lambda_{max.}$ (CHCl$_3$) 225, 400, 550 nm, $\nu_{max.}$ (CHCl$_2$CHCl$_2$) 1620 cm^{-1}] is evidently a hexachloro-polycyclic compound, possibly an extended quinone of the dihydroxy-perylene or dihydroxybinaphthyl type. It gave an intense emerald green colour in sulphuric acid but was unstable in alkaline solution.

THE HYPERICINS

Hypericin (14), $C_{30}H_{16}O_8$

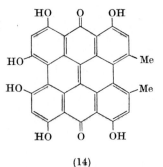

(14)

Occurrence:† in the petals, leaves and stems of numerous *Hypericum* spp. (Guttiferae), mainly in the sections *Euhypericum* and *Campylosporus*,[11, 12] and in the fresh fruit bodies of *Polystictus versicolor* (L. ex Fr.).[19]

Physical properties: dark red needles, dec. >300°, λ_{max}. (MeOH) 508, 548, 590 nm, λ_{max}. (C_5H_5N) 520, 559, 603 nm, ν_{max}. (KBr) 1590, 3450 cm^{-1}.

It has long been known that animals which have eaten *Hypericum* spp. (for example *H. perforatum*, St. John's Wort) are prone to "hypericism", a photogenic disease resulting in inflammation of the skin, oedema, fluctuation in body temperature and in extreme cases may result in death.[13] Hypericism has been observed mainly in sheep, especially in parts of Australia where *Hypericaceae* occur frequently on uncultivated pasture. The details of such "photodynamic effects" are still not clear but essentially they appear to be sensitised photo-oxidations (of protein ?) for which the requirements are light, oxygen and a sensitising substance which is usually a fluorescent pigment with a long-lived triplet state.[14] In *Hypericum* the culprit is the pigment hypericin which is usually visible in the form of black dots or streaks on petals, sepals, leaves and stems. A simple method of locating the pigment in plant tissues by "printing" onto chromatographic paper was devised‡ by Mathis and Ourisson.[11] It is of interest that hypericin appears to act as a marker substance for the beetle *Chrysolina brunsvicensis* which feeds specifically on hypericin-containing *Hypericum* plants.[108]

The structure of hypericin has been elucidated by Brockmann and his group at Göttingen.[15] Although first examined in 1830,[16] pure crystalline hypericin was not obtained until 1942.[17] It was isolated from the flowers

† See also p. 594.
‡ These authors did not distinguish between hypericin and related co-pigments as this was not of taxonomic interest. The method can be used successfully with dried herbarium material.

of *H. perforatum* by methanol extraction and precipitation with hydrochloric acid. Hypericin is sparingly soluble in most organic solvents, usually forming red solutions with a red fluorescence; it is green in aqueous sodium hydroxide and concentrated sulphuric acid, the latter solution also showing intense red fluorescence. The pigment forms a yellow hexabenzoate and an amorphous, blue, leuco-octabenzoate, but treatment with ketene[20] affords only a diacetate. Four of the six hydroxyl groups are therefore chelated to the quinone carbonyls. On zinc dust distillation hypericin gives a very small yield of a red hydrocarbon identified as anthrodianthrene (16).[17] The implication that the natural compound is an anthrodianthrone is however misleading, as the natural compound contains two C-methyl groups. It was then shown[17] that both 2,2'-dimethylhelianthrone (15) and 2,2'-dimethylnaphthodianthrone (17) undergo similar cyclisations to give small yields of (16) on zinc dust distillation, whence it follows that hypericin is either the helianthrone

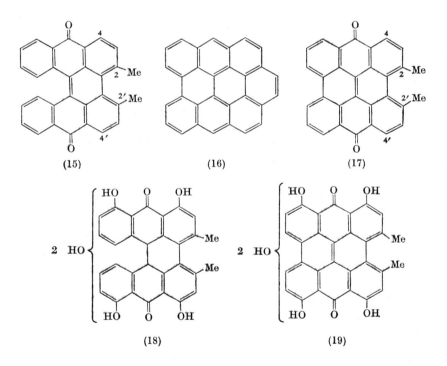

(18) or the naphthodianthrone (19). At that time it was not possible to distinguish with certainty between these two possibilities by any direct method and the problem was solved in the following way.

Brockmann and Budde[18] have demonstrated that the polycyclic structure of a polyhydroxyquinone can be easily recognised by reductive acetylation, as the electronic absorption spectrum of the product is very similar to that of the parent hydrocarbon. Moreover, it is particularly easy to distinguish between helianthrenes and naphthodianthrenes as the former are red and rapidly form yellow peroxides on exposure to light and air, while the latter are blue and are not photo-labile.[21] As vigorous reductive acetylation[22] of hypericin gave, after dehydrogenation with chloranil, a stable blue hexa-acetate† with almost the same light absorption as (20), its structure is clearly (21), and consequently

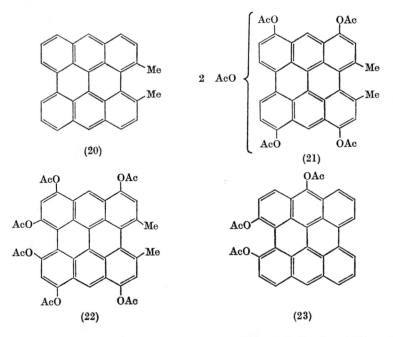

hypericin is (19). Finally structure (21) could be redefined as (22) as the near identity of the visible absorption of the hexa-acetate ($\lambda_{max.}$ 578, 625 nm) and the dimethylnaphthodianthrene (20) ($\lambda_{max.}$ 578, 627 nm) means that the acetoxyl groups in (21) are not exerting their normal bathochromic effect. This can be explained if the β-acetoxyl groups are located as shown in (22) resulting in overcrowding in that part of the molecule and some distortion of the polycyclic system. Similarly, both (20)[23] and (23)[20] absorb at shorter wavelengths than the parent hydrocarbon. As the reductive acetylation product is (22), hypericin itself must have structure (14) which has been confirmed by synthesis.[24]

When the bianthraquinone (24), obtained from 4-bromo-emodin trimethyl ether by an Ullmann reaction, was reduced with copper in

† Under milder conditions a hepta-acetate can be obtained in which only one *meso* oxygen atom has been eliminated.[20]

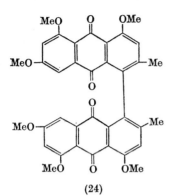

(24)

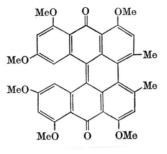

(25)

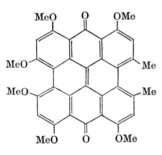

(26)

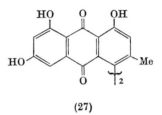

(27)

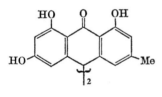

(28)

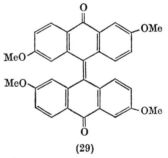

(29)

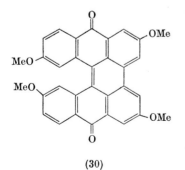

(30)

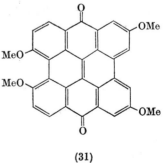

(31)

acetic acid-hydrochloric acid, in the absence of light, it gave the helianthrone (25) which was converted into the naphthodianthrone (26) by irradiation. Demethylation of this compound by heating with potassium iodide in phosphoric acid gave a product identical in all respects with hypericin.[24] Hypericin has also been obtained directly from the bianthraquinone (27) by irradiating a sulphuric acid solution in the presence of copper bronze[27] and by aerial oxidation of the bianthrone (28) in aqueous alkali.[26]

The stepwise conversion of an anthrone or anthraquinone into the highly condensed naphthodianthrone system is a facile transformation and the occurrence of these complex systems in plants is not as surprising as at first appeared. An interesting example employing a less activated starting material is the conversion of 2,6-dimethoxy-9-anthrone into a mixture containing (29), (30) and (31) simply by exposing an alkaline solution to air for several days, or by treatment with methyl iodide-silver oxide.[25]

The exact form in which hypericin occurs *in vivo* is not known. The native pigment is much more soluble than the purified, crystalline material but has the same electronic spectrum. The indications[15, 22] are that hypericin, in the plant, is bound to a solubilising moiety linked to a β-hydroxyl group(s), which is easily detached by acid treatment. Although none of the derivatives of hypericin examined showed optical activity, the bound, native, pigment may nevertheless be optically active, undergoing racemisation during the isolation and purification procedure. In this connection it is of interest that 2,2′-dimethylhelianthrone has been obtained in an optically active form with a half-life in solution of several days.[34]

Protohypericin (32), $C_{30}H_{18}O_8$
"Hyperico-dehydro-dianthrone" (33)

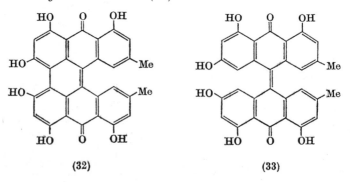

(32)   (33)

The isolation of these two precursors of hypericin from *H. hirsutum* has been briefly reported.[29] They were separated from hypericin by column chromatography of the crude pigment obtained by extraction of the flowers in the absence of light. Protohypericin, characterised as its orange hexabenzoate, gave a red, light-sensitive, helianthrene derivative on reductive acetylation, and could be converted into hypericin by irradiation in solution. It was prepared[28] later by aerial oxidation of emodin-9-anthrone in pyridine-piperidine solution, and since it differed from (25; OH in place of OMe), protohypericin must have structure (32). The quinone (33) was not obtained in crystalline form but the structure follows from its reduction with zinc and acetic acid to emodinanthrone, and its conversion by light into protohypericin and hypericin. The isolation of (33), (32) and also emodinanthrone from the flowers of *H. hirsutum*, and their transformation *in vitro* into hypericin indicates clearly the mode of biogenesis. Emodin bianthrone (28) was not detected but it is interesting that the isomer, penicilliopsin (34), is produced by the mould *Penicilliopsis clavariaeformis*.[30] Oxford and Raistrick[30] had

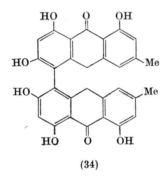

(34)

found that this orange pigment could be oxidized by air to "oxypenicilliopsin" which, on exposure to light, changed into "irradiated oxypenicilliopsin". Brockmann and Eggers[31] confirmed these transformations and identified the products as protohypericin and hypericin, respectively.

Pseudohypericin and cyclopseudohypericin

Of twenty-two *Hypericum* spp. examined by Brockmann and Sanne,[22] *H. hirsutum* contained hypericin alone and fourteen others contained a second photodynamic pigment. This became apparent when it was noted that the visible absorption of these extracts, in sulphuric acid solution, was shifted to shorter wavelengths (15–18 nm) on exposure to light, a phenomenon first observed by Pace and Mackinney.[32] The two pigments were isolated from *H. perforatum* flowers and separated by preparative ring paper chromatography,[33] and further investigation revealed that of the fourteen species mentioned, four, *H. montanum, H. inodorum, H. lobocarpum* and *H. barbatum*, contained only the new pigment, pseudo-hypericin. The compound was obtained as dark red needles having virtually the same ultraviolet-visible absorption as hypericin, and on reductive acetylation it gave a blue naphthodianthrene derivative with the same light absorption as (22). It appeared to have two C-methyl and eight hydroxyl groups, whereas the irradiation product, which was isolated, seemed to contain two C-methyl groups but only six hydroxyl groups, and on reductive acetylation it gave a red compound whose light absorption was similar to that of anthrodianthrene (16). Accordingly it must be a derivative of anthrodianthrone and pseudohypericin must be a naphthodianthrone with substituents at the 2,2'-positions capable of cyclisation. On the basis of a $C_{32}H_{20}O_{10}$ molecular formula, and the positive iodoform test given by pseudohypericin, Brockmann and Pampus[33] proposed structure (35) for this pigment, and (36) for the photo-product, named cyclopseudohypericin. As the latter has been

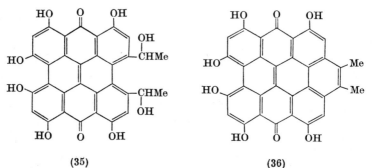

(35)   (36)

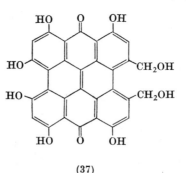

(37)

isolated from several *H. perforatum* extractions carried out in the dark, it is also a natural compound. However, it has been found recently that pseudohypericin is a C_{30} compound (possibly a mixture), and these structures are currently under revision.[107] Unpublished observations[15] suggest that *H. perforatum* contains yet another red pigment having structure (37), designated demethylcyclopseudohypericin.

Fagopyrin (38), $C_{42}H_{36}N_2O_{10}$
Protofagopyrin (39).

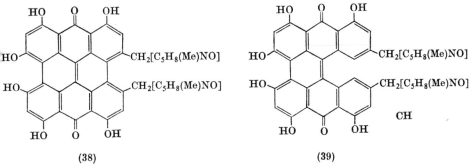

(38)   (39)

"Fagopyrism" is a photogenic disease similar to "hypericism" which is caused by the red, red-fluorescent, pigments in the flowers[35] of buckwheat (*Fagopyrum esculentum* Moench, Polygonaceae). It is not of economic importance as buckwheat is no longer fed to animals. Two of these pigments have been investigated by Brockmann and his co-workers[36] for which they cultivated half an acre of *F. esculentum*. After extensive counter current and chromatographic separation fagopyrin was eventually crystallised from phenol in dark red leaflets, $C_{42}H_{36}N_2O_{10}$. It shows the same electronic absorption as hypericin in pyridine solution, and the same green colour in sulphuric acid and methanolic alkali, and yields a hexa-*p*-nitrobenzoate. On reductive acetylation it gave a blue naphthodianthrene derivative while heating with pyridine hydrochloride afforded hypericin. Fagopyrin is clearly a hypericin derivative in which the unidentified portion, $C_{12}H_{22}N_2O_2$, seems to be linked to the β-hydroxyl groups. This moiety could also be detected by heating with potassium hydroxide in aqueous glycol but was not affected by potassium iodide in phosphoric acid. Curiously, on treatment with bromine it afforded a nitrogen-free brominated product identical with that obtained by bromination of hypericin under the same conditions. An ether linkage seems unlikely, therefore, and Brockmann[15] formulates fagopyrin as (38) with the unknown fragments attached to benzylic carbon atoms. Little is known of the nitrogeneous moieties; two C-methyl groups are present and after methylation and vigorous hydrolysis, trimethylamine is liberated.

The second pigment, protofagopyrin, has not been obtained in crystalline form, but since it gives a red, photo-labile, helianthrene derivative on reductive acetylation it is probably (39), analogous to protohypericin.[15]

Protozoan pigments. Several blue-green, brown and purple pigments occur in heterotrich ciliates,[37] some of which are fluorescent and exert a photodynamic effect, and show a general resemblance to hypericin. These include stentorin[38] (*Stentor coeruleus* Ehr.), stentorol[39] (*Stentor niger* Ehr.) and zoopurpurin[40, 41] (*Blepharisma* spp.). Stentorin is very similar to hypericin in both fluorescence and ultraviolet-visible absorption spectra, and in the spectral shifts effected by several reagents including boroacetic anhydride.[41] The two major pigments (= zoopurpurin ?) of *Blepharisma undulans*[112] are also spectroscopically very similar to hypericin and probably possess a tetra-α-hydroxynaphthodianthrone system. They have the same infrared spectra, which is like that of hypericin, but shows a strong carbonyl peak at 1740 cm^{-1}. Virtually no chemical work has been done but stentorol has been isolated[39] as an amorphous, almost black powder, [$\nu_{max.}$ (KBr) 3350, 1602 cm^{-1}], giving red solutions ($\lambda_{max.}$ 475, 520, 555 nm in CHCl$_3$) showing a red fluorescence in ultraviolet light. Addition of boro-acetate gave a green solution with a red fluorescence. Stentorol was soluble in aqueous sodium bicarbonate, and could be acetylated and benzoylated, but although it appeared to undergo reductive acetylation giving a yellow-orange product with a blue fluorescence, it was not decolourised either by dithionite or on catalytic hydrogenation.

Fringelites. Blumer[42] has discovered a remarkable series of polyhydroxy-polycyclic quinones and related hydrocarbons in a Jurassic crinoid (*Apiocrinus* sp.) found near Fringeli in Switzerland. (The anthraquinones in living crinoids were described in Chapter 5.) The lower fossil stalk and root system were intense violet and, after removing inorganic carbonates with hydrochloric acid, the hydrocarbons and less polar pigments were extracted with chloroform, the more polar material being obtained by exhaustive extraction with pyridine, and final dissolution in concentrated sulphuric acid. All the pigments give red solutions in organic solvents, and green solutions in alkali and in sulphuric acid, and can be acetylated and benzoylated. In acetic anhydride, containing sulphuric acid, they give a blue(-violet) colour reminiscent of erythroaphin derivatives. Comparison of the visible absorption spectra with those of various model compounds indicates a basic tetra-α-hydroxynaphthodianthrone structure (40) (which may be identical with fringelite H), the presence of additional β-hydroxyl groups being suggested by partial acetylation with ketene and stepwise reaction with acetic anhydride. Fringelite D, in fact, is identical with hypericin.[66]

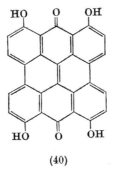

(40)

Reductive acetylation of fringelites D, E and F, and dehydrogenation with chloranil, following Brockmann's procedure, gave blue naphthodianthrene derivatives which absorbed at shorter wavelengths than the parent hydrocarbon. The normal bathochromic effect of the acetoxyl groups is therefore offset by distortion of the ring system arising from overcrowding at the 2,2'- and 7,7'-positions, the hypsochromic shifts relative to naphthodianthrene indicating a decrease in steric hindrance

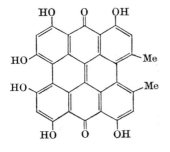

D

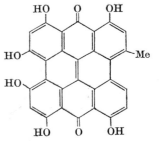

E

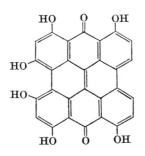

F

on passing from D to E to F. From the mass spectra of their trimethyl-silyl ethers, fringelites E and F possess, respectively, one and two methyl groups less than hypericin and the structures shown on p. 595 have been assigned accordingly.[43, 66] It should be emphasised, however, that these structural proposals are based entirely on spectroscopic data and are still considered by Blumer to be tentative.

Reductive acetylation under relatively mild conditions leads to reduction of the two right-hand benzenoid rings giving products having an electronic spectrum like that of 1,12-benzperylene, while under more vigorous conditions the left-hand rings of fringelites D and E (but not F) are also reduced giving derivatives whose spectra are consistent with a highly substituted phenanthrene structure. It is suggested that the ease of hydrogenation is related to the steric strain in the leuco-acetates formed initially. In the case of fringelite F this is relieved by formation of a 1,12-benzoperylene, but at this level of reduction there is still some overcrowding in the products from fringelites D and E which is relieved by further hydrogenation.[43]

Several additional related pigments occur in the fossil whose structures are not yet known, but zinc dust fusions of pigment mixtures gave a little anthrodianthrene suggesting the presence of further pigments alkylated in the 2,2'-positions. The fossil also contains a series of polycyclic hydrocarbons which were resolved chromatographically, ranging from anthracene to the hexahydroanthrodianthrene (42), the most abundant

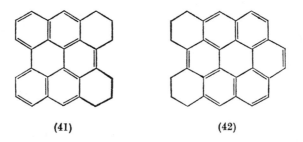

(41)   (42)

being the hexahydronaphthodianthrene (41) which has a 1,12-benzperylene-type spectrum.[44] There seems little doubt that these large hydrocarbons were formed from the pigments by a slow geochemical process. Elimination of oxygen is known[45] to occur in reducing sedimentary environments whereas the reverse process, oxidation of the hydrocarbons to the quinones, which would involve introduction of oxygen at sterically hindered positions is very unlikely at the low redox potential prevailing in the fossil.

THE APHINS

Protoaphin-fb† (43), $C_{36}H_{38}O_{16}$
Protoaphin-sl†ₗ(44), $C_{36}H_{38}O_{16}$

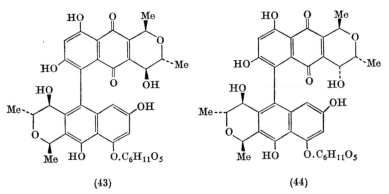

(43)   (44)

Occurrence: in the following aphids (Aphididae),[59, 60] *Aphis fabae* Scop., *A. rumicis* L., *A. philadelphi* Börner, *A. viburni* Scop., *A. evonymi* Fabr. (= *A. cognatella*), *A. corniella* H.R.L., *A. hederae* Kalt., *A. sambuci* L., *A. ilicis* Kalt., *A. farinosa* Gmelin, *Brachycaudus klugkisti* Börner, *B. rociadae* Cockerell, *Eriosoma lanigerum* Hausmann, *E. ulmi* L., *Myzus cerasi* F., *Rhopalosiphum nymphaeae* L., *Dysaphis* (formerly *Sappaphis*) *pyri* Fonsc., *Schizolachnus pineti* Fabr. (= *S. tomentosis*), *Tuberolachnus salignus* Gmelin, and in *Pineus strobi* Hartig (= *Adelges strobi*) (Adelgidae), *P. pini* (Gmelin)[98] and other adelgid species.[98] The *sl* isomer (44) has been isolated only from *Tuberolachnus salignus*, most species appear to contain the *fb* isomer (43), but adelgids possess *sl*, and both *fb* and *sl* isomers are present in *Dysaphis pyri*.[98]

Physical properties: *fb* (43) yellow prisms, $\lambda_{max.}$ (70% EtOH) 223, 273, 296, 310, 343, 357, 449 nm (log ε 4·82, 4·19, 4·06, 3·97, 3·73, 3·78, 3·69), $\nu_{max.}$ (Nujol) 3430, 1673, 1645, 1613, 1590 cm⁻¹.

sl (44) yellow microcrystalline powder, $\lambda_{max.}$ (75% EtOH) 223, 274, 297, 311, 343, 356, 450 nm (log ε 4·73, 4·14, 3·98, 3·91, 3·67, 3·72, 3·64), infl. 234 nm, $\nu_{max.}$ (Nujol) 3540, 3440, 1651, 1615, 1593 cm⁻¹.

As the protoaphins are naphthaquinones, logically these pigments should be discussed in Chapter 4. However, as they are easily transformed into extended quinones with which their chemistry is closely connected, it is convenient to consider the whole of the aphin group here. Our knowledge of these remarkable compounds is entirely due to the extensive investigations of Lord Todd and his colleagues during the last twenty years.[60, 61] They occur in at least twenty species of aphids (and are absent from many others), frequently in relatively large amounts (<3–4% live weight). Prior to the Cambridge researches, Blount[62] had isolated two pigments from aphids, including lanigerin‡ from the woolly

† The suffix is indicative of the species from which the pigment was first isolated; *fb* from *A. fabae*, *sl* from *T. salignus*.
‡ Now chrysoaphin-*fb*. The other pigment, strobinin, isolated from *Pineus strobi*, was probably erythroaphin-*sl*.

aphis (*Eriosoma lanigerum*) which was presumably the species described in 1871 by Sorby[63] as "red Aphides" found "in downy patches" on apple bark. By careful study of these and other aphids, he showed that the native pigment could undergo a number of transformations in solution giving rise to a series of related pigments, and this has been confirmed by recent work.

A protoaphin is present in the haemolymph of living insects accompanied by an enzyme which, after death, rapidly converts it into an unstable, yellow, fluorescent, xanthoaphin, which changes on keeping (or more rapidly in the presence of acid or alkali) into a slightly more stable, orange, chrysoaphin, and finally into a red, red-fluorescent, erythroaphin, the stable end-product. There are two series of stereoisomeric pigments and their inter-relationships are set out below.

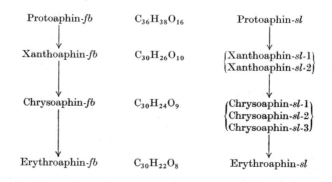

Protoaphin-*fb*	$C_{36}H_{38}O_{16}$	Protoaphin-*sl*
↓		↓
Xanthoaphin-*fb*	$C_{30}H_{26}O_{10}$	{Xanthoaphin-*sl*-1} {Xanthoaphin-*sl*-2}
↓		↓
Chrysoaphin-*fb*	$C_{30}H_{24}O_{9}$	{Chrysoaphin-*sl*-1} {Chrysoaphin-*sl*-2} {Chrysoaphin-*sl*-3}
↓		↓
Erythroaphin-*fb*	$C_{30}H_{22}O_{8}$	Erythroaphin-*sl*

To isolate protoaphins, the insects are washed off the host plant with water at 70°. This inactivates the enzymes responsible for xanthoaphin formation without doing appreciable damage to the protoaphins.[57, 58] Considerable amounts of material can be collected in seasons when infestation is heavy and in one summer more than 50 g of aphin pigment were isolated from an estimated 50 million *A. fabae* ("black fly") feeding on cultivated broad beans. Structural investigations have been carried out mainly on aphins of the *fb*-series obtained most conveniently from the bean aphid, the *sl*-pigments being isolated from the relatively large brown willow aphid, *T. salignus*.

The protoaphins[57, 58] are brownish-yellow, hygroscopic, acidic (pK_a 6·2) pigments, appreciably ionised at biological pH forming violet-red anions which are chiefly responsible for the dark colour of the insect haemolymph. A deep violet colour results on addition of sodium hydroxide suggesting the presence of additional phenolic groups. The protoaphins show redox properties, slowly decompose in solution, and yield deca-acetates and leucododeca-acetates; the latter show naphthalenic light absorption. Mild acid hydrolysis affords D-glucose and an intractable aglycone which can only be isolated under nitrogen and is convertible, by acid treatment, into the corresponding erythroaphin.

Important results were obtained by more extensive reduction, either with neutral dithionite[58] or by catalytic hydrogenolysis (uptake of 2 mols. of hydrogen) in aqueous buffer at pH 6·6.[64] After aerial re-oxidation, two products were obtained from protoaphin-fb, namely an acidic (pK_a 6·5), orange quinone (quinone A, $C_{15}H_{14}O_6$) and a colourless, glucoside (glucoside B, $C_{21}H_{26}O_{10}$) which was readily hydrolysed to D-glucose and an aglucone although the latter could not be isolated. Comparison of molecular formulae shows that the following reaction has occurred:

$$\text{protoaphin-}fb + \text{H}_2 \rightarrow \text{quinone A} + \text{glucoside B}.$$

The combined properties of the two fragments parallel those of the protoaphins and summation of their absorption spectra gives a curve almost identical with that of the original pigment. It is clear that the protoaphins consist of two parts which are effectively insulated from each other.

From its spectroscopic properties (ultraviolet-visible, infrared and n.m.r.) quinone A is undoubtedly a 5,7-dihydroxy-1,4-naphthaquinone and accordingly it afforded a monomethyl ether (diazomethane), a dimethyl ether (methyl iodide-silver oxide) and a leucotetramethyl ether. The dimethyl ether gave a monoacetate. Ten carbon atoms are now accounted for and the structure of the remainder of this molecule was largely settled by a mild chromic acid oxidation[58] which yielded D,D-(+)-dilactic acid (45) of known[65] absolute stereochemistry. Quinone A was therefore formulated[58] as (46; R = H) which takes into account the relationship of the protoaphins to the erythroaphins (see later) and a detailed analysis of the n.m.r. spectrum.[71] The alcoholic function is necessarily located as

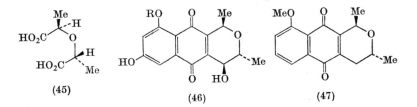

(45)   (46)   (47)

shown, the stereochemistry at that centre being determined by the coupling constant (J, 8·0 Hz) of the relevant protons which must be *trans* diaxial. This group can be removed by reduction with sodium stannite[91] to give a compound having the same dihydropyran structure as iso-eleutherin (p. 282) which has, in fact, the same stereochemistry (47) as quinone A. By dissolution in sulphuric acid quinone A can be converted into the red anhydro-quinone (48). It is worth noting that hypoiodite

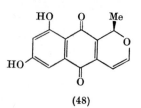

(48)

oxidation of quinone A yields iodoform (0·4 mol.) despite the absence of a MeCH(OH)– group. This must arise via oxidative breakdown of the quinone as no iodoform can be obtained from glucoside B (50) in which the heterocyclic system is fused to a benzenoid ring.[58]

Spectroscopically, glucoside B appeared to be a hydroxylated naphthalene and on oxidation with Fremy's salt (in contrast to other oxidising agents) it was smoothly converted into two naphthaquinones. One of these was a glucoside, easily hydrolysed to quinone A, and since it contained an acidic phenolic group it must be (46; R = D-glucosyl). The other quinone is still more acidic (pK_a 3·8) and gives a boro-acetate, and since its ultraviolet-visible absorption is virtually identical with that of 2,5-dihydroxy-1,4-naphthaquinone it is regarded as (49). It follows that

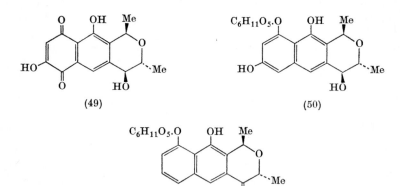

(49)   (50)

(50a)

glucoside B must be (50). The β-configuration of the sugar moiety is evident from the observation that hydrolysis is almost complete on treatment with almond emulsin whereas α-glycosidases are without effect.[67] Glucoside B has been found recently in the bright orange aphid *Aphis nerii* (and other species) along with related compounds, principally the yellow pigment, neriaphin (50a).[109]

Knowing the structure of quinone A (46; R = H) and glucoside B (50), that of protoaphin-*fb* can now be constructed bearing in mind that it gives rise to (46; R = H) and (50) on hydrogenolytic fission. Of the various ways in which these might be combined only one (43) could explain its transformation into erythroaphin (53; R = R' = H).[58] It can

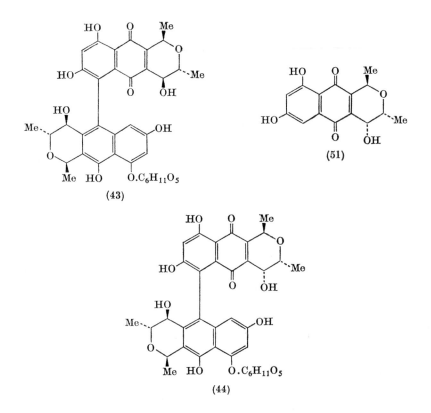

(43)

(51)

(44)

now be seen that the reductive cleavage is analogous to the fission of skyrin and other bianthraquinones although it had not been observed previously in the binaphthyl series. Clearly it is impossible for the two halves of (43) to achieve coplanarity and its behaviour as two separate chromophoric entities accounts for the spectral properties. The ultraviolet absorption of the deca-acetate is essentially a summation of the spectra of a naphthalene and a naphthaquinone. The acidic group in the protoaphin molecule is the β-hydroxyl group in the naphthaquinone moiety, and when it dissociates to the mono-anion only those peaks attributable to the naphthaquinone segment undergo a bathochromic shift.

Reductive fission of protoaphin-*sl* also gives glucoside B together with another quinone (A′) isomeric with and very similar to quinone A. As the A′ quinone also gives D,D-(+)-lactic acid (45) on oxidative degradation the two quinones differ only in their configuration at the carbon atom bearing the alcoholic group. Quinone A′ is consequently (51), the *cis*-configuration of the relevant protons being confirmed by their coupling constant (J,2 Hz). Both quinone A and A′ epimerise to a mixture of both on treatment with sodium hydroxide in the absence of air.[91] Structure (44) can therefore be assigned to protoaphin-*sl*, and it differs from the *fb*-isomer only at one asymmetric centre.[58]

It seems highly likely that the formation of the protoaphins *in vivo* involves a coupling reaction between the two halves of the molecules at some stage. This can be done *in vitro* with remarkable ease.[64] When quinone A and glucoside B are left at room temperature in aqueous solution at pH 6·6 a virtually quantitative reaction occurs giving a mixture of mainly (52) and a much smaller amount (2%) of protoaphin-*fb* (43); at 80° the yield of (43) increases to 18%. A partial synthesis of protoaphin-*sl* has also been effected by a similar coupling of quinone A′ and glucoside B.[64]

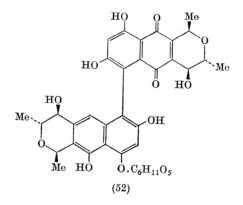

(52)

The most intriguing aspect of the chemistry of the protoaphins, namely their transformation into xanthoaphins, is discussed later (p. 612).

ERYTHROAPHINS

It is convenient, at this point, to consider the erythroaphins before discussing the intermediate xanthoaphins and chrysoaphins. The erythroaphins, being the most stable and most readily isolated aphin pigments, were the first to be examined in detail.[68, 69] They were found to be appreciably basic and advantage was taken of this in purification as they are easily extracted from chloroform with 75% sulphuric acid and, after dilution to 60%, taken back into the organic phase. The erythroaphins, $C_{30}H_{22}O_8$ [λ_{max}. (CHCl$_3$) 421, 447, 485, 520, 560, 586 nm, ν_{max}. (Nujol) 1625, 1575 cm^{-1}] are crystalline carmine-red compounds which fluoresce in solution. They show redox properties, form green alkaline solutions, and yield diacetates and leucotetra-acetates. Comparison of ultraviolet-visible spectra shows indubitably that the erythro-aphins are derivatives of 4,9-dihydroxy-3,10-perylenequinone (1); the spectra of the diacetates are almost identical with that of perylene-3,10-quinone. This accounts for twenty of the carbon atoms and four of the oxygen atoms, and explains the origin of the mellitic acid which arises on oxidation with nitric acid. The formation of 1,12-benzperylene and coronene derivatives, in addition to alkylated perylenes, on zinc dust fusion shows that alkyl groups must be attached to at least two of the "waist" positions of the chromophore. The non-aromatic part of the molecule contains four oxygen atoms and ten carbon atoms, and from the relative simplicity of the n.m.r. spectra of erythroaphin-fb derivatives the protons must be arranged symmetrically with respect to the perylene nucleus. Thus the central chromophore is flanked by a system containing five carbon atoms (including two C-methyl groups) and two oxygen atoms (which must be ethereal), and this gives rise to appreciable amounts of acetaldehyde on heating with sulphuric acid.[70] As larger fragments could not be isolated the structure of the non-aromatic part of the erythroaphins was difficult to determine until n.m.r. spectroscopy became available.

In the spectrum[71] of the leucotetramethyl ether of erythroaphin-fb the signals from the C-methyl groups appear as pairs of overlapping doublets coupled, in one case, to a quartet and in the other to a multiplet arising from a CH$_3$–CH$_A$–CH$_B$– group (J$_{AB}$ ~10 Hz). Taking all these facts into consideration, together with the derivation of the erythroaphins from the protoaphins, Lord Todd and his colleagues[70] came to the conclusion that erythroaphin-fb should be formulated as (53; R = R' = H), and the sl-isomer, which differs only at one asymmetric centre (but has a more complex n.m.r. spectrum), as (54). The structures were supported later by mass spectral studies[85] but these did not distinguish stereochemical differences. A partial synthesis[86] of erythroaphin-fb (and hence of -sl by epimerisation) was effected by oxidation of glucoside B (50) with neutral ferricyanide to give (55), readily hydrolysed by acid to erythroaphin-fb.

Structures (53) and (54) are written as perylene-3,9-quinones but the precise tautomeric structure is not clear. The n.m.r. spectrum of erythroaphin-fb shows only a single peak at τ 3·39 from the two "aromatic" protons (cf. elsinochrome A) which is consistent with structure

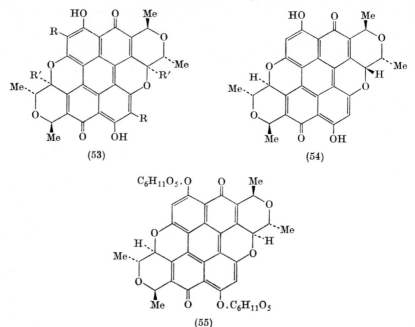

(53; R = R′ = H) and may be attributed to an averaging of quinonoid and benzenoid protons in the rapidly tautomerising system. On the other hand the electronic spectrum is more consistent with the 3,10-perylenequinone tautomer (56a; R = H). There is a marked difference between the visible absorption of perylene-3,9- and 3,10-quinones and a corresponding distinction between the two dimethyl ethers of ery-throaphin-*fb*, (56b) being dark purple while (56a; R = Me) is orange, the absorption of the latter being similar to that of the parent erythroaphin and notably lacking the intense absorption above 600 nm shown by (56b). The n.m.r. spectrum of (56a; R = Me) shows two singlets at low field (τ 3·11 and 3·69) representing the benzenoid and quinonoid protons, and two signals near τ 6 arising from two different methoxyl groups; on the other hand the spectrum of (56b) shows only one aromatic singlet (2H) at τ 3·15 and one methoxyl singlet (6H) which is consistent with the structure assigned. Methylation of 4,9-dihydroxyperylene-3,10-quinone gives only a 3,10-quinone derivative, and the formation of both (56a; R = Me) and (56b) on methylation of erythroaphin-*fb* (using methyl iodide-silver oxide-chloroform *in the dark*[99]) may be attributed to the unhindered situation of both methoxyl groups in (56b) whereas in (56a) one of these is in a crowded environment. The same argument explains the non-appearance of the isomer (56c) in the methylation reaction. Acetylation gives only (56a; R = Ac), so presumably any diacetate

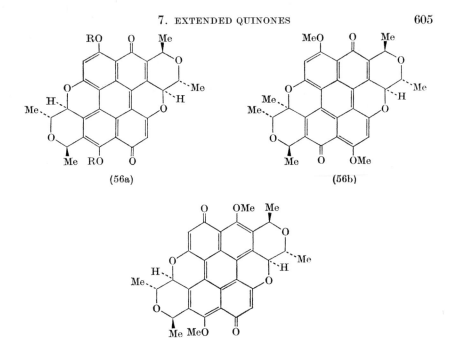

(56a)   (56b)

(56c)

corresponding to (56b) which is formed, rearranges to the more stable isomer by migration of the acetyl groups to the neighbouring *peri* oxygen functions.

During the extensive studies of the chemistry of the erythroaphins several unexpected features came to light. Whereas halogenation [70, 72] proceeds normally to give (53; R = Hal, R′ = H), amination [72] occurs not on the perylenequinone nucleus but at the central benzylic positions to give remarkably stable amino hemi-ketals (53; R = H, R′ = NR$_2''$).[70] The presence of the two "aromatic" protons and the absence of the benzylic protons (R′ in 53) in these diamino derivatives was established by the n.m.r. spectra. The complete absence of nuclear amination products is surprising (cf. 4,9-dihydroxyperylene-3,10-quinone[4]) and was further exemplified by the amination of dibromoerythroaphins which gave diaminodibromo derivatives. These reactions are considered to be nucleophilic additions to a tautomeric form of erythroaphin, such as (57a), to give an aminoquinol (57b; R′ = H) which is reoxidised in air (absorption of 1 mol. oxygen) to an aminoquinone, and the sequence is then repeated. In the amination of erythroaphin-*sl* derivatives epimerisation occurs quantitatively, the products being the more stable diamino-*fb* derivatives. This is understandable if the reaction proceeds by way of tautomers such as (57a) in which the relevant benzyl carbon has lost its asymmetry. Although side chain amination was first observed in these

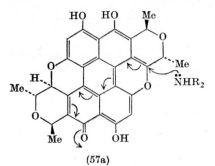

(57a)

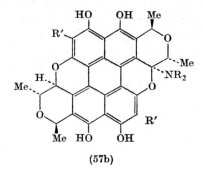

(57b)

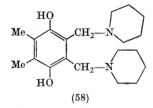

(58)

complex quinones it appears to be general and can be effected readily
with simple compounds, for example, duroquinone reacts with piperidine
at room temperature to give the diaminoquinol (58), and ultimately tri-
and tetra-piperidino derivatives.[73] In the absence of air dibromo-
erythroaphin-*fb* reacts with piperidine to give, surprisingly, the di-
piperidino-erythroaphin-*fb* (53; R = H, R′ = piperidino).[70] In this case
the aminoquinol (57b; R′ = Br), which in air oxidises to the quinone, is
considered to react in a tautomeric keto form (59) which undergoes a
base-catalysed elimination of hydrogen bromide; repetition of these

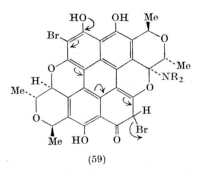

(59)

steps leads finally to (53; R = H, R′ = piperidino). The facile reductive
elimination of bromine from dibromoerythroaphins may also proceed by

elimination of hydrogen halide from keto tautomers of the leuco derivatives, for which analogies exist.[74]

Side-chain acetoxylation[70, 72] can also be effected under Thiele conditions, the reaction proceeding by way of intensely fluorescent intermediates which provide a useful colour test for erythroaphins. (The reaction is not applicable to simpler compounds such as duroquinone.) In this reaction, first observed by Blount,[62] erythroaphin is treated with acetic anhydride containing a drop of sulphuric acid to give a bright blue solution with an intense red fluorescence, attributed[99] to the oxonium cation (61) which probably arises, as indicated (60) from the quinone-methide tautomer of the diacetate. The final formation of the penta-acetate (62) occurs during work up by nucleophilic addition of acetate anion. Analogous intensely coloured green oxonium ions similar to (61) can be formed from erythroaphin derivatives under various conditions.[99] Hydrolysis and oxidation of the penta-acetate (62) leads via (57b; R′ = H, HO in place of NR₂) to hydroxyerythroaphin. Hydroxylation

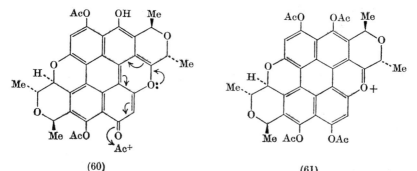

(60)   (61)

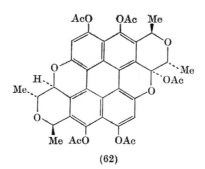

(62)

at the central benzylic positions can also be effected by periodate oxidation of the erythroaphins,[75, 76] and like the corresponding amino derivatives, reduction to the parent erythroaphin is easily effected.

It was noted above that isomerisation of *sl*- to *fb*-epimers occurs during amination, and the process can be effected by other bases and acids. Leuco derivatives are not susceptible in this way but on the other hand they undergo mutarotation in solution, especially when irradiated by ultraviolet light.[77] Both *fb*- and *sl*-derivatives are converted into the same epimeric mixture which includes a third stereochemical isomer. Thus mutarotation of the leucotetra-acetates leads to the formation of (63; R = Ac) from which a new stereoisomer (64) (erythroaphin-*tt*) can be derived, not found as yet in Nature.[78]

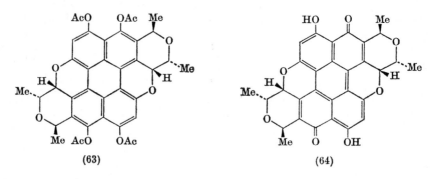

(63)   (64)

XANTHOAPHINS AND CHRYSOAPHINS

These pigments are intermediate on the pathway from the protoaphins to the erythroaphins; they include two xanthoaphins-*sl* and three chrysoaphins-*sl*. They are very labile, being easily converted by acid, base, or heat into the stable end-products. Consequently derivatives have been almost unobtainable and their structures were deduced[79] from those of the protoaphins and erythroaphins, together with the following considerations. The xanthoaphins alone, in the aphin series, show no redox properties, and are somewhat less stable than the chrysoaphins. In passing along the series xanthoaphin → chrysoaphin → erythroaphin there is a progressive bathochromic shift of the ultraviolet-visible spectra (Fig. 1), loss of the elements of water and disappearance (in the ery-throaphins) of free hydroxyl absorption near 3 μ. These changes suggest a progressive aromatisation of the central part of the molecule, pro-ceeding in fact from an anthracene to a perylene system. The electronic absorption of the xanthoaphins closely parallels that of 1,5-diacetyl-2,6-dihydroxyanthracene (65).[80] Further support for the anthracene chromophore comes from the formation of a tetrachloroxanthoaphin by chlorination of erythroaphin-*fb* in hot nitrobenzene.[79] All the spectro-scopic evidence is consistent with structure (66), formed apparently by addition of hypochlorous acid to two double bonds in dichloroery-throaphin-*fb* to which it can be reduced by zinc and acetic acid. The light absorption of (66) shows unmistakably that it has the same chromophore

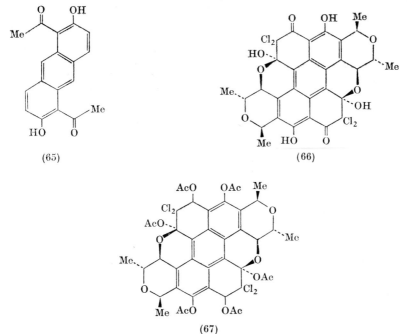

(65)   (66)

(67)

as the xanthoaphins, and reduction of the carbonyl groups with borohydride, and acetylation, is accompanied by a change in the absorption
curve to that of a polyalkylated anthracene (67).

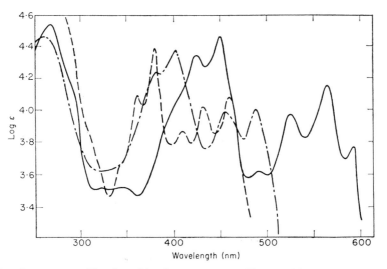

FIG. 1. ——— Xanthoaphin-*sl*-1; — · — · — · Chrysoaphin-*sl*-2; ———Ery-
throaphin-*sl*. Solvent-CHCl₃.

20

Taking all these factors into account xanthoaphin-*fb* was formulated[79] as (68) and xanthoaphin-*sl*-1 as (69). The configuration with least strain was assigned to the new asymmetric hemi-ketal centres, so that the stereochemistry at this point is determined by the stereochemistry at the asterisked carbon atoms in the precursor protoaphin. The remaining

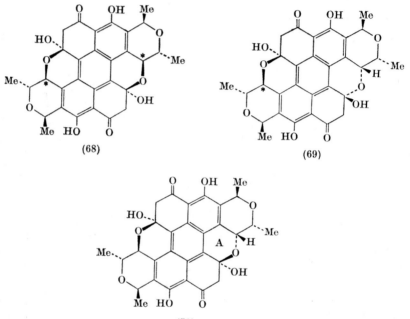

(68)

(69)

(70)

xanthoaphin-*sl*-2 can only be (70), the very low stability of this compound being attributable to the strain in ring A.[81] A model can only be constructed at the expense of some distortion of the planar anthracene system.

It now follows that chrysoaphin-*fb* is (71) which can arise by dehydration from either side of xanthoaphin-*fb* (68).[79] However the hemi-ketal groups in the xanthoaphins-*sl* (69) and (70) are not symmetrical and from these three chrysoaphins-*sl* have been isolated so far, all of which are converted into erythroaphin-*sl* on further dehydration.[76, 79, 81] Dehydration on either side of xanthoaphin-*sl*-1 (69) leads to chrysoaphin-*sl*-1 (72) and *sl*-2 (73), while the very unstable xanthoaphin-*sl*-2 (70) changes on keeping into chrysoaphin-*sl*-1 (72) and xanthoaphin-*sl*-1 (69), the strain in ring A being relieved partly by dehydration and partly by epimerisation. However, some dehydration does take place at the less-strained side of the molecule and chrysoaphin-*sl*-3 (74) can be detected in the decomposition products at low temperature. Chrysoaphin-*sl*-3 is

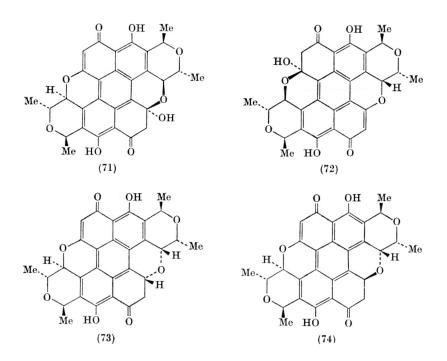

(71)   (72)

(73)   (74)

not quite as unstable as xanthoaphin-*sl*-2; relief of strain can only occur by dehydration to give erythroaphin-*sl* or by epimerisation to chryso-aphin-*sl*-2 which takes place in the presence of sodium carbonate.[81]

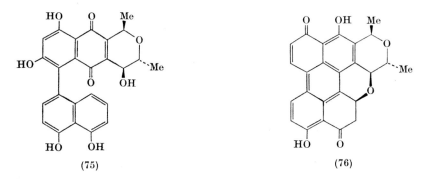

(75)   (76)

A useful confirmation of the general structure of the chrysoaphins was obtained by the synthesis of (76) which has almost the same chromophore and ultraviolet-visible absorption.[64] The orange compound (75) was prepared by coupling quinone A (46; R = H) with 1,8-dihydroxynaphthalene at room temperature (pH 6·6), and this was smoothly converted into (76) on brief heating in aqueous solution, the condensation step being analogous to that involved in xanthoaphin formation.

Protoaphin-xanthoaphin Conversion [60b]

The transformations, xanthoaphin → chrysoaphin → erythroaphin, are simply dehydrations but the initial conversion of protoaphin into xanthoaphin is much more complex and involves hydrolysis of the glucosidic link, condensation of a quinone carbonyl group with the "resorcinol ring" of the aglucone, and the formation of two hemi-ketal linkages. These reactions take place under extremely mild conditions apparently controlled by two enzyme systems. This conclusion was confirmed by the discovery[92] that extracts of *Dactynotus jaceae* (which elaborates a different group of pigments, the dactynaphins) contain a glucosidase which can bring about the initial hydrolysis, but the reaction stops at that stage as the second enzyme is not available. The first step can be effected either by acid hydrolysis or enzymically in the presence of boric acid which converts the aglucone (a 1,8-dihydroxynaphthalene) into a borate complex and so inhibits xanthoaphin formation.[84] The second stage can be brought about by more prolonged acid treatment but the yields are poor compared to the enzymic conversion. All the

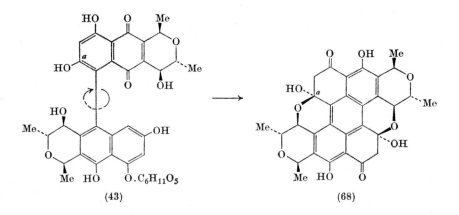

(43)   (68)

evidence suggests that the condensation step and hemiketal formation are concerted processes, the latter being reminiscent of the conversion of skyrin into pseudoskyrin derivatives (see p. 424).

A factor not mentioned hitherto is the configuration of the protoaphins with respect to the restricted rotation about the binaphthyl linkage. If, in the formula of protoaphin-*fb* (43) the lower half is considered to be in the plane of the paper, then the upper half will make an angle with that plane such that its heterocyclic ring can be directed either out of the plane towards the observer (isomer A) or into the plane away from the observer (isomer B). Xanthoaphin formation, in the case of A, requires the upper half to be rotated relative to the lower, as indicated above; in the case of B, rotation would be in the opposite sense. Considering hemi-ketal formation at C^a in (43), if this takes place through attack on C^a by the neighbouring alcoholic group it can only be from one direction, owing to restricted rotation, and it can be seen that rotation of isomer A would lead to the correct configuration at C^a in xanthoaphin-*fb* (68) whereas isomer B would lead to the opposite, and much less stable, configuration. This argument[79] suggests then, that protoaphin-*fb* is correctly represented by isomer A, and by similar reasoning protoaphin-*sl* (44) should have the same configuration and should give rise to xanthoaphin-*sl*-2 (70).[81] It is interesting to find, however, that extracts of *A. fabae* are able to convert protoaphin-*sl* (44) directly into xanthoaphin-*sl*-1 (69) which suggests that in some cases formation of the hemi-ketal furthest from the binaphthyl linkage may not be enzymically controlled.[81]

Rhodoaphin-be (77), $C_{30}H_{22}O_{10}$

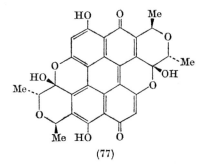

(77)

Occurrence: as (probably) a simple glycoside, heteroaphin, in the haemolymph of *Hormaphis*† *betulina*‡ (Horvath)[87, 88] and *H. spinosus* (Shimer)[87, 88] (Thelaxidae).

Physical properties: deep red crystals, $\lambda_{max.}$ (CHCl₃) 428, 454, 491, 527, 568·5, 596 nm (log ϵ 4·37, 4·50, 3·79, 4·02, 4·16, 3·70), $\nu_{max.}$ (KBr) 1630, 1570 cm⁻¹ (OH region obscured).

† Formerly *Hamamelistes*.
‡ Formerly *betulae*.

Primitive aphids of the *Hormaphis* genus contain a red, water-soluble, glycosidic pigment, heteroaphin, which is converted enzymically on the death of the insect (or by treatment with acid) into a red fluorescent aglycone. The two species listed above were first examined by MacDonald[87] who named the aglycones rhodoaphin-*be* and *sp*; these were shown later to be identical.[88] The visible absorption of rhodoaphin is virtually identical with that of dihydroxyerythroaphin-*fb* (78) and the mass spectra are identical in the high mass region (below m/e 500 the spectrum is not reproducible owing to thermal decomposition). The

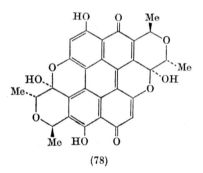

(78)

n.m.r. spectrum is remarkably simple (indicating a symmetrical structure), comprising methyl doublets coupled to methine quartets, together with a singlet from aromatic protons, all of which are consistent with an erythroaphin structure. Catalytic hydrogenation of rhodoaphin-*be* gave erythroaphin-*tt* (64) as the major product, together with smaller amounts of the *fb* and *sl*-stereoisomers. It appeared therefore that this new pigment was a dihydroxy derivative of erythroaphin-*tt*, the "unnatural" stereoisomer, as experiments with (78) showed that configuration is largely retained during hydrogenolysis. This was confirmed by brief treatment with trifluoroacetic acid which effectively epimerised rhodoaphin-*be* to dihydroxyerythroaphin-*fb*.[88]

The structure of heteroaphin is not yet known. Its visible absorption is similar to that of erythroaphin-*fb* dimethyl ether, but shifted to longer wavelengths, which suggests that it may be a simple glycoside in which the sugar is attached to one of the *peri*-hydroxyl groups.[88] Thus the native pigment in *Hormaphis* spp. is an erythroaphin derivative (69) in contrast to the *Aphididae* where erythroaphins are not present in the living insects and only appear post-mortem.

DACTYNAPHINS

Protodactynaphin-jc-1 (79), $C_{36}H_{38}O_{17} \cdot H_2O$

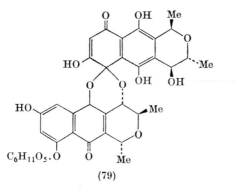

(79)

Occurrence: in the haemolymph of *Dactynotus* aphids; *D. jaceae* L.,[89] *D. cirsii* L.,[89] *D. taraxaci* Kaltenbach,[89] *D. tanaceti* L.,[89] *D. rudbeckiae* Fitch,[90] *D. ambrosiae* Thomas[90] and *D. nigrotuberculatus* Thomas Olive.[89]

Physical properties: light brown powder, $\lambda_{max.}$ (95% EtOH) 275, 295, 360(sh) nm (log ϵ 4·17, 4·05, 3·71), $\nu_{max.}$ (KBr) 3420, 1611, 1580 cm^{-1}.

This group of pigments, which appears to be confined to *Dactynotus* aphids, has been studied by Bowie and Cameron.[89, 91, 92] The native protodactynaphins ,$C_{36}H_{38}O_{17}$, are again glycosides which are converted post-mortem, or by acid treatment, into a mixture of red and yellow aglycones consisting mainly of the isomeric, interconvertible, rhodo- and xanthodactynaphins-jc-1, $C_{30}H_{28}O_{12}$, and smaller amounts of a similar pair of isomers, rhodo- and xanthodactynaphins-jc-2.[89] The jc-1 isomers are derived from protodactynaphin-jc-1 and the jc-2 isomers undoubtedly originate from a protodactynaphin-jc-2 although this has not yet been detected.[92]† Preliminary investigations[89]‡ revealed a general similarity to other members of the aphin series, for example their n.m.r. and mass spectra, and reductive acetylation of rhododactynaphin-jc-1 gave a product showing only naphthalenic absorption. However a new feature is the formation of pairs of isomeric aglycones; a red aqueous solution of rhododactynaphin-jc-1 at pH 6 quickly fades as the transformation into xanthodactynaphin-jc-1 proceeds. The reverse change is conveniently demonstrated by adsorption of xanthodactynaphin-jc-1 onto

---

† It has now been isolated.[83]

‡ Rhododactynaphin-jc-1 (= rhododactynaphin B[90]), deep red needles, dec. >300°, $\lambda_{max.}$ (95% EtOH) 277, 325(sh), 390(sh), 504, 590(sh) nm (log ϵ 4·27, 3·87, 3·25, 3·57, 3·25).

Rhododactynaphin-jc-2 (= rhododactynaphin A[90]), red needles, dec. 290°, $\lambda_{max.}$ as for jc-1, $\nu_{max.}$ (KBr) 3400, 1633, 1610 cm^{-1}.

Xanthodactynaphin-jc-1, yellow needles, dec. 250°, $\lambda_{max.}$ (95% EtOH) 232, 245(sh), 289, 326, 382 nm (log ϵ 4·23, 4·22, 4·10, 4·02, 3·86).

silica gel t.l.c. plates, followed by development some hours later when substantial conversion to rhododactynaphin-*jc*-1 can be seen. The xanthodactynaphins do not show redox properties.

The aglycones can be cleaved in two ways which give valuable information concerning the structure of these compounds. Reductive fission of both dactynaphins-*jc*-1, either with neutral dithionite or by prolonged catalytic hydrogenolysis, gave (after reoxidation) chiefly quinone A (80; R = OH), previously obtained from protoaphin-*fb*, and an unstable compound showing ultraviolet absorption similar to 2,4-dihydroxyacetophenone. As the second compound is formed by reduction of quinone A under identical conditions it is regarded as the tetralone (81), and the overall reduction is therefore $C_{30}H_{28}O_{12} + 2H_2 \rightarrow 2C_{15}H_{16}O_6$. [Reduction of rhododactynaphin-*jc*-2 gives a mixture of two C_{15} quinones (combined yield 0·9 mol.), (80; R = OH and H)].

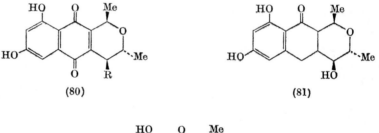

(80)   (81)

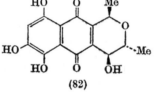

(82)

In alkaline solution rhododactynaphin-*jc*-1 is converted into the anion of xanthodactynaphin-*jc*-1 (colour change blue to yellow) and then decomposes, but this can be controlled by exclusion of air when the solution turns red, and yields two quinones one red, $C_{15}H_{14}O_7$, and one orange, $C_{15}H_{14}O_5$ (quinone A is $C_{15}H_{14}O_6$). The red compound was similar to naphthopurpurin in acidity, visible absorption, and the ease of reduction,[93] both chemically and catalytically, to quinone A (80; R = OH). Accordingly the red quinone must be (82). Confirmation was obtained by an unusual synthesis[94] in which dimethylamine was added to quinone A to give (83), and thence (82) by hydrolysis. This is a further illustration of the high reactivity of 5,7-dihydroxy-1,4-naphthaquinones at position-8, and probably another example is the formation of a purple compound, by the action of ammonia on rhododactynaphin-*jc*-2, which has the same visible absorption as (83). The orange cleavage product, $C_{15}H_{14}O_5$, had virtually the same electronic spectrum as quinone A (80; R = OH), and the n.m.r. spectrum at high field showed a marked

(83)　　　　　　　　　　　　(84)

resemblance to that of isoeleutherin (47). Accordingly, it appeared to have structure (84) which was confirmed by comparison with the compound obtained by reduction of quinone A with sodium stannite (p. 599). The same quinone can be obtained directly from the dacty-naphins by reduction in the same way.

It now seems likely that both dactynaphins comprise two units having the carbon-oxygen skeleton (85), and the remaining problem concerns their mode of attachment to each other, and the manner of their rearrangement. Various considerations indicate that rhododactynaphin-*jc*-1 contains the partial structure (86), notably the general similarity

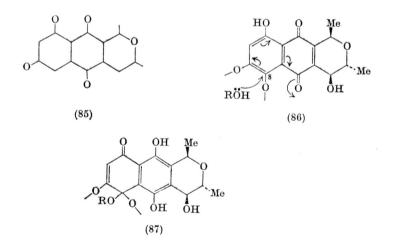

(85)　　　　　　　　　　　　(86)

(87)

between its visible absorption and that of naphthazarin while it can be inferred, *inter alia*, from the n.m.r. spectrum that the chromophore (86) is linked to the rest of the molecule by the two oxygen atoms indicated. The isomerisation of rhododactynaphin into xanthodactynaphin takes place under very mild conditions, and results in a marked change in the chromophore and the disappearance of the quinonoid system. There is no obvious intermolecular reaction of (86) to account for this which leads to the conclusion that some group located in the other half of the molecule is able to promote an intramolecular reaction. Nucleophilic attack by a hydroxyl group (ROH in 86) at position-8 of the naphthazarin nucleus is a reasonable possibility, and this would lead to the chromophoric system (87) which is consistent with the spectral data. 2,5-Dihydroxy-acetophenone provides a simple analogy, the ultraviolet absorption of which is in good agreement with the peak of longest wavelength (λ_{max}. 370 nm) in the spectrum of xanthodactynaphin-jc-1. The latter also shows λ_{max}. at 289 and 326 nm which coalesce at pH 8·5 to a more intense peak at 341 nm, and this behaviour is closely followed in the spectra of 2,4-dihydroxyacetophenone. Indeed the ultraviolet spectrum of xantho-dactynaphin in ethanol closely parallels a summation of the spectra of 2,4- and 2,5-dihydroxyacetophenone. The "other half" of the xantho-

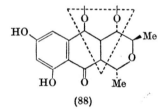

(88)

dactynaphin molecule is therefore represented as (88) taking into consideration also the n.m.r. spectrum which confirms the presence of two *meta*-coupled protons. The same partial structure (88) is also present in rhododactynaphin-jc-1 which shows the appropriate aromatic and methyl resonances in the n.m.r., and its ultraviolet-visible spectrum resembles a summation of the spectra of naphthazarin and 2,4-di-hydroxyacetophenone. A significant difference between rhodo- and xanthodactynaphin-jc-1 is that the latter is much more acidic and can be separated from the former by extraction from chloroform with buffer at pH 6. Ionisation results in a bathochromic shift (ca. 36 nm) of the band attributed to chromophore (87) and hence it is concluded that the oxygen function at C-7 is a free hydroxyl group.[91]

Only one combination of structures (87) and (88) is possible (see 89 but ignore stereochemical factors) but in order to satisfy the molecular formula for xanthodactynaphin-jc-1 an additional element of unsatura-tion is required which must not affect any of the foregoing spectral arguments. This is best positioned within the triangular area shown in (88) for which no direct experimental evidence could be obtained. Bowie and Cameron[91] therefore tentatively propose structure (89; R = OH) for xanthodactynaphin-*jc*-1. The failure of ring A in (89) to aromatise may be attributed to steric inhibition as this would tend to distort (flatten) the 1,3-dioxan ring. Other examples of cyclohexadienone systems stabilised by steric factors are known, for example, (90).[95]

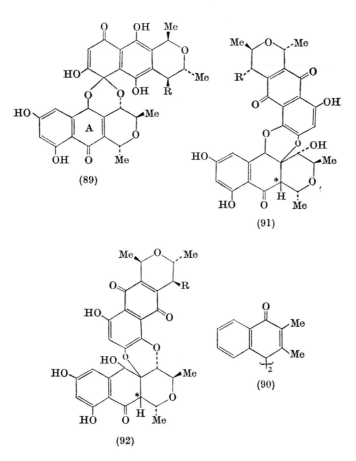

(89)

(91)

(92)

(90)

In order to arrive at the structure of the rhododactynaphin isomer it is necessary to combine (86) and (88). Two arrangements (91 and 92; R = OH) can be considered,[91] both of which are supported by a doublet (τ 6·38, J,6 Hz) in the n.m.r. spectrum which is absent from that of xanthodactynaphin-*jc*-1. This is attributed to the proton attached to the asterisked carbon atom in (91) and (92) which was confirmed by appropriate spin decoupling experiments. Either of these formulae could equally well represent rhododactynaphin-*jc*-1 but in the subsequent discussion reference will be made only to (91; R = OH).

One stereoisomer of the 1,3-dioxan ring in xanthodactynaphin is represented by (93). The two halves of the molecule are approximately at right angles and hence formation of an unstable intermediate (94) is quite feasible. Rearrangement of xantho- to rhododactynaphin can therefore be conceived as attack by the enolic hydroxyl group in (89; R = OH) (see *a* in 95) to give (94), followed by fission of one of the ketal

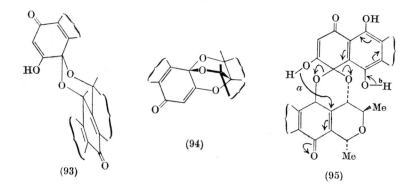

(93)   (94)   (95)

C–O bonds (*b* in 95) to give either (91 or 92; R = OH), the whole process being reversible. The formation of the two quinones (74) and (76) when rhodo- or xanthodactynaphins-*jc*-1 are treated with alkali is readily understood assuming initial abstraction of the proton attached to ring A, and subsequent electron shifts as indicated in (96). Reductive fission of the dactynaphins to give quinone A (80; R = OH) is unexceptional as only easily reducible bonds are involved. Quinone A is derived from both halves of the dactynaphin molecule and circular dichroism measurements established[91] that it had the same configuration as that derived from protoaphin-*fb*. Thus although the complete stereochemistry of the

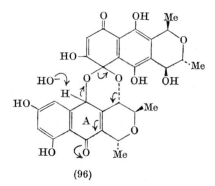

(96)

dactynaphins is not known they belong to the same series as the aphins themselves.

The dactynaphins-*jc*-2 show the same electronic absorption and chemical behaviour as the *jc*-1 isomers but contain one atom of oxygen less. The n.m.r. spectra are similar to that of isoeleutherin (47) and include signals at τ 7·0–7·5, which are not seen in the *jc*-1 spectrum. This means that a benzylic hydroxyl group in the dactynaphins-*jc*-1 is missing, and as one of these is involved in xanthodactynaphin formation rhodo- and xanthodactynaphins-*jc*-2 can be formulated as (91 or 92; R = H) and (89; R = H), respectively. Confirmation[91] was obtained by reductive fission which gave approximately equal amounts of quinone A (80; R = OH) and (84), and by alkaline treatment in the absence of air which afforded a red naphthopurpurin derivative, probably (97), and the

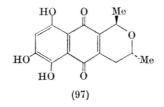

(97)

orange quinone (84). The dactynaphins-*jc*-2 thus possess the isoeleutherin (47) pyran system.[91]

If living *D. jaceae* aphids are crushed in acetone to inactivate enzymes, careful work up of the extract permits the separation of protodacty-naphins-*jc*-1 from aphinin (see p. 622) and from an unidentified colourless component with an intense yellow fluorescence in ultraviolet light.[92] By treatment with extracts of *D. jaceae*, or on acid hydrolysis, it gives rhodo- and xanthodactynaphins-*jc*-1, and D-glucose. As the enzymic hydrolysis can also be effected by extracts of *A. fabae* protodactynaphin-*jc*-1 is presumably a β-D-glucoside. The ultraviolet absorption is very similar to that of xanthodactynaphin-*jc*-1 but shifted to shorter wave-lengths. This implies that the sugar residue is attached to a phenolic group in the xanthodactynaphin molecule but not to the two acidic free hydroxyl groups in (89; R = OH) as substantial bathochromic shifts are observed when the pH is raised to 8·5. However the longwave peak at 415 nm in the proto-compound is shifted hypsochromically by 22 nm relative to that of the xantho-compound, and as this band is chiefly associated with the lower half of (89; R = OH), the sugar moiety is apparently attached to the *peri*-hydroxyl group as in (79) (and as in the protoaphins). This was confirmed by alkaline fission of protodacty-naphin-*jc*-1 to give the quinone (82) and a glucoside which afforded (84) after treatment with extracts of *D. jaceae*. Protodactynaphin-*jc*-1 is therefore regarded[92] as (79) and presumably it exists *in vivo* in equilibrium with the corresponding rhododactynaphin glucoside. This is

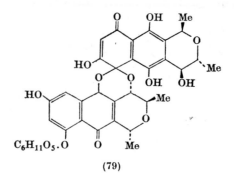

(79)

supported by an inflection at 450–550 nm in the visible absorption curve and the formation of both rhodo- and xantho-compounds on either acidic or enzymic hydrolysis. An analogous pair of glucosides corresponding to dactynaphin-*jc*-2 is probably also present, in much smaller amount, in *D. jaceae.*[92]†

APHININS

Aphinin-sm (98), $C_{36}H_{36}O_{16}$

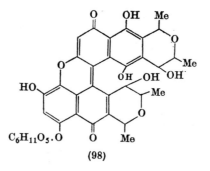

(98)

Occurrence: in the aphids *Aphis fabae* Scop.,[96] *A. sambuci* L.,[96] *A. rumicis* L.,[96] *A. farinosa* Gm.,[96] *A. corniella* HRL,[96] and many other species.[60b]

Physical properties: dark blue-green solid, $\lambda_{max.}$ 638 nm.

Aphids vary in colour from almost colourless through red and green to almost black, and there may be considerable variation in colour within one species. Aphins are characteristic of dark-coloured species but they are always accompanied by a green, water-soluble, glycosidic,

† Now confirmed.[83]

aphinin pigment. Aphinin(s) are also found in many aphin-free species including the well-known greenfly (*Macrosiphium rosae* L.), a common pest of cultivated roses. For chemical purposes aphinin-*sm* was most conveniently isolated from *A. sambuci* (parasitic on elder, *Sambucus nigra*), and was found to be very sensitive to heat, light and oxidising conditions. Only a brief report has appeared to date but Cameron and his colleagues[96, 98] have established the existence of the chromophore shown in (98) by comparison with a synthetic model.[64] Mild oxidation

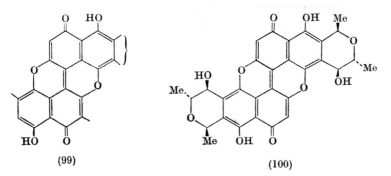

(99)   (100)

converts it into another green quinone, probably having the same chromophore (99) as xylaphin (100), a synthetic compound obtained by self-coupling of quinone A (80; R = OH).[97]

Xylindein (101), $C_{32}H_{24}O_{10}$

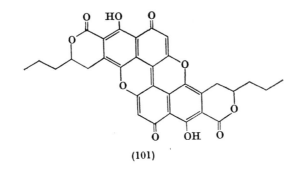

(101)

Occurrence: *Chlorociboria aeruginosa* (Nylander)† growing on dead deciduous wood[49, 50] and in laboratory culture.[51] Cultures of *Lophiostoma viridarium* Cooke.[82]

Physical properties: bronze plates, m.p. >300°, λ_{max}. (CHCl$_3$) 380, 405, 423, 603, 647 nm (log ε 4·19, 4·17, 4·06, 4·31, 4·46), ν_{max}. (KBr) 1726, 1631, 1608 cm^{-1}.

The green colour acquired by rotting wood infected with *Chlorociboria aeruginosa* has attracted attention for many years.[49, 52, 53] Oak wood stained green in this way was used at one time in making a special kind of

† Formerly described as *Chlorosplenium aeruginosum* or *Peziza aeruginosa*.

marquetry known as "Tunbridge ware", and a process for artificially colouring wood with *Chlorociboria* was patented.[54] In this case chlorophyll is not responsible for the colour and the principal pigment present, xylindein, is another rare example of a natural nitrogen-free, green, colouring matter. The compound is extremely insoluble in most organic solvents which severely handicapped early investigations, and Liebermann[49] first obtained it in crystalline form by extracting green wood with phenol, followed by crystallisation of the crude product from aqueous phenol. It is now possible to grow *C. aeruginosa* in the laboratory,[51] the pigment being conveniently extracted from the mycelium with phenol or chloroform. The first systematic chemical investigation was that of Kögl[55] who prepared a series of derivatives from which the presence of two phenolic groups, two lactone rings and an extended quinone system was deduced. Zinc dust distillation of a xylindeic acid derivative was reported to yield phenanthrene. Recently, the complete structure (101) has been elucidated independently by Todd and his co-workers,[51, 53] and by Edwards and Kale.[50]

Xylindein forms a purple dimethyl ether with diazomethane, and affords a purple diacetate under mild conditions. In these reactions, a carbonyl peak at 1625 cm^{-1} in the parent compound shifts to 1638–1640 cm^{-1} showing that two phenolic groups are present *peri* to an extended quinone system. The facile reaction with diazomethane is noteworthy. Catalytic hydrogenation affords the yellow quinol (also extracted from infected wood).[50] Zinc dust fusion of the leucotetra-acetate gave a yellowish product, $C_{30}H_{30}O_2$, whose ultraviolet spectrum very closely resembled that of *peri*-xanthenoxanthene (102; R = H); it was identified

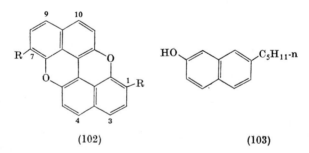

(102)   (103)

by high resolution mass spectromety[56] as a di-n-pentyl derivative, most probably (102; R = n-C_5H_{11}), or an isomer in which the side chains occupy two positions adjacent to the ether bridges. As other evidence favoured the 1,7-isomer (102; R = n-C_5H_{11}), it was synthesised for comparison and proved to be identical with the zinc degradation product.

The synthesis was effected by oxidative coupling of the naphthol (103) using copper(II) acetate in pyridine, followed by further oxidation with silver oxide in boiling benzene. It was then evident that the chromophore of xylindein must be 3,9-dihydroxy-*peri*-xanthenoxanthene-4,10-quinone (104), and indeed the visible absorption curve of xylindein dimethyl

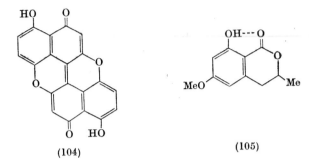

(104)   (105)

ether, and of its reduction product with lithium aluminium hydride, both resemble that of the parent *peri*-xanthenoxanthene-4,10-quinone.

Kögl[55] demonstrated that xylindein was a dilactone by treatment with sodium hydroxide to form a green tetra-sodium salt which was converted, via the di-silver salt into a dimethyl ester (xylindeic acid dimethyl ester).[50] This is confirmed by a carbonyl band at 1726 cm^{-1} in xylindein and various derivatives, and all the spectroscopic and chemical evidence indicates that the lactones are symmetrical. That the lactone carbonyl group is attached directly to the main chromophore is attested by a number of observations. When the lactone rings are opened by reduction of leucoxylindein tetramethyl ether with lithium aluminium hydride (2-CO–O–C– → 2-CH$_2$OH + 2HOC–), followed by acetylation, the product shows only a single acetate carbonyl absorption at 1727 cm^{-1}. This shows that phenolic hydroxyls are not involved in lactone formation. Furthermore, the n.m.r. spectrum of this compound reveals the presence of four benzylic acetate protons (AcOCH_2Ar) in addition to signals from four aromatic methoxyls and four aliphatic acetate groups. When the purple xylindein diacetate is reduced catalytically it forms a yellow quinol showing ν_{CO} 1770 (phenolic acetate) and 1658 (chelated lactone); further acetylation gives a tetra-acetate in which the lactone absorption is restored to 1726 cm^{-1}. Similarly, the model compound (105) shows ν_{CO} 1660 cm^{-1} which shifts to 1720 cm^{-1} when the *peri*-hydroxyl group is methylated, and leucoxylindein triacetate shows both free and chelated lactone carbonyl absorption at 1725 and 1660 cm^{-1}, respectively.

Oxidative degradation of xylindein, in various ways, gives n-butyric acid, and minor amounts of propionic and acetic acids. Clearly the C_4 acid and the C_5 side chains of the zinc dust degradation product (102; $R = n\text{-}C_5H_{11}$) are associated, and must also be incorporated into the lactone rings. It follows that two isocoumarin systems are present in xylindein, and consideration of all these factors leads to the conclusion that the natural pigment must be represented by (101).[50, 51, 53]

Although the precise structure of the tautomeric dihydroxy-*peri*-xanthenoxanthenequinone system in xylindein is not known, definite structures can be assigned to various derivatives. The quinol in which the lactone carbonyl groups are chelated (ν_{CO} 1650 cm^{-1}) must be (106). As the quinol derived from xylindein dimethyl ether shows free lactone

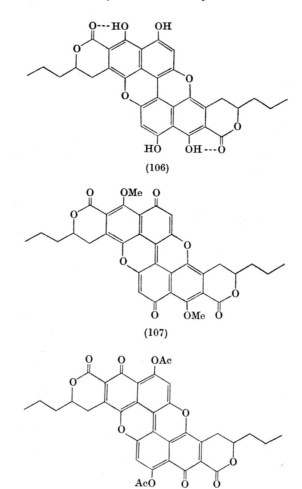

(106)

(107)

(108)

carbonyl absorption at 1720 cm^{-1} the adjacent substituent must be methoxyl, and hence xylindein dimethyl ether is (107). On the other hand the quinol obtained by catalytic hydrogenation of xylindein diacetate absorbs at 1658 cm^{-1} indicating chelation to the lactone groups, and hence the quinone diacetate should be (108) (assuming no acetyl migration during reduction).

An interesting and complicating feature of the chemistry of xylindein is the phenomenon of "auto-reduction". Kögl[55] observed that vigorous acetylation of xylindein gave a yellow tetra-acetate, and the purple dimethyl ether also gave a yellow diacetate on acetylation. The same compounds are obtained by reductive acetylation of xylindein and its dimethyl ether, respectively, and they are, in fact, derivatives of xylindein quinol which arise by an intermolecular redox reaction.[53] Whereas the yellow tetra-acetate is the sole product of reductive acetylation of xylindein, direct acetylation gives several products[50] including an orange tetra-acetate[53] which shows absorption at 470 nm not seen in the spectrum of the yellow leucotetra-acetate (106; OAc in place of OH). This implies an extension to the chromophoric system, and the compound is very probably the tetra-acetate of (109), the quinol of a bis-dehydroxylindein which could be formed in a redox reaction whereby hydrogen is transferred to another molecule of xylindein to

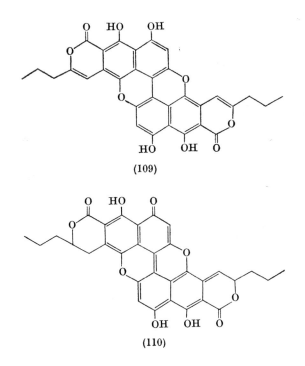

(109)

(110)

form the quinol (106).[53] A similar hydrogen transfer appears to take place when the mass spectrum is measured at 400°.[53] In the high mass region there is a group of peaks at m/e 570, 568, 566 and 564, dominated initially by that at 570 (M + 2). M + 2 peaks are not uncommon in quinone spectra but in this case the relative intensities change with time until the peak at m/e 566 (M − 4) is the strongest. This is attributed[53] to a hydrogen transfer reaction giving rise to the quinol (M + 2), and dehydro-(M − 2) and bis-dehydro-(M − 4)-xylindeins, the latter increasing in amount with time. The redox reaction may involve identical xylindein molecules or the hydrogen transfer may occur by way of a tautomeric mono-enolised form (110).

Kögl[55] thought that xylindein contained two enol-lactone groups and it is quite possible that his material, extracted from rotting wood, did contain some bis-dehydroxylindein. Xylindein isolated recently both from wood and from the cultured fungus, was not homogeneous,[50, 53] and there was evidence[53] for the presence of the mono- and/or bis-dehydro derivatives. Thus on one occasion a crude sample of the dimethyl ester of xylindeic acid was separated chromatographically into the pure

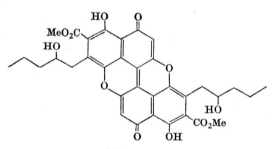

(111)

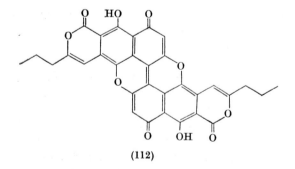

(112)

ester (111), ($\nu_{max.}$ 3400, 1730 cm^{-1}), and a second component, not completely freed from (111), which had only weak hydroxyl absorption but strong bands at 1730 cm^{-1} (ester) and 1712 cm^{-1}(dialkyl ketone), suggesting the presence of the quinone (112) or the ~~.~~ analogue. The presence of (112) in Kögl's xylindein could exp~~.~~ ydro was able to obtain a bis-semicarbazone from his sample of ~~.~~ he xylindeate.

Viopurpurin. Cultures of *Trichophyton violaceum* Sabouraud, apu~~.~~ Bodin, 1902, contains a number of pigments of which the binaphtha-quinone, xanthomegnin (113), is the major component. Two others,

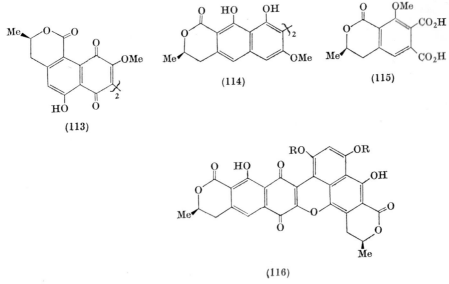

(113)

(114)

(115)

(116)

(R = H or Me)

which have a related structure are the brownish-yellow, vioxanthin (114) and the dark red quinone, viopurpurin (116).[48, 106] The quinone [$\nu_{max.}$ (KBr) 1725, 1675, 1635 cm^{-1}] could not be obtained pure. It afforded an orange triacetate, $C_{35}H_{26}O_{14}$, [$\lambda_{max.}$ (EtOH) 217, 270, 277, 425 nm] and a leucopenta-acetate [$\lambda_{max.}$ (EtOH) 286, 296, 365 nm], neither of which could be crystallised, but a red, crystalline, trimethyl ether was obtained.[106] The triacetate showed carbonyl absorption attributed to phenolic acetate (1775 cm^{-1}), a δ-lactone (1725 cm^{-1}) and a quinone function (1675 cm^{-1}), while the n.m.r. spectrum identified a methoxyl group, two different acetate groups, two different aromatic protons (τ 1·96 and 2·40) and two identical lactone rings containing the unit

$$Me-\overset{\overset{\textstyle O^-}{\vert}}{C}H-CH_2-Ar$$

found in vioxanthin. In the leucopenta-acetate the

63carbonyl absorption had disappeared and the aromatic proton almost coincided at τ 2·37. The molecular formula $C_{29}H_{18}O_{11}$ that viopurpurin is a dimeric molecule in which all the atoms have n accounted for except one atom of oxygen which must be ethereal. xidation of viopurpurin trimethyl ether with alkaline hydrogen peroxide gave the lactone (115). Of the two structures tentatively advanced, (116)[106] agrees best with the available data.

References

1. J. M. Anderson and J. Murray, *Chem. Ind.* (*London*) (1956), 376.
2. D. C. Allport and J. D. Bu'Lock, *J. Chem. Soc.* (1958), 4090.
3. D. C. Allport and J. D. Bu'Lock, *J. Chem. Soc.* (1960), 654.
4. A. Calderbank, A. W. Johnson and A. R. Todd, *J. Chem. Soc.* (1954), 1285.
5. U. Weiss, H. Flon and W. C. Burger, *Arch. Biochem. Biophys.* (1957), **69**, 311.
6. U. Weiss, H. Ziffer, T. J. Batterham, M. Blumer, W. H. L. Hackeng, H. Copier and C. A. Salemink, *Can. J. Microbiol.* (1965), **11**, 57.
7. W. H. L. Hackeng, H. Copier and C. A. Salemink, *Rec. Trav. Chim.* (1963), **82**, 322.
8. A. L. van Beverwijk, G. E. Bunschoten, C. A. Salemink, W. H. L. Hackeng and H. Copier, *Planta Med.* (1963), **11**, 407; A. E. Jenkins and A. A. Bitancourt, *Arquivos do Instituto Biologico* (*São Paulo*) (1965), **32**, 61.
9. R. J. J. Ch. Lousberg, C. A. Salemink, U. Weiss and T. J. Batterham, *J. Chem. Soc.* (*C*) (1969), 1219.
10. J. C. Overeem, A. K. Sijpesteijn and A. Fuchs, *Phytochemistry* (1967), **6**, 99.
11. C. Mathis and G. Ourisson, *Phytochemistry* (1963), **2**, 157 and references therein.
12. H. Brockmann and W. Sanne, *Chem. Ber.* (1957), **90**, 2480.
13. H. Brockmann, Jr., *Prog. Org. Chem.* (1952), **1**, 64; J. D. Biggers. *In* "The Pharmacology of Plant Phenolics" (ed. J. W. Fairbairn), (1959), p. 71. Academic Press, New York.
14. C. Reid, *Quart. Rev.* (*London*) (1958), **12**, 205.
15. H. Brockmann, *Proc. Chem. Soc.* (1957), 304; *Prog. Chem. Org. Nat. Prods.* (1957), **14**, 142.
16. A. Buchner, *Repertorium Pharm.* (1830), **24**, 217.
17. H. Brockmann, F. Pohl, K. Maier and M. N. Haschad, *Annalen* (1942), **553**, 1.
18. H. Brockmann and G. Budde, *Chem. Ber.* (1953), **86**, 432.
19. T. Shimano, K. Taki and K. Goto, *Ann. Proc. Gifu Coll. Pharm.* (1953), No. 3, 43.
20. H. Brockmann, E. H. von Falkenhausen, R. Neeff, A. Dorlars and G. Budde, *Chem. Ber.* (1951), **84**, 865.
21. H. Brockmann, E. Lindemann, K.-H. Ritter and F. Depke, *Chem. Ber.* (1950), **83**, 583.
22. H. Brockmann and W. Sanne, *Chem. Ber.* (1957), **90**, 2480.
23. H. Brockmann and R. Randebrock, *Chem. Ber.* (1951), **84**, 532.
24. H. Brockmann, F. Kluge and H. Muxfeldt, *Chem. Ber.* (1957), **90**, 2302.
25. D. W. Cameron and P. E. Schütz, *J. Chem. Soc.* (*C*) (1967), 2121.
26. H. Auterhoff and R. Sachdev, *Arch. Pharm.* (1962), **295**, 850.
27. H. Brockmann and F. Kluge, U.S.P. 2,707,704; *Chem. Abs.* (1956), **50**, 7141.

28. H. Brockmann and H. Eggers, *Chem. Ber.* (1958), **91**, 547.
29. H. Brockmann and W. Sanne, *Naturwissenschaften* (1953), **40**, 509.
30. A. E. Oxford and H. Raistrick, *Biochem. J.* (1940), **34**, 790.
31. H. Brockmann and H. Eggers, *Chem. Ber.* (1958), **91**, 81.
32. N. Pace and G. Mackinney, *J. Amer. Chem. Soc.* (1941), **63**, 2570.
33. H. Brockmann and G. Pampus, *Naturwissenschaften* (1954), **41**, 86.
34. W. Theilacker and F. Baxmann, *Naturwissenschaften* (1953), **40**, 220.
35. S. H. Wender, R. A. Gortner and O. L. Inman, *J. Amer. Chem. Soc.* (1943), **65**, 1733.
36. H. Brockmann, E. Weber and G. Pampus, *Annalen* (1952), **575**, 53.
37. H. M. Fox and G. Vevers, "The Nature of Animal Colours" (1960). Sidgwick and Jackson, London.
38. E. R. Lankester, *Quart. J. Microscop. Sci.* (1873), **13**, 139.
39. M. Barbier, E. Fauré-Fremiet and E. Lederer, *Compt. Rend.* (1956), **242**, 2182.
40. V. Arcichovskij, *Arch. Protistenk.* (1905), **6**, 227.
41. K. M. Møller, *C. R. Trav. Lab. Carlsberg* (1962), **32**, 471.
42. M. Blumer, *Mikrochemie*, 1951, **36/37**, 1048; *Nature (London)* (1960), **188**, 4756; *Science* (1965), **149**, 722.
43. M. Blumer, *Geochim. Cosmochim. Acta* (1962), **26**, 225.
44. M. Blumer, *Geochim. Cosmochim. Acta* (1962), **26**, 228; D. W. Thomas and M. Blumer, *Geochim. Cosmochim. Acta* (1964), **28**, 1467.
45. M. Blumer and G. S. Omenn, *Geochim. Cosmochim. Acta* (1961), **25**, 81.
46. K. Kumada and H. M. Hurst, *Nature (London)* (1967), **214**, 631.
47. J. H. A. Butler, D. T. Downing and R. J. Swaby, *Aust. J. Chem.* (1964), **17**, 817.
48. F. Blank, A. S. Ng and G. Just, *Canad. J. Chem.* (1966), **44**, 2873.
49. C. Liebermann, *Ber.* (1874), **7**, 1102.
50. R. L. Edwards and N. Kale, *Tetrahedron* (1965), **21**, 2095.
51. G. M. Blackburn, A. H. Neilson and Lord Todd, *Proc. Chem. Soc.* (1962), 327.
52. J. W. Döbereiner, *Schweiggers Journal* (1813), **9**, 160; L. Bley, *Arch. Pharm.* (1858), **94**, 129; M. Fordos, *Compt. Rend.* (1863), **57**, 50; M. A. Rommier, *Compt. Rend.* (1868), **66**, 108.
53. G. M. Blackburn, D. E. U. Ekong, A. H. Neilson and Lord Todd, *Chimia* (1965), **19**, 208.
54. F. T. Brooks, B.P. 24,595.
55. F. Kögl and G. von Taeuffenbach, *Annalen* (1925), **445**, 170; F. Kögl and H. Erxleben, *Annalen* (1930), **484**, 65.
56. J. H. Beynon, *Proc. Xth Colloquium Spectroscopicum Internationale, Maryland*, 1962, 764. J. H. Beynon, R. A. Saunders and A. E. Williams. "The Mass Spectra of Organic Molecules", (1968). Elsevier, Amsterdam.
57. B. R. Brown, T. Ekstrand, A. W. Johnson, S. F. MacDonald and A. R. Todd, *J. Chem. Soc.* (1952), 4925.
58. D. W. Cameron, R. I. T. Cromartie, D. G. I. Kingston and Lord Todd, *J. Chem. Soc.* (1964), 51.
59. H. Duewell, J. P. E. Human, A. W. Johnson, S. F. MacDonald and A. R. Todd, *J. Chem. Soc.* (1950), 3304.
60. (a) H. Duewell, J. P. E. Human, A. W. Johnson, S. F. MacDonald and A. R. Todd, *Nature (London)* (1948), **162**, 759; (b) D. W. Cameron and Lord Todd. *In* "Oxidative Coupling of Phenols" (eds. W. I. Taylor and A. R. Battersby) (1967). Arnold, London.
61. Lord Todd, *Chem. Brit.* (1966), **2**, 428.

62. B. K. Blount, *J. Chem. Soc.* (1936), 1034.

63. H. C. Sorby, *Quart. J. Microscop. Sci.* (1871), **11**, 352.

64. D. W. Cameron and H. W.-S. Chan, *J. Chem. Soc. (C)* (1966), 1825.

65. A. Fredga, *Tetrahedron* (1960), **8**, 126.

66. M. Blumer (1968). Personal communication.

67. D. W. Cameron and J. C. A. Craik, *J. Chem. Soc. (C)* (1968), 3068.

68. J. P. E. Human, A. W. Johnson, S. F. MacDonald and A. R. Todd, *J. Chem. Soc.* (1950), 477.

69. H. Duewell, A. W. Johnson, S. F. MacDonald and A. R. Todd, *J. Chem. Soc.* (1950), 485.

70. D. W. Cameron, R. I. T. Cromartie, Y. K. Hamied, P. M. Scott and Lord Todd, *J. Chem. Soc.* (1964), 62.

71. D. W. Cameron, R. I. T. Cromartie, Y. K. Hamied, P. M. Scott, N. Sheppard and Lord Todd, *J. Chem. Soc.* (1964), 90; D. W. Cameron, D. G. I. Kingston, N. Sheppard and Lord Todd, *J. Chem. Soc.* (1964), 98.

72. B. R. Brown, A. W. Johnson, S. F. MacDonald, J. R. Quayle and A. R. Todd, *J. Chem. Soc.* (1952), 4928.

73. D. W. Cameron, P. M. Scott and Lord Todd, *J. Chem. Soc.* (1964), 62.

74. R. H. Thomson, *Quart. Rev. (London)* (1956), **10**, 27.

75. A. W. Johnson, A. R. Todd and J. C. Watkins, *J. Chem. Soc.* (1956), 4091.

76. D. W. Cameron, R. I. T. Cromartie, Y. K. Hamied, E. Haslam, D. G. I. Kingston, Lord Todd and J. C. Watkins, *J. Chem. Soc.* (1965), 6923.

77. B. R. Brown, A. Calderbank, A. W. Johnson, J. R. Quayle and A. R. Todd, *J. Chem. Soc.* (1955), 1144.

78. D. W. Cameron, R. I. T. Cromartie, Y. K. Hamied, B. S. Joshi, P. M. Scott and Lord Todd, *J. Chem. Soc.* (1964), 72.

79. A. Calderbank, D. W. Cameron, R. I. T. Cromartie, Y. K. Hamied, E. Haslam, D. G. I. Kingston, Lord Todd and J. C. Watkins, *J. Chem. Soc.* (1964), 80.

80. D. W. Cameron, R. I. T. Cromartie, D. G. I. Kingston and G. B. V. Subramanian, *J. Chem. Soc.* (1964), 4565.

81. H. J. Banks, D. W. Cameron and J. C. A. Craik, *J. Chem. Soc. (C)* (1969), 627.

82. C. E. Stickings. Personal communication (1968).

83. H. J. Banks, Ph.D. Thesis, Cambridge (1969).

84. D. W. Cameron, H. W.-S. Chan and D. G. I. Kingston, *J. Chem. Soc.* (1965), 4363.

85. J. H. Bowie and D. W. Cameron, *J. Chem. Soc. (B)* (1966), 684.

86. D. W. Cameron, H. W.-S. Chan and E. M. Hildyard, *J. Chem. Soc. (C)* (1966), 1832.

87. S. F. MacDonald, *J. Chem. Soc.* (1954), 2378.

88. J. H. Bowie and D. W. Cameron, *J. Chem. Soc. (C)* (1967), 704.

89. J. H. Bowie anc D. W. Cameron, *J. Chem. Soc. (C)* (1967), 708.

90. U. Weiss and H. W. Altland, *Nature (London)* (1965), **207**, 1295.

91. J. H. Bowie and D. W. Cameron, *J. Chem. Soc. (C)* (1967), 712.

92. J. H. Bowie and D. W. Cameron, *J. Chem. Soc. (C)* (1967), 720.

93. J. F. Garden and R. H. Thomson, *J. Chem. Soc.* (1957), 2483.

94. H. W.-S. Chan, Ph.D. Thesis, Cambridge (1966).

95. B. R. Brown and A. H. Todd, *J. Chem. Soc.* (1963), 5564.

96. J. H. Bowie, D. W. Cameron, J. A. Findlay and J. A. K. Quartey, *Nature (London)* (1966), **210**, 395.

97. G. M. Blackburn, D. W. Cameron and H. W.-S. Chan, *J. Chem. Soc. (C)* (1966), 1836.

98. D. W. Cameron, (1968). Personal communication.

99. D. W. Cameron, H. W.-S. Chan and M. R. Thoseby, *J. Chem. Soc. (C)* (1969), 631.

100. J. D. Bu'Lock, "Essays in Biosynthesis and Microbial Development", (1967). Wiley, London; R. A. Nicolaus, *Chim. Ind. (Milan)* (1966), **48**, 341.

101. T. J. Batterham and U. Weiss, *Proc. Chem. Soc.* (1963), 89.

102. H. E. Zimmerman, *J. Amer. Chem. Soc.* (1957), **79**, 6554 and earlier papers quoted therein.

103. S. Kuyama and T. Tamura, *J. Amer. Chem. Soc.* (1957), **79**, 5725.

104. S. Kuyama and T. Tamura, *J. Amer. Chem. Soc.* (1957), **79**, 5726; S. Kuyama, *J. Org. Chem.* (1962), **27**, 939.

105. R. J. J. Ch. Lousberg, C. A. Salemink and U. Weiss, manuscript in preparation.

106. A. S. Ng, G. Just and F. Blank, *Can. J. Chem.* (1969), **47**, 1223.

107. H. Brockmann, (1969). Personal communication.

108. C. J. C. Rees, D.Phil. Thesis, University of Oxford (1966).

109. K. S. Brown, D. W. Cameron and U. Weiss, *Tetrahedron Lett.* (1969), 471.

110. R. J. J. Ch. Lousberg. Personal communication (1968).

111. R. J. J. Ch. Lousberg, Doctoral Thesis, University of Utrecht (1969).

112. M. R. Sevenants, *J. Protozool.* (1965), **12**, 240.

CHAPTER 8

Miscellaneous Quinones

This chapter deals with a miscellaneous collection of pigments which do not fit easily into any of the main groups. The first quinone considered in detail is the bacterial pigment, piloquinone, the only authentic natural phenanthraquinone apart from tanshinone I and isotanshinone I which, biogenetically, are diterpenes. Denticulatol, a pigment isolated by Chinese workers[1] from the root of *Rumex chinensis* Campd.† was considered to be a dihydroxy-methyl-9,10-phenanthraquinone isomeric with chrysophanol, a co-pigment. The structural assignment rested on the products of zinc dust distillation, namely 1-methylphenanthrene and 1-methylphenanthraquinone (formation of which was "very peculiar"), degradation with chromic acid to 3-methylphthalic acid and colour reactions. The authors concluded that denticulatol was 5,7- or 6,8-dihydroxy-1-methylphenanthraquinone but both isomers have been synthesised[48] and proved to be different from the natural quinone. In a recent investigation only hydroxyanthraquinones were found in *R. chinensis*.[68]

Piloquinone (1), $C_{21}H_{20}O_5$
4-Hydroxypiloquinone (2), $C_{21}H_{20}O_6$

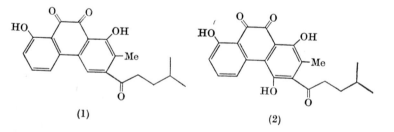

(1)   (2)

Occurrence: mycelium of *Streptomyces pilosus* Ettlinger *et al.*[2, 66]
  Physical properties: (1) brown-red needles, m.p. 176–178°, $\lambda_{max.}$ (95% EtOH) 233, 277(sh), 286, 396, 515 nm (log ε 4·50, 4·19, 4·23, 3·61, 3·74), $\nu_{max.}$ (Nujol) 1695, 1635 cm⁻¹.

  (2) red crystals, m.p. 174–176°, $\lambda_{max.}$ (95% EtOH) 235, 250, 294, 500 nm (log ε 4·38, 4·22, 3·82, 3·89), $\nu_{max.}$ (Nujol) 1710, 1608, 1596 cm⁻¹.

† For nomenclature see ref. 6.

Streptomyces pilosus elaborates a mixture of red and orange pigments and the major component, piloquinone, readily crystallises from an ether extract of the dried mycelium. It shows redox properties, and forms a diacetate, leucotetra-acetate, and a quinoxaline. Two singlets at low field in the n.m.r. spectrum show that the hydroxyl groups are strongly chelated, which is consistent with the formation of a red-fluorescent, green, solution with boro-acetic anhydride, and is confirmed by the absence of hydroxyl absorption in the 3 μ region of the infrared. The fifth oxygen atom is located in an aryl ketonic group (ν_{CO} 1695 cm^{-1}) characterised by formation of a 2,4-dinitrophenylhydrazone and reduction to an alcohol, dihydropiloquinone (ν_{OH} 3330 cm^{-1}). A Clemmensen reduction replaces CO by CH$_2$ to give deoxydihydropiloquinone and as the two hydroxyl groups in this compound are still strongly chelated (infrared, n.m.r.) they must be adjacent to the quinone carbonyl groups. The five oxygen atoms are thus defined, and the nature of the polycyclic system is evident from the phenanthrene-like ultraviolet absorption of the leucotetra-acetate of deoxydihydropiloquinone, and from the formation of phenanthrene and 2-methylphenanthrene on zinc dust distillation of the natural pigment. This degradation also indicated the position of the aromatic methyl group in piloquinone (revealed by the n.m.r. spectrum). The remaining carbon atoms comprise an isocaproyl side chain ArCOCH$_2$CH$_2$CHMe$_2$; this was established by a modified Kuhn-Roth oxidation which gave a mixture of isocaproic, isovaleric and lower fatty acids, and was confirmed by the n.m.r. and mass spectra. The latter shows a base peak at m/e 281 (M-74) corresponding to loss of a C$_5$H$_{11}$ fragment by cleavage at a in (3), and a major peak at m/e 296 corresponding to cleavage at b by way of a McLafferty rearrangement.

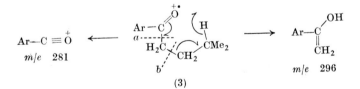

$$Ar-C \equiv \overset{+}{O}$$
$$m/e \quad 281$$

(3)

Thus piloquinone is a tetrasubstituted phenanthraquinone and the arrangement of the substituents [as in (1)] was settled by reference to the n.m.r. spectrum of the parent quinone and numerous derivatives.[3] These show the presence of four aromatic protons, three of which form an ABX system (H-5, H-6, H-7), and the fourth appears as a singlet and is therefore isolated in a pentasubstituted benzene ring. This is assigned to C-4 which is consistent with the change in chemical shift which results on Clemmensen reduction of the ketonic group and on acetylation of the hydroxyl at C-1. The orientation of the carbon side chains is in harmony with its polyketide origin (see p. 30) and the location of the methyl group *ortho* to the ketonic group is in accord with the upfield shift of its proton resonance signal when the carbonyl group is reduced to methylene.

Further support for structure (1) was obtained by oxidative degradation of piloquinone with alkaline hydrogen peroxide which gave not only the expected diphenic acid (4) but also the two isomeric lactones (5) and (6). The structures assigned to these products agree with their chemical

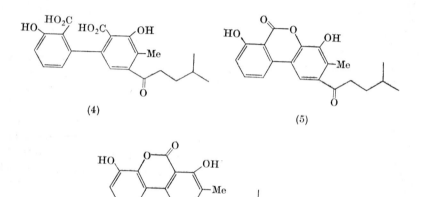

(4)   (5)

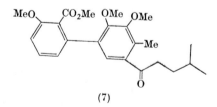

(6)

and spectroscopic properties, and it is of interest that methylation of the principal lactone (5) with diazomethane gives, according to the conditions, its mono- and dimethyl ethers, and the trimethoxy-ester (7). The trans-esterification[4] leading to (7) was effected by using methanol-chloroform as solvent, and a longer reaction time. It is well-known that

(7)

diphenic acids can be oxidized to benzocoumarins in various ways[5] but it was shown, in this case, that the acid (4) was *not* converted into the lactones (5) and (6) under the conditions used for their simultaneous formation. Lederer and his co-workers[2] consider the lactone formation to be essentially a Dakin reaction proceeding (mainly) by way of an intermediate (8) to (9), followed by hydrolysis and oxidative decarboxylation of the resulting α-keto-acid. Subsequent acidification would give

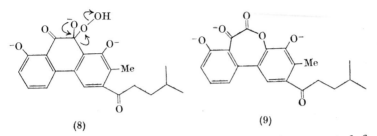

(8)   (9)

the lactone (5). Preponderant formation of (5) is accounted for by preferential nucleophilic attack on the quinone carbonyl group which is *para* to that at C-3.

4-Hydroxypiloquinone. The n.m.r. and mass spectra of this quinone establish that it has all the distinguishing features of the preceding pigment, and it differs only in the possession of an additional hydroxyl group. This is confirmed by the formation of a triacetate and a leucopentaacetate, and since there are three singlets at low field in the n.m.r. spectrum of the metabolite but no hydroxyl absorption in the 3 μ region all three phenolic groups must be strongly chelated. This suggests that hydroxypiloquinone has structure (2)[66] but the infrared absorption is anomalous.

THE ROYLEANONES

Royleanone (10; R = R' = H), $C_{20}H_{28}O_3$
Horminone (10; R = HO, R' = H), $C_{20}H_{28}O_4$
Taxoquinone (10; R = H, R' = HO), $C_{20}H_{28}O_4$
Horminone 7-acetate (10; R = OAc; R' = H), $C_{22}H_{30}O_5$
6,7-Dehydroroyleanone (11), $C_{20}H_{26}O_3$

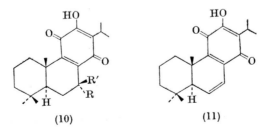

(10)   (11)

Occurrence: (10; R = H and OAc; R′ = H) and (11) in roots of *Inula royleana* D.C.[7] (Compositae) and *Salvia officinalis* L.[8] (Labiatae), leaves and stems of mixed *Plectranthus* spp.[9] (Labiatae) [(11) only in *P. parviflorus* var. *major* Willd.[9]], (10; R = OAc; R′ = H) in roots of *Salvia nemorosa* Crantz,[64] (10; R = R′ = H) and (11) in an unidentified *Plectranthus* (or *Coleon*) sp.,[62] (10; R = OH; R′ = H) in whole plants of *Horminium pyrenaicum* L.[11] (Labiatae), (10; R = R′ = H) in seeds of *Taxodium distichum* Rich[63] (Taxodiaceae).

Physical properties: (10; R = R′ = H) orange (and yellow) crystals, m.p. 181·5–183°, $[\alpha]_D$ +134° (CHCl$_3$), $\lambda_{max.}$ (CCl$_4$) 277, 283(sh), 403 nm (log ϵ 4·20, 4·18, 2·71), $\nu_{max.}$ (CHCl$_3$) 3350, 1672, 1632, 1602 cm^{-1}.

(11) orange crystals, m.p. 166·5–167°, $[\alpha]_D$ −620° (CHCl$_3$) $\lambda_{max.}$ (EtOH) 213, 245 (sh), 329, 455 nm (log ϵ 4·21, 3·89, 3·87, 2·87), $\nu_{max.}$ (CHCl$_3$) 3340, 1666, 1635, 1615(sh), 1595(sh) cm^{-1}.

(10; R = OAc; R′ = H) yellow crystals, m.p. 212–214·5°, $[\alpha]_D$ −14° (CHCl$_3$), $\lambda_{max.}$ (EtOH) 272, 407 nm (log ϵ 4·12, 2·92), $\nu_{max.}$ (CS$_2$) 3360, 1748, 1675, 1645 cm^{-1}.

(10; R = OH; R′ = H) orange plates, m.p. 172–173°, $[\alpha]_D^{20}$ −130° (CHCl$_3$), $\lambda_{max.}$ (EtOH) 217, 273, 408 nm (log ϵ 3·87, 4·10, 2·99), $\nu_{max.}$ (KBr) 3540, 3360, 1675, 1644, 1627, 1605 cm^{-1}.

(10; R = H, R′ = HO) orange needles, m.p. 212–214°, $[\alpha]_D^{28}$ +340° (CHCl$_3$), $\lambda_{max.}$ (MeOH) 276, 408 nm (log ϵ 4·08, 2·90), $\nu_{max.}$ (CHCl$_3$) 3559, 3390, 1681, 1656, 1629, 1600 cm^{-1}.

Royleanone and 6,7-dehydroroyleanone. The presence of yellow pigment in the roots of *I. royleana* was first noted by Handa *et al.*,[10] and later Edwards and his co-workers[7] identified this as a mixture of three diterpenoid quinones which they called royleanones. Complete separation of (10; R = R′ = H) and (11) is difficult.[7, 9] Royleanone is acidic (pK_a 8·5) giving a magenta solution in alkali; it absorbs one mol. of hydrogen to form a colourless product which reverts to the parent compound in air, and it affords a methyl ether. These facts, and the spectral data given above, indicate that royleanone is a hydroxybenzoquinone, and this was confirmed by reductive methylation of the methyl ether which gave a trimethoxy derivative showing benzenoid absorption. On hydrogenation, dehydroroyleanone takes up two mols. of hydrogen whence aerial re-oxidation gives royleanone. It follows that the two pigments differ by the presence of a double bond in dehydroroyleanone which is confirmed by signals from two vinylic (and one allylic) protons in the n.m.r. spectrum of its methyl ether, not present in the spectrum of royleanone. Comparison of ultraviolet visible-absorption spectra (above) shows that the double bond is in conjugation with the quinonoid chromophore, and the ultraviolet absorption of dehydroroyleanone leucotriacetate is similar to that of 1,2-dihydronaphthalene. The n.m.r. spectrum of royleanone does not extend below τ 5·95 showing that the quinone ring is fully substituted, and the partial structures (12) and (13) may be written for royleanone and dehydroroyleanone, respectively.

The remaining features of these molecules were elucidated by degradative experiments. The group R must be isopropyl as oxidation of both compounds with alkaline hydrogen peroxide yielded isobutyric acid, while more extensive breakdown of dehydroroyleanone with

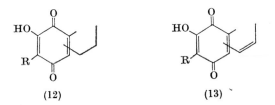

(12)   (13)

alkaline permanganate gave the di-acid (14). This could be recognised from the n.m.r. spectrum of its dimethyl ester which shows singlets from three quaternary methyl groups and another (1H) arising from an isolated proton. These facts lead to the conclusion that royleanone is

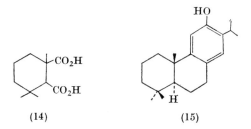

(14)   (15)

either (10; R = R' = H) or the isomer in which the hydroxyl and iso-propyl groups are interchanged. A decision in favour of (10; R = R' = H) was obtained by a synthesis [7] from ferruginol (15), which is also present in *I. royleana*. Oxidation of the phenol with acidic hydrogen peroxide gave a small yield of quinone identical with royleanone in optical rotation and other physical properties. As the absolute configuration of ferruginol is established, structures (10; R = R' = H) and (11) can be assigned to royleanone and its dehydro derivative, respectively.[7]

Horminone 7-acetate is the major pigment in the root of *I. royleana*.[7] The presence of an acetate group was evident both from the infrared and n.m.r. spectra, and a signal (1H) from a proton at τ 4·25 suggested that the acetoxyl group was secondary. Hydrolysis gave an alcohol (and acetic acid) which on dehydration with *p*-toluenesulphonic acid in boiling xylene gave 6,7-dehydroroyleanone (11). The ester group could therefore be α or β to the quinone ring, and since it can be cleaved by hydrogenolysis (or by treatment with sodium borohydride) it is clearly allylic or benzylic. This quinone is therefore 7-acetoxyroyleanone, the acetoxyl group being in the α-configuration (see below).

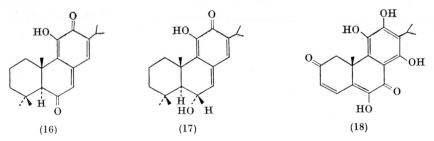

(16)   (17)   (18)

Horminone and taxoquinone. Hydrogenolysis of the hydroxyroyleanone (horminone) from *H. pyrenaicum*, or reduction with potassium boro-hydride, gave "desoxyhorminone" identical with royleanone, while the mono-acetate, formed along with the diacetate by acetylation in pyridine, is identical with 7-acetoxyroyleanone from *I. royleana*. Horminone is therefore 7-hydroxyroyleanone[11] and in fact (10; R = HO, R′ = H)[63] as the C-7 proton shows a broad signal at τ 5·23 ($W_{1/2}$, 8 Hz) corresponding to a β-equatorial proton. The epimer, taxoquinone (10; R = H, R′ = HO),[63] has an α-axial proton at C-7 distinguished by a multiplet centred at τ 5·20 ($W_{1/2}$, 20 Hz).

Traces of other unidentified hydroxyquinones are present in *H. pyrenaicum*[11] and *S. officinalis*,[8] while the seeds of *T. distichum*[63] also contain the two quinone methides taxodione (16), and taxodone (17), which show significant anti-tumour properties. Coleon B (18) (see p. 334) a yellow pigment from *Coleus ignarius* Schweinf.[65] (Labiatae), although a more highly oxidised diterpene, is not quinonoid.

THE TANSHINONES

Tanshinone I (19), $C_{18}H_{12}O_3$
Tanshinone IIA (20; R = H), $C_{19}H_{18}O_3$
Tanshinone IIB (21; R = CH₂OH), $C_{19}H_{18}O_4$
Cryptotanshinone (22), $C_{19}H_{20}O_3$
Hydroxytanshinone IIA (20; R = OH), $C_{19}H_{18}O_4$
Methyl tanshinonate (21; R = CO₂Me), $C_{20}H_{18}O_5$

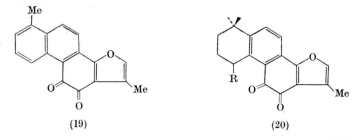

(19)   (20)

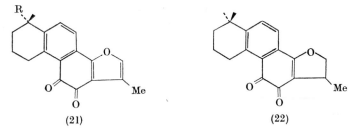

(21)   (22)

Occurrence: root of *Salvia miltiorrhiza* Bge.[12-14, 49] (Labiatae).

Physical properties: (19) brown-red needles, m.p. 233–234°, $\lambda_{max.}$ (EtOH) 244·5, 266(sh), 325, 417 nm (log ϵ 4·62, 4·31, 3·68, 3·70), $\nu_{max.}$ (KBr) 1688(sh), 1660, 1593 cm^{-1}.

(20; R = H) red needles, m.p. 198–200°, $\lambda_{max.}$ (EtOH) 225, 251, 268, 272(sh), 348, 460 nm (log ϵ 4·29, 4·30, 4·42, –, 3·24, 3·47), $\nu_{max.}$ (KBr) 1690, 1670 cm^{-1}.

(20; R = OH) red crystals, m.p. 187°, $[\alpha]_D$ 0° (CHCl$_3$), $\lambda_{max.}$ (EtOH) 222, 252, 273, 348, 462 nm, $\nu_{max.}$ (KBr) 3525, 1685, 1670 cm^{-1}.

(21; R = CH$_2$OH) red crystals, m.p. 200–204°, $[\alpha]_D^{19·8}$ −48·4° (Me$_2$CO), $\lambda_{max.}$ (EtOH) 226, 252, 269, 276(sh), 348, 465 nm, $\nu_{max.}$ (KBr) 3558, 3465, 1690, 1665, 1638 cm^{-1}.

(21; R = CO$_2$Me) red crystals, m.p. 175–176°, $[\alpha]_D$ −139° (CHCl$_3$), $\lambda_{max.}$ (EtOH) 223, 252, 269, 352, 465 nm, $\nu_{max.}$ (KBr) 1725, 1690, 1670 cm^{-1}.

(22) orange plates, m.p. 184–185°, $[\alpha]_D^{21}$ −91·4° (CHCl$_3$), $\lambda_{max.}$ (EtOH) 221, 263, 272, 290, 355, 447 nm (log ϵ 4·26, 4·47, 4·41, 3·96, 3·41, 3·48), $\nu_{max.}$ (KBr) 1680, 1648, 1620 cm^{-1}.

Tan-shen, the root of *S. miltiorrhiza*, is an ancient Chinese drug from which Nakao[12] originally isolated three red crystalline substances by alcohol extraction. These were designated tanshinones I, II and III, but II and III were not pure. Subsequently Takiura[13] showed that III was a mixture of II and another pigment, cryptotanshinone, and later he found[14] that, after removal of I, the remaining mixture could be resolved chromatographically into tanshinones IIA and IIB, and cryptotanshinone. Two further pigments, hydroxytanshinone IIA and methyl tanshinonate, have been isolated recently.[49]

Tanshinone I. This pigment was characterised as an *o*-quinone by the formation of a quinoxaline and a leucodiacetate and its structure was elucidated mainly by the work of Wessely.[15, 16] Oxidation with chromic acid gave a yellow anhydride, C$_{13}$H$_8$O$_3$, which yielded 1-methylnaphthalene after decarboxylation, and was shown, by synthesis, to have structure (23).[15] As tanshinone I is a quinone, a third six-membered ring is presumably fused to the naphthalene system at positions 3 and 4 which was confirmed independently by Takiura[18] who isolated 1 methylphenanthrene from a zinc dust distillation. Thus tanshinone I appears to be a methylphenanthraquinone, the remaining unidentified fragment, C$_3$H$_4$O, being attached to the quinonoid ring. This moiety contains a

21

C-methyl group (since tanshinone I has two) and an inert oxygen atom, and as ethylenic unsaturation appeared to be absent Wessely and Wang[15] concluded that this fragment was part of a furan ring. This leads to eight

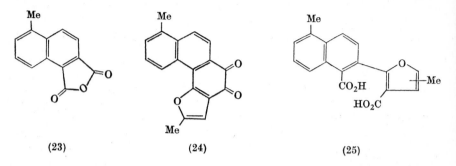

(23)   (24)   (25)

possible structures for the natural pigment of which (19) and (24) are examples. Evidence for a 3,4-quinone structure was obtained[19] by aerial oxidation in basic solution which produced a dibasic acid, $C_{18}H_{14}O_5$, easily converted into an anhydride. Decarboxylation afforded an oil, $C_{16}H_{14}O$, characterised as its red picrate, which could be further degraded by oxidation with permanganate in acetone, to 1-methylnaphthalene-6-carboxylic acid. The dibasic acid was therefore formulated as (25). Information on the position of the methyl group in the furan ring was obtained by limited ozonolysis[16] of the dimethyl ether of leucotanshinone-I. The properties of the yellow product, $C_{19}H_{18}O_4$, were consistent

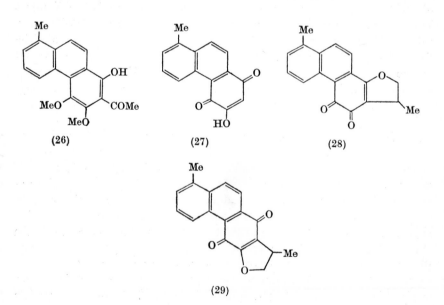

(26)   (27)   (28)

(29)

with the o-hydroxyketone structure (26) indicating that tanshinone I has structure (19) which, as Takiura[19] noted, has a close relationship to abietic acid and similar diterpenes.[50]

This structure has recently been confirmed by synthesis.[20] When the hydroxyquinone (27) was treated with β-chloro-α-methylpropionyl peroxide the initial C-alkylation was followed by spontaneous cyclisation with loss of hydrochloric acid, the quinone reacting conveniently in the tautomeric o-quinonoid form to give (28). Dehydrogenation of (28) with DDQ then gave (19) identical with tanshinone-I. In another approach[55] o-methylstyrene was added to 3-methylbenzofuran-4,7-quinone. The desired Diels-Alder adduct oxidized to the phenanthrenequinone during repeated crystallisation, and was then hydrogenated to the dihydro derivative (29), which was rearranged to the isomer (28) by dissolution in alkali and treatment with hot acid.

Cryptotanshinone.[18] Like its co-pigments, cryptotanshinone forms a quinoxaline and a leucodiacetate, and gave a yellow oil on zinc dust distillation which afforded 1-methylphenanthrene *after* dehydrogenation with sulphur (cf. tanshinone I). Although neutral, it dissolves in alcoholic sodium hydroxide giving a red solution from which a yellow hydroxy-quinone, $C_{19}H_{22}O_4$, is liberated on careful acidification. This new quinone loses the elements of water and reverts to cryptotanshinone when treated with concentrated sulphuric acid, and in this and other ways, there is a striking resemblance to the chemistry of dunnione and related compounds (see p. 214) in which a dihydrofuran or dihydropyran ring is fused to a quinone ring. Oxidation of the yellow hydroxyquinone with alkaline hydrogen peroxide yielded a volatile acid, regarded as β-hy-droxyisobutyric acid, and a crystalline dibasic acid, $C_{14}H_{16}O_4$, which was

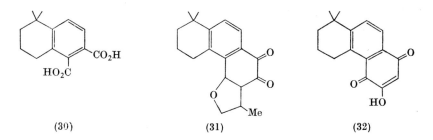

(30)   (31)   (32)

shown to be (30) since it could be decarboxylated over copper chromite to 1,1-dimethyltetralin and dehydrogenated with sulphur to 1-methyl-naphthalene-5,6-dicarboxylic acid. Consideration of all these facts leads to the conclusion that cryptotanshinone is either (22) or (31), and Takiura[18] preferred the former because of its relationship to abietic acid.

He was later shown to be correct when racemic cryptotanshinone was synthesised [20] by reaction of the hydroxyquinone (32) with β-chloro-α-methylpropionyl peroxide. An alternative synthesis [55] starts by Diels-Alder addition of the diene (33) [56] to 3-methylbenzofuran-4,7-quinone to give, exclusively, the adduct (34). After aerial oxidation to the quinone in basic solution, catalytic reduction afforded the dihydro compound (35)

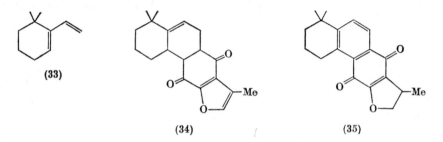

(33)   (34)   (35)

which could be rearranged to cryptotanshinone by treatment first with alkali, and then with acid.

Tanshinone IIA. It is clear from the ultraviolet-visible spectra of these pigments and their leuco derivatives that tanshinone IIA is neither a furanophenanthrenequinone like tanshinone I nor an *o*-naphthaquinone like cryptotanshinone.[21] However it is an *o*-quinone (quinoxaline formation), and since chromic acid oxidation gives the di-acid (30)[14, 17] accompanied by loss of a C_5H_4 fragment (as in the corresponding degradation of tanshinone I), structure (20a), that is, dehydrocrypto-tanshinone, seems likely by analogy. Direct evidence for this was obtained by hydrogenation [21] of IIA over palladium when two mols. of hydrogen were absorbed and reoxidation of the product then gave cryptotanshinone (isolated as the ferric chloride complex). Conversely,

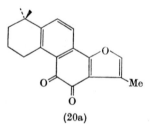

(20a)

dehydrogenation [14] of the quinoxaline of cryptotanshinone with palladium in cinnamic acid gave the quinoxaline of tanshinone IIA. Dehydrogenation of racemic cryptotanshinone, obtained synthetically, has recently been effected using DDQ which constitutes a synthesis[21] of tanshinone-IIA.

In another recent synthesis Japanese workers[53] obtained the phenanthrene (37) by standard methods starting from the tetralone (36). Curiously, the methoxyl group *peri* to the keto function in (36) was replaced by hydrogen during a Reformatsky reaction with ethyl γ-bromocrotonate. Further elaboration of (37) led to (38) from which the

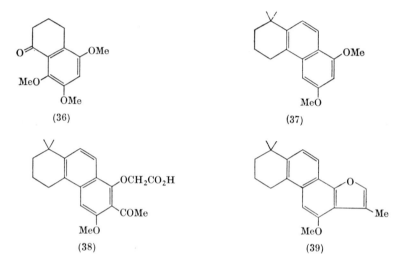

(36)　　　　　　　　　(37)

(38)　　　　　　　　　(39)

furanophenanthrene (39) was derived by refluxing with acetic anhydride and sodium acetate. After demethylation by heating with methyl magnesium iodide, the resulting phenanthrol was smoothly oxidised by air (but not by Fremy's salt) to tanshinone IIA.

Tanshinone IIB. This pigment[14] is optically active and contains one atom of oxygen more than tanshinone IIA, assigned to a hydroxyl group. All the spectra of IIA and IIB are very similar except for the 3 μ region

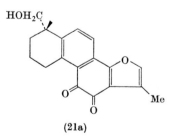

(21a)

of the infrared and a significant difference in the n.m.r. spectra where the 6-proton singlet at τ 8·81 (gem-dimethyl group) in that of IIA is replaced by a 3-proton singlet in the spectrum of IIB. Accordingly, the hydroxyl group is located as shown in (21a).[20] In the mass spectrum of IIB, the molecular ion at m/e 310 is followed by peaks at m/e 292 (M-18) and 279 (M-31) (the base peak), and the remainder of the spectrum is virtually identical with that of tanshinone IIA.

Hydroxytanshinone IIA. The ultraviolet-visible absorption curve of this pigment is superimposable on that of tanshinone IIA, and their infrared spectra are very similar in the 3–6 μ region except for a peak at 3525 cm^{-1} shown by the new quinone. It may be concluded that this $C_{19}H_{18}O_4$ molecule has the same furano-*o*-naphthaquinone chromophore as IIA with an additional hydroxyl group in the saturated part of the molecule. Reference to the n.m.r. spectrum shows clearly that this group must occupy a benzylic position as in (20b) although the compound is optically

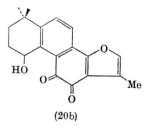

(20b)

inactive. Comparison with the n.m.r. spectrum of tanshinone IIA shows that hydroxytanshinone IIA contains exactly the same proton distribution except that the 2-proton multiplet at ca. τ 6·9 in IIA, arising from the benzylic methylene group, is replaced by a 1-proton multiplet at τ 5·62. This was confirmed by catalytic reduction over palladium which yielded, after re-oxidation, tanshinone IIA and cryptotanshinone, with hydrogenolysis of the benzylic hydroxyl group.[49]

Methyl tanshinonate was recently isolated by Kakisawa and co-workers[49] from tan-shen in 0·002% yield. The molecular formula, $C_{20}H_{18}O_5$, shows that it contains one carbon and two oxygen atoms more than tanshinone IIA, and the close similarity of the ultraviolet-visible, infrared and n.m.r. spectra again reveals a tanshinone IIA-type pigment. The presence of a methoxycarbonyl group is evident from a peak at 1725 cm^{-1} and a singlet (3H) at τ 6·34, and it is clearly located as shown in (21b) by a singlet (3H) at τ 8·38 which replaces the 6-proton singlet in the n.m.r. spectrum of tanshinone IIA arising from the *gem*-dimethyl group.

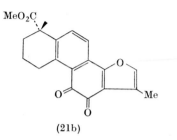

(21b)

THE ISOTANSHINONES

Isotanshinone I (40), $C_{18}H_{12}O_3$
Isotanshinone II (34), $C_{19}H_{18}O_3$
Isocryptotanshinone (35), $C_{19}H_{20}O_3$

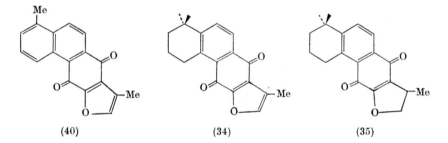

(40)   (34)   (35)

Occurrence: root of *Salvia miltiorrhiza* Bge.[67] (Labiatae).
Physical properties: (40) orange-red crystals, m.p. 219°, $\lambda_{max.}$ (EtOH) 234, 293, 346, 450 nm (log ϵ 4·58, 4·45, 3·78, 3·28), $\nu_{max.}$ (KBr) 1660, 1585 cm^{-1}.
(34) yellow crystals, m.p. 208°, $\lambda_{max.}$ (EtOH) 227, 253, 256, 303, 361 nm (log ϵ 4·02, 4·33, 4·34, 3·59, 3·63), $\nu_{max.}$ (KBr) 1655, 1585 cm^{-1}.
(35) yellow crystals, m.p. 121°, $[\alpha]_D$ +55·6° (dioxan), $\lambda_{max.}$ (EtOH) 251, 258, 299, 365 nm (log ϵ 3·97, 4·10, 3·79, 3·30), $\nu_{max.}$ (KBr) 1665, 1645 cm^{-1}.

These minor pigments had been synthesised[55] in Kakisawa's laboratory as intermediate products en route to the isomeric tanshinones before they were found[67] in tan-shen. Consequently they were quickly recognised from their spectroscopic properties, and identity was confirmed by direct comparison with synthetic samples. Comparison of the infrared and ultraviolet-visible data above with those given on p. 641 for the corresponding tanshinones shows that the new pigments are *p*-quinones. The n.m.r. spectra of appropriate pairs are virtually identical except that in the isotanshinone group the AB quartet is shifted downfield.

Miltirone (41), $C_{19}H_{22}O_2$

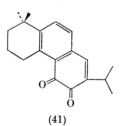

(41)

Occurrence: root of *Salvia miltiorrhiza* Bge.[69] (Labiatae).

Physical properties: red crystals, m.p. 100°, $\lambda_{max.}$ (EtOH) 260, 362, 436 nm (log ϵ 4·55, 3·36, 3·52), $\nu_{max.}$ (KBr) 1680(sh), 1660, 1635 cm⁻¹.

This is the tenth diterpenoid quinone to be isolated from tan-shen, and it differs from the rest in possessing only two atoms of oxygen. Comparison (infrared, ultraviolet-visible) with other members of the group shows that it is another *o*-quinone. All the protons are recognisable from the n.m.r. spectrum, the benzylic methylene group signal appearing typically at relatively low field (τ 6·80) due to the deshielding effect of the adjacent carbonyl group. This gives the orientation of the quinone function, the isopropyl group being obviously located as shown. Further support for structure (41) was obtained by Thiele acetylation, followed by hydrolysis and oxidation, to give the hydroxy-1,4-naphthaquinone (42) ($\nu_{max.}$ 3350, 1645 cm⁻¹) whose ultraviolet absorption was almost

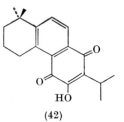

(42)

identical with that of (32). Relative to (41), the n.m.r. spectrum of the *p*-quinone (42) showed the expected changes, in particular the disappearance of the quinonoid proton doublet and the downfield shift of the AB quartet.

Tetrangulol (43), $C_{19}H_{12}O_4$

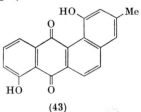

(43)

Occurrence: cultures of *Streptomyces rimosus* Sobin *et al.*[22]
Physical properties: purple needles, m.p. 198–200°, $\lambda_{max.}$ (MeOH) 225, 250(sh), 315, 425 nm (log ϵ 4·70, 4·31, 4·39, 3·85), $\nu_{max.}$ (KBr) 1640, 1625 cm^{-1}.

Tetrangulol is the first representative of a new type of natural quinone. It shows normal redox properties, and forms a diacetate, and mono- and dimethyl ethers. The absence of hydroxyl absorption in the 3 μ region and the carbonyl shifts on methylation and acetylation (to 1660–1670 cm^{-1}) indicate that both carbonyl groups are chelated. However, in contrast to other natural quinones, the ultraviolet absorption of the diacetate is like that of 1,2-benzanthraquinone and zinc dust distillation gave a crystalline hydrocarbon which was clearly a methyl-1,2-benzanthracene although the exact position of the methyl group was not determined. The n.m.r. spectra of tetrangulol and its derivatives were very informative. That of the dimethyl ether, in addition to a singlet from an aromatic methyl group, shows an AB quartet (J, 8 Hz) attributed to two *ortho*-coupled aromatic protons, two doublets (J, 2 Hz) arising from two *meta*-protons, and a more complex signal (3H), best seen in the spectrum of tetrangomycin (45) (see below), assigned to a fragment (44) by comparison with the spectrum of ϵ-rhodomycinone (p. 560). Oxidation

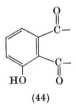

(44)

of tetrangulol with alkaline permanganate yielded benzene 1,2,3,4-tetracarboxylic acid which can only be derived from ring C of a 1,2-benzanthraquinone, and must also be the origin of the AB quartet. Consideration of all these and other data led Kunstmann and Mitscher[22] to assign structure (43) conclusively to tetrangulol. The natural occurrence of tetrangulol is of some interest in view of the well-known carcinogenic properties of benzanthracenes.

Tetrangomycin (45), $C_{19}H_{14}O_5$

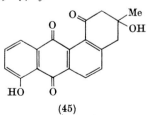

(45)

Occurrence: cultures of *Streptomyces rimosus* Sobin et al.[22]

Physical properties: yellow crystals, m.p. 182–184°, $[\alpha]_D^{25}$ +41·8° (CHCl₃), λ_{max}. (MeOH) 267, 330(sh), 400 nm (log ε 4·50, 3·46, 3·72), ν_{max}. (KBr) 3570, 3420, 1705, 1678, 1642 cm⁻¹.

On treatment with alkali tetrangomycin loses the elements of water to form its co-pigment tetrangulol. It differs from the latter in possessing a non-chelated hydroxyl group which is difficult to acetylate, a non-chelated quinone carbonyl, and a conjugated carbonyl group which absorbs at relatively high frequency (1705 cm⁻¹) and disappears on dehydration; the ultraviolet-visible spectrum is like that of a 1-hydroxy-anthraquinone. The n.m.r. spectrum shows the same AB and ABX systems as does tetrangulol, but the signals from the *meta* protons are replaced by singlets (2H) at τ 6·87 and 7·00 attributed to isolated methylene groups. All these facts are readily explained if tetrangomycin has structure (45).

Ochromycinone (46), C₁₉H₁₄O₄

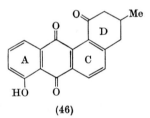

(46)

Occurrence: cultures of unidentified *Streptomyces*.[23]

Physical properties: yellow crystals, m.p. 152–153°, $[\alpha]_D^{25}$ +204·5° (CHCl₃), λ_{max}. (EtOH) 265, 405 nm (log ε 4·42, 3·55), ν_{max}. 1703, 1668, 1638 cm⁻¹.

Like the preceding quinones, zinc dust distillation of ochromycinone affords 3-methylbenz[α]anthracene which determines the carbon skeleton, assuming that no rearrangement has occurred. Both the infrared and the ultraviolet-visible spectra suggest a 1-hydroxyanthra-quinone chromophore with an additional conjugated carbonyl group. It forms a methyl ether, an acetate and, under forcing conditions, a 2,4-dinitrophenylhydrazone. On this information, four structures (47

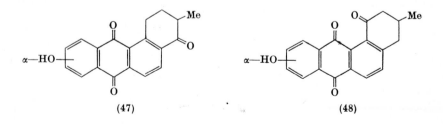

(47)   (48)

and 48) can be written for ochromycinone, all of which are in substantial agreement with the n.m.r. spectrum which confirms the substitution pattern in rings A and C and shows a doublet (3H) due to the methyl protons, and a complex pattern arising from the remaining protons of ring D. Structures (47) and (48) could be distinguished by consideration of the mass spectrum which revealed a breakdown pattern essentially that of an α-tetralone.[24] The fragmentation M–C₃H₆–CO–H (see below) was established both by the presence of the appropriate

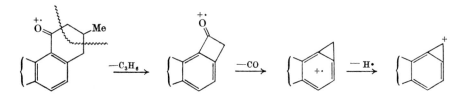

metastable peaks and by deuteriation studies. When the protons α to the carbonyl group were replaced by deuterium the fragmentation changed to M–C₃H₄D₂–CO–H. This mode of decomposition is consistent only with structures (48) of which (46) was preferred on biogenetic grounds.[23]

Phenocyclinone (49), $C_{35}H_{24}O_{14}$

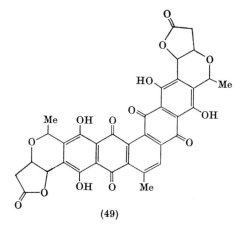

(49)

Occurrence: mycelium of *Streptomyces coelicolor* (Müller) Waksman and Henrici.[70]
Physical properties: amorphous red solid, $\lambda_{max.}$ (CHCl₃) 290, 498 nm, $\nu_{max.}$ (KBr) 1783, 1620 cm⁻¹.

Strep. coelicolor elaborates the binaphthazarin, actinorhodin (p. 302), and recently Brockmann and Christiansen[70] have described a second pigment from the same organism which is related in structure but contains a third aromatic ring. The preferred structure (49) is a derivative of the pentaphene ring system not found hitherto in natural products. The new pigment gives similar colour reactions to binaphthazarin and absorbs somewhat less than two mols. of hydrogen catalytically. Although all four hydroxyl groups are chelated (ν_{CO} 1620 → 1672 cm^{-1} on acetylation) it forms, surprisingly, an amorphous, yellow, tetramethyl ether with diazomethane.

A large part of the structure was identified when it was found that the non-aromatic part of the n.m.r. spectrum of the tetramethyl ether agreed closely with those of granaticin (p. 298), its degradation product (50) and kalafungin (51).[71]† From the relative intensities, two of these pyrano-lactone units are present. In addition there is a signal (1H) at τ 1·79

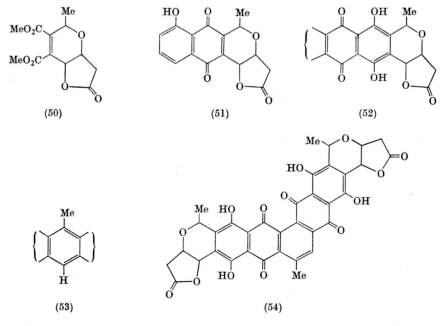

(50)   (51)   (52)

(53)   (54)

assigned to a *peri* aromatic proton and another at τ 7·14 attributed to an aromatic methyl group in a *peri*-position to a carbonyl group. Combining a pyranolactone unit with a naphthazarin nucleus leads to the part structure (52), two of which would account for 32 out of the total of 35 carbon atoms in phenocyclinone. Bearing in mind that both naphtha-zarin systems must be fully substituted, several possible formulae can now be written by introducing aromatic methine and methyl groups into a 2 × (52) structure. Of these (49) was preferred. The n.m.r. spectrum of

† From *Strep. tanashiensis.*[71] Structure determined[72] by X-ray crystallographic analysis, no other details available.

the tetramethyl ether shows that two of the methoxyl groups are in the same environment (6H singlet at τ 6·06) and the other two are in different environments (3H singlets at τ 5·88 and 5·92). This would not be the case in a linear structure centred on ring (53), but both the angular arrangement (49) and the alternative (54) with one of the pyranolactone units reversed are in agreement with the n.m.r. data. Assuming polyketide biogenesis structure (49) again seems the more likely,[70] the central carbon atoms, $CH_3-C{=}CH-$, probably originating from propionate. The tautomeric form shown (49) is also preferred as it contains the maximum number of benzenoid rings.

The n.m.r. spectrum indicates that phenocyclinone has the same relative configuration at each end of the molecule, and it must be assumed that the lactone rings are cis-fused to the adjacent heterocycle. While the absolute configuration at the asymmetric centres is not known it would be reasonable to assume that it is the same as in actinorhodin which is produced by the same organism (granaticin has the opposite configuration).

The Mitomycins†

Mitomycin A (55; R = Me; R' = H), $C_{16}H_{19}N_3O_6$
Mitomycin B (55; R = H; R' = Me), $C_{16}H_{19}O_6$
Mitomycin C (56; R = H), $C_{15}H_{18}N_4O_5$
Porfiromycin (56; R = Me), $C_{16}H_{20}N_4O_5$

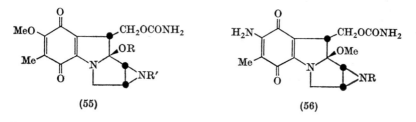

(55)   (56)

Occurrence: cultures of *Streptomyces caespitosus* Sug. & Hata,[25, 26] *Strep. verticillatus* Kriss,[27] and *Strep. ardus* DeBoer et al.[52]

† The following nomenclature has been proposed:

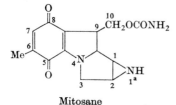

Mitosane

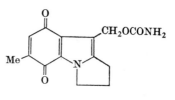

Mitosene

Physical properties: A, purple needles, m.p. 159–160°, $[\alpha]_D^{25}$ −143° (MeOH), λ_{max}. (MeOH) 218, 320, 520 nm (log ϵ 4·24, 4·02, 2·72), ν_{max}. (KBr) 3413, 3336, 1709, 1645, 1585 cm⁻¹.

B, dark purple-blue needles, dec., no m.p., $[\alpha]_D^{25}$ −835° (MeOH), λ_{max}. as for A, ν_{max}. (KBr) 3436, 3300, 3205, 1736, 1695, 1658, 1618 cm⁻¹.

C, deep purple needles, m.p. >300°,† λ_{max}. (MeOH) 217, 360, 555 nm (log ϵ 4·39, 4·36, 2·32), ν_{max}. (KBr) 3400, 3280, 1724, 1642, 1597 cm⁻¹.

Porfiromycin, dark purple crystals, m.p. 201° (dec.), $[\alpha]_D^{25}$ +275° (±55°) (MeOH), λ_{max}. as for C, ν_{max}. (Nujol) 3370, 3250, 1720, 1690, 1640, 1600 cm⁻¹.

The mitomycins were discovered originally by Japanese workers[25, 26] and proved to have marked antibiotic and antitumour activity.[28, 59] Mytomycin C is now commercially available in Japan for clinical use. Their structures, which have several unusual features, were elucidated by the Lederle group,[29, 30] who reported briefly in 1962,‡ and were supported by independent work elsewhere.[31, 32] Six antibiotics were detected in strains of *Strep. verticillatus* of which five were isolated as crystalline purple pigments[27] and four§ of these were interrelated by simple chemical transformations.[29] Mitomycin A contains two methoxyl groups, one of which can be replaced by an amino group on treatment with ammonia under very mild conditions. The resulting aminoquinone is mitomycin C. Similarly, N-methylmitomycin A, obtained by methylation with methyl iodide and sodium bicarbonate in aqueous dimethylformamide, reacts with methanolic ammonia (OMe → NH_2) to give porfiromycin which can also be obtained by methylation of mitomycin C.[46] Mitomycin B, which is isomeric with A but contains one O–Me and one N–Me group, has not been converted into or derived from any of the other pigments but identical degradation products have been obtained from both mitomycin B and porfiromycin.

The similarity of the ultraviolet-visible absorption of mitomycin A to that of 5-methoxy-2-dimethylaminobenzoquinone indicates the nature of the chromophore, and in aqueous alkali mitomycin A is converted into another compound (not isolated) whose light absorption is almost identical with that of 2,5-dihydroxy-*p*-xyloquinone. Likewise, the ultraviolet-visible absorption of mitomycin C and porfiromycin resemble that of 2,5-bis-dimethylaminobenzoquinone.

In dilute hydrochloric acid at 25°, mitomycin A is slowly converted into *apo*-mitomycin A, $C_{15}H_{17}N_3O_6$, with the liberation of one mol. of methanol.[29] This new, optically active, compound retains the methoxyl and methyl groups attached to the quinone ring of the parent compound

† Ammonium carbonate sublimes at ca. 200°.

‡ A full paper is still awaited.

§ The fifth pigment, mitiromycin, red purple needles, m.p. 124–126°, λ_{max}. (MeOH) 218, 323, 530 nm, is isomeric with mitomycins A and B, and contains 1 OMe, 1 NMe and 1 CMe group, and 2 or 3 active hydrogens, one being NH. Added in proof. See Appendix ref. 2.

(n.m.r.) and has, in addition, an aliphatic hydroxyl and an aliphatic primary amino group which show the appropriate amide and ester carbonyl absorption in the infrared spectrum of its diacetate. On further treatment with cold alkali the remaining methoxyl group was hydrolysed to give an acidic purple hydroxyquinone, $C_{14}H_{15}N_3O_6$. (The homologous blue quinone, $C_{15}H_{17}N_3O_6$, was obtained in the same way from both mitomycin B and porfiromycin.) Thus four of the six oxygen atoms in *apo*-mitomycin A have been identified and the other two are located in the uncommon group $-OCONH_2$.[29] Hydrolysis with 6N-hydrochloric acid at 25° liberated equimolecular amounts of carbon dioxide, ammonium ions and a violet compound, $C_{13}H_{14}N_2O_5$ [mitomycinone (63)[31]]; the quinone methoxyl group was also hydrolysed as the product formed a tetra-acetyl derivative. The infrared band at ca. 5·8 μ (1724 cm^{-1}) shown by *apo*-mitomycin A and all the mitomycins, but absent from the spectrum of this hydrolysis product, is therefore assigned to a carbamate carbonyl group.[33] The remaining unidentified nitrogen atom in *apo*-mitomycin A is non-basic and non-hydrolysable, and is clearly part of the aminobenzoquinone chromophore. That it is part of an indole system is evident from the near identity of the ultraviolet-visible spectrum of its alkaline hydrolysis product, $C_{14}H_{15}N_3O_6$, with that of the model compound (57). The partial structure (58) can now be written for *apo*-mitomycin A.

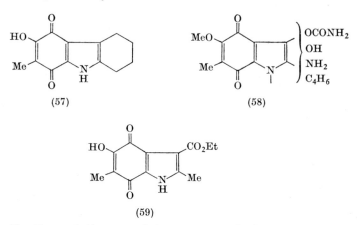

(57)   (58)

(59)

Clarification of the remaining structural details came from an examination of the product obtained by treatment of *apo*-mitomycin A with nitrous acid.[29] This optically inactive compound, $C_{15}H_{14}N_2O_6$, lacked the hydroxyl and amino groups of (58) but contained a new carbonyl group (ν_{CO} 1733 cm^{-1}). An intense band at 280 nm indicated that the new carbonyl group was in conjugation with the indolequinone chromophore, and probably attached to C-2 as the spectrum was unlike that of (59). The n.m.r. spectrum showed signals characteristic of an A_2X_2 system which is consistent with the presence of a $-CH_2-CH_2-CO-$ moiety linked by the carbonyl group to the indole nucleus at C-2 ,the other end being necessarily attached to the indole nitrogen as in (60) since a permanganate oxidation afforded β-alanine.

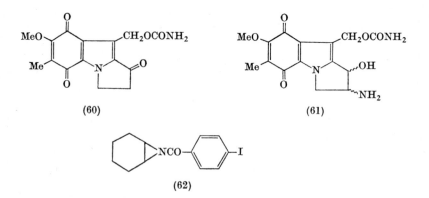

(60)   (61)

(62)

This places the $-CH_2OCONH_2$ substituent at C-3. It is now apparent that compound (60) arose by pinacolinic deamination of a 1,2-amino-alcohol function in *apo*-mitomycin A which could therefore be (61). This was sustained by the formation of an oxazolidinone on treatment of *apo*-mitomycin A with phosgene.

The formation of a 1,2-amino-alcohol structure on acid hydrolysis of mitomycin A strongly suggests that the antibiotic contains a fused aziridine ring. This receives some support[29] from (a) weak bands near 3·3 μ in all the mitomycins, possibly attributable[34] to aziridine C–H stretching vibrations; (b) a carbonyl band[35] at 1712 cm^{-1} in N-acetyl mitomycin A (prepared by treatment with 1-acetylimidazole[36]) and (c) an additional peak at 262 nm in the ultraviolet spectrum of the N-(4-iodobenzoyl) derivative in good agreement with that of compound (62). The final problem is the position of the acid-labile methoxyl group (τ 6·8) which was assigned to C-9a on the basis of the n.m.r. spectrum which shows a typical twelve-peak ABX pattern consistent with the coupling of a single proton on the asymmetric C-9 carbon atom with the dissimilar methylene protons at C-10. Mitomycin A is therefore (55; R = Me; R' = H),[29] the stereochemistry being derived from an X-ray crystallographic analysis[30] of the N-(4-bromobenzenesulphonyl) derivative which confirmed the structure proposed on chemical and spectroscopic grounds in every respect. The structures of mitomycins B and C, and porfiromycin, follow from the chemical relationships previously established and are fully consistent with their n.m.r. spectra. Structure

(61), arbitrarily assigned to *apo*-mitomycin A in the previous discussion, is also confirmed by its n.m.r. spectrum, and is the isomer to be expected by preferential attack on the more electrophilic carbon (C-1) of the aziridine ring during acid hydrolysis. The product, in fact, was a mixture of *cis*- and *trans*-isomers, and was always accompanied by a more soluble compound, probably the 1-amino-2-hydroxy isomer. Mitomycin C reacts[32] with glacial acetic acid to give the N-acetyl derivative (68) which presumably arises by initial formation of the 1-acetoxy-2-amino derivative, followed by normal acyl migration.

Independently, Japanese workers[31] arrived at a partial structure for mitomycinone, which they obtained from mitomycins A and C, and Stevens and co-workers[32] deduced the structure of mitomycin C by a detailed examination of its hydrolytic degradation products. Several interesting compounds were derived from mitomycinone (63). The

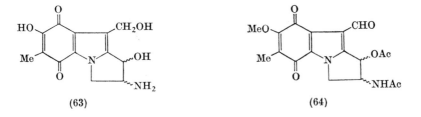

(63)   (64)

tetra-acetyl derivative gave a hexa-acetyl product on reductive acetylation, and a triacetate by selective hydrolysis of the acetoxy group at C-7 (vinylogous anhydride) with cold dilute methanolic ammonia. By treatment with cold acetic anhydride (in methanol) the N-acetyl compound was obtained and, after methylation with diazomethane, oxidation with the Jones reagent converted the primary alcohol group into an aldehyde, isolated as the diacetyl derivative (64). Direct evidence for the presence of a pyrrole ring in mitomycinone was secured by ozonolysis of the tetra-acetyl derivative which afforded the diketo-acid (65); periodate oxidation then degraded this to the di-acid

(65)   (66)

(66) which gave a positive pine splinter test and, on warming, a positive Ehrlich reaction.

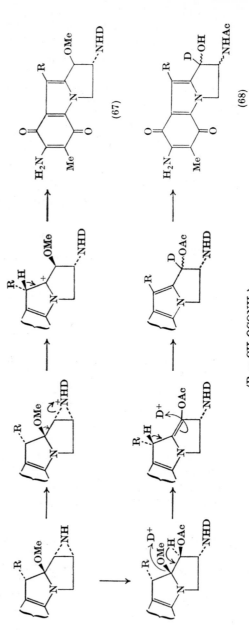

(67)

(68)

(R = CH₂OCONH₂)

In the reaction of mitomycin C with acetic acid, referred to previously, a minor product is the methoxyamine (67; H in place of D), which is the only compound so far obtained where the aziridine ring is opened and the adjacent methoxyl group retained. This is considered to involve 1,2-migration of the methoxyl function, as shown, since there is no deuterium incorporation (on carbon) when the reaction is conducted in acetic acid-d. On the other hand, the main product of the reaction (68) is deuteriated on C-1, the methoxyl group being eliminated as indicated. These reactions confirm the location of the labile methoxyl group at C-9. Another example of methoxyl elimination is seen in the catalytic hydrogenation of mitomycin B in dimethylformamide which yields the aziridinomitosene derivation (70).[37] Expulsion of the methoxyl group occurs at the quinol stage (see 69) in which the lone pair on the indole nitrogen is no longer in conjugation with the quinone system, and hence available to assist the elimination.

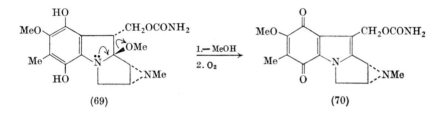

(69)   (70)

As an antibiotic the mitosene (70) is as potent as the mitomycins which suggests that this compound (or its quinol) may be the active species. It has been proposed[38] that the mitomycins owe their activity to the aziridine ring which functions as an alkylating group, analogous to other types of cytotoxic agents which are believed to attack DNA. Experiments with rat liver homogenates under both aerobic and anaerobic conditions suggest that reduction to the quinol may be a prerequisite for activation. There is some evidence[39] that mitomycin C can effect cross-linking of complementary DNA strands. In this connection it is of interest that the mitomycins contain a pyrrolizidine ring

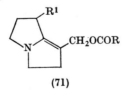

(71)

system, exemplified *par excellence* by the pyrrolizidine alkaloids.[47] Those represented by the general formula (71) cause both acute and chronic liver damage in rats, and it is considered[60] that they also act as alkylating agents, the active species arising by alkyl-oxygen fission of the allylic amino ester group. Similarly, the mitosene (54) (or its quinol) could function as an alkylating agent at C_{10} and this dual activity (C_{10} and the aziridine ring) would account for the DNA cross-linking function of the mitomycins.[59]

The tetracyclic structures of the mitomycins provide a challenge to the synthetic organic chemist which has yet to be met, but their biological activity has inspired a large amount of synthetic work in the indoloquinone and related fields.[40, 41, 61]

Streptonigrin (72), $C_{25}H_{22}N_4O_8$

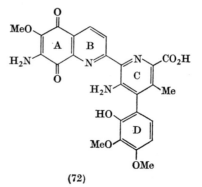

(72)

Occurrence: cultures of *Streptomyces flocculus* (Duché) Waksman & Henrici.[42]

Physical properties: dark brown plates, dec. ca. 275°, $\lambda_{max.}$ (MeOH) 245, 375 nm, $\nu_{max.}$ 3400, 3100, 1730, 1670, 1645(sh), 1610 cm^{-1}.

Like the mitomycins, the antibiotic streptonigrin is elaborated by a Streptomycete and has significant antitumour properties.[51] Evidence for the structure was put forward by Rao, Biemann and Woodward in 1963.[43]† The molecular formula was deduced by mass spectrometric comparison of the hexamethyl derivative, obtained by methylation of the leuco compound with methyl sulphate, and the hexadeuteriomethyl derivative prepared in the same manner. Alkaline hydrolysis of the former gave a pentamethyl derivative which evolved carbon dioxide when volatilised directly into the ion source of a mass spectrometer, and the spectrum showed an intense M-44 peak. Hence this compound and streptonigrin are monobasic acids.

Oxidative degradation of streptonigrin with alkaline hydrogen peroxide gave the tribasic streptonigric acid, $C_{22}H_{19}N_3O_9$, which afforded a tetramethyl derivative. Further oxidation with alkaline

† A full paper has not yet appeared.

permanganate led to the tetrabasic streptonigrinic acid, $C_{15}H_{11}N_3O_8$, characterised by its tetramethyl ester. Three of these carboxyl groups originate from substituents destroyed by oxidation, and comparison of the molecular formulae of these compounds suggested that conversion of streptonigrin into streptonigric acid eliminated a methoxyamino-quinone (or hydroxy-methoxyquinonimine) ring, and that further degradation to streptonigrinic acid removed a hydroxydimethoxy-phenyl substituent. The n.m.r. spectra of these compounds show that they all contain an aromatic methyl group and an aromatic primary amino group. Allowing for these it was deduced that the parent structure of streptonigric acid is $C_{10}H_8N_2$, that is, a bipyridyl, attached to which, in streptonigrin itself, are amino, carboxyl and methyl groups, and the substituted quinone and phenyl rings.

The structure of streptonigrinic acid was elucidated by the following reactions. The amino group was removed by treating the tetramethyl ester with nitric acid in ether, whence hydrolysis of the desamino derivative, and decarboxylation, gave 5-methyl-2,2′-bipyridyl (73).

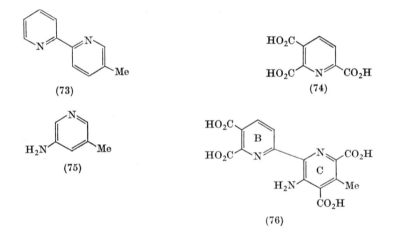

Oxidation of streptonigrinic acid with sodium hypochlorite destroyed half the molecule leaving the acid (74), and the other half was eliminated by hydrogenation over platinum, degradation of the product with alkaline permanganate, and final distillation over soda lime to yield 3-amino-5-methylpyridine (75). As two of the carboxyl groups in streptonigrinic acid presumably occupy adjacent positions as they originate from the breakdown of a quinone ring, and as the presence of two *ortho*-coupled aromatic protons is clear from an AB quartet in the n.m.r. spectrum, streptonigrinic acid must be (76), incorporating rings B and C of streptonigrin.

Turning to the larger degradation product, streptonigric acid, alkaline hydrolysis of the tetramethyl derivative, already mentioned, gave O-methylstreptonigric acid from which 2,3,4-trimethoxybenzoic acid was obtained by permanganate oxidation. This arises from the

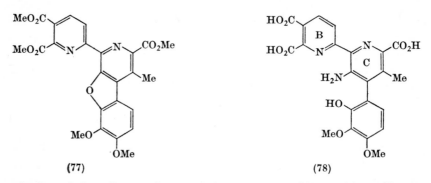

(77)   (78)

substituted phenyl group known to be present, and its position adjacent to the amino group in ring C was ascertained by treatment of the tetra-methyl derivative with nitric acid in ether. This reaction yielded the furan (77), and as the identical compound was obtained in the same way from the trideuteriomethyl ester of O-methyl streptonigric acid, the

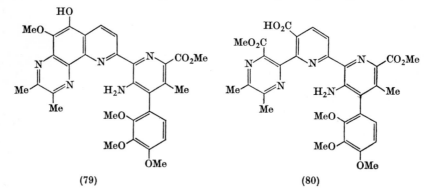

(79)   (80)

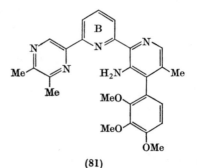

(81)

parent acid must have structure (78) with the phenolic group in ring D located as shown.

Finally, the structural arrangement of the quinone ring A in streptonigrin was arrived at in the following manner. (A quinonimine structure was excluded as prolonged methylation with methyl sulphate gave a non-acidic product containing only two new O-methyl groups.) The dimethyl derivative of streptonigrin reacted with hydroxylamine to give a mono-oxime (tautomeric with the corresponding nitrosophenol) easily reduced with dithionite to an o-diamine. This condensed with biacetyl to give a quinoxaline which must have structure (79) as oxidation with permanganate in hot pyridine resulted in cleavage of the phenolic ring and formation of the acid (80). Hydrolysis and decarboxylation of the latter gave the tetracyclic compound (81) whose n.m.r. spectrum showed, inter alia, a typical ABC pattern arising from the three adjacent protons in ring B. The substitution pattern in ring A of streptonigrin is thus defined and the complete structure is therefore (72).[43] Synthetic work[44] so far has been limited to related but much simpler molecules and no major assault on the structure has yet been announced.

Woodward and his colleagues[43] have drawn attention to the fact that streptonigrin, the mitomycins and the actinomycins (82)[45] possess the common structural unit (83), and all have marked anticancer activity.

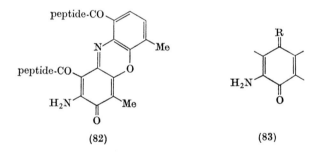

(82)   (83)

Five other streptonigrins (A_1, A_2, F, P_1 and P_2) have been isolated[57] from various strains of Strep. flocculus but no structural information is available. Valacidin,[58] an antibiotic obtained from a strain of Strep. lavendulae appears to be very similar and bruneomycin, a metabolite of Strep. albus var. bruneomycini is said[54] to be closely related.

REFERENCES

1. J. J. Chi, S. T. Hsu, M. Hu and S. Wang, J. Chinese Chem. Soc. (1947), **15**, 21.
2. B. C. Johnson, P. Cohen, J. Polonsky and E. Lederer, Nature (London) (1963), **199**, 285; J. Polonsky, B. C. Johnson, P. Cohen and E. Lederer, Bull. Soc. Chim. France, 1963, 1909.

3. A. Gaudemer, J. Polonsky and L. Alais, *Bull. Soc. Chim. France* (1963), 1918.

4. H. Bredereck, R. Sieber and L. Kamphenkel, *Chem. Ber.* (1956), **89**, 1169.

5. P. M. Brown, J. Russell, R. H. Thomson and A. G. Wylie, *J. Chem. Soc. (C)* (1968), 482 and references therein.

6. R. H. Thomson. "Naturally Occurring Quinones", (1957). 1st edit. Butterworths, London.

7. O. E. Edwards, G. Feniak and M. Los, *Can. J. Chem.* (1962), **40**, 1540.

8. C. H. Brieskorn, A. Fuchs, J. B. Bredenberg, J. D. McChesney and E. Wenkert *J. Org. Chem.* (1964), **29**, 2293

9. J. H. Gough and M. D. Sutherland, *Aust. J. Chem.* (1966), **19**, 329.

10. K. L. Handa, S. S. Chaudhary and K. S. Jamwal, *Indian J. Pharm.* (1958), **20**, 211.

11. M.-M. Janot and P. Potier, *Ann. Pharm. Franç.* (1964), **22**, 387.

12. M. Nakao, *Bull. Shanghai Sci. Inst.* (1930), **5**, No. 5; M. Nakao and T. Fukushima, *J. Pharm. Soc. Japan* (1934), **54**, 154.

13. K. Takiura, *J. Pharm. Soc. Japan* (1941), **61**, 475.

14. K. Takiura and K. Koizumi, *Chem. Pharm. Bull. (Tokyo)* (1962), **10**, 112.

15. F. von Wessely and S. Wang, *Ber.* (1940), **73**, 19.

16. F. von Wessely and A. Bauer, *Ber.* (1942), **75**, 617.

17. F. von Wessely and T. Lauterbach, *Ber.* (1942), **75**, 958.

18. K. Takiura, *J. Pharm. Soc. Japan* (1941), **61**, 482.

19. K. Takiura, *J. Pharm. Soc. Japan* (1943), **63**, 40.

20. A. C. Baillie and R. H. Thomson, *J. Chem. Soc. (C)* (1968), 48.

21. Y. Okumura, H. Kakisawa, M. Kato and Y. Hirata, *Bull. Chem. Soc. Japan* (1961), **34**, 895.

22. M. P. Kuntsmann and L. A. Mitscher, *J. Org. Chem.* (1966), **31**, 2920.

23. J. H. Bowie and A. W. Johnson, *Tetrahedron Lett.* (1967), 1449.

24. J. H. Bowie, *Aust. J. Chem.* (1966), **19**, 1619.

25. T. Hata, R. Sano, R. Sugawara, A. Matsumae, K. Kanamori, T. Shima and T. Hoshi, *J. Antibiotics, Ser. A* (1956), **9**, 141.

26. S. Wakaki, H. Marumo, K. Tomioka, G. Shimizu, E. Kato, H. Kamada, S. Kudo and Y. Fujimoto, *Antibiotics and Chemotherapy* (1958), **8**, 228.

27. D. V. Lefemine, M. Dann, F. Barbatschi, W. K. Hausmann, V. Zbinovsky, P. Monnikendam, J. Adam and N. Bohonos, *J. Amer. Chem. Soc.* (1962), **84**, 3184.

28. J. F. Collins, *Brit. Med. Bull.* (1965), **21**, 223.

29. J. S. Webb, D. B. Cosulich, J. H. Mowat, J. B. Patrick, R. W. Broschard, W. E. Meyer, R. P. Williams, C. F. Wolf, W. Fulmor, C. Pidacks and J. E. Lancaster, *J. Amer. Chem. Soc.* (1962), **84**, 3185, 3186.

30. A. Tulinsky, *J. Amer. Chem. Soc.* (1962), **84**, 3188; A. Tulinsky and J. H. van den Hende, *J. Amer. Chem. Soc.* (1967), **89**, 2905.

31. K. Uzu, Y. Harada and S. Wakaki, *Agr. Biol. Chem. (Tokyo)* (1964), **28**, 388; K. Uzu, Y. Harada, S. Wakaki and Y. Yamada, *Agr. Biol. Chem. (Tokyo)* (1964), **28**, 394.

32. C. L. Stevens, K. G. Taylor, M. E. Munk, W. S. Marshall, K. Noll, G. D. Shah, L. G. Shah and K. Uzu, *J. Med. Chem.* (1965), **8**, 1.

33. S. Pinchas and D. Ben-Ishai, *J. Amer. Chem. Soc.* (1957), **79**, 4099.

34. H. T. Hoffman, G. E. Evans and G. Glockler, *J. Amer. Chem. Soc.* (1951), **73**, 3028.

35. H. C. Brown and A. Tsukamoto, *J. Amer. Chem. Soc.* (1961), **83**, 2016.

36. G. W. Anderson and R. Paul, *J. Amer. Chem. Soc.* (1958), **80**, 4423.

37. J. B. Patrick, R. P. Williams, W. E. Meyer, W. Fulmor, D. B. Cosulich, R. W. Broschard and J. S. Webb, *J. Amer. Chem. Soc.* (1964), **86**, 1889.
38. H. S. Schwartz, J. E. Sodergren and F. S. Philips, *Science, N.Y.* (1963), **42**, 1181.
39. V. N. Iyer and W. Szybalski, *Microbiology* (1963), **50**, 355.
40. M. J. Weiss, G. S. Redin, G. R. Allen, A. C. Dornbush, H. L. Lindsay, J. F. Poletto, W. A. Remers, R. H. Roth and A. E. Sloboda, *J. Med. Chem.* (1968), **11**, 742 and previous papers; V. Carelli, M. Cardellini and F. Morlacchi, *Tetrahedron Lett.* (1967), 765.
41. V. J. Mazzolo, K. F. Bernady and R. W. Franck, *J. Org. Chem.* (1967), **32**, 486; T. Kametani, T. Yamanaka and M. Satoh, *J. Pharm. Soc. Japan* (1967), **87**, 1407; J. Auerbach and R. W. Franck, *Chem. Comm.* (1969), 991.
42. K. V. Rao and W. P. Cullen, *Antibiotics Annual* (1959–1960), 950.
43. K. V. Rao, K. Biemann and R. B. Woodward, *J. Amer. Chem. Soc.* (1963), **85**, 2532.
44. T. Kametani, K. Ogasawara, A. Kozuka and K. Nyu, *J. Pharm. Soc. Japan*, 1967, **87**, 1195 and previous papers; T. K. Liao, W. H. Nyberg and C. C. Cheng, *Angew. Chem.* (1967), **6**, 82.
45. H. Brockmann, *Prog. Chem. Org. Nat. Prods.* (1960), **18**, 1.
46. A. F. Wagner and C. O. Gitterman, *Antibiotics and Chemotherapy* (1962), **12**, 464; S. Wakaki, Y. Harada, K. Uzu, G. B. Whitfield, A. N. Wilson, A. Kalowky, E. O. Stapley, F. J. Wolf and D. E. Williams, *Antibiotics and Chemotherapy* (1962), **12**, 469.
47. F. L. Warren, *Prog. Chem. Org. Nat. Prods.* (1955), **12**, 198; (1966), **24**, 329.
48. (a) A. J. Birch and M. Kocor, unpublished work; M. V. Sargent and D. O'N. Smith, *J. Chem. Soc.* (C) (1970), 329.
49. H. Kakisawa, T. Hayashi, I. Okazaki and M. Ohashi, *Tetrahedron Lett.* (1968), 3231.
50. A. Todd, *Ann. Rept.* (1941), **38**, 209.
51. *Antibiotics and Chemotherapy* (1961), **11**, 147–189.
52. R. R. Herr, M. E. Bergy, T. E. Eble and H. K. Jahnke, "Antimicrobial Agents Annual 1960", (1961), p. 23. Plenum Press, New York.
53. H. Kakisawa, M. Tateishi and T. Kusumu, *Tetrahedron Lett.* (1968) 3783.
54. G. F. Gause, *Chem. Ind. (London)* (1966), 1506.
55. H. Kakisawa and Y. Inouye, *Chem. Comm.* (1968), 1327; *Bull. Chem. Soc. Japan* (1969), **42**, 3318.
56. H. Kakisawa and M. Ikeda, *J. Pharm. Soc. Japan* (1967), **88**, 476.
57. B.P. 1,012,684.
58. U.S.P. 2,970,943.
59. W. Szybalski and V. N. Iyer. *In* "Antibiotics, Mechanism of Action", (eds. D. Gottlieb and P. D. Shaw) (1967), Vol. 1, p. 211. Springer-Verlag, Berlin.
60. C. C. J. Culvenor, A. T. Dann and A. T. Dick, *Nature (London)* (1962), **195**, 570.
61. T. Hirata, Y. Yamada and M. Matsui, *Tetrahedron Lett.* (1969), 19, 4107; Y. Yamada, T. Hirata and M. Matsui, *Tetrahedron Lett.* (1969), 101.
62. C. H. Eugster and H. J. Leuenberger, *Palette* (1968), **27**, 25.
63. S. M. Kupchan, A. Karim and C. Marcks, *J. Amer. Chem. Soc.* (1968), **90**, 5923; *J. Org. Chem.* (1969), **34**, 3912.
64. G. F. Vlasova, A. S. Romanova, M. E. Perelson and A. I. Bankovskii, *Khim. Prirod. Soedin.* (1969), 317.
65. M. Ribi, A. Chang Sin-Ren, H. P. Küng and C. H. Eugster, *Helv. Chim. Acta* (1969), **52**, 1685.
66. M. Lounasmaa and J. Zylber, *Bull. Soc. Chim. France* (1969), 3100.

67. H. Kakisawa, T. Hayashi and T. Yamazaki, *Tetrahedron Lett.* (1969), 301.
68. A. J. Birch quoted in ref. 48(b).
69. T. Hayashi, H. Kakisawa, H.-Y. Hsu and Y. P. Chen, *Chem. Comm.* (1970), 299.
70. H. Brockmann and P. Christiansen, *Chem. Ber.* (1970), **103**, 708.
71. M. E. Bergy, *J. Antibiotics (Tokyo) Ser. A.* (1968), **21**, 454.
72. D. J. Duchamp, Abstr. Sym. "Small Angle Scattering of X-rays", p. 82. Amer. Crystallographic Assoc. Summer Meeting, 1968.

Appendix

Chapter 1

1. C. D. Snyder and H. Rapoport, "Biosynthesis of bacterial menaquinones. Origin of quinone oxygens." *Biochemistry* (1970) **9**, 2033.
2. S. Nozoe, M. Morisaki and H. Matsumoto, "Biosynthesis of helicobasidin and related compounds." *Chem. Comm.* (1970), 926.
3. Y. Yamamoto, M. Shinya and Y. Oohata, "Studies on the metabolic products of a strain of *Aspergillus fumigatus* (DH 413). IV. Biosynthesis of toluquinones and chemical structures of new metabolites" (spinulosin hydrate, spinulosin quinol hydrate, dihydrospinulosin quinol and fumigatin chlorohydrin). *Chem. Pharm. Bull.* (*Tokyo*) (1970), **18**, 561.
4. G. R. Whistance, B. S. Brown and D. R. Threlfall, "Biosynthesis of ubiquinone in non-photosynthetic Gram-negative bacteria." *Biochem. J.* (1970), **117**, 119.
5. G. R. Whistance and D. R. Threlfall, "Biosynthesis of phytoquinones." *Biochem. J.* (1970), **117**, 593.
6. A. Law, D. R. Threlfall and G. R. Whistance, "Isoprenoid quinone precursors of ubiquinone-10 (X-H$_2$) in *Aspergillus flavus*." *Biochem. J.* (1970), **117**, 799.

Chapter 2

1. D. Hausigk, "Die Spaltung von *p*-Chinonen mit Kalium-tert. butanolat." *Tetrahedron Lett.* (1970), 2447.
2. M. S. El Ezaby, T. M. Salem, A. H. Zewail and R. Issa, "Spectral studies (u.v., i.r.) of some hydroxy-derivatives of anthraquinones." *J. Chem. Soc.* (*B*) (1970), 1293.
3. K. Yamaguchi, "Spectral data of natural products", Vol. I (1970), Elsevier, Amsterdam.

Chapter 3

1. C. Tabacik and M. Hubert, "Poly-isoprénes de *Satureia montana*" (thymoquinone). *Phytochemistry* (1970), **9**, 1129.
2. R. Barr and F. L. Crane, "Comparative studies on plastoquinones. V. Changes in lipophilic chloroplast quinones during development." *Plant Physiol.* (1970), **45**, 53.
3. I. R. Peake, P. J. Dunphy and J. F. Pennock, "The chemical diversity of the plastochromanols." *Phytochemistry* (1970), **9**, 1345.
4. G. B. Cox, N. A. Newton, F. Gibson, A. M. Snoswell and J. A. Hamilton, "The function of ubiquinone in *Escherichia coli*." *Biochem. J.* (1970), **117**, 551.
5. H. Morimoto, I. Imada and G. Goto, "Photooxydation von Ubichinon-7." *Annalen* (1970), **735**, 65.
6. I. Imada, M. Watanabe, N. Matsumoto and H. Morimoto, "Metabolism of Ubiquinone-7." *Biochemistry* (1970), **9**, 2870.

7. M. Watanabe, I. Imada and H. Morimoto, "Synthesis of Ubiquinone-7 Metabolites." *Biochemistry* (1970), **9**, 2879.

8. H.-M. Cheng and J. E. Casida, "Preparation of ^{14}C- and ^{3}H-methoxyl-labelled forms of ubiquinone by photochemical O-demethylation and subsequent remethylation." *J. Labelled Compounds* (1970), **6**, 66.

9. P. Mamont, "Condensation, en milieu acide, du styrène et du diphényl-1,1-éthylène avec les ubiquinones." *Bull Soc. Chim. France* (1970), 1557.

10. P. Mamont, "Condensation, en milieu acide, du styrène avec tocoquinone-1." *Bull. Soc. Chim. France* (1970), 1564.

11. P. Mamont, "Mecanisme de l'isomérisation acide des ménaquinones, tocoquinones et ubiquinones." *Bull. Soc. Chim. France* (1970), 1568.

12. G. Read and V. M. Ruiz, "Quinone Epoxides. Part VI. The synthesis and stereochemistry of terremutin." *J. Chem. Soc. (C)* (1970), 1945.

13. J. C. Cooke and R. P. Collins, "Oosporein from *Chaetomium trilaterale.*" *Lloydia* (1970), **33**, 269.

Chapter 4

1. G. J. Kapadia, M. B. E. Fayez and M. L. Sethi, "Hennosides, the primary glycosidic constituents of henna" (lawsone precursors). *Lloydia* (1969), **32**, 523.

2. K. C. Lee and R. W. Campbell, "Nature and occurrence of juglone in *Juglans regia* L." *Hort. Science* (1969), **4**, 297.

3. A. J. Birch and V. H. Powell, "Synthesis of some polycyclic quinones through 1-methoxycyclohexa-1,3-dienes" (2,3,7-trihydroxyjuglone). *Tetrahedron Lett.* (1970), 3467.

4. H. Brockmann and A. Zeeck, "Die Konstitution von a-Rubromycin, β-Rubromycin, γ-Rubromycin und γ-*iso*-Rubromycin." *Chem. Ber.* (1970), **103**, 1709.

5. J. St. Pyrek, O. Achmatowicz and A. Zamojski, "Naphtha- and anthraquinones produced by *Streptomyces thermoviolaceus*" (granaticin and related compounds). *Abstr. IUPAC 7th Int. Sym. Chem. Nat. Prods. (Riga)* (1970), p. 604.

6. Y. Nozawa and Y. Ito, "Separation of quinone pigments from *Microsporium cookei* by thin-layer chromatography" (xanthomegnin and related compounds). *Experientia* (1970), **26**, 803.

7. J. M. Imhoff and R. Azerad, "Ménaquinones-7 et -8 d'un mutant de *Corynebacterium diphtheriae.*" *Bull. Soc. Chim. Biol.* (1970), **52**, 695.

8. T. J. Duello and J. T. Matschiner, "Identification of phylloquinone in horse liver." *Arch. Biochem. Biophys.* (1970), **138**, 640.

9. J. T. Matschiner, "Characterization of vitamin K from the contents of bovine rumen" (MK-10, 11, 12 and 13). *J. Nutrition* (1970), **100**, 190.

10. J. H. Gough and M. D. Sutherland, "Pigments of Marine Animals. Synthesis of 6-acetyl-2,3,5,7-tetrahydroxy-1,4-naphthaquinone. Its status as an echinoid pigment." *Aust. J. Chem.* (1970), **23**, 1839.

11. R. Luckner and M. Luckner, "Naphthochinonderivate aus *Drosera ramentacea* Burch. ex Harv. & Sond" [plumbagin, ramentacéone and ramenton (= 2-methylnaphthazarin)]. *Pharmazie* (1970), **25**, 261.

12. H. Ohmae and G. Katsui, "Photolysis of vitamin K_1. (V). Photolytic products in the absence of oxygen." *Vitamins* (1970), **41**, 178.

Chapter 5

1. C. F. Culberson and W. L. Culberson, "First reports of lichen substances from seven genera of lichens" (physcion). *The Bryologist* (1969), **72**, 210.
2. J. Santesson, "Some occurrences of the anthraquinone parietin in lichens." *Phytochemistry* (1970), **9**, 1565.
3. B. Franck, U. Ohnsorge and H. Flasch, "Einfache Totalsynthese von [10-^{14}C]-Endocrocin." *Tetrahedron Lett.* (1970), 3773.
4. P. J. Aucamp and C. W. Holzapfel, "Polyhydroxyanthraquinones from *Aspergillus versicolor*, *Aspergillus nidulans* and *Bipolaris* sp. Their significance in relation to biogenetic theories on aflatoxin B₁" (averufin, avermutin, averantin 1'-methyl ether, versicolorin C, bipolarin, nidurufin). *J. South African Chem. Inst.* (1970), **23**, 40.
5. A. J. Birch and V. H. Powell, "Synthesis of some polycyclic quinones through 1-methoxycyclohexa-1,3-dienes" (chrysophanol). *Tetrahedron Lett.* (1970), 3467.
6. G. C. Elsworthy, J. S. E. Holker, J. M. McKeown, J. B. Robinson and L. J. Mulheirn, "The biosynthesis of the aflatoxins" (6-deoxyversicolorin A). *Chem. Comm.* (1970), 1069.

Chapter 6

1. W. Keller-Schierlein, J. Sauerbier, U. Vogler and H. Zähner, "Aranciamycin" (a new anthracyclinone). *Helv. Chim. Acta* (1970), **53**, 779.
2. W. Keller-Schierlein and A. Müller, "The sugar component of aranciamycin: 2-0-methyl-L-rhamnose." *Experientia* (1970), **26**, 929.

Chapter 8

1. T. Hayashi and H. Kakisawa, "Hydrogenolysis of a quinonoid oxygen of alkoxynaphthaquinones" (tanshinones). *Bull. Chem. Soc. Japan* (1970), **43**, 1897.
2. G. O. Morton, G. E. Van Lear and W. Fulmor, "The structure of mitiromycin." *J. Amer. Chem. Soc.* (1970), **92**, 2588.
3. G. E. Van Lear, "Mass spectrometric studies of antibiotics. I. Mass spectra of mitomycin antibiotics." *Tetrahedron* (1970), **26**, 2587.
4. G. R. Allen and M. J. Weiss, "The mitomycin antibiotics. Synthetic studies XXIII. An attempted synthesis of 3H-pyrrolo-[1,2-a]indole-9-carboxaldehyde". *J. Heterocyclic Chem.* (1970), **7**, 193.
5. U. Hornemann and J. C. Cloyd, "Studies on the biosynthesis of the mitomycins" (abstr. only). *Lloydia* (1969), **32**, 525.

Author Index

The numbers in parentheses are reference numbers and are included to assist in locating references when authors' names are not mentioned in the text. The numbers in *italics* refer to the pages on which the references are listed.

A

Abe, J., 476 (585) 477 (586) 478 (585) *533*
Abiusso, N., 379 (79) *520*
Abraham, E. P., 115 (143) *187*
Abrahamson, S., 165 (299) *191*
Achmatowicy, O., *668*
Acton, E. M., 565 (65) *574*
Adam, J., 653 (27) 654 (27) *664*
Adams, R., 153 (255) 159 (255) *190*, 379 (68) 411 (280) 420 (315) *519, 525, 526*
Adderley, C. J. R., 111 (410) *195*
Adrian, 377 (51) *519*
Agarashi, H., 448 (408) *528*
Aghoramurthy, K., 108 (100) 112 (118) 162 (291) *185, 186, 191*
Agosti, G., 414 (267) 419 (309) *524, 526*
Ahluwalia, V. K., 368 (4) *518*
Ahmad, M. S., 389 (183) *522*
Aiyan, S. N., 137 (210) *189*
Aiyar, A. S., 28 (96) *36*
Aizawa, K., 367 (563) *532*
Akačic, B., 389 (136) 419 (136) *521*
Akagi, M., 159 (257) 160 (277, 278) 161 (277) *190*
Alais, L., 635 (3) *664*
Alcalay, W., 102 (74) *185*
Alden, R. A., 432 (461) *530*
Alexander, P., 96 (16) *183*
Algar, J., 435 (594) *533*
Ali, M. A., 460 (445) 462 (445) *529*
Ali, S. M., 3 (7) 31 (7) *33*, 223 (497) 227 (497) 228 (497) 230 (497) 237 (497) 239 (497) *364*
Allan, R. D., 165 (472) 166 (472, 473) 167 (472) *196*
Allen, C. F., 344 (445b) *363*
Allen, G. R., 660 (40) *665, 669*
Allen, M. B., 259 (389) *361*
Allport, D. C., 576 (2) 577 (2, 3) *630*
Almquist, H. J., 342 (423) *362*

Altamarano, F., 228 (131) *354*
Altland, H. W., 615 (90) *632*
Alvino, C., 18 (58) *35*
Amelotti, J. M., 341 (559) *366*
Amin, M., 470 (427) *529*
Anand, S. R., 311 (274b) 312 (274b) *358*
Anchel, M., 24 (139) *37*, 102 (58) *113* (128) 156 (262) 157 (262) *184, 186, 190,* 402 (224) *523*
Anderson, A. B., 98 (52) 99 (52) 109 (107) 110 (110) *184, 186*
Anderson, G. W., 656 (36) *664*
Anderson, H. A., 3 (8) *33*, 56 (68) *91*, 115 (137) *186*, 257 (503) 258 (326) 261 (365) 262 (326) 263 (326) 264 (326) 265 (326) 268 (369) 270 (365) *359, 360, 365*, 484 (476) 485 (476) *530*
Anderson, J. M., 576 (1) 577 (1) *630*
Anderson, R. J., 201 (447) *363*
Anderson, T., 410 (270) 412 (270) *525*
Ando, M., 39 (1) *89*
Andres, W. W., 570 (60) 572 (60) *574*
Aneshansley, D. J., 97 (469) 98 (469) *196*
Anger, V., 41 (4) *89*
Angliker, E., 403 (234) *524*
Anon., 137 (208) *189*
Anschütz, R., 125 (177) *188*
Anslow, W. K., 111 (111) 112 (111, 115) 117 (147, 148, 149, 150) *186, 187*, 435 (384) 436 (384) 437 (384) 439 (384) 494 (493) 502 (493) 504 (502) *528, 531*
Anton, R., 389 (548) 419 (548) 421 (583) 429 (548) *532, 533*
Aplin, R. T., 83 (55) *91*
Aquila, H., 115 (133) *186*
Arahata, S., 419 (545) 432 (545) 433 (545) 435 (545) 436 (545) *532*
Arai, T., 199 (66) 200 (12) 321 (489) *351, 364*
Arakawa, H., 251 (231) *357*
Arata, Y., 429 (337) 439 (337) *526*

671

22

23

Wimmer, E., 551(25) 555(25, 35) *573*
Winkler, H., 95(406) *194*
Winrow, M. J., 28(145) *37*
Winters, T. E., 537(74) *575*
Winterstein, A., 341(413) 345(433) *362*
Winzor, F. L., 57(25) *90*, 209(69) 237
 (164, 166) 254(164, 165, 172) 272(166)
 352, 355
Wirth, E., 153(254) 154(254) 159(254)
 161(254) *190*
Wirth, J. C., 311(274a, 274b) 312(274b)
 358
Wise, L. E., 203(37) *351*
Wiss, O., 69(40) *90*, 168(314) 175(354,
 357) *191, 193*, 341(413) 342(421)
 343(430) 345(433) *362*
Witanowski, W. R., 228(134) *354*
Witt, H. T., 168(434) *195*
Wittreich, P. E., 175(358) *193*, 211(72)
 352
Wolf, C. F., 544(70) *575*, 654(29) 655
 (29) 656(29) *664*
Wolf, D. E., 1(39) *34*, 168(312) 175
 (355) 177(389) 181(370) *191, 193, 194*
Wolf, F. J., 654(46) *665*
Wolff, J., 407(248) *524*
Wolff, R. E., 229(152) *354*
Wolinsky, J., 153(247) *190*
Wollmann, H., 311(279) *358*
Wolstenholme, G. E. W., 1(39) 23(76)
 34, 35
Wong, E. L., 40(78) *92*, 168(313) 175
 (358) 176(384) *191, 193, 194*, 211(71)
 345(435) *352, 362*
Wood, R. K. S., 280(237) *357*
Woods, R. J., 343(430) *362*
Woodward, R. B., 102(61) 130(417)
 195, 660(43) 663(43) *665*
Worthington, A., 107(93) *185*
Wright, D. E., 287(248) *357*, 408(458)
 450(415) 463(458) 464(458) 465(458)
 467(458) 504(415, 500) *529, 530, 531*
Würsch, J., 168(314) 175(359) *191, 193*,
 343(430) *362*
Wydler, R., 248(213) *356*
Wylie, A. G., 635(5) *664*

Y

Yakubov, G. Z., 317(488) *364*, 552(66)
 553(45) 554(66) 557(45) 558(66) 568
 (66) 570(66) *574*

Yamada, K., 475(587) *533*
Yamada, O., 316(449) *363*
Yamada, Y., 654(31) 657(31) 660(61)
 664, 665
Yamaguchi, K., 126(184) 131(193) 133
 (193) 136(193) *188, 667*
Yamaguchi, M., 265(351) 274(325b)
 359, 360
Yamaguti, K., 139(220, 223) *189*
Yamamoto, I., 423(363) 428(363) *527*
Yamamoto, K., 367(558) *532*
Yamamoto, R., 350(407) *361*
Yamanaka, T., 660(41) *665*
Yamauchi, H., 419(543) 432(543) 434
 (543) 435(575) *532, 533*
Yamazaki, M., 32(108) *36*
Yamazaki, T., 647(67) *666*
Yamomoto, T., 500(526) *531*
Yamomoto, Y., 23(78) *35*, 57(27) *90*,
 116(54) *184*, 419(545) 423(363) 428
 (363) 432(545) 433(545) 435(545) 436
 (545) 500(526) *527, 531, 532, 667*
Yanagisawa, F., 321(489) *364*
Yaoi, Y., 199(6b) *351*
Yasue, M., 325(301, 302, 303, 304, 305)
 326(301, 303) 328(301, 304) 329(301,
 303, 304) 330(305) *359*
Yates, P., 64(32) 69(32) *90*, 470(427)
 529
Yee, C. W., 368(5) 382(88) *518, 520*
Yokota, M., 389(154) 402(219) 429(219,
 327) *522, 523, 526*
Yokoyama, M., 429(355) *527*
Yoneshige, M., 389(146) 419(146) *521*
Yonezawa, Y., 462(462) *530*
Yoshida, M., 259(391) 264(342) *360, 361*
Yoshihira, K., 3(7) 31(7) *33*, 144(232)
 147(230) 148(425) 149(230) *189, 195*,
 227(156) 231(159) 233(156) 234(156,
 159) 235(156, 159) 236(156) *355*
Yoshikawa, S., 490(617) *534*
Yoshimoto, T., 57(76) *91*
Yoshino, E., 539(9) 541(9) *573*
Yosioka, I., 419(543) 432(543) 434
 (543) 435(575) *532, 533*
Young, I. G., 28(144) *37*
Youngken, H. W., 389(139) 419(139,
 300) *521, 525*

Z

Zacharov, B. V., 563(52) 566(52) *574*

Botanical Index

Zoological Index

A

Acanthaster planci, 265, 274
Adelges strobi, 597
Amblyneustes oveum, 263
Antedon bifida, 265
Anthocidaris crassispina, 263
Aphis cognatella, 597
Aphis corniella, 597, 622
Aphis evonymi, 597
Aphis fabae, 597, 598, 621, 622
Aphis farinosa, 597, 622
Aphis hederae, 597
Aphis ilici, 597
Aphis nerii, 600
Aphis philadelphi, 597
Aphis rumicis, 622
Aphis sambuci, 597, 622, 623
Aphis viburni, 597
Arbacia incisa, 263
Arbacia lixula, 257, 259, 263, 270
Arbacia pustulosa, 13, 263
Archiulus sabulosus, 94
Arenicola marina, 388
Ascaris lumbricoides var. suis, 180
Aulonopygus aculeatus, 94
Aulonopygus aculeatus barbieri, 94
Austrotachardia acaciae, 462

B

Blaps gigas, 95
Blaps lethifera, 95
Blaps mortisaga, 95, 98
Blaps mucronata, 95
Blaps requienii, 95
Blepharisma spp., 594
Blepharisma undulans, 594
Brachynus spp., 97
Brachinus crepitans, 95
Brachinus explodens, 95
Brachinus sclopeta, 95
Brachycaudus klugkisti, 597
Brachycaudus rociadae, 597
Brachyiulus unilineatus, 94

C

Callistus lunatus, 95
Cambala hubrichti, 94
Carcinas maenas, 174
Centrostephanus nitidus, 263
Centrostephanus rodgersii, 263
Ceroplastes albolineatus, 456
Chicobolus spinigerus, 94
Chlaenius vestitus, 95
Chondrocidaris gigantea, 262
Chrysolina brunsvicensis, 586
Cidaris cidaris, 262
Clivina basalis, 95
Clivina fossor, 95
Colobocentratus atratus, 263
Comatula cratera, 492, 507
Comatula pectinata, 492, 507
Coniocidaris tubaria var. impressa, 262
Cylindroiulus teutonicus, 94

D

Dactylopius coccus, 459
Dactynotus ambrosiae, 615
Dactynotus cirsii, 615
Dactynotus jaceae, 612, 615, 621, 622
Dactynotus nigrotuberculatus, 615
Dactynotus rudbeckiae, 615
Dactynotus tanaceti, 615
Dactynotus taraxaci, 615
Dendraster excentricus, 265
Diadema antillarum, 259, 263
Diadema paucispinus, 263
Diadema setosum, 263
Diaperis boleti, 95
Diaperis maculata, 95
Diploptera punctata, 95
Doratogonus annulipes, 94
Dysaphis pyri, 597

E

Echinarachnius mirabilis, 265
Echinarachnius parma, 265

Subject Index

*Principal references are given in **bold** type*

A

2-Acetoxymethylanthraquinone, 368
Actinomycins, 663
Actinorhodin, 16, **302**, 651
6-Acetyl-2,7-dihydroxyjuglone, 271
3-Acetyl-2,7-dihydroxynaphthazarin, 268
3-Acetyl-2-hydroxynaphthazarin, 273
6-Acetyl-2-hydroxynaphthazarin, 273
3-Acetyl-2-hydroxy-7-methoxy-naphthazarin, 274
3-Acetyl-2,6,7-trihydroxyjuglone, 272
6-Acetyl-2,3,7-trihydroxyjuglone, 271, 668
3-Acetyl-2,6,7-trihydroxynaphtha-zarin, 270
Adriamycin, 567
Adriamycinone, 566
Aklavin, 539
Aklavinone, 538
Alaternin, 472
-6-methyl ether, 472, 473
Alizarin, **373**, 407
biosynthesis of, 9, 11
Alizarin-1-methyl ether, 375
-2-methyl ether, 375
Alkannan, 250, **252**
Alkannin, 2, **248**
Alkannin esters, 249, 251
Allodunnione, 216, 217
Alloeleutherin, 285
Alloisoeleutherin, 285
Aloe-emodin, **399**, 419
Aloe-emodin-9-anthrone, 400
Aloe-emodin-10,10'-bianthrone, 402
Aloes, 400
Aloinoside A, 402
Aloinoside B, 401
Alternariol, 25, 152
Altersolanol A, 473
Altersolanol B, 475, 477
Amitenone, 141

Anhydro-ethylidene-3,3'-bis(2,6,7-tri-hydroxynaphthazarin), 276
Anhydrofusarin, 291
Anthracyclines, 537
Anthracyclinones, 536
distribution of, 4
stereochemistry, 547
Anthragallol, 405
-1,2-dimethyl ether, 405
-1,3-dimethyl ether, 405
-2,3-dimethyl ether, 405
-1-methyl ether, 383
-2-methyl ether, 405
-3-methyl ether, 405
trimethyl ether, 405
Anthraquinone, 12, 367
-2-aldehyde, 368
-2-carboxylic acid, 368
Anthraquinones, 368
distribution of, 1
Aphinins, 622
Aphins, 597
Aquayamycin, 570
Aranciamycin, 669
Araroba powder, 389
Arbutin, 93
Archin, 419
Archinin, 388
Ardisiaquinones, 147
Ardisiols, 147
Arenicochrome, 388
Areolatin, 451
Arnebins, 251
Aspethecin, 510
Atromentin, 140, 155, **158**, 159, 161
Atrovenetin, 215
Aurantiacin, 158
-leucodibenzoate, 159
Aurantin, 135
Aurantiogliocladin, 22, 23, **114**
biosynthesis of, 26
Aurantio-obtusin, **505**, 511
Aurantioskyrin, 497

725